FUNDAMENTALS OF PLASMA PHYSICS

This rigorous explanation of plasmas is relevant to diverse plasma applications such as controlled fusion, astrophysical plasmas, solar physics, magnetospheric plasmas, plasma thrusters, and many others. More thorough than previous texts, it exploits new, powerful mathematical techniques to develop deeper insights into plasma behavior.

The initial chapters develop the basic plasma equations from first principles and explore single particle motion with particular attention to adiabatic invariance. The author then examines the many types of plasma waves and the philosophically intriguing issue of Landau damping. Magnetohydrodynamic equilibrium and stability are then tackled with emphasis on the topological concepts of magnetic helicity and self-organization. Advanced topics follow, including magnetic reconnection, nonlinear waves, and the Fokker–Planck treatment of collisions. The book concludes by discussing unconventional plasmas such as non-neutral and dusty plasmas.

Written for beginning graduate students and advanced undergraduates, this text emphasizes the fundamental principles that apply across many different contexts. It will interest students and researchers in physics, astronomy, space physics, electrical engineering, and aeronautics.

PAUL M. BELLAN is Professor of Applied Physics at the California Institute of Technology. He received his B.Sc. from the University of Manitoba and his Ph.D. from Princeton University. During the last 30 years he has worked on many facets of experimental and theoretical plasma physics, including controlled fusion, industrial arcs, magnetospheric physics, solar physics, and astrophysics. He has designed and constructed several laboratory experimental devices including a small tokamak and several spheromaks and has worked with numerous diagnostic techniques including laser-induced fluorescence, ultra-high-speed photography, and various types of probes. He has investigated Alfvén, lower hybrid, ion acoustic, and drift waves and has had a particular interest in the phenomenon of resonance cones. Recently he has built high-power pulsed laboratory devices which produce laboratory replicas of solar prominence and astrophysical jet plasmas and is using these plasmas to explore magnetohydrodynamic self-organization, magnetic helicity, and the acceleration of high speed flows.

In 2001, Professor Bellan received the Popular Writing Award from the Solar Physics Division of the American Astronomical Society for an article he wrote for *American Scientist*. His previous book, *Spheromaks*, was published in 2000. He is a Fellow of the American Physical Society and has published over 100 papers in journals including *Physical Review Letters*, *Geophysical Research Letters*, *Physics of Plasmas*, *Nuclear Fusion*, *The Review of Scientific Instruments*, and the *Astrophysical Journal*.

FUNDAMENTALS OF PLASMA PHYSICS

PAUL M. BELLAN
California Institute of Technology

CAMBRIDGE
UNIVERSITY PRESS

University Printing House, Cambridge CB2 8BS, United Kingdom

Cambridge University Press is part of the University of Cambridge.

It furthers the University's mission by disseminating knowledge in the pursuit of education, learning and research at the highest international levels of excellence.

www.cambridge.org
Information on this title: www.cambridge.org/9780521528009

First published in hardback 2006
Paperback edition published 2008
3rd printing 2012

A catalogue record for this publication is available from the British Library

ISBN 978-0-521-82116-2 Hardback
ISBN 978-0-521-52800-9 Paperback

To my parents

Contents

Preface

This text is based on a course I have taught for many years to first-year graduate and senior-level undergraduate students at Caltech. One outcome of this experience has been the realization that although students typically decide to study plasma physics as a means towards some specific goal, they often conclude that the study of this subject has an attraction and charm of its own; in a sense the journey becomes as enjoyable as the destination. This conclusion is shared by me and I feel that a delightful aspect of plasma physics is the frequent transferability of ideas between extremely different applications so, for example, a concept developed in the context of astrophysics might suddenly become relevant to fusion or vice versa.

Applications of plasma physics are many and varied. Examples include controlled thermonuclear fusion, ionospheric physics, magnetospheric physics, solar physics, astrophysics, plasma propulsion, semiconductor processing, antimatter confinement, and metals processing. Furthermore, because plasma physics is extremely rich in both concepts and regimes, it has often served as an incubator for new ideas in applied mathematics. Concepts first developed in one of the areas listed above frequently migrate rather quickly to one or more of the other areas so it is very worthwhile to keep abreast of developments in areas of plasma physics outside of one's immediate field of interest. Dialog between plasma researchers in seemingly disconnected areas has often proved quite profitable for all concerned and it is my hope that this text will help to promote this interdisciplinary aspect.

The prerequisites for this text are a reasonable familiarity with Maxwell's equations, classical mechanics, vector algebra, vector calculus, differential equations, and complex variables – i.e., the contents of a typical undergraduate physics or engineering curriculum. Experience has shown that because of the many different applications for plasma physics, students studying plasma physics have a diversity of preparation and not all are proficient in all prerequisites. Brief derivations of many basic concepts are included to accommodate this range of preparation; these derivations are intended to assist those students who may have had little or no exposure to the concept in question and to refresh the memories of those who have had ample exposure but have forgotten the details. For example, rather than just invoke Hamilton–Lagrange methods or Laplace transforms, there is a quick derivation and then a considerable discussion showing how these concepts relate to plasma physics issues. These additional explanations make the book more self-contained and also provide a close contact with first principles.

The order of presentation and level of rigor have been chosen to establish a firm foundation and yet avoid unnecessary mathematical formalism or abstraction. In particular, the various fluid equations are derived from first principles rather than simply invoked and the consequences of the Hamiltonian nature of particle motion are emphasized early on and shown to lead to the powerful concepts of symmetry-induced constraint and adiabatic invariance. Symmetry turns out to be an essential feature of magnetohydrodynamic plasma confinement and adiabatic invariance turns out to be not only essential for understanding many types of particle motion, but also vital to many aspects of wave behavior.

The mathematical derivations have been presented with intermediate steps shown in as much detail as is reasonably possible. This occasionally leads to daunting-looking expressions, but it is my belief that it is preferable to see all the details rather than have them glossed over and then justified by an "it can be shown" statement.

The book is organized as follows: Chapters 1 to 3 lay out the foundation of the subject. Chapter 1 provides a brief introduction and overview of applications, discusses the logical framework of plasma physics, and begins the presentation by discussing Debye shielding and then showing that plasmas are quasi-neutral and nearly collisionless. Chapter 2 introduces phase-space concepts and derives the Vlasov equation and then, by taking moments of the Vlasov equation, derives the two-fluid and magnetohydrodynamic systems of equations. Chapter 2 also introduces the dichotomy between adiabatic and isothermal behavior, which is a fundamental and recurrent theme in plasma physics. Chapter 3 considers plasmas from the point of view of the behavior of a single particle and develops both exact and approximate descriptions for particle motion. In particular, Chapter 3 includes a detailed discussion of the concept of adiabatic invariance with the aim of demonstrating that this important concept is a fundamental property of all nearly periodic Hamiltonian systems and so does not have to be explained anew each time it is encountered in a different context. Chapter 3 also includes a discussion of particle motion in fixed frequency oscillatory fields; this discussion provides a foundation for later analysis of cold plasma waves and wave–particle energy transfer in warm plasma waves.

Chapters 4 to 8 discuss plasma waves; these are not only important in many practical situations, but also provide an excellent way for developing insight and intuition regarding plasma dynamics. Chapter 4 shows how linear wave dispersion relations can be deduced from systems of partial differential equations characterizing a physical system and then presents derivations for the elementary plasma waves, namely Langmuir waves, electromagnetic plasma waves, ion acoustic waves, and Alfvén waves. The beginning of Chapter 5 shows that when a plasma contains groups of particles streaming at different velocities, free energy exists that can drive an instability; the remainder of Chapter 5 then presents Landau damping and instability theory, which reveals that surprisingly strong interactions between waves and particles can lead to either wave damping or wave instability depending on the shape of the velocity distribution of the particles. Chapter 6 describes cold plasma waves in a background magnetic field and discusses the Clemmow–Mullaly–Allis diagram, an elegant categorization scheme for the large number of qualitatively different types of cold plasma waves that exist in a magnetized plasma. Chapter 7 discusses certain additional subtle and practical aspects of wave propagation including propagation in an inhomogeneous plasma and how the energy content of a wave is related to its dispersion relation. Chapter 8 begins by showing that the combination of warm plasma effects and a background magnetic field leads to the existence of the Bernstein wave, an altogether different kind of wave that has an infinite number of branches, and shows how a cold plasma wave can "mode convert" into a Bernstein wave in an inhomogeneous plasma. Chapter 8 concludes with a discussion of drift waves; these are ubiquitous low-frequency modes that have important deleterious consequences for magnetic confinement.

Chapters 9 to 12 describe plasmas from the magnetohydrodynamic point of view. Chapter 9 begins by presenting several basic magnetohydrodynamic concepts (vacuum and force-free fields, magnetic pressure and tension, frozen-in flux, and energy minimization) and then uses these concepts to develop an intuitive understanding for dynamic behavior. Chapter 9 then discusses magnetohydrodynamic equilibria and derives the Grad–Shafranov equation, an equation that depends on the existence of symmetry and characterizes three-dimensional magnetohydrodynamic equilibria. Chapter 9 ends with a discussion on accelerated magnetohydrodynamic flows such as occur in arcs, magnetoplasmadynamic thrusters, and astrophysical jets. Chapter 10 examines the stability of perfectly conducting (i.e., ideal) magnetohydrodynamic equilibria, derives the "energy principle" method for analyzing stability, discusses sausage and kink instabilities, and introduces the concepts of magnetic helicity and force-free equilibria. Chapter 11 examines magnetic helicity from a topological point of view and shows how helicity conservation and energy minimization lead to the Woltjer–Taylor model for magnetohydrodynamic self-organization. Chapter 12 departs from the ideal models presented earlier to discuss magnetic reconnection, a non-ideal behavior that permits the magnetohydrodynamic plasma to alter its topology and relax to a minimum-energy state.

Chapters 13 to 17 consist of various advanced topics. Chapter 13 considers collisions from a Fokker–Planck point of view and is essentially a revisiting of the issues in Chapter 1 using a more sophisticated analysis; the Fokker–Planck model is used to derive a more accurate model for plasma electrical resistivity and also to show the failure of Ohm's law when the electric field exceeds a critical value called the Dreicer limit. Chapter 14 considers two manifestations of wave–particle nonlinearity: (i) quasi-linear velocity-space diffusion due to weak turbulence and (ii) echoes. Quasi-linear diffusion is an extension of Landau damping to the nonlinear regime and shows how wave turbulence interacts with equilibria, while echoes validate in a dramatic fashion basic concepts underlying Landau damping and also instigate some interesting thoughts about the meaning of entropy. Chapter 15 discusses how nonlinear interactions enable energy and momentum to be transferred between waves, categorizes the large number of such wave–wave nonlinear interactions, and shows how these various interactions are all based on a few fundamental nonlinear coupling mechanisms. Chapter 16 discusses one-component plasmas (pure electron or pure ion plasmas) and shows how these plasmas have behaviors differing from conventional two-component, electron–ion plasmas. Chapter 17 discusses dusty plasmas, which are three-component plasmas (electrons, ions, and dust grains), and shows how the addition of a third component also introduces new behaviors, including the possibility of the dusty plasma condensing into a crystal. The analysis of condensation involves revisiting the Debye shielding concept and so corresponds in a sense to having the book end on the same note it started on.

Three appendices have been included: Appendix A describes an intuitive method for deriving the standard vector calculus identities, Appendix B uses vector calculus identities to provide a quick derivation of vector calculus operators in curvilinear coordinates, and Appendix C provides both a short list of physical constants and a summary of frequently used formulae with page references to the locations where these formulae were discussed.

I would like to extend my grateful appreciation to Professor Michael Brown at Swarthmore College for providing helpful feedback obtained from his using a draft version in a seminar course at Swarthmore and to Professor Roy Gould at Caltech for providing helpful insight into both the dynamics of non-neutral plasmas and the energetics of Debye shielding. I would also like to thank graduate students Deepak Kumar and Gunsu Yun for their careful scrutiny of the final drafts of the manuscript and for pointing out both ambiguities in presentation and typographical errors. In addition, I would like to thank the many students who provided useful feedback on earlier drafts of this work when it was in the form of lecture notes. Finally, I would like to acknowledge and thank my own mentors and colleagues who have introduced me to the many fascinating ideas constituting the discipline of plasma physics and the many scientists whose hard work over many decades has led to the development of this discipline.

1

Basic concepts

1.1 History of the term "plasma"

In the mid nineteenth century the Czech physiologist Jan Evangelista Purkinje introduced use of the Greek word *plasma* (meaning "formed" or "molded") to denote the clear fluid that remains after removal of all the corpuscular material in blood. About half a century later, the American scientist Irving Langmuir proposed in 1922 that the electrons, ions, and neutrals in an ionized gas could similarly be considered as corpuscular material entrained in some kind of fluid medium and called this entraining medium plasma. However it turned out that, unlike blood where there really is a fluid medium carrying the corpuscular material, there actually is no "fluid medium" entraining the electrons, ions, and neutrals in an ionized gas. Ever since, plasma scientists have had to explain to friends and acquaintances that they were not studying blood!

1.2 Brief history of plasma physics

In the 1920s and 1930s a few isolated researchers, each motivated by a specific practical problem, began the study of what is now called plasma physics. This work was mainly directed towards understanding (i) the effect of ionospheric plasma on *long-distance short-wave radio* propagation and (ii) *gaseous electron tubes* used for rectification, switching, and voltage regulation in the pre-semiconductor era of electronics. In the 1940s Hannes Alfvén developed a theory of hydromagnetic waves (now called Alfvén waves) and proposed that these waves would be important in astrophysical plasmas. In the early 1950s large-scale plasma physics based *magnetic fusion energy* research started simultaneously in the USA, Britain, and the then Soviet Union. Since this work was an offshoot of thermonuclear weapon research, it was initially classified but, because of scant progress in each country's effort and the realization that controlled fusion research was unlikely to be of military value, all three countries declassified their efforts in 1958 and have co-operated since. Many other countries now participate in fusion research as well.

Fusion progress was slow through most of the 1960s, but by the end of that decade the empirically developed Russian *tokamak* configuration began producing plasmas with parameters far better than the lackluster results of the previous two decades. By the 1970s and 1980s many tokamaks with progressively improved performance were constructed and at the end of the twentieth century fusion break-even had nearly been achieved in tokamaks. International agreement was reached in the early twenty-first century to build the International Thermonuclear Experimental Reactor (ITER), a break-even tokamak designed to produce 500 megawatts of fusion output power. Non-tokamak approaches to fusion have also been pursued with varying degrees of success; many involve magnetic confinement schemes related to that used in tokamaks. In contrast to fusion schemes based on magnetic confinement, inertial confinement schemes were also developed in which high-power lasers or similarly intense power sources bombard millimeter-diameter pellets of thermonuclear fuel with ultra-short, extremely powerful pulses of strongly focused directed energy. The intense incident power causes the pellet surface to ablate and, in so doing, act like a rocket exhaust pointing radially outwards from the pellet. The resulting radially inwards force compresses the pellet adiabatically, making it both denser and hotter; with sufficient adiabatic compression, fusion ignition conditions are predicted to be achieved.

Simultaneously with the fusion effort, there has been an equally important and extensive study of space plasmas. Measurements of near-Earth space plasmas, such as the aurora and the ionosphere, have been obtained by ground-based instruments since the late nineteenth century. Space plasma research was greatly stimulated when it became possible to use spacecraft to make routine *in situ* plasma measurements of the Earth's *magnetosphere*, the *solar wind*, and the magnetospheres of other planets. Additional interest has resulted from ground-based and spacecraft measurements of topologically complex, dramatic structures sometimes having explosive dynamics in the *solar corona*. Using radio telescopes, optical telescopes, Very Long Baseline Interferometry, and most recently the Hubble and Spitzer spacecraft, large numbers of *astrophysical jets* shooting out from magnetized objects such as stars, active galactic nuclei, and black holes have been observed. Space plasmas often behave in a manner qualitatively similar to laboratory plasmas, but have a much grander scale.

Since the 1960s an important effort has been directed towards using plasmas for *space propulsion*. Plasma thrusters have been developed ranging from small *ion thrusters* for spacecraft attitude correction to powerful *magnetoplasmadynamic thrusters* that – given an adequate power supply – could be used for interplanetary missions. Plasma thrusters are now in use on some spacecraft and are under serious consideration for new and more ambitious spacecraft designs.

Starting in the late 1980s a new application of plasma physics appeared – *plasma processing* – a critical aspect of the fabrication of the tiny, complex integrated circuits used in modern electronic devices. This application is now of great economic importance.

In the 1980s investigations began on *non-neutral plasmas*; these mimic the equations of incompressible hydrodynamics and so provide a compelling analog computer for problems in incompressible hydrodynamics. Another application of non-neutral plasmas is as a means to store large quantities of positrons. In the 1990s studies began on *dusty plasmas*. Dust grains immersed in a plasma can become electrically charged and then act as an additional charged particle species. Because dust grains are massive compared to electrons or ions and can be charged to varying amounts, new physical behavior occurs that is sometimes an extension of what happens in a regular plasma and sometimes altogether new. Both non-neutral and dusty plasmas can also form bizarre, strongly coupled collective states where the plasma resembles a solid (e.g., forms quasi-crystalline structures).

In addition to the above activities there have been continuing investigations of industrially relevant plasmas such as *arcs*, *plasma torches*, and *laser plasmas*. In particular, approximately 40% of the steel manufactured in the United States is recycled in huge electric arc furnaces capable of melting over 100 tons of scrap steel in a few minutes. Plasma displays are used for flat-panel televisions and of course there are naturally occurring terrestrial plasmas such as *lightning*.

1.3 Plasma parameters

Three fundamental parameters[1] characterize a plasma:

1. the particle density n (measured in particles per cubic meter),
2. the temperature T of each species (usually measured in eV, where 1 eV=11 605 K),
3. the steady-state magnetic field B (measured in Tesla).

A host of subsidiary parameters (e.g., Debye length, Larmor radius, plasma frequency, cyclotron frequency, thermal velocity) can be derived from these three fundamental parameters. For partially ionized plasmas, the fractional ionization and cross-sections of neutrals are also important.

[1] In older plasma literature, density and magnetic fields are often expressed in cgs units, i.e., densities are given in particles per cubic centimeter, and magnetic fields are given in Gauss. Since the 1990s there has been general agreement to use SI units when possible. SI units have the distinct advantage that electrical units are in terms of familiar quantities such as amps, volts, and ohms and so a model prediction in SI units can much more easily be compared to the results of an experiment than a prediction given in cgs units.

1.4 Examples of plasmas

1.4.1 Non-fusion terrestrial plasmas

It takes considerable resources and skill to make a hot, fully ionized plasma and so, except for the specialized fusion plasmas, most terrestrial plasmas (e.g., arcs, neon signs, fluorescent lamps, processing plasmas, welding arcs, and lightning) have electron temperatures of a few eV and, for reasons given later, have ion temperatures that are colder, often at room temperature. These "everyday" plasmas usually have no imposed steady-state magnetic field and do not produce significant self-magnetic fields. Typically, these plasmas are weakly ionized and dominated by collisional and radiative processes. Densities in these plasmas range from 10^{14} to $10^{22}\,\mathrm{m}^{-3}$ (for comparison, the density of air at STP is $2.7\times 10^{25}\,\mathrm{m}^{-3}$).

1.4.2 Fusion-grade terrestrial plasmas

Using carefully designed, expensive, and often large plasma confinement systems together with high heating power and obsessive attention to purity, fusion researchers have succeeded in creating fully ionized hydrogen or deuterium plasmas which attain temperatures ranging from tens of eV to *tens of thousands* of eV. In typical magnetic confinement devices (e.g., tokamaks, stellarators, reversed field pinches, mirror devices) an externally produced 1–10 T magnetic field of carefully chosen geometry is imposed on the plasma. Magnetic confinement devices generally have densities in the range 10^{19}–$10^{21}\,\mathrm{m}^{-3}$. Plasmas used in inertial fusion are much more dense; the goal is to attain for a brief instant densities one or two orders of magnitude larger than solid density ($\sim 10^{28}\,\mathrm{m}^{-3}$).

1.4.3 Space plasmas

The parameters of these plasmas cover an enormous range. For example, the density of space plasmas varies from $10^{6}\,\mathrm{m}^{-3}$ in interstellar space to $10^{20}\,\mathrm{m}^{-3}$ in the solar atmosphere. Most of the astrophysical plasmas that have been investigated have temperatures in the range of 1–100 eV and these plasmas are usually fully ionized.

1.5 Logical framework of plasma physics

Plasmas are complex and exist in a wide variety of situations differing by many orders of magnitude. An important situation where plasmas do not normally exist is ordinary human experience. Consequently, people do not have the sort of intuition for plasma behavior that they have for solids, liquids, or gases. Although plasma behavior seems non- or counter-intuitive at first, with suitable effort a good intuition for plasma behavior can be developed. This intuition can be helpful for making initial predictions about plasma behavior in a new situation, because plasmas have the remarkable property of being extremely *scalable*; i.e., the same qualitative phenomena often occur in plasmas differing by many orders of magnitude. Plasma physics is usually not a precise science. It is rather a web of overlapping points of view, each modeling a limited range of behavior. Understanding of plasmas is developed by studying these various points of view, all the while keeping in mind the linkages between the points of view.

Plasma dynamics is determined by the *self-consistent interaction between electromagnetic fields and statistically large numbers of charged particles* as shown schematically in Fig. 1.1. In principle, the time evolution of a plasma can be calculated as follows:

1. given the trajectory $\mathbf{x}_j(t)$ and velocity $\mathbf{v}_j(t)$ of each and every particle j, the electric field $\mathbf{E}(\mathbf{x},t)$ and magnetic field $\mathbf{B}(\mathbf{x},t)$ can be evaluated using Maxwell's equations and simultaneously;
2. given the instantaneous electric and magnetic fields $\mathbf{E}(\mathbf{x},t)$ and $\mathbf{B}(\mathbf{x},t)$, the forces on each and every particle j can be evaluated using the Lorentz equation and then used to update the trajectory $\mathbf{x}_j(t)$ and velocity $\mathbf{v}_j(t)$ of each particle.

While this approach is conceptually easy to understand, it is normally impractical to implement because of the extremely large number of particles and, to a lesser extent, because of the complexity of the electromagnetic field. To gain a practical understanding, we therefore do not attempt to evaluate the entire complex behavior all at once but, instead, study plasmas by considering specific phenomena. For each phenomenon under immediate consideration, appropriate simplifying approximations are made, leading to a more tractable problem and hopefully revealing the essence of what is going on. A situation where a certain set of approximations is valid and provides a self-consistent description is called a regime. There are a number of general categories of simplifying approximations, namely:

1. *Approximations involving the electromagnetic field:*
   (a) assuming the magnetic field is zero (unmagnetized plasma);
   (b) assuming there are no inductive electric fields (electrostatic approximation);

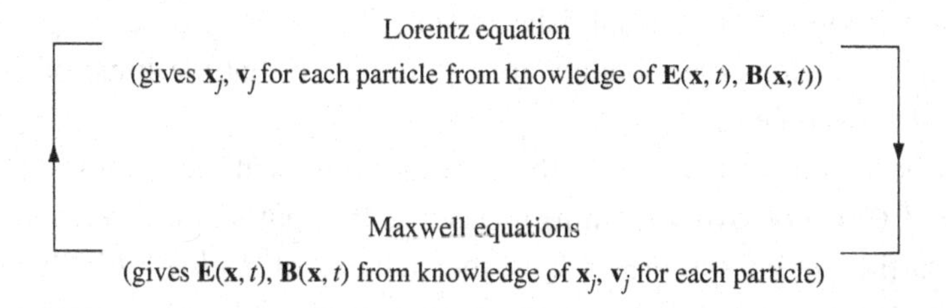

Fig. 1.1 Interrelation between Maxwell's equations and the Lorentz equation.

(c) neglecting the displacement current in Ampère's law (suitable for phenomena having characteristic velocities much slower than the speed of light);
(d) assuming that all magnetic fields are produced by conductors external to the plasma;
(e) various assumptions regarding geometric symmetry (e.g., spatially uniform, uniform in a particular direction, azimuthally symmetric about an axis).

2. *Approximations involving the particle description:*
   (a) Averaging of the Lorentz force over some sub-group of particles:
      i. Vlasov theory: average over all particles of a given species (electrons or ions) having the same velocity at a given location and characterize the plasma using the distribution function $f_\sigma(\mathbf{x}, \mathbf{v}, t)$, which gives the density of particles of species σ having velocity $\mathbf{v}$ at position $\mathbf{x}$ at time t;
      ii. Two-fluid theory: average velocities over all particles of a given species at a given location and characterize the plasma using the species density $n_\sigma(\mathbf{x}, t)$, mean velocity $\mathbf{u}_\sigma(\mathbf{x}, t)$, and pressure $P_\sigma(\mathbf{x}, t)$ defined relative to the species mean velocity.
      iii. Magnetohydrodynamic theory: average momentum over all particles of all species and characterize the plasma using the center-of-mass density $\rho(\mathbf{x}, t)$, center-of-mass velocity $\mathbf{U}(\mathbf{x}, t)$, and pressure $P(\mathbf{x}, t)$ defined relative to the center-of-mass velocity.
   (b) Assumptions about time (e.g., assume the phenomenon under consideration is fast or slow compared to some characteristic frequency of the particles such as the cyclotron frequency).
   (c) Assumptions about space (e.g., assume the scale length of the phenomenon under consideration is large or small compared to some characteristic plasma length such as the cyclotron radius).
   (d) Assumptions about velocity (e.g., assume the phenomenon under consideration is fast or slow compared to the thermal velocity $v_{T\sigma}$ of a particular species σ).

The large number of possible permutations and combinations that can be constructed from the above list means that there will be a large number of regimes. Since developing an intuitive understanding requires making approximations of the sort listed above and since these approximations lack an obvious hierarchy, it is not clear where to begin. In fact, as sketched in Fig. 1.2, the models for particle motion (Vlasov, two-fluid, MHD) involve a circular argument. Wherever we start on this circle, we are always forced to take at least one new concept on trust and hope that its validity will be established later. The reader is encouraged to refer to Fig. 1.2 as its various components are examined so that the logic of this circle will eventually become clear.

Because the argument is circular, the starting point is at the author's discretion, and for good (but not overwhelming) reasons, this author has decided that the optimum starting point on Fig. 1.2 is the subject of *Debye shielding*. Debye concepts, the Rutherford model for how charged particles scatter from each other,

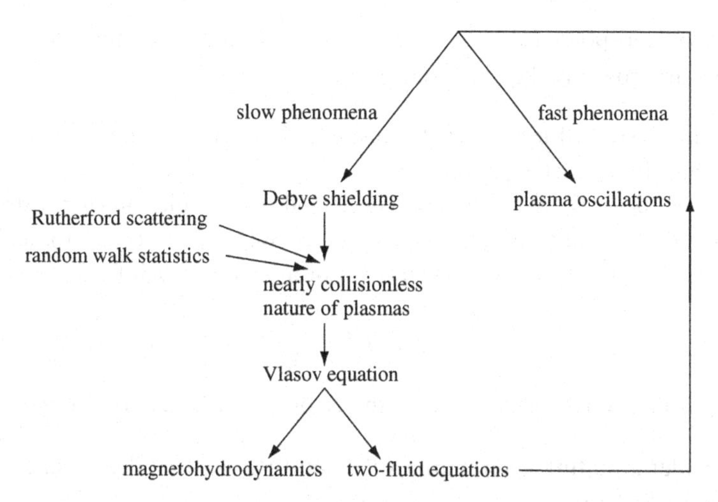

Fig. 1.2 Hierarchy of models of plasmas showing circular nature of logic.

and some elementary statistics will be combined to construct an argument showing that plasmas are *weakly collisional*. We will then discuss *phase-space* concepts and introduce the *Vlasov equation* for the phase-space density. Averages of the Vlasov equation will provide *two-fluid equations* and also the *magnetohydrodynamic* (MHD) equations. Having established this framework, we will then return to study features of these points of view in more detail, often tying up loose ends that occurred in our initial derivation of the framework. Somewhat separate from the study of Vlasov, two-fluid, and MHD equations (which all attempt to give a self-consistent picture of the plasma) is the study of *single particle orbits in prescribed fields*. This provides useful intuition on the behavior of a typical particle in a plasma, and can provide important inputs or constraints for the self-consistent theories.

1.6 Debye shielding

We begin our study of plasmas by examining Debye shielding, a concept originating from the theory of liquid electrolytes (Debye and Hückel 1923). Consider a finite-temperature plasma consisting of a statistically large number of electrons and ions and assume that the ion and electron densities are initially equal and spatially uniform. As will be seen later, the ions and electrons need not be in thermal equilibrium with each other, and so the ions and electrons will be allowed to have separate temperatures denoted by T_i, T_e.

Since the ions and electrons have random thermal motion, thermally induced perturbations about the equilibrium will cause small, transient spatial variations of the electrostatic potential ϕ. In the spirit of circular argument the following assumptions are now invoked without proof:

1. The plasma is assumed to be nearly collisionless so that collisions between particles may be neglected to first approximation.
2. Each species, denoted as σ, may be considered as a "fluid" having a density n_σ, a temperature T_σ, a pressure $P_\sigma = n_\sigma \kappa T_\sigma$ (κ is Boltzmann's constant), and a mean velocity $\mathbf{u}_\sigma$ so that the collisionless equation of motion for each fluid is

$$m_\sigma \frac{\mathrm{d}\mathbf{u}_\sigma}{\mathrm{d}t} = q_\sigma \mathbf{E} - \frac{1}{n_\sigma} \nabla P_\sigma, \tag{1.1}$$

where m_σ is the particle mass, q_σ is the charge of a particle, and $\mathbf{E}$ is the electric field.

Now consider a perturbation with a sufficiently *slow* time dependence to allow the following assumptions:

1. The inertial term $\sim \mathrm{d}/\mathrm{d}t$ on the left-hand side of Eq. (1.1) is negligible and may be dropped.
2. Inductive electric fields are negligible so the electric field is almost entirely electrostatic, i.e., $\mathbf{E} \sim -\nabla\phi$.
3. All temperature gradients are smeared out by thermal particle motion so that the temperature of each species is spatially uniform.
4. The plasma remains in thermal equilibrium throughout the perturbation (i.e., it can always be characterized by a temperature).

Invoking these approximations, Eq. (1.1) reduces to

$$0 \approx -n_\sigma q_e \nabla\phi - \kappa T_\sigma \nabla n_\sigma, \tag{1.2}$$

a simple balance between the force due to the electrostatic electric field and the force due to the isothermal pressure gradient. Equation (1.2) is readily solved to give the *Boltzmann relation*

$$n_\sigma = n_{\sigma 0} \exp(-q_\sigma \phi / \kappa T_\sigma), \tag{1.3}$$

where $n_{\sigma 0}$ is a constant. It is important to emphasize that the Boltzmann relation results from the assumption that the perturbation is *very slow;* if this is not the case, then inertial effects, inductive electric fields, or temperature gradient effects will cause the plasma to have a completely different behavior from the Boltzmann relation. Situations exist where this "slowness" assumption is valid for electron dynamics but not for ion dynamics, in which case the Boltzmann condition will apply only to the electrons but not to the ions (the converse situation does not normally occur because ions, being heavier, are always more sluggish than electrons and so it is only possible for a phenomenon to appear slow to electrons but not to ions).

Let us now imagine slowly inserting a single additional particle (so-called "test" particle) with charge q_T into an initially unperturbed, spatially uniform, neutral plasma. To keep the algebra simple, we define the origin of our coordinate system to be at the location of the test particle. Before insertion of the test particle, the plasma potential was $\phi = 0$ everywhere because the ion and electron densities were spatially uniform and equal, but now the ions and electrons will be perturbed because of their interaction with the test particle. Particles having the same polarity as q_T will be slightly repelled whereas particles of opposite polarity will be slightly attracted. The slight displacements resulting from these repulsions and attractions will result in a small, but finite, potential in the plasma. This potential will be the superposition of the test particle's own potential and the potential of the plasma particles that have moved slightly in response to the test particle.

This slight displacement of plasma particles is called *shielding* or *screening* of the test particle because the displacement tends to reduce the effectiveness of the test particle field. To see this, suppose the test particle is a positively charged ion. When immersed in the plasma it will attract nearby electrons and repel nearby ions; the net result is an effectively negative charge cloud surrounding the test particle. An observer located far from the test particle and its surrounding cloud would see the combined potential of the test particle and its associated cloud. Because the cloud has the opposite polarity to the test particle, the cloud potential will partially cancel (i.e., *shield* or *screen*) the test particle potential.

Screening is calculated using Poisson's equation with the source terms being the test particle and its associated cloud. The cloud contribution is determined using the Boltzmann relation for the particles that participate in the screening. This is a "self-consistent" calculation for the potential because the shielding cloud is affected by its self-potential.

Thus, Poisson's equation becomes

$$\nabla^2 \phi = -\frac{1}{\varepsilon_0}\left[q_T \delta(\mathbf{r}) + \sum_\sigma n_\sigma(\mathbf{r}) q_\sigma \right], \tag{1.4}$$

where the term $q_T \delta(\mathbf{r})$ on the right-hand side represents the charge density due to the test particle and the term $\sum_\sigma n_\sigma(\mathbf{r}) q_\sigma$ represents the charge density of all plasma particles that participate in the screening (i.e., everything except the test particle). Before the test particle was inserted $\sum_{\sigma=i,e} n_\sigma(\mathbf{r}) q_\sigma$ vanished because the plasma was assumed to be initially neutral.

Since the test particle was inserted slowly, the plasma response will be Boltzmann-like and we may substitute for $n_\sigma(\mathbf{r})$ using Eq. (1.3). Furthermore, because the perturbation due to a single test particle is infinitesimal, we can safely assume that $|q_\sigma \phi| \ll \kappa T_\sigma$, in which case Eq. (1.3) becomes simply $n_\sigma / n_{\sigma 0} \approx 1 - q_\sigma \phi / \kappa T_\sigma$ and so Eq. (1.4) becomes

$$\nabla^2 \phi = -\frac{1}{\varepsilon_0} \left[q_T \delta(\mathbf{r}) + \left(1 - \frac{q_e \phi}{\kappa T_e}\right) n_{e0} q_e + \left(1 - \frac{q_i \phi}{\kappa T_i}\right) n_{i0} q_i \right]. \tag{1.5}$$

The assumption of initial neutrality means that $n_{e0} q_e + n_{i0} q_i = 0$, in which case Eq. (1.5) reduces to

$$\nabla^2 \phi - \frac{1}{\lambda_D^2} \phi = -\frac{q_T}{\varepsilon_0} \delta(\mathbf{r}), \tag{1.6}$$

where the effective Debye length is defined by

$$\frac{1}{\lambda_D^2} = \sum_\sigma \frac{1}{\lambda_\sigma^2} \tag{1.7}$$

and the species Debye length λ_σ is

$$\lambda_\sigma^2 = \frac{\varepsilon_0 \kappa T_\sigma}{n_{\sigma 0} q_\sigma^2}. \tag{1.8}$$

The second term on the left-hand side of Eq. (1.6) is just the negative of the shielding cloud charge density. The summation in Eq. (1.7) is over all species that participate in the shielding. Since ions cannot move fast enough to keep up with an electron test charge, which would be moving at the nominal electron thermal velocity, the shielding of electrons is only by other electrons, whereas the shielding of ions is by both ions and electrons.

Equation (1.6) can be solved using standard mathematical techniques (cf. assignments) to give

$$\phi(\mathbf{r}) = \frac{q_T}{4\pi\varepsilon_0 r} \mathrm{e}^{-r/\lambda_D}; \tag{1.9}$$

this is sometimes called the Yukawa potential. For $r \ll \lambda_D$ the potential $\phi(\mathbf{r})$ is identical to the potential of a test particle in vacuum, whereas for $r \gg \lambda_D$ the test charge is completely screened by its surrounding shielding cloud. The nominal radius of the shielding cloud is λ_D. Because the test particle is completely screened for $r \gg \lambda_D$, the total shielding cloud charge is equal in magnitude to the charge on the test particle and opposite in sign. This test-particle/shielding-cloud analysis makes sense only if there is a macroscopically large number of plasma particles in the shielding cloud; i.e., the analysis makes sense only if $4\pi n_0 \lambda_D^3 / 3 \gg 1$. This will be seen later to be the condition for the plasma to be nearly collisionless and so validate assumption 1 at the top of p. 8.

In order for shielding to be a relevant issue, the Debye length must be small compared to the overall dimensions of the plasma, because otherwise no point in the plasma could be outside the shielding cloud. Finally, it should be realized that *any* particle could have been construed as being "the" test particle and so we conclude that the time-averaged effective potential of any selected particle in the plasma is given by Eq. (1.9). (From a statistical point of view, a selected particle is no longer assumed to have a random thermal velocity, in which case its effective potential results from the summation of the non-random, static potential associated with its own charge and a screening potential, which is the time-average of a rapidly changing potential associated with the fleeting, random thermal motions of all other particles.)

1.7 Quasi-neutrality

The Debye shielding analysis assumed that the plasma was initially neutral, i.e., that the initial electron and ion densities were equal. We now demonstrate that if the Debye length is a microscopic length, then it is indeed an excellent assumption that plasmas remain extremely close to neutrality, while not being exactly neutral. It is found that the electrostatic electric field associated with any reasonable configuration is easily produced by having only a tiny deviation from perfect neutrality. This tendency to be *quasi-neutral* occurs because a conventional plasma does not have sufficient internal energy to become substantially non-neutral over distances exceeding a Debye length (there do exist non-neutral plasmas that violate this concept, but these involve rotation of plasma in a background magnetic field, which effectively plays the neutralizing role of ions in a conventional plasma).

To prove the assertion that plasmas tend to be quasi-neutral, we consider an initially neutral plasma with temperature T and calculate the largest radius sphere that could spontaneously become depleted of electrons due to thermal fluctuations. Let r_{max} be the radius of this presumed sphere. Complete depletion (i.e., maximum non-neutrality) would occur if a random thermal fluctuation caused all the electrons originally in the sphere to vacate the volume of the sphere and move to its surface. The electrons would have to come to rest on the surface of the presumed sphere because if they did not, they would still have available kinetic energy, which could then be used to move out towards an even larger radius, violating the assumption that the sphere was the largest radius sphere that could become fully depleted of electrons. This situation is of course extremely artificial and likely to be so rare as to be essentially negligible because it requires all the electrons to be moving radially relative to some origin. In reality, the electrons would be moving in random directions.

When the electrons exit the sphere they leave behind an equal number of ions. The remnant ions produce a radial electric field, which pulls the electrons back towards the center of the sphere. One way of calculating the energy stored in this system is to calculate the work done by the electrons as they leave the sphere and collect on the surface, but a simpler way is to calculate the energy stored in the electrostatic electric field produced by the ions remaining in the sphere. This electrostatic energy did not exist when the electrons were initially in the sphere and balanced the ion charge and so it must be equivalent to the work done by the electrons on leaving the sphere.

The energy density of an electric field is $\varepsilon_0 E^2/2$ and, because of the spherical symmetry assumed here, the electric field produced by the remnant ions must be in the radial direction. The ion charge in a sphere of radius r is $Q = 4\pi n e r^3/3$ and so, after all the electrons have vacated the sphere, the electric field at radius r is therefore $E_r = Q/4\pi\varepsilon_0 r^2 = ner/3\varepsilon_0$. The electrostatic field energy in the ion-filled, electron-depleted sphere of radius r_{max} is thus

$$W = \int_0^{r_{max}} \frac{\varepsilon_0 E_r^2}{2} 4\pi r^2 \mathrm{d}r = \pi r_{max}^5 \frac{2n_e^2 e^2}{45\varepsilon_0}. \tag{1.10}$$

Equating this potential energy to the initial electron thermal kinetic energy, $W_{kinetic}$, gives

$$\pi r_{max}^5 \frac{2n_e^2 e^2}{45\varepsilon_0} = \frac{3}{2} n\kappa T \times \frac{4}{3}\pi r_{max}^3, \tag{1.11}$$

which may be solved to give

$$r_{max}^2 = 45 \frac{\varepsilon_0 \kappa T}{n_e e^2} \tag{1.12}$$

so that $r_{max} \simeq 7\lambda_D$.

Thus, the largest spherical volume that could spontaneously become depleted of electrons has a radius of a few Debye lengths. Since the probability of all electrons in a sphere simultaneously having radially outwards velocities at some instant is essentially zero, this electron depletion of a sphere is extremely unlikely and so regions of non-neutrality will be much smaller than given by Eq. (1.12). We conclude that *plasma is quasi-neutral over scale lengths much larger than the Debye length.* When a biased electrode such as a wire probe is inserted into a plasma, the plasma screens the field due to the potential on the electrode in the same way that the test charge potential was screened. The screening region is called the *sheath,* which is a region of non-neutrality having an extent of the order of a Debye length.

1.8 Small- vs. large-angle collisions in plasmas

We now consider what happens to the momentum and energy of a test particle of charge q_T and mass m_T that is injected with velocity $\mathbf{v}_T$ into a plasma. This test particle will make a sequence of random collisions with the plasma particles (called "field" particles and denoted by subscript F); these collisions will alter both the momentum and the energy of the test particle. A collision is characterized by a deflection (also called scattering) of the test particle by an angle θ from its original direction of motion. The scattering angle θ depends on a number of parameters, including the test and target particle masses, the speed of the test particle relative to the target particle, and whether the collision is a bull's-eye collision or merely a grazing collision. Determining this functional dependence is called the Rutherford scattering problem, a standard problem in classical mechanics.

Solution of the Rutherford scattering problem in the center-of-mass frame shows (Assignment 1, this chapter) that the scattering angle θ is given by

$$\tan\left(\frac{\theta}{2}\right) = \frac{q_T q_F}{4\pi\varepsilon_0 b \mu v_0^2} \sim \frac{\text{Coulomb interaction energy}}{\text{kinetic energy}}, \tag{1.13}$$

where $\mu^{-1} = m_T^{-1} + m_F^{-1}$ is the reduced mass, b is the impact parameter shown in Fig. 1.3, and v_0 is the initial relative velocity. It is useful to separate scattering events (i.e., collisions) into two approximate categories, namely (1) large-angle collisions where $\pi/2 \leq \theta \leq \pi$ and (2) small-angle (grazing) collisions where $\theta \ll \pi/2$.

Let us denote $b_{\pi/2}$ as the impact parameter for 90° collisions; from Eq. (1.13) this is

$$b_{\pi/2} = \frac{q_T q_F}{4\pi\varepsilon_0 \mu v_0^2} \tag{1.14}$$

and corresponds to the radius of the inner (small) shaded circle in Fig. 1.3. Large-angle scatterings will occur if the test particle is incident anywhere within this

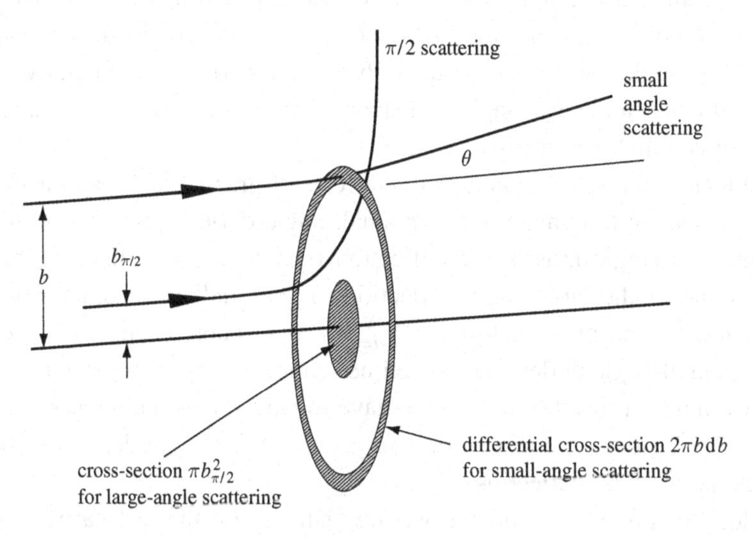

Fig. 1.3 Differential scattering cross-sections for large and small deflections.

circle and so the total cross-section for *all* large-angle collisions is

$$\begin{aligned}\sigma_{\text{large}} &\approx \pi b_{\pi/2}^2 \\ &= \pi\left(\frac{q_T q_F}{4\pi\varepsilon_0 \mu v_0^2}\right)^2.\end{aligned} \tag{1.15}$$

Grazing (small-angle) collisions occur when the test particle impinges *outside* the shaded circle and so *occur much more frequently* than large-angle collisions. Although each grazing collision does not deflect the test particle very much, there are far more grazing collisions than large-angle collisions and so it is important to compare the *cumulative* effect of grazing collisions with the *cumulative* effect of large-angle collisions.

To make matters even more complicated, the effective cross-section of grazing collisions depends on impact parameter, since the larger b is, the smaller the deflection. To take this weighting of impact parameters into account, the area outside the shaded circle is subdivided into a set of concentric annuli, called differential cross-sections. If the test particle impinges on the differential cross-section having radius between b and $b+\mathrm{d}b$, then the test particle will be deflected by an angle lying between $\theta(b)$ and $\theta(b+\mathrm{d}b)$ as determined by Eq. (1.13). The area of the differential cross-section is $2\pi b \mathrm{d}b$, which is therefore the effective cross-section for scattering between $\theta(b)$ and $\theta(b+\mathrm{d}b)$. Because the azimuthal angle about the direction of incidence is random, the simple average of N small-angle deflections vanishes, i.e., $N^{-1}\sum_{i=1}^{N}\theta_i = 0$, where θ_i is the deflection due to the ith collision and N is a large number.

Random walk statistics must therefore be used to describe the cumulative effect of a large number of small-angle deflections and so we will use the *square* of the deflection angle, i.e., θ_i^2, as the quantity for comparing the cumulative effects of small (grazing) and large-angle collisions. Thus, scattering (i.e., a sequence of random deflections) is a diffusive process.

To compare the respective cumulative effects of grazing and large-angle collisions, we calculate how many random small-angle deflections must occur to be equivalent to a single large-angle deflection (i.e., $\theta_{\text{large}}^2 \approx 1$); here we pick the nominal value of the large-angle deflection to be 1 radian. In other words, we ask what must N be in order to have $\sum_{i=1}^{N}\theta_i^2 \approx 1$, where each θ_i represents an individual small-angle deflection (sometimes called a "scattering event"). Equivalently, we may ask what time t do we have to wait for the cumulative effect of the grazing collisions on a test particle to give an effective deflection equivalent to a single large-angle deflection.

To calculate this, let us imagine we are "sitting" on the test particle. In this test particle frame the field particles approach the test particle with the velocity

v_{rel}, and so the apparent flux of field particles is $\Gamma = n_F v_{rel}$, where v_{rel} is the relative velocity between the test and field particles. The number of small-angle scattering events in time t for impact parameters between b and $b+\mathrm{d}b$ is $\Gamma t 2\pi b \mathrm{d}b$ and so the time required for the cumulative effect of small-angle collisions to be equivalent to a large-angle collision is given by

$$1 \approx \sum_{i=1}^{N} \theta_i^2 = \Gamma t \int 2\pi b \mathrm{d}b [\theta(b)]^2. \tag{1.16}$$

The definitions of scattering theory show (see Assignment 9) that $\sigma\Gamma = t^{-1}$, where σ is the cross-section for an event and t is the time one has to wait for the event to occur. Substituting for Γt in Eq. (1.16) gives the cross-section σ^* for the cumulative effect of grazing collisions to be equivalent to a single large-angle scattering event,

$$\sigma^* = \int 2\pi b \mathrm{d}b [\theta(b)]^2. \tag{1.17}$$

The appropriate lower limit for the integral in Eq. (1.17) is $b_{\pi/2}$, since impact parameters smaller than this value produce large-angle collisions. What should the upper limit of the integral be? We recall from our Debye discussion that the field of the scattering center is *screened out* for distances greater than λ_D. Hence, small-angle collisions occur *only* for impact parameters in the range $b_{\pi/2} < b < \lambda_D$ because the scattering potential is non-existent for distances larger than λ_D.

For small-angle collisions, Eq. (1.13) gives

$$\theta(b) = \frac{q_T q_F}{2\pi\varepsilon_0 \mu v_0^2 b} \tag{1.18}$$

so Eq. (1.17) becomes

$$\sigma^* = \int_{b_{\pi/2}}^{\lambda_D} 2\pi b \mathrm{d}b \left(\frac{q_T q_F}{2\pi\varepsilon_0 \mu v_0^2 b} \right)^2 \tag{1.19}$$

or

$$\sigma^* = 8 \ln \left(\frac{\lambda_D}{b_{\pi/2}} \right) \sigma_{\text{large}}. \tag{1.20}$$

Thus, the cross-section σ^* will significantly exceed σ_{large} if $\lambda_D / b_{\pi/2} \gg 1$. Since $b_{\pi/2} = 1/2n\lambda_D^2$, the condition $\lambda_D \gg b_{\pi/2}$ is equivalent to $n\lambda_D^3 \gg 1$, which is just the criterion for there to be a large number of particles in a sphere having a radius λ_D (a so-called *Debye sphere*). This was the condition for the Debye shielding cloud argument to make sense. We conclude that the criterion for an ionized gas to behave as a plasma (i.e., Debye shielding is important and grazing collisions dominate large-angle collisions) is the condition $n\lambda_D^3 \gg 1$. For most plasmas $n\lambda_D^3$ is a large number with natural logarithm of order 10; typically, when making rough estimates of σ^*, one uses $\ln(\lambda_D/b_{\pi/2}) \approx 10$. The reader may have developed a concern about the seeming arbitrary nature of the choice of $b_{\pi/2}$ as the "dividing line" between large-angle and grazing collisions. This arbitrariness is of no consequence since the logarithmic dependence means that any other choice having the same order of magnitude for the "dividing line" would give essentially the same result.

By substituting for $b_{\pi/2}$, the cross-section can be rewritten as

$$\sigma^* = \frac{1}{2\pi}\left(\frac{q_T q_F}{\varepsilon_0 \mu v_0^2}\right)^2 \ln\left(\frac{\lambda_D}{b_{\pi/2}}\right). \tag{1.21}$$

Thus, σ^* decreases approximately as the fourth power of the relative velocity. In a hot plasma where v_0 is large, σ^* will be *very small* and so scattering by Coulomb collisions is often much less important than other phenomena. A useful way to decide whether Coulomb collisions are important is to compare the collision frequency $\nu = \sigma^* n v$ with the frequency of other effects, or equivalently the mean free path of collisions, $l_{mfp} = 1/\sigma^* n$, with the characteristic length of other effects. If the collision frequency is small, or the mean free path is large (in comparison to other effects), collisions may be neglected to first approximation, in which case the plasma under consideration is called a collisionless or "ideal" plasma. The effective Coulomb cross-section σ^* and its related parameters ν and l_{mfp} can be used to evaluate transport properties such as electrical resistivity, mobility, and diffusion.

1.9 Electron and ion collision frequencies

One of the fundamental physical constants influencing plasma behavior is the ion to electron mass ratio. The large value of this ratio often causes electrons and ions to experience qualitatively distinct dynamics. In some situations, one species may determine the essential character of a particular plasma behavior while the other species has little or no effect. Let us now examine how mass ratio affects:

1. momentum change (scattering) of a given incident particle due to collision between
   (a) like particles (i.e., electron-electron or ion-ion collisions, denoted *ee* or *ii*),
   (b) unlike particles (i.e., electrons scattering from ions denoted *ei* or ions scattering from electrons denoted *ie*);
2. kinetic energy change (scattering) of a given incident particle due to collisions between like or unlike particles.

Momentum scattering is characterized by the time required for collisions to deflect the incident particle by an angle $\pi/2$ from its initial direction or, more commonly, by the inverse of this time, called the collision frequency. The momentum scattering collision frequencies are denoted as $\nu_{ee}, \nu_{ii}, \nu_{ei}, \nu_{ie}$ for the various possible interactions between species and the corresponding times as τ_{ee}, etc. Energy scattering is characterized by the time required for an incident particle to transfer all its kinetic energy to the target particle. Energy transfer collision frequencies are denoted respectively by $\nu_{Eee}, \nu_{Eii}, \nu_{Eei}, \nu_{Eie}$.

We now show that these frequencies separate into categories having *three distinct orders of magnitude* with relative scalings $1 : (m_i/m_e)^{1/2} : m_i/m_e$. In order to estimate the orders of magnitude of the collision frequencies we assume the incident particle is "typical" for its species and so take its incident velocity to be the species thermal velocity, $v_{T\sigma} = (2\kappa T_\sigma/m_\sigma)^{1/2}$. While this is reasonable for a rough estimate, it should be realized that, because of the v^{-4} dependence in σ^*, a more careful averaging over all particles in the thermal distribution will differ. This careful averaging is quite intricate and will be deferred to Chapter 13.

We normalize all collision frequencies to ν_{ee}, and for further simplification assume that the ion and electron temperatures are of the same order of magnitude. First consider ν_{ei}: the reduced mass for *ei* collisions is the same as for *ee* collisions (except for a factor of 2, which we neglect) and the relative velocity is the same – hence, we conclude that $\nu_{ei} \sim \nu_{ee}$. Now consider ν_{ii}: because the temperatures were assumed equal, $\sigma^*_{ii} \approx \sigma^*_{ee}$ and so the collision frequencies will differ only because of the different velocities in the expression $\nu = n\sigma v$. The ion thermal velocity is lower by an amount $(m_e/m_i)^{1/2}$, giving $\nu_{ii} \approx (m_e/m_i)^{1/2}\nu_{ee}$.

Care is required when calculating ν_{ie}. Strictly speaking, this calculation should be done in the center-of-mass frame and then transformed back to the lab frame, but an easy way to estimate ν_{ie} using lab-frame calculations is to note that momentum is conserved in a collision so that in the lab frame $m_i\Delta\mathbf{v}_i = -m_e\Delta\mathbf{v}_e$, where Δ means the change in a quantity as a result of the collision. If the collision of an ion head-on with a stationary electron is taken as an example, then the electron bounces off forward with twice the ion's velocity (corresponding to a specular reflection of the electron in a frame where the ion is stationary); this gives $\Delta\mathbf{v}_e = 2\mathbf{v}_i$ and $|\Delta\mathbf{v}_i| / |\mathbf{v}_i| = 2m_e/m_i$. Thus, in order to have $|\Delta\mathbf{v}_i| / |\mathbf{v}_i|$ of order unity, it is necessary to have m_i/m_e head-on collisions of an ion with electrons, whereas in order to have $|\Delta\mathbf{v}_e| / |\mathbf{v}_e|$ of order unity it is only necessary to have one collision of an electron with an ion. Hence, $\nu_{ie} \sim (m_e/m_i)\nu_{ee}$.

Now consider energy changes in collisions. If a moving electron makes a head-on collision with an electron at rest, then the incident electron stops (loses all its momentum and energy), while the originally stationary electron flies off with the same momentum and energy that the incident electron had. A similar picture holds for an ion hitting an ion. Thus, like-particle collisions transfer energy at the same rate as momentum, so $\nu_{Eee} \sim \nu_{ee}$ and $\nu_{Eii} \sim \nu_{ii}$.

Table 1.1

~ 1	$\sim (m_e/m_i)^{1/2}$	$\sim m_e/m_i$
ν_{ee}	ν_{ii}	ν_{ie}
ν_{ei}	ν_{Eii}	ν_{Eei}
ν_{Eee}		ν_{Eie}

Interspecies collisions are more complicated. Consider an electron hitting a stationary ion head-on. Because the ion is massive, it barely recoils and the electron reflects with a velocity nearly equal in magnitude to its incident velocity. Thus, the change in electron momentum is $-2m_e\mathbf{v}_e$. From conservation of momentum, the momentum of the recoiling ion must be $m_i\mathbf{v}_i = 2m_e\mathbf{v}_e$. The energy transferred to the ion in this collision is $m_i v_i^2/2 = 4(m_e/m_i)m_e v_e^2/2$. Thus, an electron has to make $\sim m_i/m_e$ such collisions in order to transfer all its energy to ions. Hence, $\nu_{Eei} = (m_e/m_i)\nu_{ee}$.

Similarly, if an incident ion hits an electron at rest the electron will fly off with twice the incident ion velocity (in the center-of-mass frame, the electron is reflecting from the ion). The electron gains energy $m_e v_i^2/2$ so that, again, $\sim m_i/m_e$ collisions are required for the ion to transfer all its energy to electrons. The orders of magnitudes of collision frequencies are summarized in Table 1.1.

Although collisions are typically unimportant for fast, transient processes, they may eventually determine many properties of a given plasma. The wide disparity of collision frequencies shows that one has to be careful when determining which collisional process is relevant to a given phenomenon. Perhaps the best way to illustrate how collisions must be considered is by an example, such as the following:

Suppose half the electrons in a plasma initially have a directed velocity $\mathbf{v}_0$ while the other half of the electrons and all the ions are initially at rest. This may be thought of as a high-density beam of electrons passing through a cold plasma. On the fast (i.e., ν_{ee}) time scale the beam electrons will:

(i) Collide with the stationary electrons and share their momentum and energy so that after a time of order ν_{ee}^{-1} the beam will become indistinguishable from the background electrons. Since momentum must be conserved, the combined electrons will have a mean velocity $\mathbf{v}_0/2$.

(ii) Collide with the stationary ions, which will act as nearly fixed scattering centers so that the beam electrons will scatter in direction but not transfer significant energy to the ions.

Both of the above processes will randomize the velocity distribution of the electrons until this distribution becomes Maxwellian (the maximum entropy distribution); the Maxwellian will be centered about the average velocity discussed in (i) above.

On the very slow ν_{Eei} time scale (down by a factor m_i/m_e), the electrons will transfer momentum to the ions, so on this time scale the electrons will share their momentum with the ions, in which case the electrons will slow down and the ions will speed up until the average electron velocity and the average ion velocity become identical. Similarly, the electrons will share energy with the ions, in which case the ions will heat up while the electrons will cool.

If, instead, a beam of ions were injected into the plasma, the ion beam would thermalize and share momentum with the background ions on the intermediate ν_{ii} time scale, and then only share momentum and energy with the electrons on the very slow ν_{Eie} time scale.

This collisional sharing of momentum and energy and thermalization of velocity distribution functions to make Maxwellians is the process by which thermodynamic equilibrium is achieved. Collision frequencies vary as $T^{-3/2}$ and so, for hot plasmas, collision processes are often slower than many other phenomena. Since collisions are the means by which thermodynamic equilibrium is achieved, plasmas are typically *not in thermodynamic equilibrium*, although some components of the plasma may be in a partial equilibrium (for example, the electrons may be in thermal equilibrium with each other but not with the ions). Hence, thermodynamically based descriptions of the plasma are often inappropriate. It is not unusual, for example, to have a plasma where the electron and ion temperatures differ by more than an order of magnitude. This can occur when one species or the other has been subject to heating and the plasma lifetime is shorter than the interspecies energy equilibration time, $\sim \nu_{Eei}^{-1}$.

1.10 Collisions with neutrals

If a plasma is weakly ionized then collisions with neutrals must be considered. These collisions differ fundamentally from collisions between charged particles because now the interaction forces are short-range (unlike the long-range Coulomb interaction) and so the neutral can be considered simply as a hard body with cross-section of the order of its actual geometrical size. All atoms have radii of the order of 10^{-10} m so the typical neutral cross-section is $\sigma_{\rm neut} \sim 3 \times 10^{-20}$ m^2. When a particle hits a neutral it can simply scatter with no change in the internal energy of the neutral; this is called *elastic scattering*. It can also transfer energy to the structure of the neutral and so cause an internal change in the neutral; this is called *inelastic scattering*. Inelastic scattering includes ionization and excitation of atomic level transitions (with accompanying optical radiation).

Another process can occur when ions collide with neutrals – the incident ion can capture an electron from the neutral and become neutralized while simultaneously ionizing the original neutral. This process, called *charge exchange*, is used for producing energetic neutral beams. In this process a high-energy beam of ions is injected into a gas of neutrals, captures electrons, and exits as a high-energy beam of neutrals.

Because ions have approximately the same mass as neutrals, ions rapidly exchange energy with neutrals and tend to be in thermal equilibrium with the neutrals if the plasma is weakly ionized. As a consequence, ions are typically cold in weakly ionized plasmas, because the neutrals are in thermal equilibrium with the walls of the container.

1.11 Simple transport phenomena

1. Electrical resistivity. When a uniform electric field $\mathbf{E}$ exists in a plasma, the electrons and ions are accelerated in opposite directions creating a difference between the average velocities of the two species, i.e., a relative average velocity, $\mathbf{u}_{\rm rel} = \mathbf{u}_e - \mathbf{u}_i$, of electrons relative to ions. The creation of this relative average velocity due to oppositely directed $\mathbf{E}$ field acceleration of electrons and ions competes with the simultaneous dissipation of the relative average velocity due to interspecies collisions (this dissipation of relative average velocity is known as "drag"). Balancing these competing forces on an average electron gives

$$0 = -e\mathbf{E} - \nu_{ei} m_e \mathbf{u}_{\rm rel}, \tag{1.22}$$

since the drag force on the average electron is $\nu_{ei} m_e (\mathbf{u}_e - \mathbf{u}_i)$. However, since the electric current is $\mathbf{J} = -n_e e \mathbf{u}_{\rm rel}$, Eq. (1.22) can be rewritten as

$$\mathbf{E} = \eta \mathbf{J}, \tag{1.23}$$

where

$$\eta = \frac{m_e \nu_{ei}}{n_e e^2} \tag{1.24}$$

is the plasma electrical resistivity. Substituting $\nu_{ei} = \sigma^* n_i v_{Te}$ and noting from quasi-neutrality that $Z n_i = n_e$, the plasma electrical resistivity is

$$\eta = \frac{Z e^2}{2\pi m_e \varepsilon_0^2 v_{Te}^3} \ln\left(\frac{\lambda_D}{b_{\pi/2}}\right), \tag{1.25}$$

from which we see that resistivity is independent of density, proportional to $T_e^{-3/2}$, and also proportional to the ion charge Z. This expression for the resistivity is only approximate since we did not properly average over the electron velocity distribution (a more accurate expression, differing by a factor of order unity, will be derived in Chapter 13). Resistivity resulting from grazing collisions between electrons and ions as given by Eq. (1.25) is known as Spitzer resistivity (Spitzer and Harm 1953). It should be emphasized that although this discussion assumes existence of a *uniform* electric field in the plasma, a uniform field will not exist in what naively appears to be the most obvious geometry, namely a plasma between two parallel plates charged to different potentials. This is because Debye shielding will concentrate virtually all the potential drop into thin sheaths adjacent to the electrodes, resulting in near-zero electric field inside the plasma. A practical way to obtain a uniform electric field is to create the field by induction so that there are no electrodes that can be screened out.

2. Diffusion and ambipolar diffusion. Standard random walk arguments show that particle diffusion coefficients scale as $D \sim (\Delta x)^2/\tau$, where Δx is the characteristic step size in the random walk and τ is the time between steps. This can also be expressed as $D \sim v_T^2/\nu$, where $\nu = \tau^{-1}$ is the collision frequency and $v_T = \Delta x/\tau = \nu \Delta x$ is the thermal velocity. Since the random step size for particle collisions is the mean free path and the time between steps is the inverse of the collision frequency, the electron diffusion coefficient in an unmagnetized plasma scales as

$$D_e = \nu_e l_{mfp,e}^2 = \frac{\kappa T_e}{m_e \nu_e}, \tag{1.26}$$

where $\nu_e = \nu_{ee} + \nu_{ei} \sim \nu_{ee}$ is the 90° scattering rate for electrons and $l_{mfp,e} = \sqrt{\kappa T_e / m_e \nu_e^2}$ is the electron mean free path. Similarly, the ion diffusion coefficient in an unmagnetized plasma is

$$D_i = \nu_i l_{mfp,i}^2 = \frac{\kappa T_i}{m_i \nu_i}, \tag{1.27}$$

where $\nu_i = \nu_{ii} + \nu_{ie} \sim \nu_{ii}$ is the effective ion collision frequency. The electron diffusion coefficient is typically much larger than the ion diffusion coefficient in an unmagnetized plasma (it is the other way around for diffusion across a magnetic field in a magnetized plasma where the step size is the Larmor radius). However, if the electrons in an unmagnetized plasma did in fact diffuse across a density gradient at a rate two orders of magnitude faster than the ions, the ions would be left behind and the plasma would no longer be quasi-neutral. What actually happens is that the electrons try to diffuse faster than the ions, but in so doing set up a slight charge separation. This charge separation establishes a so-called "*ambipolar*" electrostatic electric field oriented to retard the electrons and accelerate the ions, i.e., decrease the outward electron flux and increase the outward ion flux. The amount of charge separation self-regulates to produce an ambipolar electric field having just the right magnitude to equalize the outward electron and ion fluxes. This results in an effective diffusion, called the ambipolar diffusion, which is less than the electron rate, but greater than the ion rate. Equation 1.22 shows that an electric field establishes an average electron momentum $m_e \mathbf{u}_e = -e\mathbf{E}/\nu_e$, where ν_e is the rate at which the average electron loses momentum owing to collisions with ions or neutrals. Electron-electron collisions are excluded from this calculation because an "average electron" cannot lose momentum owing to collisions with all the other electrons, because the other electrons have on average the same momentum as this average electron. Since the electric field cannot impart momentum to the plasma as a whole, the momentum imparted to ions must be equal and opposite, so $m_i\mathbf{u}_i = e\mathbf{E}/\upsilon_e$. Because diffusion in the presence of a density gradient produces an electron flux, $-D_e\nabla n_e$, the net electron flux resulting from the combination of an electric field and a diffusion across a density gradient is

$$\mathbf{\Gamma}_e = n_e\mu_e\mathbf{E} - D_e\nabla n_e, \tag{1.28}$$

where

$$\mu_e = -\frac{e}{m_e\upsilon_e} \tag{1.29}$$

is called the electron mobility. Similarly, the net ion flux is

$$\mathbf{\Gamma}_i = n_i\mu_i\mathbf{E} - D_i\nabla n_i, \tag{1.30}$$

where

$$\mu_i = \frac{e}{m_i\upsilon_i} \tag{1.31}$$

is the ion mobility. In order to maintain quasi-neutrality, the electric field automatically self-adjusts to give $\mathbf{\Gamma}_e = \mathbf{\Gamma}_i = \mathbf{\Gamma}_{ambipolar}$ and $n_i = n_e = n$; this ambipolar electric field is

$$\begin{aligned}\mathbf{E}_{ambipolar} &= \frac{(D_e - D_i)}{(\mu_e - \mu_i)}\nabla \ln n \\ &\simeq \frac{D_e}{\mu_e}\nabla \ln n \\ &= \frac{\kappa T_e}{e}\nabla \ln n.\end{aligned} \tag{1.32}$$

Substitution for $\mathbf{E}$ gives the ambipolar diffusion to be

$$\mathbf{\Gamma}_{ambipolar} = -\left(\frac{\mu_e D_i - D_e\mu_i}{\mu_e - \mu_i}\right)\nabla n, \tag{1.33}$$

so the ambipolar diffusion coefficient is

$$\begin{aligned}D_{ambipolar} &= \frac{\mu_e D_i - D_e\mu_i}{\mu_e - \mu_i} \\ &= \frac{\dfrac{D_i}{\mu_i} - \dfrac{D_e}{\mu_e}}{\dfrac{1}{\mu_i} - \dfrac{1}{\mu_e}} \\ &= \frac{D_i\dfrac{m_i\upsilon_i}{e} + D_e\dfrac{m_e\upsilon_e}{e}}{\dfrac{m_i\upsilon_i}{e} + \dfrac{m_e\upsilon_e}{e}} \\ &\simeq \frac{\kappa(T_i + T_e)}{m_i\upsilon_i},\end{aligned} \tag{1.34}$$

where Eqs. (1.26) and (1.27) have been used as well as the relation $\upsilon_i \sim (m_e/m_i)^{1/2}\upsilon_e$. If the electrons are much hotter than the ions, then for a given ion temperature, the ambipolar diffusion scales as T_e/m_i. Ambipolar diffusion is thus somewhat like the situation of a small child tugging on his/her parent (the energy of the small child is like the electron temperature, the parental mass is like the ion mass, and the tension in the arm that accelerates the parent and decelerates the child is like the ambipolar electric field); the resulting motion (parent and child move together faster than the parent would like and slower than the child would like) is analogous to electrons being retarded and ions being accelerated by the ambipolar electric field in such a way as to maintain quasi-neutrality.

1.12 A quantitative perspective

Relevant physical constants are

$$\begin{aligned} e &= 1.6 \times 10^{-19}\ \mathrm{C} \\ m_e &= 9.1 \times 10^{-31}\ \mathrm{kg} \\ m_p/m_e &= 1836 \\ \varepsilon_0 &= 8.85 \times 10^{-12}\ \mathrm{F\,m^{-1}} \end{aligned}$$

The temperature is measured in units of electron volts, so that $\kappa = 1.6 \times 10^{-19}\ \mathrm{J\,V^{-1}}$; i.e., $\kappa = e$. Thus, the Debye length is

$$\begin{aligned} \lambda_D &= \sqrt{\frac{\varepsilon_0 \kappa T}{ne^2}} \\ &= \sqrt{\frac{\varepsilon_0}{e}}\sqrt{\frac{T_{eV}}{n}} \\ &= 7.4 \times 10^3 \sqrt{\frac{T_{eV}}{n}}\ \mathrm{m}. \end{aligned} \tag{1.35}$$

We will assume that the typical velocity is related to the temperature by

$$\frac{1}{2}mv^2 = \frac{3}{2}\kappa T. \tag{1.36}$$

For electron-electron scattering $\mu = m_e/2$ so that the small-angle scattering cross-section is

$$\begin{aligned} \sigma^* &= \frac{1}{2\pi}\left(\frac{e^2}{\varepsilon_0 m v^2/2}\right)^2 \ln\left(\lambda_D/b_{\pi/2}\right) \\ &= \frac{1}{2\pi}\left(\frac{e^2}{3\varepsilon_0 \kappa T}\right)^2 \ln \Lambda, \end{aligned} \tag{1.37}$$

Table 1.2 *Comparison of parameters for a wide variety of plasmas.*

	n	T	λ_D	$n\lambda_D{}^3$	$\ln\Lambda$	ν_{ee}	l_{mfp}	L
units	m^{-3}	eV	m			s^{-1}	m	m
Solar corona (loops)	10^{15}	100	10^{-3}	10^7	19	10^2	10^5	10^8
Solar wind (near Earth)	10^7	10	10	10^9	25	10^{-5}	10^{11}	10^{11}
Magnetosphere (tail lobe)	10^4	10	10^2	10^{11}	28	10^{-8}	10^{14}	10^8
Ionosphere	10^{11}	0.1	10^{-2}	10^4	14	10^2	10^3	10^5
Mag. fusion (tokamak)	10^{20}	10^4	10^{-4}	10^7	20	10^4	10^4	10
Inertial fusion (imploded)	10^{31}	10^4	10^{-10}	10^2	8	10^{14}	10^{-7}	10^{-5}
Lab plasma (dense)	10^{20}	5	10^{-6}	10^3	9	10^8	10^{-2}	10^{-1}
Lab plasma (diffuse)	10^{16}	5	10^{-4}	10^5	14	10^4	10^1	10^{-1}

where

$$\begin{aligned}\Lambda &= \frac{\lambda_D}{b_{\pi/2}} \\ &= \sqrt{\frac{\varepsilon_0 \kappa T}{ne^2}}\frac{4\pi\varepsilon_0 mv^2/2}{e^2} \\ &= 6\pi n\lambda_D^3 \end{aligned} \tag{1.38}$$

is typically a very large number corresponding to there being a macroscopically large number of particles in a sphere having a radius equal to a Debye length; different authors will have slightly different numerical coefficients, depending on how they identify velocity with temperature. This difference is of no significance because of the logarithmic dependence.

The collision frequency is $\nu = \sigma^* nv$, so

$$\begin{aligned}\nu_{ee} &= \frac{n}{2\pi}\left(\frac{e^2}{3\varepsilon_0\kappa T}\right)^2\sqrt{\frac{3\kappa T}{m_e}}\ln\Lambda \\ &= \frac{e^{5/2}}{2\times 3^{3/2}\pi\varepsilon_0^2 m_e^{1/2}}\frac{n\ln\Lambda}{T_{eV}^{3/2}} \\ &= 4\times 10^{-12}\frac{n\ln\Lambda}{T_{eV}^{3/2}}. \end{aligned} \tag{1.39}$$

For most plasmas $\ln\Lambda$ lies in the range 8–25; for situations where only a simple order of magnitude estimate of the collision frequency is required it is usually sufficient to assume $\ln\Lambda \sim 10$.

Table 1.2 lists nominal parameters for several plasmas of interest and shows these plasmas have an enormous range of densities, temperatures, scale lengths, mean free paths, and collision frequencies. The crucial issue is the ratio of the mean free path to the characteristic scale length.

Arc plasmas and magnetoplasmadynamic thrusters are in the category of dense lab plasmas; these plasmas are very collisional (the mean free path is much smaller than the characteristic scale length). The plasmas used in semiconductor processing and many research plasmas are in the diffuse lab plasma category; these plasmas are collisionless. It is possible to make both collisional and collisionless lab plasmas, and in fact if there are large temperature or density gradients it is possible to have both collisional and collisionless behavior in the same device.

1.13 Assignments

1. Rutherford scattering. This assignment involves deriving the Rutherford scattering formula using a geometrical analysis, which exploits the symmetry of the scattering trajectory.

   (a) Show that the equation of motion in the center-of-mass frame is

$$\mu \ddot{\mathbf{r}} = \frac{q_1 q_2}{4\pi\varepsilon_0 r^2}\hat{r}.$$

   By taking the time derivative of $\mathbf{r} \times \dot{\mathbf{r}}$ show that the angular momentum $\mathbf{L} = \mu \mathbf{r} \times \dot{\mathbf{r}}$ is a constant of the motion. The calculations will be done in the center-of-mass frame using a cylindrical coordinate system r, ϕ, z with origin at the scattering center and the z axis chosen to be parallel to $\mathbf{L}$. Show that the particle does not move in the z direction (hint: consider the dot product between $\dot{\mathbf{r}}$ and $\mathbf{L}$) so the particle is confined to the $z = 0$ plane. Let θ be the scattering angle, and let b be the impact parameter as indicated in Fig. 1.4. Also, define a Cartesian coordinate

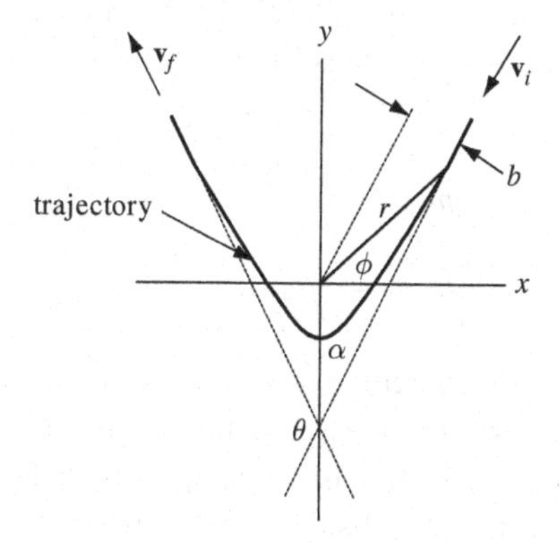

Fig. 1.4 Geometry of scattering in center-of-mass frame. Scattering center is at the origin and θ is the scattering angle. Note symmetries of velocities before and after scattering.

system x, y so that $y = r \sin\phi$, etc.; these Cartesian coordinates are also shown in Fig. 1.4.

(b) Show that $|\mathbf{L}| = \mu b v_\infty = \mu r^2 \dot{\phi}$ so $\dot{\phi} = b v_\infty / r^2$.

(c) Let $\mathbf{v}_i$ and $\mathbf{v}_f$ be the initial and final velocities as shown in Fig. 1.4. Since energy is conserved during scattering, the magnitudes of these two velocities must be the same, i.e., $|\mathbf{v}_i| = |\mathbf{v}_f| = v_\infty$. From the symmetry of the figure it is seen that the x component of velocity at infinity is the same before and after the collision, even though it is altered during the collision. However, the y component of the velocity reverses direction as a result of the collision. Let Δv_y be the net change in the y velocity over the entire collision. Express Δv_y in terms of v_{yi}, the y component of $\mathbf{v}_i$.

(d) Using the y component of the equation of motion, obtain a relationship between $\mathrm{d}v_y$ and $\mathrm{d}\cos\phi$. (Hint: it is useful to use conservation of angular momentum to eliminate $\mathrm{d}t$ in favor of $\mathrm{d}\phi$.) Let ϕ_i and ϕ_f be the initial and final values of ϕ. By integrating $\mathrm{d}v_y$, calculate Δv_y over the entire collision. How is ϕ_f related to ϕ_i and to α (refer to figure)?

(e) How is v_{yi} related to ϕ_i and v_∞? How is θ related to α? Use the expressions for Δv_y obtained in parts (c) and (d) above to obtain the Rutherford scattering formula

$$\tan\left(\frac{\theta}{2}\right) = \frac{q_1 q_2}{4\pi\varepsilon_0 \mu b v_\infty^2}.$$

What is the scattering angle for grazing (small-angle collisions) and how does this small-angle scattering relate to the initial center-of-mass kinetic energy and to the potential energy at distance b? For grazing collisions how does b relate to the distance of closest approach? What impact parameter gives 90° scattering?

2. One-dimensional scattering relations. The separation of collision types according to m_e/m_i can also be understood by considering how the combination of conservation of momentum and of energy constrain certain properties of collisions. Suppose a particle with mass m_1 and incident velocity v_1 makes a head-on collision with a stationary target particle having mass m_2. The conservation equations for momentum and energy can be written as

$$\begin{aligned} m_1 v_1 &= m_1 v_1' + m_2 v_2' \\ \frac{1}{2} m_1 v_1^2 &= \frac{1}{2} m_1 v_1'^2 + \frac{1}{2} m_2 v_2'^2, \end{aligned}$$

where prime refers to the value after the collision. By eliminating v_1' between these two equations obtain v_2' as a function of v_1. Use this to construct an expression showing the ratio $m_2 v_2'^2 / m_1 v_1^2$, i.e., the fraction of the incident particle energy transferred to the target particle as a result of the collision. How does this fraction depend on m_1/m_2 when m_1/m_2 is equal to unity, very large, or very small? If m_1/m_2 is very large or very small how many collisions are required to transfer approximately all of the incident particle energy to target particles?

3. Some basic facts you should know. Memorize the value of ε_0 (or else arrange for the value to be close at hand). What is the value of Boltzmann's constant when temperatures are measured in electron volts? What is the density of the air you are breathing, measured in particles per cubic meter? What is the density of particles in solid copper, measured in particles per cubic meter? What is room temperature, expressed in electron volts? What is the ionization potential (in eV) of a hydrogen atom? What is the mass of an electron and of an ion (in kilograms)? What is the strength of the Earth's magnetic field at your location, expressed in Tesla? What is the strength of the magnetic field produced by a straight wire carrying 1 ampere as measured by an observer located 1 meter from the wire and what is the direction of the magnetic field? What is the relationship between Tesla and Gauss, between particles per cubic centimeter and particles per cubic meter? What is magnetic flux? If a circular loop of wire with a break in it links a magnetic flux of 29.83 Wb, which increases at a constant rate to a flux of 30.83 Wb, in one second, what voltage appears across the break?
4. Solve Eq. (1.6) the "easy" way using Gauss' law to show that the solution of

$$\nabla^2\phi = -\frac{1}{\varepsilon_0}\delta(\mathbf{r})$$

is

$$\phi = \frac{1}{4\pi\varepsilon_0 r}.$$

Show that this implies

$$\nabla^2\frac{1}{4\pi r} = -\delta(\mathbf{r}) \tag{1.40}$$

is a representation for the delta function. Then, use spherical polar coordinates and symmetry to show the Laplacian reduces to

$$\nabla^2\phi = \frac{1}{r^2}\frac{\partial}{\partial r}\left(r^2\frac{\partial\phi}{\partial r}\right).$$

Explicitly calculate $\nabla^2(1/r)$ and then reconcile your result with Eq. (1.40). Using these results guess that the solution to Eq. (1.6) has the form

$$\phi = \frac{g(r)}{4\pi\varepsilon_0 r}.$$

Substitute this guess into Eq. (1.6) to obtain a differential equation for g that is trivial to solve.
5. Solve Eq. (1.6) for $\phi(\mathbf{r})$ using a more general method that illustrates several important mathematical techniques and formalisms. Begin by defining the 3-D Fourier transform

$$\tilde{\phi}(\mathbf{k}) = \int \mathrm{d}\mathbf{r}\phi(\mathbf{r})\mathrm{e}^{-\mathrm{i}\mathbf{k}\cdot\mathbf{r}}, \tag{1.41}$$

in which case the inverse transform is

$$\phi(\mathbf{r}) = \frac{1}{(2\pi)^3}\int \mathrm{d}\mathbf{k}\tilde{\phi}(\mathbf{k})\mathrm{e}^{\mathrm{i}\mathbf{k}\cdot\mathbf{r}} \tag{1.42}$$

and note that the Dirac delta function can be expressed as

$$\delta(\mathbf{r}) = \frac{1}{(2\pi)^3}\int \mathrm{d}\mathbf{k}\mathrm{e}^{\mathrm{i}\mathbf{k}\cdot\mathbf{r}}. \tag{1.43}$$

Now multiply Eq. (1.6) by $\exp(-\mathrm{i}\mathbf{k}\cdot\mathbf{r})$ and then integrate over all $\mathbf{r}$, i.e., operate with $\int \mathrm{d}\mathbf{r}$. The term involving ∇^2 is integrated by parts, which effectively replaces the ∇ operator with $\mathrm{i}\mathbf{k}$.
Show that the Fourier transform of the potential is

$$\tilde{\phi}(\mathbf{k}) = \frac{q_T}{\varepsilon_0(k^2+\lambda_D^{-2})} \tag{1.44}$$

and use this in Eq. (1.42).
Because of spherical symmetry use spherical polar coordinates for the k-space integral. The only fixed direction is the $\mathbf{r}$ direction so choose the polar axis of the $\mathbf{k}$ coordinate system to be parallel to $\mathbf{r}$. Thus $\mathbf{k}\cdot\mathbf{r} = kr\alpha$, where $\alpha = \cos\theta$ and θ is the polar angle. Also, $\mathrm{d}\mathbf{k} = -\mathrm{d}\phi k^2 \mathrm{d}\alpha \mathrm{d}k$, where ϕ is the azimuthal angle. What are the limits of the respective ϕ, α, and k integrals? In answering this, you should first obtain an integral of the form

$$\phi(\mathbf{r}) \sim \int_{\phi=?}^{?}\mathrm{d}\phi\int_{\alpha=?}^{?}\mathrm{d}\alpha\int_{k=?}^{?}k^2\mathrm{d}k\times(?), \tag{1.45}$$

where the limits and the integrand with appropriate coefficients are specified (i.e., replace all the question marks and $\sim$ by the correct quantities). Upon evaluation of the ϕ and α integrals Eq. (1.45) becomes an even function of k so that the range of integration can be extended to $-\infty$ providing the overall integral is multiplied by 1/2. Realizing that $\sin kr = \mathrm{Im}\,[\mathrm{e}^{ikr}]$, derive an expression of the general form

$$\phi(\mathbf{r}) \sim \mathrm{Im}\int_{-\infty}^{\infty} k\mathrm{d}k\frac{\mathrm{e}^{ikr}}{f(k^2)} \tag{1.46}$$

but specify the coefficient and exact form of $f(k^2)$. Explain why the integration contour (which is along the real k axis) can be completed in the upper half complex k-plane. Complete the contour in the upper half-plane and show that the integrand has a single pole in the upper half-plane at $k = ?$ Use the method of residues to obtain $\phi(\mathbf{r})$.

6. Make sure you know how to evaluate quickly $\mathbf{A}\times(\mathbf{B}\times\mathbf{C})$ and $(\mathbf{A}\times\mathbf{B})\times\mathbf{C}$. A useful mnemonic, which works for both forms, is: "both forms = middle times (other two dotted together) – outer times (other two dotted together)," where outer refers to the outer vector of the parentheses (furthest from the center of the triad) and middle refers to the middle vector in the triad of vectors; see Appendix A.
7. Particle trajectory integration scheme (Birdsall and Langdon 1985). In this assignment you will develop a simple, but powerful, "leap-frog" numerical integration scheme. This is a type of "implicit" numerical integration scheme. This numerical scheme can

later be used to evaluate particle orbits in time-dependent fields having complex topology. These calculations can be considered as numerical experiments used in conjunction with the analytic theory we will develop. This combined analytical/numerical approach provides a deeper insight into charged particle dynamics than analysis alone.
Brief note on Implicit vs. Explicit numerical integration schemes
Suppose it is desired to use numerical methods to integrate the equation

$$\frac{\mathrm{d}y}{\mathrm{d}t} = f(y(t), t).$$

Unfortunately, since $y(t)$ is the sought-after quantity, we do not know what to use in the right-hand side for $y(t)$. A naive choice would be to use the previous value of y in the right-hand side to get a scheme of the form

$$\frac{y_{\text{new}} - y_{\text{old}}}{\Delta t} = f(y_{\text{old}}, t),$$

which may be solved to give

$$y_{\text{new}} = y_{\text{old}} + \Delta t\, f(y_{\text{old}}, t).$$

Simple and appealing as this is, it does not work since it is numerically unstable. However, if we use the following scheme we will get a stable result:

$$\frac{y_{\text{new}} - y_{\text{old}}}{\Delta t} = f((y_{\text{new}} + y_{\text{old}})/2, t). \tag{1.47}$$

In other words, we have used the average of the new and the old values of y in the right-hand side. This makes sense because the right-hand side is a function evaluated at time t, whereas $y_{\text{new}} = y(t + \Delta t/2)$ and $y_{\text{old}} = y(t - \Delta t/2)$. If these last quantities are Taylor expanded, it is seen that to lowest order $y(t) = [y(t + \Delta t/2) + y(t - \Delta t/2)]/2$. Since y_{new} occurs on both sides of the equation we will have to solve some sort of equation, or invert some sort of matrix, to get y_{new}.
Start with

$$m\frac{\mathrm{d}\mathbf{v}}{\mathrm{d}t} = q(\mathbf{E} + \mathbf{v} \times \mathbf{B}).$$

Define the angular cyclotron frequency vector $\boldsymbol{\Omega} = q\mathbf{B}/m$ and the normalized electric field $\boldsymbol{\Sigma} = q\mathbf{E}/m$ so that the above equation becomes

$$\frac{\mathrm{d}\mathbf{v}}{\mathrm{d}t} = \boldsymbol{\Sigma} + \mathbf{v} \times \boldsymbol{\Omega}. \tag{1.48}$$

Using the implicit scheme of Eq. (1.47), show that Eq. (1.48) becomes

$$\mathbf{v}_{\rm new} + \mathbf{A}\times\mathbf{v}_{\rm new} = \mathbf{C},$$

where $\mathbf{A} = \boldsymbol{\Omega}\Delta t/2$ and $\mathbf{C} = \mathbf{v}_{\rm old} + \Delta t\ (\boldsymbol{\Sigma} + \mathbf{v}_{\rm old}\times\boldsymbol{\Omega}/2)$. By first dotting the above equation with $\mathbf{A}$ and then crossing it with $\mathbf{A}$ show that the new value of velocity is given by

$$\mathbf{v}_{\rm new} = \frac{\mathbf{C} + \mathbf{A}\mathbf{A}\cdot\mathbf{C} - \mathbf{A}\times\mathbf{C}}{1+A^2}.$$

The new position is simply given by

$$\mathbf{x}_{\rm new} = \mathbf{x}_{\rm old} + \mathbf{v}_{\rm new}dt.$$

The above two equations can be used to solve charged particle motion in complicated, 3-D, time-dependent fields. Use this particle integrator to calculate and plot the trajectory of an electron moving in crossed electric and magnetic fields where the non-vanishing components are $E_x = 1\ \mathrm{V\,m^{-1}}$ and $B_z = 1$ T. Try varying the field strengths and polarities, and also try ions instead of electrons.

8. Use the leap-frog numerical integration scheme to demonstrate the Rutherford scattering problem:

   (i) Define a characteristic length for this problem to be the impact parameter for a 90° degree scattering angle, $b_{\pi/2}$. A reasonable choice for the characteristic velocity is v_∞. What is the characteristic time?
   (ii) Define a Cartesian coordinate system such that the z axis is parallel to the incident relative velocity vector $\mathbf{v}_\infty$ and goes through the scattering center. Let the impact parameter be in the y direction so that the incident particle is traveling in the $y-z$ plane. Establish a plotting frame spanning the region $-50 \leq z/b_{\pi/2} \leq 50$ and $-50 \leq y/b_{\pi/2} \leq 50$.
   (iii) Set the magnetic field to be zero, and let the electric field be

   $$\mathbf{E} = -\nabla\phi,$$

   where $\phi = ?$ so $E_x = ?$, etc.
   (iv) By using $r^2 = x^2 + y^2 + z^2$ calculate the electric field at each particle position, and so determine the particle trajectory.
   (v) Demonstrate that the scattering is indeed at 90° when $b = b_{\pi/2}$. What happens when b is much larger or much smaller than $b_{\pi/2}$? What happens when q_1, q_2 have the same or opposite signs?
   (vi) Have your code draw the relevant theoretical scattering angle θ_s and show that the numerical result is in agreement with the theoretical prediction.

9. Collision relations. Show that $\sigma n_t l_{mfp} = 1$, where σ is the cross-section for a collision, n_t is the density of target particles, and l_{mfp} is the mean free path. Show also that the collision frequency is given by $v = \sigma n_t v$, where v is the velocity of the incident particle. Calculate the electron–electron collision frequency for the following plasmas: fusion ($n \sim 10^{20}\ \mathrm{m^{-3}}$, $T \sim 10$ keV), partially ionized discharge plasma ($n \sim 10^{16}\ \mathrm{m^{-3}}$, $T \sim 10$ eV). At what temperature does the conductivity of plasma equal that of copper, and of steel? Assume $Z = 1$.

10. Cyclotron motion. Suppose that a particle is immersed in a uniform magnetic field $\mathbf{B} = B\hat{z}$ and there is no electric field. Suppose that at $t = 0$ the particle's initial position is at $\mathbf{x} = 0$ and its initial velocity is $\mathbf{v} = v_0\hat{x}$. Using the Lorentz equation, calculate the particle position and velocity as a function of time (be sure to take initial conditions into account). What is the direction of rotation for ions and for electrons (right handed or left handed with respect to the magnetic field)? If you had to make up a mnemonic for the sense of ion rotation, would it be Lions or Rions? Now, repeat the analysis but this time with an electric field $\mathbf{E} = \hat{x}E_0 \cos(\omega t)$. What happens in the limit where $\omega \to \Omega$, where $\Omega = qB/m$ is the cyclotron frequency? Assume that the particle is a proton and that $B = 1\,\text{T}$, $v_0 = 10^5\,\text{m s}^{-1}$, and compare your results with direct numerical solution of the Lorentz equation. Use $E_0 = 10^4\,V\,\text{m}^{-1}$ for the electric field.
11. Space-charge-limited current. When a metal or metal oxide is heated to high temperatures it emits electrons from its surface. This process, called *thermionic emission,* is the basis of vacuum tube technology and is also essential when high currents are drawn from electrodes in a plasma. The electron-emitting electrode is called a *cathode*, while the electrode to which the electrons flow is called an *anode.* An idealized configuration is shown in Fig. 1.5. This configuration can operate in two regimes: (i) the *temperature-limited* regime, where the current is determined by the thermionic emission capability of the cathode, and (ii) the *space-charge-limited* regime, where the current is determined by a buildup of electron density in the region between cathode and anode (inter-electrode region). Let us now discuss this space-charge-limited regime: if the current is small, then the number of electrons required to carry the current is small and so the inter-electrode region is nearly a vacuum, in which case the electric field in this region will be nearly uniform and be given by $E = V/d$, where V is the anode–cathode potential difference and d is the anode–cathode separation. This electric field will accelerate the electrons from anode to cathode. However, if the current is large, there will be a significant electron density in the inter-electrode

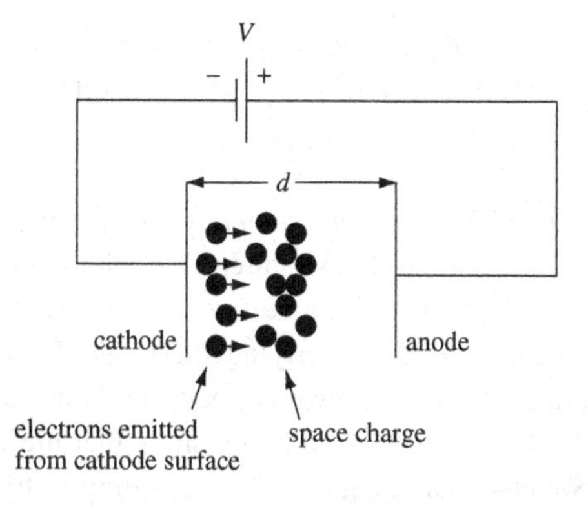

Fig. 1.5 Electron cloud accelerated from cathode to anode encounters space charge of previously emitted electrons.

region. This "space charge" will create a localized depression in the potential (since electrons have negative charge).
The result is that the electric field will be reduced in the region near the cathode. If the space charge is sufficiently large, the electric field at the cathode vanishes. In this situation, attempting to increase the current by increasing the number of electrons ejected by the cathode will not succeed because an increase in current (which will give an increase in space charge) will produce a repulsive electric field that prevents additional electrons from leaving the cathode. Let us now calculate the space-charge-limited current and relate it to our discussion on Debye shielding. The current density in this system is

$$J = -n(x)ev(x) = \text{a negative constant.}$$

Since potential is undefined with respect to a constant, let us choose this constant so that the cathode potential is zero, in which case the anode potential is V_0. Assuming electrons leave the cathode with zero velocity, show that the dependence of electron velocity on position is given by

$$v(x) = \sqrt{\frac{2eV(x)}{m_e}}.$$

Show that the above two equations, plus Poisson's equation, can be combined to give the following differential equation for the potential:

$$\frac{\mathrm{d}^2 V}{\mathrm{d}x^2} - \lambda V^{-1/2} = 0,$$

where $\lambda = \varepsilon_0^{-1} |J| \sqrt{m_e/2e}$. By multiplying this equation with the integrating factor $\mathrm{d}V/\mathrm{d}x$ and using the space-charge-limited boundary condition $E = 0$ at $x = 0$, solve for $V(x)$. By rearranging the expression for $V(x)$ show that the space-charge-limited current is

$$J = \frac{4}{9}\varepsilon_0 \sqrt{\frac{2e}{m_e}} \frac{V^{3/2}}{d^2}.$$

This is called the Child–Langmuir space charge limited current. For reference the temperature-limited current is given by the Richardson–Dushman law,

$$J = AT^2 e^{-\phi_0/\kappa T},$$

where the coefficient A and the *work function* ϕ_0 are properties of the cathode material, while T is the cathode temperature. Thus, the actual cathode current will be whichever is the smaller of the above two expressions. Show there is a close relationship between the physics underlying the Child–Langmuir law and Debye shielding (hint: characterize the electron velocity as being a thermal velocity and its energy as being a thermal energy and then show that the inter-electrode spacing corresponds to ?). Suppose that a cathode was operating in the space-charge-limited regime and that some positively charged ions were placed in the inter-electrode region. What would happen to the space charge – would it be possible to draw more or less current from the cathode? Suppose the entire inter-electrode region were filled with plasma with electron temperature T_e. What would be the appropriate value of d and how much current could be drawn from the cathode (assuming it were sufficiently hot)? Does this give you any ideas on why high current switch tubes (called ignitrons) use plasma to conduct the current?

2

The Vlasov, two-fluid, and MHD models of plasma dynamics

2.1 Overview

We begin this chapter by developing the concept of conservation of particles in phase-space and then use this concept as the basis for establishing the three main models of plasma dynamics, namely Vlasov theory, two-fluid theory, and magnetohydrodynamics (MHD). The Vlasov model is the most detailed and characterizes plasma dynamics by following the temporal evolution of electron and ion velocity distribution functions. The two-fluid model is intermediate in complexity and approximates plasma as a system of mutually interacting, finite-pressure electron and ion fluids. The MHD model is the least detailed and approximates plasma as a single, finite-pressure, electrically conducting fluid. The question of which of these models to use when analyzing a given situation is essentially a matter of selecting the best tool for the task and furthermore, just as a mechanic might alternate between using a screwdriver and a pair of pliers for a specific task, it is often advantageous to alternate between these models when analyzing a specific problem. As we develop these three models, we will also take the opportunity to explore some immediate and important fundamental consequences of these models, most notably the strong dependence of a collisionless plasma on its past history (Vlasov model) and the freezing of magnetic flux into the arbitrarily moving frame of a perfectly conducting plasma (MHD). The chapter concludes with an examination of some applications having a close association with the derivation of these models; these are classical transport of plasma across a magnetic field, sheaths at the edge of a plasma bounded by a conducting wall, and Langmuir probes, a simple method for diagnosing plasmas.

2.2 Phase-space

Consider a particle moving in a one-dimensional space and let the position of the particle be $x = x(t)$ and the velocity of the particle be $v = v(t)$. A way to visualize

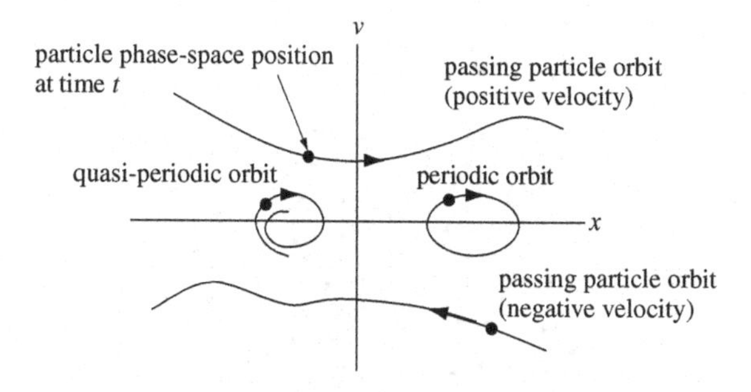

Fig. 2.1 Phase-space showing different types of possible particle orbits.

the x and v trajectories simultaneously is to plot these trajectories parametrically on a two-dimensional graph, where the horizontal coordinate is given by $x(t)$ and the vertical coordinate is given by $v(t)$. This $x - v$ plane is called *phase-space.* The trajectory (or orbit) of several particles can be represented as a set of curves in phase-space as shown in Fig. 2.1. Examples of a few qualitatively different phase-space orbits are shown in Fig. 2.1.

Particles in the upper half-plane always move to the right, since they have a positive velocity, while those in the lower half-plane always move to the left. Particles having exact periodic motion (e.g., $x = A\cos[\omega t]$, $v = -\omega A\sin[\omega t]$) alternate between moving to the right and the left and so describe an ellipse in phase-space. Particles with nearly periodic (quasi-periodic) motions will have near-ellipses or spiral orbits. A particle that does not reverse direction is called a passing particle, while a particle confined to a certain region of phase-space (e.g., a particle with periodic motion) is called a trapped particle.

2.3 Distribution function and Vlasov equation

At any given time, each particle has a specific position and velocity. We can therefore characterize the instantaneous configuration of a large number of particles by specifying the density of particles at each point x, v in phase-space. The function prescribing the instantaneous density of particles in phase-space is called the *distribution function* and is denoted by $f(x, v, t)$. Thus, $f(x, v, t)\mathrm{d}x\mathrm{d}v$ is the number of particles at time t having positions in the range between x and $x+\mathrm{d}x$ and velocities in the range between v and $v+\mathrm{d}v$. As time progresses, the particle motion and acceleration causes the number of particles in these x and v ranges to change and so f will change. This temporal evolution of f gives a description of the system more detailed than a fluid description, but less detailed than following the trajectory of each individual particle. Using the evolution of f to characterize

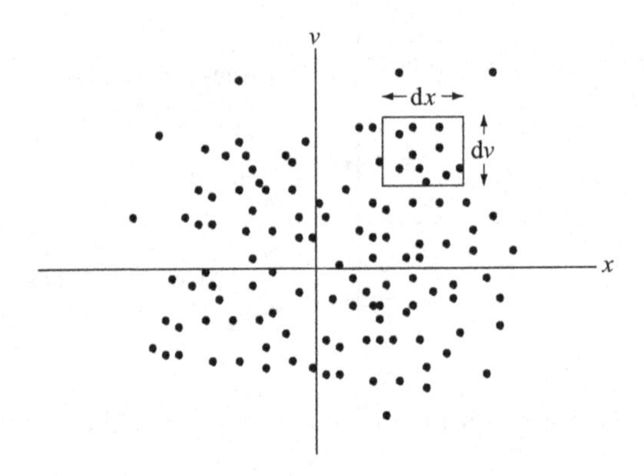

Fig. 2.2 A box within phase-space having width dx and height dv.

the system does not keep track of the trajectories of individual particles, but rather characterizes classes of particles having the same x, v.

Now consider the rate of change of the number of particles inside a small box in phase-space, such as is shown in Fig. 2.2. Defining $a(x, v, t)$ to be the acceleration of a particle, it is seen that the particle flux in the horizontal direction is fv and the particle flux in the vertical direction is fa. Thus, the particle fluxes into the four sides of the box are:

1. flux into left side of box is $f(x, v, t)v\mathrm{d}v$,
2. flux into right side of box is $-f(x+\mathrm{d}x, v, t)v\mathrm{d}v$,
3. flux into bottom of box is $f(x, v, t)a(x, v, t)\mathrm{d}x$,
4. flux into top of box is $-f(x, v+\mathrm{d}v, t)a(x, v+\mathrm{d}v, t)\mathrm{d}x$.

The number of particles in the box is $f(x, v, t)\mathrm{d}x\mathrm{d}v$ so that the rate of change of the number of particles in the box is

$$\begin{aligned}\frac{\partial f(x, v, t)}{\partial t}\mathrm{d}x\mathrm{d}v = & -f(x+\mathrm{d}x, v, t)v\mathrm{d}v + f(x, v, t)v\mathrm{d}v \\ & -f(x, v+\mathrm{d}v, t)a(x, v+\mathrm{d}v, t)\mathrm{d}x \\ & +f(x, v, t)a(x, v, t)\mathrm{d}x\end{aligned} \tag{2.1}$$

or, on Taylor expanding the quantities on the right-hand side, we obtain the one-dimensional *Vlasov equation*,

$$\frac{\partial f}{\partial t} + v\frac{\partial f}{\partial x} + \frac{\partial}{\partial v}(af) = 0. \tag{2.2}$$

It is straightforward to generalize Eq. (2.2) to three dimensions and so obtain the three-dimensional Vlasov equation,

$$\frac{\partial f}{\partial t} + \mathbf{v}\cdot\frac{\partial f}{\partial \mathbf{x}} + \frac{\partial}{\partial \mathbf{v}}\cdot(\mathbf{a}f) = 0. \tag{2.3}$$

Because **x**, **v** are independent quantities in phase-space, the spatial derivative term has the commutation property,

$$\mathbf{v}\cdot\frac{\partial f}{\partial \mathbf{x}} = \frac{\partial}{\partial \mathbf{x}}\cdot(\mathbf{v}f). \tag{2.4}$$

The particle acceleration is given by the Lorentz force

$$\mathbf{a} = \frac{q}{m}(\mathbf{E}+\mathbf{v}\times\mathbf{B}). \tag{2.5}$$

Because $(\mathbf{v}\times\mathbf{B})_i = v_j B_k - v_k B_j$ is independent of v_i, the term $\partial(\mathbf{v}\times\mathbf{B})_i/\partial v_i$ vanishes so that even though the Lorentz acceleration **a** is velocity-dependent, it nevertheless commutes with the vector velocity derivative as

$$\mathbf{a}\cdot\frac{\partial f}{\partial \mathbf{v}} = \frac{\partial}{\partial \mathbf{v}}\cdot(\mathbf{a}f). \tag{2.6}$$

Because of this commutation property the Vlasov equation can also be written as

$$\frac{\partial f}{\partial t}+\mathbf{v}\cdot\frac{\partial f}{\partial \mathbf{x}}+\mathbf{a}\cdot\frac{\partial f}{\partial \mathbf{v}} = 0. \tag{2.7}$$

If we "sit on top of" a particle that has a phase-space trajectory $\mathbf{x} = \mathbf{x}(t)$, $\mathbf{v} = \mathbf{v}(t)$ and measure the distribution function as we move along with the particle, the observed rate of change of the distribution function will be $\mathrm{d}f(\mathbf{x}(t), \mathbf{v}(t), t)/\mathrm{d}t$, where the $\mathrm{d}/\mathrm{d}t$ means that the derivative is measured in the moving frame. Because $\mathrm{d}\mathbf{x}/\mathrm{d}t = \mathbf{v}$ and $\mathrm{d}\mathbf{v}/\mathrm{d}t = \mathbf{a}$, this observed rate of change is

$$\left(\frac{\mathrm{d}f(\mathbf{x}(t), \mathbf{v}(t), t)}{\mathrm{d}t}\right)_{\text{orbit}} = \frac{\partial f}{\partial t}+\mathbf{v}\cdot\frac{\partial f}{\partial \mathbf{x}}+\mathbf{a}\cdot\frac{\partial f}{\partial \mathbf{v}} = 0. \tag{2.8}$$

Thus, the distribution function f as measured when moving along a particle trajectory (orbit) is *constant.* This gives a powerful method for finding solutions to the Vlasov equation. Since f is a constant when measured in a frame following an orbit, we can choose f to depend on *any quantity* that is constant along the orbit (Jeans 1915, Watson 1956).

For example, if the energy E of particles is constant along their orbits then $f = f(E)$ is a solution to the Vlasov equation. On the other hand, if both the energy and the momentum p are constant along particle orbits, then any distribution function with the functional dependence $f = f(E, p)$ is a solution to the Vlasov equation. Depending on the situation at hand, the energy and/or momentum may or may not be constant along an orbit and so whether or not $f = f(E, p)$ is a solution to the Vlasov equation depends on the specific problem under consideration. However, there always exists at least one constant of the motion for any trajectory because, just like every human being has an invariant birthday, the initial conditions of a particle trajectory are invariant along this trajectory. As a simple example, consider a situation where there is no electromagnetic field so that $\mathbf{a} = 0$, in which case the particle trajectories are simply $\mathbf{x}(t) = \mathbf{x}_0 + \mathbf{v}_0 t$, $\mathbf{v}(t) = \mathbf{v}_0$, where $\mathbf{x}_0, \mathbf{v}_0$ are the initial position and velocity. Let us check to see whether $f(\mathbf{x}_0)$ is a solution to the Vlasov equation. By writing $\mathbf{x}_0 = \mathbf{x}(t) - \mathbf{v}_0 t$ so $f(\mathbf{x}_0) = f(\mathbf{x}(t) - \mathbf{v}_0 t)$ we observe that indeed $f = f(\mathbf{x}_0)$ is a solution, since

$$\frac{\partial f}{\partial t} + \mathbf{v} \cdot \frac{\partial f}{\partial \mathbf{x}} + \mathbf{a} \cdot \frac{\partial f}{\partial \mathbf{v}} = -\mathbf{v}_0 \cdot \frac{\partial f}{\partial \mathbf{x}} + \mathbf{v} \cdot \frac{\partial f}{\partial \mathbf{x}} = 0. \tag{2.9}$$

2.4 Moments of the distribution function

Let us count the particles in the shaded vertical strip in Fig. 2.3. The number of particles in this strip is the number of particles lying between x and $x + \mathrm{d}x$, where x is the location of the left-hand side of the strip and $x + \mathrm{d}x$ is the location of the right-hand side. The number of particles in the strip is equivalently defined as $n(x, t)\mathrm{d}x$, where $n(x)$ is the *density* of particles at x. Thus we see that $\int f(x, v)\mathrm{d}v = n(x)$; the transition from a phase-space description (i.e., x, v are independent variables) to a conventional description (i.e., only x is an independent variable) involves "integrating out" the velocity dependence to obtain a quantity (e.g., density) depending only on position. Since the number of particles is finite, and since f is a positive quantity, f must vanish as $v \to \pm\infty$.

Another way of viewing f is to consider $f(x, v, t)/n(x, t)$ as the probability that a randomly selected particle at position x has the velocity v at time t. Using this point of view, we see that averaging over the velocities of all particles at x gives the mean velocity $u(x, t) = \int \mathrm{d}v\, v f(x, v, t)/n(x, t)$. Similarly, multiplying $f(x, v, t)/n(x, t)$ by $mv^2/2$ and integrating over velocity will give an expression for the mean kinetic energy of all the particles. This procedure of multiplying f by various powers of v and then integrating over velocity is called *taking moments of the distribution function.*

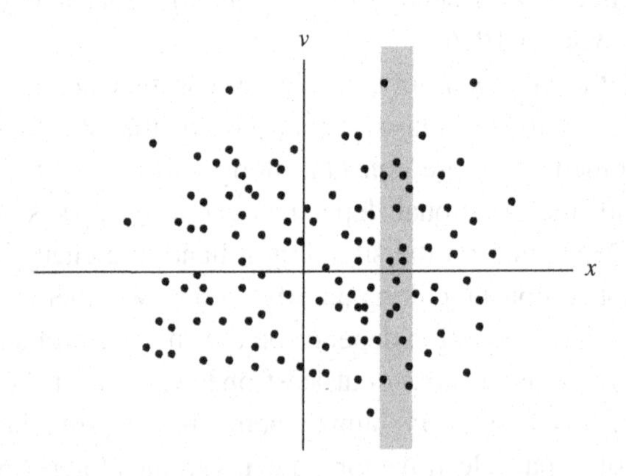

Fig. 2.3 Moments give weighted averages of the particles in the shaded vertical strip.

It is straightforward to generalize this "moment-taking" to three-dimensional problems simply by taking integrals over three-dimensional velocity space. Thus, in three dimensions the density becomes

$$n(\mathbf{x}, t) = \int f(\mathbf{x}, \mathbf{v}, t) d\mathbf{v} \tag{2.10}$$

and the mean velocity becomes

$$\mathbf{u}(\mathbf{x}, t) = \frac{\int \mathbf{v} f(\mathbf{x}, \mathbf{v}, t) d\mathbf{v}}{n(\mathbf{x}, t)}. \tag{2.11}$$

2.4.1 Treatment of collisions in the Vlasov equation

It was shown in Section 1.8 that the cumulative effect of grazing collisions dominates the cumulative effect of the more infrequently occurring large-angle collisions. In order to see how collisions affect the Vlasov equation, let us now temporarily imagine that the grazing collisions are replaced by an equivalent sequence of abrupt large scattering angle encounters as shown in Fig. 2.4. Two particles involved in a collision do not significantly change their positions during the course of a collision, but they do substantially change their velocities. For example, a particle making a head-on collision with an equal mass stationary particle will stop after the collision, while the target particle will assume the velocity of the incident particle. If we draw the detailed phase-space trajectories characterized by a collision between two particles we see that each particle has a sudden change in its vertical coordinate (i.e., velocity) but no change in its horizontal coordinate (i.e., position). The collision-induced velocity jump occurs very fast so that if the phase-space trajectories were recorded with a "movie

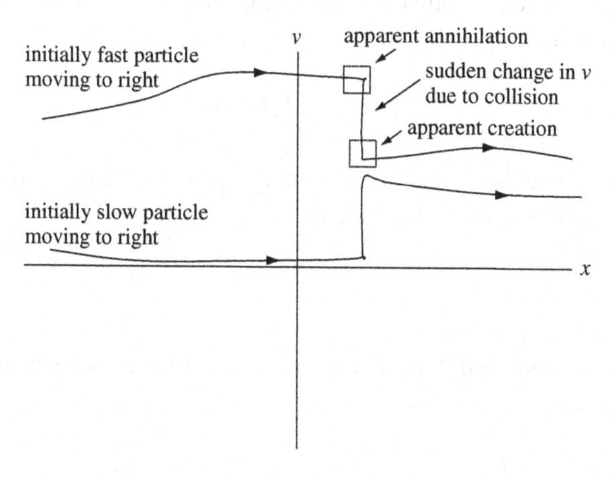

Fig. 2.4 Detailed view of collisions causing "jumps" in phase-space.

camera" having insufficient framing rate to catch the details of the jump, the resulting movie would show particles being spontaneously created or annihilated within given volumes of phase-space (e.g., within the boxes shown in Fig. 2.4).

The details of these individual jumps in phase-space are complicated and yet of little interest since all we really want to know is the cumulative effect of many collisions. It is therefore both efficient and sufficient to follow the trajectories on the slow time scale while accounting for the apparent "creation" or "annihilation" of particles by inserting a *collision operator* on the right-hand side of the Vlasov equation. In the example shown here it is seen that when a particle is apparently "created" in one box, another particle must be simultaneously "annihilated" in another box at the same x coordinate but a different v coordinate (of course, what is actually happening is that a single particle is suddenly moving from one box to the other). This coupling of the annihilation and creation rates in different boxes constrains the form of the collision operator. We will not attempt to derive collision operators in this chapter but will simply discuss the constraints on these operators. From a more formal point of view, collisions are characterized by constrained sources and sinks for particles in phase-space and inclusion of collisions in the Vlasov equation causes the Vlasov equation to assume the form

$$\frac{\partial f_\sigma}{\partial t} + \frac{\partial}{\partial \mathbf{x}} \cdot (\mathbf{v} f_\sigma) + \frac{\partial}{\partial \mathbf{v}} \cdot (\mathbf{a} f_\sigma) = \sum_\alpha C_{\sigma\alpha}(f_\sigma), \tag{2.12}$$

where $C_{\sigma\alpha}(f_\sigma)$ is the rate of change of f_σ due to collisions of species σ with species α.

The constraints that must be satisfied by the collision operator $C_{\sigma\alpha}(f_\sigma)$ are as follows:

- Conservation of particles – Collisions cannot change the total number of particles at a particular location so

$$\int d\mathbf{v} C_{\sigma\alpha}(f_\sigma) = 0. \tag{2.13}$$

- Conservation of momentum – Collisions between particles of the same species cannot change the total momentum of that species so

$$\int d\mathbf{v} m_\sigma \mathbf{v} C_{\sigma\sigma}(f_\sigma) = 0 \tag{2.14}$$

while collisions between different species must conserve the total momentum of both species together so

$$\int d\mathbf{v} m_i \mathbf{v} C_{ie}(f_i) + \int d\mathbf{v} m_e \mathbf{v} C_{ei}(f_e) = 0. \tag{2.15}$$

- Conservation of energy – Collisions between particles of the same species cannot change the total energy of that species so

$$\int d\mathbf{v} m_\sigma v^2 C_{\sigma\sigma}(f_\sigma) = 0 \tag{2.16}$$

while collisions between different species must conserve the total energy of both species together so

$$\int d\mathbf{v} m_i v^2 C_{ie}(f_i) + \int d\mathbf{v} m_e v^2 C_{ei}(f_e) = 0. \tag{2.17}$$

2.5 Two-fluid equations

Instead of just taking moments of the distribution function f itself, moments will now be taken of the entire Vlasov equation to obtain a set of partial differential equations relating the mean quantities $n(\mathbf{x})$, $\mathbf{u}(\mathbf{x})$, etc. We begin by integrating the Vlasov equation, Eq. (2.12), over velocity for each species. This first and simplest step in the procedure is called taking the "zeroth" moment, since the operation of multiplying by unity can be considered as multiplying the entire Vlasov equation by $\mathbf{v}$ raised to the power zero. Multiplying the Vlasov equation by unity and then integrating over velocity gives

$$\int \left[\frac{\partial f_\sigma}{\partial t} + \frac{\partial}{\partial \mathbf{x}} \cdot (\mathbf{v} f_\sigma) + \frac{\partial}{\partial \mathbf{v}} \cdot (\mathbf{a} f_\sigma) \right] d\mathbf{v} = \sum_\alpha \int C_{\sigma\alpha}(f_\sigma) d\mathbf{v}. \tag{2.18}$$

The velocity integral commutes with both the time and space derivatives on the left-hand side because $\mathbf{x}$, $\mathbf{v}$, and t are independent variables, while the third term on the left-hand side is the volume integral of a divergence in velocity space. Gauss' theorem (i.e., $\int_{vol} d\mathbf{x} \nabla \cdot \mathbf{Q} = \int_{sfc} d\mathbf{s} \cdot \mathbf{Q}$) gives f_σ evaluated on a surface at $v = \infty$. However, because $f_\sigma \to 0$ as $v \to \infty$, this surface integral in velocity space vanishes. Using Eqs. (2.10), (2.11), and (2.13), we see that Eq. (2.18) becomes the species *continuity equation*

$$\frac{\partial n_\sigma}{\partial t} + \nabla \cdot (n_\sigma \mathbf{u}_\sigma) = 0. \tag{2.19}$$

Now let us multiply Eq. (2.12) by $\mathbf{v}$ and integrate over velocity to take the "first moment,"

$$\int \mathbf{v} \left[\frac{\partial f_\sigma}{\partial t} + \frac{\partial}{\partial \mathbf{x}} \cdot (\mathbf{v} f_\sigma) + \frac{\partial}{\partial \mathbf{v}} \cdot (\mathbf{a} f_\sigma) \right] d\mathbf{v} = \sum_\alpha \int \mathbf{v} C_{\sigma\alpha}(f_\sigma) d\mathbf{v}. \tag{2.20}$$

This may be rearranged in a more tractable form by:

(i) "pulling" both the time and space derivatives out of the velocity integral,
(ii) writing $\mathbf{v} = \mathbf{v}'(\mathbf{x}, \mathbf{t}) + \mathbf{u}(\mathbf{x}, t)$, where $\mathbf{v}'(\mathbf{x}, t)$ is the *random* part of a given velocity, i.e., that part of the velocity that differs from the mean (note that $\mathbf{v}$ is independent of both $\mathbf{x}$ and t but $\mathbf{v}'$ is not; also $\mathrm{d}\mathbf{v} = \mathrm{d}\mathbf{v}'$),
(iii) integrating by parts in 3-D velocity space on the acceleration term and using

$$\left(\frac{\partial \mathbf{v}}{\partial \mathbf{v}}\right)_{ij} = \delta_{ij}.$$

After performing these manipulations, the first moment of the Vlasov equation becomes

$$\frac{\partial (n_\sigma \mathbf{u}_\sigma)}{\partial t} + \frac{\partial}{\partial \mathbf{x}} \cdot \int (\mathbf{v}'\mathbf{v}' + \mathbf{v}'\mathbf{u}_\sigma + \mathbf{u}_\sigma \mathbf{v}' + \mathbf{u}_\sigma \mathbf{u}_\sigma) f_\sigma \mathrm{d}\mathbf{v}'$$
$$- \frac{q_\sigma}{m_\sigma} \int (\mathbf{E} + \mathbf{v} \times \mathbf{B}) f_\sigma \mathrm{d}\mathbf{v}' = -\frac{1}{m_\sigma} \mathbf{R}_{\sigma\alpha}, \quad (2.21)$$

where $\mathbf{R}_{\sigma\alpha}$ is the net frictional drag force due to collisions of species σ with species α. Note that $\mathbf{R}_{\sigma\sigma} = 0$, since a species cannot exert a net drag force on itself. This is because, just like one cannot pull oneself up by one's own bootstraps, the totality of electrons cannot cause a force that changes the center-of-mass velocity of the totality of electrons. The frictional terms have the form

$$\mathbf{R}_{ei} = \nu_{ei} m_e n_e (\mathbf{u}_e - \mathbf{u}_i), \quad (2.22)$$

$$\mathbf{R}_{ie} = \nu_{ie} m_i n_i (\mathbf{u}_i - \mathbf{u}_e), \quad (2.23)$$

so that in the ion frame the drag on electrons is simply the total electron momentum $m_e n_e \mathbf{u}_e$ measured in this frame multiplied by the rate ν_{ei} at which this momentum is destroyed by collisions with ions. This form for frictional drag force has the following properties: (i) $\mathbf{R}_{ei} + \mathbf{R}_{ie} = 0$, showing that the plasma cannot exert a frictional drag force on itself, (ii) friction causes the faster species to be slowed down by the slower species, and (iii) there is no friction between species if both have the same mean velocity.

Equation (2.21) can be further simplified by factoring $\mathbf{u}$ out of the velocity integrals and recalling that by definition $\int \mathbf{v}' f_\sigma \mathrm{d}\mathbf{v}' = 0$. Thus, Eq. (2.21) reduces to

$$m_\sigma \left[\frac{\partial (n_\sigma \mathbf{u}_\sigma)}{\partial t} + \frac{\partial}{\partial \mathbf{x}} \cdot (n_\sigma \mathbf{u}_\sigma \mathbf{u}_\sigma)\right] = n_\sigma q_\sigma (\mathbf{E} + \mathbf{u}_\sigma \times \mathrm{B}) - \frac{\partial}{\partial \mathbf{x}} \cdot \overleftrightarrow{\mathbf{P}}_\sigma - \mathbf{R}_{\sigma\alpha}, \quad (2.24)$$

where the *pressure tensor* $\overleftrightarrow{\mathbf{P}}$ is defined by

$$\overleftrightarrow{\mathbf{P}}_\sigma = m_\sigma \int \mathbf{v}'\mathbf{v}' f_\sigma \mathrm{d}\mathbf{v}'.$$

If f_σ is an isotropic function of $\mathbf{v}'$, then the off-diagonal terms in $\overleftrightarrow{\mathbf{P}}_\sigma$ vanish and the three diagonal terms are identical. In this case, it is useful to define the diagonal terms to be the *scalar pressure* P_σ, i.e.,

$$\begin{aligned} P_\sigma &= m_\sigma \int v'_x v'_x f_\sigma \mathrm{d}\mathbf{v}' = m_\sigma \int v'_y v'_y f_\sigma \mathrm{d}\mathbf{v}' = m_\sigma \int v'_z v'_z f_\sigma \mathrm{d}\mathbf{v}' \\ &= \frac{m_\sigma}{3} \int \mathbf{v}' \cdot \mathbf{v}' f_\sigma \mathrm{d}\mathbf{v}'. \end{aligned} \tag{2.25}$$

Equation (2.25) defines pressure for a three-dimensional isotropic system. However, we will often deal with systems of reduced dimensionality, i.e., systems with just one or two dimensions. Equation (2.25) can therefore be generalized to these other cases by introducing the general N-dimensional definition for scalar pressure

$$P_\sigma = \frac{m_\sigma}{N} \int \mathbf{v}' \cdot \mathbf{v}' f_\sigma \mathrm{d}^N\mathbf{v}' = \frac{m_\sigma}{N} \int \sum_{j=1}^{N} v_j'^2 f_\sigma \mathrm{d}^N\mathbf{v}', \tag{2.26}$$

where $\mathbf{v}'$ is the N-dimensional random velocity.

It is important to emphasize that assuming isotropy is done largely for mathematical convenience and that in real systems the distribution function is often quite anisotropic. Collisions, being randomizing, drive the distribution function towards isotropy, while competing processes simultaneously drive it towards anisotropy. Thus, each situation must be considered individually in order to determine whether there is sufficient collisionality to make f isotropic. Because fully ionized hot plasmas often have insufficient collisions to make f isotropic, the oft-used assumption of isotropy is an oversimplification, which may or may not be acceptable depending on the phenomenon under consideration.

On expanding the derivatives on the left-hand side of Eq. (2.24), it is seen that two of the terms combine to give $\mathbf{u}$ times Eq. (2.19). After removing this embedded continuity equation, Eq. (2.24) reduces to

$$n_\sigma m_\sigma \frac{\mathrm{d}\mathbf{u}_\sigma}{\mathrm{d}t} = n_\sigma q_\sigma (\mathbf{E} + \mathbf{u}_\sigma \times \mathbf{B}) - \nabla P_\sigma - \mathbf{R}_{\sigma\alpha}, \tag{2.27}$$

where the operator $\mathrm{d}/\mathrm{d}t$ is defined to be the *convective derivative*

$$\frac{\mathrm{d}}{\mathrm{d}t} = \frac{\partial}{\partial t} + \mathbf{u}_\sigma \cdot \nabla, \tag{2.28}$$

which characterizes the temporal rate of change seen by an observer *moving with the mean fluid velocity* $\mathbf{u}_\sigma$ of species σ. An everyday example of the convective term would be the apparent temporal increase in density of automobiles seen by a motorcyclist who enters a traffic jam of stationary vehicles and is not impeded by the traffic jam.

At this point in the procedure it becomes evident that a certain pattern recurs for each successive moment of the Vlasov equation. When we took the zeroth moment, an equation for the density $\int f_\sigma \mathrm{d}\mathbf{v}$ resulted, but this also introduced a term involving the next higher moment, namely the mean velocity $\sim \int \mathbf{v} f_\sigma \mathrm{d}\mathbf{v}$. Then, when we took the first moment to get an equation for the velocity, an equation was obtained containing a term involving the next higher moment, namely the pressure $\sim \int \mathbf{v}\mathbf{v} f_\sigma \mathrm{d}\mathbf{v}$. Thus, each time we take a moment of the Vlasov equation, an equation for the moment we want is obtained, but because of the $\mathbf{v}\cdot\nabla f$ term in the Vlasov equation, a next higher moment also appears. Thus, moment-taking never leads to a closed system of equations; there will always be a "loose end," a highest moment for which there is no determining equation. Some sort of ad hoc closure procedure must always be invoked to terminate this chain (as seen below, typical closures involve invoking adiabatic or isothermal assumptions). Another feature of taking moments is that each higher moment embeds terms that contain complete lower moment equations multiplied by some factor. Algebraic manipulation can identify these lower moment equations and eliminate them to give a simplified higher moment equation.

Let us now take the second moment of the Vlasov equation. Unlike the zeroth and first moments, the dimensionality of the system now enters explicitly so the more general pressure definition given by Eq. (2.26) will be used. Multiplying the Vlasov equation by $m_\sigma v^2/2$ and integrating over velocity gives

$$\left\{\begin{array}{l} \dfrac{\partial}{\partial t}\displaystyle\int \dfrac{m_\sigma v^2}{2} f_\sigma \mathrm{d}^N\mathbf{v} \\ +\dfrac{\partial}{\partial \mathbf{x}}\cdot\displaystyle\int \dfrac{m_\sigma v^2}{2}\mathbf{v} f_\sigma \mathrm{d}^N\mathbf{v} \\ +q_\sigma \displaystyle\int \dfrac{v^2}{2}\dfrac{\partial}{\partial \mathbf{v}}\cdot(\mathbf{E}+\mathbf{v}\times\mathbf{B}) f_\sigma \mathrm{d}^N\mathbf{v} \end{array}\right\} = \sum_\alpha \int m_\sigma \frac{v^2}{2} C_{\sigma\alpha} f_\sigma \mathrm{d}^N\mathbf{v}. \tag{2.29}$$

We consider each term of this equation separately as follows:

1. The time derivative term becomes

$$\frac{\partial}{\partial t}\int \frac{m_\sigma v^2}{2} f_\sigma \mathrm{d}^N\mathbf{v} = \frac{\partial}{\partial t}\int \frac{m_\sigma(\mathbf{v}'+\mathbf{u}_\sigma)^2}{2} f_\sigma \mathrm{d}^N\mathbf{v}' = \frac{\partial}{\partial t}\left(\frac{NP_\sigma}{2}+\frac{m_\sigma n_\sigma u_\sigma^2}{2}\right).$$

2. Again using $\mathbf{v}=\mathbf{v}'+\mathbf{u}_\sigma$ the space derivative term becomes

$$\frac{\partial}{\partial \mathbf{x}}\cdot\int \frac{m_\sigma v^2}{2}\mathbf{v} f_\sigma \mathrm{d}^N\mathbf{v} = \nabla\cdot\left(\mathbf{Q}_\sigma + \frac{2+N}{2}P_\sigma \mathbf{u}_\sigma + \frac{m_\sigma n_\sigma u_\sigma^2}{2}\mathbf{u}_\sigma\right),$$

where $\mathbf{Q}_\sigma = \int \frac{m_\sigma v'^2}{2}\mathbf{v}' f_\sigma \mathrm{d}^N\mathbf{v}$ is called the *heat flux*.

3. On integrating by parts, the acceleration term becomes

$$q_\sigma \int \frac{v^2}{2} \frac{\partial}{\partial \mathbf{v}} \cdot [(\mathbf{E}+\mathbf{v}\times\mathbf{B}) f_\sigma] \mathrm{d}^N\mathbf{v} = -q_\sigma \int \mathbf{v}\cdot\mathbf{E} f_\sigma \mathrm{d}\mathbf{v} = -q_\sigma n_\sigma \mathbf{u}_\sigma \cdot \mathbf{E}.$$

4. The collision term becomes (using Eq. (2.16))

$$\sum_\alpha \int m_\sigma \frac{v^2}{2} C_{\sigma\alpha} f_\sigma \mathrm{d}\mathbf{v} = \int_{\alpha\neq\sigma} m_\sigma \frac{v^2}{2} C_{\sigma\alpha} f_\sigma \mathrm{d}\mathbf{v} = -\left(\frac{\partial W}{\partial t}\right)_{E\sigma\alpha},$$

where $(\partial W/\partial t)_{E\sigma\alpha}$ is the rate at which species σ collisionally transfers energy to species α.

Combining the above four relations, Eq. (2.29) becomes

$$\frac{\partial}{\partial t}\left(\frac{NP_\sigma}{2} + \frac{m_\sigma n_\sigma u_\sigma^2}{2}\right) + \nabla\cdot\left(\mathbf{Q}_\sigma + \frac{2+N}{2} P_\sigma \mathbf{u}_\sigma + \frac{m_\sigma n_\sigma u_\sigma^2}{2}\mathbf{u}_\sigma\right) - q_\sigma n_\sigma \mathbf{u}_\sigma \cdot \mathbf{E}$$
$$= -\left(\frac{\partial W}{\partial t}\right)_{E\sigma\alpha}. \tag{2.30}$$

This equation can be simplified by invoking two mathematical identities, the first being

$$\frac{\partial}{\partial t}\left(\frac{m_\sigma n_\sigma u_\sigma^2}{2}\right) + \nabla\cdot\left(\frac{m_\sigma n_\sigma u_\sigma^2}{2}\mathbf{u}_\sigma\right) = n_\sigma\left(\frac{\partial}{\partial t} + \mathbf{u}_\sigma\cdot\nabla\right)\frac{m_\sigma u_\sigma^2}{2} = n_\sigma \frac{\mathrm{d}}{\mathrm{d}t}\left(\frac{m_\sigma u_\sigma^2}{2}\right). \tag{2.31}$$

The second identity is obtained by dotting the equation of motion with $\mathbf{u}_\sigma$ and is

$$n_\sigma m_\sigma \left[\frac{\partial}{\partial t}\left(\frac{u_\sigma^2}{2}\right) + \mathbf{u}_\sigma \cdot \left(\nabla\left(\frac{u_\sigma^2}{2}\right) - \mathbf{u}_\sigma \times \nabla \times \mathbf{u}_\sigma\right)\right]$$
$$= n_\sigma q_\sigma \mathbf{u}_\sigma \cdot \mathbf{E} - \mathbf{u}_\sigma \cdot \nabla P_\sigma - \mathbf{R}_{\sigma\alpha} \cdot \mathbf{u}_\sigma$$

or

$$n_\sigma \frac{\mathrm{d}}{\mathrm{d}t}\left(\frac{m_\sigma u_\sigma^2}{2}\right) = n_\sigma q_\sigma \mathbf{u}_\sigma \cdot \mathbf{E} - \mathbf{u}_\sigma \cdot \nabla P_\sigma - \mathbf{R}_{\sigma\alpha} \cdot \mathbf{u}_\sigma. \tag{2.32}$$

Inserting Eqs. (2.31) and (2.32) in Eq. (2.30) gives the energy evolution equation

$$\frac{N}{2}\frac{\mathrm{d}P_\sigma}{\mathrm{d}t} + \frac{2+N}{2} P \nabla\cdot\mathbf{u}_\sigma = -\nabla\cdot\mathbf{Q}_\sigma + \mathbf{R}_{\sigma\alpha}\cdot\mathbf{u}_\sigma - \left(\frac{\partial W}{\partial t}\right)_{E\sigma\alpha}. \tag{2.33}$$

The first term on the right-hand side represents the heat flux, the second term gives the frictional heating of species σ due to frictional drag on species α, while the last term on the right-hand side gives the rate at which species σ collisionally transfers energy to other species. Although Eq. (2.33) is complicated, two important limiting situations become evident if we let t_{char} be the characteristic time scale for a given phenomenon and l_{char} be its characteristic length scale. A characteristic velocity

$V_{ph} \sim l_{char}/t_{char}$ may then be defined for the phenomenon and so, replacing temporal derivatives by t_{char}^{-1} and spatial derivatives by l_{char}^{-1} in Eq. (2.33), it is seen that the two limiting situations are:

1. *Isothermal limit* – The heat flux term dominates all other terms, in which case the temperature becomes spatially uniform. This occurs if (i) $v_{T\sigma} \gg V_{ph}$ since the ratio of the left-hand side terms to the heat flux term is $\sim V_{ph}/v_{T\sigma}$ and (ii) the collisional terms are small enough to be ignored.
2. *Adiabatic limit* – The heat flux terms and the collisional terms are small enough to be ignored compared to the left-hand side terms; this occurs when $V_{ph} \gg v_{T\sigma}$. Adiabatic is a Greek word meaning "impassable," and is used here to denote that no heat is flowing, i.e., the volume under consideration is thermally isolated from the outside world.

Both of these limits make it possible to avoid solving for $\mathbf{Q}_\sigma$, which involves the third moment, and so both the adiabatic and isothermal limit provide a closure to the moment equations.

The energy equation may be greatly simplified in the adiabatic limit by rearranging the continuity equation to give

$$\nabla \cdot \mathbf{u}_\sigma = -\frac{1}{n_\sigma}\frac{\mathrm{d}n_\sigma}{\mathrm{d}t} \tag{2.34}$$

and then substituting this expression into the left-hand side of Eq. (2.33) to obtain

$$\frac{1}{P_\sigma}\frac{\mathrm{d}P_\sigma}{\mathrm{d}t} = \frac{\gamma}{n_\sigma}\frac{\mathrm{d}n_\sigma}{\mathrm{d}t}, \tag{2.35}$$

where

$$\gamma = \frac{N+2}{N}. \tag{2.36}$$

Equation (2.35) implies

$$\frac{\mathrm{d}}{\mathrm{d}t}\left(\frac{P_\sigma}{n_\sigma^\gamma}\right) = 0 \tag{2.37}$$

so

$$\frac{P_\sigma}{n_\sigma^\gamma} = \text{constant } \textit{in the frame of the moving plasma.} \tag{2.38}$$

This constitutes a derivation of adiabaticity based on geometry and statistical mechanics rather than on thermodynamic arguments.

2.5.1 Entropy of a distribution function

Collisions cause the distribution function to tend towards a simple final state characterized by having the maximum entropy for the given constraints (e.g., fixed total energy). To see this, we provide a brief discussion of entropy and show how it relates to a distribution function.

Suppose we throw two dice, labeled A and B, and let R denote the result of a throw. Thus R ranges from 2 through 12. The complete set of (A, B) combinations that gives these R's is listed below:

$R=2 \Longleftrightarrow$(1,1)
$R=3 \Longleftrightarrow$(1,2),(2,1)
$R=4 \Longleftrightarrow$(1,3),(3,1),(2,2)
$R=5 \Longleftrightarrow$(1,4),(4,1),(2,3),(3,2)
$R=6 \Longleftrightarrow$(1,5),(5,1),(2,4),(4,2),(3,3)
$R=7 \Longleftrightarrow$(1,6),(6,1),(2,5),(5,2),(3,4),(4,3)
$R=8 \Longleftrightarrow$(2,6),(6,2),(3,5),(5,3),(4,4)
$R=9 \Longleftrightarrow$(3,6),(6,3),(4,5),(5,4)
$R=10 \Longleftrightarrow$(4,6),(6,4),(5,5)
$R=11 \Longleftrightarrow$(5,6),(6,5)
$R=12 \Longleftrightarrow$(6,6)

There are six (A, B) pairs that give $R=7$, but only one pair for $R=2$ and only one pair for $R=12$. Thus, there are six microscopic states (distinct (A, B) pairs) corresponding to $R=7$ but only one microscopic state corresponding to each of $R=2$ or $R=12$. Thus, we *know more* about the microscopic state of the system if $R=2$ or 12 than if $R=7$. We define the entropy S to be the *natural logarithm of the number of microscopic states corresponding to a given macroscopic state.* Thus, for the dice, the entropy would be the natural logarithm of the number of (A, B) pairs that correspond to a given R. The entropy for $R=2$ or $R=12$ would be zero since $S=\ln(1)=0$, while the entropy for $R=7$ would be $S=\ln(6)$ since there were six different ways of obtaining $R=7$.

If the dice were to be thrown a statistically large number of times the most likely result for any throw is $R=7$; this is the macroscopic state with the largest number of microscopic states. Since any of the possible microscopic states is an equally likely outcome, the most likely macroscopic state after a large number of dice throws is the macroscopic state with the highest entropy.

Now consider a situation more closely related to the concept of a distribution function. In order to do this we first pose the following simple problem: suppose we have a pegboard with $\mathcal{N}$ holes, labeled $h_1, h_2, \ldots, h_{\mathcal{N}}$, and we also have $\mathcal{N}$ pegs labeled by $p_1, p_2, \ldots, p_{\mathcal{N}}$. What is the number of ways of putting all the pegs in all the holes? Starting with hole h_1, we have a choice of $\mathcal{N}$ different pegs, but when we get to hole h_2 there are now only $\mathcal{N}-1$ pegs remaining so that there are now only $\mathcal{N}-1$ choices. Using this argument for subsequent holes, we see there are $\mathcal{N}!$ ways of putting all the pegs in all the holes.

Let us complicate things further. Suppose that we arrange the holes in $\mathcal{M}$ groups. Say group G_1 has the first 10 holes, group G_2 has the next 19 holes, group G_3 has the next 4 holes and so on, up to group $\mathcal{M}$. We will use f to denote the number of holes in a group, thus $f(1)=10$, $f(2)=19$, $f(3)=4$, etc. The number of ways of arranging pegs within a group is just the factorial of the number of pegs in the group, e.g., the number of ways of arranging the pegs within group 1 is just 10! and so in general the number of ways of arranging the pegs in the jth group is $[f(j)]!$.

Let us denote C as the number of ways of putting all the pegs in all the groups *without caring* about the internal arrangement within groups. The number of ways of putting the pegs in all the groups *caring* about the internal arrangements in all the groups is $C \times f(1)! \times f(2)! \times \ldots f(\mathcal{M})!$, but this is just the number of ways of putting all the pegs in all the holes, i.e.,

$$C \times f(1)! \times f(2)! \times \ldots f(\mathcal{M})! = \mathcal{N}!$$

or

$$C = \frac{\mathcal{N}!}{f(1)! \times f(2)! \times \ldots f(\mathcal{M})!}.$$

Now C is just the number of microscopic states corresponding to the macroscopic state of the prescribed grouping $f(1) = 10$, $f(2) = 19$, $f(3) = 4$, etc. so the entropy is just $S = \ln\, C$ or

$$\begin{aligned} S &= \ln\left(\frac{\mathcal{N}!}{f(1)! \times f(2)! \times \ldots f(\mathcal{M})!}\right) \\ &= \ln \mathcal{N}! - \ln f(1)! - \ln f(2)! - \ldots - \ln f(\mathcal{M})!. \end{aligned} \tag{2.39}$$

Stirling's formula shows that the large-argument asymptotic limit of the factorial function is

$$\lim_{k \to \text{large}} \ln k! = k \ln k - k. \tag{2.40}$$

Noting that $f(1) + f(2) + \ldots f(\mathcal{M}) = N$ the entropy becomes

$$\begin{aligned} S &= \mathcal{N} \ln \mathcal{N} - f(1) \ln f(1) - f(2) \ln f(2) - \ldots - f(M) \ln f(M) \\ &= \mathcal{N} \ln \mathcal{N} - \sum_{j=1}^{\mathcal{M}} f(j) \ln f(j). \end{aligned} \tag{2.41}$$

The constant $\mathcal{N} \ln \mathcal{N}$ is often dropped, giving

$$S = -\sum_{j=1}^{\mathcal{M}} f(j) \ln f(j). \tag{2.42}$$

If j is made into a continuous variable, say $j \to v$ so that $f(v)\mathrm{d}v$ is the number of items in the group labeled by v, then the entropy can be written as

$$S = -\int \mathrm{d}v f(v) \ln f(v). \tag{2.43}$$

By now, it is obvious that f could be the velocity distribution function, in which case $f(v)\mathrm{d}v$ is just the number of particles in the group having velocity between v and $v+\mathrm{d}v$. Since the peg groups correspond to different velocity range coordinates, having more dimensions just means having more groups and so for three dimensions the entropy generalizes to

$$S = -\int \mathrm{d}\mathbf{v}\, f(\mathbf{v}) \ln f(\mathbf{v}). \tag{2.44}$$

If the distribution function depends on position as well, this corresponds to still more peg groups, and so a distribution function that depends on both velocity and position will have the entropy

$$S = -\int \mathrm{d}\mathbf{x} \int \mathrm{d}\mathbf{v}\, f(\mathbf{x}, \mathbf{v}) \ln f(\mathbf{x}, \mathbf{v}). \tag{2.45}$$

2.5.2 *Effect of collisions on entropy*

The highest entropy state is the most likely state of the system because the highest entropy state has the highest number of microscopic states corresponding to the macroscopic state. Collisions (or other forms of randomization) will take some initial prescribed microscopic state and scramble the phase-space positions of the particles, thereby transforming the system to a different microscopic state. This new state could in principle be any microscopic state, but is most likely to be a member of the class of microscopic states belonging to the highest entropy macroscopic state. Thus, any randomization process such as collisions will cause the system to evolve towards the macroscopic state having the maximum entropy.

An important shortcoming of this argument is that it neglects any conservation relations that have to be satisfied. Thus, a qualification must be added to the argument. Randomizing processes will scramble the system to attain the state of maximum entropy *consistent* with any constraints placed on the system. Examples of such constraints would be the requirements that the total system energy and the total number of particles must be conserved. We therefore reformulate the problem as: given an isolated system with $\mathcal{N}$ particles in a fixed volume V and initial average energy per particle $\langle E \rangle$, what is the maximum entropy state consistent with conservation of energy and conservation of number of particles?

This is a variational problem because the goal is to maximize S subject to the constraint that both $\mathcal{N}$ and $\mathcal{N}\langle E\rangle$ are fixed. The method of Lagrange multipliers can then be used to take into account these constraints (see Assignment 2 for a derivation of the Lagrange multiplier method). Using this method the variational problem becomes

$$\delta S - \lambda_1 \delta \mathcal{N} - \lambda_2 \delta(\mathcal{N}\langle E\rangle) = 0, \tag{2.46}$$

where λ_1 and λ_2 are as-yet undetermined Lagrange multipliers. The number of particles is

$$\mathcal{N} = V \int f \mathrm{d}v. \tag{2.47}$$

The energy of an individual particle is $E = mv^2/2$, where v is the velocity measured in the rest frame of the center of mass of the entire collection of $\mathcal{N}$ particles. Thus, the total kinetic energy of all the particles in this rest frame is

$$\mathcal{N}\langle E\rangle = V \int \frac{mv^2}{2} f(v) \mathrm{d}v \tag{2.48}$$

and so the variational problem becomes

$$\delta \int \mathrm{d}v \left(f \ln f - \lambda_1 V f - \lambda_2 V \frac{mv^2}{2} f \right) = 0. \tag{2.49}$$

Incorporating the volume V into the Lagrange multipliers, and factoring out the coefficient δf, this becomes

$$\int \mathrm{d}v \, \delta f \left(1 + \ln f - \lambda_1 - \lambda_2 \frac{mv^2}{2} \right) = 0. \tag{2.50}$$

Since δf is arbitrary, the integrand must vanish, giving

$$\ln f = \lambda_2 \frac{mv^2}{2} - \lambda_1, \tag{2.51}$$

where the "1" has been incorporated into λ_1.

The maximum entropy distribution function of an isolated, energy and particle conserving system is therefore

$$f = \lambda_1 \exp(-\lambda_2 mv^2/2); \tag{2.52}$$

this is called a Maxwellian distribution function. We will often assume that the plasma is locally Maxwellian so that $\lambda_1 = \lambda_1(\mathbf{x}, t)$ and $\lambda_2 = \lambda(\mathbf{x}, t)$. We define the temperature to be

$$\kappa T_\sigma(\mathbf{x}, t) = \frac{1}{\lambda_2(\mathbf{x}, t)}, \tag{2.53}$$

where Boltzmann's factor κ allows temperature to be measured in various units. The normalization factor is set to be

$$\lambda_1(\mathbf{x}, t) = n(\mathbf{x}, t)\left(\frac{m_\sigma}{2\pi\kappa T_\sigma(\mathbf{x}, t)}\right)^{N/2}, \tag{2.54}$$

where N is the dimensionality (1, 2, or 3) so that $\int f_\sigma(\mathbf{x}, \mathbf{v}, t)\mathrm{d}^N\mathbf{v} = n_\sigma(\mathbf{x}, t)$. Because the kinetic energy of individual particles was defined in terms of velocities measured in the rest frame of the center of mass of the complete system of particles, if this center of mass is moving in the lab frame with a velocity $\mathbf{u}_\sigma$, then in the lab frame the Maxwellian will have the form

$$f_\sigma(\mathbf{x}, \mathbf{v}, t) = n_\sigma\left(\frac{m_\sigma}{2\pi\kappa T_\sigma}\right)^{N/2} \exp\left(-m_\sigma(\mathbf{v}-\mathbf{u}_\sigma)^2/2\kappa T_\sigma\right). \tag{2.55}$$

2.5.3 Relation between pressure and Maxwellian

The scalar pressure has a simple relation to the generalized Maxwellian as seen by recasting Eq. (2.26) as

$$\begin{aligned}
P_\sigma &= \frac{m_\sigma}{N}\int \mathbf{v}'\cdot\mathbf{v}' f_\sigma \mathrm{d}^N\mathbf{v}' \\
&= \frac{n_\sigma m_\sigma}{N}\left(\frac{m_\sigma}{2\pi\kappa T_\sigma}\right)^{N/2}\int (\mathbf{v}')^2 \exp\left(-m_\sigma (\mathbf{v}')^2/2\kappa T_\sigma\right)\mathrm{d}^N\mathbf{v}' \\
&= -\frac{n_\sigma m_\sigma}{N}\left(\frac{\alpha}{\pi}\right)^{N/2}\frac{\mathrm{d}}{\mathrm{d}\alpha}\int e^{-\alpha v'^2}\mathrm{d}^N\mathbf{v}', \text{ defining } \alpha = m_\sigma/2\kappa T_\sigma \\
&= -\frac{n_\sigma m_\sigma}{N}\left(\frac{\alpha}{\pi}\right)^{N/2}\frac{\mathrm{d}}{\mathrm{d}\alpha}\left(\frac{\alpha}{\pi}\right)^{-N/2} \\
&= n_\sigma \kappa T_\sigma,
\end{aligned} \tag{2.56}$$

which is just the ideal gas law. Thus, the definitions that have been proposed for pressure and temperature are consistent with everyday notions for these quantities.

Despite this correspondence to familiar concepts, we must be careful not to become overconfident regarding the descriptive power of the fluid point of view because weaknesses exist in this point of view. For example, as discussed on p. 46 neither the adiabatic nor the isothermal approximation is appropriate when $V_{ph} \sim v_{T\sigma}$. The fluid description breaks down in this situation and, as will be seen in a later chapter, the Vlasov description must be used in this situation. Furthermore, the distribution function is Maxwellian only if there are sufficient collisions or some other randomizing process. Because of the weak collisionality of a plasma, this is very often not the case. In particular, since the collision frequency scales as v^{-3}, fast particles take much longer to become Maxwellian than slow particles. It is not at all unusual for a real plasma to be in a state where the low-velocity particles have reached a Maxwellian distribution whereas the fast particles form a non-Maxwellian "tail."

We now summarize the two-fluid equations:

- continuity equation for each species

$$\frac{\partial n_\sigma}{\partial t} + \nabla \cdot (n_\sigma \mathbf{u}_\sigma) = 0; \tag{2.57}$$

- equation of motion for each species

$$n_\sigma m_\sigma \frac{d\mathbf{u}_\sigma}{dt} = n_\sigma q_\sigma (\mathbf{E} + \mathbf{u}_\sigma \times \mathbf{B}) - \nabla P_\sigma - \mathbf{R}_{\sigma\alpha}; \tag{2.58}$$

- equation of state for each species

Regime	Equation of state	Name
$V_{ph} \gg v_{T\sigma}$	$P_\sigma \sim n_\sigma^\gamma$	adiabatic
$V_{ph} \ll v_{T\sigma}$	$P_\sigma = n_\sigma \kappa T_\sigma$, $T_\sigma = \text{constant}$	isothermal

- Maxwell's equations

$$\nabla \times \mathbf{E} = -\frac{\partial \mathbf{B}}{\partial t} \tag{2.59}$$

$$\nabla \times \mathbf{B} = \mu_0 \sum_\sigma n_\sigma q_\sigma \mathbf{u}_\sigma + \mu_0 \varepsilon_0 \frac{\partial \mathbf{E}}{\partial t} \tag{2.60}$$

$$\nabla \cdot \mathbf{B} = 0 \tag{2.61}$$

$$\nabla \cdot \mathbf{E} = \frac{1}{\varepsilon_0} \sum_\sigma n_\sigma q_\sigma. \tag{2.62}$$

2.6 Magnetohydrodynamic equations

Particle motion in the two-fluid system was described by the individual species' mean velocities $\mathbf{u}_e$, $\mathbf{u}_i$ and by the pressures $\overleftrightarrow{\mathbf{P}}_e$, $\overleftrightarrow{\mathbf{P}}_i$, which provide an accounting for the mean square of the random deviation of the velocity from its average value. Magnetohydrodynamics is an alternate description of the plasma where, instead of using $\mathbf{u}_e$, $\mathbf{u}_i$ to describe mean motion, two new velocity variables that are a linear combination of $\mathbf{u}_e$, $\mathbf{u}_i$ are used. As will be seen below, this means a slightly different definition for pressure must also be used.

The new velocity-like variables are (i) the current density

$$\mathbf{J} = \sum_\sigma n_\sigma q_\sigma \mathbf{u}_\sigma, \tag{2.63}$$

which is essentially the relative velocity between ions and electrons, and (ii) the center-of-mass velocity

$$\mathbf{U} = \frac{1}{\rho}\sum_\sigma m_\sigma n_\sigma \mathbf{u}_\sigma, \tag{2.64}$$

where

$$\rho = \sum_\sigma m_\sigma n_\sigma \tag{2.65}$$

is the total mass density. Magnetohydrodynamics is primarily concerned with low-frequency, long-wavelength, *magnetic* behavior of the plasma.

2.6.1 MHD continuity equation

Multiplying Eq. (2.19) by m_σ and summing over species gives the MHD continuity equation

$$\frac{\partial \rho}{\partial t} + \nabla \cdot (\rho \mathbf{U}) = 0. \tag{2.66}$$

2.6.2 MHD equation of motion

To obtain an equation of motion we take the first moment of the Vlasov equation, then multiply by m_σ and sum over species to obtain

$$\frac{\partial}{\partial t}\sum_\sigma m_\sigma \int \mathbf{v} f_\sigma d\mathbf{v} + \frac{\partial}{\partial \mathbf{x}} \cdot \sum_\sigma \int m_\sigma \mathbf{v}\mathbf{v} f_\sigma d\mathbf{v} + \sum_\sigma q_\sigma \int \mathbf{v} \frac{\partial}{\partial \mathbf{v}} \cdot [(\mathbf{E} + \mathbf{v} \times \mathbf{B}) f_\sigma] = 0; \tag{2.67}$$

the right-hand side is zero since $\mathbf{R}_{ei} + \mathbf{R}_{ie} = 0$, i.e., the total plasma cannot exert drag on itself. We now define random velocities relative to $\mathbf{U}$ (rather than to $\mathbf{u}_\sigma$ as was the case for the two-fluid equations) so that the second term can be written as

$$\sum_\sigma \int m_\sigma \mathbf{v}\mathbf{v} f_\sigma d\mathbf{v} = \sum_\sigma \int m_\sigma (\mathbf{v}' + \mathbf{U})(\mathbf{v}' + \mathbf{U}) f_\sigma d\mathbf{v} = \sum_\sigma \int m_\sigma \mathbf{v}'\mathbf{v}' f_\sigma d\mathbf{v} + \rho \mathbf{U}\mathbf{U}, \tag{2.68}$$

where $\sum_\sigma \int m_\sigma \mathbf{v}' f_\sigma d\mathbf{v} = 0$ has been used to eliminate terms linear in $\mathbf{v}'$. The MHD pressure tensor is now defined in terms of the random velocities relative to $\mathbf{U}$ and is given by

$$\overleftrightarrow{\mathbf{P}}^{\mathrm{MHD}} = \sum_\sigma \int m_\sigma \mathbf{v}'\mathbf{v}' f_\sigma d\mathbf{v}. \tag{2.69}$$

We insert Eqs. (2.68) and (2.69) in Eq. (2.67), integrate by parts on the acceleration term, and perform the summation over species to obtain the MHD equation of motion

$$\frac{\partial(\rho\mathbf{U})}{\partial t} + \nabla\cdot(\rho\mathbf{U}\mathbf{U}) = \left(\sum_\sigma n_\sigma q_\sigma\right)\mathbf{E} + \mathbf{J}\times\mathbf{B} - \nabla\cdot\overleftrightarrow{\mathbf{P}}^{\mathrm{MHD}}. \tag{2.70}$$

Magnetohydrodynamics is typically used to describe phenomena having spatial scales large enough for the plasma to be essentially neutral, i.e., $\sum_\sigma n_\sigma q_\sigma$ is assumed to be so small that the $(\sum_\sigma n_\sigma q_\sigma)\,\mathbf{E}$ term can be dropped from Eq. (2.70). Just as in the two-fluid situation, the left-hand side of Eq. (2.70) contains a factor times the MHD continuity equation, since the left-hand side can be expanded as

$$\begin{aligned}\frac{\partial(\rho\mathbf{U})}{\partial t} + \nabla\cdot(\rho\mathbf{U}\mathbf{U}) &= \underbrace{\left[\frac{\partial\rho}{\partial t} + \nabla\cdot(\rho\mathbf{U})\right]}_{\text{zero}}\mathbf{U} + \rho\frac{\partial\mathbf{U}}{\partial t} + \rho\mathbf{U}\cdot\nabla\mathbf{U} \\ &= \rho\left(\frac{\partial\mathbf{U}}{\partial t} + \mathbf{U}\cdot\nabla\mathbf{U}\right).\end{aligned} \tag{2.71}$$

Using Eq. (2.71) in Eq. (2.70) leads to the standard form for the MHD equation of motion,

$$\rho\frac{\mathrm{D}\mathbf{U}}{\mathrm{D}t} = \mathbf{J}\times\mathbf{B} - \nabla\cdot\overleftrightarrow{\mathbf{P}}^{\mathrm{MHD}}, \tag{2.72}$$

where

$$\frac{\mathrm{D}}{\mathrm{D}t} = \frac{\partial}{\partial t} + \mathbf{U}\cdot\nabla \tag{2.73}$$

is the convective derivative defined using the center-of-mass velocity. Scalar approximations of the MHD pressure tensor will be postponed until after discussing the MHD Ohm's law and its implications.

2.6.3 MHD Ohm's law

Equation (2.72) provides one equation relating $\mathbf{J}$ and $\mathbf{U}$; let us now find the other one. In order to do this, consider the two-fluid electron equation of motion,

$$m_e\frac{d\mathbf{u}_e}{dt} = -e(\mathbf{E} + \mathbf{u}_e\times\mathbf{B}) - \frac{1}{n_e}\nabla(n_e\kappa T_e) - \nu_{ei}m_e(\mathbf{u}_e - \mathbf{u}_i). \tag{2.74}$$

In MHD we are interested in low-frequency phenomena with large spatial scales. If the characteristic time scale of the phenomenon is long compared to the electron cyclotron motion, then the electron inertia term $m_e d\mathbf{u}_e/dt$ can be dropped since it is small compared to the magnetic force term $-e(\mathbf{u}_e \times \mathbf{B})$. This assumption is reasonable for velocities perpendicular to $\mathbf{B}$, but can be a poor approximation for the velocity component parallel to $\mathbf{B}$, since parallel velocities do not provide a magnetic force. Since $\mathbf{u}_e - \mathbf{u}_i = -\mathbf{J}/n_e e$ and $\mathbf{u}_i \simeq \mathbf{U}$, Eq. (2.74) reduces to the generalized Ohm's law

$$\mathbf{E}+\mathbf{U}\times\mathbf{B}-\frac{1}{n_e e}\mathbf{J}\times\mathbf{B}+\frac{1}{n_e e}\nabla\left(n_e\kappa T_e\right)=\eta\mathbf{J}. \tag{2.75}$$

The term $-\mathbf{J}\times\mathbf{B}/n_e e$ on the left-hand side of Eq. (2.75) is called the *Hall term* and can be neglected in either of the following two cases:

1. The pressure term in the MHD equation of motion, Eq. (2.72) is negligible compared to the other two terms, which therefore must balance, giving

$$|\mathbf{J}| \sim \omega\rho|\mathbf{U}|/|\mathbf{B}|;$$

here $\omega \sim \mathrm{D}/\mathrm{D}t$ is the characteristic frequency of the phenomenon. In this case comparison of the Hall term with the $\mathbf{U}\times\mathbf{B}$ term shows that the Hall term is small by a factor $\sim \omega/\omega_{ci}$, where $\omega_{ci} = q_i B/m_i$ is the ion cyclotron frequency. Thus dropping the Hall term is justified for phenomena having characteristic frequencies small compared to ω_{ci}.
2. The electron–ion collision frequency is large compared to the electron cyclotron frequency $\omega_{ce} = q_e B/m_e$, in which case the Hall term may be dropped since it is small by a factor $\omega_{ce}/\upsilon_{ei}$ compared to the right-hand side resistive term $\eta\mathbf{J} = (m_e \nu_{ei}/n_e e^2)\mathbf{J}$.

From now on, when using MHD it will be assumed that one of these conditions is true and Hall terms will be dropped (if Hall terms are retained, the system is called Hall MHD). Typically, Eq. (2.75) will not be used directly; instead its curl will be used to provide the *induction* equation

$$-\frac{\partial\mathbf{B}}{\partial t}+\nabla\times(\mathbf{U}\times\mathbf{B})-\frac{1}{n_e e}\nabla n_e\times\nabla\kappa T_e=\nabla\times\left(\frac{\eta}{\mu_0}\nabla\times\mathbf{B}\right). \tag{2.76}$$

Usually the density gradient is parallel to the temperature gradient so that the thermal electromotive force term $(n_e e)^{-1}\,\nabla n_e \times \nabla\kappa T_e$ can be dropped, in which case the induction equation reduces to

$$-\frac{\partial\mathbf{B}}{\partial t}+\nabla\times(\mathbf{U}\times\mathbf{B})=\nabla\times\left(\frac{\eta}{\mu_0}\nabla\times\mathbf{B}\right). \tag{2.77}$$

The thermal term is often simply ignored in the MHD Ohm's law, which is then written as

$$\mathbf{E}+\mathbf{U}\times\mathbf{B}=\eta\mathbf{J}; \tag{2.78}$$

this is only acceptable providing we intend to take the curl and providing $\nabla n_e \times \nabla \kappa T_e \simeq 0$.

2.6.4 *Ideal MHD and frozen-in flux*

If the resistive term $\eta\mathbf{J}$ is so small as to be negligible compared to the other terms in Eq. (2.78), then the plasma is said to be *ideal* or *perfectly conducting*. From the Lorentz transformation of electromagnetic theory we realize that $\mathbf{E}+\mathbf{U}\times\mathbf{B}=\mathbf{E}'$, where $\mathbf{E}'$ is the electric field observed in the frame moving with velocity $\mathbf{U}$. This implies that the magnetic flux in ideal plasmas is time-invariant in the frame moving with velocity $\mathbf{U}$, because otherwise Faraday's law would imply the existence of an electric field in the moving frame. The frozen-in flux concept is the essential "bed-rock" concept underlying ideal MHD. While this concept is often an excellent approximation, it must be kept in mind that the concept becomes invalid in situations when any one of the electron inertia, electron pressure, or Hall terms becomes important and leads to different, more complex behavior.

The frozen-in flux concept is frequently expressed in a slightly different form as a frozen-in *field* concept, i.e., it is often said that magnetic field lines are frozen into the plasma so that the plasma and magnetic field lines move together as an ensemble. While this point of view can be quite intuitive and useful, it contains some ambiguity because an ideal plasma can actually move across magnetic field lines in certain situations. These situations are such that magnetic flux is preserved within the plasma even though it is moving across field lines. Discussion of this subtlety will be deferred to Section 3.5.5.

A formal proof of the frozen-in flux property will now be established by direct calculation of the rate of change of the magnetic flux through a surface $S(t)$ bounded by a material line $C(t)$, i.e., a closed contour that moves with the plasma. This magnetic flux is

$$\Phi(t)=\int_{S(t)}\mathbf{B}(\mathbf{x},t)\cdot \mathrm{d}\mathbf{s} \tag{2.79}$$

and the flux changes with respect to time due to either (i) the explicit time dependence of $\mathbf{B}(t)$ or (ii) changes in the surface $S(t)$ resulting from plasma motion. The rate of change of flux is thus

$$\frac{\mathrm{D}\Phi}{\mathrm{D}t}=\lim_{\delta t\to 0}\left(\frac{\int_{S(t+\delta t)}\mathbf{B}(\mathbf{x},t+\delta t)\cdot \mathrm{d}\mathbf{s}-\int_{S(t)}\mathbf{B}(\mathbf{x},t)\cdot \mathrm{d}\mathbf{s}}{\delta t}\right). \tag{2.80}$$

The displacement of a segment $\mathbf{dl}$ of the bounding contour C is $\mathbf{U}\delta t$, where $\mathbf{U}$ is the velocity of this segment. The incremental change in surface area due to this displacement is $\Delta S = \mathbf{U}\delta t \times \mathbf{dl}$. The rate of change of flux can thus be expressed as

$$\begin{aligned}
\frac{\mathrm{D}\Phi}{\mathrm{D}t} &= \lim_{\delta t\to 0} \frac{\int_{S(t+\delta t)} \left(\mathbf{B}+\delta t\frac{\partial \mathbf{B}}{\partial t}\right)\cdot \mathbf{ds} - \int_{S(t)} \mathbf{B}\cdot\mathbf{ds}}{\delta t} \\
&= \lim_{\delta t\to 0} \frac{\int_{S(t)} \left(\mathbf{B}+\delta t\frac{\partial \mathbf{B}}{\partial t}\right)\cdot \mathbf{ds} + \oint_C \mathbf{B}\cdot\mathbf{U}\delta t\times\mathbf{dl} - \int_{S(t)} \mathbf{B}\cdot\mathbf{ds}}{\delta t} \\
&= \int_{S(t)} \frac{\partial \mathbf{B}}{\partial t}\cdot\mathbf{ds} + \oint_C \mathbf{B}\cdot\mathbf{U}\times\mathbf{dl} \\
&= \int_{S(t)} \left[\frac{\partial \mathbf{B}}{\partial t} + \nabla\times(\mathbf{B}\times\mathbf{U})\right]\cdot\mathbf{ds}.
\end{aligned} \tag{2.81}$$

Thus, if

$$\frac{\partial \mathbf{B}}{\partial t} = \nabla\times(\mathbf{U}\times\mathbf{B}) \tag{2.82}$$

then

$$\frac{\mathrm{D}\Phi}{\mathrm{D}t} = 0 \tag{2.83}$$

so that the magnetic flux linked by any closed material line is constant. Therefore, magnetic flux is frozen into an ideal plasma because Eq. (2.77) reduces to Eq. (2.82) if $\eta = 0$. Equation (2.82) is called the ideal MHD induction equation.

2.6.5 MHD equations of state

Double adiabatic laws

A procedure analogous to that which led to Eq. (2.35) gives the MHD adiabatic relation

$$\frac{P^{MHD}}{\rho^\gamma} = const., \tag{2.84}$$

where again $\gamma = (N+2)/N$ and N is the number of dimensions of the system. It was shown in the previous section that magnetic flux is conserved in the plasma frame. This means that, as shown in Fig. 2.5, a tube of plasma initially occupying the same volume as a magnetic flux tube is constrained to evolve in such a way that $\int \mathbf{B}\cdot\mathbf{ds}$ stays constant over the plasma tube cross-section. For a flux tube of infinitesimal cross-section, the magnetic field is approximately uniform

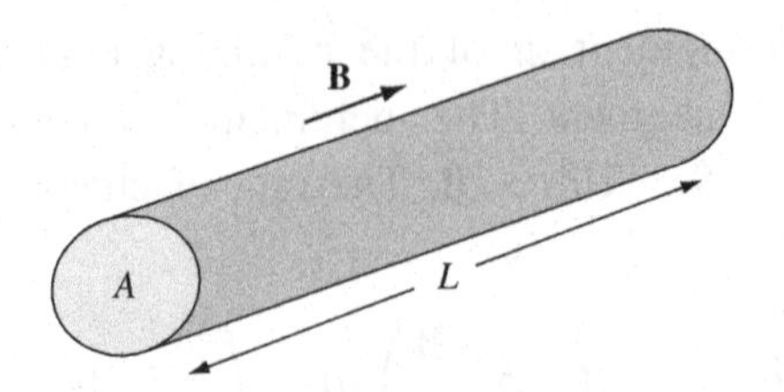

Fig. 2.5 Magnetic flux tube with flux $\Phi = BA$.

over the cross-section and we may write this as $BA = const.$, where A is the cross-sectional area.

Let us define two temperatures for this magnetized plasma, namely $T_\perp$, the temperature corresponding to motions perpendicular to the magnetic field, and $T_\parallel$, the temperature corresponding to motions parallel to the magnetic field. If for some reason (e.g., anisotropic heating or compression) the temperature develops an anisotropy such that $T_\perp \neq T_\parallel$ and if collisions are infrequent, this anisotropy will persist for a long time, since collisions are the means by which the two temperatures equilibrate. Thus, rather than assuming that the MHD pressure is fully isotropic, we consider the less restrictive situation where the MHD pressure tensor is given by

$$\overleftrightarrow{\mathbf{P}}^{MHD} = \begin{bmatrix} P_\perp & 0 & 0 \\ 0 & P_\perp & 0 \\ 0 & 0 & P_\parallel \end{bmatrix} = P_\perp \overleftrightarrow{\mathbf{I}} + (P_\parallel - P_\perp)\hat{B}\hat{B}. \tag{2.85}$$

The first two coordinates (x, y-like) in the above matrix refer to the directions perpendicular to the local magnetic field $\mathbf{B}$ and the third coordinate (z-like) refers to the direction parallel to $\mathbf{B}$. The tensor expression on the right-hand side is equivalent (here $\overleftrightarrow{\mathbf{I}}$ is the unit tensor) but allows for arbitrary, curvilinear geometry. We now develop separate adiabatic relations for the perpendicular and parallel directions:

- Parallel direction: here the number of dimensions is $N = 1$ so that $\gamma = 3$ and so the adiabatic law gives

$$\frac{P_\parallel^{1D}}{\rho_{1D}^3} = const., \tag{2.86}$$

where ρ_{1D} is the *one-dimensional* mass density; i.e., $\rho_{1D} \sim 1/L$, where L is the length along the one-dimension, e.g., along the length of the flux tube in Fig. 2.5. The three-dimensional mass density ρ, which has been used implicitly until now, has the proportionality $\rho \sim 1/LA$, where A is the cross-section of the flux tube; similarly the three-dimensional pressure has the proportionality $P_\parallel \sim \rho T_\parallel$. However, we must be careful to realize that $P_\parallel^{1D} \sim \rho_{1D} T_\parallel$ so, using $BA = const.$, Eq. (2.86) can be recast as

$$const. = \frac{P_\parallel^{1D}}{\rho_{1D}^3} \sim \frac{\rho_{1D} T_\parallel}{\rho_{1D}^3} \sim T_\parallel L^2 \sim \underbrace{\left(\frac{1}{LA}\right) T_\parallel}_{P_\parallel} \underbrace{(LA)^3}_{\rho^{-3}} B^2 \tag{2.87}$$

or

$$\frac{P_\parallel B^2}{\rho^3} = const. \tag{2.88}$$

- Perpendicular direction: here the number of dimensions is $N = 2$ so that $\gamma = 2$ and the adiabatic law gives

$$\frac{P_\perp^{2D}}{\rho_{2D}^2} = const., \tag{2.89}$$

where $P_\perp^{2D}$ is the 2-D perpendicular pressure, and has dimensions of energy per unit area, while ρ_{2D} is the 2-D mass density and has dimensions of mass per unit area. Thus, $\rho_{2D} \sim 1/A$ so $P_\perp^{2D} \sim \rho_{2D} T_\perp \sim T_\perp / A$, in which case Eq. (2.89) can be rewritten as

$$const. = \frac{P_\perp^{2D}}{\rho_{2D}^2} \sim T_\perp A \sim \left(\frac{1}{LA}\right) T_\perp \frac{LA}{B} \tag{2.90}$$

or

$$\frac{P_\perp}{\rho B} = const. \tag{2.91}$$

Equations (2.88) and (2.91) are called the double adiabatic or CGL laws after Chew, Goldberger, and Low (1956) who first developed these laws.

Single adiabatic law

If collisions are sufficiently frequent to equilibrate the perpendicular and parallel temperatures, then the pressure tensor becomes fully isotropic and the dimensionality of the system is $N = 3$ so that $\gamma = 5/3$. There is now just one pressure and one temperature and the adiabatic relation becomes

$$\frac{P}{\rho^{5/3}} = const. \tag{2.92}$$

2.6.6 MHD approximations for Maxwell's equations

The various assumptions contained in MHD lead to a simplifying approximation of Maxwell's equations. In particular, the assumption of charge neutrality in MHD makes Poisson's equation superfluous because Poisson's equation prescribes the relationship between non-neutrality and the electrostatic component of the electric field. The assumption of charge neutrality also has implications for the current density. To see this, the two-fluid continuity equation is multiplied by q_σ and then summed over species to obtain the charge conservation equation

$$\frac{\partial}{\partial t}\left(\sum n_\sigma q_\sigma\right) + \nabla \cdot \mathbf{J} = 0. \tag{2.93}$$

Thus, charge neutrality implies

$$\nabla \cdot \mathbf{J} = 0. \tag{2.94}$$

Let us now consider Ampère's law

$$\nabla \times \mathbf{B} = \mu_0 \mathbf{J} + \mu_0 \varepsilon_0 \frac{\partial \mathbf{E}}{\partial t}. \tag{2.95}$$

Taking the divergence gives

$$\nabla \cdot \mathbf{J} + \varepsilon_0 \frac{\partial \nabla \cdot \mathbf{E}}{\partial t} = 0, \tag{2.96}$$

which is equivalent to Eq. (2.93) if Poisson's equation is invoked.

Finally, MHD is restricted to phenomena having characteristic velocities V_{ph} slow compared to the speed of light in vacuum, $c = (\varepsilon_0 \mu_0)^{-1/2}$. Again, t_{char} is assumed to represent the characteristic time scale for a given phenomenon and l_{char} is assumed to represent the corresponding characteristic length scale so that $V_{ph} \sim l_{char}/t_{char}$. Faraday's equation gives the scaling

$$\nabla \times \mathbf{E} = -\frac{\partial \mathbf{B}}{\partial t} \Longrightarrow E \sim B l_{char}/t_{char}. \tag{2.97}$$

On comparing the magnitude of the displacement current term in Eq. (2.95) to the left-hand side it is seen that

$$\frac{\mu_0 \varepsilon_0 \left|\dfrac{\partial \mathbf{E}}{\partial t}\right|}{|\nabla \times \mathbf{B}|} \sim \frac{c^{-2} E/t_{char}}{B/l_{char}} \sim \left(\frac{V_{ph}}{c}\right)^2. \tag{2.98}$$

Thus, if $V_{ph} \ll c$ the displacement current term can be dropped from Amperè's law resulting in the so-called "pre-Maxwell" form

$$\nabla \times \mathbf{B} = \mu_0 \mathbf{J}. \tag{2.99}$$

The divergence of Eq. (2.99) gives Eq. (2.94) so it is unnecessary to specify Eq. (2.94) separately.

2.7 Summary of MHD equations

We may now summarize the MHD equations:

1. Mass conservation

$$\frac{\partial \rho}{\partial t} + \nabla \cdot (\rho \mathbf{U}) = 0. \tag{2.100}$$

2. Equation of state and associated equation of motion
   (a) Single adiabatic regime, collisions equilibrate perpendicular and parallel temperatures so that pressure and temperature are both isotropic,

$$\frac{P}{\rho^{5/3}} = const., \tag{2.101}$$

   and the equation of motion is

$$\rho \frac{\mathrm{D}\mathbf{U}}{\mathrm{D}t} = \mathbf{J} \times \mathbf{B} - \nabla \mathbf{P}. \tag{2.102}$$

   (b) Double adiabatic regime, the collision frequency is insufficient to equilibrate perpendicular and parallel temperatures so that

$$\frac{P_{\parallel} B^2}{\rho^3} = const., \qquad \frac{P_{\perp}}{\rho B} = const. \tag{2.103}$$

   and the equation of motion is

$$\rho \frac{\mathrm{D}\mathbf{U}}{\mathrm{D}t} = \mathbf{J} \times \mathbf{B} - \nabla \cdot \left[P_{\perp} \overleftrightarrow{I} + (P_{\parallel} - P_{\perp}) \hat{B} \hat{B} \right]. \tag{2.104}$$

3. Faraday's law

$$\nabla \times \mathbf{E} = -\frac{\partial \mathbf{B}}{\partial t}. \tag{2.105}$$

4. Amperè's law

$$\nabla \times \mathbf{B} = \mu_0 \mathbf{J}. \tag{2.106}$$

5. Ohm's law

$$\mathbf{E} + \mathbf{U} \times \mathbf{B} = \eta \mathbf{J}. \tag{2.107}$$

These equations provide a self-consistent description of phenomena that satisfy all the various assumptions we have made, namely:

(i) The plasma is charge-neutral since characteristic lengths are much longer than a Debye length;
(ii) The characteristic velocity of the phenomenon under consideration is slow compared to the speed of light;
(iii) The pressure and density gradients are parallel, so there is no electrothermal EMF;
(iv) The time scale is long compared to both the electron and ion cyclotron periods.

Even though these assumptions are self-consistent, they may not accurately portray a real plasma and so MHD models, while intuitively appealing, must be used with caution.

Finally, it is worth mentioning that MHD plasmas can be categorized yet another way, namely according to the relative importance of the magnetic force $\mathbf{J}\times\mathbf{B}$ compared to the hydrodynamic force $\nabla\cdot\overleftrightarrow{\mathbf{P}}$. If the magnetic force is negligible compared to the hydrodynamic force, then there is not much point in using MHD because in this case the system of equations is simply classical hydrodynamics. Thus, the only non-trivial MHD situations are where the magnetic and hydrodynamic forces are of comparable importance or where the magnetic force is much more important than the hydrodynamic force. Using Amperè's law, Eq. (2.106), the nominal ratio of the hydrodynamic force to the magnetic force is defined as

$$\beta = \frac{P}{B^2/2\mu_0} \sim \frac{\nabla\cdot\overleftrightarrow{\mathbf{P}}}{\mathbf{J}\times\mathbf{B}} \sim \frac{P/L}{B^2/2\mu_0 L}; \tag{2.108}$$

the characteristic gradient scale length L is assumed to be comparable for both types of forces and so cancels out in the comparison. Low-β plasmas are those where $B^2/2\mu_0$ is much larger than P so the hydrodynamic force is negligible compared to the magnetic force, whereas $\beta = \mathcal{O}(1)$ plasmas are those where the magnetic and hydrodynamic forces are comparable.

2.8 Classical transport

Consider a cylindrical coordinate system $\{r, \theta, z\}$ with a uniform steady-state magnetic field $\mathbf{B} = B_z\hat{z}$ and suppose that an azimuthally symmetric, finite-pressure plasma exists in the region $r \leq a$ and is surrounded by vacuum. Since the right-hand side of Eq. (2.105) must vanish in steady state, if an equilibrium electric field exists, it must be electrostatic, i.e., be of the form $\mathbf{E} = -\nabla\phi$. Because of the assumed symmetry, this electrostatic electric field would also have to be azimuthally symmetric. The last two considerations mean that the θ component of the electric field would have to be zero since $E_\theta = -r^{-1}\partial\phi/\partial\theta$ and θ derivatives of all physical quantities must vanish because of the assumed steady-state azimuthal symmetry. Because $E_\theta = 0$, the θ component of the MHD Ohm's law, Eq. (2.107), is simply

$$-U_r B_z = \eta J_\theta. \tag{2.109}$$

This leads to the immediate and important conclusion that if the plasma is perfectly conducting (i.e., $\eta = 0$) then $U_r = 0$ and so the plasma will not be able to move across the magnetic field. In this case, one can say that the plasma is frozen to the field lines, but it should be noted that this behavior relies on the existence of azimuthal symmetry.

We now consider the more general situation where the resistivity is finite, in which case there will be a finite radial motion of the plasma across the magnetic field, i.e., $U_r = -\eta J_\theta / B_z$. The azimuthal current J_θ can be determined from the steady-state r component of the MHD equation of motion, Eq. (2.104),

$$0 = J_\theta B_z - \frac{\partial P_\perp}{\partial r} \tag{2.110}$$

so, combining Eqs. (2.109) and (2.110) gives

$$U_r = -\frac{\eta}{B_z^2} \frac{\partial P_\perp}{\partial r}. \tag{2.111}$$

This gives an azimuthally symmetric, radially outward motion proportional to resistivity (i.e., to electron–ion collisions) and the radial pressure gradient, but inversely proportional to the square of the magnetic field strength. Hence, transport of plasma across magnetic fields is driven by pressure gradients, is allowed to happen if the plasma is not perfectly conducting, and is reduced if the magnetic field is strong. This process is called "classical transport." The concept of a plasma being frozen to an azimuthally symmetric flux tube is thus valid if the magnetic field is sufficiently strong and the plasma is a sufficiently good conductor for classical transport to be negligible. Using $P_\perp = n(\kappa T_{e\perp} + \kappa T_{i\perp})$ and $\eta = m_e \nu_{ei} / ne^2$, the classical transport flux $\Gamma = nU_r$ can be expressed as

$$\Gamma_r = -D_{classical} \frac{\partial n}{\partial r}, \tag{2.112}$$

where

$$D_{classical} = \frac{m_e \nu_{ei} (\kappa T_{e\perp} + \kappa T_{i\perp})}{e^2 B_z^2} \tag{2.113}$$

is called the classical particle diffusion coefficient. This coefficient can equivalently be written as

$$D_{classical} = \left(1 + \frac{T_{i\perp}}{T_{e\perp}}\right) \frac{r_{Le}^2}{\tau_{ei}}, \tag{2.114}$$

where $r_{Le} = \sqrt{\kappa T_e / m_e \omega_{ce}^2}$ is the nominal electron-cyclotron radius and $\omega_{ce} = eB/m_e$ is the electron cyclotron frequency. Like any diffusion coefficient, the classical diffusion coefficient has the dimensions of (step-size)^2/(time-step). It is intrinsically ambipolar because it has been derived in the context of MHD where the plasma is intrinsically neutral. An examination of the individual ion and electron cross-field diffusions would show that the ion diffusion coefficient is much larger than the electron diffusion coefficient, but then a resulting ambipolar electric field would be established, which would increase the outward electron flux and decrease the outward ion flux so as to produce the ambipolar result given here. Since $r_{Le}^2 \sim T/B$ and $\tau_{ei} \sim T^{3/2}$ it is seen that $D_{classical} \sim 1/T^{1/2}B^2$ and so is very small for hot plasmas in strong magnetic fields.

2.9 Sheath physics and Langmuir probe theory

Let us now turn attention back to Vlasov theory and discuss an immediate practical application of this theory. The properties of collisionless Vlasov equilibria can be combined with Poisson's equation to develop a model for the potential in the steady-state transition region between a plasma and a conducting wall; this region is known as a sheath and is important in many practical situations. The sheath is non-neutral and its width is of the order of a Debye length. The exact sheath potential profile must be solved numerically because of the transcendental nature of the relevant equations, but a useful approximate solution can be obtained by a simple analytic argument, which will now be presented. Sheath physics is of particular importance for interpreting the current-voltage characteristics of Langmuir probes, small metal wires inserted into low-temperature plasmas for diagnostic purposes. Biasing a Langmuir probe at a sequence of voltages and then measuring the resulting current provides a simple way to gauge both the plasma density and the electron temperature.

The model presented here is the simplest possible model for sheaths and Langmuir probes and so is one dimensional. The geometry, sketched in Fig. 2.6, idealizes the Langmuir probe as a metal wall located at $x = 0$ and biased to a potential ϕ_{probe}; this geometry could also be used to describe an actual biased metal wall at $x = 0$ in a two-dimensional plasma. The plasma is assumed to be collisionless and unmagnetized and to have an ambipolar potential ϕ_{plasma} that differs from the laboratory reference potential (so-called ground potential) because of a difference in the diffusion rates of electrons and ions out of the plasma. The plasma is assumed to extend into the semi-infinite left-hand half-plane, $-\infty < x < 0$. If $\phi_{probe} = \phi_{plasma}$, then neither electrons nor ions will be accelerated or decelerated on leaving the plasma and so each species will strike the probe (or wall) at a rate given by its respective thermal velocity. Since $m_e \ll m_i$, the electron thermal velocity greatly exceeds the ion thermal velocity. Thus, for $\phi_{probe} = \phi_{plasma}$ the electron flux to the probe (or wall) greatly exceeds the ion flux and so the current collected by the probe (or wall) will be negative.

Now consider what happens to this electron flow if the probe (or wall) is biased *negative* with respect to the plasma as shown in Fig. 2.6. To simplify the

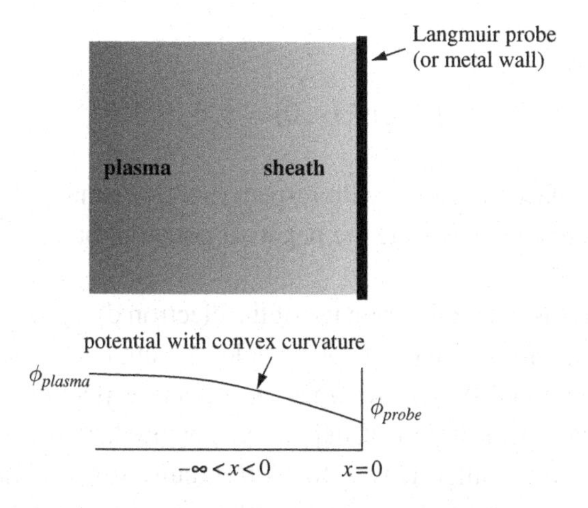

Fig. 2.6 Sketch of sheath. Top: ions are accelerated in the sheath towards the probe (wall) at $x = 0$ whereas electrons are repelled. Bottom: convex curvature of sheath requires $n_i(x)$ always to be greater than $n_e(x)$.

notation, a bar will be used to denote a potential measured relative to the plasma potential, i.e.,

$$\bar{\phi}(x) = \phi(x) - \phi_{plasma}. \tag{2.115}$$

The bias potential imposed on the probe (or wall) will be shielded out by the plasma within a distance of the order of the Debye length; this region is the sheath. The relative potential $\bar{\phi}(x)$ varies within the sheath and has the two limiting behaviors:

$$\begin{aligned} \lim_{x \to 0} \bar{\phi}(x) &= \phi_{probe} - \phi_{plasma} \\ \lim_{|x| \gg \lambda_D} \bar{\phi}(x) &= 0. \end{aligned} \tag{2.116}$$

Inside the plasma, i.e., for $|x| \gg \lambda_D$, it is assumed that the electron distribution function is Maxwellian with temperature T_e. Since the distribution function depends only on constants of the motion, the one-dimensional electron velocity distribution function must depend only on the electron energy $mv^2/2 + q_e\bar{\phi}(x)$, a constant of the motion, and so must be of the form

$$f_e(v, x) = \frac{n_0}{\sqrt{2\pi\kappa T_e/m_e}} \exp\left(-\left(\frac{mv^2/2 + q_e\bar{\phi}(x)}{\kappa T_e}\right)\right) \tag{2.117}$$

in order to be Maxwellian when $|x| \gg \lambda_D$.

The electron density is

$$n_e(x) = \int_{-\infty}^{\infty} \mathrm{d}v f_e(v, 0) = n_0 \mathrm{e}^{-q_e \bar{\phi}(x)/\kappa T_e}. \tag{2.118}$$

When the probe is biased negative with respect to the plasma, only those electrons with sufficient energy to overcome the negative potential barrier will be collected by the probe.

The ion dynamics is not a mirror image of the electron dynamics. This is because a repulsive potential prevents passage of particles having insufficient initial energy to climb over a potential barrier whereas an attractive potential allows passage of all particles entering a region of depressed potential. Particle density is thus reduced compared to the inlet density for *both* repulsive and attractive potentials but for different reasons. As shown in Eq. (2.118) a repulsive potential reduces the electron density exponentially (this is essentially the Boltzmann analysis developed in the theory of Debye shielding). Suppose the ions are cold and enter a region of attractive potential with velocity u_0. Flux conservation shows that $n_0 u_0 = n_i(x) u_i(x)$ and since the ions accelerate to higher velocity when falling down the attractive potential, the ion density must decrease. Thus the electron density scales as $\exp(-\left|q_e \bar{\phi}\right| / \kappa T_e)$ and so decreases upon approaching the wall in response to what is a repulsive potential for electrons whereas the ion density scales as $1/u_i(x)$ and *also* decreases upon approaching the wall in response to what is an attractive potential for ions.

It is now important to recall that the probe was assumed to be biased negatively with respect to the plasma. Since quasi-neutrality within the plasma mandates that the electric field must vanish inside the plasma, the potential must have a downward slope on going from the plasma to the probe and the derivative of this slope must also be downward. This means that the potential $\bar{\phi}$ must have a convex curvature, i.e., a negative second derivative as indicated in Fig. 2.6 (bottom). However, the one-dimensional Poisson's equation

$$\frac{\mathrm{d}^2 \bar{\phi}}{\mathrm{d}x^2} = -\frac{e}{\varepsilon_0} \left(n_i(x) - n_e(x)\right) \tag{2.119}$$

shows that in order for $\bar{\phi}$ to have a negative second derivative, it is necessary to have $n_i(x) > n_e(x)$ everywhere. This condition will now be used to estimate the inflow velocity of the ions at the location where they enter the sheath from the bulk plasma.

Ion energy conservation gives

$$\frac{1}{2} m_i u^2(x) + e\bar{\phi}(x) = \frac{1}{2} m_i u_0^2, \tag{2.120}$$

which can be solved to give

$$u(x) = \sqrt{u_0^2 - 2e\bar{\phi}(x)/m_i}. \tag{2.121}$$

Using the ion flux conservation relation $n_0u_0 = n_i(x)u_i(x)$ the local ion density is found to be

$$n_i(x) = \frac{n_0}{\left(1 - 2e\bar{\phi}(x)/mu_0^2\right)^{1/2}}. \tag{2.122}$$

The convexity requirement $n_i(x) > n_e(x)$ implies

$$\left(1 - 2e\bar{\phi}(x)/m_iu_0^2\right)^{-1/2} > \mathrm{e}^{e\bar{\phi}(x)/\kappa T_e}. \tag{2.123}$$

Bearing in mind that $\bar{\phi}(x)$ is negative, this can be rearranged as

$$1 + \frac{2e|\bar{\phi}(x)|}{m_iu_0^2} < \mathrm{e}^{2e|\bar{\phi}(x)|/\kappa T_e} \tag{2.124}$$

or

$$\frac{2e|\bar{\phi}(x)|}{m_iu_0^2} < \frac{2e|\bar{\phi}(x)|}{\kappa T_e} + \frac{1}{2}\left(\frac{2e|\bar{\phi}(x)|}{\kappa T_e}\right)^2 + \frac{1}{3!}\left(\frac{2e|\bar{\phi}(x)|}{\kappa T_e}\right)^3 + \cdots \tag{2.125}$$

which can only be satisfied for arbitrary $|\bar{\phi}(x)|$ if

$$u_0^2 > \kappa T_e/m_i. \tag{2.126}$$

Thus, in order to be consistent with the assumption that the probe is more negative than the plasma to keep $\mathrm{d}^2\bar{\phi}/\mathrm{d}x^2$ negative and hence $\bar{\phi}$ convex, it is necessary to have the ions enter the region of non-neutrality with a velocity slightly larger than the so-called "ion acoustic" velocity $c_s = \sqrt{\kappa T_e/m_i}$.

The ion current collected by the probe is given by the ion flux times the probe area, i.e.,

$$\begin{aligned} I_i &= n_0u_0q_iA \\ &= n_0c_seA. \end{aligned} \tag{2.127}$$

The electron current density incident on the probe is

$$\begin{aligned} J_e(x) &= \int_0^\infty \mathrm{d}v q_e v f_e(v, 0) \\ &= \frac{n_0q_e\mathrm{e}^{-q_e\bar{\phi}(x)/\kappa T_e}}{\sqrt{\pi 2\kappa T_e/m_e}} \int_0^\infty \mathrm{d}v\, v\mathrm{e}^{-mv^2/2\kappa T_e} \\ &= n_0q_e\sqrt{\frac{\kappa T_e}{2\pi m_e}}\mathrm{e}^{-e|\bar{\phi}(x)|/\kappa T_e} \end{aligned} \tag{2.128}$$

and so the electron current collected at the probe is

$$I_e = -n_0 eA\sqrt{\frac{\kappa T_e}{2\pi m_e}}e^{-e|\bar{\phi}(x)|/\kappa T_e}. \tag{2.129}$$

Thus, the combined electron and ion current collected by the probe is

$$\begin{aligned} I &= I_i + I_e \\ &= n_0 c_s eA - n_0 eA\sqrt{\frac{\kappa T_e}{2\pi m_e}}\mathrm{e}^{-e|\bar{\phi}(x)|/\kappa T_e}. \end{aligned} \tag{2.130}$$

The electron and ion currents cancel each other when

$$\sqrt{\frac{2\kappa T_e}{m_i}} = \sqrt{\frac{\kappa T_e}{2\pi m_e}}\mathrm{e}^{-e|\bar{\phi}(x)|/\kappa T_e} \tag{2.131}$$

i.e., when

$$\begin{aligned} e|\bar{\phi}_{probe}|/\kappa T_e &= \ln\sqrt{\frac{m_i}{4\pi m_e}} \\ &= 2.5 \text{ for hydrogen.} \end{aligned} \tag{2.132}$$

This can be expressed as

$$\phi_{probe} = \phi_{plasma} - \frac{\kappa T_e}{e}\ln\sqrt{\frac{m_i}{4\pi m_e}} \tag{2.133}$$

and shows that when the probe potential is more negative than the plasma potential by an amount $T_e \ln\sqrt{m_i/4\pi m_e}$, where T_e is expressed in electron volts, then no current flows to the probe. This potential is called the floating potential, because an insulated object immersed in the plasma will always charge up to the floating potential since this is the potential at which no current flows to the object.

These relationships can be used as a simple diagnostic for the plasma density and electron temperature. If a probe is biased with a large negative potential, then no electrons are collected but an ion flux is collected. The collected current is called the ion saturation current and is given by $I_{sat} = n_0 c_s eA$. The ion saturation current is then subtracted from all subsequent measurements giving the electron current $I_e = I - I_{sat} = n_0 eA\left(\kappa T_e/2\pi m_e\right)^{1/2}\exp\left(-e|\bar{\phi}(x)|/\kappa T_e\right)$. The slope of a logarithmic plot of I_e versus ϕ gives $1/\kappa T_e$ and so can be used to measure the electron temperature. Once the electron temperature is known, c_s can be calculated. The plasma density can then be calculated from the ion saturation current measurement and knowledge of the probe area. Langmuir probe measurements are simple to implement but are not very precise, typically having an uncertainty of a factor of two or more.

2.10 Assignments

1. Prove Stirling's formula. To do this first show

$$\ln N! = \ln 1 + \ln 2 + \ln 3 + \ldots \ln N$$
$$= \sum_{j=1}^{N} \ln j.$$

Now assume N is large and, using a graphical argument, show that the above expression can be expressed as an integral

$$\ln N! \approx \int_{?}^{?} h(x)\mathrm{d}x,$$

where the form of $h(x)$ and the limits of integration are to be provided. Evaluate the integral and obtain Stirling's formula

$$\ln N! \simeq N \ln N - N \text{ for large } N.$$

This is an effective way of calculating the values of factorials of large numbers. Check the accuracy of Stirling's formula by evaluating the left- and right-hand sides of Stirling's formula numerically and plot the results for $N = 1, 10, 100, 1000, 10^4$ and higher if possible.

2. Variational calculus and Lagrange multipliers. Many physical problems involve finding the function $f(x)$ that maximizes or minimizes integrals of the form

$$I = \int_{x_1}^{x_2} w(f(x), x)\mathrm{d}x \text{ or} \tag{2.134a}$$

$$I = \int_{x_1}^{x_2} w(f(x), \mathrm{d}f/\mathrm{d}x, x)\mathrm{d}x \tag{2.134b}$$

and there may or may not be restrictions on the value of $f(x_1)$ and $f(x_2)$. Entropy is of the form given in Eq. (2.134a), since we may identify $w = -f \ln f$, $x_1 = -\infty$, and $x_2 = +\infty$ and replace the dummy variable x by the dummy variable v so

$$S = -\int_{-\infty}^{\infty} f(v) \ln f(v)\, \mathrm{d}v.$$

The total kinetic energy of a distribution of particles is also of the form given by Eq. (2.134a), since we may again use v as the dummy variable and let $w = mv^2 f(v)/2$. The Lagrange method for mechanics (to be discussed in Chapter 3) involves integrals of the form given in Eq. (2.134b). For now, let us consider the specific problem of determining the function $f(x)$ that minimizes I as given in Eq. (2.134a). We assume that such an f exists and call this minimizing function $f_{\min}$. We then consider the neighboring function

$$f(x) = f_{\min}(x) + \varepsilon\eta(x), \tag{2.135}$$

where ε is much smaller than unity and $\eta(x)$ is an *arbitrary* function of order unity. We may then write Eq. (2.134a) as

$$I=\int_{x_1}^{x_2} w(f_{\min}(x)+\varepsilon\eta(x),x)\mathrm{d}x. \tag{2.136}$$

Because ε is small, Taylor expansion of I to second order in ε gives

$$I=\int_{x_1}^{x_2}\left\{w(f_{\min}(x),x)+\varepsilon\eta(x)\left(\frac{\partial w}{\partial f}\right)_{f=f_{\min}}+\frac{1}{2}(\varepsilon\eta(x))^2\left(\frac{\partial^2 w}{\partial f^2}\right)_{f=f_{\min}}\right\}\mathrm{d}x. \tag{2.137}$$

(a) Define

$$I_{\min}=\int_{x_1}^{x_2} w(f_{\min}(x),x)\mathrm{d}x \tag{2.138}$$

and consider $I-I_{\min}$, where I is given by Eq. (2.136). If $f_{\min}$ is indeed the function that minimizes I, explain that $I-I_{\min}$ must be positive if ε is not zero.

(b) Let the term that is first order in ε in Eq. (2.137) be called δI and the term that is second order in ε be called $\delta^2 I$ so

$$\delta I=\varepsilon\int_{x_1}^{x_2}\eta(x)\left(\frac{\partial w}{\partial f}\right)_{f=f_{\min}}\mathrm{d}x \tag{2.139a}$$

$$\delta^2 I=\frac{\varepsilon^2}{2}\int_{x_1}^{x_2}(\eta(x))^2\left(\frac{\partial^2 w}{\partial f^2}\right)_{f=f_{\min}}\mathrm{d}x. \tag{2.139b}$$

Show that δI dominates $\delta^2 I$ if ε is made sufficiently small and then show that the sign of ε could then be chosen to make $I-I_{\min}$ negative unless we insist that $\delta I=0$. Since ε is finite and $\eta(x)$ is arbitrary, argue that insisting $\delta I=0$ requires having

$$\left(\frac{\partial w}{\partial f}\right)_{f=f_{\min}}=0, \tag{2.140}$$

in which case

$$I-I_{\min}=\delta^2 I. \tag{2.141}$$

Since $I-I_{\min}$ must be positive and $\eta(x)$ is arbitrary what sign must $\partial^2 w/\partial f^2$ have? How is the function that minimizes I determined (hint: consider Eq. (2.140))?

(c) Now suppose that some sort of integral constraint exists, so the problem becomes one of finding the function f that minimizes I subject to the constraint

$$K=\int_{x_1}^{x_2} k(f(x),x)\mathrm{d}x=const. \tag{2.142}$$

By inserting $f=f_{\min}+\varepsilon\eta(x)$ into the constraint equation, show that

$$\delta K=\varepsilon\int_{x_1}^{x_2}\eta(x)\frac{\partial k}{\partial f}\,\mathrm{d}x=0, \tag{2.143}$$

where $\partial k/\partial f$ is understood to be evaluated at $f=f_{\min}$.

(d) Since $\eta(x)$ is arbitrary, choose it to be

$$\eta(x) = \varepsilon\delta(x-a) + \mu\varepsilon\delta(x-b), \tag{2.144}$$

where $x_1 < a < x_2$, $x_1 < b < x_2$, $a \neq b$, μ is an arbitrary constant of order unity, and the delta function used here is defined specifically as

$$\delta(x) = \left\{ \left| \begin{array}{ll} \dfrac{1}{\varepsilon} & \text{if } |x| < \varepsilon/2 \\ 0 & \text{if } |x| \geq \varepsilon/2. \end{array} \right| \right. \tag{2.145}$$

Show that these definitions are consistent with the requirement that $\eta(x)$ is of order unity and arbitrary.

(e) By using Eq. (2.144) in Eqs. (2.139a) and (2.143) obtain the coupled equations

$$\left(\frac{\partial w}{\partial f}\right)_{x=a} + \mu \left(\frac{\partial w}{\partial f}\right)_{x=b} = 0$$
$$\left(\frac{\partial k}{\partial f}\right)_{x=a} + \mu \left(\frac{\partial k}{\partial f}\right)_{x=b} = 0, \tag{2.146}$$

where f, if it appears explicitly, is understood to be f_{min}, i.e., the f that minimizes I subject to the constraint $K = const.$

(f) Define the constant λ to be

$$\lambda = \frac{\left(\dfrac{\partial w}{\partial f}\right)_{x=a}}{\left(\dfrac{\partial k}{\partial f}\right)_{x=a}} \tag{2.147}$$

and use Eqs. (2.146) to prove that

$$\frac{\left(\dfrac{\partial w}{\partial f}\right)_{x=b}}{\left(\dfrac{\partial k}{\partial f}\right)_{x=b}} = \lambda. \tag{2.148}$$

(g) Using the property that b is arbitrary show that

$$\frac{\partial w}{\partial f} - \lambda\frac{\partial k}{\partial f} = 0 \tag{2.149}$$

must be true for any x.

(h) Use the preceding results to explain why finding the f that minimizes I subject to the constraint $K = const.$ is equivalent to finding the f that minimizes $I - \lambda K$ where λ is a constant. The constant λ is called a Lagrange multiplier.

(i) Show that if there are two constraints, say $K = const.$ and $L = const.$, then minimization of I subject to these two constraints is equivalent to minimizing $I - \lambda_1 K - \lambda_2 L$, where λ_1 and λ_2 are two independent constants. (Hint: do this in two stages).

(j) Show that maximizing S is equivalent to minimizing $-S$.

3. Suppose that a group of N particles with charge q_σ and mass m_σ are located in an electrostatic potential $\phi(\mathbf{x})$. What is the maximum entropy distribution function for this situation (give a derivation)?
4. Prove that

$$\int_{-\infty}^{\infty} \mathrm{d}x\, \mathrm{e}^{-ax^2} = \sqrt{\frac{\pi}{a}}. \tag{2.150}$$

   Hint: consider the integral

$$\int_{-\infty}^{\infty} \mathrm{d}x\, \mathrm{e}^{-x^2} \int_{-\infty}^{\infty} \mathrm{d}y\, \mathrm{e}^{-y^2} = \int_{-\infty}^{\infty} \int_{-\infty}^{\infty} \mathrm{d}x\mathrm{d}y\, \mathrm{e}^{-(x^2+y^2)},$$

   and note that $\mathrm{d}x\mathrm{d}y$ is an element of area. Then, instead of using Cartesian coordinates for the integral over area, use cylindrical coordinates and express all quantities in the double integral in cylindrical coordinates.
5. Evaluate the integrals

$$\int_{-\infty}^{\infty} \mathrm{d}x\, x^2 \mathrm{e}^{-ax^2}, \int_{-\infty}^{\infty} \mathrm{d}x\, x^4 \mathrm{e}^{-ax^2}.$$

   Hint: take the derivative of both sides of Eq. (2.150) with respect to a.
6. Suppose that a particle starts at time $t = t_0$ with velocity $\mathbf{v}_0$ at location $\mathbf{x}_0$ and is located in a uniform, constant magnetic field $\mathbf{B} = B\hat{z}$. There is no electric field. Calculate its position and velocity as a function of time. Make sure your solution satisfies the initial conditions on both velocity and position. Be careful to treat motion parallel to the magnetic field as well as perpendicular. Express your answer in vector form as much as possible; use the subscripts $\parallel, \perp$ to denote directions parallel and perpendicular to the magnetic field, and use $\omega_c = qB/m$ to denote the cyclotron frequency. Show that $f(\mathbf{x}_0)$ is a solution of the Vlasov equation.
7. Thermal force (Braginskii 1965). If there is a temperature gradient then because of the temperature dependence of collisions, there turns out to be a subtle additional drag force, which is proportional to ∇T. To find this force, suppose a temperature gradient exists in the x direction, and consider the frictional drag on electrons passing a point $x = x_0$. The electrons moving to the right (positive velocity) at x_0 have traveled without collision from the point $x_0 - l_{mfp}$, where the temperature was $T(x_0 - l_{mfp})$, while those moving to the left (negative velocity) will have come collisionlessly from the point $x_0 + l_{mfp}$ where the temperature is $T(x_0 + l_{mfp})$. Suppose that both electrons and ions have no mean velocity at x_0; i.e., $u_e = u_i = 0$. Show that the total drag force on all the electrons at x_0 is

$$R_{thermal} = -2m_e n_e l_{mfp} \frac{\partial}{\partial x} \left(\nu_{ei} v_{Te} \right).$$

   Normalize the collision frequency, thermal velocity, and mean free paths to their values at $x = x_0$, where $T = T_0$; e.g., $v_{Te}(T) = v_{Te0}(T/T_0)^{1/2}$. By writing $\partial/\partial x = (\partial T/\partial x)\, \partial/\partial T$ and using these normalized values show that

$$\mathbf{R}_{thermal} = -2n_e \kappa \nabla T_e.$$

A more accurate treatment that does a proper averaging over velocities gives

$$\mathbf{R}_{thermal} = -0.71 n_e \kappa \nabla T_e.$$

8. MHD with neutrals. Suppose a plasma is partially (perhaps weakly) ionized so that besides moment equations for ions and electrons there will also be moment equations for neutrals. Now the constraints will be different, since ionization and recombination will genuinely create and annihilate both plasma particles and neutrals. Construct a set of constraint equations on the collision operators, which now include ionization and recombination as well as scattering. Take the zeroth and first moment of the three Vlasov equations for ions, electrons, and neutrals and show that the continuity equation is formally the same as before, i.e.,

$$\frac{\partial \rho}{\partial t} + \nabla \cdot (\rho \mathbf{U}) = 0,$$

providing ρ refers to the total mass density of the entire fluid (electrons, ions, and neutrals) and $\mathbf{U}$ refers to the center-of-mass velocity of the entire fluid. Show also that the equation of motion is formally the same as before, provided the pressure refers to the pressure of the entire configuration:

$$\rho \frac{\mathrm{D}\mathbf{U}}{\mathrm{D}t} = -\nabla P + \mathbf{J} \times \mathbf{B}.$$

Show that Ohm's law will be the same as before, providing the plasma is sufficiently collisional so that the Hall term can be dropped, and so Ohm's law is

$$\mathbf{E} + \mathbf{U} \times \mathbf{B} = \eta \mathbf{J}$$

Explain how the neutral component of the plasma gets accelerated by the $\mathbf{J} \times \mathbf{B}$ force – this must happen since the inertial part of the equation of motion (i.e., $\rho \mathrm{D}\mathbf{U}/\mathrm{D}t$) includes the acceleration on neutrals. Assume electron temperature gradients are parallel to electron density gradients so that the electrothermal force can be ignored.

9. MHD Heat Transport Equation. Define the MHD N-dimensional pressure

$$P = \frac{1}{N} \sum_\sigma m_\sigma \int \mathbf{v}' \cdot \mathbf{v}' f_\sigma \mathrm{d}^N \mathbf{v},$$

where $\mathbf{v}' = \mathbf{v} - \mathbf{U}$ and N is the number of dimensions of motion (e.g., if motion in only one dimension is considered, then $N = 1$, and $\mathbf{v}, \mathbf{U}$ are one dimensional, etc.). Also define the isotropic MHD heat flux

$$\mathbf{q} = \sum_\sigma m_\sigma \int \mathbf{v}' \frac{v'^2}{2} f \mathrm{d}^N \mathbf{v}.$$

(i) By taking the second moment of the Vlasov equation for each species (i.e., use $v^2/2$) and summing over species show

$$\frac{N}{2} \frac{\mathrm{D}P}{\mathrm{D}t} + \frac{N+2}{2} P \nabla \cdot \mathbf{U} = -\nabla \cdot \mathbf{q} + \mathbf{J} \cdot (\mathbf{E} + \mathbf{U} \times \mathbf{B}).$$

Table 2.1

	a	B	n	T_e, T_i	$t \sim \dfrac{a^2}{D}$
	(m)	(T)	(m^{-3})	(eV)	(s)
solar corona loop	10^6	10^{-2}	10^{15}	100	?
fusion-level tokamak	2	5	10^{20}	10^4	?
nominal lab discharge	0.05	0.1	10^{18}	$5-10$	?

Hints:

(a) Prove that $\mathbf{U}\cdot\left(m_\sigma \sum \int \mathbf{v}'\mathbf{v}' f_\sigma \mathrm{d}^N\mathbf{v}\right) = P\mathbf{U}$ assuming that f_σ is isotropic.
(b) What happens to $\sum_\sigma m_\sigma \int v^2 C_{\sigma\alpha} f_\sigma \mathrm{d}^N\mathbf{v}$?
(c) Using the momentum and continuity equations prove that

$$\frac{\partial}{\partial t}\left(\frac{\rho U^2}{2}\right) + \nabla\cdot\left(\frac{\rho U^2}{2}\mathbf{U}\right) = -\mathbf{U}\cdot\nabla P + \mathbf{U}\cdot(\mathbf{J}\times\mathbf{B}).$$

(ii) Using the continuity equation and Ohm's law show that

$$\frac{N}{2}\frac{\mathrm{D}P}{\mathrm{D}T} - \frac{N+2}{2}\frac{P}{\rho}\frac{\mathrm{D}\rho}{\mathrm{D}t} = -\nabla\cdot\mathbf{q} + \eta J^2.$$

Show that if both the heat flux term $-\nabla\cdot\mathbf{q}$ and the Ohmic heating term ηJ^2 can be ignored, then the pressure and density are related by the adiabatic condition $P \sim \rho^\gamma$, where $\gamma = (N+2)/N$. By assuming $\mathrm{D}/\mathrm{D}t \sim \omega$ and $\nabla \sim k$ show that the dropping of these two right-hand terms is equivalent to assuming $\omega \gg \nu_{ei}$ and $\omega/k \gg v_T$. Explain why the phenomenon should be isothermal if $\omega/k \ll v_T$.

10. Classical diffusion. Show by dimensional analysis that the nominal distance diffused in a time t is given by $x^2 \sim Dt$ (do this by combining Fick's law $\Gamma = -D\nabla n$ and the continuity equation). Using the parameters given in Table 2.1 for a solar corona loop, a tokamak with fusion-level parameters, and a nominal laboratory discharge, calculate the time required for the plasma to diffuse out of the configuration if the diffusion is classical. In other words, fill in the last column of the table using Eq. (2.114) to calculate the nominal time (or range of times) to diffuse a distance a, where a is the nominal radius of the configuration. Are these times long or short compared to what is of interest?
11. Sketch the current collected by a Langmuir probe as a function of the bias voltage and indicate the ion saturation current, the exponentially changing electron current, the floating potential, and the plasma potential. Calculate the ion saturation current collected by a 1 cm long, 0.25 mm diameter probe immersed in a 5 eV argon plasma that has a density $n = 10^{16}$ m^{-3}. Calculate the electron saturation current also (i.e., the current when the probe is at the plasma potential). What is the offset of the floating potential relative to the plasma potential?

3

Motion of a single plasma particle

3.1 Motivation

Single particle motion in neutral gases is trivial – particles move in straight lines until they hit other particles or the wall. Because of this simplicity, there is no point in keeping track of the details of single particle motion in a neutral gas and instead a statistical averaging of this motion suffices; this averaging shows that neutral gases have Maxwellian velocity distributions and are in a local thermodynamic equilibrium. In contrast, plasma particles are nearly collisionless and typically have complex trajectories that are strongly affected by both electric and magnetic fields.

As discussed in the previous chapter, the velocity distribution in a plasma will become Maxwellian when enough collisions have occurred to maximize the entropy. However, since collisions occur infrequently in hot plasmas, many important phenomena have time scales shorter than the time required for the plasma velocity distribution to become Maxwellian. A collisionless model is thus required to characterize these fast phenomena. In these situations randomization does not occur, entropy is conserved, the distribution function need not be Maxwellian, and the plasma is not in thermodynamic equilibrium. Thermodynamic concepts therefore do not apply, and the plasma is instead characterized by concepts from classical mechanics such as momentum or energy conservation of individual particles. In these collisionless situations the complex details of single particle dynamics are not washed out by collisions but instead persist and influence the macroscopic scale. As an example, the cyclotron resonance of a single particle can be important at the macroscopic scale in a collisionless plasma. This chapter examines various aspects of single particle motion and shows how these aspects can influence the macroscopic properties of a plasma.

Furthermore, study of single particle dynamics has a very direct relevance to Vlasov theory because, as shown in Section 2.3, *any* function constructed from constants of single particle motion is a valid solution of the collisionless Vlasov equation. Thus, knowledge of single particle dynamics provides a "repertoire" of constants of the motion from which solutions to the Vlasov equation suitable for various situations can be constructed.

Finally, the study of single particle motion develops valuable intuition regarding wave–particle interactions and identifies certain unusual situations, such as stochastic or non-adiabatic particle motion, that are beyond the descriptive capability of fluid models.

3.2 Hamilton–Lagrange formalism vs. Lorentz equation

Two mathematically equivalent formalisms describe charged particle dynamics, namely (i) the Lorentz equation

$$m\frac{\mathrm{d}\mathbf{v}}{\mathrm{d}t} = q(\mathbf{E}+\mathbf{v}\times\mathbf{B}) \tag{3.1}$$

and (ii) Hamiltonian–Lagrangian dynamics.

The two formalisms are complementary: the Lorentz equation is intuitive and suitable for approximate methods, whereas the more abstract Hamiltonian–Lagrangian formalism exploits time and space symmetries. A brief review of the Hamiltonian–Lagrangian formalism follows, emphasizing aspects relevant to dynamics of charged particles.

The starting point is to postulate the existence of a function L, called the Lagrangian, which:

1. contains *all* information about the particle dynamics for a given situation;
2. depends only on generalized coordinates $Q_i(t)$, $\dot{Q}_i(t)$ appropriate to the problem;
3. possibly has an explicit dependence on time t.

If such a function $L(Q_i(t), \dot{Q}_i(t), t)$ exists, then information on particle dynamics is retrieved by manipulation of the *action integral*

$$S = \int_{t_1}^{t_2} L(Q_i(t), \dot{Q}_i(t), t)\mathrm{d}t. \tag{3.2}$$

This manipulation is based on d'Alembert's principle of least action. According to this principle, one considers the infinity of possible trajectories a particle *could* follow to get from its initial position $Q_i(t_1)$ to its final position $Q_i(t_2)$, and postulates that the trajectory *actually* followed is the one that results in the *lowest* value of S. Thus, the value of S must be minimized (note that S here is action, and not entropy as in the previous chapter). Minimizing S does not give the actual trajectory directly, but rather gives equations of motion, which can be solved to give the actual trajectory.

Minimizing S is accomplished by considering an arbitrary nearby *alternative* trajectory $Q_i(t)+\delta Q_i(t)$ having the *same* beginning and end points as the actual trajectory, i.e., $\delta Q_i(t_1)=\delta Q_i(t_2)=0$. In order to make the variational argument more precise, δQ_i is expressed as

$$\delta Q_i(t)=\epsilon\eta_i(t), \tag{3.3}$$

where ϵ is an arbitrarily adjustable scalar assumed to be small so that $\epsilon^2<\epsilon$ and $\eta_i(t)$ is a function of t that vanishes when $t=t_1$ or $t=t_2$ but is otherwise arbitrary. Calculating δS to second order in ϵ gives

$$\begin{aligned}
\delta S &= \int_{t_1}^{t_2} L(Q_i+\delta Q_i,\ \dot{Q}_i+\delta\dot{Q}_i,\ t)\mathrm{d}t-\int_{t_1}^{t_2} L(Q_i,\dot{Q}_i,t)\ \mathrm{d}t \\
&= \int_{t_1}^{t_2} L(Q_i+\epsilon\eta_i,\ \dot{Q}_i+\epsilon\dot{\eta}_i,\ t)\mathrm{d}t-\int_{t_1}^{t_2} L(Q_i,\dot{Q}_i,t)\ \mathrm{d}t \\
&= \int_{t_1}^{t_2}\left(\epsilon\eta_i\frac{\partial L}{\partial Q_i}+\frac{(\epsilon\eta_i)^2}{2}\frac{\partial^2 L}{\partial Q_i^2}+\epsilon\dot{\eta}_i\frac{\partial L}{\partial \dot{Q}_i}+\frac{(\epsilon\dot{\eta}_i)^2}{2}\frac{\partial^2 L}{\partial \dot{Q}_i^2}\right)\mathrm{d}t.
\end{aligned} \tag{3.4}$$

Suppose the trajectory $Q_i(t)$ is the one that minimizes S. Any other trajectory must lead to a higher value of S and so δS must be positive for any finite value of ϵ. If ϵ is chosen to be sufficiently small, then the absolute values of the terms of order ϵ^2 in Eq. (3.4) will be smaller than the absolute values of the terms linear in ϵ. The sign of ϵ could then be chosen to make δS negative, but this would violate the requirement that δS must be positive. The only way out of this dilemma is to insist that the sum of the terms linear in ϵ in Eq. (3.4) vanishes so $\delta S\sim\epsilon^2$ and is therefore always positive as required. Insisting that the sum of terms linear in ϵ vanishes implies

$$0=\int_{t_1}^{t_2}\left(\eta_i\frac{\partial L}{\partial Q_i}+\dot{\eta}_i\frac{\partial L}{\partial \dot{Q}_i}\right)\mathrm{d}t. \tag{3.5}$$

Using $\dot{\eta}_i=\mathrm{d}\eta_i/\mathrm{d}t$ the above expression may be integrated by parts to obtain

$$\begin{aligned}
0 &= \int_{t_1}^{t_2}\left(\eta_i\frac{\partial L}{\partial Q_i}+\frac{\mathrm{d}\eta_i}{\mathrm{d}t}\frac{\partial L}{\partial \dot{Q}_i}\right)\mathrm{d}t \\
&= \left[\eta_i\frac{\partial L}{\partial \dot{Q}_i}\right]_{t_1}^{t_2}+\int_{t_1}^{t_2}\left\{\eta_i\frac{\partial L}{\partial Q_i}-\eta_i\frac{\mathrm{d}}{\mathrm{d}t}\left(\frac{\partial L}{\partial \dot{Q}_i}\right)\right\}\mathrm{d}t.
\end{aligned} \tag{3.6}$$

Since $\eta_i(t_{1,2})=0$, the integrated term vanishes and since η_i was an arbitrary function of t, the coefficient of η_i in the integrand must vanish, yielding *Lagrange's equation*

$$\frac{\mathrm{d}P_i}{\mathrm{d}t}=\frac{\partial L}{\partial Q_i}, \tag{3.7}$$

where the *canonical momentum* P_i is defined as

$$P_i = \frac{\partial L}{\partial \dot{Q}_i}. \tag{3.8}$$

Equation (3.7) shows that if L does *not* depend on a particular generalized coordinate Q_j, then $\mathrm{d}P_j/\mathrm{d}t = 0$, in which case the canonical momentum P_j is a *constant of the motion;* the coordinate Q_j is called a *cyclic* or *ignorable* coordinate. This is a very powerful and profound result. Saying that the Lagrangian function does not depend on a coordinate is equivalent to saying that the system is *symmetric* in that coordinate or translationally invariant with respect to that coordinate. The quantities P_j and Q_j are called conjugate and action has the dimensions of the product of these quantities.

Hamilton extended this formalism by introducing a new function related to the Lagrangian. This new function, called the Hamiltonian, provides further useful information and is defined as

$$H \equiv \left(\sum_i P_i \dot{Q}_i\right) - L. \tag{3.9}$$

Partial derivatives of H with respect to P_i and to Q_i give Hamilton's equations

$$\dot{Q}_i = \frac{\partial H}{\partial P_i} \qquad \dot{P}_i = -\frac{\partial H}{\partial Q_i}, \tag{3.10}$$

which are equations of motion having a close relation to phase-space concepts. The time derivative of the Hamiltonian is

$$\frac{\mathrm{d}H}{\mathrm{d}t} = \sum_i \frac{\mathrm{d}P_i}{\mathrm{d}t}\dot{Q}_i + \sum_i P_i \frac{\mathrm{d}\dot{Q}_i}{\mathrm{d}t} - \left(\sum_i \frac{\partial L}{\partial Q_i}\dot{Q}_i + \sum_i \frac{\partial L}{\partial \dot{Q}_i}\frac{\mathrm{d}\dot{Q}_i}{\mathrm{d}t} + \frac{\partial L}{\partial t}\right) = -\frac{\partial L}{\partial t}. \tag{3.11}$$

This shows that if L does not explicitly depend on time, i.e., $\partial L/\partial t = 0$, the Hamiltonian H is a *constant of the motion.* As will be shown later, H corresponds to the energy of the system, so if $\partial L/\partial t = 0$, the energy is a constant of the motion. Thus, energy is conjugate to time in analogy to canonical momentum being conjugate to position (note that energy $\times$ time also has the units of action). If the Lagrangian does not explicitly depend on time, then the system can be thought of as being "symmetric" with respect to time, or "translationally" invariant with respect to time.

The Lagrangian for a charged particle in an electromagnetic field is

$$L = \frac{mv^2}{2} + q\mathbf{v}\cdot\mathbf{A}(\mathbf{x}, t) - q\phi(\mathbf{x}, t); \tag{3.12}$$

the validity of Eq. (3.12) will now be established by showing that it generates the Lorentz equation when inserted into Lagrange's equation. Since no symmetry is assumed, there is no reason to use any special coordinate system and so ordinary Cartesian coordinates will be used as the canonical coordinates, in which case Eq. (3.8) gives the canonical momentum as

$$\mathbf{P} = m\mathbf{v} + q\mathbf{A}(\mathbf{x},t). \tag{3.13}$$

The left-hand side of Eq. (3.7) becomes

$$\frac{d\mathbf{P}}{dt} = m\frac{d\mathbf{v}}{dt} + q\left(\frac{\partial \mathbf{A}}{\partial t} + \mathbf{v}\cdot\nabla\mathbf{A}\right), \tag{3.14}$$

while the right-hand side of Eq. (3.7) becomes

$$\begin{aligned}\frac{\partial L}{\partial \mathbf{x}} &= q\nabla\left(\mathbf{v}\cdot\mathbf{A}\right) - q\nabla\phi = q\left(\mathbf{v}\cdot\nabla\mathbf{A} + \mathbf{v}\times\nabla\times\mathbf{A}\right) - q\nabla\phi \\ &= q\left(\mathbf{v}\cdot\nabla\mathbf{A} + \mathbf{v}\times\mathbf{B}\right) - q\nabla\phi. \end{aligned} \tag{3.15}$$

Equating the above two expressions gives the Lorentz equation, where the electric field is defined as $\mathbf{E} = -\partial\mathbf{A}/\partial t - \nabla\phi$ in accordance with Faraday's law. This proves that Eq. (3.12) is mathematically equivalent to the Lorentz equation when used with the principle of least action.

The Hamiltonian associated with this Lagrangian is, in Cartesian coordinates,

$$\begin{aligned} H &= \mathbf{P}\cdot\mathbf{v} - L \\ &= \frac{mv^2}{2} + q\phi \\ &= \frac{(\mathbf{P} - q\mathbf{A}(\mathbf{x},t))^2}{2m} + q\phi(\mathbf{x},t), \end{aligned} \tag{3.16}$$

where the last line is the form more suitable for use with Hamilton's equations, i.e., $H = H(\mathbf{x}, \mathbf{P}, t)$. Equation (3.16) also shows that H is, as promised, the particle energy. If generalized coordinates are used, the energy can be written in a general form as $E = H(Q, P, t)$. Equation (3.11) showed that even though both Q and P depend on time, the energy depends on time only if H explicitly depends on time. Thus, in a situation where H does not explicitly depend on time, the energy would have the form $E = H(Q(t), P(t)) = const.$

It is important to realize that both canonical momentum and energy depend on the reference frame. For example, a bullet fired in an airplane in the direction opposite to the airplane motion, and with a speed equal to the airplane's speed, has a large energy as measured in the airplane frame, but zero energy as measured by an observer on the ground. A more subtle example (of importance to later analysis of waves and Landau damping) occurs when $\mathbf{A}$ and/or ϕ have a wave-like dependence, e.g., $\phi(\mathbf{x},t) = \phi(\mathbf{x} - \mathbf{v}_{ph}t)$, where $\mathbf{v}_{ph}$ is the wave phase velocity. This potential is *time-dependent* in the lab frame and so the associated Lagrangian has an explicit dependence on time in the lab frame, which implies that *energy is not a constant of the motion in the lab frame*. In contrast, ϕ is *time-independent* in the wave frame and so the energy is a *constant of the motion in the wave frame*. Existence of a constant of the motion reduces the complexity of the system of equations and typically makes it possible to integrate at least one equation in closed form. Thus, it is advantageous to analyze the system in the frame having the most constants of the motion.

3.3 Adiabatic invariant of a pendulum

Perfect symmetry is never attained in reality. This leads to the practical question of how constants of the motion behave when space and/or time symmetries are "good," but not perfect. Does the utility of constants of the motion collapse abruptly when the slightest non-symmetrical blemish rears its ugly head, does the utility decay gracefully, or does something completely different happen? To answer these questions, we begin by considering the problem of a small-amplitude pendulum having a time-dependent, but *slowly changing* resonant frequency $\omega(t)$. Since $\omega^2 = g/l$, the time-dependence of the frequency might result from either a slow change in the gravitational acceleration g or else from a slow change in the pendulum length l. In both cases the pendulum equation of motion will be

$$\frac{\mathrm{d}^2x}{\mathrm{d}t^2} + \omega^2(t)x = 0. \tag{3.17}$$

This equation cannot be solved exactly for arbitrary $\omega(t)$ but if a modest restriction is put on $\omega(t)$ the equation can be solved *approximately* using the WKB method (Wentzel 1926, Kramers 1926, Brillouin 1926). This method is based on the hypothesis that the solution for a time-dependent frequency is likely to be a generalization of the constant-frequency solution

$$x = \mathrm{Re}\left[A\exp(\mathrm{i}\omega t)\right], \tag{3.18}$$

where this generalization is postulated to be of the form

$$x(t) = \mathrm{Re}\left[A(t)\mathrm{e}^{\mathrm{i}\int^t \omega(t')\mathrm{d}t'}\right]. \tag{3.19}$$

Here $A(t)$ is an amplitude function determined as follows: calculate the first derivative of Eq. (3.19),

$$\frac{\mathrm{d}x}{\mathrm{d}t} = \mathrm{Re}\left[\mathrm{i}\omega A\mathrm{e}^{\mathrm{i}\int^t \omega(t')\mathrm{d}t'} + \frac{\mathrm{d}A}{\mathrm{d}t}\mathrm{e}^{\mathrm{i}\int^t \omega(t')\mathrm{d}t'}\right], \tag{3.20}$$

then the second derivative

$$\frac{\mathrm{d}^2x}{\mathrm{d}t^2} = \mathrm{Re}\left[\left(\mathrm{i}\frac{\mathrm{d}\omega}{\mathrm{d}t}A + 2\mathrm{i}\omega\frac{\mathrm{d}A}{\mathrm{d}t} - \omega^2 A + \frac{\mathrm{d}^2A}{\mathrm{d}t^2}\right)\mathrm{e}^{\mathrm{i}\int^t \omega(t')\mathrm{d}t'}\right], \tag{3.21}$$

and insert this last result into Eq. (3.17) which reduces to

$$\mathrm{i}\frac{\mathrm{d}\omega}{\mathrm{d}t}A+2\mathrm{i}\omega\frac{\mathrm{d}A}{\mathrm{d}t}+\frac{\mathrm{d}^2A}{\mathrm{d}t^2}=0, \tag{3.22}$$

since the terms involving ω^2 cancel exactly. To proceed further, we make an assumption – the validity of which is to be checked later – that the time dependence of $\mathrm{d}A/\mathrm{d}t$ is *sufficiently slow* to allow dropping the last term in Eq. (3.22) relative to the middle term. The two terms that remain in Eq. (3.22) can then be rearranged as

$$\frac{1}{\omega}\frac{\mathrm{d}\omega}{\mathrm{d}t}=-\frac{2}{A}\frac{\mathrm{d}A}{\mathrm{d}t}, \tag{3.23}$$

which has the exact solution

$$A(t)\sim\frac{1}{\sqrt{\omega(t)}}. \tag{3.24}$$

The assumption of slowness is thus at least self-consistent, for if $\omega(t)$ is indeed slowly changing, Eq. (3.24) shows that $A(t)$ will also be slowly changing and the dropping of the last term in Eq. (3.22) is justified. The slowness requirement can be quantified by assuming that the frequency has an exponential dependence

$$\omega(t)=\omega_0\mathrm{e}^{\alpha t}. \tag{3.25}$$

Thus,

$$\alpha=\frac{1}{\omega}\frac{\mathrm{d}\omega}{\mathrm{d}t} \tag{3.26}$$

is a measure of how fast the frequency is changing compared to the frequency itself. Hence, dropping the last term in Eq. (3.22) is legitimate if

$$\alpha\ll 4\omega_0 \tag{3.27}$$

or

$$\frac{1}{\omega}\frac{\mathrm{d}\omega}{\mathrm{d}t}\ll 4\omega. \tag{3.28}$$

In other words, if Eq. (3.28) is satisfied, then the fractional change of the pendulum period per period is small.

Equation (3.24) indicates that when ω is time-dependent, the pendulum amplitude is not constant and so the pendulum energy is *not* conserved. It turns out that what *is* conserved is the *action integral*

$$S=\oint v\mathrm{d}x, \tag{3.29}$$

where the integration is over one period of oscillation. This integral can also be written in terms of time as

$$S = \int_{t_0}^{t_0+\tau} v\frac{\mathrm{d}x}{\mathrm{d}t}\mathrm{d}t, \tag{3.30}$$

where t_0 is a time when x is at an instantaneous maximum and τ, the period of a complete cycle, is defined as the interval between two successive times when $\mathrm{d}x/\mathrm{d}t = 0$ and $\mathrm{d}^2x/\mathrm{d}t^2$ has the same sign (e.g., for a pendulum, t_0 would be a time when the pendulum has swung all the way to the right and so is reversing its velocity while τ is the time one has to wait for this to happen again). To show that action is conserved, Eq. (3.29) can be integrated by parts as

$$\begin{aligned} S &= \int_{t_0}^{t_0+\tau} \left[\frac{\mathrm{d}}{\mathrm{d}t}\left(x\frac{\mathrm{d}x}{\mathrm{d}t}\right) - x\frac{\mathrm{d}^2x}{\mathrm{d}t^2}\right]\mathrm{d}t \\ &= \left[x\frac{\mathrm{d}x}{\mathrm{d}t}\right]_{t_0}^{t_0+\tau} - \int_{t_0}^{t_0+\tau} x\frac{\mathrm{d}^2x}{\mathrm{d}t^2}\mathrm{d}t \\ &= \int_{t_0}^{t_0+\tau} \omega^2 x^2 \mathrm{d}t, \end{aligned} \tag{3.31}$$

where (i) the integrated term has vanished by virtue of the definitions of t_0 and τ, and (ii) Eq. (3.17) has been used to substitute for $\mathrm{d}^2x/\mathrm{d}t^2$. Equations (3.19) and (3.24) can be combined to give

$$x(t) = x(t_0)\sqrt{\frac{\omega(t_0)}{\omega(t)}}\cos\left(\int_{t_0}^{t} \omega(t')\mathrm{d}t'\right) \tag{3.32}$$

so Eq. (3.31) becomes

$$\begin{aligned} S &= \int_{t_0}^{t_0+\tau} \omega(t')^2 \left\{x(t_0)\sqrt{\frac{\omega(t_0)}{\omega(t')}}\cos\left(\int_{t_0}^{t'} \omega(t'')\mathrm{d}t''\right)\right\}^2 \mathrm{d}t' \\ &= [x(t_0)]^2\,\omega(t_0)\int_{t_0}^{t_0+\tau} \omega(t')\cos^2\left(\int_{t_0}^{t'} \omega(t'')\mathrm{d}t''\right)\mathrm{d}t' \\ &= [x(t_0)]^2\,\omega(t_0)\int_0^{2\pi} \mathrm{d}\xi\cos^2\xi = \pi\,[x(t_0)]^2\,\omega(t_0) = const., \end{aligned} \tag{3.33}$$

where $\xi = \int_{t_0}^{t'} \omega(t'')\mathrm{d}t''$ and $\mathrm{d}\xi = \omega(t')\mathrm{d}t'$. Equation (3.29) shows that S is the area in phase-space enclosed by the trajectory $\{x(t), v(t)\}$ and Eq. (3.33) shows that for a slowly changing pendulum frequency, *this area is a constant of the motion*. Since the average energy of the pendulum scales as $\sim [\omega(t)x(t)]^2$, we see from Eq. (3.24) that the ratio

$$\frac{\text{energy}}{\text{frequency}} \sim \omega(t)x^2(t) \sim S \sim const. \tag{3.34}$$

The ratio in Eq. (3.34) is the classical equivalent of the quantum number N of a simple harmonic oscillator because in quantum mechanics the energy E of a simple harmonic oscillator is related to the frequency by the relation $E/\hbar\omega = N + 1/2$.

This analysis clearly applies to *any* dynamical system having an equation of motion of the form of Eq. (3.17). Hence, if the dynamics of plasma particles happens to be of this form, then S can be added to our repertoire of constants of the motion.

3.4 Extension of WKB method to general adiabatic invariant

Action has the dimensions of (canonical momentum) × (canonical coordinate) so we may anticipate that for general Hamiltonian systems, the action integral given in Eq. (3.29) is not an invariant because v is *not*, in general, proportional to P. We postulate that the general form for the action integral is

$$S = \oint P\mathrm{d}Q, \tag{3.35}$$

where the integral is over one period of the periodic motion and P, Q are the relevant canonical momentum–coordinate conjugate pair. The proof of adiabatic invariance used for Eq. (3.29) does not work directly for Eq. (3.35); we now present a slightly more involved proof to show that Eq. (3.35) is indeed the more general form of adiabatic invariant.

Let us define the radius vector in the $Q-P$ plane to be $\mathbf{R} = (Q, P)$ and define unit vectors in the Q and P directions by $\hat{Q}$ and $\hat{P}$; these definitions are shown in Fig. 3.1. Furthermore, we define the z direction as being normal to the $Q-P$

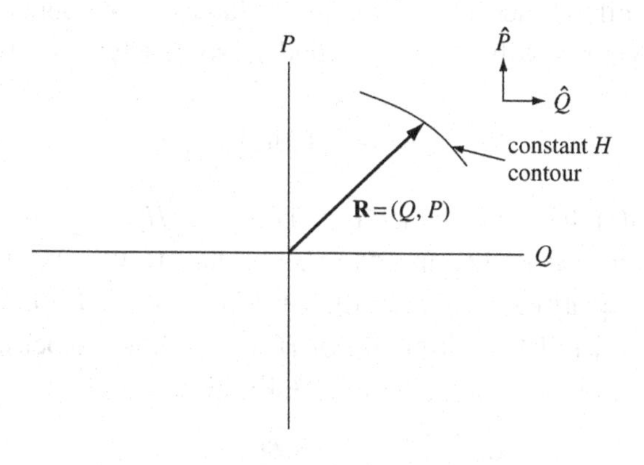

Fig. 3.1 Q – P plane.

plane; thus, the unit vector $\hat{z}$ is "out of the paper," i.e., $\hat{z} = \hat{Q} \times \hat{P}$. Hamilton's equations (i.e., $\dot{P} = -\partial H/\partial Q$, $\dot{Q} = \partial H/\partial P$) may be written in vector form as

$$\frac{\mathrm{d}\mathbf{R}}{\mathrm{d}t} = -\hat{z} \times \nabla H, \tag{3.36}$$

where

$$\nabla = \hat{Q}\frac{\partial}{\partial Q} + \hat{P}\frac{\partial}{\partial P} \tag{3.37}$$

is the gradient operator in the $Q-P$ plane. Equation (3.36) shows that the phase-space "velocity" $\mathrm{d}\mathbf{R}/\mathrm{d}t$ is *orthogonal* to ∇H. Hence, $\mathbf{R}$ *stays on a level contour of* H. If H is constant, then, in order for the motion to be periodic, the path along this level contour must circle around and join itself, like a road of *constant elevation* around the rim of a mountain (or a crater). If H is not constant, but slowly changing in time, the contour will circle around and *nearly* join itself.

Equation (3.36) can be inverted by crossing it with $\hat{z}$ to give

$$\nabla H = \hat{z} \times \frac{\mathrm{d}\mathbf{R}}{\mathrm{d}t}. \tag{3.38}$$

For periodic and near-periodic motions, $\mathrm{d}\mathbf{R}/\mathrm{d}t$ is always in the same sense (always clockwise or always counterclockwise). Thus, Eq. (3.38) shows that an "observer" following the path would always see that H is increasing on the left-hand side of the path and decreasing on the right-hand side (or vice versa). For clarity, the origin of the $Q-P$ plane is redefined to be at a local maximum or minimum of H. Hence, near the extremum H must have the Taylor expansion

$$H(P,Q) = H_{extremum} + \frac{P^2}{2}\left[\frac{\partial^2 H}{\partial P^2}\right]_{P=0,Q=0} + \frac{Q^2}{2}\left[\frac{\partial^2 H}{\partial Q^2}\right]_{P=0,Q=0}, \tag{3.39}$$

where $\left[\partial^2 H/\partial P^2\right]_{P=0,Q=0}$ and $\left[\partial^2 H/\partial Q^2\right]_{P=0,Q=0}$ are either both positive (valley) or both negative (hill). Since H is assumed to have a slow dependence on time, these second derivatives will be time-dependent so that Eq. (3.39) has the form

$$H = \alpha(t)\frac{P^2}{2} + \beta(t)\frac{Q^2}{2}, \tag{3.40}$$

where $\alpha(t)$ and $\beta(t)$ have the same sign. The term $H_{extremum}$ in Eq. (3.39) has been dropped because it is just an additive constant to the energy and does not affect Hamilton's equations. From Eq. (3.36) the direction of rotation of $\mathbf{R}$ is seen to be counterclockwise if the extremum of H is a hill, and clockwise if a valley.

Hamilton's equations operating on Eq. (3.40) give

$$\frac{\mathrm{d}P}{\mathrm{d}t} = -\beta Q, \quad \frac{\mathrm{d}Q}{\mathrm{d}t} = \alpha P. \tag{3.41}$$

These equations do not directly generate the simple harmonic oscillator equation because of the time dependence of α, β. However, if we define the auxiliary variable

$$\tau = \int^{t} \beta(t')\mathrm{d}t' \tag{3.42}$$

then

$$\frac{\mathrm{d}}{\mathrm{d}t} = \frac{\mathrm{d}\tau}{\mathrm{d}t}\frac{\mathrm{d}}{\mathrm{d}\tau} = \beta\frac{\mathrm{d}}{\mathrm{d}\tau}$$

so Eq. (3.41) becomes

$$\frac{\mathrm{d}P}{\mathrm{d}\tau} = -Q, \quad \frac{\mathrm{d}Q}{\mathrm{d}\tau} = \frac{\alpha}{\beta}P. \tag{3.43}$$

Substituting for Q in the right-hand equation using the left-hand equation gives

$$\frac{\mathrm{d}^2 P}{\mathrm{d}\tau^2} + \frac{\alpha}{\beta}P = 0, \tag{3.44}$$

which is a simple harmonic oscillator with $\omega^2(\tau) = \alpha(\tau)/\beta(\tau)$. The action integral may be rewritten as

$$S = \int P\frac{\mathrm{d}Q}{\mathrm{d}\tau}\mathrm{d}\tau, \tag{3.45}$$

where the integral is over one period of the motion. Using Eq. (3.43) and following the same procedure as was used with Eq. (3.32) and Eq. (3.33), this becomes

$$S = \int P^2\frac{\alpha}{\beta}\mathrm{d}\tau = \lambda^2\int\left[\left(\frac{\alpha(\tau')}{\beta(\tau')}\right)^{1/2}\cos^2\left(\int^{\tau'}(\alpha/\beta)^{1/2}\mathrm{d}\tau''\right)\right]\mathrm{d}\tau', \tag{3.46}$$

where λ is a constant dependent on initial conditions. By introducing the orbit phase $\phi = \int^{\tau}(\alpha/\beta)^{1/2}\mathrm{d}\tau$, Eq. (3.46) becomes

$$S = \lambda^2\int_0^{2\pi}\mathrm{d}\phi\cos^2\phi = const. \tag{3.47}$$

Thus, the general action integral is indeed an adiabatic invariant. This proof is of course only valid in the vicinity of an extremum of H, i.e., only where H can be adequately represented by Eq. (3.40).

3.4.1 *Proof for the general adiabatic invariant*

We now develop a proof for the general adiabatic invariant. This proof is not restricted to small oscillations (i.e., being near an extremum of H) as was the previous discussion. Let the Hamiltonian depend on time via a slowly changing parameter $\lambda(t)$, so that $H = H(P, Q, \lambda(t))$. From Eq. (3.16) the energy is given by

$$E(t) = H(P, Q, \lambda(t)) \tag{3.48}$$

and, in principle, this relation can be inverted to give $P = P(E(t), Q, \lambda(t))$. Suppose a particle is executing nearly periodic motion in the $Q - P$ plane. We define the turning point Q_{tp} as a position where $\mathrm{d}Q/\mathrm{d}t = 0$. Since Q is oscillating, there will be a turning point associated with Q having its maximum value and a turning point associated with Q having its minimum value. From now on let us only consider turning points where Q has its maximum value, that is, we only consider the turning points on the right-hand side of the nearly periodic trajectories in the $Q - P$ plane shown in Fig. 3.2.

If the motion is periodic, then the turning point for the $N + 1^{\text{th}}$ period will be the same as the turning point for the Nth period, but if the motion is only nearly periodic, there will be a slight difference as shown in Fig. 3.2. This difference can be characterized by making the turning point a function of time so $Q_{tp} = Q_{tp}(t)$. This function is only defined for the times when $\mathrm{d}Q/\mathrm{d}t = 0$. When the motion is not exactly periodic, this turning point is such that $Q_{tp}(t + \tau) \neq Q_{tp}(t)$, where τ is the time interval required for the particle to go from the first turning point to the next turning point. The action integral is over one entire period of oscillation starting from a right-hand turning point and then going to the next right-hand turning point (cf. Fig. 3.2) and so can be written as

$$\begin{aligned} S &= \oint P \mathrm{d}Q \\ &= \int_{Q_{tp}(t)}^{Q_{tp}(t+\tau)} P \mathrm{d}Q. \end{aligned} \tag{3.49}$$

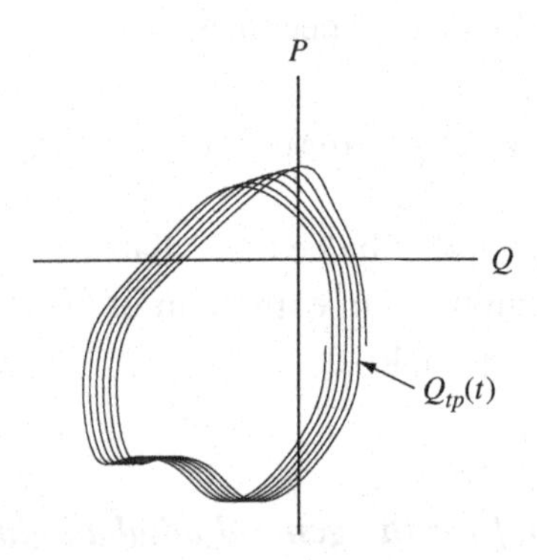

Fig. 3.2 Nearly periodic-phase space trajectory for slowly changing Hamiltonian. The turning point $Q_{tp}(t)$ is where Q is at its maximum.

From Eq. (3.16) it is seen that P/m is not, in general, the velocity and so the velocity $\mathrm{d}Q/\mathrm{d}t$ is not, in general, proportional to P. Thus, the turning points are not necessarily at the locations where P vanishes, and in fact P need not change sign during a period. However, S still corresponds to the area of phase-space enclosed by one period of the phase-space trajectory.

We can now calculate

$$\begin{aligned}\frac{\mathrm{d}S}{\mathrm{d}t} &= \frac{\mathrm{d}}{\mathrm{d}t}\oint P\mathrm{d}Q = \frac{\mathrm{d}}{\mathrm{d}t}\int_{Q_{tp}(t)}^{Q_{tp}(t+\tau)} P(E(t), Q, \lambda(t))\mathrm{d}Q \\ &= \left[P\frac{\mathrm{d}Q}{\mathrm{d}t}\right]_{Q_{tp}(t)}^{Q_{tp}(t+\tau)} + \int_{Q_{tp}(t)}^{Q_{tp}(t+\tau)} \left(\frac{\partial P}{\partial t}\right)_Q \mathrm{d}Q \\ &= \int_{Q_{tp}(t)}^{Q_{tp}(t+\tau)} \left[\left(\frac{\partial P}{\partial E}\right)_{Q,\lambda}\frac{\mathrm{d}E}{\mathrm{d}t} + \left(\frac{\partial P}{\partial \lambda}\right)_{Q,E}\frac{\mathrm{d}\lambda}{\mathrm{d}t}\right]\mathrm{d}Q.\end{aligned} \tag{3.50}$$

Because $\mathrm{d}Q/\mathrm{d}t = 0$ at the turning point, the integrated term vanishes and so there is no contribution from motion of the turning point. From Eq. (3.48) we have

$$1 = \frac{\partial H}{\partial P}\left(\frac{\partial P}{\partial E}\right)_{Q,\lambda} \tag{3.51}$$

and

$$0 = \frac{\partial H}{\partial P}\left(\frac{\partial P}{\partial \lambda}\right)_{Q,E} + \frac{\partial H}{\partial \lambda} \tag{3.52}$$

so Eq. (3.50) becomes

$$\frac{\mathrm{d}S}{\mathrm{d}t} = \oint \left(\frac{\partial H}{\partial P}\right)^{-1}\left[\frac{\mathrm{d}E}{\mathrm{d}t} - \frac{\partial H}{\partial \lambda}\frac{\mathrm{d}\lambda}{\mathrm{d}t}\right]\mathrm{d}Q. \tag{3.53}$$

From Eq. (3.48) we have

$$\frac{\mathrm{d}E}{\mathrm{d}t} = \frac{\partial H}{\partial P}\frac{\mathrm{d}P}{\mathrm{d}t} + \frac{\partial H}{\partial Q}\frac{\mathrm{d}Q}{\mathrm{d}t} + \frac{\partial H}{\partial \lambda}\frac{\mathrm{d}\lambda}{\mathrm{d}t} = \frac{\partial H}{\partial \lambda}\frac{\mathrm{d}\lambda}{\mathrm{d}t}, \tag{3.54}$$

since the first two terms canceled due to Hamilton's equations. Substitution of Eq. (3.54) into Eq. (3.53) gives $\mathrm{d}S/\mathrm{d}t = 0$, completing the proof of adiabatic invariance. No assumption has been made here that P, Q are close to the values associated with an extremum of H.

This proof seems too neat, because it has established adiabatic invariance simply by careful use of the chain rule, and by taking partial derivatives. However, this observation reveals the underlying essence of adiabaticity, namely it is the differentiability of H, P with respect to λ from one period to the next and the

Hamilton nature of the system, which together provide the conditions for the adiabatic invariant to exist. If the motion had been such that after one cycle the motion had changed so drastically that taking a derivative of H or P with respect to λ would not make sense, then the adiabatic invariant would not exist.

3.5 Drift equations

We show in this section that it is possible to deduce intuitive and quite accurate analytic solutions for the velocity (drift) of charged particles in arbitrarily complicated electric and magnetic fields provided the fields are *slowly changing in both space and time* (this requirement is essentially the slowness requirement for adiabatic invariance). Drift solutions are obtained by solving the Lorentz equation

$$m\frac{d\mathbf{v}}{dt} = q\left(\mathbf{E}+\mathbf{v}\times\mathbf{B}\right) \tag{3.55}$$

iteratively, taking advantage of the assumed separation of scales between fast and slow motions.

3.5.1 Simple $\mathbf{E}\times\mathbf{B}$ *and force drifts*

Before developing the general method for analyzing drifts, a simple example illustrating the basic idea will now be discussed. This example consists of an ion starting at rest in a spatially uniform magnetic field $\mathbf{B} = B\hat{z}$ and a spatially uniform electric field $\mathbf{E} = E\hat{y}$. The origin is defined to be at the ion's starting position and both electric and magnetic fields are constant in time. The assumed spatial uniformity and time-independence of the fields represent the extreme limit of assuming that the fields are slowly changing in space and time.

Because the magnetic force $q\mathbf{v}\times\mathbf{B}$ is perpendicular to $\mathbf{v}$, the magnetic force does no work and so only the electric field can change the ion's energy (this can be seen by dotting Eq. (3.55) with $\mathbf{v}$). Also, because all fields are uniform and static the electric field can be expressed as $\mathbf{E} = -\nabla\phi$, where $\phi = -Ey$ is an electrostatic potential. Since the ion lowers its potential energy $q\phi$ on moving to larger y, motion in the positive y direction corresponds to the ion "falling downhill." Since the ion starts from rest at $y = 0$, where $\phi = 0$, its total energy $W = mv^2/2 + q\phi$ is initially zero. Furthermore, the time-independence of the fields implies that W must remain zero for all time. Because the kinetic energy $mv^2/2$ is positive-definite, the ion can only attain finite kinetic energy if it falls downhill, i.e., moves into regions of positive y. If for any reason the ion y-coordinate becomes zero at some later time, then at such a time the ion would again have to have $\mathbf{v} = 0$ because $W = mv^2 - qEy = 0$.

When the ion begins moving, it is acted on primarily by the electric force $qE\hat{y}$ because the magnetic force $q\mathbf{v}\times\mathbf{B}$ is negligible at small velocities. The electric force accelerates the ion in the y direction so the ion develops a positive v_y and also moves towards larger positive y as it "falls downhill" in the potential. As it develops a positive v_y, the ion starts to experience a magnetic force $qv_y\hat{y}\times B\hat{z} = v_y qB\hat{x}$, which accelerates the ion in the positive x direction causing the ion to develop in addition a positive v_x. The trajectory now becomes curved as the ion veers in the x direction while moving towards larger y. The positive v_x continues to increase and as a consequence a new magnetic force $qv_x\hat{x}\times B\hat{z} = -v_x\, qB\hat{y}$ develops and, being in the negative y direction, this increasing magnetic force counteracts the steady electric force, eventually causing the ion to decelerate in the y direction. The velocity v_y now decreases and ultimately reverses so that the ion starts to head in the negative y direction back towards $y=0$. As a consequence of the reversal of v_y, the magnetic force $qv_y\hat{y}\times B\hat{z}$ will become negative and so the ion will also decelerate in the x direction. Moving with negative v_y means the ion is going uphill in the electrostatic potential and when it reaches $y=0$, its potential energy must go back to zero. As noted above, the ion must come to rest at this point, because its total energy is always zero. Because the x velocity was never negative, the result of all this is that the ion makes a net positive displacement in the x direction. The whole process then repeats with the result that the ion keeps advancing in x while making a sequence of semi-circles in which v_y oscillates in polarity while v_x is never negative. The ion consequently moves like a leap-frog, which bounces up and down in the y direction while continuously advancing in the x direction. If an electron had been used instead of an ion, the sign of both the electric and magnetic forces would have reversed and the electron would have been confined to regions where $y\leq 0$. However, the net displacement would also be in the positive x direction (this is easily seen by repeating the above argument using an electron).

If an ion starts with a finite rather than a zero velocity, it will execute cyclotron (also called Larmor) orbits, which take the ion into regions of both positive and negative y. However, the ion will have a larger gyroradius in its $y>0$ orbit segment than in its $y<0$ orbit segment, resulting again in an average drift to the right as shown in Fig. 3.3. Electrons have larger gyroradii in the $y<0$ portions of their orbit, but have a counterclockwise rotation so electrons *also* drift to the right. The magnitude of this steady drift is easily calculated by assuming the existence of a constant perpendicular drift velocity in the Lorentz equation, and then averaging out the cyclotron motion:

$$0 = \mathbf{E} + \langle\mathbf{v}\rangle\times\mathbf{B}. \tag{3.56}$$

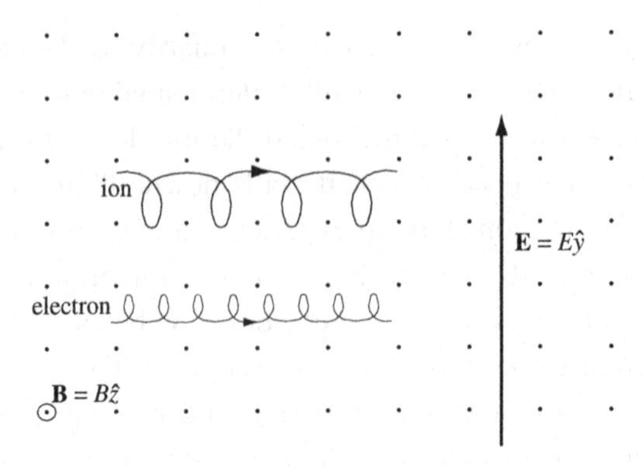

Fig. 3.3 $\mathbf{E}\times\mathbf{B}$ drifts for particles having finite initial energy.

This may be solved to give the average drift velocity

$$\mathbf{v}_E = \langle \mathbf{v} \rangle = \frac{\mathbf{E}\times\mathbf{B}}{B^2}. \tag{3.57}$$

This steady drift, called the $\mathbf{E}\times\mathbf{B}$ drift (pronounced "E cross B"), is independent of both the particle's polarity and its initial velocity. One way of interpreting this behavior is to recall that according to the theory of special relativity the electric field $\mathbf{E}'$ observed in a frame moving with velocity $\mathbf{u}$ is $\mathbf{E}' = \mathbf{E} + \mathbf{u}\times\mathbf{B}$ and so Eq. (3.56) is simply a statement that a particle drifts in such a way to ensure that the electric field seen in its own frame vanishes. The $\mathbf{E}\times\mathbf{B}$ drift analysis can be easily generalized to describe the effect on a charged particle of *any* force orthogonal to $\mathbf{B}$ by simply making the replacement $\mathbf{E}\to\mathbf{F}/q$ in the Lorentz equation. Thus, any spatially uniform, temporally constant force orthogonal to $\mathbf{B}$ will cause a drift

$$\mathbf{v}_F \equiv \langle \mathbf{v} \rangle = \frac{\mathbf{F}\times\mathbf{B}}{qB^2}. \tag{3.58}$$

Equations (3.57) and (3.58) lead to two counter-intuitive and important conclusions:

1. A steady-state electric field perpendicular to a magnetic field does not drive currents in a plasma, but instead causes a bulk motion of the entire plasma across the magnetic field with the velocity $\mathbf{v}_E$.
2. A steady-state *force* (e.g., gravity, centrifugal force, etc.) perpendicular to the magnetic field causes oppositely directed motions for electrons and ions and so drives a cross-field current

$$\mathbf{J}_F = \sum_\sigma n_\sigma \frac{\mathbf{F}\times\mathbf{B}}{B^2}. \tag{3.59}$$

3.5.2 Drifts in slowly changing arbitrary fields

We now consider charged particle motion in arbitrarily complicated but slowly changing fields subject to the following restrictions:

1. The time variation is so slow that the fields can be considered as approximately constant during each cyclotron period of the motion.
2. The fields vary so *gradually* in space that they are nearly uniform over the spatial extent of any single complete cyclotron orbit.
3. The electric and magnetic fields are related by Faraday's law $\nabla \times \mathbf{E} = -\partial \mathbf{B}/\partial t$.
4. $E/B \ll c$ so that relativistic effects are unimportant (otherwise there would be a problem with v_E becoming faster than c).

In this more general situation a charged particle will gyrate about **B**, stream parallel to **B**, have $\mathbf{E} \times \mathbf{B}$ drifts across **B**, and may also have force-based drifts. The analysis is based on the assumption that all these various motions are well separated (i.e., easily distinguishable from each other); this assumption is closely related to the requirement that the fields vary slowly and also to the concept of adiabatic invariance.

The assumed separation of scales is expressed by decomposing the particle motion into a fast, oscillatory component – the gyromotion – and a slow component obtained by averaging out the gyromotion. As sketched in Fig. 3.4, the particle's position and velocity are each decomposed into two terms

$$\mathbf{x}(t) = \mathbf{x}_{gc}(t) + \mathbf{r}_L(t), \quad \mathbf{v}(t) = \frac{\mathrm{d}\mathbf{x}}{\mathrm{d}t} = \mathbf{v}_{gc}(t) + \mathbf{v}_L(t), \tag{3.60}$$

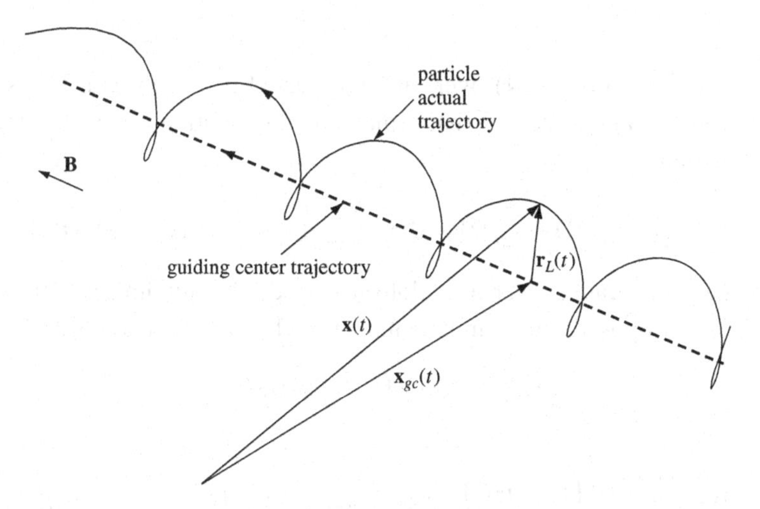

Fig. 3.4 Drift in an arbitrarily complicated field.

where $\mathbf{r}_L(t)$, $\mathbf{v}_L(t)$ give the *fast* gyration of the particle in a cyclotron orbit and $\mathbf{x}_{gc}(t)$, $\mathbf{v}_{gc}(t)$ are the *slowly* changing motion of the *guiding center* obtained after averaging out the cyclotron motion. Ignoring any time dependence of the fields for now, the magnetic field seen by the particle can be written as

$$\begin{aligned}\mathbf{B}(\mathbf{x}(t)) &= \mathbf{B}(\mathbf{x}_{gc}(t)+\mathbf{r}_L(t)) \\ &= \mathbf{B}(\mathbf{x}_{gc}(t))+(\mathbf{r}_L(t)\cdot\nabla)\,\mathbf{B}.\end{aligned} \tag{3.61}$$

Because $\mathbf{B}$ was assumed to be nearly uniform over the cyclotron orbit, it is sufficient to keep only the first term in the Taylor expansion of the magnetic field. The electric field may be expanded in a similar fashion.

After insertion of these Taylor expansions for the non-uniform electric and magnetic fields, the Lorentz equation becomes

$$\begin{aligned} m\frac{\mathrm{d}\left[\mathbf{v}_{gc}(t)+\mathbf{v}_L(t)\right]}{\mathrm{d}t} &= q\left[\mathbf{E}(\mathbf{x}_{gc}(t))+(\mathbf{r}_L(t)\cdot\nabla)\,\mathbf{E}\right] \\ &\quad +q\left[\mathbf{v}_{gc}(t)+\mathbf{v}_L(t)\right]\times\left[\mathbf{B}(\mathbf{x}_{gc}(t))+(\mathbf{r}_L(t)\cdot\nabla)\,\mathbf{B}\right]. \end{aligned} \tag{3.62}$$

The gyromotion (i.e., the fast cyclotron motion) is defined to be the solution of the equation

$$m\frac{\mathrm{d}\mathbf{v}_L(t)}{\mathrm{d}t} = q\mathbf{v}_L(t)\times\mathbf{B}(\mathbf{x}_{gc}(t)); \tag{3.63}$$

subtracting this fast motion equation from Eq. (3.62) leaves

$$\begin{aligned} m\frac{\mathrm{d}\mathbf{v}_{gc}(t)}{\mathrm{d}t} &= q\left[\mathbf{E}(\mathbf{x}_{gc}(t))+(\mathbf{r}_L(t)\cdot\nabla)\,\mathbf{E}\right] \\ &\quad +q\left\{\mathbf{v}_{gc}(t)\times\left[\mathbf{B}(\mathbf{x}_{gc}(t))+(\mathbf{r}_L(t)\cdot\nabla)\,\mathbf{B}\right]+\mathbf{v}_L(t)\times(\mathbf{r}_L(t)\cdot\nabla)\,\mathbf{B}\right\}. \end{aligned} \tag{3.64}$$

Let us now average Eq. (3.64) over one gyroperiod in which case terms *linear* in gyromotion average to zero. What remains is an equation describing the slow quantities, namely

$$m\frac{\mathrm{d}\mathbf{v}_{gc}(t)}{\mathrm{d}t} = q\left\{\mathbf{E}(\mathbf{x}_{gc}(t))+\mathbf{v}_{gc}(t)\times\mathbf{B}(\mathbf{x}_{gc}(t))+\langle\mathbf{v}_L(t)\times(\mathbf{r}_L(t)\cdot\nabla)\,\mathbf{B}\rangle\right\}, \tag{3.65}$$

where $\langle\rangle$ means averaged over a cyclotron period. The guiding center velocity can now be decomposed into components perpendicular and parallel to $\mathbf{B}$,

$$\mathbf{v}_{gc}(t) = \mathbf{v}_{\perp gc}(t)+v_{\parallel gc}(t)\hat{B} \tag{3.66}$$

so that

$$\frac{\mathrm{d}\mathbf{v}_{gc}(t)}{\mathrm{d}t} = \frac{\mathrm{d}\mathbf{v}_{\perp gc}(t)}{\mathrm{d}t}+\frac{\mathrm{d}\left(v_{\parallel gc}(t)\hat{B}\right)}{\mathrm{d}t} = \frac{\mathrm{d}\mathbf{v}_{\perp gc}(t)}{\mathrm{d}t}+\frac{\mathrm{d}v_{\parallel gc}(t)}{\mathrm{d}t}\hat{B}+v_{\parallel gc}(t)\frac{\mathrm{d}\hat{B}}{\mathrm{d}t}. \tag{3.67}$$

Denoting the distance along the magnetic field by s, the derivative of the magnetic field unit vector can be written, to lowest order, as

$$\frac{\mathrm{d}\hat{B}}{\mathrm{d}t} = \frac{\partial \hat{B}}{\partial s}\frac{\mathrm{d}s}{\mathrm{d}t} = v_{\parallel gc}\hat{B}\cdot\nabla\hat{B}, \tag{3.68}$$

so Eq. (3.65) becomes

$$\begin{aligned} m\left[\frac{\mathrm{d}\mathbf{v}_{\perp gc}(t)}{\mathrm{d}t} + \frac{\mathrm{d}v_{\parallel gc}(t)}{\mathrm{d}t}\hat{B} + v_{\parallel gc}^2\hat{B}\cdot\nabla\hat{B}\right] &= q\mathbf{E}(\mathbf{x}_{gc}(t)) \\ &\quad + q\mathbf{v}_{gc}(t)\times\mathbf{B}(\mathbf{x}_{gc}(t)) \\ &\quad + q\left\langle \mathbf{v}_L(t)\times(\mathbf{r}_L(t)\cdot\nabla)\,\mathbf{B}\right\rangle. \end{aligned} \tag{3.69}$$

The component of this equation along $\mathbf{B}$ is

$$m\frac{\mathrm{d}v_{\parallel gc}(t)}{\mathrm{d}t} = q\left[E_{\parallel}(\mathbf{x}_{gc}(t)) + \left\langle \mathbf{v}_L(t)\times(\mathbf{r}_L(t)\cdot\nabla)\,\mathbf{B}\right\rangle_{\parallel}\right] \tag{3.70}$$

while the component perpendicular to $\mathbf{B}$ is

$$m\left[\frac{\mathrm{d}\mathbf{v}_{\perp gc}(t)}{\mathrm{d}t} + v_{\parallel gc}^2\hat{B}\cdot\nabla\hat{B}\right] = q\left[\begin{array}{l} \mathbf{E}_{\perp}(\mathbf{x}_{gc}(t)) \\ +\mathbf{v}_{gc}(t)\times\mathbf{B}(\mathbf{x}_{gc}(t)) \\ +\left\langle \mathbf{v}_L(t)\times(\mathbf{r}_L(t)\cdot\nabla)\,\mathbf{B}\right\rangle_{\perp} \end{array}\right]. \tag{3.71}$$

Equation (3.71) is of the generic form

$$m\frac{\mathrm{d}\mathbf{v}_{\perp gc}}{\mathrm{d}t} = \mathbf{F}_{\perp} + q\mathbf{v}_{gc}\times\mathbf{B}. \tag{3.72}$$

where

$$\begin{aligned} \mathbf{F}_{\perp} = {} & q\left[\mathbf{E}_{\perp}(\mathbf{x}_{gc}(t)) + \left\langle \mathbf{v}_L(t)\times(\mathbf{r}_L(t)\cdot\nabla)\,\mathbf{B}\right\rangle_{\perp}\right] \\ & -mv_{\parallel gc}^2\hat{B}\cdot\nabla\hat{B}. \end{aligned} \tag{3.73}$$

Equation (3.72) is solved iteratively based on the assumption that $\mathbf{v}_{\perp gc}$ has a slow time dependence. In the first iteration, the time dependence is neglected altogether so that the left-hand side of Eq. (3.72) is set to zero to obtain the "first guess" for the perpendicular drift to be

$$\mathbf{v}_{\perp gc} \simeq \mathbf{v}_F \equiv \frac{\mathbf{F}_{\perp}\times\mathbf{B}}{qB^2}.$$

Next, $\mathbf{v}_p$ is defined to be a correction to this first guess, where $\mathbf{v}_p$ is assumed small and incorporates effects due to any time dependence of $\mathbf{v}_{\perp gc}$. To determine $\mathbf{v}_p$, we write $\mathbf{v}_{\perp gc} = \mathbf{v}_F + \mathbf{v}_p$ so, to second order Eq. (3.72) becomes

$$m\frac{\mathrm{d}(\mathbf{v}_F + \mathbf{v}_p)}{\mathrm{d}t} = \mathbf{F}_{\perp} + q(\mathbf{v}_F + \mathbf{v}_p)\times\mathbf{B}. \tag{3.74}$$

In accordance with the slowness condition, it is assumed that $|\mathrm{d}\mathbf{v}_p/\mathrm{d}t| \ll |\mathrm{d}\mathbf{v}_F/\mathrm{d}t|$ so Eq. (3.74) becomes

$$0 = -m\frac{\mathrm{d}\mathbf{v}_F}{\mathrm{d}t} + q\mathbf{v}_p \times \mathbf{B}. \tag{3.75}$$

Crossing this equation with $\mathbf{B}$ gives the general *polarization drift*

$$\mathbf{v}_p = -\frac{m}{qB^2}\frac{\mathrm{d}\mathbf{v}_F}{\mathrm{d}t} \times \mathbf{B}. \tag{3.76}$$

The most important example of the polarization drift is when $\mathbf{v}_F$ is the $\mathbf{E}\times\mathbf{B}$ drift in a uniform, constant magnetic field so that

$$\begin{aligned}\mathbf{v}_p &= -\frac{m}{qB^2}\frac{\mathrm{d}}{\mathrm{d}t}\left(\frac{\mathbf{E}\times\mathbf{B}}{B^2}\right)\times\mathbf{B} \\ &= \frac{m}{qB^2}\frac{\mathrm{d}\mathbf{E}}{\mathrm{d}t}.\end{aligned} \tag{3.77}$$

We now evaluate the middle term in Eq. (3.73); this term is called the "grad B" force and is

$$\mathbf{F}_{\nabla B} = q\left\langle \mathbf{v}_L(t) \times (\mathbf{r}_L(t)\cdot\nabla)\,\mathbf{B}\right\rangle. \tag{3.78}$$

To simplify the algebra for the averaging over cyclotron orbits, a local Cartesian coordinate system is used with x axis in the direction of the gyrovelocity at $t=0$ and z axis in the direction of the magnetic field *at the gyrocenter*. Thus, the Larmor orbit velocity has the form

$$\mathbf{v}_L(t) = v_L\left[\hat{x}\cos\omega_c t - \hat{y}\sin\omega_c t\right], \tag{3.79}$$

where

$$\omega_c = \frac{qB}{m} \tag{3.80}$$

is the cyclotron frequency and the Larmor orbit position has the form

$$\mathbf{r}_L(t) = \frac{v_L}{\omega_c}\left[\hat{x}\sin\omega_c t + \hat{y}\cos\omega_c t\right]. \tag{3.81}$$

Inserting the above two expressions in Eq. (3.78) gives

$$\mathbf{F}_{\nabla B} = q\frac{v_L^2}{\omega_c}\left\langle \left[\hat{x}\cos\omega_c t - \hat{y}\sin\omega_c t\right] \times \left(\left[\hat{x}\sin\omega_c t + \hat{y}\cos\omega_c t\right]\cdot\nabla\right)\mathbf{B}\right\rangle. \tag{3.82}$$

Noting that $\langle \sin^2 \omega_c t \rangle = \langle \cos^2 \omega_c t \rangle = 1/2$ while $\langle \sin(\omega_c t) \cos(\omega_c t) \rangle = 0$, this reduces to

$$\begin{aligned}
\mathbf{F}_{\nabla B} &= \frac{q v_L^2}{2\omega_c} \left[\hat{x} \times \frac{\partial \mathbf{B}}{\partial y} - \hat{y} \times \frac{\partial \mathbf{B}}{\partial x} \right] \\
&= \frac{m v_L^2}{2B} \left[\hat{x} \times \frac{\partial (B_y \hat{y} + B_z \hat{z})}{\partial y} - \hat{y} \times \frac{\partial (B_x \hat{x} + B_z \hat{z})}{\partial x} \right] \\
&= \frac{m v_L^2}{2B} \left[\hat{z} \left(\frac{\partial B_y}{\partial y} + \frac{\partial B_x}{\partial x} \right) - \hat{y} \frac{\partial B_z}{\partial y} - \hat{x} \frac{\partial B_z}{\partial x} \right].
\end{aligned} \tag{3.83}$$

But from $\nabla \cdot \mathbf{B} = 0$, it is seen that $\dfrac{\partial B_y}{\partial y} + \dfrac{\partial B_x}{\partial x} = -\dfrac{\partial B_z}{\partial z}$, so the "grad B" force is

$$\mathbf{F}_{\nabla B} = -\frac{m v_L^2}{2B} \nabla B, \tag{3.84}$$

where the approximation $B_z \simeq B$ has been used, since the magnetic field direction is mainly in the $\hat{z}$ direction.

Let us now define

$$\mathbf{F}_c = -m v_{\parallel gc}^2 \hat{B} \cdot \nabla \hat{B} \tag{3.85}$$

and consider this force. Suppose the magnetic field lines are curved and consider a particular point on a specific field line. Define, as shown in Fig. 3.5, a two-dimensional cylindrical coordinate system (R, ϕ) with origin at the field line

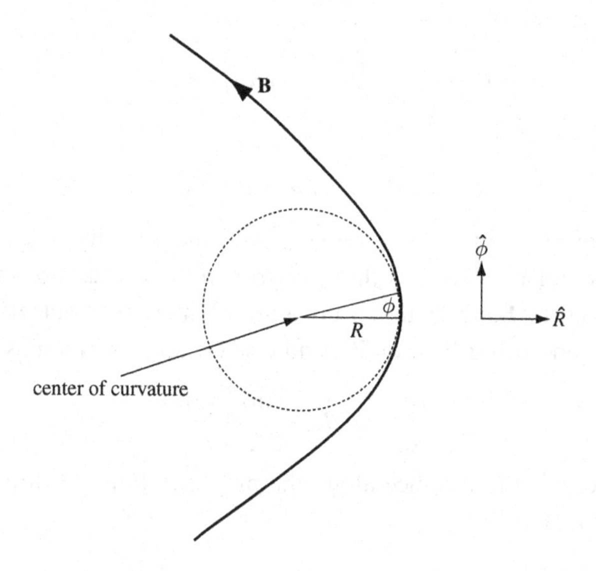

Fig. 3.5 Local cylindrical coordinate system defined by curved magnetic field, $\hat{\phi} = \hat{B}$.

center of curvature for this specific point and lying in the plane of the field line at this point. The radial position of the chosen point in this cylindrical coordinate system is the local radius of curvature of the field line and, since $\hat{\phi} = \hat{B}$, it is seen that $\hat{B} \cdot \nabla \hat{B} = \hat{\phi} \cdot \nabla \, \hat{\phi} = -\hat{R}/R$. The force associated with curvature of a field line,

$$\mathbf{F}_c = \frac{mv_{\parallel gc}^2 \hat{R}}{R}, \tag{3.86}$$

is therefore just the centrifugal force resulting from the motion along the curve of the particle's guiding center.

The drifts can be summarized as

$$\mathbf{v}_{\perp gc} = \mathbf{v}_E + \mathbf{v}_{\nabla B} + \mathbf{v}_c + \mathbf{v}_p, \tag{3.87}$$

where:

1. the "**E** cross **B**" drift is

$$\mathbf{v}_E = \frac{\mathbf{E} \times \mathbf{B}}{B^2}; \tag{3.88}$$

2. the "grad B" drift is

$$\mathbf{v}_{\nabla B} = -\frac{mv_L^2}{2qB^3} \nabla B \times \mathbf{B}; \tag{3.89}$$

3. the "curvature" drift is

$$\mathbf{v}_c = -\frac{mv_{\parallel gc}^2}{qB^2} \hat{B} \cdot \nabla \hat{B} \times \mathbf{B} = \frac{1}{qB^2} \left(\frac{mv_{\parallel gc}^2 \hat{R}}{R} \right) \times \mathbf{B}; \tag{3.90}$$

4. the "polarization" drift is

$$\mathbf{v}_p = -\frac{m}{qB^2} \left[\frac{\mathrm{d}}{\mathrm{d}t} (\mathbf{v}_E + \mathbf{v}_{\nabla B} + \mathbf{v}_c) \right] \times \mathbf{B}. \tag{3.91}$$

3.5.3 μ conservation

We now imagine being in a frame moving with the velocity $\mathbf{v}_{\perp gc}$; in this frame the only perpendicular velocity is the cyclotron velocity (Larmor motion). Since $\mathbf{v}_{\perp gc}$ is orthogonal to **B**, the parallel equation of motion is not affected by this change of frame and using Eqs. (3.70) and (3.84) can be written as

$$m \frac{\mathrm{d}v_\parallel}{\mathrm{d}t} = qE_\parallel - \frac{mv_L^2}{2B} \frac{\partial B}{\partial s}, \tag{3.92}$$

where, as before, s is the distance along the magnetic field. Multiplication by $v_\parallel$ gives an energy relation

$$\frac{\mathrm{d}}{\mathrm{d}t} \left(\frac{mv_\parallel^2}{2} \right) = qE_\parallel v_\parallel - \frac{mv_L^2}{2B} v_\parallel \frac{\partial B}{\partial s}. \tag{3.93}$$

The perpendicular force defined in Eq. (3.73) does not exist in this moving frame because it has been "transformed away" by the change of frames. Also, recall that it was assumed that the characteristic scale lengths of $\mathbf{E}$ and $\mathbf{B}$ are large compared to the gyroradius (Larmor radius). However, if the magnetic field has an *absolute time derivative*, Faraday's law dictates that there must be an inductive electric field, i.e., an electric field where $\oint \mathbf{E}\cdot \mathbf{dl}$ is finite. Because $\oint \mathbf{E}\cdot \mathrm{d}\mathbf{l} \neq 0$, the force provided by an inductive field is *non-conservative* and so is intrinsically different from the conservative force provided by the static electric field previously discussed. The consequences of this inductive field and its non-conservative force must therefore be taken into account explicitly.

To understand the effect of an inductive electric field on a specific particle, we dot the Lorentz equation with $\mathbf{v}$ to obtain

$$\frac{\mathrm{d}}{\mathrm{d}t}\left(\frac{mv_\parallel^2}{2}+\frac{mv_L^2}{2}\right)=qv_\parallel E_\parallel+q\mathbf{v}_\perp\cdot\mathbf{E}_\perp, \tag{3.94}$$

where $\mathbf{v}_\perp$ is the vector Larmor orbit velocity and $v_L=\sqrt{\mathbf{v}_\perp.\mathbf{v}_\perp}$ is its scalar magnitude. Subtracting Eq. (3.93) from (3.94) gives

$$\frac{\mathrm{d}}{\mathrm{d}t}\left(\frac{mv_L^2}{2}\right)=q\mathbf{v}_\perp\cdot\mathbf{E}_\perp+\frac{mv_L^2}{2B}v_\parallel\frac{\partial B}{\partial s}. \tag{3.95}$$

Integration of Faraday's law over the cross-section of the Larmor orbit gives

$$\int \mathrm{d}\mathbf{s}\cdot\nabla\times\mathbf{E}=-\int \mathrm{d}\mathbf{s}\cdot\frac{\partial \mathbf{B}}{\partial t} \tag{3.96}$$

or

$$\oint \mathrm{d}\mathbf{l}\cdot\mathbf{E}=-\pi r_L^2\frac{\partial B}{\partial t}, \tag{3.97}$$

where it has been assumed that the magnetic field is changing sufficiently slowly for the orbit radius to be approximately constant during each orbit.

Equation (3.95) involves the local electric field $\mathbf{E}_\perp$ but Eq. (3.97) only gives the line integral of the electric field. This line integral can still be used if Eq. (3.95) is averaged over a cyclotron period. The critical term is the time average over the

Larmor orbit of $q\mathbf{v}_\perp \cdot \mathbf{E}_\perp$ (which gives the rate at which the perpendicular electric field does work on the particle),

$$\begin{aligned} < q\mathbf{v}_\perp \cdot \mathbf{E}_\perp >_{orbit} &= \frac{\omega_c}{2\pi}\int \mathrm{d}t\, q\mathbf{v}_\perp \cdot \mathbf{E}_\perp \\ &= -\frac{q\omega_c}{2\pi}\oint \mathrm{d}\mathbf{l}\cdot\mathbf{E}_\perp \\ &= \frac{q\omega_c}{2} r_L^2 \frac{\partial B}{\partial t}. \end{aligned} \tag{3.98}$$

The substitution $\mathbf{v}_\perp \mathrm{d}t = -\mathrm{d}\mathbf{l}$ has been used and the minus sign is invoked because particle motion is diamagnetic (e.g., ions have a left-handed orbit, whereas in Stokes' theorem $\mathrm{d}\mathbf{l}$ is assumed to be a right-handed line element). Averaging of Eq. (3.95) gives

$$\left\langle \frac{\mathrm{d}}{\mathrm{d}t}\left(\frac{mv_L^2}{2}\right)\right\rangle = \frac{mv_L^2}{2B}\frac{\partial B}{\partial t} + \frac{mv_L^2}{2B}v_\parallel \frac{\partial B}{\partial s} = \frac{mv_L^2}{2B}\frac{\mathrm{d}B}{\mathrm{d}t}, \tag{3.99}$$

where $\mathrm{d}B/\mathrm{d}t = \partial B/\partial t + v_\parallel \partial B/\partial s$ is the *total derivative* of the average magnetic field experienced by the particle over a Larmor orbit. Defining the Larmor orbit kinetic energy as $W_\perp = mv_L^2/2$, Eq. (3.99) can be rewritten as

$$\frac{1}{W_\perp}\frac{\mathrm{d}W_\perp}{\mathrm{d}t} = \frac{1}{B}\frac{\mathrm{d}B}{\mathrm{d}t}, \tag{3.100}$$

which has the solution

$$\frac{W_\perp}{B} \equiv \mu = const. \tag{3.101}$$

for magnetic fields that can be changing in *both* time and space. In plasma physics terminology, μ is called the "first adiabatic" invariant, and the invariance of μ shows that the ratio of the kinetic energy of gyromotion to gyrofrequency is a conserved quantity. The derivation assumed the magnetic field changed sufficiently slowly for the instantaneous field strength $B(t)$ during an orbit to differ only slightly from the orbit-averaged field strength $\langle B\rangle$, i.e., $|B(t) - \langle B\rangle| \ll \langle B\rangle$.

3.5.4 Relation of μ conservation to other conservation relations

μ conservation is both of fundamental importance and a prime example of the adiabatic invariance of the action integral associated with a periodic motion. The μ conservation concept unites several seemingly disparate points of view, namely:

1. *Conservation of magnetic moment of a particle* – According to electromagnetic theory[1] the magnetic moment m of a current loop is $\mathsf{m} = IA$, where I is the current flowing in

[1] For example, see p. 186 of (Jackson 1998).

the loop and A is the area enclosed by the loop. Because a gyrating particle traces out a circular orbit at the frequency $\omega_c/2\pi$ and has a charge q, it effectively constitutes a current loop having $I = q\omega_c/2\pi$ and cross-sectional area $A = \pi r_L^2$. Thus, the magnetic moment of the gyrating particle is

$$\mathsf{m} = \left(\frac{q\omega_c}{2\pi}\right)\pi r_L^2 = \frac{mv_L^2}{2B} = \mu \tag{3.102}$$

and so the magnetic moment m is an adiabatically conserved quantity.

2. *Conservation of magnetic flux enclosed by gyro-orbit* – Because the magnetic flux Φ enclosed by the gyro-orbit is

$$\Phi = B\pi r_L^2 = \left(\frac{2m\pi}{q^2}\right)\mu, \tag{3.103}$$

μ conservation further implies conservation of the magnetic flux enclosed by a gyro-orbit. This is consistent with the concept that magnetic flux is frozen into the plasma, since if the field is made stronger, the field lines squeeze together in such a way that the density of field lines per unit area increases in proportion to the field strength. As shown in Fig. 3.6, the particle orbit area contracts in inverse proportion to the field strength so after a compression of field, the particle orbit links the same number of field lines as before the compression.

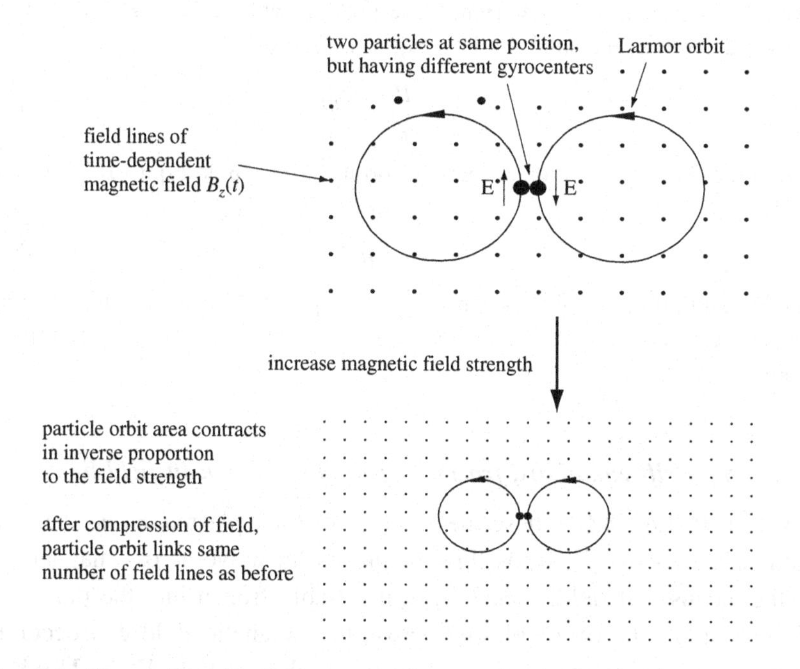

Fig. 3.6 Illustration showing how conservation of flux linked by an orbit is equivalent to frozen-in field; also, increasing magnetic field results in magnetic compression.

3. *Hamiltonian point of view (cylindrical geometry with azimuthal symmetry)* – Define a cylindrical coordinate system (r, θ, z) with z axis along the axis of rotation of the gyrating particle. Since $B_z = r^{-1}\partial(rA_\theta)/\partial r$, the vector potential is $A_\theta = rB_z/2$. The velocity vector is $\mathbf{v} = \dot{r}\hat{r} + r\dot{\theta}\hat{\theta} + \dot{z}\hat{z}$ and the Lagrangian is

$$L = \frac{m}{2}\left(\dot{r}^2 + r^2\dot{\theta}^2 + \dot{z}^2\right) + qr\dot{\theta}A_\theta - q\phi \tag{3.104}$$

so the canonical angular momentum is

$$P_\theta = mr^2\dot{\theta} + qrA_\theta = mr^2\dot{\theta} + qr^2B_z/2. \tag{3.105}$$

Since particles are diamagnetic, $\dot{\theta} = -\omega_c$. Because of the azimuthal symmetry, P_θ will be a constant of the motion and so

$$const. = P_\theta = -mr^2\omega_c + qr^2B/2 = -\frac{mv_\theta^2}{2\omega_c} = -\frac{m}{q}\mu. \tag{3.106}$$

This shows that constancy of canonical angular momentum is equivalent to μ conservation. It is important to realize that constancy of angular momentum due to perfect axisymmetry is a much more restrictive assumption than the slowness assumption used for adiabatic invariance.

4. *Adiabatic gas law* – The pressure associated with gyrating particles has dimensionality $N = 2$, i.e., $P = (m/2)\int \mathbf{v}' \cdot \mathbf{v}' f d^2v$, where $\mathbf{v}' = v_x\hat{x} + v_y\hat{y}$ and the $x - y$ plane is the plane of the gyration. Also, the density for a two-dimensional system has units of particles/area, i.e., $n \sim 1/A$. Hence, the pressure will scale as $P \sim v_{T\perp}^2/A$. Since $\gamma = (N+2)/N = 2$, the adiabatic law, Eq. (2.38), gives

$$const. \sim \frac{P}{n^2} \sim \frac{v_{T\perp}^2}{A}A^2; \tag{3.107}$$

but from the flux conservation property of orbits, $A \sim 1/B$ so Eq. (3.107) becomes

$$\frac{P}{n^2} \sim \frac{v_{T\perp}^2}{B}, \tag{3.108}$$

which is again proportional to μ, since $v_{T\perp}^2$ is proportional to the mean perpendicular thermal energy, i.e., the average of the gyrational energies of the individual particles making up the fluid.

3.5.5 Drift equations, frozen-in flux, and frozen-in field lines

The $\mathbf{E} \times \mathbf{B}$ drift, Eq. (3.57), describes motion of bulk plasma across a magnetic field and so shows that it is possible for plasma to move across magnetic field lines. This means that field lines are not invariably frozen into the plasma as is sometimes stated. Motion of plasma across the magnetic field can occur if no constraint exists preventing establishment of an electric field $\mathbf{E} = -\mathbf{U} \times \mathbf{B}$. The plasma can move in the direction perpendicular to the magnetic field provided the motion is such as to preserve the magnetic flux linked by the plasma.

If the magnetic field is static, then the electric field is constrained to be a potential electric field. As an example of a plasma moving across a static magnetic field consider a cylindrical plasma immersed in a static magnetic field $\mathbf{B} = B\hat{z}$ and a radial static electric field $\mathbf{E} = E_r\hat{r}$. The plasma will rotate with a velocity $\mathbf{U} = -\hat{\theta}E_r/B$ and so will be moving across the static magnetic field lines. This motion violates the statement that plasma is frozen to magnetic field lines, but is consistent with the statement that magnetic flux is frozen into the frame of the plasma. This azimuthal rotation of a plasma across a static magnetic field is quite common in laboratory plasmas and occurs spontaneously because of the development of an ambipolar radial electric field established by the difference in the outward radial diffusion rates of electrons and ions. Symmetry considerations or boundary conditions can also constrain the electric field. For example, if the configuration is azimuthally symmetric so $\partial/\partial\theta = 0$, then no azimuthal electrostatic electric field $E_\theta = -r^{-1}\partial\phi/\partial\theta$ is allowed, in which case an ideal plasma cannot move in the radial direction across a constant magnetic field $\mathbf{B} = B\hat{z}$.

If the magnetic field is time-dependent, then an inductive electric field will exist. The magnetic field lines can now be construed as moving, and the drift motion due to the inductive electric field will be such as to keep the plasma attached to the moving field lines as illustrated in Fig. 3.6. This motion will also be such as to preserve the flux linked by the plasma. Thus, a reasonable way of stating the frozen-in field concept is to say that plasma is frozen to moving magnetic field lines, but the converse is not necessarily true, i.e., field lines are not necessarily frozen to a moving plasma. These latter situations involve establishment of a static electric field, i.e., $\mathbf{E} = -\mathbf{U}\times\mathbf{B}$, where $\nabla\times\mathbf{E} = 0$, in which case the plasma moves across the magnetic field but the magnetic field does not change and so the magnetic field lines cannot be construed as moving. If the motion of the plasma across the magnetic field is such that $\nabla\times(\mathbf{U}\times\mathbf{B})$ is not zero, then there will have to be a finite $\partial\mathbf{B}/\partial t$, and so the field lines will move. The motion of the field lines will be such as to preserve the flux linked by any element of plasma.

3.5.6 Magnetic mirrors - a consequence of μ conservation

Consider a charged particle moving in a static, but spatially non-uniform magnetic field. The non-uniformity is such that the field strength varies in the direction of the field line so $\partial B/\partial s \neq 0$, where s is the distance along a field line. Such a field cannot be straight because if it were and so had the form $\mathbf{B} = B_z(z)\hat{z}$, it would necessarily have a non-zero divergence, i.e., it would have $\nabla\cdot\mathbf{B} = \partial B_z/\partial z \neq 0$. Because magnetic fields must have zero divergence, the magnetic field vector must have another component besides B_z and this other component must be spatially non-uniform in order to contribute to the divergence. Hence the field must be curved if the field strength varies along the direction of the field.

This curvature is easy to see by sketching field lines, as shown in Fig. 3.7. The density of field lines is proportional to the strength of the magnetic field and so a gradient of field strength along the field means that the field lines squeeze together as the field becomes stronger. Because magnetic field lines have zero divergence they are endless and so must bend as they squeeze together. This means that if $\partial B_z/\partial z \neq 0$ there must also be a field transverse to the initial direction of the magnetic field, i.e., a field in the x or y direction. In a cylindrically symmetric system, this transverse field must be a radial field as indicated by the vector decomposition $\mathbf{B} = B_z\hat{z} + B_r\hat{r}$ in Fig. 3.7.

The magnetic field is assumed to be static so that $\nabla \times \mathbf{E} = 0$, in which case $\mathbf{E} = -\nabla\phi$ and Eq. (3.92) can be written as

$$m\frac{\mathrm{d}v_\parallel}{\mathrm{d}t} = -q\frac{\partial \phi}{\partial s} - \mu\frac{\partial B}{\partial s}. \tag{3.109}$$

Multiplying Eq. (3.109) by $v_\parallel$ gives

$$\frac{\mathrm{d}}{\mathrm{d}t}\left[\frac{mv_\parallel^2}{2} + q\phi + \mu B\right] = 0, \tag{3.110}$$

assuming that the electrostatic potential is also constant in time. Time integration gives

$$\frac{mv_\parallel^2}{2} + q\phi(s) + \mu B(s) = const. \tag{3.111}$$

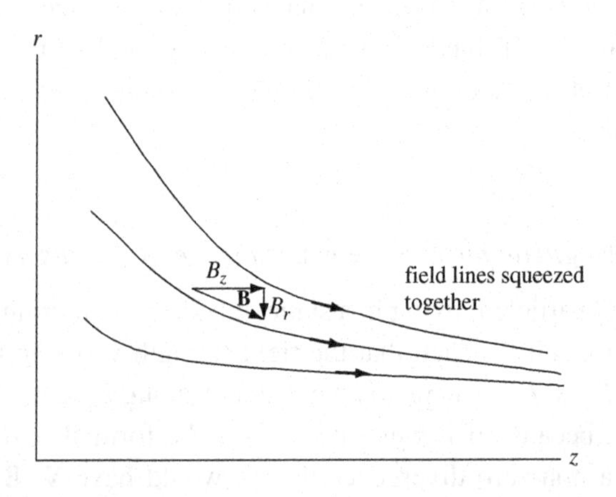

Fig. 3.7 Field lines squeezing together when **B** has a gradient. B field is stronger on the right than on the left because density of field lines is larger on the right.

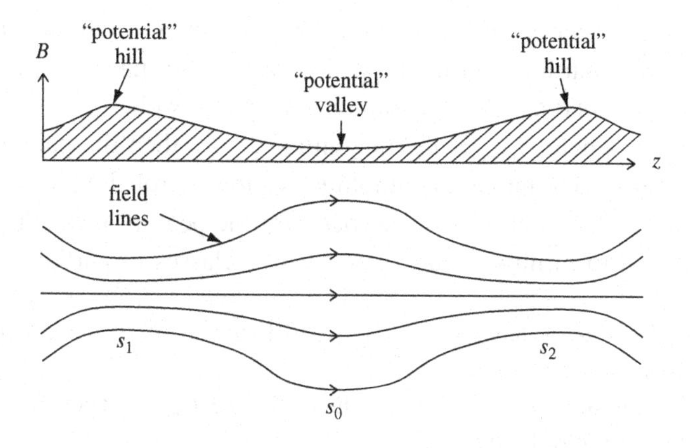

Fig. 3.8 Magnetic mirror.

Thus, $\mu B(s)$ acts as an effective potential energy since it adds to the electrostatic potential energy $q\phi(s)$. This property has the consequence that if $B(s)$ has a minimum with respect to s as shown in Fig. 3.8, then μB acts as an effective potential well that can trap particles. A magnetic trap of this sort can be produced by two axially separated coaxial coils. On each field line $B(s)$ has maxima at the locations s_1 and s_2 near the coils, a minimum at location s_0 between the coils, and $B(s)$ tends to zero as $s \to \pm\infty$. To focus attention on magnetic trapping, suppose now no electrostatic potential exists so Eq. (3.111) reduces to

$$\frac{mv_\parallel^2}{2} + \mu B(s) = const. \tag{3.112}$$

Now consider a particle with parallel velocity $v_{\parallel 0}$ located at the well minimum s_0 at time $t = 0$. Evaluating Eq. (3.112) at $s = s_0$, $t = 0$, and then again when the particle is at some arbitrary position s gives

$$\frac{mv_\parallel^2(s)}{2} + \mu B(s) = \frac{mv_{\parallel 0}^2}{2} + \mu B(s_0) = \frac{m\left(v_{\parallel 0}^2 + v_{\perp 0}^2\right)}{2} = W_0, \tag{3.113}$$

where W_0 is the particle's total kinetic energy at $t = 0$. Solving Eq. (3.113) for $v_\parallel(s)$ gives

$$v_\parallel(s) = \pm\sqrt{\frac{2}{m}\left[W_0 - \mu B(s)\right]}. \tag{3.114}$$

If $\mu B(s) = W_0$ at some position s, then $v_\parallel(s)$ must vanish at this position, in which case the particle must reverse its direction of motion just like a pendulum reversing direction when its velocity goes through zero. This velocity reversal corresponds to a reflection of the particle and so this configuration is called a magnetic *mirror*. A particle can be trapped between two magnetic mirrors; such a configuration is called a magnetic trap or a magnetic well.

If $W_0 > \mu B_{\max}$, where $B_{\max}$ is the magnitude at $s_{1,2}$ then the parallel velocity does not go to zero at the maximum amplitude of the mirror field. In this case the particle does not reflect, but instead escapes over the peak of the $\mu B(s)$ potential hill and travels out to infinity. Thus, there are two classes of particles:

1. trapped particles – these have $W_0 < \mu B_{\max}$ and bounce back and forth between the mirrors of the magnetic well,
2. untrapped (or passing) particles – these have $W_0 > \mu B_{\max}$ and are retarded at the potential hills but not reflected.

Since $\mu = mv_{\perp 0}^2/2B_{\min}$ and $W_0 = mv_0^2/2$, the criterion for trapping can be written as

$$\frac{B_{\min}}{B_{\max}} < \frac{v_{\perp 0}^2}{v_0^2}. \tag{3.115}$$

Let us define θ as the angle the velocity vector makes with respect to the magnetic field at s_0, i.e., $\sin\theta = v_{\perp 0}/v_0$, and also define

$$\theta_{trap} = \sin^{-1}\sqrt{\frac{B_{\min}}{B_{\max}}}. \tag{3.116}$$

Thus, as shown in Fig. 3.9, all particles with $\theta > \theta_{trap}$ are trapped, while all particles with $\theta < \theta_{trap}$ are untrapped. Suppose at $t = 0$ the particle velocity distribution at s_0 is isotropic. After a time interval long enough for all untrapped particles to have escaped the trap, there will be no particles in the $\theta < \theta_{trap}$ region of velocity space. The velocity distribution will thus be zero for $\theta < \theta_{trap}$; such a distribution function is called a *loss-cone* distribution function.

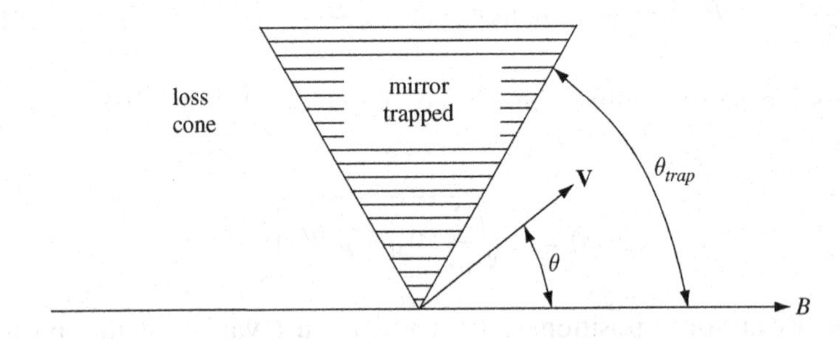

Fig. 3.9 Loss-cone velocity distribution. Particles with velocity angle $\theta > \theta_{trap}$ are mirror trapped, others are lost.

3.5.7 $\mathcal{J}$, the second adiabatic invariant

Trapped particles have periodic motion in the magnetic well. Thus, applying the concept of adiabatic invariance presented in Section 3.4.1, the quantity

$$\mathcal{J} = \oint P_{\parallel}\mathrm{d}s = \oint (mv_{\parallel} + qA_{\parallel})\mathrm{d}s \tag{3.117}$$

will be an invariant if:

1. any time dependence of the well shape is slow compared to the bounce frequency of the trapped particle,
2. any spatial inhomogeneities of the well magnetic field are so gradual that the particle's bounce trajectory changes by only a small amount from one bounce to the next.

To determine the circumstances where $A_{\parallel} \neq 0$, we use Coulomb gauge (i.e., assume $\nabla \cdot \mathbf{A} = 0$) and at any given location define a local Cartesian coordinate system with z axis parallel to the local field. From Ampère's law it is seen that

$$[\nabla \times (\nabla \times \mathbf{A})]_z = -\nabla^2 A_z = \mu_0 J_z \tag{3.118}$$

so A_z is finite only if there is a *current parallel to the magnetic field*. Because J_z acts as the source term in a Poisson-like partial differential equation for A_z, the parallel current need not be at the same location as A_z. If there are no currents parallel to the magnetic field anywhere then $A_{\parallel} = 0$, and in this case the second adiabatic invariant reduces to

$$\mathcal{J} = m \oint v_{\parallel}\mathrm{d}s. \tag{3.119}$$

Having a current flow along the magnetic field corresponds to a more complicated magnetic topology. The axial current produces an associated azimuthal magnetic field, which links the axial magnetic field resulting in a helical twist. This more complicated situation of finite magnetic helicity will be discussed in a later chapter.

3.5.8 Consequences of $\mathcal{J}$-invariance

Just as μ-invariance was related to the perpendicular CGL adiabatic invariant discussed in Section 2.6.5, $\mathcal{J}$-invariance is closely related to the parallel CGL adiabatic invariant also discussed in Section 2.6.5. To see this, recall that density in a one-dimensional system has dimensions of particles per unit length, i.e., $n_{1D} \sim 1/L$, and pressure in a one-dimensional system has dimensions of kinetic energy per unit length, i.e., $P_{1D} \sim v_\parallel^2/L$. For parallel motion the number of dimensions is $N=1$ so $\gamma=(N+2)/N=3$ and the fluid adiabatic relation is

$$const. \sim \frac{P_{1D}}{n_{1D}^3} \sim \frac{v_\parallel^2/L}{L^{-3}} \sim \left(v_\parallel L\right)^2, \tag{3.120}$$

which is a simplified form of Eq. (3.119) since Eq. (3.119) has the scaling $\mathcal{J} \sim v_\parallel L = const.$

$\mathcal{J}$-invariance combined with mirror trapping/detrapping is the basis of an acceleration mechanism proposed by Fermi (1954) as a means for accelerating cosmic-ray particles to ultra-relativistic velocities. The Fermi mechanism works as follows: consider a particle initially trapped in a magnetic mirror. This particle has an initial angle in velocity space $\theta > \theta_{trap}$; both θ and θ_{trap} are measured when the particle is at the mirror minimum. Now suppose the distance between the magnetic mirrors is slowly reduced so that the bounce distance L of the mirror-trapped particle slowly decreases. This would typically occur by reducing the axial separation between the coils producing the magnetic mirror field. Because $\mathcal{J} \sim v_\parallel L$ is invariant, the particle's parallel velocity increases on each successive bounce as L slowly decreases. This steady increase in $v_\parallel$ means that the velocity angle θ decreases. Eventually, θ becomes smaller than θ_{trap} whereupon the particle becomes detrapped and escapes from one end of the mirror with a large parallel velocity. This mechanism provides a slow pumping to very high energy, followed by a sudden and automatic ejection of the energetic particle.

3.5.9 *The third adiabatic invariant*

Consider a particle bouncing back and forth in either of the two geometries shown in Fig. (3.10). In Fig. 3.10(a), the magnetic field is produced by a single

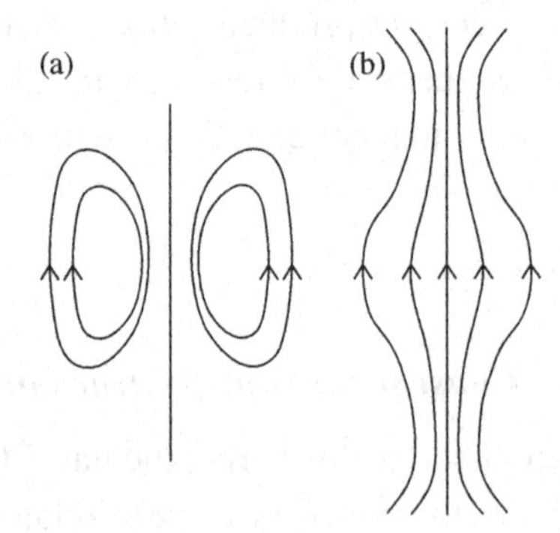

Fig. 3.10 Magnetic field lines relevant to discussion of third adiabatic invariant: (a) field lines always have same curvature (dipole field), (b) field lines have both concave and convex curvature (mirror field).

magnetic dipole and the field lines always have convex curvature, i.e., the radius of curvature is always on the inside of the field lines. The field decreases in magnitude with increasing distance from the dipole.

In Fig. 3.10(b), the field is produced by two coils and has convex curvature near the mirror minimum and concave curvature in the vicinity of the coils. On defining a cylindrical coordinate system (r, θ, z) with z axis coaxial with the coils, it is seen that in the region between the two coils where the field bulges out, the field strength is a decreasing function of r, i.e., $\partial B/\partial r < 0$, whereas in the plane of each coil the opposite is true. Thus, in the mirror minimum, both the centrifugal and grad B forces are radially outward, whereas the opposite is true near the coils.

In both Figs. 3.10(a) and (b) a particle moving along the field line can be mirror-trapped because in both cases the field has a minimum flanked by two maxima. However, for Fig. 3.10(a), the particle will have grad B and curvature drifts always in the same azimuthal sense, whereas for Fig. 3.10(b) the azimuthal direction of these drifts will depend on whether the particle is in a region of concave or convex curvature. Thus, in addition to the mirror bouncing motion, much slower curvature and grad B drifts also exist, directed along the field binormal (i.e., the direction orthogonal to both the field and its radius of curvature). These higher-order drifts may alternate sign during the mirror bouncing. The binormally directed displacement made by a particle during its ith complete period τ of mirror bouncing is

$$\delta \mathbf{r}_j = \int_0^\tau \mathbf{v} \mathrm{d}t, \tag{3.121}$$

where τ is the mirror bounce period and $\mathbf{v}$ is the sum of the curvature and grad B drifts experienced in the course of a mirror bounce. This displacement is due to the cumulative effect of the curvature and grad B drifts experienced during one complete period of bouncing between the magnetic mirrors. The average velocity associated with this slow drifting may be defined as

$$\langle \mathbf{v} \rangle = \frac{1}{\tau} \int_0^\tau \mathbf{v} \mathrm{d}t. \tag{3.122}$$

Let us calculate the action associated with a sequence of $\delta \mathbf{r}_j$. This action is

$$S = \sum_j [m \langle \mathbf{v} \rangle + q\mathbf{A}]_j \cdot \delta \mathbf{r}_j, \tag{3.123}$$

where the quantity in square brackets is evaluated on the line segment $\delta \mathbf{r}_j$. If the $\delta \mathbf{r}_j$ are small then this can be converted into an action "integral" for the path traced out by the $\delta \mathbf{r}_j$. If the $\delta \mathbf{r}_j$ are sufficiently small to behave as differentials, then we may write them as $\mathbf{dr}_{bounce}$ and express the summation as an action integral

$$S = \int [m \langle \mathbf{v} \rangle + q\mathbf{A}] \cdot \mathbf{dr}_{bounce}, \tag{3.124}$$

where it must be remembered that $\langle \mathbf{v} \rangle$ is the *bounce-averaged* velocity. The quantity $m \langle \mathbf{v} \rangle + q\mathbf{A}$ is just the canonical momentum associated with the effective motion along the sequence of line segments $\delta \mathbf{r}_j$. The vector $\mathbf{r}_{bounce}$ is a vector pointing from the origin to the particle's location at successive bounces and so is the generalized coordinate associated with the bounce-averaged velocity. If the motion resulting from $\langle \mathbf{v} \rangle$ is periodic, we expect S to be an adiabatic invariant. The first term in Eq. (3.124) will be of the order of $mv_{drift} 2\pi r$, where r is the radius of the trajectory described by the $\delta \mathbf{r}_j$. The second term is just $q\Phi$, where Φ is the magnetic flux enclosed by the trajectory. We compare the ratio of these two terms to obtain

$$\frac{\int m \langle \mathbf{v} \rangle \cdot \mathbf{dr}}{\int q\mathbf{A} \cdot \mathbf{dr}} \sim \frac{mv_{drift} 2\pi r}{qB\pi r^2} \sim \frac{v_{drift}}{\omega_c r} \sim \frac{r_L^2}{r^2}, \tag{3.125}$$

where we have used $v_{\nabla B} \sim v_c \sim v_\perp^2/\omega_c r \sim \omega_c r_L^2/r$. Thus, if the Larmor radius is much smaller than the characteristic scale length of the field, the magnetic flux term dominates the action integral and adiabatic invariance corresponds to the particle *staying on a constant flux surface* as its orbit evolves following the various curvature and grad B drifts. This third adiabatic invariant is much more fragile than $\mathcal{J}$, which in turn was more fragile than μ, because the analysis here is based on the rather strong assumption that the curvature and grad B drifts are small enough for the $\delta \mathbf{r}_j$ to trace out a nearly periodic orbit.

3.6 Relation of drift equations to the double adiabatic MHD equations

The derivation of the MHD Ohm's law involved dropping the Hall term (see p. 55) and the basis for dropping this term was assuming $\omega \ll \omega_{ci}$, where ω is the characteristic rate of change of the electromagnetic field. The derivation of the single particle drift equations involved essentially the same assumption (i.e., the motion was slow compared to $\omega_{c\sigma}$). Thus, if the characteristic rate of change of the electromagnetic field is slow compared to ω_{ci}, both the MHD and the single particle drift equations ought to be equally valid descriptions of the plasma dynamics. If so, there should be some sort of a correspondence principle relating these ostensibly different points of view. Preliminary evidence supporting this hypothesis was the observation that the single particle adiabatic invariants μ and $\mathcal{J}$ were respectively related to the perpendicular and parallel double adiabatic MHD equations. It thus seems reasonable to expect additional connections between the drift equations and the double adiabatic MHD equations.

In fact, an approximate derivation of the double adiabatic MHD equations can be obtained by summing the currents associated with the various particle drifts – providing one additional effect, *diamagnetic current,* is added to this sum. Diamagnetic current is a peculiar concept because it is a consequence of the macroscopic phenomenon of pressure gradients and so has no meaning in the context of a single particle description.

In order to establish this microscopic–macroscopic relationship we begin by recalling from electromagnetic theory[2] that a magnetic material with density $\mathbf{M}$ of magnetic dipole moments per unit volume has an associated magnetization current

$$\mathbf{J}_M = \nabla \times \mathbf{M}. \tag{3.126}$$

The magnitude of the magnetic moment of a charged particle in a magnetic field was shown in Section 3.5.4 to be μ. Since a magnetic dipole is represented by a vector pointing in the direction of the magnetic field on the axis of the current loop constituting the dipole, the vector magnetic moment of a charged particle gyrating in a magnetic field will be $\mathbf{m} = -\mu\hat{B}$. The minus sign corresponds to cyclotron motion being diamagnetic, i.e., the magnetic field resulting from cyclotron rotation opposes the original field in which the particle is rotating. For example, an individual ion placed in a magnetic field $\mathbf{B} = B\hat{z}$ rotates in the negative θ direction, and so the current associated with the ion motion creates a magnetic field pointing in the $-\hat{z}$ direction inside the ion orbit.

Suppose there exists a large number or ensemble of particles with density n_σ and mean magnetic moment $\bar{\mu}_\sigma$. The density of magnetic moments, or magnetization density, of this ensemble is

$$\mathbf{M} = -\sum_\sigma n_\sigma \bar{\mu}_\sigma \hat{B} = -\sum_\sigma n_\sigma \left\langle \frac{m_\sigma v_\perp^2}{2B} \right\rangle \hat{B} = -\frac{P_\perp \hat{B}}{B}, \tag{3.127}$$

where $\langle\rangle$ denotes averaging over the velocity distribution and Eq. (2.26) has been used. Inserting Eq. (3.127) into Eq. (3.126) shows that this ensemble of charged particles in a magnetic field has a *diamagnetic current*

$$\mathbf{J}_M = -\nabla \times \left(\frac{P_\perp \hat{B}}{B} \right). \tag{3.128}$$

[2] For example, see p. 192 of (Jackson 1998).

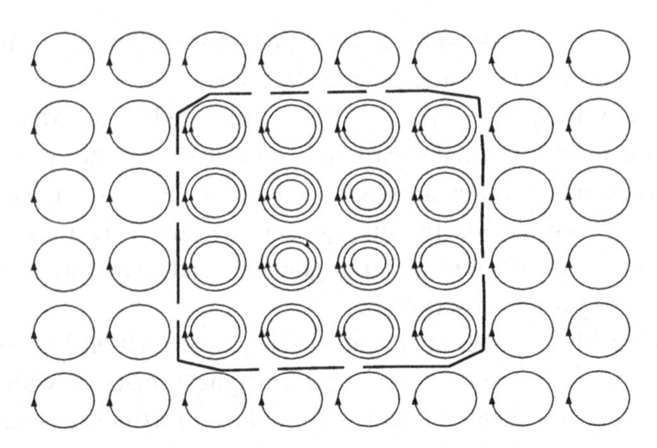

Fig. 3.11 Gradient of magnetized particles gives apparent current as observed on dashed line.

Figure 3.11 shows the physical origin of $\mathbf{J}_M$. Here, a collection of ions all rotate clockwise in a magnetic field pointing out of the page. The azimuthally directed current on the dashed curve is the sum of contributions from (i) particles with guiding centers located one Larmor radius *inside* the dashed curve and (ii) particles with guiding centers located one Larmor radius *outside* the dashed curve. From the point of view of an observer located on the dashed curve, the inside particles (group (i)) constitute a clockwise current, whereas the outside (group (ii)) particles constitute a counterclockwise current. If there are unequal numbers of inside and outside particles (indicated here by concentric circles inside the dashed curve), then the two opposing currents do not cancel and a net macroscopic current appears to flow around the dashed curve, even though *no* actual particles flow around the dashed curve. Inequality of the numbers of inside and outside particles corresponds to a density gradient and so we see that a radial density gradient of gyrating particles gives a net macroscopic azimuthal current. Similarly, if there is a radial temperature gradient, the velocities of the inner and outer groups differ, resulting again in an apparent macroscopic azimuthal current. The combination of density and temperature gradients is such that the net macroscopic current depends on the pressure gradient as given by Eq. (3.128).

Taking diamagnetic current into account is critical for establishing a correspondence between the single particle drifts and the MHD equations, and having recognized this, we are now in a position to derive this correspondence. In order for the derivation to be tractable yet non-trivial, it will be assumed that the magnetic field is time-independent, but the electric field will be allowed to depend on time. It is also assumed that the *dominant* cross-field particle motion is the $\mathbf{v}_E = \mathbf{E} \times \mathbf{B}/B^2$ drift; this assumption is consistent with the hierarchy of particle drifts (i.e., polarization drift is a higher-order correction to $\mathbf{v}_E$).

Because both species have the same $\mathbf{v}_E$, no macroscopic current results from $\mathbf{v}_E$, and so all cross-field currents must result from the other, smaller drifts, namely $\mathbf{v}_{\nabla B}$, $\mathbf{v}_c$, and $\mathbf{v}_p$. Let us now add the magnetization current to the currents associated with these other drifts to obtain the total macroscopic current

$$\mathbf{J}_{\text{total}} = \mathbf{J}_M + \mathbf{J}_{\nabla B} + \mathbf{J}_c + \mathbf{J}_p = \mathbf{J}_M + \sum_\sigma n_\sigma q_\sigma \left(\mathbf{u}_{\nabla B,\sigma} + \mathbf{u}_{c,\sigma} + \mathbf{u}_{p,\sigma} \right), \tag{3.129}$$

where $\mathbf{J}_{\nabla B}$, $\mathbf{J}_c$, $\mathbf{J}_p$ are currents due to grad B, curvature, and polarization drifts respectively and $\mathbf{u}_{\nabla B,\sigma}$, $\mathbf{u}_{c,\sigma}$, and $\mathbf{u}_{p,\sigma}$ are the mean (i.e., fluid) velocities associated with these drifts. These currents are explicitly:

1. grad B current

$$\begin{aligned}\mathbf{J}_{\nabla B} &= \sum_\sigma n_\sigma q_\sigma \mathbf{u}_{\nabla B,\sigma} \\ &= -\sum_\sigma \frac{m_\sigma n_\sigma q_\sigma \left\langle v_{\perp\sigma}^2 \right\rangle}{2B} \frac{\nabla B \times \mathbf{B}}{q_\sigma B^2} = -P_\perp \frac{\nabla B \times \mathbf{B}}{B^3};\end{aligned} \tag{3.130}$$

2. curvature current

$$\begin{aligned}\mathbf{J}_c &= \sum_\sigma n_\sigma q_\sigma \mathbf{u}_{c,\sigma} \\ &= -\sum_\sigma n_\sigma q_\sigma m_\sigma \left\langle v_{\parallel\sigma}^2 \right\rangle \frac{\hat{B} \cdot \nabla \hat{B} \times \mathbf{B}}{q_\sigma B^2} = -P_\parallel \frac{\hat{B} \cdot \nabla \hat{B} \times \mathbf{B}}{B^2};\end{aligned} \tag{3.131}$$

3. polarization current

$$\mathbf{J}_p = \sum_\sigma n_\sigma q_\sigma \mathbf{u}_{p,\sigma} = \sum_\sigma n_\sigma q_\sigma \left(\frac{m_\sigma}{q_\sigma B^2} \frac{\mathrm{d}\mathbf{E}_\perp}{\mathrm{d}t} \right) = \frac{\rho}{B^2} \frac{\mathrm{d}\mathbf{E}_\perp}{\mathrm{d}t}. \tag{3.132}$$

Because the magnetic field was assumed to be constant, the time derivative of $\mathbf{v}_E$ is the only contributor to the polarization drift current.

The total magnetic force is

$$\begin{aligned}\mathbf{J}_{total} \times \mathbf{B} &= \left(\mathbf{J}_M + \mathbf{J}_{\nabla B} + \mathbf{J}_c + \mathbf{J}_p \right) \times \mathbf{B} \\ &= \begin{bmatrix} -\nabla \times \left(\dfrac{P_\perp \hat{B}}{B} \right) - P_\perp \dfrac{\nabla B \times \mathbf{B}}{B^3} \\ -P_\parallel \dfrac{\hat{B} \cdot \nabla \hat{B} \times \mathbf{B}}{B^2} + \dfrac{\rho}{B^2} \dfrac{\mathrm{d}\mathbf{E}}{\mathrm{d}t} \end{bmatrix} \times \mathbf{B}.\end{aligned} \tag{3.133}$$

The grad B current cancels *part* of the magnetization current as follows:

$$\nabla\times\left(\frac{P_\perp\hat{B}}{B}\right)+P_\perp\frac{\nabla B\times\mathbf{B}}{B^3}=\left[\nabla\left(\frac{1}{B}\right)\times P_\perp\hat{B}+\frac{1}{B}\nabla\times\left(P_\perp\hat{B}\right)\right]$$
$$+P_\perp\frac{\nabla B\times\mathbf{B}}{B^3}$$
$$=\frac{1}{B}\nabla\times\left(P_\perp\hat{B}\right)=\frac{P_\perp}{B}\nabla\times\hat{B}+\frac{\nabla P_\perp\times\hat{B}}{B} \quad (3.134)$$

so

$$\mathbf{J}_{total}\times\mathbf{B}=-\left[P_\perp\nabla\times\hat{B}+\nabla P_\perp\times\hat{B}+P_\parallel\hat{B}\cdot\nabla\hat{B}\times\hat{B}-\frac{\rho}{B}\frac{\mathrm{d}\mathbf{E}}{\mathrm{d}t}\right]\times\hat{B}. \quad (3.135)$$

The first term on the right-hand side of Eq. (3.135) can be recast using the vector identity

$$\nabla\left(\frac{\hat{B}\cdot\hat{B}}{2}\right)=0=\hat{B}\cdot\nabla\hat{B}+\hat{B}\times\nabla\times\hat{B} \quad (3.136)$$

and the electric field in the last term of Eq. (3.135) can be replaced using $\mathbf{E}=-\mathbf{U}\times\mathbf{B}$ to give

$$\mathbf{J}_{total}\times\mathbf{B}=-\left(P_\perp-P_\parallel\right)\hat{B}\cdot\nabla\hat{B}+\nabla_\perp P_\perp-\frac{\rho}{B}\frac{\mathrm{d}\left(\mathbf{U}\times\mathbf{B}\right)}{\mathrm{d}t}\times\hat{B}. \quad (3.137)$$

The relation $\left[\hat{B}\cdot\nabla\hat{B}\right]_\perp=\hat{B}\cdot\nabla\hat{B}$ has been used here; this relation follows from Eq. (3.136). Finally, it is observed that

$$\left[\nabla\cdot\left(\hat{B}\hat{B}\right)\right]_\perp=\left[\left(\nabla\cdot\hat{B}\right)\hat{B}+\hat{B}\cdot\nabla\hat{B}\right]_\perp=\hat{B}\cdot\nabla\hat{B} \quad (3.138)$$

so

$$\left(P_\perp-P_\parallel\right)\hat{B}\cdot\nabla\hat{B}=\left(P_\perp-P_\parallel\right)\left[\nabla\cdot\left(\hat{B}\hat{B}\right)\right]_\perp=\left[\nabla\cdot\left\{\left(P_\perp-P_\parallel\right)\hat{B}\hat{B}\right\}\right]_\perp. \quad (3.139)$$

Furthermore,

$$\frac{\rho}{B}\frac{\mathrm{d}\left(\mathbf{U}\times\mathbf{B}\right)}{\mathrm{d}t}\times\hat{B}=-\left[\rho\frac{\mathrm{d}\mathbf{U}}{\mathrm{d}t}\right]_\perp, \quad (3.140)$$

since it has been assumed the magnetic field is time-independent. Inserting these last two results in Eq. (3.137) gives

$$\mathbf{J}_{total}\times\mathbf{B}=\left[\nabla\cdot\left\{\left(P_\parallel-P_\perp\right)\hat{B}\hat{B}\right\}\right]_\perp+\nabla_\perp P_\perp+\left[\rho\frac{\mathrm{d}\mathbf{U}}{\mathrm{d}t}\right]_\perp \quad (3.141)$$

or

$$\left[\rho\frac{\mathrm{d}\mathbf{U}}{\mathrm{d}t}\right]_\perp=\left[\mathbf{J}_{total}\times\mathbf{B}-\nabla\cdot\left\{P_\perp\overleftrightarrow{\mathbf{I}}+\left(P_\parallel-P_\perp\right)\hat{B}\hat{B}\right\}\right]_\perp, \quad (3.142)$$

which is just the perpendicular component of the double adiabatic MHD equation of motion. This demonstrates that if diamagnetic current is taken into account, the drift equations for phenomena with characteristic frequencies ω much smaller than ω_{ci} and the double adiabatic MHD equations are equivalent descriptions of plasma dynamics. This analysis also shows one has to be extremely careful when invoking single particle concepts to explain macroscopic behavior, because if diamagnetic effects are omitted, erroneous conclusions can result.

The justification for the designation "polarization current" results from comparing this current to the current flowing through a parallel plate capacitor with dielectric ε. The capacitance of the parallel plate capacitor is $C = \varepsilon A/d$, where A is the cross-sectional area of the capacitor plates and d is the gap between the plates. The charge on the capacitor is $Q = CV$, where V is the voltage across the capacitor plates. The current through the capacitor is $I = \mathrm{d}Q/\mathrm{d}t$ so

$$I = C\frac{\mathrm{d}V}{\mathrm{d}t} = \frac{\varepsilon A}{d}\frac{\mathrm{d}V}{\mathrm{d}t}. \tag{3.143}$$

Because the electric field between the plates is $E = V/d$ and the current density is $J = I/A$, Eq. (3.143) can be expressed as

$$J = \varepsilon\frac{\mathrm{d}E}{\mathrm{d}t}, \tag{3.144}$$

which gives the alternating current density in a medium with dielectric ε. If this is compared to the polarization current

$$\mathbf{J}_p = \frac{\rho}{B^2}\frac{\mathrm{d}\mathbf{E}_\perp}{\mathrm{d}t} \tag{3.145}$$

it is seen that the plasma acts like a dielectric medium in the direction perpendicular to the magnetic field and has an effective dielectric constant given by ρ/B^2.

The polarization current is intimately related to the acceleration of macroscopic volumes of plasma and also to the details of how an $\mathbf{E}\times\mathbf{B}$ drift becomes established. This is shown schematically in Fig. 3.12(a), where a cube of perfectly conducting plasma is placed at rest between two capacitor plates located in a magnetic field $\mathbf{B} = B\hat{z}$. When the capacitor plates are connected to a source of EMF as in Fig. 3.12(b) so that the upper capacitor plate becomes negatively charged and the lower capacitor plate becomes positively charged, a vertical electric field E_y is established. This causes the electrons and the ions to have an $\mathbf{E}\times\mathbf{B}$ drift to the right as sketched in Fig. 3.12(b) and so the bulk plasma moves to the right with the velocity $\mathbf{U} = \mathbf{E}\times\mathbf{B}/B^2 = \hat{x}E_y/B$. The slight upward motion of the ions and the slight downward motion of the electrons polarizes the cube so a positive surface charge develops on the upper surface of the cube and a negative surface charge develops on the lower surface. The surface polarization creates an electric field opposing the vacuum field due to the capacitor plates and so

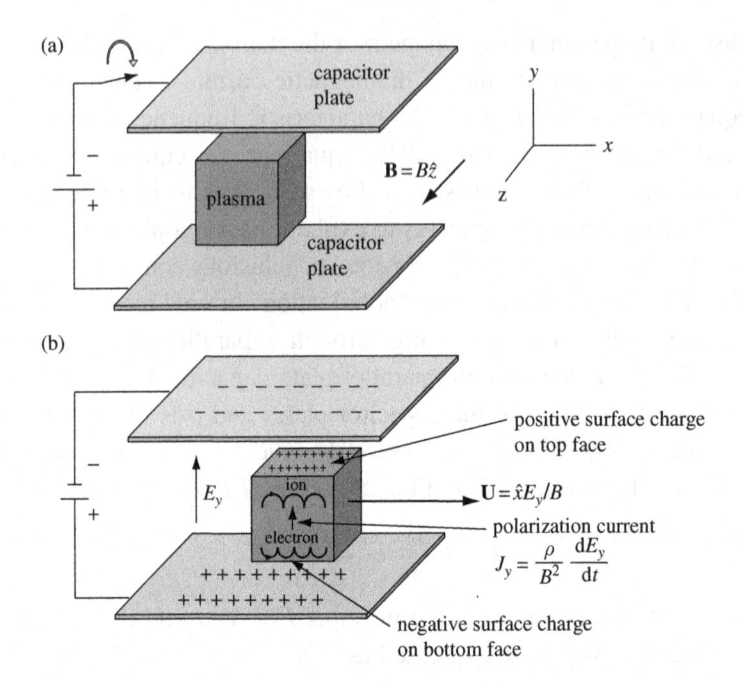

Fig. 3.12 (a) A plasma cube is initially at rest between two capacitor plates located in a magnetic field $\mathbf{B} = B\hat{z}$. A source of EMF is connected to the capacitor plates so as to establish a vertical electric field E_y. (b) The $E \times B$ motion of ions causes a positive surface charge to develop on the top of the cube while the corresponding electron $E \times B$ motion creates a negative surface charge on the bottom of the cube. This surface charge is a polarization charge and reduces the strength of the applied electric field. The polarization current establishes the polarization charge.

the net electric field is reduced relative to the vacuum level, just as the electric field inside a capacitor is reduced by the polarization of the dielectric in the capacitor. The polarization current is the means by which these surface charges are established. The polarization current $\mathbf{J}_p$ flows transiently in the y direction as the cube is accelerated from rest to its steady-state velocity, and this acceleration corresponds to the temporal increase of E_y. This entire process can be equivalently characterized using the MHD point of view, in which case the acceleration of the cube to the right is interpreted as resulting from the cross-product between the polarization current $\mathbf{J}_p$ and the magnetic field $\mathbf{B}$, since

$$\rho\frac{\mathrm{d}\mathbf{U}}{\mathrm{d}t} = \mathbf{J}_p \times \mathbf{B} = \frac{\rho}{B^2}\frac{\mathrm{d}\mathbf{E}_\perp}{\mathrm{d}t} \times \mathbf{B} = \rho\frac{\mathrm{d}}{\mathrm{d}t}\left(\frac{\mathbf{E}\times\mathbf{B}}{B^2}\right). \tag{3.146}$$

The importance of symmetry becomes apparent if one changes the geometrical arrangement to cylindrical geometry $\{r, \phi, z\}$ with $\mathbf{B} = B\hat{z}$ and attempts to have radial acceleration of an annular shell of perfectly conducting plasma, i.e., a plasma shell defined by $a < r < b$, where a is the inner radius and b is the outer radius. The annular shell corresponds to letting $x \to r$ and $y \to \phi$ when relating to the situation sketched in Fig. 3.12. Let us imagine the annular shell is decomposed into N equal angular sectors, so that the first sector lies between $\phi = 0$ and $\phi = 2\pi/N$, etc. Viewed individually, each annulus sector should behave in a manner similar to the cube sketched in Fig. 3.12 and develop a surface polarization charge on its ϕ faces, corresponding to the development of an azimuthal electric field E_ϕ. For example, radial motion of the first sector would cause negative polarity to develop on the $\phi = 0$ surface and positive polarity to develop on the $\phi = 2\pi/N$ surface. However, the next sector (i.e., the sector lying between $\phi = 2\pi/N$ and $\phi = 4\pi/N$) would have a negative surface charge at $\phi = 2\pi/N$, which would *cancel* the positive surface charge on the first sector. The azimuthal symmetry essentially short-circuits the development of a finite electrostatic E_ϕ and so prevents the plasma from moving in the radial direction. Thus, whether or not a plasma can move across a magnetic field is intimately related to the symmetry of the configuration. Moving across a magnetic field requires establishment of a polarization surface charge and associated electrostatic field and this cannot occur if prohibited by considerations of symmetry.

3.7 Non-adiabatic motion in symmetric geometry

Adiabatic behavior occurs when temporal or spatial changes in the electromagnetic field from one cyclical orbit to the next are sufficiently gradual to be effectively continuous and differentiable (i.e., analytic). Thus, adiabatic behavior corresponds to situations where variations of the electromagnetic field are sufficiently gradual to be characterized by the techniques of calculus (differentials, limits, Taylor expansions, etc.).

Non-adiabatic particle motion occurs when this is not so. It is therefore no surprise that it is usually not possible to construct analytic descriptions of non-adiabatic particle motion. However, there exist certain special situations where non-adiabatic motion can be described analytically. Using these special cases as a guide, it is possible to develop an understanding for what happens when motion is non-adiabatic.

One special situation is when the electromagnetic field is *geometrically symmetric* with respect to some coordinate Q_j, in which case the symmetry makes it possible to develop analytic descriptions of non-adiabatic motion. This is because symmetry in Q_j causes the canonical momentum P_j to be an exact constant of the motion. The critical feature is that P_j remains constant no matter how drastically the field changes in time or space because Lagrange's equation $\dot{P}_j = -\partial L/\partial Q_j$ has no limitations on the rate at which changes can occur. In effect, being symmetric trumps being non-analytic. The absolute invariance of P_j when $\partial L/\partial Q_j = 0$ reduces the number of equations and allows a partial or sometimes even a complete solution of the motion. Solutions to symmetric problems give valuable insight regarding the more general situation of being both non-adiabatic and asymmetric.

Two closely related examples of non-adiabatic particle motion will now be analyzed: (i) sudden temporal and (ii) sudden spatial reversal of the polarity of an azimuthally symmetric magnetic field having no azimuthal component. The most general form of such a field can be written in cylindrical coordinates (r, θ, z) as

$$\mathbf{B} = \frac{1}{2\pi}\nabla\psi(r, z, t) \times \nabla\theta; \tag{3.147}$$

a field of this form is called poloidal. Rather than using $\hat{\theta}$ explicitly, the form $\nabla\theta$ has been used because $\nabla\theta$ is better suited for use with the various identities of vector calculus (e.g., $\nabla \times \nabla\theta = 0$) and leads to greater algebraic clarity. The relationship between $\nabla\theta$ and $\hat{\theta}$ is seen by simply taking the gradient:

$$\nabla\theta = \left(\hat{r}\frac{\partial}{\partial r} + \frac{\hat{\theta}}{r}\frac{\partial}{\partial \theta} + \hat{z}\frac{\partial}{\partial z}\right)\theta = \frac{\hat{\theta}}{r}. \tag{3.148}$$

Equation (3.147) automatically satisfies $\nabla \cdot \mathbf{B} = 0$ (by virtue of the vector identity $\nabla \cdot (\mathbf{G} \times \mathbf{H}) = \mathbf{H} \cdot \nabla \times \mathbf{G} - \mathbf{G} \cdot \nabla \times \mathbf{H}$), has no θ component, and is otherwise arbitrary, since ψ is arbitrary. As shown in Fig. 3.13, the magnetic flux linking a circle of radius r with center at axial position z is

$$\begin{aligned}\int \mathbf{B} \cdot \mathrm{d}\mathbf{s} &= \int_0^r 2\pi r \mathrm{d}r \hat{z} \cdot \left[\frac{1}{2\pi}\nabla\psi(r, z, t) \times \nabla\theta\right] \\ &= \int_0^r \mathrm{d}r \frac{\partial\psi(r, z, t)}{\partial r} = \psi(r, z, t) - \psi(0, z, t).\end{aligned} \tag{3.149}$$

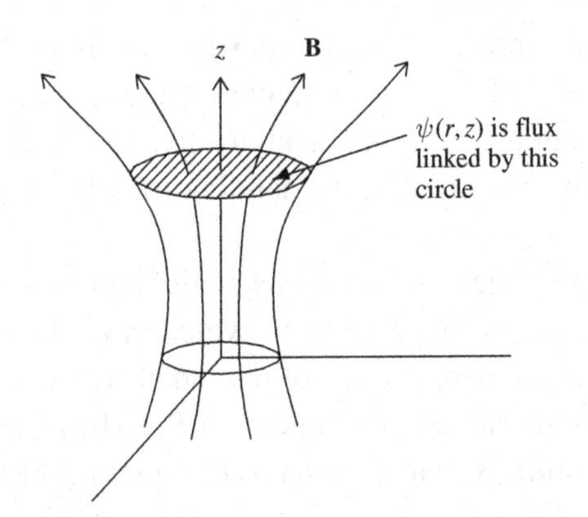

Fig. 3.13 Azimuthally symmetric flux surface.

However,

$$B_r(r, z, t) = -\frac{1}{2\pi r}\frac{\partial\psi}{\partial z} \tag{3.150}$$

and since $\nabla\cdot\mathbf{B} = 0$, B_r must vanish at $r = 0$, and so $\partial\psi/\partial z = 0$ on the symmetry axis $r = 0$.

Thus ψ is constant along the symmetry axis $r = 0$; for convenience we choose this constant to be zero. Hence, $\psi(r, z, t)$ is precisely the magnetic flux enclosed by a circle of radius r at axial location z. We can also use the vector potential $\mathbf{A}$ to calculate the magnetic flux through the same circle and obtain

$$\int \mathbf{B}\cdot d\mathbf{s} = \int \nabla\times\mathbf{A}\cdot d\mathbf{s} = \oint \mathbf{A}\cdot d\mathbf{l} = \int_0^{2\pi} A_\theta r d\theta = 2\pi r A_\theta. \tag{3.151}$$

This shows that the flux ψ and the vector potential A_θ are related by

$$\psi(r, z, t) = 2\pi r A_\theta. \tag{3.152}$$

No other component of vector potential is required to determine the magnetic field and so we may set $\mathbf{A} = A_\theta(r, z, t)\hat{\theta}$.

The current $\mathbf{J} = \mu_0^{-1}\nabla\times\mathbf{B}$ producing this magnetic field is purely azimuthal as can be seen by considering the r and z components of $\nabla\times\mathbf{B}$. The actual current density is

$$\begin{aligned} J_\theta &= \mu_0^{-1} r\nabla\theta\cdot\nabla\times\mathbf{B} \\ &= \mu_0^{-1} r\nabla\cdot(\mathbf{B}\times\nabla\theta) \\ &= -\frac{r}{2\pi\mu_0}\nabla\cdot\left(\frac{1}{r^2}\nabla\psi\right) \\ &= -\frac{r}{2\pi\mu_0}\left[\frac{\partial}{\partial r}\left(\frac{1}{r^2}\frac{\partial\psi}{\partial r}\right) + \frac{1}{r^2}\frac{\partial^2\psi}{\partial z^2}\right], \end{aligned} \tag{3.153}$$

a Poisson-like equation. Since no current loops can exist at infinity, the field prescribed by Eq. (3.147) must be produced by a set of coaxial coils having various finite radii r and various finite axial positions z.

The axial magnetic field is

$$B_z = \frac{1}{2\pi r}\frac{\partial\psi}{\partial r}. \tag{3.154}$$

Near $r = 0$, ψ can always be Taylor expanded as

$$\psi(r, z) = 0 + r\frac{\partial\psi(r = 0, z)}{\partial r} + \frac{r^2}{2}\frac{\partial^2\psi(r = 0, z)}{\partial r^2} + \ldots \tag{3.155}$$

Suppose $\partial\psi/\partial r$ is non-zero at $r = 0$, i.e., $\psi \sim r$ near $r = 0$. If this were the case, then the first term in the right-hand side of the last line of Eq. (3.153) would become infinite and so lead to an infinite current density at $r = 0$. Such a result is non-physical and so we require that the first non-zero term in the Taylor expansion of ψ about $r = 0$ to be the r^2 term.

Every field line looping through the inside of a current loop also loops back in the reverse direction on the outside, so there is no net magnetic flux at infinity. This means ψ must vanish at infinity and so as r increases, ψ increases from its value of zero at $r = 0$ to some maximum value ψ_{max} at $r = r_{max}$, and then slowly decreases back to zero as $r \to \infty$. As seen from Eq. (3.154) this corresponds to B_z being positive for $r < r_{max}$ and negative for $r > r_{max}$. A contour plot of the $\psi(r, z)$ flux surfaces and a plot of $\psi(r, z = 0)$ versus r is shown in Fig. 3.14.

In this cylindrical coordinate system the Lagrangian, Eq. (3.12), has the form

$$L = \frac{m}{2}\left(\dot{r}^2 + r^2\dot{\theta}^2 + \dot{z}^2\right) + qr\dot{\theta}A_\theta - q\phi(r, z, t). \tag{3.156}$$

Since θ is an ignorable coordinate, the canonical angular momentum is a constant of the motion, i.e.,

$$P_\theta = \frac{\partial L}{\partial \dot{\theta}} = mr^2\dot{\theta} + qrA_\theta = const. \tag{3.157}$$

or, in terms of flux,

$$P_\theta = mr^2\dot{\theta} + \frac{q}{2\pi}\psi(r, z, t) = const. \tag{3.158}$$

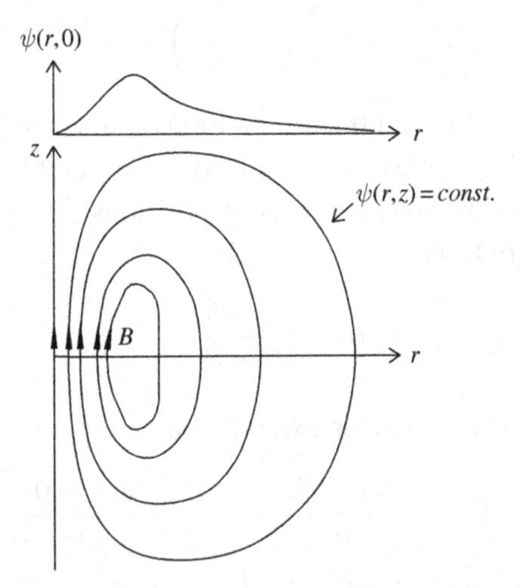

Fig. 3.14 Contour plot of flux surfaces.

Thus, the Hamiltonian is

$$
\begin{aligned}
H &= \frac{m}{2}\left(\dot{r}^2 + r^2\dot{\theta}^2 + \dot{z}^2\right) + \phi(r,z,t) \\
&= \frac{m}{2}\left(\dot{r}^2 + \dot{z}^2\right) + \frac{\left(P_\theta - q\psi(r,z,t)/2\pi\right)^2}{2mr^2} + \phi(r,z,t) \\
&= \frac{m}{2}\left(\dot{r}^2 + \dot{z}^2\right) + \chi(r,z,t),
\end{aligned}
\tag{3.159}
$$

where

$$
\chi(r,z,t) = \frac{1}{2m}\left[\frac{P_\theta - q\psi(r,z,t)/2\pi}{r}\right]^2 \tag{3.160}
$$

is an *effective* potential. For purposes of plotting, the effective potential can be written in a dimensionless form as

$$
\frac{\chi(r,z,t)}{\chi_0} = \left(\frac{\dfrac{2\pi P_\theta}{q\psi_0} - \dfrac{\psi(r,z,t)}{\psi_0}}{r/L}\right)^2, \tag{3.161}
$$

where L is some reference scale length, ψ_0 is some arbitrary reference value for the flux, and $\chi_0 = q\psi_0^2/8\pi^2L^2m$. For simplicity we have set $\phi(r,z,t) = 0$, since this term gives the motion of a particle in a readily understood, two-dimensional electrostatic potential.

Suppose that for times $t < t_1$ the coil currents are constant, in which case the associated magnetic field and flux are also constant. Since the Lagrangian does not explicitly depend on time, the energy H is a constant of the motion. Hence, there are two constants of the motion, H and P_θ. Consider now a particle located initially on the midplane $z = 0$ with $r < r_{\max}$. The particle motion depends on the sign of $q\psi/P_\theta$ and so we consider each polarity separately.

1. $q\psi/P_\theta$ is *positive*. If $2\pi|P_\theta| < |q\psi_{\max}|$ there exists a location inside $r_{\max}$ where

$$
P_\theta = \frac{q}{2\pi}\psi \tag{3.162}
$$

and there exists a location outside $r_{\max}$ where this equality holds as well. χ vanishes at these two points, which are also local minima of χ because χ is positive-definite. The top plot in Fig. 3.15 shows a nominal $\psi(r)/\psi_0$ flux profile and the middle plot shows the corresponding $\chi(r)/\chi_0$; the z and t dependences are suppressed from the arguments for clarity. There exists a maximum of χ between the two minima. We consider a particle initially located in one of the two minima of χ. If $H < \chi_{\max}$ the

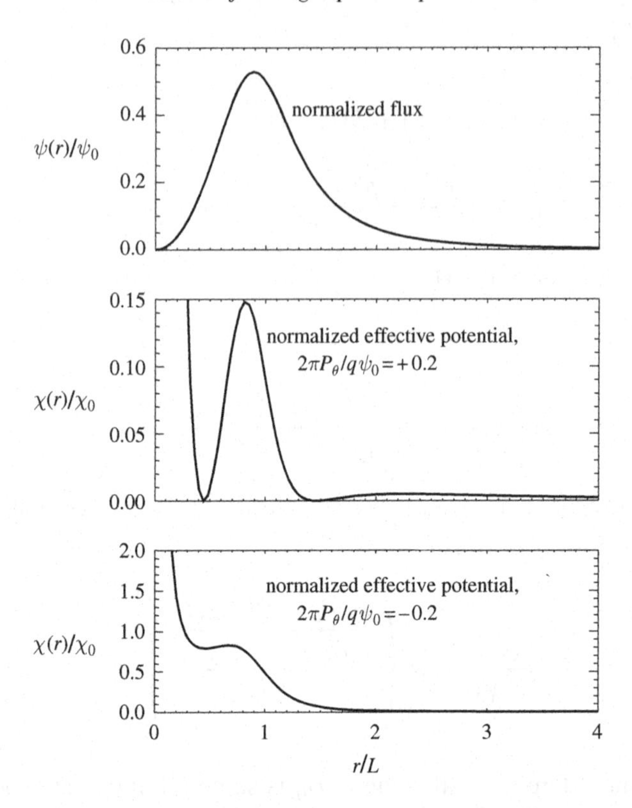

Fig. 3.15 Specific example (with z dependence suppressed) showing ψ and χ relationship: top is plot of function $\psi(r)/\psi_0 = (r/L)^2/(1+(r/L)^6)$, middle and bottom plots show corresponding normalized effective potential for $2\pi P_\theta/q\psi_0 = +0.2$ and $2\pi P_\theta/q\psi_0 = -0.2$. Both middle and bottom plots have a minimum at $r/L \simeq 0.45$; middle plot also has a minimum at $r/L \simeq 1.4$. The two minima in the middle plot occur when $\chi(r)/\chi_0 = 0$ but the single minimum in the bottom plot occurs at a finite value of $\chi(r)/\chi_0$, indicating that an axis-encircling particle must have finite energy.

particle will be confined to an effective potential well centered about the flux surface defined by Eq. (3.162). From Eq. (3.158) the angular velocity is

$$\dot{\theta} = \frac{1}{mr^2}\left(P_\theta - \frac{q\psi}{2\pi}\right). \tag{3.163}$$

The sign of $\dot{\theta}$ reverses periodically as the particle bounces back and forth in the χ potential well. This corresponds to localized gyromotion as shown in Fig. 3.16 (left).

2. $q\psi/P_\theta$ is *negative*. In this case χ can never vanish, because $P_\theta - q\psi/2\pi$ never vanishes. Nevertheless, it is still possible for χ to have a minimum and, hence, a potential well. This possibility can be seen by setting $\partial\chi/\partial r = 0$, which occurs when

$$\left(P_\theta - \frac{q}{2\pi}\psi\right)\frac{\partial}{\partial r}\left(\frac{P_\theta - q\psi/2\pi}{r}\right) = 0. \tag{3.164}$$

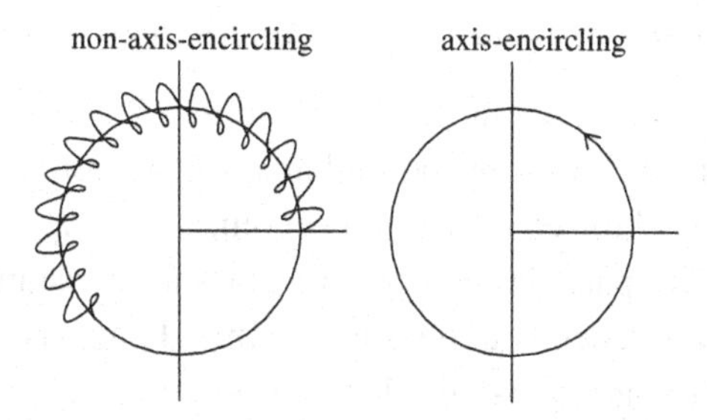

Fig. 3.16 Localized gyromotion associated with particle bouncing in effective potential well.

Equation (3.164) can be satisfied by having

$$\frac{\partial}{\partial r}\left(\frac{P_\theta - \dfrac{q}{2\pi}\psi}{r}\right) = 0, \tag{3.165}$$

which implies

$$P_\theta = -\frac{qr^2}{2\pi}\frac{\partial}{\partial r}\left(\frac{\psi}{r}\right). \tag{3.166}$$

Recall that ψ had a maximum, $\psi \sim r^2$ near $r = 0$, and also $\psi \to 0$ as $r \to \infty$. Thus $\psi/r \sim r$ for small r and $\psi/r \to 0$ for $r \to \infty$ so ψ/r also has a maximum; this maximum is located at an r somewhat inside of the maximum of ψ. Equation (3.166) can only be valid at points inside of this maximum; otherwise the assumption of opposite signs for P_θ and ψ would be incorrect. Furthermore, Eq. (3.166) can only be satisfied if $|P_\theta|$ is not too large, because the right-hand side of Eq. (3.166) has a maximum value. If all these conditions are satisfied, then χ will have a non-zero minimum as shown in the bottom plot of Fig. 3.15.

A particularly simple example of this behavior occurs if Eq. (3.166) is satisfied near the $r = 0$ axis (i.e., where $\psi \sim r^2$) so that this equation becomes simply

$$P_\theta = -\frac{q}{2\pi}\psi, \tag{3.167}$$

which is just the opposite of Eq. (3.162). Substituting in Eq. (3.163) we see that $\dot{\theta}$ now *never* changes sign; i.e., the particle is *axis-encircling* as sketched in Fig. 3.16 (right). The Larmor radius of this axis-encircling particle is just the radius of the minimum of the potential well, the radius where Eq. (3.166) holds. The azimuthal kinetic energy of the particle corresponds to the height of the minimum of χ in the bottom plot of Fig. 3.15.

3.7.1 Temporal reversal of magnetic field-energy gain

Armed with this information about axis-encircling and non-axis-encircling particles, we now examine the strongly non-adiabatic situation where a coil current starts at $I = I_0$, is reduced to zero, and then becomes $I = -I_0$, so that all fields and fluxes reverse sign. The particle energy will not stay constant for this situation because the Lagrangian depends explicitly on time. However, since symmetry is maintained, P_θ *must* remain constant. Thus, a non-axis-encircling particle (with radial location determined by Eq. (3.162)) will change to an axis-encircling particle if a minimum exists for χ when the sign of ψ is reversed. If such a minimum does exist and if the initial radius was near the axis where $\psi \sim r^2$, then comparison of Eqs. (3.162) and (3.167) shows that the particle will have the same radius after the change of sign as before. The particle will gain energy during the field reversal by an amount corresponding to the finite value of the minimum of χ for the axis-encircling case.

This process can also be considered from the point of view of particle drifts: initially, the non-axis-encircling particle is frozen to a constant ψ surface (flux surface). When the coil current starts to decrease, the maximum value of the flux correspondingly decreases. The constant ψ contours on the *inside* of ψ_{max} move outwards towards the location of ψ_{max} where they are annihilated. Likewise, the contours outside of ψ_{max} move inwards to ψ_{max} where they are also annihilated.

To the extent that the $\mathbf{E} \times \mathbf{B}$ drift is a valid approximation, its effect is to keep the particle attached to a surface of constant flux. This can be seen by integrating Faraday's law over the area of a circle of radius r to obtain $\int \mathrm{d}s \cdot \nabla \times \mathbf{E} = -\int \mathrm{d}s \cdot \partial \mathbf{B}/\partial t$ and then invoking Stokes' theorem to give

$$E_\theta 2\pi r = -\frac{\partial \psi}{\partial t}. \tag{3.168}$$

The theta component of $\mathbf{E} + \mathbf{v} \times \mathbf{B} = 0$ is

$$E_\theta + v_z B_r - v_r B_z = 0 \tag{3.169}$$

and from (3.147), $B_r = -(2\pi r)^{-1}\, \partial\psi/\partial z$ and $B_z = -(2\pi r)^{-1}\, \partial\psi/\partial r$. Combination of Eqs. (3.168) and (3.169) thus gives

$$\frac{\partial \psi}{\partial t} + v_r \frac{\partial \psi}{\partial r} + v_z \frac{\partial \psi}{\partial z} = 0. \tag{3.170}$$

Because $\psi(r(t), z(t), t)$ is the flux measured in the frame of a particle moving with trajectory $r(t)$ and $z(t)$, Eq. (3.170) shows that the $\mathbf{E} \times \mathbf{B}$ drift maintains the particle on a surface of constant ψ, i.e., the $\mathbf{E} \times \mathbf{B}$ drift is such as to maintain $d\psi/dt = 0$, where d/dt means time derivative as measured in the particle frame.

The implication of this attachment of the particle to a surface of constant ψ can be appreciated by making an analogy to the motion of people initially located on the beach of a volcanic island that is slowly sinking into the sea. In order to avoid being drowned as the island sinks, the people will move towards the mountain top to stay at a constant height above the sea. The location of ψ_{max} here corresponds to the mountain top and the particles trying to stay on surfaces of constant ψ correspond to people trying to stay at constant altitude. A particle initially located at some location away from the "mountain top" ψ_{max} moves towards ψ_{max} if the overall level of all the ψ surfaces is sinking. The reduction of ψ as measured at a fixed position will create the azimuthal electric field given by Eq. (3.168) and this electric field will, as shown by Eqs. (3.169) and (3.170), cause an $\mathbf{E}\times\mathbf{B}$ drift, which convects each particle in just such a way as to stay on a constant ψ contour.

The $\mathbf{E}\times\mathbf{B}$ drift approximation breaks down when $\mathbf{B}$ becomes zero, i.e., when ψ changes polarity. This breakdown corresponds to a breakdown of the adiabatic approximation. If ψ changes polarity before a particle reaches ψ_{max}, the particle becomes axis-encircling. The extra energy associated with being axis-encircling is obtained when $\psi \simeq 0$ but $\partial\psi/\partial t \neq 0$ so that there is an electric field E_θ, but no magnetic field. Finite E_θ and no magnetic field results in a simple theta acceleration of the particle. Thus, when ψ reverses polarity the particle is accelerated azimuthally and develops finite kinetic energy. After ψ has changed polarity the magnitude of ψ increases and the adiabatic approximation again becomes valid. Because the polarity is reversed, increase of the magnitude of ψ is now analogous to creating an ever deepening crater. Particles again try to stay on constant flux surfaces as dictated by Eq. (3.170) and as the crater deepens, the particles have to move away from ψ_{min} to stay at the same altitude. When the reversed flux attains the same magnitude as the original flux, the flux surfaces have the same shape as before. However, the particles are now axis-encircling and have the extra kinetic energy obtained at field reversal.

3.7.2 *Spatial reversal of field-cusps*

Suppose two solenoids with constant currents are arranged coaxially with their magnetic fields opposing each other as shown in Fig. 3.17(a). Since the solenoid currents are constant, the Lagrangian does not depend explicitly on time, in which case energy is a constant of the motion. Because of the geometrical arrangement, the flux function is antisymmetric in z , where $z = 0$ defines the midplane between the two solenoids.

Consider a particle injected with initial velocity $\mathbf{v} = v_{z0}\hat{z}$ at $z = -L, r = a$. Since this particle has no initial $v_\perp$, it simply streams along a magnetic field line. However, when the particle approaches the cusp region, the magnetic field lines start to curve causing the particle to develop both curvature and grad B drifts perpendicular to the magnetic field. When the particle approaches the $z = 0$ plane, the drift approximation breaks down because $B \to 0$ and so the particle's motion becomes non-adiabatic (cf. Fig. 3.17(a)).

Although the particle trajectory is very complex in the vicinity of the cusp, it is still possible to determine whether the particle will cross into the positive z half-plane, i.e., cross the cusp. Such an analysis is possible because two constants of the motion exist, namely P_θ and H. The energy

$$H = \frac{P_r^2}{2m} + \frac{P_z^2}{2m} + \frac{\left(P_\theta - \frac{q}{2\pi}\psi(r,z)\right)^2}{2mr^2} = const. \tag{3.171}$$

can be evaluated using

$$P_\theta = \left[mr^2\dot{\theta} + \frac{q}{2\pi}\psi\right]_{\text{initial}} = \frac{q}{2\pi}\psi_0, \tag{3.172}$$

since initially $\dot{\theta} = 0$. Here

$$\psi_0 = \psi(r = a, z = -L) \tag{3.173}$$

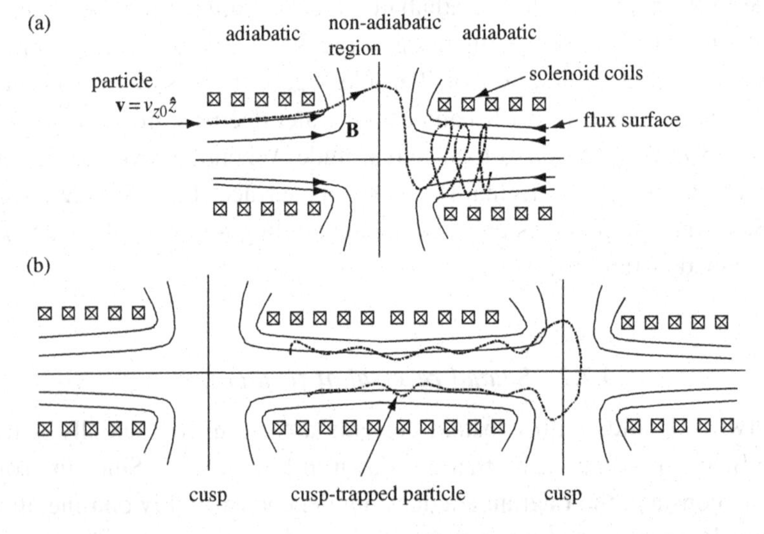

Fig. 3.17 (a) Cusp field showing trajectory for particle with sufficient initial energy to penetrate the cusp; (b) two cusps used as magnetic trap to confine particles.

is the flux at the particle's initial position. Inserting initial values of all quantities in Eq. (3.171) gives

$$H = \frac{mv_{z0}^2}{2} \tag{3.174}$$

and so Eq. (3.171) becomes

$$\begin{aligned}\frac{mv_{z0}^2}{2} &= \frac{mv_r^2}{2} + \frac{mv_z^2}{2} + \frac{\left(\frac{q}{2\pi}\right)^2 (\psi_0 - \psi(r,z))^2}{2mr^2} \\ &= \frac{mv_r^2}{2} + \frac{mv_z^2}{2} + \frac{mv_\theta^2}{2}.\end{aligned} \tag{3.175}$$

The extent to which a particle penetrates the cusp can be easily determined if the particle starts close enough to $r = 0$ so that the flux may be approximated as $\psi \sim r^2$. Specifically, the flux will be $\psi = B_{z0}\pi r^2$, where B_{z0} is the on-axis magnetic field in the $z \ll 0$ region. The canonical momentum is simply $P_\theta = q\psi/2\pi = qB_{z0}a^2/2$, since the particle started as non-axis-encircling.

Suppose the particle penetrates the cusp and arrives at some region where again $\psi \sim r^2$. Since the particle is now axis-encircling, the relation between canonical momentum and flux is $P_\theta = -q\psi/2\pi = -q(-B_{z0}\pi r^2)/2\pi = qB_{z0}r^2/2$ from which it is concluded that $r = a$. Thus, *if* the particle is able to move across the cusp, it becomes an axis-encircling particle with the *same* radius $r = a$ it originally had when it was non-axis-encircling. The minimum energy an axis-encircling particle can have is when it is purely axis-encircling, i.e., has $v_r = 0$ and $v_z = 0$. Thus, in order for the particle to cross the cusp and reach a location where it becomes purely axis-encircling, the particle's initial energy must satisfy

$$\frac{mv_{z0}^2}{2} \geq \frac{m(\omega_c a)^2}{2} \tag{3.176}$$

or simply

$$v_{z0} \geq \omega_c a. \tag{3.177}$$

If v_{z0} is too small to satisfy this relation, the particle reflects from the cusp and returns back to the negative z half-plane. Plasma confinement schemes have been designed based on particles reflecting from cusps as shown in Fig. 3.17(b). Here a particle is trapped between two cusps and, so long as its parallel energy is insufficient to violate Eq. (3.177), the particle is confined between the two cusps.

Cusps have also been used to trap relativistic electron beams in mirror fields (Kribel *et al.* 1974, Hudgings *et al.* 1978). In this scheme an additional opposing solenoid is added to one end of a magnetic mirror so as to form a cusp outside the mirror region. A relativistic electron beam is injected through the cusp into the mirror. The beam changes from non-axis-encircling into axis-encircling on passing through the cusp as in Fig. 3.17(a). If energy is conserved, the beam is not trapped because the beam will reverse its trajectory and bounce back out of the mirror. However, if axial energy is removed from the beam once it is in the mirror, then the motion will not be reversible and the beam will be trapped. Removal of beam axial energy has been achieved by having the beam collide with neutral particles or by having the beam induce currents in a resistive wall.

3.7.3 Stochastic motion in large-amplitude, low-frequency waves

The particle drifts ($\mathbf{E}\times\mathbf{B}$, polarization, etc.) were derived using an iteration scheme based on the assumption that spatial changes in the electric and magnetic fields are sufficiently gradual to allow Taylor expansions of the fields about their values at the gyrocenter.

We now examine a situation where the fields change gradually in space relative to the initial gyro-orbit dimensions, but the fields also pump energy into the particle motion so the size of the gyro-orbit eventually increases to the point that the smallness assumption fails. To see how this might occur, consider motion of a particle in an electrostatic wave

$$\mathbf{E} = \hat{y}k\phi\sin(ky-\omega t) \tag{3.178}$$

propagating in a plasma immersed in a uniform magnetic field $\mathbf{B} = B\hat{z}$. The wave frequency is much lower than the cyclotron frequency of the particle in question. This $\omega \ll \omega_c$ condition indicates that, in principle, the drift equations can be used and so according to these equations, the charged particle should have both an $\mathbf{E}\times\mathbf{B}$ drift

$$\mathbf{v}_E = \frac{\mathbf{E}\times\mathbf{B}}{B^2} = \hat{x}\frac{k\phi}{B}\sin(ky-\omega t) \tag{3.179}$$

and a polarization drift

$$\mathbf{v}_p = \frac{m}{qB^2}\frac{\mathrm{d}\mathbf{E}_\perp}{\mathrm{d}t} = \hat{y}\frac{mk\phi}{qB^2}\frac{\mathrm{d}}{\mathrm{d}t}\sin(ky-\omega t). \tag{3.180}$$

If the wave amplitude is infinitesimal, the spatial displacements associated with $\mathbf{v}_E$ and $\mathbf{v}_p$ are negligible and so the guiding center value of y may be used in the right-hand side of Eq. (3.180) to obtain

$$\mathbf{v}_p = -\hat{y}\frac{\omega mk\phi}{qB^2}\cos(ky-\omega t). \tag{3.181}$$

Equations (3.179) and (3.181) show that the combined $\mathbf{v}_E$ and $\mathbf{v}_p$ particle drift motion results in an elliptical trajectory.

Now suppose the wave amplitude becomes so large that the particle is displaced significantly from its initial position. Since the polarization drift is in the y direction, there will be a substantial displacement in the y direction. Thus, the right side of Eq. (3.180) should be construed as $\sin(ky(t)-\omega t)$ so, taking into account the time dependence of y on the right-hand side, Eq. (3.180) becomes

$$\mathbf{v}_p = \hat{y}\frac{mk\phi}{qB^2}\frac{\mathrm{d}}{\mathrm{d}t}\sin(ky-\omega t) = \hat{y}\frac{mk\phi}{qB^2}\left(k\frac{\mathrm{d}y}{\mathrm{d}t}-\omega\right)\cos(ky-\omega t). \tag{3.182}$$

However, $\mathrm{d}y/\mathrm{d}t = v_p$, since $\mathbf{v}_p$ is the motion in the y direction. Equation (3.182) becomes an implicit equation for $\mathbf{v}_p$ and may be solved to give

$$\mathbf{v}_p = -\hat{y}\frac{\omega mk\phi}{qB^2}\frac{\cos(ky-\omega t)}{[1-\alpha\cos(ky-\omega t)]}, \tag{3.183}$$

where

$$\alpha = \frac{mk^2\phi}{qB^2} \tag{3.184}$$

is a non-dimensional measure of the wave amplitude (McChesney, Stern, and Bellan 1987, White, Chen and Lin 2002).

If $\alpha > 1$, the denominator in Eq. (3.183) vanishes when $ky-\omega t = \cos^{-1}\alpha^{-1}$ and this vanishing denominator would result in an infinite polarization drift. However, the derivation of the polarization drift was based on the assumption that the time derivative of the polarization drift was negligible compared to the time derivative of $\mathbf{v}_E$, i.e., it was explicitly assumed $\mathrm{d}\mathbf{v}_p/\mathrm{d}t \ll \mathrm{d}\mathbf{v}_E/\mathrm{d}t$. Clearly, this assumption fails when $\mathbf{v}_p$ becomes infinite and so the iteration scheme used to derive the particle drifts fails. What is happening is that when $\alpha \sim 1$, the particle displacement due to polarization drift becomes $\sim k^{-1}$, i.e., the displacement of the particle from its gyrocenter becomes of the order of a wavelength. In such a situation it is incorrect to represent the particle's actual location by its gyrocenter because the particle experiences the wave field at its actual location, not at its gyrocenter. Because the wave field is significantly different at two locations separated by $\sim k^{-1}$, it is essential to use the wave field evaluated at the actual particle location rather than at the gyrocenter.

Direct numerical integration of the Lorentz equation in this large-amplitude limit shows that when α exceeds unity, particle motion becomes chaotic and cannot be described by analytic formulae. Onset of chaotic motion resembles heating of the particles since chaos and heating both broaden the velocity distribution function. However, chaotic heating is not a true heating because entropy is not increased – the motion is deterministic and not random. Nevertheless, this chaotic (or stochastic) heating is indistinguishable for practical purposes from ordinary collisional thermalization of directed motion.

An alternate way of looking at this issue is to consider the Lorentz equations for two initially adjacent particles, denoted by subscripts 1 and 2, that are in a wave electric field and a uniform, steady-state magnetic field (Stasiewicz, Lundin and Marklund 2000). The respective Lorentz equations of the two particles are

$$\frac{\mathrm{d}\mathbf{v}_1}{\mathrm{d}t} = \frac{q}{m}[\mathbf{E}(\mathbf{x}_1,t)+\mathbf{v}_1\times\mathbf{B}]$$
$$\frac{\mathrm{d}\mathbf{v}_2}{\mathrm{d}t} = \frac{q}{m}[\mathbf{E}(\mathbf{x}_2,t)+\mathbf{v}_2\times\mathbf{B}]. \tag{3.185}$$

Subtracting these two equations gives an equation for the difference between the velocities of the two particles, $\delta\mathbf{v}=\mathbf{v}_1-\mathbf{v}_2$, in terms of the difference $\delta\mathbf{x}=\mathbf{x}_1-\mathbf{x}_2$ in their positions, i.e.,

$$\frac{\mathrm{d}\delta\mathbf{v}}{\mathrm{d}t} = \frac{q}{m}[\delta\mathbf{x}\cdot\nabla\mathbf{E}+\delta\mathbf{v}\times\mathbf{B}]. \tag{3.186}$$

The difference velocity is related to the difference in positions by $\mathrm{d}\delta\mathbf{x}/\mathrm{d}t=\delta\mathbf{v}$. Let y be the direction in which the electric field is non-uniform, i.e., with this choice of coordinate system $\mathbf{E}$ depends only on the y direction. To simplify the algebra, define $\mathcal{E}_x = qE_x/m$ and $\mathcal{E}_y = qE_y/m$ so the components of Eq. (3.186) transverse to the magnetic field are

$$\delta\ddot{x} = \delta y\frac{\partial\mathcal{E}_x}{\partial y}+\omega_c\delta\dot{y}$$
$$\delta\ddot{y} = \delta y\frac{\partial\mathcal{E}_y}{\partial y}-\omega_c\delta\dot{x}. \tag{3.187}$$

Now take the time derivative of the lower equation to obtain

$$\delta\dddot{y} = \delta\dot{y}\frac{\partial\mathcal{E}_y}{\partial y}+\delta y\frac{\partial}{\partial y}\left(\frac{\mathrm{d}\mathcal{E}_y}{\mathrm{d}t}\right)-\omega_c\delta\ddot{x} \tag{3.188}$$

and then substitute for $\delta\ddot{x}$ giving

$$\delta\dddot{y} = \delta\dot{y}\frac{\partial\mathcal{E}_y}{\partial y}+\delta y\frac{\partial}{\partial y}\left(\frac{\mathrm{d}\mathcal{E}_y}{\mathrm{d}t}\right)-\omega_c\left(\delta y\frac{\partial\mathcal{E}_x}{\partial y}+\omega_c\delta\dot{y}\right). \tag{3.189}$$

This can be rearranged as

$$\delta\dddot{y}+\omega_c^2\left(1-\frac{1}{\omega_c^2}\frac{\partial\mathcal{E}_y}{\partial y}\right)\delta\dot{y} = -\omega_c\delta y\frac{\partial\mathcal{E}_x}{\partial y}+\delta y\frac{\partial}{\partial y}\left(\frac{\mathrm{d}\mathcal{E}_y}{\mathrm{d}t}\right). \tag{3.190}$$

Consider the right-hand side of the equation as being a forcing term for the left-hand side. If $\omega_c^{-2}\partial\mathcal{E}_y/\partial y < 1$, then the left-hand side is a simple harmonic oscillator equation in the variable $\delta\dot{y}$. However, if $\omega_c^{-2}\partial\mathcal{E}_y/\partial y$ exceeds unity, then the left-hand side becomes an equation with solutions that grow exponentially in time. Thus, if two particles are initially separated by the infinitesimal distance

δy and if $\omega_c^{-2}\partial\mathcal{E}_y/\partial y < 1$, the separation distance between the two particles will undergo harmonic oscillations, but if $\omega_c^{-2}\partial\mathcal{E}_y/\partial y > 1$ the separation distance will exponentially diverge with time. It is seen that α corresponds to $\omega_c^{-2}\partial\mathcal{E}_y/\partial y$ for a sinusoidal wave. Exponential growth of the separation distance between two particles that are initially arbitrarily close together is called stochastic behavior.

3.8 Particle motion in small-amplitude oscillatory fields

Suppose a small-amplitude electromagnetic field exists in a plasma that in addition has a large, uniform, steady-state magnetic field and no steady-state electric field. The fields can thus be written as

$$\begin{aligned}\mathbf{E} &= \mathbf{E}_1(\mathbf{x}, t) \\ \mathbf{B} &= \mathbf{B}_0 + \mathbf{B}_1(\mathbf{x}, t),\end{aligned} \tag{3.191}$$

where the subscript 1 denotes the small-amplitude oscillatory quantities and the subscript 0 denotes large, uniform equilibrium quantities. A typical particle in this plasma will develop an oscillatory motion

$$\mathbf{x}(t) = \langle\mathbf{x}(t)\rangle + \delta\mathbf{x}(t), \tag{3.192}$$

where $\langle\mathbf{x}(t)\rangle$ is the particle's time-averaged position and $\delta\mathbf{x}(t)$ is the instantaneous deviation from this average position. If the amplitudes of $\mathbf{E}_1(\mathbf{x}, t)$ and $\mathbf{B}_1(\mathbf{x}, t)$ are sufficiently small, then the fields at the particle position can be approximated as

$$\begin{aligned}\mathbf{E}(\langle\mathbf{x}(t)\rangle + \delta\mathbf{x}(t), t) &\simeq \mathbf{E}_1(\langle\mathbf{x}(t)\rangle, t) \\ \mathbf{B}(\langle\mathbf{x}(t)\rangle + \delta\mathbf{x}(t), t) &\simeq \mathbf{B}_0 + \mathbf{B}_1(\langle\mathbf{x}(t)\rangle, t).\end{aligned} \tag{3.193}$$

This is the opposite limit from what was considered in Section 3.7.3. The Lorentz equation reduces in this small-amplitude limit to

$$m\frac{\mathrm{d}\mathbf{v}}{\mathrm{d}t} = q\left[\mathbf{E}_1(\langle\mathbf{x}\rangle, t) + \mathbf{v}\times(\mathbf{B}_0 + \mathbf{B}_1(\langle\mathbf{x}(t)\rangle, t))\right]. \tag{3.194}$$

Since the oscillatory fields are small, the resulting particle velocity will also be small (unless there is a resonant response as would happen at the cyclotron frequency). If the particle velocity is small, then the term $\mathbf{v}\times\mathbf{B}_1(\mathbf{x},t)$ is of second-order smallness, whereas $\mathbf{E}_1$ and $\mathbf{v}\times\mathbf{B}_0$ are of first-order smallness. The $\mathbf{v}\times\mathbf{B}_1(\mathbf{x}, t)$ term is thus insignificant compared to the other two terms on the right-hand side and therefore can be discarded so that the Lorentz equation reduces to

$$m\frac{\mathrm{d}\mathbf{v}}{\mathrm{d}t} = q\left[\mathbf{E}_1(\langle\mathbf{x}\rangle, t) + \mathbf{v}\times\mathbf{B}_0\right], \tag{3.195}$$

a linear differential equation for $\mathbf{v}$. Since $\delta\mathbf{x}$ is assumed to be so small that it can be ignored, the average brackets will be omitted from now on and the first order electric field will simply be written as $\mathbf{E}_1(\mathbf{x}, t)$, where $\mathbf{x}$ can be interpreted as being either the actual or the average position of the particle.

The oscillatory electric field can be decomposed into Fourier modes, each having time dependence $\sim \exp(-\mathrm{i}\omega t)$ and, since Eq. (3.195) is linear, the particle response to a field $\mathbf{E}_1(\mathbf{x}, t)$ is just the linear superposition of its response to each Fourier mode. Thus, it is appropriate to consider motion in a single Fourier mode of the electric field, say

$$\mathbf{E}_1(\mathbf{x}, t) = \tilde{\mathbf{E}}(\mathbf{x}, \omega) \exp(-\mathrm{i}\omega t). \tag{3.196}$$

If initial conditions are ignored for now, the particle motion can be found by simply assuming that the particle velocity also has the time dependence $\exp(-\mathrm{i}\omega t)$, in which case the Lorentz equation becomes

$$-\mathrm{i}\omega m\mathbf{v} = q\left[\tilde{\mathbf{E}}(\mathbf{x}) + \mathbf{v}\times\mathbf{B}_0\right], \tag{3.197}$$

where a factor $\exp(-\mathrm{i}\omega t)$ is implicitly assumed for all terms and also an ω argument is implicitly assumed for $\tilde{\mathbf{E}}$. Equation (3.197) is a vector equation of the form

$$\mathbf{v} + \mathbf{v}\times\mathbf{A} = \mathbf{C}, \tag{3.198}$$

where

$$\begin{aligned} \mathbf{A} &= \frac{\omega_c}{\mathrm{i}\omega}\hat{z}, \\ \omega_c &= \frac{qB_0}{m}, \\ \mathbf{C} &= \frac{\mathrm{i}q}{\omega m}\tilde{\mathbf{E}}(\mathbf{x}), \end{aligned} \tag{3.199}$$

and the z axis has been chosen to be in the direction of $\mathbf{B}_0$. Equation (3.198) can be solved for $\mathbf{v}$ by first dotting with $\mathbf{A}$ to obtain

$$\mathbf{A}\cdot\mathbf{v} = \mathbf{C}\cdot\mathbf{A} \tag{3.200}$$

and then crossing with $\mathbf{A}$ to obtain

$$\mathbf{v}\times\mathbf{A} + \mathbf{A}\mathbf{A}\cdot\mathbf{v} - \mathbf{v}A^2 = \mathbf{C}\times\mathbf{A}. \tag{3.201}$$

Substituting for $\mathbf{A}\cdot\mathbf{v}$ using Eq. (3.200) and for $\mathbf{v}\times\mathbf{A}$ using Eq. (3.198) gives

$$\mathbf{v} = \frac{\mathbf{C} + \mathbf{A}\mathbf{A}\cdot\mathbf{C} - \mathbf{C}\times\mathbf{A}}{1+A^2} = C_\parallel\hat{z} + \frac{\mathbf{C}_\perp}{1+A^2} + \frac{\mathbf{A}\times\mathbf{C}}{1+A^2}, \tag{3.202}$$

where $\mathbf{C}$ has been split into parallel and perpendicular parts relative to $\mathbf{B}_0$ and $\mathbf{AA}\cdot\mathbf{C}=A^2C_{\|}\hat{z}$ has been used. On substituting for $\mathbf{A}$ and $\mathbf{C}$ this becomes

$$\mathbf{v}=\frac{\mathrm{i}q}{\omega m}\left[\tilde{E}_{\|}(\mathbf{x})\hat{z}+\frac{\tilde{\mathbf{E}}_{\perp}(\mathbf{x})}{1-\omega_c^2/\omega^2}-\frac{\mathrm{i}\omega_c}{\omega}\frac{\hat{z}\times\tilde{\mathbf{E}}(\mathbf{x})}{1-\omega_c^2/\omega^2}\right]\mathrm{e}^{-\mathrm{i}\omega t}. \tag{3.203}$$

The third term on the right-hand side is a generalization of the $\mathbf{E}\times\mathbf{B}$ drift, since for $\omega\ll\omega_c$ this term reduces to the $\mathbf{E}\times\mathbf{B}$ drift. Similarly, the middle term on the right-hand side is a generalization of the polarization drift, since for $\omega\ll\omega_c$ this term reduces to the polarization drift. The first term on the right-hand side, the parallel quiver velocity, does not involve the magnetic field B_0. This non-dependence on magnetic field is to be expected because no magnetic force results from motion parallel to a magnetic field. In fact, if the magnetic field were zero, then the second term would add to the first and the third term would vanish, giving a three-dimensional unmagnetized quiver velocity $\mathbf{v}=\mathrm{i}q\tilde{\mathbf{E}}(\mathbf{x})/\omega m$.

If the electric field is in addition decomposed into spatial Fourier modes with dependence $\sim\exp(\mathrm{i}\mathbf{k}\cdot\mathbf{x})$, then the velocity for a typical mode will be

$$\mathbf{v}(\mathbf{x},t)=\frac{\mathrm{i}q}{\omega m}\left[\tilde{E}_{\|}\hat{z}+\frac{\tilde{\mathbf{E}}_{\perp}}{1-\omega_c^2/\omega^2}-\frac{\mathrm{i}\omega_c}{\omega}\frac{\hat{z}\times\tilde{\mathbf{E}}}{1-\omega_c^2/\omega^2}\right]\mathrm{e}^{\mathrm{i}\mathbf{k}\cdot\mathbf{x}-\mathrm{i}\omega t}. \tag{3.204}$$

The convention of a negative coefficient for ω and a positive coefficient for $\mathbf{k}$ has been adopted to give waves propagating in the positive $\mathbf{x}$ direction. Equation (3.204) will later be used as the starting point for calculating wave-generated plasma currents.

3.9 Wave–particle energy transfer

3.9.1 *"Average velocity"*

Anyone who has experienced delay in a traffic jam knows that it is usually impossible to make up for the lost time by going faster after escaping from the traffic jam. To see why, define α as the fraction of the total trip length in the traffic jam, v_s as the slow (traffic jam) speed, and v_f as the fast speed (out of traffic jam). It is tempting, but wrong, to say that the average velocity is $(1-\alpha)v_f+\alpha v_s$ because

$$\text{average velocity of a trip}=\frac{\text{total distance}}{\text{total time}}. \tag{3.205}$$

Since the fast-portion duration is $t_f=(1-\alpha)L/v_f$ while the the slow-portion duration is $t_s=\alpha L/v_s$, the average velocity of the complete trip is

$$v_{avg}=\frac{L}{(1-\alpha)L/v_f+\alpha L/v_s}=\frac{1}{(1-\alpha)/v_f+\alpha/v_s}. \tag{3.206}$$

Thus, if $v_s \ll v_f$ then $v_{avg} \simeq v_s/\alpha$. The average velocity is almost entirely determined by the slow-portion velocity and as verified by sad experience in traffic jams, once v_s becomes infinitesimal, it is impossible to boost v_{avg} by increasing v_f.

3.9.2 Motion of particles in a sawtooth potential

The exact motion of a particle in a sinusoidal potential can be solved using elliptic integrals, but the obtained solution is implicit, i.e., the solution is expressed in the form of time as a function of position. While exact, the implicit nature of this solution obscures the essential physics. In order to shed some light on the underlying physics, we will first consider particle motion in the somewhat artificial situation of the periodic sawtooth-shaped potential shown in Fig. 3.18 and then later will consider particle motion in a more realistic, but harder to analyze, sinusoidal potential.

A particle in the downward-sloping portion of the sawtooth potential experiences a constant acceleration $+a$ and when in the upward portion experiences a constant acceleration $-a$. The goal here is to determine the average velocity of a group of particles injected with an initial velocity v_0 into the system. Care is required when using the word "average" because this word has two meanings depending on whether one is referring to the average velocity of a single particle or the average velocity of a group of particles. The average velocity of a single particle is defined by Eq. (3.205) whereas the average velocity of a group of particles is defined as the sum of the velocities of all the particles in the group divided by the number of particles in the group.

The average velocity of any given individual particle depends on where the particle was injected. Consider the four particles denoted as A, B, C, and D in Fig. 3.18 as representatives of the various possibilities for injection location. Particle A is injected at a potential maximum, particle C is injected at a potential minimum, particle B is injected half-way on the downslope, and particle D is injected half-way on the upslope.

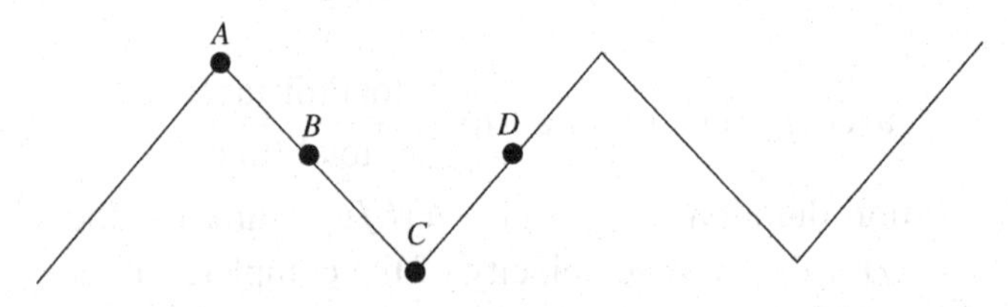

Fig. 3.18 Initial positions of particles A, B, C, and D. All are injected with same initial velocity v_0 moving to the right.

The average velocity for each of these four representative particles will now be evaluated:

Particle A – Let the distance between maximum and minimum potential be d. Let $x=0$ be the location of the minimum so the injection point is at $x=-d$. Thus, the trajectory on the downslope is

$$x(t)=-d+v_0t+at^2/2 \tag{3.207}$$

and the time for particle A to go from its injection point to the potential minimum is found by setting $x(t)=0$ giving

$$t_{down}^A=\frac{v_0}{a}\left(-1+\sqrt{1+2\delta}\right), \tag{3.208}$$

where $\delta=ad/v_0^2$ is the normalized acceleration. When particle A reaches the next potential peak, it again has velocity v_0 and if the time and space origins are reset to be at the new peak, the trajectory will be

$$x(t)=v_0t-at^2/2. \tag{3.209}$$

The negative time when the particle is at the preceding potential minimum is found from

$$-d=v_0t-at^2/2. \tag{3.210}$$

Solving for this negative time and then calculating the time increment to go from the minimum to the maximum shows that this time is the same as going from the maximum to the minimum, i.e., $t_{down}^A=t_{up}^A$. Thus, the average velocity for particle A is

$$v_{avg}^A=\frac{da/v_0}{-1+\sqrt{1+2\delta}}. \tag{3.211}$$

The average velocity of particle A is thus always *faster* than its injection velocity.

Particle C – Now let $x=0$ be the location of maximum potential and $x=-d$ be the point of injection so the particle trajectory is

$$x(t)=-d+v_0t-at^2/2 \tag{3.212}$$

and the time to get to $x=0$ is

$$t_{up}^C=\frac{v_0}{a}\left(1-\sqrt{1-2\delta}\right). \tag{3.213}$$

From symmetry it is seen that the time to go from the maximum to the minimum will be the same so the average velocity will be

$$v_{avg}^{C} = \frac{ad/v_0}{1 - \sqrt{1 - 2\delta}}. \tag{3.214}$$

Because particle C is always on a potential hill relative to its injection position, its average velocity is always *slower* than its injection velocity.

Particles B and D – Particle B can be considered as first traveling in a potential well and then in a potential hill, while the reverse is the case for particle D. For the potential well portion, the forces are the same as for particle A, but the distances are half as much, so the time to traverse the potential well portion is

$$t_{well} = \frac{2v_0}{a}\left(-1 + \sqrt{1+\delta}\right). \tag{3.215}$$

Similarly, the time required to traverse the potential hill portion will be

$$t_{hill} = \frac{2v_0}{a}\left(1 - \sqrt{1-\delta}\right) \tag{3.216}$$

so the average velocity for particles B and D will be

$$v_{avg}^{B,D} = \frac{ad/v_0}{\sqrt{1+\delta} - \sqrt{1-\delta}}. \tag{3.217}$$

These particles move *slower* than the injection velocity, but the effect is second order in δ.

The average velocity of the four particles will be

$$\begin{aligned} v_{avg} &= \frac{1}{4}\left(v_{avg}^{A} + v_{avg}^{B} + v_{avg}^{C} + v_{avg}^{D}\right) \\ &= \frac{ad}{4v_0}\left[\frac{1}{-1+\sqrt{1+2\delta}} + \frac{1}{1-\sqrt{1-2\delta}} + \frac{2}{\sqrt{1+\delta}-\sqrt{1-\delta}}\right]. \end{aligned} \tag{3.218}$$

If δ is small this expression can be approximated as

$$\begin{aligned} v_{avg} &\simeq \frac{ad}{4\delta v_0}\left[\frac{1}{1-\delta/2+\delta^2/2} + \frac{1}{1+\delta/2+\delta^2/2} + \frac{2}{1+\delta^2/8}\right] \\ &= \frac{ad}{2\delta v_0}\left[\frac{1+\delta^2/2}{1+3\delta^2/4} + \frac{1}{1+\delta^2/8}\right] \\ &\simeq v_0\left[1 - 3\delta^2/16\right] \end{aligned} \tag{3.219}$$

so that the average velocity of the four representative particles is *smaller* than the injection velocity. This effect is second order in δ and means that a group of particles injected at random locations with identical velocities into a sawtooth periodic potential will, on average, be slowed down.

3.9.3 Slowing down, energy conservation, and average velocity

The sawtooth potential analysis above shows that it is necessary to be very careful about what is meant by energy and average velocity. Each particle individually conserves energy and regains its injection velocity when it returns to the phase at which it was injected. However, the average velocities of the particles are not the same as the injection velocities. Particle A has an average velocity higher than its injection velocity whereas particles B, C, and D have average velocities smaller than their respective injection velocities. The average velocity of all the particles is less than the injection velocity so that the average kinetic energy of the particles is reduced relative to the injection kinetic energy. Thus the average velocity of a group of particles slows down in a periodic potential, yet paradoxically individual particles do not lose energy. The energy that appears to be missing is contained in the instantaneous potential energy of the individual particles.

3.9.4 Wave–particle energy transfer in a sinusoidal wave

The calculation will now be redone for the physically more relevant situation where a group of particles interacts with a sinusoidal wave. As a prerequisite for doing this calculation it must first be recognized that two distinct classes of particles exist, namely those that are trapped in the wave and those that are not. The trajectories of trapped particles differ in a substantive way from those of untrapped particles, but for low-amplitude waves the number of trapped particles is so small as to be of no consequence. It will therefore be assumed that the wave amplitude is sufficiently small that the trapped particles can be ignored.

Particle energy is conserved in the wave frame but not in the lab frame because the particle Hamiltonian is time-independent in the wave frame but not in the lab frame. Since each additional conserved quantity reduces the number of equations to be solved, it is advantageous to calculate the particle dynamics in the wave frame, and then transform back to the lab frame.

The analysis in Section 3.9.2 of particle motion in a sawtooth potential showed that randomly phased groups of particles have their average velocity slowed down, i.e., the average velocity of the group tends towards zero as observed in the frame of the sawtooth potential. If the sawtooth potential were moving with respect to the lab frame, the sawtooth potential would appear as a propagating wave in the lab frame. A lab-frame observer would see the particle velocities tending to come to rest in the sawtooth frame, i.e., the lab-frame average of the particle velocities would tend to converge towards the velocity with which the sawtooth frame moves in the lab frame.

The quantitative motion of a particle in a one-dimensional wave potential $\phi(x, t) = \phi_0 \cos(kx - \omega t)$ will now be analyzed in some detail. This situation corresponds to a particle being acted on by a wave traveling in the positive x direction with phase velocity ω/k. It is assumed that there is no magnetic field so the equation of motion is simply

$$\frac{\mathrm{d}v}{\mathrm{d}t} = \frac{qk\phi_0}{m}\sin(kx - \omega t). \tag{3.220}$$

At $t = 0$ the particle's position is $x = x_0$ and its velocity is $v = v_0$. The wave phase at the particle location is defined to be $\psi = kx - \omega t$. This is a more convenient variable than x and so the differential equation for x will be transformed into a corresponding differential equation for ψ. Using ψ as the dependent variable corresponds to transforming to the wave frame, i.e., the frame moving with the phase velocity ω/k, and makes it possible to take advantage of the wave-frame energy being a constant of the motion. The equations are less cluttered with minus signs if a slightly modified phase variable $\theta = kx - \omega t - \pi$ is used.

The respective first and second derivatives of θ are

$$\frac{\mathrm{d}\theta}{\mathrm{d}t} = kv - \omega \tag{3.221}$$

and

$$\frac{\mathrm{d}^2\theta}{\mathrm{d}t^2} = k\frac{\mathrm{d}v}{\mathrm{d}t}. \tag{3.222}$$

Substitution of Eq. (3.222) into Eq. (3.220) gives

$$\frac{\mathrm{d}^2\theta}{\mathrm{d}t^2} + \frac{k^2 q\phi_0}{m}\sin\theta = 0. \tag{3.223}$$

By defining the *bounce* frequency

$$\omega_b^2 = \frac{k^2 q\phi_0}{m} \tag{3.224}$$

and the dimensionless bounce-normalized time

$$\tau = \omega_b t, \tag{3.225}$$

Eq. (3.223) reduces to the pendulum-like equation

$$\frac{\mathrm{d}^2\theta}{\mathrm{d}\tau^2} + \sin\theta = 0. \tag{3.226}$$

Upon multiplying by the integrating factor $2\mathrm{d}\theta/\mathrm{d}\tau$, Eq. (3.226) becomes

$$\frac{\mathrm{d}}{\mathrm{d}\tau}\left[\left(\frac{\mathrm{d}\theta}{\mathrm{d}\tau}\right)^2 - 2\cos\theta\right] = 0. \tag{3.227}$$

This integrates to give

$$\left(\frac{\mathrm{d}\theta}{\mathrm{d}\tau}\right)^2 - 2\cos\theta = \lambda = const. \tag{3.228}$$

which indicates the expected energy conservation in the wave frame. The value of λ is determined by two initial conditions, namely the wave-frame injection velocity

$$\left(\frac{\mathrm{d}\theta}{\mathrm{d}\tau}\right)_{\tau=0} = \frac{1}{\omega_b}\left(\frac{\mathrm{d}\theta}{\mathrm{d}t}\right)_{t=0} = \frac{kv_0 - \omega}{\omega_b} \equiv \alpha \tag{3.229}$$

and the wave-frame injection phase

$$\theta_{\tau=0} = kx_0 - \pi \equiv \theta_0. \tag{3.230}$$

Inserting these initial values in the left-hand side of Eq. (3.228) gives

$$\lambda = \alpha^2 - 2\cos\theta_0. \tag{3.231}$$

Except for a constant factor,

- λ is the total wave-frame energy,
- $(\mathrm{d}\theta/\mathrm{d}\tau)^2$ is the wave-frame kinetic energy,
- $-2\cos\theta$ is the wave-frame potential energy.

If $-2 < \lambda < 2$, then the particle is trapped in a specific wave trough and oscillates back and forth in this trough. However, if $\lambda > 2$, the particle is untrapped and travels continuously in the same direction, speeding up when traversing a potential valley and slowing down when traversing a potential hill.

Attention will now be restricted to untrapped particles with kinetic energy greatly exceeding potential energy. For these particles

$$\alpha^2 \gg 2, \tag{3.232}$$

which corresponds to considering small-amplitude waves, since $\alpha \sim \omega_b^{-1}$ and $\omega_b \sim \sqrt{\phi_0}$.

We wish to determine how these untrapped particles exchange energy with the wave. To accomplish this, the lab-frame kinetic energy must be expressed in terms of wave-frame quantities. From Eqs. (3.221) and (3.225) the lab-frame velocity is

$$v = \frac{1}{k}\left(\omega + \frac{\mathrm{d}\theta}{\mathrm{d}t}\right) = \frac{\omega_b}{k}\left(\frac{\omega}{\omega_b} + \frac{\mathrm{d}\theta}{\mathrm{d}\tau}\right) \tag{3.233}$$

so that the lab-frame kinetic energy can be expressed as

$$W = \frac{1}{2}mv^2 = \frac{m\omega_b^2}{2k^2}\left[\left(\frac{\omega}{\omega_b}\right)^2 + 2\frac{\omega}{\omega_b}\frac{\mathrm{d}\theta}{\mathrm{d}\tau} + \left(\frac{\mathrm{d}\theta}{\mathrm{d}\tau}\right)^2\right]. \tag{3.234}$$

Substituting for $(\mathrm{d}\theta/\mathrm{d}\tau)^2$ using Eq. (3.228) gives

$$W = \frac{m\omega_b^2}{2k^2}\left[\left(\frac{\omega}{\omega_b}\right)^2 + 2\frac{\omega}{\omega_b}\frac{\mathrm{d}\theta}{\mathrm{d}\tau} + \lambda + 2\cos\theta\right]. \tag{3.235}$$

Since wave–particle energy transfer is of interest, attention is now focused on the *changes* in the lab-frame particle kinetic energy and so we consider

$$\begin{aligned}\frac{\mathrm{d}W}{\mathrm{d}t} &= \frac{m\omega_b^3}{k^2}\left[\frac{\omega}{\omega_b}\frac{\mathrm{d}^2\theta}{\mathrm{d}\tau^2} - \frac{\mathrm{d}\theta}{\mathrm{d}\tau}\sin\theta\right] \\ &= -\frac{m\omega_b^3}{k^2}\sin\theta\left[\frac{\omega}{\omega_b} + \frac{\mathrm{d}\theta}{\mathrm{d}\tau}\right],\end{aligned} \tag{3.236}$$

where Eq. (3.226) has been used. To proceed further, it is necessary to obtain the time dependence of both $\sin\theta$ and $\mathrm{d}\theta/\mathrm{d}t$.

Solving Eq. (3.228) for $\mathrm{d}\theta/\mathrm{d}\tau$ and assuming $\alpha \gg 1$ (corresponding to untrapped particles) gives

$$\begin{aligned}\frac{\mathrm{d}\theta}{\mathrm{d}\tau} &= \pm\sqrt{\lambda + 2\cos\theta} \\ &= \pm\sqrt{\alpha^2 + 2(\cos\theta - \cos\theta_0)} \\ &= \alpha\left(1 + \frac{2(\cos\theta - \cos\theta_0)}{\alpha^2}\right)^{1/2} \\ &\simeq \alpha + \frac{\cos\theta - \cos\theta_0}{\alpha}.\end{aligned} \tag{3.237}$$

This expression is valid for both positive and negative α, i.e., for particles going in either direction in the wave frame. The first term in the last line of Eq. (3.237) gives the velocity the particle would have if there were *no* wave (unperturbed orbit) while the second term gives the *perturbation* due to the small-amplitude wave. The particle orbit $\theta(\tau)$ is now solved iteratively. To lowest order (i.e., dropping terms of order α^{-2}) the particle velocity is

$$\frac{\mathrm{d}\theta}{\mathrm{d}\tau} = \alpha \tag{3.238}$$

and so the rate at which energy is transferred from the wave to the particles is

$$\begin{aligned}\frac{\mathrm{d}W}{\mathrm{d}t} &= -\frac{m\omega_b^3}{k^2}\sin\theta\left[\frac{\omega}{\omega_b} + \alpha\right] \\ &\simeq -\frac{m\omega_b^2 v_0}{k}\sin\theta.\end{aligned} \tag{3.239}$$

Integration of Eq. (3.238) gives the unperturbed orbit solution

$$\theta(\tau) = \theta_0 + \alpha\tau. \tag{3.240}$$

This first approximation is then substituted back into Eq. (3.237) to get the corrected form

$$\frac{d\theta}{d\tau} = \alpha + \frac{\cos(\theta_0 + \alpha\tau) - \cos\theta_0}{\alpha}, \tag{3.241}$$

which may be integrated to give the corrected phase

$$\theta(\tau) = \theta_0 + \alpha\tau + \frac{\sin(\theta_0 + \alpha\tau) - \sin\theta_0}{\alpha^2} - \frac{\tau}{\alpha}\cos\theta_0. \tag{3.242}$$

From Eq. (3.242) we may write

$$\sin\theta = \sin[(\theta_0 + \alpha\tau) + \Delta(\tau)], \tag{3.243}$$

where

$$\Delta(\tau) = \frac{\sin(\theta_0 + \alpha\tau) - \sin\theta_0}{\alpha^2} - \frac{\tau}{\alpha}\cos\theta_0 \tag{3.244}$$

is the "perturbed-orbit" correction to the phase. If consideration is restricted to times where $\tau \ll |\alpha|$, the phase correction $\Delta(\tau)$ will be small. This restriction corresponds to

$$(\omega_b t)^2 \ll |kv_0 - \omega| t, \tag{3.245}$$

which means that the number of wave peaks the particle passes greatly exceeds the number of bounce times. Since bounce frequency is proportional to wave amplitude, this condition will be satisfied for all finite times for an infinitesimal amplitude wave. Because Δ is assumed to be small, Eq. (3.243) may be expanded as

$$\sin\theta = \sin(\theta_0 + \alpha\tau)\cos\Delta + \sin\Delta\cos(\theta_0 + \alpha\tau) \simeq \sin(\theta_0 + \alpha\tau) + \Delta\cos(\theta_0 + \alpha\tau) \tag{3.246}$$

so that Eq. (3.239) becomes

$$\frac{dW}{dt} = -\frac{m\omega_b^2 v_0}{k}\left[\sin(\theta_0 + \alpha\tau) + \Delta\cos(\theta_0 + \alpha\tau)\right]. \tag{3.247}$$

The wave-to-particle energy transfer rate depends on the particle initial position. This is analogous to the earlier sawtooth potential analysis where it was shown that whether a particle gains or loses average velocity depends on its injection phase. It is now assumed that there exist many particles with *evenly spaced* initial positions and then an averaging will be performed over all these particles, which corresponds to averaging over all initial injection phases. Denoting such averaging by $\langle\rangle$ gives

$$\left\langle\frac{\mathrm{d}W}{\mathrm{d}t}\right\rangle = -\frac{m\omega_b^2 v_0}{k}\langle\Delta\cos(\theta_0+\alpha\tau)\rangle$$
$$= -\frac{m\omega_b^2 v_0}{k}\left\langle\left[\frac{\sin(\theta_0+\alpha\tau)-\sin\theta_0}{\alpha^2}-\frac{\tau}{\alpha}\cos\theta_0\right]\cos(\theta_0+\alpha\tau)\right\rangle. \quad (3.248)$$

Using the identities

$$\langle\sin(\theta_0+\alpha\tau)\cos(\theta_0+\alpha\tau)\rangle = 0$$
$$\langle\sin\theta_0\cos(\theta_0+\alpha\tau)\rangle = -\frac{1}{2}\sin\alpha\tau$$
$$\langle\cos\theta_0\cos(\theta_0+\alpha\tau)\rangle = \frac{1}{2}\cos\alpha\tau \quad (3.249)$$

the wave-to-particle energy transfer rate becomes

$$\left\langle\frac{\mathrm{d}W}{\mathrm{d}t}\right\rangle = -\frac{m\omega_b^2 v_0}{2k}\left(\frac{\sin\alpha\tau}{\alpha^2}-\frac{\tau}{\alpha}\cos\alpha\tau\right) = \frac{m\omega_b^2 v_0}{2k}\frac{\mathrm{d}}{\mathrm{d}\alpha}\left(\frac{\sin\alpha\tau}{\alpha}\right). \quad (3.250)$$

At this point it is recalled that one representation for a delta function is

$$\delta(z) = \lim_{N\to\infty}\frac{\sin(Nz)}{\pi z} \quad (3.251)$$

so that for $|\alpha\tau| \gg 1$, Eq. (3.250) becomes

$$\left\langle\frac{\mathrm{d}W}{\mathrm{d}t}\right\rangle = \frac{\pi m\omega_b^2 v_0}{2k}\frac{\mathrm{d}}{\mathrm{d}\alpha}\delta(\alpha). \quad (3.252)$$

Since $\delta(z)$ has an infinite positive slope just to the left of $z=0$ and an infinite negative slope just to the right of $z=0$, the derivative of the delta function consists of a positive spike just to the left of $z=0$ and a negative spike just to the right of $z=0$. Furthermore, $\alpha=(kv_0-\omega)/\omega_b$ is slightly positive for particles moving a little faster than the wave-phase velocity and slightly negative for particles moving a little slower. Thus $\langle\mathrm{d}W/\mathrm{d}t\rangle$ is large and positive for particles moving slightly slower than the wave, while it is large and negative for particles moving slightly faster. If the number of particles moving slightly slower than the wave equals the number moving slightly faster, the energy gained by the slightly slower particles is equal and opposite to that gained by the slightly faster particles.

However, if the number of slightly slower particles *differs* from the number of slightly faster particles, there will be a net transfer of energy from wave to particles or vice versa. Specifically, if there are more slow particles than fast particles, there will be a transfer of energy to the particles. This energy must come from the wave and a more complete analysis (cf. Section 5.3) will show that the wave *damps*. The direction of energy transfer depends critically on the slope of the distribution function in the vicinity of $v = \omega/k$, since this slope determines the ratio of slightly faster to slightly slower particles.

We now consider a large number of particles with an initial one-dimensional distribution function $f(v_0)$ and calculate the net wave-to-particle energy transfer rate averaged over all particles. Since $f(v_0)\mathrm{d}v_0$ is the probability that a particle had its initial velocity between v_0 and $v_0 + \mathrm{d}v_0$, the energy transfer rate averaged over all particles is

$$
\begin{aligned}
\left\langle \frac{\mathrm{d}W_{\mathrm{total}}}{\mathrm{d}t} \right\rangle &= \int \mathrm{d}v_0 f(v_0) \frac{\pi m \omega_b^2 v_0}{2k} \frac{\mathrm{d}}{\mathrm{d}\alpha}\delta(\alpha) \\
&= \frac{\pi m \omega_b^3}{2k^2} \int \mathrm{d}v_0 f(v_0) v_0 \frac{\mathrm{d}}{\mathrm{d}v_0}\delta\left(\frac{kv_0 - \omega}{\omega_b}\right) \\
&= \frac{\pi m \omega_b^4}{2k^3} \int \mathrm{d}v_0 f(v_0) v_0 \frac{\mathrm{d}}{\mathrm{d}v_0}\delta\left(v_0 - \frac{\omega}{k}\right) \\
&= -\frac{\pi m \omega_b^4}{2k^3} \left[\frac{\mathrm{d}}{\mathrm{d}v_0}\left(f(v_0)v_0\right)\right]_{v_0=\omega/k}.
\end{aligned}
\tag{3.253}
$$

If the distribution function has the Maxwellian form $f \sim \exp(-v_0^2/v_T^2)$, where v_T is the thermal velocity, and if $\omega/k \gg v_T$ then

$$
\begin{aligned}
\left[\frac{\mathrm{d}}{\mathrm{d}v_0}\left(f(v_0)v_0\right)\right]_{v_0=\omega/k} &= \left[v_0 \frac{\mathrm{d}}{\mathrm{d}v_0}\left(f(v_0)\right) + f(v_0)\right]_{v_0=\omega/k} \\
&= \left[-2\frac{v_0^2}{v_T^2} f(v_0) + f(v_0)\right]_{v_0=\omega/k}
\end{aligned}
\tag{3.254}
$$

showing that the derivative of f is the dominant term. Hence, Eq. (3.253) becomes

$$
\left\langle \frac{\mathrm{d}W_{\mathrm{total}}}{\mathrm{d}t} \right\rangle = -\frac{\pi m \omega_b^4 \omega}{2k^4} \left[\frac{\mathrm{d}}{\mathrm{d}v_0}\left(f(v_0)\right)\right]_{v_0=\omega/k}. \tag{3.255}
$$

Substituting for the bounce frequency using Eq. (3.224) this becomes

$$
\left\langle \frac{\mathrm{d}W_{\mathrm{total}}}{\mathrm{d}t} \right\rangle = -\frac{\pi m \omega}{2k^2} \left(\frac{qk\phi_0}{m}\right)^2 \left[\frac{\mathrm{d}}{\mathrm{d}v_0}\left(f(v_0)\right)\right]_{v_0=\omega/k}. \tag{3.256}
$$

Thus, particles *gain* kinetic energy at the expense of the wave if the distribution function has negative slope in the range $v \sim \omega/k$. This process is called Landau damping and will be examined in Section 5.3 from the wave viewpoint.

3.10 Assignments

1. A charged particle starts from *rest* in combined static fields $\mathbf{E} = E\hat{y}$ and $\mathbf{B} = B\hat{z}$, where $E/B \ll c$ and c is the speed of light. Calculate and plot its exact trajectory (do this both analytically and numerically).
2. Calculate (qualitatively and numerically) the trajectory of a particle starting from rest at $x = 0,\ y = 5a$ in combined E and B fields, where $\mathbf{E} = E_0\hat{x}$ and $\mathbf{B} = \hat{z}B_0 y/a$. What happens to μ conservation on the line $y = 0$? Sketch the motion showing both the Larmor motion and the guiding center motion. Explain the particle motion in terms of $E \times B$, ∇B, and polarization drifts and note that the particle can never move into the $x < 0$ region because of energy conservation considerations.
3. Calculate the motion of a particle in the steady-state electric field produced by a line charge λ along the z axis and a steady-state magnetic field $\mathbf{B} = B_0\hat{z}$. Obtain an approximate solution using drift theory and also obtain a solution using Hamilton–Lagrange theory. Hint: for the drift theory show that the electric field has the form $\mathbf{E} = \hat{r}\lambda/2\pi r$. Assume that λ is small for approximate solutions.
4. Consider the magnetic field produced by a toroidal coil system; this coil consists of a single wire threading the hole of a torus (donut) N times with the N turns evenly arranged around the circumference of the torus. Use Ampère's law to show that the magnetic field is in the toroidal direction and has the form $B = \mu_0 NI/2\pi r$, where N is the total number of turns in the coil and I is the current through the turn. What are the drifts for a particle having finite initial velocities both parallel and perpendicular to this toroidal field?
5. Show that of all the standard drifts ($\mathbf{E} \times \mathbf{B}$, ∇B, curvature, polarization) only the polarization drift causes a change in the particle energy. Hint: consider what happens when the following equation is dotted with $\mathbf{v}$:

$$m\frac{\mathrm{d}\mathbf{v}}{\mathrm{d}t} = \mathbf{F} + \mathbf{v} \times \mathbf{B}.$$

6. Use the numerical Lorentz solver to calculate the motion of a charged particle in a uniform magnetic field $\mathbf{B} = B\hat{z}$ and an electric field given by Eq. (3.178). Compare the motion to the predictions of drift theory ($\mathbf{E} \times \mathbf{B}$, polarization). Describe the motion for cases where $\alpha \ll 1$, $\alpha \simeq 1$, and $\alpha \gg 1$, where $\alpha = mk^2\phi/qB^2$. Describe what happens when α becomes of order unity.
7. A "magnetic mirror" field in cylindrical coordinates r, θ, z can be expressed as $\mathbf{B} = (2\pi)^{-1}\nabla\psi \times \nabla\theta$, where $\psi = B_0\pi r^2(1 + (z/L)^2)$ and L is a characteristic length. Sketch by hand the field line pattern in the r, z plane and write out the components of $\mathbf{B}$. What are appropriate characteristic lengths, times, and velocities for an electron in this configuration? Use $r = (x^2 + y^2)^{1/2}$ and numerically integrate the orbit of an electron starting at $x = 0,\ y = L,\ z = 0$ with initial velocity $v_x = 0$ and initial v_y, v_z of the order of the characteristic velocity (try different values). Simultaneously plot the motion in the $z - y$ plane and in the $x - y$ plane. What interesting phenomena can be observed (e.g., reflection)? Does the electron stay on a constant ψ contour?

8. Consider the motion of a charged particle in the magnetic field

$$\mathbf{B} = \frac{1}{2\pi} \nabla\psi(r, z, t) \times \nabla\theta,$$

where

$$\psi(r, z, t) = B_{\min} \pi r^2 \left[1 + 2\lambda \frac{\zeta^2}{\zeta^4 + 1} \right]$$

and

$$\zeta = \frac{z}{L(t)}.$$

Show by explicit evaluation of the flux derivatives and also by plotting contours of constant flux that this is an example of a magnetic mirror field with minimum axial field $B_{\min}$ when $z = 0$ and maximum axial field $(1+\lambda)B_{\min}$ at $z = L(t)$. By making $L(t)$ a slowly decreasing function of time show that the magnetic mirrors slowly move together. Using numerical techniques to integrate the equation of motion, demonstrate Fermi acceleration of a particle when the mirrors move slowly together. Do not forget the electric field associated with the time-changing magnetic field (this electric field is closely related to the time derivative of $\psi(r, z, t)$; use Faraday's law). Plot the velocity-space angle at $z = 0$ for each bounce between mirrors and show that the particle becomes detrapped when this angle decreases below $\theta_{trap} = \sin^{-1}\left(1/\sqrt{1+\lambda}\right)$.

9. Consider a point particle bouncing with nominal velocity v between a stationary wall and a second wall that is approaching the first wall with speed u. Calculate the change in speed of the particle after it bounces from the moving wall (hint: do this first in the frame of the moving wall, and then translate back to the lab frame). Calculate τ_b, the time for the particle to make one complete bounce between the walls if the nominal distance between walls is L. Calculate ΔL, the change in L during one complete bounce and show that if $u \ll v$, then Lv is a conserved quantity. By considering collisionless particles bouncing in a cube that is slowly shrinking self-similarly in three dimensions show that $PV^{5/3}$ is constant where $P = n\kappa T$, n is the density of the particles, and T is the average kinetic energy of the particles. What happens if the shrinking is not self-similar (hint: consider the effect of collisions and see discussion in Bellan (2004a)).

10. Using numerical techniques to integrate the equation of motion, illustrate how a charged particle changes from being non-axis-encircling to axis-encircling when a magnetic field $\mathbf{B} = (2\pi)^{-1}\nabla\psi(r, z, t) \times \nabla\theta$ reverses polarity at $t = 0$. For simplicity use $\psi = B(t)\pi r^2$, i.e., a uniform magnetic field. To make the solution as general as possible, normalize time to the cyclotron frequency by defining $\tau = \omega_c t$, and set $B(\tau) = \tanh\tau$ to represent a polarity-reversing field. Normalize lengths to some reference length L and normalize velocities to $\omega_c L$. Show that the canonical angular momentum is conserved. Hint: do not forget to include the inductive electric field associated with the time-dependent magnetic field.

11. Consider a cusp magnetic field given by $\mathbf{B} = (2\pi)^{-1}\nabla\psi(r,z) \times \nabla\theta$, where the flux function

$$\psi(r,z) = B\pi r^2 \frac{z}{\sqrt{1+z^2/a^2}}$$

is antisymmetric in z. Plot the surfaces of constant flux. Using numerical techniques to integrate the equation of motion, demonstrate that a particle incident at $z \ll -a$ and $r = r_0$ with incident velocity $\mathbf{v} = v_{z0}\hat{z}$ will reflect from the cusp if $v_{z0} < r_0\omega_c$, where $\omega_c = qB/m$.

12. Consider the motion of a charged particle starting from rest in a simple one-dimensional electrostatic wave field,

$$m\frac{\mathrm{d}^2 x}{\mathrm{d}t^2} = -q\nabla\phi(x,t),$$

where $\phi(x,t) = \bar{\phi}\cos(kx - \omega t)$. How large does $\bar{\phi}$ have to be to give trapping of particles that start from rest? Demonstrate this trapping threshold numerically.

13. Prove Equation (3.219).

14. Prove that

$$\delta(z) = \lim_{N\to\infty} \frac{\sin(Nz)}{\pi z}$$

is a valid representation for the delta function.

15. As sketched in Fig. 3.19, a current loop (radius r, current I) is located in the $x-y$ plane; the loop's axis defines the z axis of the coordinate system, so that the center of the loop is at the origin. The loop is immersed in a non-uniform magnetic field $\mathbf{B}$ produced by external coils and oriented so that the magnetic-field lines converge symmetrically about the z axis. The current I is small and does not significantly modify $\mathbf{B}$. Consider the following three circles: the current loop, a circle of radius b coaxial with the loop but with center at $z = -L/2$, and a circle of radius a with

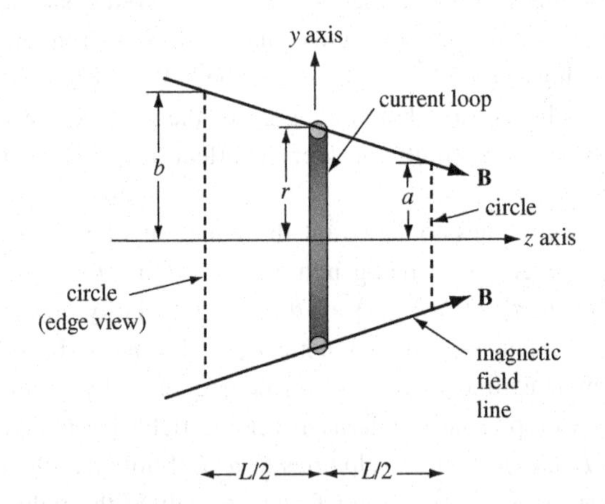

Fig. 3.19 Non-uniform magnetic field acting on a current loop.

center at $z = +L/2$. The radii a and b are chosen so as to intercept the field lines that intercept the current loop (see figure). Assume the figure is somewhat exaggerated so that B_z is approximately uniform over each of the three circular surfaces and so one may ignore the radial dependence of B_z and therefore express $B_z = B_z(z)$.

(a) Note that $r = (a+b)/2$. What is the force (magnitude and direction) on the current loop expressed in terms of I, $B_z(0)$, a, b, and L only? (Hint: use the field line slope to give a relationship between B_r and B_z at the loop radius.)
(b) For each of the circles and the current loop, express the magnetic flux enclosed in terms of B_z at the respective entity and the radius of the entity. What is the relationship between the B_z's at these three entities?
(c) By combining the results of parts (a) and (b) above and taking the limit $L \to 0$, show that the force on the loop can be expressed in terms of a derivative of B_z.

4

Elementary plasma waves

4.1 General method for analyzing small-amplitude waves

All plasma phenomena can be described by combining Maxwell's equations with the Lorentz force equation where the latter is represented by the Vlasov, the two-fluid, or the MHD approximation. The subject of linear plasma waves provides a good introduction to the study of plasma phenomena because linear waves are relatively simple to analyze and yet demonstrate many essential features of plasma behavior.

Linear analysis, a straightforward method applicable to any set of partial differential equations describing a physical system, reveals the physical system's simplest non-trivial, self-consistent dynamical behavior. In the context of plasma dynamics, the method is as follows:

1. By making appropriate physical assumptions, the general Maxwell–Lorentz system of equations is reduced to the simplest set of equations characterizing the phenomena under consideration.
2. An equilibrium solution is determined for this set of equations. The equilibrium might be trivial such that densities are uniform, the plasma is neutral, and all velocities are zero. However, less trivial equilibria could also be invoked where there are density gradients or flow velocities. Equilibrium quantities are designated by the subscript 0, indicating "zero-order" in smallness.
3. If $\{f(\mathbf{x}, t), g(\mathbf{x}, t), h(\mathbf{x}, t), \ldots\}$ constitutes the set of dependent variables and a specific, physically allowed perturbation is prescribed for one of these variables, then solving the system of differential equations will give the responses of all the other dependent variables to this prescribed perturbation. For example, suppose that a perturbation ϵf_1 is prescribed for the dependent variable f where $\epsilon \ll 1$ so that f becomes

$$f = f_0 + \epsilon f_1 \tag{4.1}$$

and $\epsilon\,|f_1| \ll |f_0|$ for all $\mathbf{x}$ and t. The system of differential equations gives the functional dependence of the other variables on f and so, for example, would give $g = g(f) =$

$g(f_0+\epsilon f_1)$. Since the functional dependence of g on f is in general nonlinear, Taylor expansion gives $g = g_0+\epsilon g_1+\epsilon^2 g_2+\epsilon^3 g_3+\ldots$ The ϵ's are, from now on, considered implicit and so the variables are written as

$$\begin{aligned} f &= f_0+f_1 \\ g &= g_0+g_1+g_2+\ldots \\ h &= h_0+h_1+h_2+\ldots \end{aligned} \tag{4.2}$$

and it is assumed that the order of magnitude of f_1 is smaller than the order of magnitude of f_0 by a factor ϵ, etc. The smallness of the perturbation is an assumption that obviously must be satisfied in the real situation being modeled. Note that there are no f_j terms with $j \geq 2$ because the perturbation to f was prescribed as being f_1.

4. Each partial differential equation is rewritten with all dependent quantities expanded to first order as in Eq. (4.2). For example, the two-fluid continuity equation becomes

$$\frac{\partial(n_0+n_1)}{\partial t}+\nabla\cdot[(n_0+n_1)(\mathbf{u}_0+\mathbf{u}_1)] = 0. \tag{4.3}$$

By assumption, equilibrium quantities satisfy

$$\frac{\partial n_0}{\partial t}+\nabla\cdot(n_0\mathbf{u}_0) = 0. \tag{4.4}$$

The essence of linearization consists of subtracting the equilibrium equation (e.g., Eq. (4.4)) from the expanded equation (e.g., Eq. (4.3)). For this example such a subtraction yields

$$\frac{\partial n_1}{\partial t}+\nabla\cdot(n_1\mathbf{u}_0+n_0\mathbf{u}_1+n_1\mathbf{u}_1) = 0. \tag{4.5}$$

The nonlinear term $n_1\mathbf{u}_1$, which is a product of *two* first order quantities, is discarded because it is of order ϵ^2 whereas all the other terms are of order ϵ. What remains is called the linearized equation, i.e., the equation that consists of only first-order terms. For the example here, the linearized equation would be

$$\frac{\partial n_1}{\partial t}+\nabla\cdot(n_1\mathbf{u}_0+n_0\mathbf{u}_1) = 0. \tag{4.6}$$

The linearized equation is in a sense the differential of the original equation.

Before engaging in a methodical study of the large variety of waves that can propagate in a plasma, a few special cases of fundamental importance will first be examined.

4.2 Two-fluid theory of unmagnetized plasma waves

The simplest plasma waves are those described by two-fluid theory in an unmagnetized plasma, i.e., a plasma that has no equilibrium magnetic field. The theory for these waves also applies to magnetized plasmas in the special situation where all fluid motions are strictly parallel to the equilibrium magnetic field because fluid flowing along a magnetic field experiences no $\mathbf{u}\times\mathbf{B}$ force and so behaves as if there were no magnetic field.

The two-fluid equation of motion relevant to an unmagnetized plasma is

$$m_\sigma n_\sigma \frac{\mathrm{d}\mathbf{u}_\sigma}{\mathrm{d}t} = q_\sigma n_\sigma \mathbf{E} - \nabla P_\sigma \tag{4.7}$$

and these simple plasma waves are found by linearizing about an equilibrium where $\mathbf{u}_{\sigma 0} = 0$, $\mathbf{E}_0 = 0$, and $\nabla P_{\sigma 0} = 0$. The linearized form of Eq. (4.7) is then

$$m_\sigma n_{\sigma 0} \frac{\partial \mathbf{u}_{\sigma 1}}{\partial t} = q_\sigma n_{\sigma 0} \mathbf{E}_1 - \nabla P_{\sigma 1}. \tag{4.8}$$

The electric field can be expressed as

$$\mathbf{E} = -\nabla\phi - \frac{\partial \mathbf{A}}{\partial t}, \tag{4.9}$$

a form that automatically satisfies Faraday's law. The vector potential $\mathbf{A}$ is undefined with respect to a gauge since $\mathbf{B} = \nabla\times(\mathbf{A}+\nabla\psi) = \nabla\times\mathbf{A}$. It is convenient to choose ψ so as to have $\nabla\cdot\mathbf{A} = 0$. This is called Coulomb gauge and causes the divergence of Eq. (4.9) to give Poisson's equation so that charge density provides the only source term for the electrostatic potential ϕ. Since Eq. (4.9) is linear to begin with, its linearized form is just

$$\mathbf{E}_1 = -\nabla\phi_1 - \frac{\partial \mathbf{A}_1}{\partial t}. \tag{4.10}$$

4.2.1 Electrostatic (compressional) waves

These waves are characterized by having finite $\nabla\cdot\mathbf{u}_1$ and are variously called compressional, electrostatic, or longitudinal waves. The first step in the analysis is to take the divergence of Eq. (4.8) to obtain

$$m_\sigma n_{\sigma 0} \frac{\partial \nabla\cdot\mathbf{u}_{\sigma 1}}{\partial t} = -q_\sigma n_{\sigma 0}\nabla^2\phi_1 - \nabla^2 P_{\sigma 1}. \tag{4.11}$$

Because Eq. (4.11) involves three variables (i.e., $\mathbf{u}_{\sigma 1}$, ϕ_1, $P_{\sigma 1}$) two more equations are required to provide a complete description. One of these additional equations is the linearized continuity equation

$$\frac{\partial n_{\sigma 1}}{\partial t} + n_0 \nabla\cdot\mathbf{u}_{\sigma 1} = 0, \tag{4.12}$$

which, after substitution into Eq. (4.11), gives

$$m_\sigma \frac{\partial^2 n_{\sigma 1}}{\partial t^2} = q_\sigma n_{\sigma 0}\nabla^2\phi_1 + \nabla^2 P_{\sigma 1}. \tag{4.13}$$

For adiabatic processes the pressure and density are related by

$$\frac{P_\sigma}{n_\sigma^\gamma} = const., \tag{4.14}$$

where $\gamma = (N+2)/N$ and N is the dimensionality of the system, whereas for isothermal processes

$$\frac{P_\sigma}{n_\sigma} = const. \tag{4.15}$$

The same formalism can therefore be used for both isothermal and adiabatic processes by using Eq. (4.14) for both and then simply setting $\gamma = 1$ if the process is isothermal. Linearization of Eq. (4.14) gives

$$\frac{P_{\sigma 1}}{P_{\sigma 0}} = \gamma \frac{n_{\sigma 1}}{n_{\sigma 0}} \tag{4.16}$$

so Eq. (4.13) becomes

$$m_\sigma \frac{\partial^2 n_{\sigma 1}}{\partial t^2} = q_\sigma n_{\sigma 0} \nabla^2 \phi_1 + \gamma \kappa T_{\sigma 0} \nabla^2 n_{\sigma 1}, \tag{4.17}$$

where $P_{\sigma 0} = n_{\sigma 0} \kappa T_{\sigma 0}$ has been used.

Although this system of linear equations could be solved by the formal method of Fourier transforms, we instead take the shortcut of simply assuming that the linear perturbation happens to be a single Fourier mode. Thus, it is assumed that *all* linearized dependent variables have the wave-like dependence

$$n_{\sigma 1} \sim \exp(i\mathbf{k}\cdot\mathbf{x} - i\omega t), \quad \phi_1 \sim \exp(i\mathbf{k}\cdot\mathbf{x} - i\omega t), \quad \text{etc.} \tag{4.18}$$

so that $\nabla \to i\mathbf{k}$ and $\partial/\partial t \to -i\omega$. Equation (4.17) therefore reduces to the algebraic equation

$$m_\sigma \omega^2 n_{\sigma 1} = q_\sigma n_{\sigma 0} k^2 \phi_1 + \gamma \kappa T_{\sigma 0} k^2 n_{\sigma 1}, \tag{4.19}$$

which may be solved for $n_{\sigma 1}$ to give

$$n_{\sigma 1} = \frac{q_\sigma n_{\sigma 0}}{m_\sigma} \frac{k^2 \phi_1}{(\omega^2 - \gamma k^2 \kappa T_{\sigma 0}/m_\sigma)}. \tag{4.20}$$

Poisson's equation provides another relation between ϕ_1 and $n_{\sigma 1}$, namely

$$k^2 \phi_1 = \frac{1}{\varepsilon_0} \sum_\sigma n_{\sigma 1} q_\sigma. \tag{4.21}$$

Equation (4.20) is substituted into Poisson's equation to give

$$k^2 \phi_1 = \sum_\sigma \frac{n_{\sigma 0} q_\sigma^2}{\varepsilon_0 m_\sigma} \frac{k^2 \phi_1}{(\omega^2 - \gamma k^2 \kappa T_{\sigma 0}/m_\sigma)}, \tag{4.22}$$

which may be rearranged as

$$\left[1-\sum_{\sigma}\frac{\omega_{p\sigma}^{2}}{(\omega^{2}-\gamma k^{2}\kappa T_{\sigma 0}/m_{\sigma})}\right]\phi_{1}=0, \tag{4.23}$$

where

$$\omega_{p\sigma}^{2}\equiv\frac{n_{\sigma 0}q_{\sigma}^{2}}{\varepsilon_{0}m_{\sigma}} \tag{4.24}$$

is the square of the *plasma frequency* of species σ. A useful way to recast Eq. (4.23) is

$$(1+\chi_{e}+\chi_{i})\phi_{1}=0, \tag{4.25}$$

where

$$\chi_{\sigma}=-\frac{\omega_{p\sigma}^{2}}{(\omega^{2}-\gamma k^{2}\kappa T_{\sigma 0}/m_{\sigma})} \tag{4.26}$$

is called the susceptibility of species σ. In Eq. (4.25) the "1" comes from the "vacuum" part of Poisson's equation (i.e., the left-hand side term $\nabla^{2}\phi$) while the susceptibilities give the respective contributions of each species to the right-hand side of Poisson's equation. This formalism follows that of dielectrics where the displacement vector is $\mathbf{D}=\varepsilon\mathbf{E}$ and the dielectric constant is $\varepsilon=1+\chi$, where χ is a susceptibility.

Equation (4.25) shows that if $\phi_{1}\neq 0$, the quantity $1+\chi_{e}+\chi_{i}$ must vanish. In other words, in order to have a non-trivial normal mode it is necessary to have

$$1+\chi_{e}+\chi_{i}=0. \tag{4.27}$$

This is called a dispersion relation and prescribes a functional relation between ω and k. The dispersion relation can be considered as the determinant-like equation for the eigenvalues $\omega(k)$ of the system of equations.

The normal modes can be identified by noting that Eq. (4.26) has two limiting behaviors depending on how the wave-phase velocity compares to $\sqrt{\kappa T_{\sigma 0}/m_{\sigma}}$, a quantity that is of the order of the thermal velocity. These limiting behaviors are:

1. Adiabatic regime: $\omega/k\gg\sqrt{\kappa T_{\sigma 0}/m_{\sigma}}$ and $\gamma=(N+2)/N$. Because plane waves are one-dimensional perturbations (i.e., the plasma is compressed in the $\hat{k}$ direction only), $N=1$ so that $\gamma=3$. Hence the susceptibility has the limiting form

$$\begin{aligned}\chi_{\sigma}&=-\frac{\omega_{p\sigma}^{2}}{\omega^{2}(1-\gamma k^{2}\kappa T_{\sigma 0}/m_{\sigma}\omega^{2})}\\&\simeq-\frac{\omega_{p\sigma}^{2}}{\omega^{2}}\left(1+3\frac{k^{2}}{\omega^{2}}\frac{\kappa T_{\sigma 0}}{m_{\sigma}}\right)\\&=-\frac{1}{k^{2}\lambda_{D\sigma}^{2}}\frac{k^{2}}{\omega^{2}}\frac{\kappa T_{\sigma}}{m_{\sigma}}\left(1+3\frac{k^{2}}{\omega^{2}}\frac{\kappa T_{\sigma 0}}{m_{\sigma}}\right).\end{aligned} \tag{4.28}$$

2. Isothermal regime: $\omega/k \ll \sqrt{\kappa T_{\sigma 0}/m_\sigma}$ and $\gamma = 1$. Here the susceptibility has the limiting form

$$\chi_\sigma = \frac{\omega_{p\sigma}^2}{k^2\kappa T_{\sigma 0}/m_\sigma} = \frac{1}{k^2\lambda_{D\sigma}^2}. \tag{4.29}$$

Figure 4.1 shows a plot of χ_σ versus $\omega/k\sqrt{\kappa T_{\sigma 0}/m_\sigma}$. The isothermal and adiabatic susceptibilities are seen to be substantially different and, in particular, do not coalesce when $\omega/k\sqrt{\kappa T_{\sigma 0}/m_\sigma} \to 1$. This non-coalescence as $\omega/k\sqrt{\kappa T_{\sigma 0}/m_\sigma} \to 1$ indicates that the fluid description, while valid in both the adiabatic and isothermal limits, fails in the vicinity of $\omega/k \sim \sqrt{\kappa T_{\sigma 0}/m_\sigma}$. As will be seen later, the more accurate Vlasov description must be used in the $\omega/k \sim \sqrt{\kappa T_{\sigma 0}/m_\sigma}$ regime.

Since the ion-to-electron mass ratio is large, ions and electrons typically have thermal velocities differing by at least one and sometimes two orders of magnitude. Furthermore, ion and electron temperatures often differ, again allowing substantially different electron and ion thermal velocities. Three different situations can occur in a typical plasma depending on how the wave-phase velocity compares to thermal velocities. These situations are:

1. Case where $\omega/k \gg \sqrt{\kappa T_{e0}/m_e}, \sqrt{\kappa T_{i0}/m_i}$.
   Here both electrons and ions are adiabatic and the dispersion relation becomes

$$1 - \frac{\omega_{pe}^2}{\omega^2}\left(1 + 3\frac{k^2}{\omega^2}\frac{\kappa T_{e0}}{m_e}\right) - \frac{\omega_{pi}^2}{\omega^2}\left(1 + 3\frac{k^2}{\omega^2}\frac{\kappa T_{i0}}{m_i}\right) = 0. \tag{4.30}$$

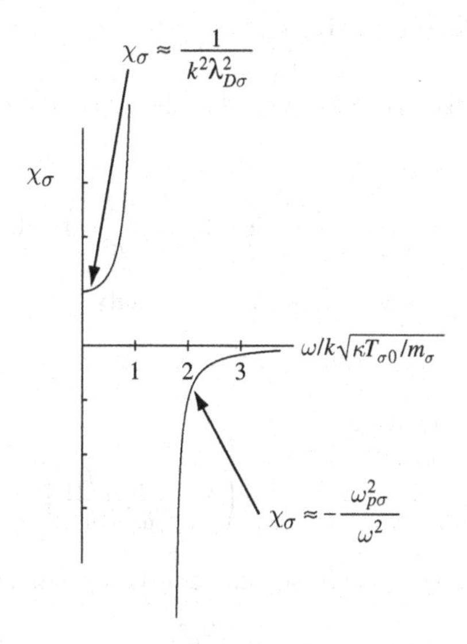

Fig. 4.1 Susceptibility χ_σ as a function of $\omega/k\sqrt{\kappa T_{\sigma 0}/m_\sigma}$.

Since $\omega_{pe}^2/\omega_{pi}^2 = m_i/m_e$ the ion contribution can be dropped, and the dispersion becomes

$$1 - \frac{\omega_{pe}^2}{\omega^2}\left(1 + 3\frac{k^2}{\omega^2}\frac{\kappa T_{e0}}{m_e}\right) = 0. \tag{4.31}$$

To lowest order, the solution of this equation is simply $\omega^2 = \omega_{pe}^2$. An iterative solution may be obtained by substituting this lowest order solution into the thermal term, which, by assumption, is a small correction because $\omega/k \gg \sqrt{\kappa T_{e0}/m_e}$. This gives the standard form for the high-frequency, electrostatic, unmagnetized plasma wave

$$\omega^2 = \omega_{pe}^2 + 3k^2\frac{\kappa T_{e0}}{m_e}. \tag{4.32}$$

This most basic of plasma waves is called the electron plasma wave, the Langmuir wave (Langmuir 1928), or the Bohm–Gross wave (Bohm and Gross 1949).

2. Case where $\omega/k \ll \sqrt{\kappa T_{e0}/m_e}, \sqrt{\kappa T_{i0}/m_i}$.
Here both electrons and ions are isothermal and the dispersion becomes

$$1 + \sum_\sigma \frac{1}{k^2\lambda_{D\sigma}^2} = 0. \tag{4.33}$$

This has no frequency dependence, and is just the Debye shielding derived in Chapter 1. Thus, when $\omega/k \ll \sqrt{\kappa T_{e0}/m_e}, \sqrt{\kappa T_{i0}/m_i}$ the plasma approaches the steady-state limit and screens out any applied perturbation. This limit shows why ions cannot provide Debye shielding for electrons, because if the test particle were chosen to be an electron, its nominal speed would be the electron thermal velocity. From the point of view of ions, this fast-moving electron test particle would zoom through with a phase velocity $\omega/k \sim v_{Te}$ and so violate the requirement that $\omega/k \ll \sqrt{\kappa T_{i0}/m_i}$ for the ions to be able to respond; the ions would not be able to move fast enough to do any shielding (Wang, Joyce, and Nicholson 1981).

3. Case where $\sqrt{\kappa T_{i0}/m_i} \ll \omega/k \ll \sqrt{\kappa T_{e0}/m_e}$.
Here the ions act adiabatically whereas the electrons act isothermally so that the dispersion becomes

$$1 + \frac{1}{k^2\lambda_{De}^2} - \frac{\omega_{pi}^2}{\omega^2}\left(1 + 3\frac{k^2}{\omega^2}\frac{\kappa T_{i0}}{m_i}\right) = 0. \tag{4.34}$$

It is conventional to define the "ion acoustic" velocity

$$c_s^2 = \omega_{pi}^2\lambda_{De}^2 = \kappa T_e/m_i \tag{4.35}$$

so that Eq. (4.34) can be recast as

$$\omega^2 = \frac{k^2c_s^2}{1 + k^2\lambda_{De}^2}\left(1 + 3\frac{k^2}{\omega^2}\frac{\kappa T_{i0}}{m_i}\right). \tag{4.36}$$

Since $\omega/k \gg \sqrt{\kappa T_{i0}/m_i}$, this may be solved iteratively by first assuming $T_{i0} = 0$ giving

$$\omega^2 = \frac{k^2c_s^2}{1 + k^2\lambda_{De}^2}. \tag{4.37}$$

This is the most basic form for the *ion acoustic wave* dispersion and in the limit $k^2\lambda_{De}^2 \gg 1$, becomes simply $\omega^2 = c_s^2/\lambda_{De}^2 = \omega_{pi}^2$. To obtain the next higher order of precision for the ion acoustic dispersion, Eq. (4.37) may be used to eliminate k^2/ω^2 from the ion thermal term of Eq. (4.36) giving

$$\omega^2 = \frac{k^2 c_s^2}{1+k^2\lambda_{De}^2} + 3k^2\frac{\kappa T_{i0}}{m_i}. \tag{4.38}$$

For self-consistency, it is necessary to have $c_s^2 \gg \kappa T_{i0}/m_i$; if this were not true, the ion acoustic wave would become $\omega^2 = 3k^2\kappa T_{i0}/m_i$, which would violate the assumption that $\omega/k \gg \sqrt{\kappa T_{i0}/m_i}$. The condition $c_s^2 \gg \kappa T_{i0}/m_i$ is the same as $T_e \gg T_i$ so ion acoustic waves can only propagate when the electrons are much hotter than the ions. This issue will be explored in more depth later when ion acoustic waves are re-examined from the Vlasov point of view.

4.2.2 Electromagnetic (incompressible) waves

The compressional waves discussed in the previous section were obtained by taking the divergence of Eq. (4.8). An arbitrary vector field $\mathbf{V}$ can always be decomposed into a gradient of a potential and a solenoidal part, i.e., it can always be written as $\mathbf{V} = \nabla\psi + \nabla\times\mathbf{Q}$, where ψ and $\mathbf{Q}$ can be determined from $\mathbf{V}$. The $\nabla\psi$ term has zero curl and so describes a conservative field, whereas the solenoidal term $\nabla\times\mathbf{Q}$ has zero divergence and describes a non-conservative field. Because we have chosen to use Coulomb gauge in our analysis of waves, the $-\nabla\phi$ term on the right-hand side of Eq. (4.9) is the only conservative field; the $-\partial\mathbf{A}/\partial t$ term is the solenoidal or non-conservative field.

Waves involving finite $\mathbf{A}$ have coupled electric and magnetic fields and are a generalization of vacuum electromagnetic waves such as light or radio waves. These finite $\mathbf{A}$ waves are variously called electromagnetic, transverse, or incompressible waves. Since no electrostatic potential is involved, $\nabla\cdot\mathbf{E} = 0$ and the plasma remains neutral. Because $\mathbf{A} \neq 0$, these waves involve electric currents.

Since the electromagnetic waves are solenoidal, the $-\nabla\phi$ term in Eq. (4.8) is superfluous and can be eliminated by taking the curl of Eq. (4.8) giving

$$\frac{\partial}{\partial t}\nabla\times(m_\sigma n_\sigma \mathbf{u}_{\sigma 1}) = -q_\sigma n_\sigma \frac{\partial \mathbf{B}_1}{\partial t}. \tag{4.39}$$

To obtain an equation involving currents, Eq. (4.39) is integrated with respect to time, multiplied by q_σ/m_σ, and then summed over species to give

$$\nabla\times\mathbf{J}_1 = -\varepsilon_0\omega_p^2\mathbf{B}_1, \tag{4.40}$$

where

$$\omega_p^2 = \sum_\sigma \omega_{p\sigma}^2. \tag{4.41}$$

However, Ampère's law can be written in the form

$$\mathbf{J}_1 = \frac{1}{\mu_0}\nabla\times\mathbf{B}_1 - \varepsilon_0\frac{\partial \mathbf{E}_1}{\partial t}, \tag{4.42}$$

which, after substitution into Eq. (4.40), gives

$$\nabla\times\left(\nabla\times\mathbf{B}_1 - \frac{1}{c^2}\frac{\partial \mathbf{E}_1}{\partial t}\right) = -\frac{\omega_p^2}{c^2}\mathbf{B}_1. \tag{4.43}$$

Using the vector identity $\nabla\times(\nabla\times\mathbf{Q}) = \nabla(\nabla\cdot\mathbf{Q}) - \nabla^2\mathbf{Q}$ and Faraday's law this becomes

$$\nabla^2\mathbf{B}_1 = \frac{1}{c^2}\frac{\partial^2\mathbf{B}_1}{\partial t^2} + \frac{\omega_p^2}{c^2}\mathbf{B}_1. \tag{4.44}$$

In the limit of no plasma, $\omega_p^2 \to 0$, so that Eq. (4.44) reduces to the standard vacuum electromagnetic wave. If it is assumed that $\mathbf{B}_1 \sim \exp(\mathrm{i}\mathbf{k}\cdot\mathbf{x} - \mathrm{i}\omega t)$, Eq. (4.44) becomes the electromagnetic, unmagnetized plasma wave dispersion

$$\omega^2 = \omega_p^2 + k^2c^2. \tag{4.45}$$

Waves satisfying Eq. (4.45) are often used to measure plasma density. Such a measurement can be accomplished two ways:

1. Cutoff method
   If $\omega^2 < \omega_p^2$ then k^2 becomes negative, the wave does not propagate, and only exponentially growing or decaying spatial behavior occurs (such behavior is called evanescent). If the wave is excited by an antenna driven by a fixed-frequency oscillator, the boundary condition that the wave field does not diverge at infinity means that only waves that decay away from the antenna exist. Thus, the field is localized near the antenna and there is no wave-like behavior. This is called *cutoff*. When the oscillator frequency is raised above ω_p, the wave starts to propagate so that a receiver located some distance away will abruptly start to pick up the wave. By scanning the transmitter frequency and noting the frequency at which the wave starts to propagate, ω_p^2 is determined, giving a direct, unambiguous measurement of the plasma density.

2. Phase shift method
   Here the oscillator frequency is set to be well above cutoff so that the wave is always propagating. The dispersion relation is solved for k and the phase delay $\Delta\phi$ of the wave through the plasma is measured by interferometric fringe-counting. The total phase delay through a length L of plasma is

$$\phi = \int_0^L k\mathrm{d}x = \frac{1}{c}\int_0^L \left[\omega^2 - \omega_p^2\right]^{1/2}\mathrm{d}x \simeq \frac{\omega}{c}\int_0^L\left[1 - \frac{\omega_p^2}{2\omega^2}\right]\mathrm{d}x \tag{4.46}$$

   so that the phase delay due to the presence of plasma is

$$\Delta\phi = -\frac{1}{2\omega c}\int_0^L \omega_{p_e}^2\mathrm{d}x = -\frac{e^2}{2\omega c m_e\varepsilon_0}\int_0^L n\mathrm{d}x. \tag{4.47}$$

Thus, measurement of the phase shift $\Delta\phi$ due to the presence of plasma can be used to measure the average density along L; this density is called the *line-averaged density*.

4.3 Low-frequency magnetized plasma: Alfvén waves

4.3.1 Overview of Alfvén waves

We now consider low-frequency waves propagating in a *magnetized* plasma, i.e., a plasma immersed in a uniform, constant magnetic field $\mathbf{B} = B_0\hat{z}$. By low frequency, it is meant that the wave frequency ω is much smaller than the ion cyclotron frequency ω_{ci}. Several types of waves exist in this frequency range; certain of these involve electric fields having a purely electrostatic character (i.e., $\nabla\times\mathbf{E} = 0$), whereas others involve electric fields having an inductive character (i.e., $\nabla\times\mathbf{E} \neq 0$). Faraday's law, $\nabla\times\mathbf{E} = -\partial\mathbf{B}/\partial t$, shows that if the electric field is electrostatic the magnetic field must be constant, whereas inductive electric fields must have an associated time-dependent magnetic field.

We now further restrict attention to a specific category of these $\omega \ll \omega_{ci}$ modes, namely the Alfvén waves. These waves are the normal modes of MHD, involve magnetic perturbations, and have characteristic velocities of the order of the Alfvén velocity $v_A = B/\sqrt{\mu_0\rho}$. The existence of such modes is not surprising if one considers that ordinary neutral gas sound waves have a velocity $c_{sound} = \sqrt{\gamma P/\rho}$ and the magnetic stress tensor scales as $\sim B^2/\mu_0\rho$ so that Alfvén-type velocities will result if P is replaced by $B^2/2\mu_0$. Two distinct kinds of Alfvén modes exist and to complicate matters these are called a variety of names by different authors. One mode, variously called the fast mode, the compressional mode, or the magnetosonic mode resembles a sound wave and involves compression and rarefaction of magnetic field lines; this mode has a finite B_{z1}. The other mode, variously called the Alfvén mode, the shear mode, the torsional mode, or the slow mode, involves twisting, shearing, or plucking motions; this mode has $B_{z1} = 0$. This latter mode appears in distinct β-dependent versions when modeled using two-fluid or Vlasov theory; these are respectively called the inertial Alfvén wave and the kinetic Alfvén wave.

Both Faraday's law and the pre-Maxwell Ampère's law are fundamental to Alfvén wave dynamics. The system of linearized equations thus is

$$\nabla\times\mathbf{E}_1 = -\frac{\partial\mathbf{B}_1}{\partial t} \tag{4.48}$$

$$\nabla\times\mathbf{B}_1 = \mu_0\mathbf{J}_1. \tag{4.49}$$

If the dependence of $\mathbf{J}_1$ on $\mathbf{E}_1$ can be determined, then the combination of Ampère's law and Faraday's law provides a complete self-consistent description of the coupled fields $\mathbf{E}_1, \mathbf{B}_1$ and hence describes the normal modes. From a mathematical point of view, specifying $\mathbf{J}_1(\mathbf{E}_1)$ means that there are as many equations as dependent variables in Eqs. (4.48) and (4.49). The relationship between $\mathbf{J}_1$ and $\mathbf{E}_1$ is determined by the Lorentz equation or some generalization thereof (e.g., drift equations, Vlasov equation, fluid equation of motion). The analysis of solutions to Eqs. (4.48) and (4.49) will proceed in three steps. First, we will consider the zero-pressure MHD approximation, next, the finite-pressure MHD approximation, and finally, the finite-pressure two-fluid approximation. MHD is a simpler description than two-fluid theory because MHD essentially ignores parallel dynamics. Instead of attempting a description of the parallel dynamics, the MHD approximation invokes what is in effect an ad hoc closure relation for the parallel current density in order to maintain overall charge neutrality.

4.3.2 Zero-pressure MHD model

The zero-pressure MHD model reveals the fundamental nature of the Alfvén modes and shows that the essence of these MHD modes comes from the polarization drift associated with a time-dependent perpendicular electric field, namely

$$\mathbf{u}_{\sigma,polarization} = \frac{m_\sigma}{q_\sigma B^2}\frac{\mathrm{d}\mathbf{E}_\perp}{\mathrm{d}t}; \tag{4.50}$$

this was discussed in the derivation of Eq. (3.77). The polarization drift results in a polarization current

$$\begin{aligned} \mathbf{J}_\perp &= \sum n_\sigma q_\sigma \mathbf{u}_{\sigma,polarization} \\ &= \frac{\rho}{B^2}\frac{\mathrm{d}\mathbf{E}_\perp}{\mathrm{d}t}, \end{aligned} \tag{4.51}$$

where $\rho = \sum n_\sigma m_\sigma$ is the mass density. This can be recast as

$$\begin{aligned} \frac{\mathrm{d}\mathbf{E}_\perp}{\mathrm{d}t} &= \frac{B^2}{\mu_0\rho}\mu_0\mathbf{J}_\perp \\ &= v_A^2\left(\nabla\times\mathbf{B}_1\right)_\perp, \end{aligned} \tag{4.52}$$

where

$$v_A^2 = \frac{B^2}{\mu_0\rho} \tag{4.53}$$

is the Alfvén velocity. Equation (4.48) and the linearized version of Eq. (4.52) give the two basic coupled equations governing these Alfvén modes, namely

$$\frac{\partial\mathbf{B}_1}{\partial t} = -\nabla\times\mathbf{E}_1 \tag{4.54a}$$

$$\frac{\partial\mathbf{E}_{\perp 1}}{\partial t} = v_A^2\left(\nabla\times\mathbf{B}_1\right)_\perp. \tag{4.54b}$$

The omission of an equation for the parallel component of Ampère's law is an essential property of the MHD model. Unlike all other dependent variables, which appear in two equations and so have to be solved for self-consistently, the only place where J_z appears in MHD is in the $\nabla\cdot\mathbf{J}=0$ equation and so J_z is solved for in MHD as a sort of "after-thought" using the relation $J_z = -\int \mathrm{d}z \nabla_\perp\cdot\mathbf{J}_\perp$.

A critical feature of MHD related to this omission of a dynamical prescription for J_z results from dotting the linearized ideal Ohm's law

$$\mathbf{E}_1 + \mathbf{U}_1 \times \mathbf{B} = 0 \tag{4.55}$$

with $\hat{z}$ to obtain $E_{z1} = 0$. Thus, in the MHD limit, it is assumed there is no parallel electric field so there is no means for accelerating particles in the parallel direction and thus no means for attaining the parallel particle velocities associated with the existence of a parallel current. Even though MHD does not provide the dynamics for establishing a parallel current, MHD nevertheless requires the existence of a parallel current in order to satisfy $\nabla\cdot\mathbf{J}=0$. Thus, according to the MHD approximation, a parallel current spontaneously comes into existence without any parallel electric field to drive this current. Equation (4.55) also implies that the plasma velocity for these modes is just the $\mathbf{E}\times\mathbf{B}$ drift, i.e., $\mathbf{U}_1 = \mathbf{E}_1\times\mathbf{B}/B^2$.

We now proceed with the derivation of the two zero-pressure MHD modes. Since E_{z1} is zero, the electric field is only in the perpendicular direction. The components of the electric field, the magnetic field, and the ∇ operator can thus be expressed as

$$\begin{aligned}
\mathbf{E}_1 &= \mathbf{E}_{\perp 1} \\
\mathbf{B}_1 &= \mathbf{B}_{\perp 1} + B_{z1}\hat{z} \\
\nabla &= \nabla_\perp + \hat{z}\frac{\partial}{\partial z},
\end{aligned} \tag{4.56}$$

where $\perp$ refers to the x and y components of a vector or a vector operator so, for example, $\nabla_\perp = \hat{x}\partial/\partial x + \hat{y}\partial/\partial y$. The curl operators appearing in Eqs. (4.54) can therefore be expanded as

$$\begin{aligned}
\nabla\times\mathbf{E}_1 &= \left(\nabla_\perp + \hat{z}\frac{\partial}{\partial z}\right)\times\mathbf{E}_{\perp 1} \\
&= \nabla_\perp\times\mathbf{E}_{\perp 1} + \hat{z}\times\frac{\partial\mathbf{E}_{\perp 1}}{\partial z}
\end{aligned} \tag{4.57}$$

and

$$\begin{aligned}
(\nabla\times\mathbf{B}_1)_\perp &= \left(\left(\nabla_\perp + \hat{z}\frac{\partial}{\partial z}\right)\times\left(\mathbf{B}_{\perp 1} + B_{z1}\hat{z}\right)\right)_\perp \\
&= \nabla_\perp B_{z1}\times\hat{z} + \hat{z}\times\frac{\partial\mathbf{B}_{\perp 1}}{\partial z},
\end{aligned} \tag{4.58}$$

where it should be noted that both $\nabla_\perp \times \mathbf{E}_{\perp 1}$ and $\nabla_\perp \times \mathbf{B}_{\perp 1}$ are in the z direction. Using Eqs. (4.57) and (4.58) in Eqs. (4.54) gives the basic pair of coupled equations for MHD waves, namely

$$\frac{\partial \mathbf{B}_1}{\partial t} = -\nabla_\perp \times \mathbf{E}_{\perp 1} - \hat{z} \times \frac{\partial \mathbf{E}_{\perp 1}}{\partial z} \tag{4.59a}$$

$$\frac{\partial \mathbf{E}_{\perp 1}}{\partial t} = v_A^2 \left(\nabla_\perp B_{z1} \times \hat{z} + \hat{z} \times \frac{\partial \mathbf{B}_{\perp 1}}{\partial z} \right). \tag{4.59b}$$

Slow or Alfvén mode (mode where $B_{z1} = 0$)

In this case $\mathbf{B}_1 = \mathbf{B}_{\perp 1}$ and Eqs. (4.59) become

$$\frac{\partial \mathbf{B}_{\perp 1}}{\partial t} = -\hat{z} \times \frac{\partial \mathbf{E}_{\perp 1}}{\partial z} \tag{4.60a}$$

$$\frac{\partial \mathbf{E}_{\perp 1}}{\partial t} = v_A^2 \hat{z} \times \frac{\partial \mathbf{B}_{\perp 1}}{\partial z} \tag{4.60b}$$

with the condition

$$\nabla_\perp \times \mathbf{E}_{\perp 1} = 0, \tag{4.61}$$

which is obtained from the z component of Eq. (4.59a). Equations (4.60) can be rewritten as

$$\frac{\partial \mathbf{B}_{\perp 1}}{\partial t} = -\frac{\partial}{\partial z} (\hat{z} \times \mathbf{E}_{\perp 1}) \tag{4.62a}$$

$$\frac{\partial}{\partial t} (\hat{z} \times \mathbf{E}_{\perp 1}) = -v_A^2 \frac{\partial \mathbf{B}_{\perp 1}}{\partial z}, \tag{4.62b}$$

which leads to a wave equation in the coupled variables $\hat{z} \times \mathbf{E}_{\perp 1}$ and $\mathbf{B}_{\perp 1}$. In particular, taking a second time derivative of Eq. (4.62a) and then substituting for $\partial/\partial t\, (\hat{z} \times \mathbf{E}_{\perp 1})$ using Eq. (4.62b), gives the wave equation for the slow mode (Alfvén mode),

$$\frac{\partial^2 \mathbf{B}_{\perp 1}}{\partial t^2} = v_A^2 \frac{\partial^2 \mathbf{B}_{\perp 1}}{\partial z^2}. \tag{4.63}$$

This is the mode originally derived by Alfvén (1943). The condition given by Eq. (4.61) corresponds to setting $B_{z1} = 0$ and implies that $\mathbf{E}_{\perp 1}$ can be expressed as the gradient of a potential. This further implies that the velocity $\mathbf{U}_1 = \mathbf{E}_1 \times \mathbf{B}/B^2$ is incompressible since $\nabla \cdot \mathbf{U}_1 = \nabla \cdot \left(\mathbf{E}_1 \times \mathbf{B}/B^2 \right) = B^{-2} \mathbf{B} \cdot \nabla \times \mathbf{E}_1 = B^{-2} \mathbf{B} \cdot \nabla_\perp \times \mathbf{E}_{\perp 1} = 0$.

4.3.3 Fast mode (mode where $B_{z1} \neq 0$)

In this case only the z component of Eq. (4.59a) is used and after crossing Eq. (4.59b) with $\hat{z}$, Eqs. (4.59) become

$$\frac{\partial B_{z1}}{\partial t} = -\hat{z} \cdot \nabla_\perp \times \mathbf{E}_{\perp 1} = -\nabla \cdot (\mathbf{E}_{\perp 1} \times \hat{z}) \tag{4.64a}$$

$$\frac{\partial}{\partial t}\mathbf{E}_{\perp 1} \times \hat{z} = v_A^2 \left(\nabla_\perp B_{z1} \times \hat{z} + \hat{z} \times \frac{\partial \mathbf{B}_{\perp 1}}{\partial z} \right) \times \hat{z} = v_A^2 \left(-\nabla_\perp B_{z1} + \frac{\partial \mathbf{B}_{\perp 1}}{\partial z} \right). \tag{4.64b}$$

Taking a time derivative of Eq. (4.64a) and then using Eq. (4.64b) to substitute for $\partial/\partial t\,(\mathbf{E}_{\perp 1} \times \hat{z})$ gives

$$\begin{aligned} \frac{\partial^2 B_{z1}}{\partial t^2} &= -v_A^2 \nabla \cdot \left(-\nabla_\perp B_{z1} + \frac{\partial \mathbf{B}_{\perp 1}}{\partial z} \right) \\ &= v_A^2 \nabla_\perp^2 B_{z1} - v_A^2 \frac{\partial}{\partial z} (\nabla \cdot \mathbf{B}_{\perp 1}). \end{aligned} \tag{4.65}$$

However, using $\nabla \cdot \mathbf{B}_1 = 0$ it is seen that

$$\nabla \cdot \mathbf{B}_{\perp 1} = -\frac{\partial B_{z1}}{\partial z} \tag{4.66}$$

and so the fast wave equation becomes

$$\frac{\partial^2 B_{z1}}{\partial t^2} = v_A^2 \nabla^2 B_{z1}. \tag{4.67}$$

4.3.4 Comparison of the two modes

The slow mode Eq. (4.63) involves only z derivatives and so has a dispersion relation

$$\omega^2 = k_z^2 v_A^2, \tag{4.68}$$

whereas the fast mode involves the ∇^2 operator and so has the dispersion relation

$$\omega^2 = k^2 v_A^2. \tag{4.69}$$

The slow mode is incompressible and has $B_{z1} = 0$ so its perturbed magnetic field is entirely orthogonal to the equilibrium field. Thus, the slow mode magnetic perturbation corresponds to a twisting or plucking of the equilibrium field. The fast mode has $B_{z1} \neq 0$, which corresponds to a compression of the equilibrium field as shown in Fig. 4.2; since the plasma is frozen to the magnetic field, the plasma is also compressed.

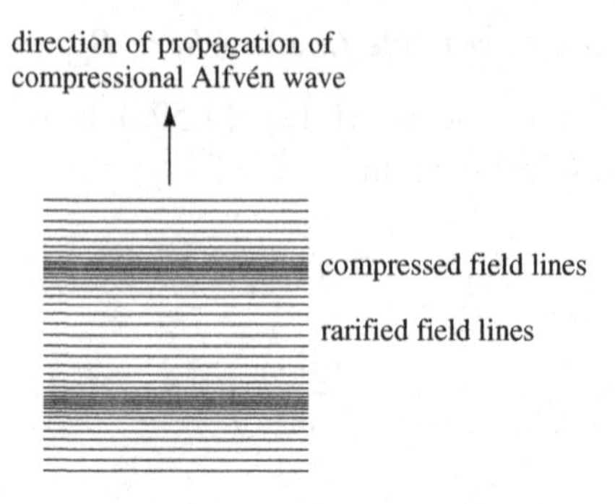

Fig. 4.2 Compressional Alfvén waves.

4.3.5 Finite-pressure analysis of MHD waves

If the pressure is allowed to be finite, then the two modes become coupled and an acoustic mode appears. Using the vector identity $\nabla B^2 = 2(\mathbf{B}\cdot\nabla\mathbf{B}+\mathbf{B}\times\nabla\times\mathbf{B})$ the $\mathbf{J}\times\mathbf{B}$ force in the MHD equation of motion can be written as

$$\mathbf{J}\times\mathbf{B} = \frac{1}{\mu_0}(\nabla\times\mathbf{B})\times\mathbf{B} = -\nabla\left(\frac{B^2}{2\mu_0}\right)+\frac{1}{\mu_0}\mathbf{B}\cdot\nabla\mathbf{B}. \tag{4.70}$$

The MHD equation of motion thus becomes

$$\rho\frac{\mathrm{D}\mathbf{U}}{\mathrm{D}t} = -\nabla\left(P+\frac{B^2}{2\mu_0}\right)+\frac{1}{\mu_0}\mathbf{B}\cdot\nabla\mathbf{B}. \tag{4.71}$$

Linearizing this equation about a stationary equilibrium, where the pressure and the density are uniform and constant, gives

$$\rho\frac{\partial\mathbf{U}_1}{\partial t} = -\nabla\left(P_1+\frac{\mathbf{B}\cdot\mathbf{B}_1}{\mu_0}\right)+\frac{1}{\mu_0}\mathbf{B}\cdot\nabla\mathbf{B}_1, \tag{4.72}$$

where unsubscripted dependent variables are equilibrium (zero-order) quantities. The curl of the linearized ideal MHD Ohm's law,

$$\mathbf{E}_1+\mathbf{U}_1\times\mathbf{B} = 0, \tag{4.73}$$

gives the induction equation

$$\frac{\partial\mathbf{B}_1}{\partial t} = \nabla\times(\mathbf{U}_1\times\mathbf{B}), \tag{4.74}$$

while the linearized continuity equation

$$\frac{\partial\rho_1}{\partial t}+\rho\nabla\cdot\mathbf{U}_1 = 0 \tag{4.75}$$

together with the equation of state

$$\frac{P_1}{P} = \gamma \frac{\rho_1}{\rho} \tag{4.76}$$

give

$$\frac{\partial P_1}{\partial t} = -\gamma P \nabla \cdot \mathbf{U}_1. \tag{4.77}$$

To obtain an equation involving $\mathbf{U}_1$ only, we take the time derivative of Eq. (4.72) and use Eqs. (4.74) and (4.77) to eliminate the time derivatives of P_1 and $\mathbf{B}_1$. This gives

$$\rho \frac{\partial^2 \mathbf{U}_1}{\partial t^2} = -\nabla \left(-\gamma P \nabla \cdot \mathbf{U}_1 + \frac{1}{\mu_0} \mathbf{B} \cdot \nabla \times (\mathbf{U}_1 \times \mathbf{B}) \right) + \frac{1}{\mu_0} (\mathbf{B} \cdot \nabla) \nabla \times (\mathbf{U}_1 \times \mathbf{B}). \tag{4.78}$$

This can be simplified using the identity $\nabla \cdot (\mathbf{a} \times \mathbf{b}) = \mathbf{b} \cdot \nabla \times \mathbf{a} - \mathbf{a} \cdot \nabla \times \mathbf{b}$ so that

$$\mathbf{B} \cdot \nabla \times (\mathbf{U}_1 \times \mathbf{B}) = \nabla \cdot [(\mathbf{U}_1 \times \mathbf{B}) \times \mathbf{B}] = -B^2 \nabla \cdot \mathbf{U}_{1\perp}. \tag{4.79}$$

Furthermore,

$$\mathbf{B} \cdot \nabla = B \frac{\partial}{\partial z} = \mathrm{i} k_z B. \tag{4.80}$$

Using these relations Eq. (4.78) becomes

$$\frac{\partial^2 \mathbf{U}_1}{\partial t^2} = \nabla \left(c_s^2 \nabla \cdot \mathbf{U}_1 + v_A^2 \nabla \cdot \mathbf{U}_{1\perp} \right) + \mathrm{i} k_z v_A^2 \nabla \times (\mathbf{U}_1 \times \hat{z}), \tag{4.81}$$

where $c_s^2 = \gamma P / \rho$ is the velocity of the *conventional sound wave* in a gas (this wave is not to be confused with the ion acoustic wave, which is a result of two-fluid analysis and which has a velocity $c_s^2 = \kappa T_e / m_i$). To proceed further we take either the divergence or the curl of this equation to obtain expressions for compressional or incompressible motions.

4.3.6 MHD compressional (fast) mode

Here we take the divergence of Eq. (4.81) to obtain

$$\frac{\partial^2 \nabla \cdot \mathbf{U}_1}{\partial t^2} = \nabla^2 \left(c_s^2 \nabla \cdot \mathbf{U}_1 + v_A^2 \nabla \cdot \mathbf{U}_{1\perp} \right) \tag{4.82}$$

or

$$\omega^2 \nabla \cdot \mathbf{U}_1 = \left(k_\perp^2 + k_z^2 \right) \left(c_s^2 \nabla \cdot \mathbf{U}_1 + v_A^2 \nabla \cdot \mathbf{U}_{1\perp} \right). \tag{4.83}$$

On the other hand, if Eq. (4.81) is operated on with $\nabla_\perp = \nabla - ik_z\hat{z}$ we obtain

$$\frac{\partial^2 \nabla_\perp \cdot U_1}{\partial t^2} = \nabla_\perp^2 \left(c_s^2 \nabla \cdot \mathbf{U}_1 + v_A^2 \nabla \cdot \mathbf{U}_{1\perp} \right) + k_z^2 v_A^2 \hat{z} \cdot \nabla \times (\mathbf{U}_1 \times \hat{z}). \tag{4.84}$$

Using

$$\begin{aligned} \hat{z} \cdot \nabla \times (\mathbf{U}_1 \times \hat{z}) &= \nabla \cdot ((\mathbf{U}_1 \times \hat{z}) \times \hat{z}) \\ &= -\nabla \cdot \mathbf{U}_{1\perp} \end{aligned} \tag{4.85}$$

Eq. (4.84) becomes

$$\omega^2 \nabla_\perp \cdot \mathbf{U}_1 = k_\perp^2 \left(c_s^2 \nabla \cdot \mathbf{U}_1 + v_A^2 \nabla \cdot \mathbf{U}_{1\perp} \right) + k_z^2 v_A^2 \nabla \cdot \mathbf{U}_{1\perp}. \tag{4.86}$$

Equations (4.83) and (4.86) constitute two coupled equations in the variables $\nabla \cdot \mathbf{U}_1$ and $\nabla \cdot \mathbf{U}_{1\perp}$, namely

$$\begin{aligned} \left(\omega^2 - k^2 c_s^2\right) \nabla \cdot \mathbf{U}_1 - k^2 v_A^2 \nabla \cdot \mathbf{U}_{1\perp} &= 0 \\ k_\perp^2 c_s^2 \nabla \cdot \mathbf{U}_1 + \left(k^2 v_A^2 - \omega^2\right) \nabla \cdot \mathbf{U}_{1\perp} &= 0. \end{aligned} \tag{4.87}$$

These coupled equations have the determinant

$$\left(\omega^2 - k^2 c_s^2\right)\left(k^2 v_A^2 - \omega^2\right) + k^2 v_A^2 k_\perp^2 c_s^2 = 0, \tag{4.88}$$

which can be rearranged as a fourth order polynomial in ω,

$$\omega^4 - \omega^2 k^2 \left(v_A^2 + c_s^2\right) + k^2 k_z^2 v_A^2 c_s^2 = 0, \tag{4.89}$$

having roots

$$\omega^2 = \frac{k^2\left(v_A^2 + c_s^2\right) \pm \sqrt{k^4\left(v_A^2 + c_s^2\right)^2 - 4k^2 k_z^2 v_A^2 c_s^2}}{2}. \tag{4.90}$$

This dispersion relation has the following limiting forms

$$\left.\begin{aligned} &\omega^2 = k_\perp^2 \left(v_A^2 + c_s^2\right) \\ &\text{or} \\ &\omega^2 = 0 \end{aligned}\right\} \text{ if } k_z = 0 \tag{4.91}$$

and

$$\left.\begin{aligned} &\omega^2 = k_z^2 v_A^2 \\ &\text{or} \\ &\omega^2 = k_z^2 c_s^2 \end{aligned}\right\} \text{ if } k_\perp^2 = 0; \tag{4.92}$$

the $\omega^2 = k_\perp^2 \left(v_A^2 + c_s^2\right)$ mode existing in the $k_z = 0$ limit is called the magnetosonic mode and corresponds to choosing the plus sign in Eq. (4.90).

4.3.7 MHD shear (slow) mode

It is now assumed that $\nabla\cdot\mathbf{U}_1 = 0$ and taking the curl of Eq. (4.81) gives

$$\begin{aligned}\frac{\partial^2\nabla\times\mathbf{U}_1}{\partial t^2} &= v_A^2\nabla\times\nabla\times\left(\frac{\partial\mathbf{U}_1}{\partial z}\times\hat{z}\right)\\ &= v_A^2\nabla\times\left(\frac{\partial\mathbf{U}_1}{\partial z}\underbrace{\nabla\cdot\hat{z}}_{\text{zero}}+\hat{z}\cdot\nabla\frac{\partial\mathbf{U}_1}{\partial z}-\underbrace{\hat{z}\nabla\cdot\frac{\partial\mathbf{U}_1}{\partial z}}_{\text{zero}}-\frac{\partial\mathbf{U}_1}{\partial z}\cdot\underbrace{\nabla\hat{z}}_{\text{zero}}\right)\\ &= v_A^2\frac{\partial^2}{\partial z^2}\nabla\times\mathbf{U}_1,\end{aligned} \tag{4.93}$$

where the vector identity $\nabla\times(\mathbf{F}\times\mathbf{G}) = \mathbf{F}\nabla\cdot\mathbf{G}+\mathbf{G}\cdot\nabla\mathbf{F}-\mathbf{G}\nabla\cdot\mathbf{F}-\mathbf{F}\cdot\nabla\mathbf{G}$ has been used to obtain the second line.

Equation (4.93) reduces to the slow wave dispersion relation Eq. (4.68). The associated spatial behavior is such that $\nabla\times\mathbf{U}_1 \neq 0$, and the mode is unaffected by existence of finite pressure. The perturbed magnetic field is orthogonal to the equilibrium field, i.e., $\mathbf{B}_1\cdot\mathbf{B} = 0$, since it has been assumed that $\nabla\cdot\mathbf{U}_1 = 0$ and since finite $\mathbf{B}_1\cdot\mathbf{B}$ corresponds to finite $\nabla\cdot\mathbf{U}_1$. Since $\nabla\times\mathbf{U}_1$ is the fluid vorticity, the slow mode involves propagation of vorticity.

4.3.8 Limitations of the MHD model

The MHD model ignores parallel electron dynamics and so has a shear mode dispersion $\omega^2 = k_z^2 v_A^2$ that has no dependence on $k_\perp$. Some authors interpret this as a license to allow arbitrarily large $k_\perp$, in which case a shear mode could be localized to a single field line. However, the two-fluid model of the shear mode does have a dependence on $k_\perp$, which becomes important when $k_\perp$ becomes large. The nature of the two-fluid shear mode depends on how the parallel phase velocity of the wave compares to the electron and ion thermal velocities, i.e., on how $\omega/k_z \sim v_A$ compares to v_{Te} and v_{Ti}. Since it is possible to have (i) $v_A \ll v_{Ti}, v_{Te}$, (ii) $v_{Ti} \ll v_A \ll v_{Te}$, or (iii) $v_{Ti}, v_{Te} \ll v_A$, three distinct two-fluid regimes can exist and, as will be shown below, the plasma β determines which is the relevant regime for a given plasma.

MHD also predicts a sound wave that is identical to the ordinary hydrodynamic sound wave of an unmagnetized gas. The perpendicular behavior of this sound wave is consistent with the two-fluid model because both two-fluid and MHD perpendicular motions involve compressional behavior associated with having finite B_{z1}. However, the parallel behavior of the MHD sound wave is problematic because E_{z1} is assumed to be identically zero in MHD. According to the two-fluid model, any parallel acceleration requires a parallel electric field. The two-fluid B_{z1} mode is decoupled from the two-fluid E_{z1} mode so that the two-fluid B_{z1} mode is both compressional and has no parallel motion associated with it.

The MHD analysis makes no restriction on the electron to ion temperature ratio and predicts that a sound wave will exist when $T_e = T_i$. In contrast, the two-fluid model shows that sound waves can only exist when $T_e \gg T_i$ because only in this regime is it possible to have $\kappa T_i/m_i \ll \omega^2/k_z^2 \ll \kappa T_e/m_e$ and so have inertial behavior for ions and kinetic behavior for electrons.

Various paradoxes develop in the MHD treatment of the shear mode but not in the two-fluid description. These paradoxes illustrate the limitations of the MHD description of a plasma and show that MHD results must be treated with caution for the shear (slow) mode. MHD provides an adequate description for the fast (compressional) mode.

4.4 Two-fluid model of Alfvén modes

We now examine these modes from a two-fluid point of view. An important point that will be demonstrated is that the two-fluid point of view shows that the shear mode has three distinct forms depending on how v_A^2 compares to the ion and electron thermal velocities. The Alfvén velocity could be slower than both electron and ion thermal velocities, faster than the ion thermal velocity while slower than the electron thermal velocity, or faster than both the electron and ion thermal velocities. Which of these situations occurs depends upon the ratio of hydrodynamic pressure to magnetic pressure. This ratio is defined for each species σ as

$$\beta_\sigma = \frac{n\kappa T_\sigma}{B^2/\mu_0}; \tag{4.94}$$

the subscript σ is not used if electrons and ions have the same temperature. β_i measures the ratio of ion thermal velocity to the Alfvén velocity since

$$\frac{v_{Ti}^2}{v_A^2} = \frac{\kappa T_i/m_i}{B^2/nm_i\mu_0} = \beta_i. \tag{4.95}$$

Thus, $v_{Ti} \ll v_A$ corresponds to $\beta_i \ll 1$. Magnetic forces dominate hydrodynamic forces in a low-β plasma, whereas in a high-β plasma the opposite is true.

The ratio of electron thermal velocity to Alfvén velocity is

$$\frac{v_{Te}^2}{v_A^2} = \frac{\kappa T_e/m_e}{B^2/nm_i\mu_0} = \frac{m_i}{m_e}\beta_e. \tag{4.96}$$

Thus, $v_{Te}^2 \gg v_A^2$ when $\beta_e \gg m_e/m_i$ and $v_{Te}^2 \ll v_A^2$ when $\beta_e \ll m_e/m_i$. Shear Alfvén wave physics is different in the $\beta_e \gg m_e/m_i$ and $\beta_e \ll m_e/m_i$ regimes, which therefore must be investigated separately. MHD ignores this β_σ dependence and so oversimplifies these waves.

The MHD derivation used the polarization drift to give a relationship between $\mathbf{J}_{\perp 1}$ and $\mathbf{E}_{\perp 1}$ but was ambiguous regarding the relationship between J_{z1} and E_{z1}. The two-fluid model is based on the linearized equation of motion

$$m_\sigma n \frac{\partial \mathbf{u}_{\sigma 1}}{\partial t} = n q_\sigma \left(\mathbf{E}_1 + \mathbf{u}_{\sigma 1} \times \mathbf{B}\right) - \nabla \cdot \mathbf{P}_{\sigma 1}; \tag{4.97}$$

this equation gives essentially the same physics as MHD in the perpendicular direction, but not in the parallel direction. As in MHD, charge neutrality is assumed so that $n_i = n_e = n$. Also, to have increased generality, the pressure is allowed to be anisotropic so

$$\nabla \cdot \mathbf{P}_{\sigma 1} = \nabla \cdot \begin{bmatrix} P_{\sigma \perp 1} & 0 & 0 \\ 0 & P_{\sigma \perp 1} & 0 \\ 0 & 0 & P_{\sigma z 1} \end{bmatrix} = \nabla_\perp P_{\sigma \perp 1} + \hat{z} \frac{\partial P_{\sigma z 1}}{\partial z}. \tag{4.98}$$

Because quasi-neutrality is assumed, the perpendicular current given by two-fluid theory is essentially identical to MHD and consists of ion polarization drift and diamagnetic drift. The existence of these drifts can be seen by invoking the $\omega \ll \omega_{c\sigma}$ assumption so that the left-hand side of Eq. (4.97) can be neglected to first approximation, in which case the perpendicular component of Eq. (4.97) can be expressed as

$$\mathbf{u}_{\sigma 1} \times \mathbf{B} \simeq -\mathbf{E}_{\perp 1} + \nabla_\perp P_{\sigma \perp 1} / n q_\sigma. \tag{4.99}$$

Crossing with $\mathbf{B}$ to solve for $\mathbf{u}_{\sigma \perp 1}$ gives

$$\mathbf{u}_{\sigma \perp 1} = \frac{\mathbf{E}_1 \times \mathbf{B}}{B^2} - \frac{\nabla P_{\sigma \perp 1} \times \mathbf{B}}{n q_\sigma B^2}, \tag{4.100}$$

where the first term is the single-particle $\mathbf{E} \times \mathbf{B}$ drift and the second term is the diamagnetic drift, a fluid effect. Because $\mathbf{u}_{\sigma \perp 1}$ is time-dependent the polarization drift, given in Eq. (4.50), should also appear. Polarization drift results from solving the perpendicular equation of motion to first order in $\omega / \omega_{c\sigma}$. Recalling that the form of the single-particle polarization drift is $\mathbf{v}_p = \left(m_\sigma / q_\sigma B^2\right) \mathrm{d}\mathbf{E}_{\perp 1} / \mathrm{d}t$ and then using $\mathbf{E}_{\perp 1} - \nabla_\perp P_{\sigma \perp 1} / n q_\sigma$ for the fluid model being considered here instead of just $\mathbf{E}_{\perp 1}$ for single particles (cf. right-hand side of Eq. (4.99)), the fluid polarization drift is obtained. With the inclusion of this higher order correction, the perpendicular fluid motion becomes

$$\mathbf{u}_{\sigma \perp 1} = \frac{\mathbf{E}_1 \times \mathbf{B}}{B^2} - \frac{\nabla P_{\sigma \perp 1} \times \mathbf{B}}{n q_\sigma B^2} + \frac{m_\sigma}{q_\sigma B^2} \frac{\mathrm{d}\mathbf{E}_{\perp 1}}{\mathrm{d}t} - \frac{m_\sigma}{n q_\sigma^2 B^2} \nabla_\perp \frac{\mathrm{d}P_{\sigma \perp 1}}{\mathrm{d}t}. \tag{4.101}$$

The $\mathrm{d}\mathbf{E}_{\perp 1} / \mathrm{d}t$ and $\mathrm{d}P_{\sigma \perp 1} / \mathrm{d}t$ terms are respectively smaller than the corresponding $\mathbf{E}_1$ and $P_{\sigma \perp 1}$ terms by the ratio $\omega / \omega_{c\sigma}$ and so may be ignored when electron and ion fluid velocities are considered separately because $\omega \ll \omega_{c\sigma}$ by assumption. However, when calculating the perpendicular current, i.e., $\mathbf{J}_{\perp 1} = \sum nq_\sigma \mathbf{u}_{\sigma\perp 1}$, the electron and ion $\mathbf{E}\times\mathbf{B}$ drift terms cancel each other so that the polarization terms become the leading terms involving the electric field. Because of the mass in the numerator, the ion polarization drift is much larger than the electron polarization drift. Thus, the perpendicular current comes from ion polarization drift and diamagnetic current

$$\mu_0 \mathbf{J}_{\perp 1} = \frac{\mu_0 n m_i \dot{\mathbf{E}}_{\perp 1}}{B^2} - \sum_\sigma \frac{\nabla P_{\sigma\perp 1}\times\mathbf{B}}{B^2} = \frac{1}{v_A^2}\dot{\mathbf{E}}_{\perp 1} - \frac{\mu_0 \nabla P_{\perp 1}\times\mathbf{B}}{B^2}, \tag{4.102}$$

where $P_{\perp 1} = \sum P_{\sigma\perp 1}$ and the dot on top of $\mathbf{E}_{\perp 1}$ denotes time derivative. The term involving $\dot{P}_{\perp 1}$ has been dropped because it is small by ω/ω_c compared to the $P_{\perp 1}$ term. Equation (4.102) is essentially the same as what one would obtain by crossing the MHD equation of motion with $\mathbf{B}$ and assuming that the MHD velocity $\mathbf{U}$ is given by the $\mathbf{E}\times\mathbf{B}$ drift. Thus, the two-fluid perpendicular dynamics is essentially the same as MHD perpendicular dynamics.

We now reconsider equations of the form given by Eqs. (4.57) and (4.58), but do so without assuming that E_{z1} is zero. Thus, all vectors are decomposed into components parallel and perpendicular to the equilibrium magnetic field, e.g., $\mathbf{E}_1 = \mathbf{E}_{\perp 1} + E_{z1}\hat{z}$. The ∇ operator is similarly decomposed into components parallel to and perpendicular to the equilibrium magnetic field, i.e., $\nabla = \nabla_\perp + \hat{z}\partial/\partial z$, and all quantities are assumed to be proportional to $f(x,y)\exp(\mathrm{i}k_z z - \mathrm{i}\omega t)$. Faraday's law can then be written as

$$\nabla_\perp \times \mathbf{E}_{\perp 1} + \nabla_\perp \times E_{z1}\hat{z} + \hat{z}\frac{\partial}{\partial z}\times\mathbf{E}_{\perp 1} = -\frac{\partial}{\partial t}\left(\mathbf{B}_{\perp 1} + B_{z1}\hat{z}\right), \tag{4.103}$$

which has a parallel component

$$\hat{z}\cdot\nabla_\perp \times \mathbf{E}_{\perp 1} = \mathrm{i}\omega B_{z1} \tag{4.104}$$

and a perpendicular component

$$\left(\nabla_\perp E_{z1} - \mathrm{i}k_z \mathbf{E}_{\perp 1}\right)\times\hat{z} = \mathrm{i}\omega \mathbf{B}_{\perp 1}. \tag{4.105}$$

Similarly, Ampère's law can be decomposed into a parallel component

$$\hat{z}\cdot\nabla_\perp \times \mathbf{B}_{\perp 1} = \mu_0 J_{z1} \tag{4.106}$$

and a perpendicular component

$$\left(\nabla_\perp B_{z1} - \mathrm{i}k_z \mathbf{B}_{\perp 1}\right)\times\hat{z} = \mu_0 \mathbf{J}_{\perp 1}. \tag{4.107}$$

Unlike the MHD model, here we have taken into account both finite E_{z1} and the parallel component of Ampère's law. Substituting the perpendicular current given by Eq. (4.102) into Eq. (4.107) gives

$$\left(\nabla_\perp B_{z1} - \mathrm{i}k_z \mathbf{B}_{\perp 1}\right) \times \hat{z} = -\frac{\mathrm{i}\omega}{v_A^2}\mathbf{E}_{\perp 1} - \frac{\mu_0 \nabla P_1 \times \hat{z}}{B} \tag{4.108}$$

or, after rearrangement,

$$\nabla_\perp \left(B_{z1} + \frac{\mu_0 P_{\perp 1}}{B} \right) \times \hat{z} - \mathrm{i}k_z \mathbf{B}_{\perp 1} \times \hat{z} = -\frac{\mathrm{i}\omega}{v_A^2}\mathbf{E}_{\perp 1}. \tag{4.109}$$

Equations (4.105) and (4.109) can be considered as two coupled equations involving $\mathbf{E}_{\perp 1}$ and $\mathbf{B}_{\perp 1}$. After crossing Eq. (4.105) with $\hat{z}$, they can also be written as two coupled equations involving $\mathbf{E}_{\perp 1}$ and $\mathbf{B}_{\perp 1} \times \hat{z}$, namely

$$\mathrm{i}\omega \mathbf{B}_{\perp 1} \times \hat{z} - \mathrm{i}k_z \mathbf{E}_{\perp 1} = -\nabla_\perp E_{z1} \tag{4.110a}$$

$$-\mathrm{i}k_z \mathbf{B}_{\perp 1} \times \hat{z} + \frac{\mathrm{i}\omega}{v_A^2}\mathbf{E}_{\perp 1} = -\nabla_\perp \left(B_{z1} + \frac{\mu_0 P_{\perp 1}}{B} \right) \times \hat{z}. \tag{4.110b}$$

These may now be solved algebraically for $\mathbf{E}_{\perp 1}$ and $\mathbf{B}_{\perp 1} \times \hat{z}$ to obtain

$$\mathbf{E}_{\perp 1} = \frac{\mathrm{i}k_z \nabla_\perp E_{z1} + \mathrm{i}\omega \nabla_\perp \left(B_{z1} + \mu_0 P_{\perp 1}/B\right) \times \hat{z}}{\omega^2/v_A^2 - k_z^2} \tag{4.111a}$$

$$\mathbf{B}_{\perp 1} \times \hat{z} = \frac{\mathrm{i}\left(\omega/v_A^2\right) \nabla_\perp E_{z1} + \mathrm{i}k_z \nabla_\perp \left(B_{z1} + \mu_0 P_{\perp 1}/B\right) \times \hat{z}}{\omega^2/v_A^2 - k_z^2}. \tag{4.111b}$$

Equations (4.104) and (4.106), the respective parallel components of Faraday's and Ampère's laws, can be written as

$$\nabla \cdot (\mathbf{E}_{\perp 1} \times \hat{z}) = \mathrm{i}\omega B_{z1}, \tag{4.112a}$$

$$\nabla \cdot (\mathbf{B}_{\perp 1} \times \hat{z}) = \mu_0 J_{z1}, \tag{4.112b}$$

so substituting for $\mathbf{E}_{\perp 1} \times \hat{z}$ and $\mathbf{B}_{\perp 1} \times \hat{z}$ gives

$$\nabla \cdot \left(\frac{\mathrm{i}k_z \nabla_\perp E_{z1} \times \hat{z} - \mathrm{i}\omega \nabla_\perp \left(B_{z1} + \mu_0 P_{\perp 1}/B\right)}{\omega^2/v_A^2 - k_z^2} \right) = \mathrm{i}\omega B_{z1} \tag{4.113a}$$

$$\nabla \cdot \left(\frac{\mathrm{i}\left(\omega/v_A^2\right) \nabla_\perp E_{z1} + \mathrm{i}k_z \nabla_\perp \left(B_{z1} + \mu_0 P_{\perp 1}/B\right) \times \hat{z}}{\omega^2/v_A^2 - k_z^2} \right) = \mu_0 J_{z1}. \tag{4.113b}$$

Because $\nabla \cdot \left(\nabla_\perp E_{z1} \times \hat{z}\right) = 0$ and $\nabla \cdot \left(\nabla_\perp \left(B_{z1} + \mu_0 P_{\perp 1}/B\right) \times \hat{z}\right) = 0$ by virtue of the vector identity $\nabla \cdot (\nabla_\perp \psi \times \hat{z}) = \nabla \cdot (\nabla \psi \times \hat{z}) = \hat{z} \cdot \nabla \times \nabla \psi = 0$, these equations simplify to

$$\nabla \cdot \left(\frac{\nabla_\perp \left(B_{z1} + \mu_0 P_{\perp 1}/B\right)}{\omega^2/v_A^2 - k_z^2} \right) = -B_{z1} \tag{4.114a}$$

$$\nabla \cdot \left(\frac{\mathrm{i}\omega \nabla_\perp E_{z1}}{\omega^2/v_A^2 - k_z^2} \right) = v_A^2 \mu_0 J_{z1}. \tag{4.114b}$$

Equation (4.114a) is essentially the equation for the compressional mode, but some further effort is required to relate $P_{\perp 1}$ to B_{z1} in order to establish this. Equation (4.114b) is essentially the equation for the shear mode, but here some further effort is required to establish the relationship between J_{z1} and E_{z1}.

An important property distinguishing these modes is whether or not the motion in the perpendicular direction involves compression. To demonstrate this, we take the divergence of Eq. (4.100) (the small polarization drift terms included in the extended form given in Eq. (4.101) are ignored since these polarization terms are higher order in $\omega/\omega_{c\sigma}$ and are important only when calculating perpendicular current). The result is

$$\nabla \cdot \mathbf{u}_{\sigma\perp 1} = \frac{1}{B^2} \mathbf{B} \cdot \nabla \times \mathbf{E}_1 = -\frac{1}{B^2} \mathbf{B} \cdot \frac{\partial \mathbf{B}_1}{\partial t} = \frac{\mathrm{i}\omega}{B} B_{z1}. \tag{4.115}$$

Thus, modes where $B_{z1} = 0$ involve no compression in the perpendicular direction; these are the shear modes. The modes where B_{z1} is finite are the compressional modes (finite B_{z1} leads to finite $\nabla \cdot \mathbf{u}_{\sigma\perp 1}$, which leads to finite $P_{\perp 1}$ as the plasma is squeezed in the perpendicular direction).

4.4.1 Two-fluid slow (shear) modes

The two-fluid shear modes fall into three categories depending on how the Alfvén velocity compares to the electron and ion thermal velocities. These three regimes are shown schematically in Fig. 4.3. The shear modes have $B_{z1} = 0$ so $\nabla \cdot \mathbf{u}_{\sigma\perp 1} = 0$. The parallel component of the linearized equation of motion, Eq. (4.97), is

$$n m_\sigma \frac{\partial u_{\sigma z1}}{\partial t} = n q_\sigma E_{z1} - \gamma_\sigma \kappa T_{\|\sigma} \frac{\partial n_{\sigma 1}}{\partial z}, \tag{4.116}$$

where $P_{\sigma\|1} = \gamma_\sigma n_{\sigma 1} \kappa T_{\|\sigma}$ has been used. We choose $\gamma = 1$ if the motion is isothermal and $\gamma_\sigma = 3$ if the motion is adiabatic since the compression, being parallel, is one-dimensional. The isothermal case corresponds to $\omega^2/k_z^2 \ll \kappa T_\sigma/m_\sigma$ and vice versa for the adiabatic case.

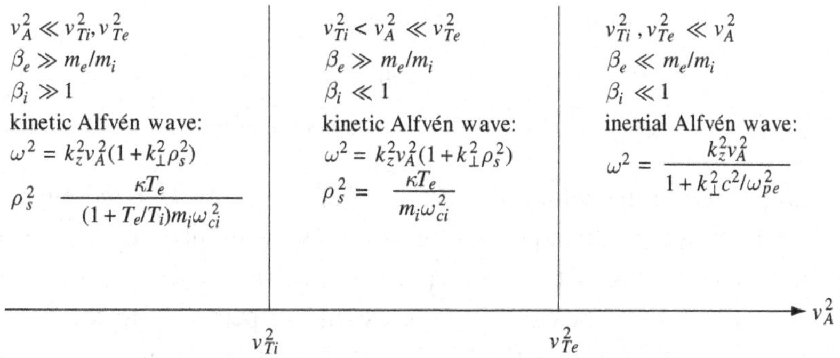

$v_A^2 \ll v_{Ti}^2, v_{Te}^2$	$v_{Ti}^2 < v_A^2 \ll v_{Te}^2$	$v_{Ti}^2, v_{Te}^2 \ll v_A^2$
$\beta_e \gg m_e/m_i$	$\beta_e \gg m_e/m_i$	$\beta_e \ll m_e/m_i$
$\beta_i \gg 1$	$\beta_i \ll 1$	$\beta_i \ll 1$
kinetic Alfvén wave:	kinetic Alfvén wave:	inertial Alfvén wave:
$\omega^2 = k_z^2 v_A^2(1+k_\perp^2\rho_s^2)$	$\omega^2 = k_z^2 v_A^2(1+k_\perp^2\rho_s^2)$	$\omega^2 = \dfrac{k_z^2 v_A^2}{1+k_\perp^2 c^2/\omega_{pe}^2}$
$\rho_s^2 \quad \dfrac{\kappa T_e}{(1+T_e/T_i) m_i \omega_{ci}^2}$	$\rho_s^2 = \dfrac{\kappa T_e}{m_i \omega_{ci}^2}$	

v_{Ti}^2 $\qquad$ v_{Te}^2 $\qquad$ v_A^2

Fig. 4.3 The two-fluid shear mode dispersion relation depends on how v_A^2 compares to the electron and ion thermal velocities and this comparison depends on the electron and ion β's.

Because $\nabla \cdot \mathbf{u}_{\sigma\perp 1} = 0$ the continuity equation is

$$\frac{\partial n_{\sigma 1}}{\partial t} + \frac{\partial}{\partial z}\left(n u_{\sigma z1}\right) = 0. \tag{4.117}$$

Taking the time derivative of Eq. (4.116) and invoking Eq. (4.117) gives

$$\frac{\partial^2 u_{\sigma z1}}{\partial t^2} - \gamma_\sigma \frac{\kappa T_\sigma}{m_\sigma}\frac{\partial^2 u_{\sigma z1}}{\partial z^2} = \frac{q_\sigma}{m_\sigma}\frac{\partial E_{z1}}{\partial t}, \tag{4.118}$$

which is similar to electron plasma wave and ion acoustic wave dynamics except it has *not* been assumed that E_{z1} is electrostatic.

Invoking the assumption that all quantities are of the form $f(x, y)\exp(\mathrm{i}k_z z - \mathrm{i}\omega t)$ Eq. (4.118) can be solved to give

$$u_{\sigma z1} = \frac{\mathrm{i}\omega q_\sigma}{m_\sigma}\frac{E_{z1}}{\omega^2 - \gamma_\sigma k_z^2 \kappa T_\sigma/m_\sigma} \tag{4.119}$$

and so the sought-after relation between parallel current and parallel electric field is

$$\mu_0 J_{z1} = \frac{\mathrm{i}\omega}{c^2}E_{z1}\sum_\sigma \frac{\omega_{p\sigma}^2}{\omega^2 - \gamma_\sigma k_z^2 \kappa T_\sigma/m_\sigma}. \tag{4.120}$$

This provides the prescription for J_{z1} needed in the right-hand side of Eq. (4.114b) and contains the parallel dynamics physics omitted from the MHD description. Substitution of Eq. (4.120) into the right-hand side of Eq. (4.114b) gives the differential equation for the shear wave

$$\nabla \cdot \left(\frac{\nabla_\perp E_{z1}}{\omega^2/v_A^2 - k_z^2}\right) - \frac{v_A^2}{c^2}E_{z1}\sum_\sigma \frac{\omega_{p\sigma}^2}{\omega^2 - \gamma_\sigma k_z^2 \kappa T_\sigma/m_\sigma} = 0. \tag{4.121}$$

On replacing $\nabla_\perp \to i\mathbf{k}_\perp$, Eq. (4.121) becomes

$$\frac{k_\perp^2}{\omega^2 - k_z^2 v_A^2} + \frac{\omega_{pe}^2}{c^2}\frac{1}{\omega^2 - \gamma_e k_z^2 \kappa T_e/m_e} + \frac{\omega_{pi}^2}{c^2}\frac{1}{\omega^2 - \gamma_i k_z^2 \kappa T_i/m_i} = 0. \quad (4.122)$$

We are interested in the solution to Eq. (4.122) where $\omega^2/k_z^2 \sim v_A^2$ but need to take into account three different possibilities for how v_A^2 might compare to electron and ion thermal velocities, namely (i) $v_A^2 \gg v_{Te}^2, v_{Ti}^2$, (ii) $v_{Te}^2 \gg v_A^2 \gg v_{Ti}^2$, or (iii) $v_{Te}^2, v_{Ti}^2 \gg v_A^2$. These three possibilities correspond respectively to the following β situations: (i) $\beta_e \ll m_e/m_i$, (ii) $m_i\beta_e/m_e \gg 1 \gg \beta_i$, and (iii) $\beta_i \gg 1$. These three situations are shown schematically in Fig. 4.3 and will now be discussed in detail.

In situation (i) where $\omega^2/k_z^2 \gg \kappa T_e/m_e$, the second term dominates the third term since $\omega_{pe}^2 \gg \omega_{pi}^2$ and so Eq. (4.122) can be recast as

$$\omega^2 = \frac{k_z^2 v_A^2}{1 + k_\perp^2 c^2/\omega_{pe}^2}, \quad (4.123)$$

which is called the inertial Alfvén wave (IAW). The quantity c/ω_{pe} is called the electron collisionless skin depth. If $k_\perp^2 c^2/\omega_{pe}^2$ is not too large, then ω/k_z is of the order of the Alfvén velocity and the condition $\omega^2 \gg k_z^2 \kappa T_e/m_e$ corresponds to $v_A^2 \gg \kappa T_e/m_e$ or

$$\beta_e = \frac{n\kappa T_e}{B^2/\mu_0} \ll \frac{m_e}{m_i}. \quad (4.124)$$

Thus, inertial Alfvén wave shear modes exist only in the ultra-low β regime where $\beta_e \ll m_e/m_i$. The inertial Alfvén wave is called a cold plasma wave because its dispersion relation does not depend on temperature.

In situation (ii) where $\kappa T_i/m_i \ll \omega^2/k_z^2 \ll \kappa T_e/m_e$, Eq. (4.122) can be recast as

$$\frac{k_\perp^2}{\omega^2 - k_z^2 v_A^2} - \frac{\omega_{pe}^2}{c^2}\frac{1}{k_z^2 \kappa T_e/m_e} + \frac{\omega_{pi}^2}{c^2}\frac{1}{\omega^2} = 0. \quad (4.125)$$

Because ω^2 appears in the respective denominators of two distinct terms, Eq. (4.125) is fourth order in ω^2 and so describes two distinct modes. Let us suppose that the mode in question is much faster than the ion acoustic velocity, i.e., $\omega^2/k_z^2 \gg \kappa T_e/m_i$. In this case the last term can be dropped and the remaining two terms can be rearranged to give

$$\omega^2 = k_z^2 v_A^2 \left(1 + \frac{k_\perp^2}{v_A^2}\frac{\kappa T_e}{m_e}\frac{c^2}{\omega_{pe}^2}\right); \quad (4.126)$$

this is called the kinetic Alfvén wave (KAW). By defining

$$\rho_s^2 = \frac{1}{v_A^2} \frac{\kappa T_e}{m_e} \frac{c^2}{\omega_{pe}^2} = \frac{\kappa T_e}{m_i \omega_{ci}^2} \tag{4.127}$$

as a fictitious ion Larmor radius calculated using the electron temperature instead of the ion temperature, the kinetic Alfvén wave (KAW) dispersion relation can be expressed more succinctly as

$$\omega^2 = k_z^2 v_A^2 \left(1 + k_\perp^2 \rho_s^2\right). \tag{4.128}$$

If $k_\perp^2 \rho_s^2$ is not too large, then ω/k_z is again of the order of v_A and so the condition $\omega^2 \ll k_z^2 \kappa T_e/m_e$ corresponds to having $\beta_e \gg m_e/m_i$. The condition $\omega^2/k_z^2 \gg \kappa T_e/m_i$, which was also assumed, corresponds to assuming that $\beta_e \ll 1$. Thus, the KAW dispersion relation Eq. (4.128) is valid in the regime $m_e/m_i \ll \beta_e \ll 1$.

Let us now consider situation (iii) where $\omega^2/k_z^2 \ll \kappa T_i/m_i,\ \kappa T_e/m_e$. In this case Eq. (4.122) again reduces to

$$\omega^2 = k_z^2 v_A^2 \left(1 + k_\perp^2 \rho_s^2\right) \tag{4.129}$$

but now ρ_s^2 is defined as

$$\rho_s^2 = \frac{\omega_{pi}^2}{\omega_{ci}^2} \left(\frac{1}{\lambda_{De}^2} + \frac{1}{\lambda_{Di}^2} \right)^{-1} = \frac{\kappa T_e}{\left(1 + T_e/T_i\right) m_i \omega_{ci}^2}. \tag{4.130}$$

This situation would describe shear modes in a high-β plasma (ion thermal velocity faster than Alfvén velocity).

To summarize: the shear mode has $B_{z1} = 0$, $E_{z1} \neq 0$, $J_{z1} \neq 0$, and exists as a cold plasma inertial Alfvén wave for $\beta_e \ll m_e/m_i$ and as one of two types of warm plasma kinetic Alfvén waves for $\beta_e \gg m_e/m_i$. These are shown in Fig. 4.3. The shear mode involves incompressible perpendicular motion, i.e., $\nabla \cdot \mathbf{u}_{\sigma\perp 1} = i\mathbf{k}_\perp \cdot \mathbf{u}_{\sigma\perp 1} = 0$, which means that $\mathbf{k}_\perp$ is orthogonal to $\mathbf{u}_{\sigma\perp 1}$. For example, in Cartesian geometry, this means that if $\mathbf{u}_{\sigma\perp 1}$ is in the x direction, then $\mathbf{k}_\perp$ must be in the y direction, while in cylindrical geometry, this means that if $\mathbf{u}_{\sigma\perp 1}$ is in the θ direction, then $\mathbf{k}_\perp$ must be in the r direction. The inertial Alfvén wave is known as a cold plasma wave because its dispersion relation does not depend on temperature (such a mode would exist even in the limit of a cold plasma). The kinetic Alfvén wave depends on the plasma having finite temperature and is therefore called a warm plasma wave. The shear mode parallel dynamics is related to parallel propagating ion acoustic modes since both shear Alfvén waves and parallel propagating ion acoustic modes involve parallel motion driven by finite E_{z1}; this parallel physics is missing from the MHD description.

4.4.2 Two-fluid compressional modes

The compressional mode has $B_{z1} \neq 0$ and $E_{z1} = 0$. Having $E_{z1} = 0$ means $J_{z1} = 0$ and, since there is no parallel motion, the linearized continuity equation becomes

$$\frac{\partial n_{\sigma 1}}{\partial t} + n\nabla \cdot \mathbf{u}_{\sigma\perp 1} = 0. \tag{4.131}$$

Using Eq. (4.115) this may be expressed as

$$\frac{\partial n_{\sigma 1}}{\partial t} - \frac{n}{B}\frac{\partial B_{z1}}{\partial t} = 0, \tag{4.132}$$

which may be integrated with respect to time to give

$$\frac{n_{\sigma 1}}{n} = \frac{B_{z1}}{B}. \tag{4.133}$$

Assuming an adiabatic response for this perpendicular compression gives

$$\frac{P_{\perp 1}}{P} = \gamma\frac{n_1}{n} = \gamma\frac{B_{z1}}{B}. \tag{4.134}$$

Substitution for $P_{\perp 1}$ in Eq. (4.114a) gives

$$\nabla_\perp \cdot \left(\frac{(v_A^2 + c_s^2)}{\omega^2 - k_z^2 v_A^2} \nabla_\perp B_{z1} \right) + B_{z1} = 0, \tag{4.135}$$

where

$$c_s^2 = \gamma\kappa\frac{T_e + T_i}{m_i}. \tag{4.136}$$

On replacing $\nabla_\perp \to \mathbf{i}\mathbf{k}_\perp$, Eq. (4.135) becomes the compressional mode dispersion relation

$$\frac{-k_\perp^2\left(v_A^2 + c_s^2\right)}{\omega^2 - k_z^2 v_A^2} + 1 = 0 \tag{4.137}$$

or

$$\omega^2 = k^2 v_A^2 + k_\perp^2 c_s^2, \tag{4.138}$$

where $k^2 = k_z^2 + k_\perp^2$.

4.5 Assignments

1. Plot frequency versus wavenumber for both the electron plasma wave and the ion acoustic wave in an unmagnetized argon plasma, which has $n = 10^{18}\ \mathrm{m}^{-3}$, $T_e = 10\,\mathrm{eV}$, and $T_i = 1\,\mathrm{eV}$.

2. Let $\Delta\phi$ be the difference between the phase shift a 632.8 nm helium–neon laser beam experiences on traversing a given length of vacuum and on traversing the same length of plasma. What is $\Delta\phi$ when the laser beam passes through 10 cm of plasma having a density of $n = 10^{22}$ m^3? How could this be used as a density diagnostic?
3. Prove that the electrostatic plasma wave $\omega^2 = \omega_{pe}^2 + 3k^2\kappa T_e/m_e$ can also be written as

$$\omega^2 = \omega_{pe}^2(1 + 3k^2\lambda_{De}^2)$$

and show over what range of $k^2\lambda_{De}^2$ the dispersion is valid. Plot the dispersion $\omega(k)$ versus k for both negative and positive k. Next, plot on the same graph the electromagnetic dispersion $\omega^2 = \omega_{pe}^2 + k^2c^2$ and show the limits of validity. Plot the ion acoustic dispersion $\omega^2 = k^2c_s^2/(1 + k^2\lambda_{De}^2)$ on this graph showing its region of validity. Finally, plot the ion acoustic dispersion with a finite ion temperature. Show the limits of validity of the ion acoustic dispersion.
4. Physical picture of plasma oscillations. Suppose that a plasma is cold and initially neutral. Consider a spherical volume of this plasma and imagine that a thin shell of electrons at spherical radius r having thickness δr moves radially outward by a distance equal to its thickness. Suppose further that the ions are infinitely massive and cannot move. What is the total ion charge acting on the electrons (consider the charge density and volume of the ions left behind when the electron shell is moved out)? What is the electric field due to these ions? By considering the force due to this electric field on an individual electron in the shell, show that the entire electron shell will execute simple harmonic motion at the frequency ω_{pe}. If the ions had finite mass how would you expect the problem to be modified (hint: consider the reduced mass)?
5. Suppose that an MHD plasma immersed in a uniform magnetic field $\mathbf{B} = B_0\hat{z}$ has an oscillating electric field $\tilde{\mathbf{E}}_\perp$, where $\perp$ means in the direction perpendicular to $\hat{z}$. What is the polarization current associated with $\tilde{\mathbf{E}}_\perp$? By substituting this polarization current into the MHD approximation of Ampère's law, find a relationship between $\partial\tilde{\mathbf{E}}_\perp/\partial t$ and a spatial operator on $\tilde{\mathbf{B}}$. Use Faraday's law to obtain a similar relationship between $\partial\tilde{\mathbf{B}}_\perp/\partial t$ and a spatial operator on $\tilde{\mathbf{E}}$. Consider a mode where $\tilde{E}_x(z,t)$ and $\tilde{B}_y(z,t)$ are the only finite components and derive a wave equation. Do the same for the pair $\tilde{E}_y(x,t)$ and $\tilde{B}_z(x,t)$. Which mode is the compressional mode and which is the shear mode?

5

Streaming instabilities and the Landau problem

5.1 Overview

The previous chapter demonstrated the existence of several types of plasma waves while Section 3.9 showed that particles can interchange energy with waves. The present chapter is devoted to establishing an important theory governing collisionless plasma waves, namely the theory of Landau damping. This theory characterizes the energy exchange that can occur between particles and waves when the wave-phase velocity is of the order of the particle thermal velocity. The Landau theory thus fills in the missing link between the adiabatic and isothermal susceptibilities discussed on p. 151. The first part of this chapter develops some preparatory ideas by showing that complex frequencies, i.e., wave instabilities, can develop if there are suitably arranged particle beams in a plasma. If one then considers a particle velocity distribution $f(v)$ as a collection of particle beams having different velocities, it should come as no surprise that waves with complex frequencies can occur. The next and main part of the chapter develops the detailed theory of Landau damping and summarizes the results in the form of a special function, called the "plasma dispersion function". The last part of the chapter examines the Penrose criterion, a method for determining the stability properties of a given velocity distribution function $f(v)$.

5.2 Streaming instabilities

The electrostatic dispersion relation for a zero-temperature plasma is simply

$$1-\sum_{\sigma}\frac{\omega_{p\sigma}^2}{\omega^2}=0, \tag{5.1}$$

indicating that a *spatially-independent* oscillation at the plasma frequency

$$\omega_p=\sqrt{\omega_{pe}^2+\omega_{pi}^2} \tag{5.2}$$

is a normal mode of a cold plasma. Once started, such an oscillation would persist indefinitely because no dissipative mechanism exists to quench it. On the other hand, the oscillation would have to be initiated by some source, because no available free energy exists from which the oscillation could draw on to start spontaneously.

We now consider a slightly different situation where cold electrons or ions stream at some spatially uniform initial velocity instead of being at rest in equilibrium. In the special situation where electrons and ions stream at the same velocity, the center of mass would also move at this velocity. One could then simply move to the center-of-mass frame where both species would be stationary and so, as argued in the previous paragraph, an oscillation would not start spontaneously. However, in the more general situation where the electrons and ions stream at *different* velocities, then both species have kinetic energy in the center-of-mass frame. This free energy could drive an instability.

In order to determine how such an instability could occur, the situation where each species has a specified equilibrium streaming velocity $\mathbf{u}_{\sigma 0}$ will now be examined. The linearized equation of motion, continuity equation, and Poisson's equation respectively become

$$\frac{\partial \mathbf{u}_{\sigma 1}}{\partial t} + \mathbf{u}_{\sigma 0} \cdot \nabla \mathbf{u}_{\sigma 1} = -\frac{q_\sigma}{m_\sigma} \nabla \phi_1, \tag{5.3}$$

$$\frac{\partial n_{\sigma 1}}{\partial t} + \mathbf{u}_{\sigma 0} \cdot \nabla n_{\sigma 1} = -n_{\sigma 0} \nabla \cdot \mathbf{u}_{\sigma 1}, \tag{5.4}$$

and

$$\nabla^2 \phi_1 = -\frac{1}{\varepsilon_0} \sum_\sigma q_\sigma n_{\sigma 1}. \tag{5.5}$$

As before, all first-order dependent variables are assumed to vary as $\exp(i\mathbf{k}\cdot\mathbf{x} - i\omega t)$. Combining the equation of motion and the continuity equation gives

$$n_{\sigma 1} = n_{\sigma 0} \frac{k^2}{(\omega - \mathbf{k}\cdot\mathbf{u}_{\sigma 0})^2} \frac{q_\sigma}{m_\sigma} \phi_1. \tag{5.6}$$

Substituting this into Eq. (5.5) gives the dispersion relation

$$1 - \sum_\sigma \frac{\omega_{p\sigma}^2}{(\omega - \mathbf{k}\cdot\mathbf{u}_{\sigma 0})^2} = 0, \tag{5.7}$$

which is just like the susceptibility for stationary cold species except that here ω is replaced by the Doppler-shifted frequency $\omega_{Doppler} = \omega - \mathbf{k}\cdot\mathbf{u}_{\sigma 0}$.

Two examples of streaming instability will now be considered: (i) equal densities of positrons and electrons streaming past each other with equal and opposite velocities, and (ii) electrons streaming past stationary ions.

Positron–electron streaming instability

The positron/electron example, while difficult to realize in practice, is worth analyzing because it reveals essential features of the streaming instability with a minimum of mathematical effort. The equilibrium positron and electron densities are assumed equal so as to have charge neutrality. Since electrons and positrons have identical mass, the positron plasma frequency ω_{pp} is the same as the electron plasma frequency ω_{pe}. Let $\mathbf{u}_0$ be the electron stream velocity and $-\mathbf{u}_0$ be the positron stream velocity. Defining $z = \omega/\omega_{pe}$ and $\lambda = \mathbf{k}\cdot\mathbf{u}_0/\omega_{pe}$, Eq. (5.7) reduces to

$$1 = \frac{1}{(z-\lambda)^2} + \frac{1}{(z+\lambda)^2}, \tag{5.8}$$

a quartic equation in z. Because of the symmetry, no odd powers of z appear and Eq. (5.8) becomes

$$z^4 - 2z^2(\lambda^2+1) + \lambda^4 - 2\lambda^2 = 0, \tag{5.9}$$

which may be solved for z^2 to give

$$z^2 = (\lambda^2+1) \pm \sqrt{4\lambda^2+1}. \tag{5.10}$$

Each choice of the $\pm$ sign gives two roots for z. If $z^2 > 0$, then the two roots are real, equal in magnitude, and opposite in sign. On the other hand, if $z^2 < 0$, then the two roots are *pure imaginary,* equal in magnitude, and opposite in sign. Recalling that $\omega = \omega_{pe}z$ and that the perturbation varies as $\exp(i\mathbf{k}\cdot\mathbf{x} - i\omega t)$, it is seen that the positive imaginary root corresponds to instability; i.e., to a perturbation that *grows* exponentially in time.

Hence the condition for instability is $z^2 < 0$. Because only the choice of minus sign in Eq. (5.10) allows this possibility, we select the minus sign and determine that the condition for instability is

$$\sqrt{4\lambda^2+1} > \lambda^2 + 1, \tag{5.11}$$

which corresponds to

$$0 < \lambda < \sqrt{2}. \tag{5.12}$$

The maximum growth rate is found by maximizing the right-hand side of Eq. (5.10) with the minus sign chosen. Taking the derivative with respect to λ and setting $\mathrm{d}z/\mathrm{d}\lambda = 0$ to find the maximum, gives

$$2z\frac{\mathrm{d}z}{\mathrm{d}\lambda} = 2\lambda - \frac{4\lambda}{\sqrt{4\lambda^2+1}} = 0 \tag{5.13}$$

or $\lambda = \sqrt{3}/2$. Substituting this most unstable λ back into Eq. (5.10) (with the minus sign selected, since this choice gives the potentially unstable root) gives the maximum growth rate to be $y = 1/2$, where $z = x + iy$.

Changing back to physical variables, it is seen that onset of instability occurs when

$$ku_0 < \sqrt{2}\omega_{pe}, \tag{5.14}$$

and the maximum growth rate occurs when

$$ku_0 = \frac{\sqrt{3}}{2}\omega_{pe}, \tag{5.15}$$

in which case

$$\omega = \mathrm{i}\frac{\omega_{pe}}{2}. \tag{5.16}$$

Figure 5.1 plots the normalized instability growth rate $\mathrm{Im}\, z$ as a function of λ; both onset and maximum growth rate are indicated. Since the instability has a pure imaginary frequency it is called a purely growing mode. Because the growth rate is of the order of magnitude of the plasma frequency, the instability grows extremely quickly.

Electron–ion streaming instability

Now consider the more realistic situation where electrons stream with velocity $\mathbf{v}_0$ through a background of stationary neutralizing ions. The dispersion relation in this case is

$$1 - \frac{\omega_{pi}^2}{\omega^2} - \frac{\omega_{pe}^2}{(\omega - \mathbf{k}\cdot\mathbf{u}_0)^2} = 0, \tag{5.17}$$

which can be recast in non-dimensional form by defining $z = \omega/\omega pe$, $\epsilon = m_e/m_i$, and $\lambda = \mathbf{k}\cdot\mathbf{u}_0/\omega_{pe}$, giving

$$1 = \frac{\epsilon}{z^2} + \frac{1}{(z-\lambda)^2}. \tag{5.18}$$

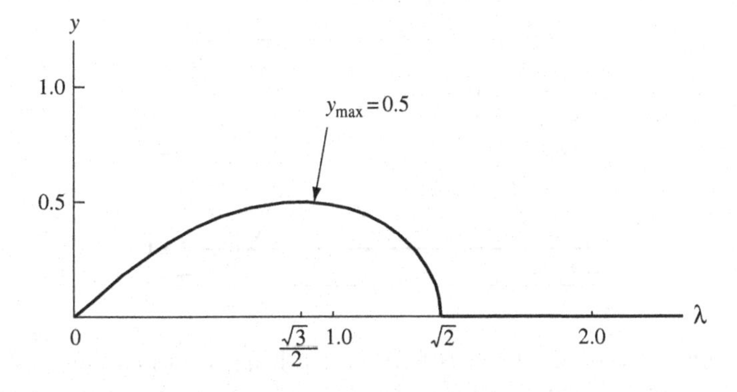

Fig. 5.1 Normalized growth rate vs. normalized wavenumber.

The value of λ at which onset of instability occurs is found by plotting the right-hand side (RHS) of Eq. (5.18) versus z and comparing this to the left-hand side, which always has the value of unity. This plot is shown in the sequence Fig. 5.2(a)–(c). The first term ϵ/z^2 diverges at $z=0$, while the second term diverges at $z=\lambda$. Between $z=0$ and $z=\lambda$, the right-hand side of Eq. (5.18) has a minimum. If the value of the right-hand side at this minimum is below unity, as in Fig. 5.2(a), there will be two places between $z=0$ and $z=\lambda$ where the right-hand side of Eq. (5.18) equals unity. For $z>\lambda$, there is always one and only one place where the right-hand side equals unity and similarly for $z<0$ there is one and only one place where the right-hand side equals unity. Thus, when the minimum of the right-hand side is below unity, Eq. (5.18) has four real roots. If λ is decreased to some critical value, then, as shown in Fig. 5.2(b), the two roots located between 0 and λ coalesce. If λ is decreased still more, then, as shown in Fig. 5.2(c), the minimum of the right-hand side is above unity so there are only two real roots (those in the regions $z>\lambda$ and $z<0$). In this latter case the other two roots of this quartic equation must be complex.

Because a quartic equation must be expressible in the form

$$(z-z_1)(z-z_2)(z-z_3)(z-z_4)=0, \tag{5.19}$$

and because the coefficients of Eq. (5.18) are real, the two complex roots must be complex conjugates of each other. To see this, suppose the complex roots are

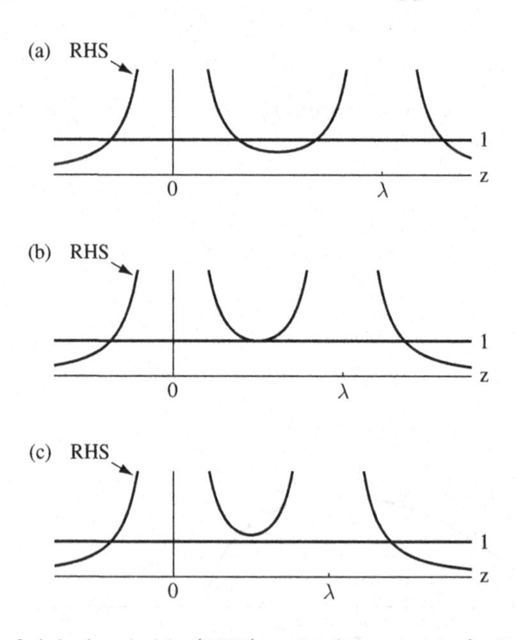

Fig. 5.2 Plot of right-hand side (RHS) and left-hand side (unity) of Eq. (5.18) for sequence of λ values. (a) Four real roots, (b) middle two roots coalesce, (c) two real roots only.

z_1 and z_2 and the real roots are z_3 and z_4. The product of the first two factors in Eq. (5.19) is $z^2 - (z_1 + z_2)z + z_1 z_2$; if the complex roots are not complex conjugates of each other then this product will contain complex coefficients and, when multiplied with the product of the terms involving the real roots, will result in an equation that contains complex coefficients. However, Eq. (5.18) has only real coefficients so the two complex roots must be complex conjugates of each other. The complex root with positive imaginary part will give rise to instability.

Thus, when the minimum of the right-hand side of Eq. (5.18) is greater than unity, two of the roots become complex, and one of these complex roots gives instability. The onset of instability occurs when the minimum of the right-hand side of Eq. (5.18) equals unity. Straightforward analysis (cf. assignments) shows this occurs when

$$\lambda = (1 + \epsilon^{1/3})^{3/2}, \tag{5.20}$$

i.e., instability starts when

$$\mathbf{k} \cdot \mathbf{u}_0 = \omega_{pe} \left[1 + \left(\frac{m_e}{m_i} \right)^{1/3} \right]^{3/2}. \tag{5.21}$$

The maximum growth rate of the instability may be found by solving Eq. (5.18) for λ and then simplifying the resulting expression using ϵ as a small parameter. The details of this are worked out in the assignments, where it is demonstrated that the maximum growth rate is

$$\max \omega_i \simeq \frac{\sqrt{3}}{2} \left(\frac{m_e}{2m_i} \right)^{1/3} \omega_{pe}, \tag{5.22}$$

which occurs when

$$\mathbf{k} \cdot \mathbf{u}_0 \simeq \omega_{pe}. \tag{5.23}$$

As before, this is a very fast-growing instability, about one order of magnitude smaller than the electron plasma frequency.

Streaming instabilities are a reason why certain simple proposed methods for attaining thermonuclear fusion will not work. These methods involve shooting an energetic deuterium beam at an oppositely directed energetic tritium beam with the expectation that collisions between the two beams would produce fusion reactions. However, such a system is extremely unstable with respect to the two-stream instability. This instability typically has a growth rate much faster than the fusion reaction rate and so will destroy the beams before significant fusion reactions can occur.

5.3 The Landau problem

A plasma wave behavior of great philosophical interest and of great practical importance can now be investigated. Before doing so, we recall three seemingly disconnected results obtained thus far, namely:

1. When the exchange of energy between charged particles and a simple one-dimensional electrostatic wave with dependence $\sim \exp(ikx - i\omega t)$ was considered, the particles were categorized into two general classes, trapped and untrapped, and it was found that untrapped particles tended to be dragged towards the wave phase velocity. Thus, untrapped particles moving slower than the wave gain kinetic energy, whereas untrapped particles moving faster lose kinetic energy. This has the consequence that if there are more slow than fast particles, the particles gain net kinetic energy overall and this gain presumably comes at the expense of the wave. Conversely, if there are more fast than slow particles, net energy flows from the particles to the wave.
2. When electrostatic plasma waves in an unmagnetized, uniform, stationary plasma were considered, it was found that wave behavior was characterized by a dispersion relation $1+\chi_e(\omega, k)+\chi_i(\omega, k)=0$, where $\chi_\sigma(\omega, k)$ is the susceptibility of each species σ. As sketched in Fig. 4.1 these susceptibilities had simple limiting forms when $\omega/k \ll \sqrt{\kappa T_{\sigma 0}/m_\sigma}$ (isothermal limit) and when $\omega/k \gg \sqrt{\kappa T_{\sigma 0}/m_\sigma}$ (adiabatic limit), but the fluid analysis failed when $\omega/k \sim \sqrt{\kappa T_{\sigma 0}/m_\sigma}$ and the susceptibilities became undefined.
3. When the behavior of interacting beams of particles was considered, it was found that under certain conditions a fast-growing instability would develop.

The analysis of the Landau problem, to be presented in the remainder of this chapter, will show that these three results are both interrelated and part of a larger picture.

5.3.1 Attempt to solve the linearized Vlasov–Poisson system of equations using Fourier analysis

The method for manipulating fluid equations to find wave solutions was as follows: (i) the relevant fluid equations were linearized, (ii) a perturbation $\sim \exp(i\mathbf{k}\cdot\mathbf{x} - i\omega t)$ was assumed, (iii) the system of partial differential equations was transformed into a system of algebraic equations, and then finally (iv) the roots of the determinant of the system of algebraic equations provided the dispersion relations that characterized the various wave solutions.

It seems reasonable to use this method again in order to investigate waves from the Vlasov point of view. However, it will be seen that this approach *fails* and that, instead, a more complicated Laplace transform technique must be used. However, once the underlying difference between the Laplace and Fourier transform techniques has been identified, it is possible to go back and "patch up" the Fourier technique. Although perhaps not entirely elegant, this patching approach turns out to be a reasonable compromise in that it incorporates both the simplicity of the Fourier method and the correct mathematics/physics of the Laplace method.

The Fourier method will now be presented and, to highlight how this method fails, the simplest relevant example will be considered, namely a one-dimensional, unmagnetized plasma with a stationary Maxwellian equilibrium. The ions are assumed to be so massive as to be immobile and the ion density is assumed to equal the electron equilibrium density. The electrostatic electric field $E = -\partial\phi/\partial x$ is therefore zero in equilibrium because there is charge neutrality in equilibrium. Since ions do not move there is no need to track ion dynamics. Thus, all perturbed quantities refer to electrons and so it is redundant to label these with a subscript "e." In order to have a well-defined, physically meaningful problem, the equilibrium electron velocity distribution is assumed to be Maxwellian, i.e.,

$$f_0(v) = n_0 \frac{1}{\pi^{1/2} v_T} \mathrm{e}^{-v^2/v_T^2}, \tag{5.24}$$

where $v_T \equiv \sqrt{2\kappa T/m}$.

The one-dimensional, unmagnetized Vlasov equation is

$$\frac{\partial f}{\partial t} + v\frac{\partial f}{\partial x} - \frac{q}{m}\frac{\partial \phi}{\partial x}\frac{\partial f}{\partial v} = 0 \tag{5.25}$$

and linearization of this equation gives

$$\frac{\partial f_1}{\partial t} + v\frac{\partial f_1}{\partial x} - \frac{q}{m}\frac{\partial \phi_1}{\partial x}\frac{\partial f_0}{\partial v} = 0. \tag{5.26}$$

Because the Vlasov equation describes evolution in phase-space, v is an *independent* variable just like x and t. Assuming a normal mode dependence $\sim \exp(\mathrm{i}kx - \mathrm{i}\omega t)$, Eq. (5.26) becomes

$$-\mathrm{i}(\omega - kv)f_1 - \mathrm{i}k\phi_1 \frac{q}{m}\frac{\partial f_0}{\partial v} = 0, \tag{5.27}$$

which gives

$$f_1 = -\frac{k}{(\omega - kv)}\frac{q}{m}\frac{\partial f_0}{\partial v}\phi_1. \tag{5.28}$$

The electron density perturbation is

$$n_1 = \int_{-\infty}^{\infty} f_1 \mathrm{d}v = -\frac{q}{m}\phi_1 \int_{-\infty}^{\infty} \frac{k}{(\omega - kv)}\frac{\partial f_0}{\partial v}\mathrm{d}v, \tag{5.29}$$

a relationship between n_1 and ϕ_1. Another relationship between n_1 and ϕ_1 is Poisson's equation

$$\frac{\partial^2 \phi_1}{\partial x^2} = -\frac{n_1 q}{\varepsilon_0}. \tag{5.30}$$

Replacing $\partial/\partial x$ by ik, Eq. (5.30) becomes

$$k^2 \phi_1 = \frac{n_1 q}{\varepsilon_0}. \tag{5.31}$$

Combining Eqs. (5.29) and (5.31) gives the dispersion relation

$$1 + \frac{q^2}{k^2 m \varepsilon_0} \int_{-\infty}^{\infty} \frac{k}{(\omega - kv)} \frac{\partial f_0}{\partial v} \mathrm{d}v = 0. \tag{5.32}$$

This can be written more elegantly by substituting for f_0 using Eq. (5.24), defining the non-dimensional particle velocity $\xi = v/v_T$, and the non-dimensional phase velocity $\alpha = \omega/kv_T$ to give

$$1 - \frac{1}{2k^2\lambda_D^2} \frac{1}{\pi^{1/2}} \int_{-\infty}^{\infty} \mathrm{d}\xi \frac{1}{(\xi - \alpha)} \frac{\partial}{\partial \xi} \mathrm{e}^{-\xi^2} = 0 \tag{5.33}$$

or

$$1 + \chi = 0, \tag{5.34}$$

where the electron susceptibility is

$$\chi = -\frac{1}{2k^2\lambda_D^2} \frac{1}{\pi^{1/2}} \int_{-\infty}^{\infty} \mathrm{d}\xi \frac{1}{(\xi - \alpha)} \frac{\partial}{\partial \xi} \mathrm{e}^{-\xi^2}. \tag{5.35}$$

In contrast to the earlier two-fluid wave analysis, where in effect the zeroth, first, and second moments of the Vlasov equation were combined (continuity equation, equation of motion, and equation of state), here only the Vlasov equation is involved. Thus the Vlasov equation contains all the information of the moment equations and more. The Vlasov method therefore seems a simpler and more direct way for calculating the susceptibilities than the fluid method, except for a serious difficulty: the integral in Eq. (5.35) is mathematically ill-defined because the denominator vanishes when $\xi = \alpha$ (i.e., when $\omega = kv_T$). Because it is not clear how to deal with this singularity, the ξ integral cannot be evaluated and the Fourier method fails. This is essentially the same as the problem encountered in fluid analysis when ω/k became comparable to $\sqrt{\kappa T/m}$.

5.3.2 Landau method: Laplace transforms

Landau (1946) argued that the Fourier problem presented above is ill-posed and showed that the linearized Vlasov–Poisson problem should be treated as an *initial-value* problem, rather than as a normal mode problem. The initial-value point of view is conceptually related to the analysis of single particle motion in sawtooth or sine waves. Before presenting the Landau analysis of the linearized Vlasov–Poisson problem, certain important features of Laplace transforms will now be reviewed.

The Laplace transform of a function $\psi(t)$ is defined as

$$\tilde{\psi}(p) = \int_0^\infty \psi(t) \mathrm{e}^{-pt} \mathrm{d}t \tag{5.36}$$

and can be considered as a "half of a Fourier transform" since the time integration starts at $t = 0$ rather than $t = -\infty$. Caution is required regarding the convergence of this integral for situations where $\psi(t)$ contains exponentially growing terms.

Suppose such exponentially growing terms exist. As $t \to \infty$, the fastest growing term, say $\exp(\gamma t)$, will dominate all other terms contributing to $\psi(t)$. The integral in Eq. (5.36) will then diverge as $t \to \infty$, unless a *restriction* is imposed on the real part of p. In particular, if it is *required* that Re $p > \gamma$, then the decaying $\exp(-pt)$ factor will always overwhelm the growing $\exp(\gamma t)$ factor so that the integral in Eq. (5.36) will converge. These issues of convergence are ignored in Fourier transforms where it is implicitly assumed that the function being transformed has neither exponentially growing terms (which diverge at $t = \infty$) nor exponentially decaying terms (which diverge at $t = -\infty$).

Thus, the integral transform in Eq. (5.36) is defined *only* for Re $p > \gamma$. To emphasize this restriction, Eq. (5.36) is rewritten as

$$\tilde{\psi}(p) = \int_0^\infty \psi(t) \mathrm{e}^{-pt} \mathrm{d}t, \quad \mathrm{Re}\ p > \gamma, \tag{5.37}$$

where γ is the fastest growing exponential term contained in $\psi(t)$. Since p is typically complex, Eq. (5.37) means that $\tilde{\psi}(p)$ is *only* defined in that part of the complex p-plane lying to the *right* of γ as sketched in Fig. 5.3(a). Whenever $\tilde{\psi}(p)$ is used, one must be very careful to avoid venturing outside the region in p-space where $\tilde{\psi}(p)$ is defined (this restriction will later become an important issue).

To construct an inverse transform, consider the integral

$$g(t) = \int_C \mathrm{d}p\, \tilde{\psi}(p) \mathrm{e}^{pt}. \tag{5.38}$$

This integral is ambiguously defined for now because the integration contour C is unspecified. However, whatever integration contour is ultimately selected must not venture into regions where $\tilde{\psi}(p)$ is undefined. Thus, an allowed integration path must have Re $p > \gamma$. Substitution of Eq. (5.37) into Eq. (5.38) and interchanging the order of integration gives

$$g(t) = \int_0^\infty \mathrm{d}t' \int_C \mathrm{d}p\, \psi(t') \mathrm{e}^{p(t-t')}, \quad \mathrm{Re}\ p > \gamma. \tag{5.39}$$

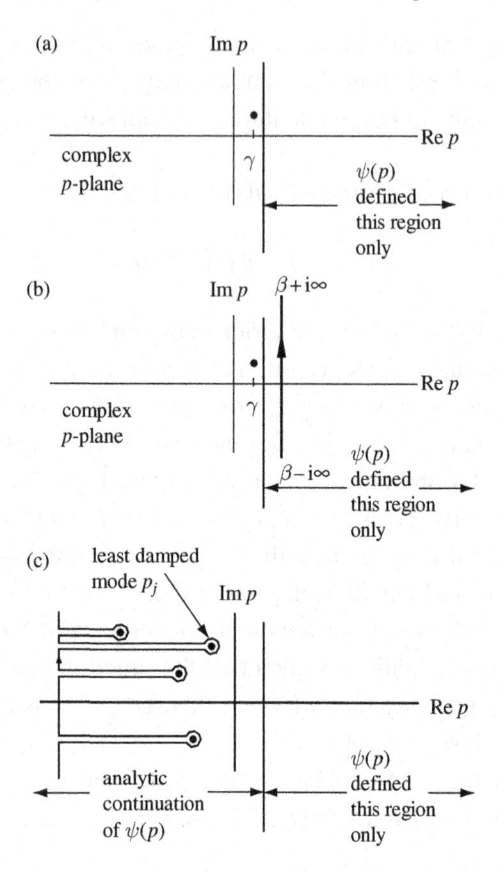

Fig. 5.3 Contours in complex p-plane.

A useful integration path C for the p integral will now be determined. Recall from the theory of Fourier transforms that the Dirac delta function can be expressed as

$$\delta(t) = \frac{1}{2\pi}\int_{-\infty}^{\infty} \mathrm{d}\omega\, \mathrm{e}^{\mathrm{i}\omega t}, \tag{5.40}$$

which is an integral along the real ω axis so that ω is always real. The integration path for Eq. (5.39) will now be chosen such that the real part of p stays constant, say at a value β that is larger than γ, while the imaginary part of p goes from $-\infty$ to ∞. This path is shown in Fig. 5.3(b), and is called the Bromwich contour.

For this choice of path, Eq. (5.39) becomes

$$\begin{aligned} g(t) &= \int_0^{\infty} \mathrm{d}t' \int_{\beta-\mathrm{i}\infty}^{\beta+\mathrm{i}\infty} \mathrm{d}(p_r+\mathrm{i}p_i)\, \psi(t')\mathrm{e}^{(p_r+\mathrm{i}p_i)(t-t')} \\ &= \mathrm{i}\int_0^{\infty} \mathrm{d}t' \mathrm{e}^{\beta(t-t')}\psi(t') \int_{-\infty}^{\infty} \mathrm{d}p_i\, \mathrm{e}^{\mathrm{i}p_i(t-t')} \end{aligned}$$

$$= 2\pi \mathrm{i} \int_0^\infty \mathrm{d}t' \mathrm{e}^{\beta(t-t')} \psi(t') \delta(t-t')$$
$$= 2\pi \mathrm{i} \psi(t), \tag{5.41}$$

where Eq. (5.40) has been used. Thus, $\psi(t) = (2\pi \mathrm{i})^{-1} g(t)$ and so the inverse of the Laplace transform is

$$\psi(t) = \frac{1}{2\pi \mathrm{i}} \int_{\beta - \mathrm{i}\infty}^{\beta + \mathrm{i}\infty} \mathrm{d}p \, \psi(p) \mathrm{e}^{pt}, \quad \beta > \gamma. \tag{5.42}$$

Before returning to physics, recall another peculiarity of Laplace transforms, namely the transformation procedure for derivatives. The Laplace transform of $\mathrm{d}\psi/\mathrm{d}t$ may be simplified by integrating by parts to give

$$\int_0^\infty \mathrm{d}t \frac{\mathrm{d}\psi}{\mathrm{d}t} \mathrm{e}^{-pt} = \left[\psi(t)\mathrm{e}^{-pt}\right]_0^\infty + p \int_0^\infty \mathrm{d}t \, \psi(t) \mathrm{e}^{-pt} = p\tilde{\psi}(p) - \psi(0). \tag{5.43}$$

Unlike Fourier transforms, here the *initial value* forms part of the transform. Thus, Laplace transforms contain information about the initial value and so should be better suited than Fourier transforms for investigating initial value problems. The importance of the initial value was also evident in the Chapter 3 analysis of particle motion in sawtooth or sine wave potentials.

The requisite mathematical tools are now in hand for investigating the Vlasov–Poisson system and its dependence on initial value. To obtain extra insights with little additional effort, the analysis is extended to the more general situation of a three-dimensional plasma where ions are allowed to move. Again, electrostatic waves are considered, and it is assumed that the equilibrium plasma is stationary, spatially uniform, neutral, and unmagnetized.

The equilibrium velocity distribution of each species is assumed to be a three-dimensional Maxwellian distribution function

$$f_{\sigma 0}(\mathbf{v}) = n_{\sigma 0} \left(\frac{m_\sigma}{2\pi \kappa T_\sigma} \right)^{3/2} \exp\left(-m_\sigma v^2 / 2\kappa T_\sigma\right). \tag{5.44}$$

The equilibrium electric field is assumed to be zero so that the equilibrium potential is a constant chosen to be zero. It is further assumed that at $t = 0$ there exists a small perturbation of the distribution function and that this perturbation evolves in time so that at later times

$$f_\sigma(\mathbf{x}, \mathbf{v}, t) = f_{\sigma 0}(\mathbf{v}) + f_{\sigma 1}(\mathbf{x}, \mathbf{v}, t). \tag{5.45}$$

The linearized Vlasov equation for each species is therefore

$$\frac{\partial f_{\sigma 1}}{\partial t} + \mathbf{v} \cdot \nabla f_{\sigma 1} - \frac{q_\sigma}{m_\sigma} \nabla \phi_1 \cdot \frac{\partial f_{\sigma 0}}{\partial \mathbf{v}} = 0. \tag{5.46}$$

All perturbed quantities are assumed to have the spatial dependence $\sim \exp(\mathrm{i}\mathbf{k}\cdot\mathbf{x})$; this is equivalent to Fourier transforming in space. Equation (5.46) becomes

$$\frac{\partial f_{\sigma 1}}{\partial t} + \mathrm{i}\mathbf{k}\cdot\mathbf{v}f_{\sigma 1} - \frac{q_\sigma}{m_\sigma}\phi_1 \mathrm{i}\mathbf{k}\cdot\frac{\partial f_{\sigma 0}}{\partial \mathbf{v}} = 0. \tag{5.47}$$

Laplace transforming in time gives

$$(p+\mathrm{i}\mathbf{k}\cdot\mathbf{v})\tilde{f}_{\sigma 1}(\mathbf{v},p) - f_{\sigma 1}(\mathbf{v},0) - \frac{q_\sigma}{m_\sigma}\tilde{\phi}_1(p)\mathrm{i}\mathbf{k}\cdot\frac{\partial f_{\sigma 0}}{\partial \mathbf{v}} = 0, \tag{5.48}$$

which may be solved for $\tilde{f}_{\sigma 1}(\mathbf{v},p)$ to give

$$\tilde{f}_{\sigma 1}(\mathbf{v},p) = \frac{1}{(p+\mathrm{i}\mathbf{k}\cdot\mathbf{v})}\left[f_{\sigma 1}(\mathbf{v},0) + \frac{q_\sigma}{m_\sigma}\tilde{\phi}_1(p)\mathrm{i}\mathbf{k}\cdot\frac{\partial f_{\sigma 0}}{\partial \mathbf{v}}\right]. \tag{5.49}$$

This is similar to Eq. (5.28), except that now the Laplace variable p occurs instead of the Fourier variable $-\mathrm{i}\omega$ and also the initial value $f_{\sigma 1}(\mathbf{v},0)$ appears. As before, Poisson's equation can be written as

$$\nabla^2\phi_1 = -\frac{1}{\varepsilon_0}\sum_\sigma q_\sigma n_{\sigma 1} = -\frac{1}{\varepsilon_0}\sum_\sigma q_\sigma \int \mathrm{d}^3v f_{\sigma 1}(\mathbf{x},\mathbf{v},t). \tag{5.50}$$

Replacing $\nabla \to \mathrm{i}\mathbf{k}$ and Laplace transforming with respect to time, Poisson's equation becomes

$$k^2\tilde{\phi}_1(p) = \frac{1}{\varepsilon_0}\sum_\sigma q_\sigma \int \mathrm{d}^3v \tilde{f}_{\sigma 1}(\mathbf{v},p). \tag{5.51}$$

Substitution of Eq. (5.49) into the right-hand side of Eq. (5.51) gives

$$k^2\tilde{\phi}_1(p) = \frac{1}{\varepsilon_0}\sum_\sigma q_\sigma \int \mathrm{d}^3v \left\{\frac{f_{\sigma 1}(\mathbf{v},0) + \dfrac{q_\sigma}{m_\sigma}\tilde{\phi}_1(p)\mathrm{i}\mathbf{k}\cdot\dfrac{\partial f_{\sigma 0}}{\partial \mathbf{v}}}{(p+\mathrm{i}\mathbf{k}\cdot\mathbf{v})}\right\}, \tag{5.52}$$

which is similar to Eq. (5.32) except that $-\mathrm{i}\omega \to p$ and the initial value appears. Equation (5.52) may be solved for $\tilde{\phi}_1(p)$ to give

$$\tilde{\phi}_1(p) = \frac{N(p)}{D(p)}, \tag{5.53}$$

where the numerator is

$$N(p) = \frac{1}{k^2\varepsilon_0}\sum_\sigma q_\sigma \int \mathrm{d}^3v \frac{f_{\sigma 1}(\mathbf{v},0)}{(p+\mathrm{i}\mathbf{k}\cdot\mathbf{v})} \tag{5.54}$$

and the denominator is

$$D(p) = 1 - \frac{1}{k^2}\sum_\sigma \frac{q_\sigma^2}{\varepsilon_0 m_\sigma}\int \mathrm{d}^3v \frac{\mathrm{i}\mathbf{k}\cdot\dfrac{\partial f_{\sigma 0}}{\partial \mathbf{v}}}{(p+\mathrm{i}\mathbf{k}\cdot\mathbf{v})}. \tag{5.55}$$

Note that the denominator is similar to Eq. (5.32). All that has to be done now is take the inverse Laplace transform of Eq. (5.53) to obtain

$$\phi_1(t) = \frac{1}{2\pi\mathrm{i}} \int_{\beta-\mathrm{i}\infty}^{\beta+\mathrm{i}\infty} \mathrm{d}p \frac{N(p)}{D(p)} \mathrm{e}^{pt}, \tag{5.56}$$

where β is chosen to be larger than the fastest growing exponential term in $N(p)/D(p)$.

This is an exact formal solution to the problem. However, because of the complexity of $N(p)$ and $D(p)$ it is impossible to evaluate the integral in Eq. (5.56). Nevertheless, it turns out to be feasible to evaluate the long-time asymptotic limit of this integral and, for practical purposes, this is a sufficient answer to the problem.

5.3.3 The relationship between poles, exponential functions, and analytic continuation

Before evaluating Eq. (5.56), it is useful to examine the relationship between exponentially growing/decaying functions, Laplace transforms, poles, residues, and analytic continuation. This relationship is demonstrated by considering the exponential function

$$f(t) = \mathrm{e}^{qt}, \tag{5.57}$$

where q is a complex constant. If the real part of q is positive, then the amplitude of $f(t)$ is exponentially growing, whereas if the real part of q is negative, the amplitude of $f(t)$ is exponentially decaying. Now, calculate the Laplace transform of $f(t)$; it is

$$\tilde{f}(p) = \int_0^\infty \mathrm{e}^{(q-p)t} \mathrm{d}t = \frac{1}{p-q}, \quad \text{defined only for Re } p > \text{Re } q. \tag{5.58}$$

Let us examine the Bromwich contour integral for $\tilde{f}(p)$ and temporarily call this integral $F(t)$; evaluation of $F(t)$ ought to yield $F(t) = f(t)$. Thus, we define

$$F(t) = \frac{1}{2\pi\mathrm{i}} \int_{\beta-\mathrm{i}\infty}^{\beta+\mathrm{i}\infty} \mathrm{d}p \tilde{f}(p) \mathrm{e}^{pt}, \quad \beta > \text{Re } q. \tag{5.59}$$

If the Bromwich contour could be closed in the left-hand p-plane, the integral could easily be evaluated using the method of residues but closure of the contour to the left is forbidden because of the restriction that $\beta > \text{Re } q$. This annoyance may be overcome by constructing a new function $\hat{f}(p)$ that:

1. equals $\tilde{f}(p)$ in the region $\beta > \text{Re } q$,
2. is *also* defined in the region $\beta < \text{Re } q$, and
3. is analytic.

Integration of $\hat{f}(p)$ along the Bromwich contour gives the same result as does integration of $\tilde{f}(p)$ along the same contour because the two functions are *identical* along this contour (cf. stipulation (1) above). Thus, it is seen that

$$F(t) = \frac{1}{2\pi \mathrm{i}} \int_{\beta-\mathrm{i}\infty}^{\beta+\mathrm{i}\infty} \mathrm{d}p \hat{f}(p) \mathrm{e}^{pt}, \tag{5.60}$$

but now there is no restriction on which part of the p-plane may be used. So long as the end points are kept fixed and no poles are crossed, the path of integration of an analytic function can be arbitrarily deformed. This is because the difference between the original path and a deformed path is a closed contour, which integrates to zero if it does not enclose any poles. Because $f(\hat{p}) \to 0$ at the endpoints $\beta \pm \mathrm{i}\infty$, the integration path of $\hat{f}(p)$ can be deformed into the left-hand plane as long as $\hat{f}(p)$ remains analytic (i.e., does not jump over any poles or branch cuts). How can this magic function $\hat{f}(p)$ be constructed?

The answer is simple: we *define* a function $\hat{f}(p)$ having the identical functional form as $\tilde{f}(p)$, but *without* the restriction that Re $p >$ Re q. Thus, the analytic continuation of

$$\tilde{f}(p) = \frac{1}{p-q}, \quad \text{defined only for } \mathrm{Re}\, p > \mathrm{Re}\, q, \tag{5.61}$$

is simply

$$\hat{f}(p) = \frac{1}{p-q}, \quad \text{defined for all } p, \text{ provided } \hat{f}(p) \text{ remains analytic.} \tag{5.62}$$

The Bromwich contour can now be deformed into the left-hand plane as shown in Fig. 5.4. Because $\exp(pt) \to 0$ for positive t and negative Re p, the integration contour can be closed by an arc that goes to the left (cf. Fig. 5.4) into the region where Re $p \to -\infty$. The resulting contour encircles the pole at $p = q$ and so the integral can be evaluated using the method of residues as follows:

$$F(t) = \frac{1}{2\pi \mathrm{i}} \oint \frac{1}{p-q} \mathrm{e}^{pt} \mathrm{d}p = \lim_{p \to q} 2\pi \mathrm{i}(p-q) \left[\frac{1}{2\pi \mathrm{i}(p-q)} \mathrm{e}^{pt} \right] = \mathrm{e}^{qt}. \tag{5.63}$$

This simple example shows that while the Bromwich contour formally gives the inverse Laplace transform of $\tilde{f}(p)$, the Bromwich contour by itself does not allow use of the method of residues, since the poles of interest are located in the left-hand complex p-plane where $\tilde{f}(p)$ is undefined. However, analytic continuation of $f(p)$ allows deformation of the Bromwich contour into the formerly forbidden area, and then the inverse transform may be easily evaluated using the method of residues.

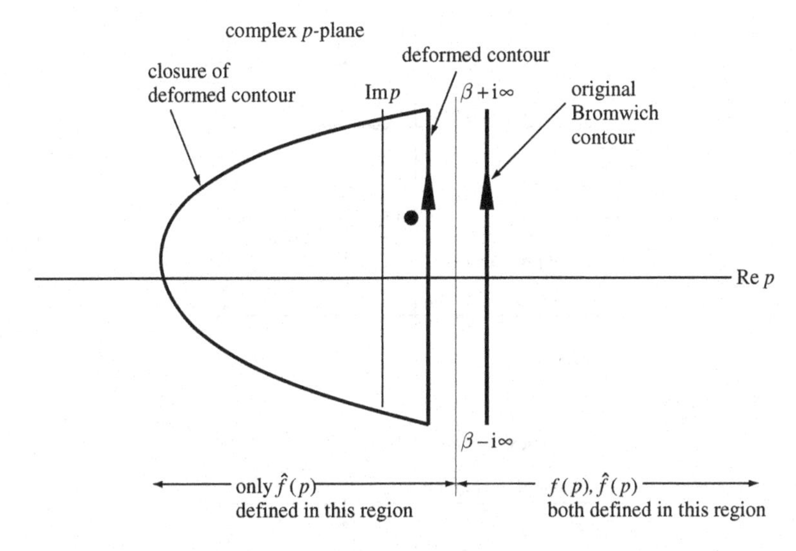

Fig. 5.4 Bromwich contour.

5.3.4 *Asymptotic long-time behavior of the potential oscillation*

We now return to the more daunting problem of evaluating Eq. (5.56). As in the simple example above, the goal is to close the contour to the left but, because the functions $N(p)$ and $D(p)$ are not defined for Re $p < \gamma$, this is not immediately possible. It is first necessary to construct analytic continuations of $N(p)$ and $D(p)$ that extend the definition of these functions into regions of negative Re p. As in the simple example, the desired analytic continuations may be constructed by taking the same formal expressions as obtained before, but now extending the definition to the entire p-plane with the proviso that *the functions remain analytic* as the region of definition is pushed leftwards in the p-plane.

Consider first construction of an analytic continuation for the function $N(p)$. This function can be written as

$$N(p) = \frac{1}{k^2\varepsilon_0}\sum_\sigma q_\sigma \int_{-\infty}^{\infty} \mathrm{d}v_\parallel \frac{F_{\sigma 1}(v_\parallel, 0)}{(p+\mathrm{i}kv_\parallel)} = \frac{1}{\mathrm{i}k^3\varepsilon_0}\sum_\sigma q_\sigma \int_{-\infty}^{\infty} \mathrm{d}v_\parallel \frac{F_{\sigma 1}(v_\parallel, 0)}{(v_\parallel - \mathrm{i}p/k)}. \quad (5.64)$$

Here, $\parallel$ means in the $\mathbf{k}$ direction, and the parallel component of the initial value of the perturbed distribution function has been defined as

$$F_{\sigma 1}(v_\parallel, 0) = \int \mathrm{d}^2\mathbf{v}_\perp f_{\sigma 1}(\mathbf{v}, 0). \quad (5.65)$$

The integrand in Eq. (5.54) has a pole at $v_\parallel = \mathrm{i}p/k$. Let us assume that $k > 0$ (the coordinate system can always be defined so that this is so). Before we construct an analytic continuation, Re p is restricted to be greater than γ so that the pole $v_\parallel = \mathrm{i}p/k$ is in the *upper half* of the complex $v_\parallel$-plane as shown in Fig. 5.5(a).

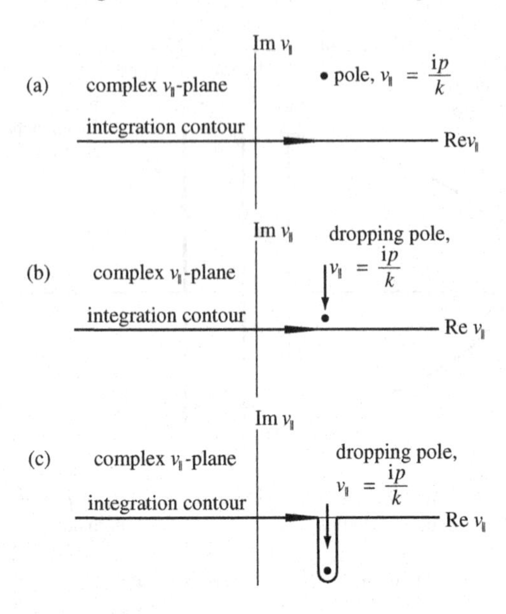

Fig. 5.5 Complex $v_\parallel$-plane.

When $N(p)$ is analytically continued to the left-hand region, the definition of $N(p)$ is extended to allow Re p to become less than γ and even negative. As shown in Fig. 5.5(b), decreasing Re p means that the pole at $v_\parallel = \mathrm{i}p/k$ in Eq. (5.54) drops from its initial location in the upper half $v_\parallel$-plane towards the lower half $v_\parallel$-plane. A critical question now arises: how should we arrange this construction when Re p passes through zero? If the pole is allowed to jump from being above the path of $v_\parallel$ integration (which is along the real $v_\parallel$ axis) to being below, the function $N(p)$ will *not* be analytic because it will have a discontinuous jump of $2\pi\mathrm{i}$ times the residue associated with the pole. Since it was stipulated that $N(p)$ must be analytic, the pole cannot be allowed to jump over the $v_\parallel$ contour of integration. Instead, the prescription proposed by Landau will be used, which is to *deform* the $v_\parallel$ contour as Re p becomes negative so that the contour *always* lies below the pole; this deformation is shown in Fig. 5.5(c).

$D(p)$ involves a similar integration along the real $v_\parallel$ axis. It also has a pole that is initially in the upper half-plane when Re $p > 0$, but then drops to being below the axis as Re p is allowed to become negative. Thus, analytic continuation of $D(p)$ is also constructed by deforming the path of the $v_\parallel$ integration so that the contour always lies below the pole.

Equipped with these suitably constructed analytic continuations of $N(p)$ and $D(p)$ into the left-hand p-plane, evaluation of Eq. (5.56) can now be undertaken. As shown in the simple example, it is computationally advantageous to deform the Bromwich contour into the left-hand p-plane. The deformed contour evaluates to the same result as the original Bromwich contour (provided the deformation does not jump over any poles) and this evaluation may be accomplished via the method of residues. In the general case where $N(p)/D(p)$ has several poles in the left-hand p-plane, then, as shown in Fig. 5.3(c), the contour may be deformed so that the vertical portion is pushed to the far left, except where there is a pole p_j; the contour "snags" around each pole p_j as shown in Fig. 5.3(c). For $\mathrm{Re}\, p \to -\infty$, the numerator $N(p) \to 0$, while the denominator $D(p) \to 1$. Since $\exp(pt) \to 0$ for $\mathrm{Re}\ p \to -\infty$ and positive t, the left-hand vertical line does not contribute to the integral and Eq. (5.56) simply consists of the sum of the residues of all the poles, i.e.,

$$\phi_1(t) = \sum_j \lim_{p \to p_j} \left[(p - p_j) \frac{N(p)}{D(p)} \mathrm{e}^{pt} \right]. \tag{5.66}$$

Where do the poles p_j come from? Upon examining Eq. (5.66), it is clear that poles could come either from (i) $N(p)$ having an explicit pole, i.e., $N(p)$ contains a term $\sim 1/(p - p_j)$, or (ii) from $D(p)$ containing a factor $\sim (p - p_j)$, i.e., p_j is a root of the equation $D(p) = 0$. The integrand in Eq. (5.64) has a pole in the $v_\parallel$-plane; this pole is "used up" as a residue upon performing the $v_\parallel$ integration, and so does not contribute a pole to $N(p)$. The only other possibility is that the initial value $F_{\sigma 1}(v_\parallel, 0)$ somehow provides a pole, but $F_{\sigma 1}(v_\parallel, 0)$ is a physical quantity with a bounded integral i.e., $\int F_{\sigma 1}(v_\parallel, 0)\mathrm{d}v_\parallel$ is finite and so cannot contribute a pole in $N(p)$. It is therefore concluded that all poles in $N(p)/D(p)$ must come from the roots (also called zeros) of $D(p)$.

The problem can be simplified by deciding to be content with a *less* than complete solution. Instead of attempting to calculate $\phi_1(t)$ for all positive times (i.e., all the poles p_j contribute to the solution), we restrict ourselves to the less burdensome problem of finding the long-time asymptotic behavior of $\phi_1(t)$. Because each term in Eq. (5.66) has a factor $\exp(p_j t)$, the least damped term (i.e., the term with pole furthest to the right in Fig. 5.3(c)) will dominate all the other terms at large t. Hence, in order to find the long-time asymptotic behavior, all that is required is to find the root p_j having the largest real part.

The problem is thus reduced to finding the roots of $D(p)$; this requires performing the $v_\parallel$ integration sketched in Fig. 5.5(c). Before doing this, it is convenient to integrate out the perpendicular velocity dependence from $D(p)$ so that

$$\begin{aligned} D(p) &= 1 - \frac{1}{k^2} \sum_\sigma \frac{q_\sigma^2}{\varepsilon_0 m_\sigma} \int \mathrm{d}^3 v \frac{\mathrm{i}\mathbf{k} \cdot \dfrac{\partial f_{\sigma 0}}{\partial \mathbf{v}}}{(p + \mathrm{i}\mathbf{k} \cdot \mathbf{v})} \\ &= 1 - \frac{1}{k^2} \sum_\sigma \frac{q_\sigma^2}{\varepsilon_0 m_\sigma} \int_{-\infty}^{\infty} \mathrm{d}v_\parallel \frac{\dfrac{\partial F_{\sigma 0}}{\partial v_\parallel}}{(v_\parallel - \mathrm{i}p/k)}. \end{aligned} \tag{5.67}$$

Thus, the relation $D(p) = 0$ can be written in terms of susceptibilities as

$$D(p) = 1 + \chi_i + \chi_e = 0, \tag{5.68}$$

since the quantities being summed in Eq. (5.67) are essentially the electron and ion perturbations associated with the oscillation, and $D(p)$ is the Laplace transform analog of the Fourier transform of Poisson's equation. In the special case where the equilibrium distribution function is Maxwellian, the susceptibilities can be written in a standardized form as

$$\begin{aligned}
\chi_\sigma &= -\frac{1}{2k^2\lambda_{D\sigma}^2}\frac{1}{\pi^{1/2}}\int_{-\infty}^{\infty} \mathrm{d}\xi \frac{1}{(\xi - \mathrm{i}p/kv_{T\sigma})}\frac{\partial}{\partial \xi}\exp(-\xi^2) \\
&= \frac{1}{k^2\lambda_{D\sigma}^2}\left[\frac{1}{\pi^{1/2}}\int_{-\infty}^{\infty} \mathrm{d}\xi \frac{(\xi - \mathrm{i}p/kv_{T\sigma} + \mathrm{i}p/kv_{T\sigma})}{(\xi - \mathrm{i}p/kv_{T\sigma})}\exp(-\xi^2)\right] \\
&= \frac{1}{k^2\lambda_{D\sigma}^2}\left[1 + \frac{1}{\pi^{1/2}}\alpha\int_{-\infty}^{\infty} \mathrm{d}\xi \frac{\exp(-\xi^2)}{(\xi - \alpha)}\right] \\
&= \frac{1}{k^2\lambda_{D\sigma}^2}\left[1 + \alpha Z(\alpha)\right],
\end{aligned} \tag{5.69}$$

where $\alpha = \mathrm{i}p/kv_{T\sigma}$, and the last line introduces the *plasma dispersion* function $Z(\alpha)$ defined as

$$Z(\alpha) \equiv \frac{1}{\pi^{1/2}}\int_{-\infty}^{\infty} \mathrm{d}\xi \frac{\exp(-\xi^2)}{(\xi - \alpha)}, \tag{5.70}$$

where the ξ integration path is under the dropped pole.

5.3.5 Evaluation of the plasma dispersion function

If the pole corresponding to the fastest growing (i.e., least damped) mode turns out to have dropped well below the real axis (corresponding to Re p being large and negative), the fastest growing mode would be highly damped. We argue that this does not happen because there ought to be a correspondence between the Vlasov and fluid models in regimes where both are valid. Since the fluid model indicated the existence of undamped plasma waves when ω/k was much larger than the thermal velocity, the Vlasov model should predict nearly the same wave in this regime. The fluid wave model had no damping and so any damping introduced by the Vlasov model should be weak in order to maintain an approximate correspondence between fluid and Vlasov models. The Vlasov solution corresponding to the fluid mode can therefore have a pole only slightly below the real axis, i.e., only slightly negative. In this case, it is only necessary to analytically continue the definition of $N(p)/D(p)$ *slightly* into the negative p-plane. Thus, the pole in Eq. (5.70) drops only slightly below the real axis as shown in Fig. 5.6.

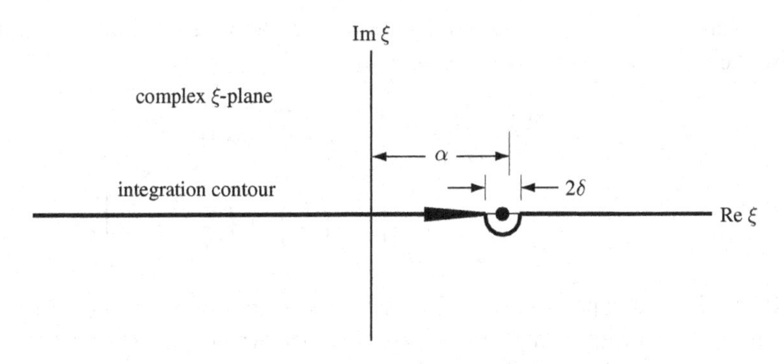

Fig. 5.6 Contour for evaluating plasma dispersion function.

The ξ integration contour can therefore be divided into three portions, namely (i) from $\xi = -\infty$ to $\xi = \alpha - \delta$, just to the left of the pole; (ii) a counterclockwise semi-circle of radius δ half-way around and *under* the pole (cf. Fig. 5.6); and (iii) a straight line from $\alpha + \delta$ to $+\infty$. The sum of the straight line segments (i) and (iii) in the limit $\delta \to 0$ is called the *principle part* of the integral and is denoted by a "P" in front of the integral sign. The semi-circle portion is *half* a residue and so makes a contribution that is just πi times the residue (rather than the standard 2πi for a complete residue). Hence, the plasma dispersion function for a pole slightly below the real axis is

$$Z(\alpha) = \frac{1}{\pi^{1/2}}\left[\mathrm{P}\int_{-\infty}^{\infty} \mathrm{d}\xi \frac{\exp(-\xi^2)}{(\xi - \alpha)}\right] + \mathrm{i}\pi^{1/2}\exp(-\alpha^2), \tag{5.71}$$

where P means principle part of the integral. Equation (5.71) prescribes how to evaluate ill-defined integrals of the type we first noted in Eq. (5.32).

There are two important limiting situations for $Z(\alpha)$, namely $|\alpha| \gg 1$ (corresponding to the adiabatic fluid limit since $\omega/k \gg v_{T\sigma}$) and $|\alpha| \ll 1$ (corresponding to the isothermal fluid limit since $\omega/k \ll v_{T\sigma}$). Asymptotic evaluations of $Z(\alpha)$ are possible in both cases and are found as follows:

1. $\alpha \gg 1$ case.
   Here, it is noted that the factor $\exp(-\xi^2)$ contributes significantly to the integral only when ξ is of order unity or smaller. In the important part of the integral where this exponential term is finite, $|\alpha| \gg \xi$. In this region of ξ the other factor in the integrand can be expanded as

$$\frac{1}{(\xi - \alpha)} = -\frac{1}{\alpha}\left(1 - \frac{\xi}{\alpha}\right)^{-1} = -\frac{1}{\alpha}\left[1 + \frac{\xi}{\alpha} + \left(\frac{\xi}{\alpha}\right)^2 + \left(\frac{\xi}{\alpha}\right)^3 + \left(\frac{\xi}{\alpha}\right)^4 + \ldots\right]. \tag{5.72}$$

The expansion is carried to fourth order because of numerous cancelations that eliminate several of the lower order terms. Substitution of Eq. (5.72) into the integral in

Eq. (5.71) and noting that all odd terms in Eq. (5.72) do not contribute to the integral because the rest of the integrand is even gives

$$P\frac{1}{\pi^{1/2}}\int_{-\infty}^{\infty}\mathrm{d}\xi\frac{\exp(-\xi^2)}{(\xi-\alpha)} = -\frac{1}{\alpha}\frac{1}{\pi^{1/2}}\int_{-\infty}^{\infty}\mathrm{d}\xi\exp(-\xi^2) \times\left[1+\left(\frac{\xi}{\alpha}\right)^2+\left(\frac{\xi}{\alpha}\right)^4+\ldots\right]. \tag{5.73}$$

The "P" has been dropped from the right-hand side of Eq. (5.73) because there is no longer any problem with a singularity. These Gaussian-type integrals may be evaluated by taking successive derivatives with respect to a of the Gaussian

$$\frac{1}{\pi^{1/2}}\int\mathrm{d}\xi\exp(-a\xi^2) = \frac{1}{a^{1/2}} \tag{5.74}$$

and then setting $a=1$. Thus,

$$\frac{1}{\pi^{1/2}}\int\mathrm{d}\xi\,\xi^2\exp(-\xi^2) = \frac{1}{2},\quad \frac{1}{\pi^{1/2}}\int\mathrm{d}\xi\,\xi^4\exp(-\xi^2) = \frac{3}{4} \tag{5.75}$$

so Eq. (5.73) becomes

$$P\frac{1}{\pi^{1/2}}\int_{-\infty}^{\infty}\mathrm{d}\xi\frac{\exp(-\xi^2)}{(\xi-\alpha)} = -\frac{1}{\alpha}\left[1+\frac{1}{2\alpha^2}+\frac{3}{4\alpha^4}+\ldots\right]. \tag{5.76}$$

In summary, for $|\alpha|\gg 1$, the plasma dispersion function has the asymptotic form

$$Z(\alpha) = -\frac{1}{\alpha}\left[1+\frac{1}{2\alpha^2}+\frac{3}{4\alpha^4}+\ldots\right]+\mathrm{i}\pi^{1/2}\exp(-\alpha^2). \tag{5.77}$$

2. $|\alpha|\ll 1$ case.
In order to evaluate the principle part integral in this regime, the variable $\eta=\xi-\alpha$ is introduced so that $\mathrm{d}\eta=\mathrm{d}\xi$. The integral may be evaluated as follows:

$$\begin{aligned}P\frac{1}{\pi^{1/2}}\int_{-\infty}^{\infty}\mathrm{d}\xi\frac{\exp(-\xi^2)}{(\xi-\alpha)} &= \frac{1}{\pi^{1/2}}P\int_{-\infty}^{\infty}\mathrm{d}\eta\frac{\mathrm{e}^{-\eta^2-2\alpha\eta-\alpha^2}}{\eta}\\ &= \frac{\mathrm{e}^{-\alpha^2}}{\pi^{1/2}}P\int_{-\infty}^{\infty}\mathrm{d}\eta\frac{\mathrm{e}^{-\eta^2}}{\eta}\left[\begin{array}{c}1-2\alpha\eta+\dfrac{(-2\alpha\eta)^2}{2!}\\ +\dfrac{(-2\alpha\eta)^3}{3!}+\ldots\end{array}\right]\\ &= -2\alpha\frac{\mathrm{e}^{-\alpha^2}}{\pi^{1/2}}\int_{-\infty}^{\infty}\mathrm{d}\eta\,\mathrm{e}^{-\eta^2}\left[1+\frac{2\eta^2\alpha^2}{3}+\ldots\right]\\ &= -2\alpha\left(1-\alpha^2+\ldots\right)\left(1+\frac{\alpha^2}{3}+\ldots\right)\\ &= -2\alpha\left(1-\frac{2\alpha^2}{3}+\ldots\right),\end{aligned} \tag{5.78}$$

where in the third line all odd terms from the second line integrated to zero due to their symmetry. Thus, for $\alpha \ll 1$, the plasma dispersion function has the asymptotic limit

$$Z(\alpha) = -2\alpha\left(1 - \frac{2\alpha^2}{3} + \ldots\right) + \mathrm{i}\pi^{1/2}\exp\left(-\alpha^2\right). \tag{5.79}$$

5.3.6 Landau damping of electron plasma waves

The plasma susceptibilities given by Eq. (5.69) can now be evaluated. For $|\alpha| \gg 1$, using Eq. (5.77), and introducing the "frequency" $\omega = \mathrm{i}p$ so that $\alpha = \omega/kv_{T\sigma}$ and $\alpha_i = \omega_i/kv_{T\sigma}$ the susceptibility is seen to be

$$\begin{aligned}
\chi_\sigma &= \frac{1}{k^2\lambda_{D\sigma}^2}\left\{1 + \alpha\left[-\frac{1}{\alpha}\left(1 + \frac{1}{2\alpha^2} + \frac{3}{4\alpha^4} + \ldots\right) + \mathrm{i}\pi^{1/2}\exp\left(-\alpha^2\right)\right]\right\} \\
&= \frac{1}{k^2\lambda_{D\sigma}^2}\left\{-\left(\frac{1}{2\alpha^2} + \frac{3}{4\alpha^4} + \ldots\right) + \mathrm{i}\alpha\pi^{1/2}\exp\left(-\alpha^2\right)\right\} \\
&= -\frac{\omega_{p\sigma}^2}{\omega^2}\left(1 + 3\frac{k^2}{\omega^2}\frac{\kappa T_\sigma}{m_\sigma} + \ldots\right) + \mathrm{i}\frac{\omega}{kv_{T\sigma}}\frac{\pi^{1/2}}{k^2\lambda_{D\sigma}^2}\exp\left(-\omega^2/k^2v_{T\sigma}^2\right).
\end{aligned} \tag{5.80}$$

Thus, if the root is such that $|\alpha| \gg 1$, the equation for the poles $D(p) = 1 + \chi_i + \chi_e = 0$ becomes

$$\begin{aligned}
1 &- \frac{\omega_{pe}^2}{\omega^2}\left(1 + 3\frac{k^2}{\omega^2}\frac{\kappa T_e}{m_e} + \ldots\right) + \mathrm{i}\frac{\omega}{kv_{Te}}\frac{\pi^{1/2}}{k^2\lambda_{De}^2}\exp\left(-\omega^2/k^2v_{Te}^2\right) \\
&- \frac{\omega_{pi}^2}{\omega^2}\left(1 + 3\frac{k^2}{\omega^2}\frac{\kappa T_i}{m_i} + \ldots\right) + \mathrm{i}\frac{\omega}{kv_{Ti}}\frac{\pi^{1/2}}{k^2\lambda_{Di}^2}\exp\left(-\omega^2/k^2v_{Ti}^2\right) = 0.
\end{aligned} \tag{5.81}$$

This expression is similar to the previously obtained fluid dispersion relation, Eq. (4.32), but contains additional imaginary terms that did not exist in the fluid dispersion. Furthermore, Eq. (5.81) is not actually a dispersion relation. Instead, it is to be understood as the equation for the roots of $D(p)$. These roots determine the poles in $N(p)/D(p)$ producing the least damped oscillations resulting from some prescribed initial perturbation of the distribution function. Since $\omega_{pe}^2/\omega_{pi}^2 = m_i/m_e$, and in general $v_{Ti} \ll v_{Te}$, both the real and imaginary parts of the ion terms are much smaller than the corresponding electron terms. On dropping the ion terms, the expression becomes

$$1 - \frac{\omega_{pe}^2}{\omega^2}\left(1 + 3\frac{k^2}{\omega^2}\frac{\kappa T_e}{m_e} + \ldots\right) + \mathrm{i}\frac{\omega}{kv_{Te}}\frac{\pi^{1/2}}{k^2\lambda_{De}^2}\exp\left(-\omega^2/k^2v_{Te}^2\right) = 0. \tag{5.82}$$

Recalling that $\omega = \mathrm{i}p$ is complex, we write $\omega = \omega_r + \mathrm{i}\omega_i$ and then proceed to find the complex ω that is the root of Eq. (5.82). Although it would not be particularly difficult to substitute $\omega = \omega_r + i\omega_i$ into Eq. (5.82) and then manipulate the coupled real and imaginary parts of this equation to solve for ω_r and ω_i, it is better to take this analysis as an opportunity to introduce a more general way for solving equations of this sort.

Equation (5.82) can be written as

$$D(\omega_r + i\omega_i) = D_r(\omega_r + i\omega_i) + iD_i(\omega_r + i\omega_i) = 0, \tag{5.83}$$

where D_r is the part of D that does not explicitly contain i and D_i is the part that does explicitly contain i. Thus,

$$D_r = 1 - \frac{\omega_{pe}^2}{\omega^2}\left(1 + 3\frac{k^2}{\omega^2}\frac{\kappa T_e}{m_e} + \ldots\right), \quad D_i = \frac{\omega}{kv_{Te}}\frac{\pi^{1/2}}{k^2\lambda_{De}^2}\exp\left(-\omega^2/k^2v_{Te}^2\right). \tag{5.84}$$

Since the oscillation has been assumed to be weakly damped, $\omega_i \ll \omega_r$ and so Eq. (5.83) can be Taylor expanded in the small quantity ω_i,

$$D_r(\omega_r) + i\omega_i\left(\frac{dD_r}{d\omega}\right)_{\omega=\omega_r} + i\left[D_i(\omega_r) + i\omega_i\left(\frac{dD_i}{d\omega}\right)_{\omega=\omega_r}\right] = 0. \tag{5.85}$$

Since $\omega_i \ll \omega_r$, the real part of Eq. (5.85) is

$$D_r(\omega_r) \simeq 0. \tag{5.86}$$

Balancing the two imaginary terms in Eq. (5.85) gives

$$\omega_i = -\frac{D_i(\omega_r)}{\dfrac{dD_r}{d\omega}}. \tag{5.87}$$

Thus, Eqs. (5.86) and (5.84) give the real part of the frequency as

$$\omega_r^2 = \omega_{pe}^2\left(1 + 3\frac{k^2}{\omega_r^2}\frac{\kappa T_e}{m_e}\right) \simeq \omega_{pe}^2\left(1 + 3k^2\lambda_{De}^2\right), \tag{5.88}$$

while Eqs. (5.87) and (5.84) give the imaginary part of the frequency, called the *Landau damping*, as

$$\begin{aligned}\omega_i &= -\sqrt{\frac{\pi}{8}}\frac{\omega_{pe}}{k^3\lambda_{De}^3}\exp\left(-\omega^2/k^2v_{T\sigma}^2\right) \\ &= -\sqrt{\frac{\pi}{8}}\frac{\omega_{pe}}{k^3\lambda_{De}^3}\exp\left[-\left(1 + 3k^2\lambda_{De}^2\right)/2k^2\lambda_{De}^2\right].\end{aligned} \tag{5.89}$$

Since the least damped oscillation goes as $\exp(pt) = \exp(-i\omega t) = \exp(-i(\omega_r + i\omega_i)t) = \exp(-i\omega_r t + \omega_i t)$ and Eq. (5.89) gives a negative ω_i, this is indeed a damping. It is interesting to note that while Landau damping was proposed theoretically by Landau in 1949, it took sixteen years before Landau damping was verified experimentally (Malmberg and Wharton 1964).

What is meant by weak damping vs. strong damping? In order to calculate ω_i it was assumed that ω_i is small compared to ω_r suggesting perhaps that ω_i is unimportant. However, even though small, ω_i can be important, because the factor 2π affects the real and imaginary parts of the wave phase differently. Suppose for example that the imaginary part of the frequency is $1/2\pi \sim 1/6$ the magnitude of the real part. This ratio is surely small enough to justify the Taylor expansion used in Eq. (5.85) and also to justify the assumption that the pole p_j corresponding to this mode is only slightly to the left of the imaginary p axis. Let us calculate how much the wave is attenuated in one period $\tau = 2\pi/\omega_r$. This attenuation will be $\exp(-|\omega_i|\tau) = \exp(-2\pi/6) \sim \exp(-1) \sim 0.3$. Thus, the wave amplitude decays to one third of its original value in just one period, which is certainly important.

5.3.7 Power relationships

It is premature to calculate the power associated with wave damping, because we do not yet know how to add up all the energy in the wave. Nevertheless, if we are willing to assume temporarily that the wave energy is entirely in the wave electric field (it turns out there is also energy in coherent particle motion – to be discussed in Chapter 14), it is seen that the power being lost from the wave electric field is

$$
\begin{aligned}
P_{wavelost} &\sim \frac{\mathrm{d}}{\mathrm{d}t}\left\langle \frac{\varepsilon_0 E_{wave}^2}{2} \right\rangle \sim \frac{\mathrm{d}}{\mathrm{d}t}\left[\frac{\varepsilon_0 |E_{wave}^2|}{4} \exp\left(-2|\omega_i|t\right) \right] = -\frac{|\omega_i| \varepsilon_0 E_{wave}^2}{2} \\
&= \sqrt{\frac{\pi}{8}} \frac{\omega_{pe}}{2k^3\lambda_{De}^3} \exp\left(-\omega^2/k^2 v_{T\sigma}^2\right) \varepsilon_0 E_{wave}^2,
\end{aligned}
\tag{5.90}
$$

where $\langle E_{wave}^2 \rangle = |E_{wave}|^2 \langle \cos(kx - \omega t) \rangle = |E_{wave}|^2/2$ has been used. However, in Section 3.8, it was shown that the energy gained by untrapped resonant particles in a wave is

$$
\begin{aligned}
P_{partgain} &= \frac{-\pi m \omega}{2k^2} \left(\frac{qE_{wave}}{m} \right)^2 \left[\frac{\mathrm{d}}{\mathrm{d}v_0} f(v_0) \right]_{v_0 = \omega/k} \\
&= \frac{-\pi m \omega}{2k^2} \left(\frac{qE_{wave}}{m} \right)^2 \left[\frac{\mathrm{d}}{\mathrm{d}v_0} \left\{ \left(\frac{m}{2\pi\kappa T} \right)^{1/2} n_0 \exp\left(-\frac{mv^2}{2\kappa T} \right) \right\} \right]_{v_0 = \omega/k} \\
&= \frac{\pi m \omega}{2k^2} \left(\frac{qE_{wave}}{m} \right)^2 \left(\frac{m}{2\pi\kappa T} \right)^{1/2} \left(\frac{m}{\kappa T} \frac{\omega}{k} \right) n_0 \exp\left(-\frac{\omega^2}{k^2 v_{T\sigma}^2} \right);
\end{aligned}
\tag{5.91}
$$

using $\omega \sim \omega_{pe}$ this is seen to be the same as Eq. (5.90) except for a factor of two. We shall see later that this factor of two comes from the fact that the wave electric field actually contains *half* the energy of the electron plasma wave, with the other half in coherent particle motion, so the true power loss rate is really twice that given in Eq. (5.90).

5.3.8 Landau damping for ion acoustic waves

Ion acoustic waves resulted from a two-fluid analysis in the regime where the wave-phase velocity was intermediate between the electron and ion thermal velocities. In this situation the electrons behave isothermally and the ions behave adiabatically. This suggests there might be another root of $D(p)$ if $|\alpha_e| \ll 1$ and $|\alpha_i| \gg 1$ or equivalently $v_{Ti} \ll \omega/k \ll v_{Te}$. From Eqs. (5.69) and (5.79), the susceptibility for $|\alpha| \ll 1$ is found to be

$$\begin{aligned}\chi_\sigma &= \frac{1}{k^2\lambda_{D\sigma}^2}\left[1+\alpha Z(\alpha)\right] \\ &= \frac{1}{k^2\lambda_{D\sigma}^2}\left\{1-2\alpha^2\left(1-\frac{2\alpha^2}{3}+\ldots\right)+\mathrm{i}\alpha\pi^{1/2}\exp\left(-\alpha^2\right)\right\} \\ &\simeq \frac{1}{k^2\lambda_{D\sigma}^2}+\mathrm{i}\frac{\alpha}{k^2\lambda_{D\sigma}^2}\pi^{1/2}\exp\left(-\alpha^2\right).\end{aligned} \tag{5.92}$$

Using Eq. (5.92) for the electron susceptibility and Eq. (5.80) for the ion susceptibility gives

$$\begin{aligned}D(\omega) = 1 &+ \frac{1}{k^2\lambda_{De}^2}+\mathrm{i}\frac{\omega}{kv_{Te}}\frac{\pi^{1/2}}{k^2\lambda_{De}^2}\exp\left(-\omega^2/k^2v_{Te}^2\right) \\ &-\frac{\omega_{pi}^2}{\omega^2}\left(1+3\frac{k^2}{\omega^2}\frac{\kappa T_i}{m_i}+\ldots\right)+\mathrm{i}\frac{\omega}{kv_{Ti}}\frac{\pi^{1/2}}{k^2\lambda_{Di}^2}\exp\left(-\omega^2/k^2v_{Ti}^2\right).\end{aligned} \tag{5.93}$$

On applying the Taylor expansion technique discussed in conjunction with Eqs. (5.86) and (5.87) we find that ω_r is the root of

$$D_r(\omega_r) = 1+\frac{1}{k^2\lambda_{De}^2}-\frac{\omega_{pi}^2}{\omega_r^2}\left(1+3\frac{k^2}{\omega^2}\frac{\kappa T_i}{m_i}\right)=0, \tag{5.94}$$

i.e.,

$$\omega_r^2 = \frac{k^2c_s^2}{1+k^2\lambda_{De}^2}\left(1+3\frac{k^2}{\omega_r^2}\frac{\kappa T_i}{m_i}\right) \simeq \frac{k^2c_s^2}{1+k^2\lambda_{De}^2}+3k^2\frac{\kappa T_i}{m_i}. \tag{5.95}$$

Here, as in the two-fluid analysis of ion acoustic waves, $c_s^2 = \omega_{pi}^2\lambda_{De}^2 = \kappa T_e/m_i$ has been defined. The imaginary part of the frequency is found to be

$$\begin{aligned}
\omega_i &\simeq -\frac{D_i(\omega_r)}{\mathrm{d}D_r/\mathrm{d}\omega} \\
&= -\frac{\omega\pi^{1/2}}{k^3}\left[\frac{\dfrac{1}{\lambda_{De}^2 v_{Te}}\exp\left(-\omega^2/k^2v_{Te}^2\right)+\dfrac{1}{\lambda_{Di}^2 v_{Ti}}\exp\left(-\omega^2/k^2v_{Ti}^2\right)}{2\omega_{pi}^2/\omega^3}\right] \\
&= -\frac{\omega^4}{k^3c_s^3}\sqrt{\frac{\pi}{8}}\left[\sqrt{\frac{m_e}{m_i}}+\left(\frac{T_e}{T_i}\right)^{3/2}\exp\left(-\omega^2/k^2v_{Ti}^2\right)\right] \\
&= -\frac{|\omega_r|}{\left(1+k^2\lambda_{De}^2\right)^{3/2}}\sqrt{\frac{\pi}{8}}\left[\sqrt{\frac{m_e}{m_i}}+\left(\frac{T_e}{T_i}\right)^{3/2}\exp\left(-\frac{T_e/2T_i}{1+k^2\lambda_{De}^2}-\frac{3}{2}\right)\right]. \qquad (5.96)
\end{aligned}$$

The dominant Landau damping comes from the ions, since the electron Landau damping term has the small factor $\sqrt{m_e/m_i}$. If $T_e \gg T_i$ the ion term also becomes small because $x^{3/2}\exp(-x) \to 0$ as x becomes large. Hence, strong ion Landau damping occurs when T_i approaches T_e and so ion acoustic waves can only propagate without extreme attenuation if the plasma has $T_e \gg T_i$. Landau damping of ion acoustic waves was first observed experimentally by Wong, Motley and D'Angelo (1964).

5.3.9 The Plemelj formula

The Landau method showed that the correct way to analyze problems that lead to ill-defined integrals such as Eq. (5.35) is to pose the problem as an initial value problem rather than as a steady-state situation. The essential result of the Landau method can be summarized by the Plemelj formula

$$\lim_{\varepsilon\to 0}\frac{1}{\xi - a \mp \mathrm{i}|\varepsilon|} = \mathrm{P}\frac{1}{\xi - a} \pm \mathrm{i}\pi\delta(\xi - a), \qquad (5.97)$$

which is a prescription showing how to deal with singular integrands of the form appearing in the plasma dispersion function. From now on, instead of repeating the lengthy Laplace transform analysis, we instead will use the less cumbersome, but formally incorrect, Fourier method and then invoke Eq. (5.97) as a "patch" to resolve any ambiguities regarding integration contours.

5.4 The Penrose criterion

The analysis so far showed that electrostatic plasma waves are subject to Landau damping, a collisionless attenuation proportional to $[\partial f/\partial v]_{v=\omega/k}$, and that this damping is consistent with the calculation of power input to particles by an electrostatic wave. Since a Maxwellian distribution function has a negative slope, its associated Landau damping is always a true wave damping. This is consistent with the physical picture developed in the single particle analysis, which showed that energy is transferred from wave to particles if there are more slow than fast particles in the vicinity of the wave phase velocity. What happens if there is a non-Maxwellian distribution function, in particular one where there are more fast particles than slow particles in the vicinity of the wave-phase velocity, i.e., $[\partial f/\partial v]_{v=\omega/k} > 0$? Because $f(v) \to 0$ as $|v| \to \infty$, f can only have a positive slope for a finite range of positive velocity and can only have a negative slope for a finite range of negative velocity; in particular, positive slopes of the distribution function must always be located to the left of a localized maximum in $f(v)$ in the $v > 0$ velocity-space region. A localized maximum in $f(v)$ corresponds to a beam of fast particles superimposed on a (possible) background of particles having a monotonically decreasing $f(v)$. Can the Landau damping process be run in reverse and so provide Landau growth, i.e., wave instability? The answer is yes. We will now discuss a criterion due to Penrose (1960) that shows how strong a beam must be to give Landau instability.

The procedure used to derive Eq. (5.32) is repeated, giving

$$1 + \frac{q^2}{k^2 m\varepsilon_0} \int_{-\infty}^{\infty} \frac{k}{(\omega - kv)} \frac{\partial f_0}{\partial v} \mathrm{d}v = 0, \tag{5.98}$$

which may be recast as

$$k^2 = Q(z), \tag{5.99}$$

where

$$Q(z) = \frac{q^2}{m\varepsilon_0} \int_{-\infty}^{\infty} \frac{1}{(v - z)} \frac{\partial f_0}{\partial v} \mathrm{d}v \tag{5.100}$$

is a complex function of the complex variable $z = \omega/k$. The wavenumber k is assumed to be a positive real quantity and the Plemelj formula will be used to resolve the ambiguity due to the singularity in the integrand.

The left-hand side of Eq. (5.99) is, by assumption, always real and positive for any choice of k. A solution of this equation can therefore always be found if $Q(z)$ is simultaneously pure real and positive. The actual magnitude of $Q(z)$ does not matter, since the magnitude of k^2 can be adjusted to match the magnitude of $Q(z)$.

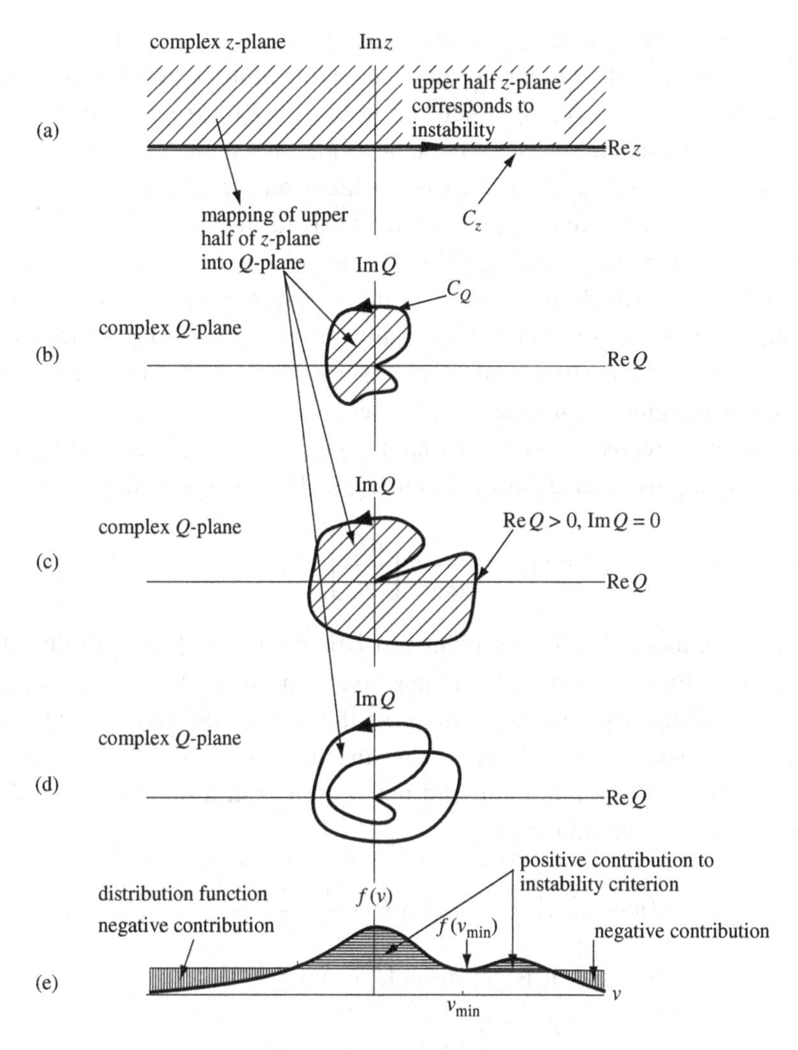

Fig. 5.7 Penrose criterion: (a)–(d) mappings, (e) instability criterion.

The function $Q(z)$ may be interpreted as a mapping from the complex z-plane to the complex Q-plane. Because solutions of Eq. (5.99) giving instability are those for which $\mathrm{Im}\,\omega > 0$, the upper half of the complex z-plane corresponds to instability and the real z axis represents the dividing line between stability and instability. Let us consider a straight-line contour C_z parallel to the real z axis, and slightly above. As shown in Fig. 5.7(a) this contour can be prescribed as $z = z_r + i\delta$, where δ is a small constant and z_r ranges from $-\infty$ to $+\infty$.

The function $Q(z) \to 0$ when $z \to \pm\infty$ and so, as z is moved along the C_z contour, the corresponding path C_Q traced in the Q-plane must start at the origin and end at the origin. Furthermore, since Q can be evaluated using the Plemelj formula, it is seen that Q is finite for all z on the path C_z. Thus, C_Q is a continuous finite curve starting at the Q-plane origin and ending at this same origin as shown by the various possible mappings sketched in Figs. 5.7(b), (c), and (d).

The upper half z-plane maps to the area inside the curve C_Q. If C_Q is of the form shown in Fig. 5.7(b), then $Q(z)$ never takes on a positive real value for z being in the upper half z-plane; thus a curve of this form cannot give a solution to Eq. (5.99) corresponding to an instability. However, curves of the form sketched in Figs. 5.7 (c) and (d) do have $Q(z)$ taking on positive real values and so do correspond to unstable solutions. Marginally unstable situations correspond to where C_Q crosses the positive real Q axis, since C_Q is a mapping of C_z which was the set of marginally unstable frequencies.

Let us therefore focus attention on what happens when C_Q crosses the positive real Q axis. Using the Plemelj formula on Eq. (5.100) it is seen that

$$\operatorname{Im} Q = \frac{q^2}{m\varepsilon_0}\pi\left[\frac{\partial f_0}{\partial v}\right]_{v=\omega_r/k} \tag{5.101}$$

and, on moving along C_Q from a point just below the real Q axis to just above the real Q axis, $\operatorname{Im} Q$ goes from being negative to positive. Thus, $[\partial f_0/\partial v]_{v=\omega_r/k}$ changes from being negative to positive, so that on the positive real Q axis f_0 is a *minimum* at some value $v = v_{\min}$ (here the subscript "min" means the value of v for which f_0 is at a minimum and not v itself is at a minimum). A Taylor expansion about this minimum gives

$$f(v) = f\left[v_{\min} + (v - v_{\min})\right] = f(v_{\min}) + 0 + \frac{(v - v_{\min})^2}{2} f''(v_{\min}) + \ldots \tag{5.102}$$

Since $f(v_{\min})$ is a constant, it is permissible to write

$$\frac{\partial f_0}{\partial v} = \frac{\partial}{\partial v}\left[f_0(v) - f_0(v_{\min})\right]. \tag{5.103}$$

This innocuous insertion of $f(v_{\min})$ makes it easy to integrate Eq. (5.100) by parts

$$\begin{aligned} Q(z = v_{\min}) &= \frac{q^2}{m\varepsilon_0}\left\{ \mathrm{P}\int_{-\infty}^{\infty} \mathrm{d}v \frac{\dfrac{\partial}{\partial v}\left[f(v) - f(v_{\min})\right]}{(v - v_{\min})} + \mathrm{i}\pi\left[\frac{\partial}{\partial v} f(v)\right]_{v=v_{\min}} \right\} \\ &= \frac{q^2}{m\varepsilon_0}\mathrm{P}\int_{-\infty}^{\infty} \mathrm{d}v \frac{1}{(v - v_{\min})^2}\left[f(v) - f(v_{\min})\right] \\ &= \frac{q^2}{m\varepsilon_0}\int_{-\infty}^{\infty} \mathrm{d}v \frac{1}{(v - v_{\min})^2}\left[0 + \frac{(v - v_{\min})^2}{2} f''(v_{\min}) + \ldots\right]; \end{aligned} \tag{5.104}$$

in the second line, advantage has been taken of the imaginary part being zero by assumption, and in the third line the "P" for principle part has been dropped because there is no longer a singularity at $v = v_{\min}$. In fact, since the leading term of $f(v) - f(v_{\min})$ is proportional to $(v - v_{\min})^2$, this qualifying "P" can also be dropped from the second line. The requirement for marginal instability can be summarized as: $f(v)$ has a minimum at $v = v_{\min}$, and the value of Q is positive, i.e.,

$$Q(v_{\min}) = \frac{q^2}{m\varepsilon_0}\int_{-\infty}^{\infty} \mathrm{d}v \left[\frac{f(v) - f(v_{\min})}{(v - v_{\min})^2}\right] > 0. \tag{5.105}$$

This is just a weighted measure of the strength of the bump in f located to the right of the minimum as shown in Fig. 5.7(e). The hatched areas with horizontal lines make positive contributions to Q, while the hatched areas with vertical lines make negative contributions. These contributions are weighted according to how far they are from $v_{\min}$ by the factor $(v - v_{\min})^{-2}$.

The Penrose criterion extends the two-stream instability analysis to an arbitrary distribution function containing finite temperature beams.

5.5 Assignments

1. Show that the electrostatic dispersion relation for electrons streaming with velocity $\mathbf{v}_0$ through stationary ions is

$$1 - \frac{\omega_{pi}^2}{\omega^2} - \frac{\omega_{pe}^2}{(\omega - \mathbf{k}\cdot\mathbf{v}_0)^2} = 0.$$

(a) Show that instability begins when

$$\left(\frac{\mathbf{k}\cdot\mathbf{v}_0}{\omega_{pe}}\right)^2 < \left[1 + \left(\frac{m_e}{m_i}\right)^{1/3}\right]^3.$$

(b) Split the frequency into its real and imaginary parts so that $\omega = \omega_r + \mathrm{i}\omega_i$. Show that the instability has maximum growth rate

$$\frac{\omega_i}{\omega_{pe}} = \frac{\sqrt{3}}{2}\left(\frac{m_e}{2m_i}\right)^{1/3}.$$

What is the value of kv_0/ω_{pe} when the instability has maximum growth rate? Sketch the dependence of ω_i/ω_{pe} on kv_0/ω_{pe}. (Hint: define non-dimensional variables $\epsilon = m_e/m_i$, $z = \omega/\omega_{pe}$, and $\lambda = kv_0/\omega_{pe}$. Let $z = x + \mathrm{i}y$ and look for the maximum y satisfying the dispersion. A particularly neat way to solve the dispersion is to solve the dispersion for the imaginary part of λ, which of course is zero, since by assumption k is real. Take advantage of the assumption $\epsilon \ll 1$ to find a relatively simple expression involving y. Maximize y with respect to x and then find the respective values of x, y, and λ at this point of maximum y.)

2. Prove the Plemelj formula.
3. Suppose

$$\mathbf{E}(\mathbf{x}, t) = \int \tilde{\mathbf{E}}(\mathbf{k}) \mathrm{e}^{\mathrm{i}\mathbf{k}\cdot\mathbf{x}-\mathrm{i}\omega(\mathbf{k})t} \mathbf{dk}, \tag{5.106}$$

where

$$\omega = \omega_r(\mathbf{k}) + \mathrm{i}\omega_i(\mathbf{k})$$

is determined by an appropriate dispersion relation. Assuming $\mathbf{E}(\mathbf{x}, t)$ is a real quantity, show, by comparing Eq. (5.106) to its complex conjugate, that $\omega_r(\mathbf{k})$ must always be an odd function of $\mathbf{k}$, while ω_i must always be an even function of $\mathbf{k}$. (Hint: note that the left-hand side of Eq. (5.106) is real by assumption, and so the right-hand side must also be real. Take the complex conjugate of both sides and replace the dummy variable of integration $\mathbf{k}$ by $-\mathbf{k}$ so that $\mathbf{dk} \to -\mathbf{dk}$ and the $\pm\infty$ limits of integration are also interchanged.)
4. Plot the real and imaginary parts of the plasma dispersion function. Plot the real and imaginary parts of the susceptibilities.
5. Is it possible to have a propagating electrostatic plasma wave that has $k\lambda_{De} \gg 1$. Hint: consider Landau damping.
6. Plot the potential versus time in units of the real period of an electron plasma wave for various values of $\omega/k\sqrt{\kappa T_e/m_e}$ showing the onset of Landau damping.
7. Plot ω_i/ω_r for ion acoustic waves for various values of T_e/T_i and show that these waves have strong ion Landau damping when the ion temperature approaches the electron temperature.
8. Landau instability for ion acoustic waves. Plasmas with $T_e > T_i$ support propagation of ion acoustic waves; these waves are Landau damped by both electrons and ions. However, if there is a sufficiently strong current J flowing in the plasma giving a relative streaming velocity $u_0 = J/ne$ between the ions and electrons, the Landau damping can operate in reverse, and give a Landau growth. This can be seen by moving to the ion frame, in which case the electrons appear as an offset Gaussian. If the offset is large enough it will be possible to have $[\partial f_e/\partial v]_{v\sim c_s} > 0$, giving more fast than slow particles at the wave-phase velocity. Now, since f_e is a Gaussian with its center shifted to be at u_0, show that if $u_0 > \omega/k$ the portion of f_e immediately to the left of $u_0$will have positive slope and so lead to instability. These qualitative ideas can easily be made quantitative, by considering a 1-D equilibrium where the ion distribution is

$$f_{i0} = \frac{n_0}{\pi^{1/2} v_{Ti}} \mathrm{e}^{-v^2/v_{Ti}^2}$$

and the drifting electron distribution is

$$f_{e0} = \frac{n_0}{\pi^{1/2} v_{Te}} \mathrm{e}^{-(v-u_0)^2/v_{Te}^2}.$$

The ion susceptibility will be the same as before, but to determine the electron susceptibility we must reconsider the linearized Vlasov equation

$$\frac{\partial f_{e1}}{\partial t} + v\frac{\partial f_{e1}}{x} - \frac{q_e}{m_e}\frac{\partial \phi_1}{\partial x}\frac{\partial}{\partial v}\left[\frac{n_0}{\pi^{1/2} v_{Te}} \mathrm{e}^{-(v-u_0)^2/v_{Te}^2}\right] = 0.$$

This equation can be simplified by defining $v' = v - u_0$. Show that the electron susceptibility becomes

$$\chi_e = \frac{1}{k^2 \lambda_{De}^2}\left[1 + \alpha Z(\alpha)\right],$$

where now $\alpha = (\omega - ku_0)/kv_{Te}$. Suppose $T_e \gg T_i$ so that the electron Landau damping term dominates. Show that if $u_0 > \omega_r/k$ the electron imaginary term will reverse sign and give instability.

9. Suppose a current I flows in a long cylindrical plasma of radius a, density n, ion mass m_i for which $T_e \gg T_i$. Write a criterion for ion acoustic instability in terms of an appropriate subset of these parameters. Suppose a cylindrical mercury plasma with $T_e = 1 - 2$ eV, $T_i = 0.1$ eV, diameter 2.5 cm, carries a current of $I = 0.35$ amps. At what density would an ion acoustic instability be expected to develop? Does this configuration remind you of an everyday object? Hint: there are several hanging from the ceiling of virtually every classroom.

6

Cold plasma waves in a magnetized plasma

6.1 Overview

Chapter 4 showed that finite temperature is responsible for the lowest order dispersive terms in both electron plasma waves (dispersion $\omega^2 = \omega_p^2 + 3k^2\kappa T_e/m_e$) and ion acoustic waves (dispersion $\omega^2 = k^2 c_s^2/(1 + k^2\lambda_{De}^2)$). Furthermore, finite temperature was shown in Chapter 5 to be essential to Landau damping and instability.

Chapter 4 also contained a derivation of the electromagnetic plasma wave (dispersion $\omega^2 = \omega_{pe}^2 + k^2c^2$) and of the inertial Alfvén wave (dispersion $\omega^2 = k_z^2 v_A^2/(1 + k_x^2 c^2/\omega_{pe}^2)$), both of which had no dependence on temperature. To distinguish waves that depend on temperature from waves that do not, the terminology "cold plasma wave" and "hot plasma wave" is used. A cold plasma wave is a wave having a temperature-independent dispersion relation so that the temperature could be set to zero without changing the wave, whereas a hot plasma wave has a temperature-dependent dispersion relation. Thus, hot and cold do not refer to a "temperature" of the wave, but rather to the wave's dependence or lack thereof on plasma temperature. Generally speaking, cold plasma waves are just the consequence of a large number of particles having identical Hamiltonian–Lagrangian dynamics whereas hot plasma waves involve different groups of particles having different dynamics because they have different initial velocities. Thus, hot plasma waves involve statistical mechanical or thermodynamic considerations. The general theory of cold plasma waves in a uniform magnetized plasma is presented in this chapter and hot plasma waves will be discussed in later chapters.

6.2 Redundancy of Poisson's equation in electromagnetic mode analysis

When electrostatic waves were examined in Chapter 4 it was seen that the plasma response to the wave electric field could be expressed as a sum of susceptibilities, where the susceptibility of each species was proportional to the density perturbation of that species. Combining the susceptibilities with Poisson's equation gave a dispersion relation. However, because electric fields can also be generated inductively, electrostatic waves are not the only type of wave. Inductive electric fields result from time-dependent currents, i.e., from charged particle acceleration, and do not involve density perturbations. As an example, the electromagnetic plasma wave involved inductive rather than electrostatic electric fields. The inertial Alfvén wave involved inductive electric fields in the parallel direction and electrostatic electric fields in the perpendicular direction.

One might expect that a procedure analogous to the previous derivation of electrostatic susceptibilities could be used to derive inductive "susceptibilities," which would then be used to construct dispersion relations for inductive modes. It turns out that such a procedure not only gives dispersion relations for inductive modes, but also includes the electrostatic modes. Thus, it turns out to be unnecessary to analyze electrostatic modes separately. The main reason for investigating electrostatic modes separately as done earlier is pedagogical – it is easier to understand a simpler system. To see why electrostatic modes are automatically included in an electromagnetic analysis, consider the interrelationship between Poisson's equation, Ampère's law, and a charge-weighted summation of the two-fluid continuity equation,

$$\nabla \cdot \mathbf{E} = -\frac{1}{\varepsilon_0}\sum_\sigma n_\sigma q_\sigma, \tag{6.1}$$

$$\nabla \times \mathbf{B} = \mu_0 \mathbf{J} + \varepsilon_0 \mu_0 \frac{\partial \mathbf{E}}{\partial t}, \tag{6.2}$$

$$\sum_\sigma q_\sigma \left[\frac{\partial n_\sigma}{\partial t} + \nabla \cdot (n_\sigma \mathbf{v}_\sigma)\right] = \left[\frac{\partial}{\partial t}\sum_\sigma n_\sigma q_\sigma\right] + \nabla \cdot \mathbf{J} = 0. \tag{6.3}$$

The divergence of Eq. (6.2) gives

$$\nabla \cdot \mathbf{J} + \varepsilon_0 \frac{\partial}{\partial t} \nabla \cdot \mathbf{E} = 0 \tag{6.4}$$

and substituting Eq. (6.3) gives

$$\frac{\partial}{\partial t}\left[-\sum_\sigma n_\sigma q_\sigma + \varepsilon_0 \nabla \cdot \mathbf{E}\right] = 0, \tag{6.5}$$

which is just the time derivative of Poisson's equation. Integrating Eq. (6.5) shows

$$-\sum_\sigma n_\sigma q_\sigma + \varepsilon_0 \nabla \cdot \mathbf{E} = const. \tag{6.6}$$

Poisson's equation, Eq. (6.1), thus provides an *initial condition*, which fixes the value of the constant in Eq. (6.6). Since all small-amplitude perturbations are assumed to have the phase dependence $\exp(\mathrm{i}\mathbf{k}\cdot\mathbf{x}-\mathrm{i}\omega t)$ and therefore behave as a single Fourier mode, the $\partial/\partial t$ operator in Eq. (6.5) is replaced by $-\mathrm{i}\omega$, in which case the constant in Eq. (6.6) is automatically set to zero, making a separate consideration of Poisson's equation redundant. In summary, the Fourier-transformed Ampère's law effectively embeds Poisson's equation and so a discussion of waves based solely on currents describes inductive and electrostatic modes as well as modes, such as the inertial Alfvén wave, that involve a mixture of inductive and electrostatic electric fields.

6.3 Dielectric tensor

Section 3.8 showed that a single particle immersed in a constant, uniform equilibrium magnetic field $\mathbf{B}=B_0\hat{z}$ and subject to a *small-amplitude* wave with electric field $\sim\exp(\mathrm{i}\mathbf{k}\cdot\mathbf{x}-\mathrm{i}\omega t)$ has the velocity

$$\tilde{\mathbf{v}}_\sigma=\frac{\mathrm{i}q_\sigma}{\omega m_\sigma}\left[\tilde{E}_z\hat{z}+\frac{\tilde{\mathbf{E}}_\perp}{1-\omega_{c\sigma}^2/\omega^2}-\frac{\mathrm{i}\omega_{c\sigma}}{\omega}\frac{\hat{z}\times\tilde{\mathbf{E}}}{1-\omega_{c\sigma}^2/\omega^2}\right]\mathrm{e}^{\mathrm{i}\mathbf{k}\cdot\mathbf{x}-\mathrm{i}\omega t}. \tag{6.7}$$

The tilde ˜ denotes a small-amplitude oscillatory quantity with space-time dependence $\exp(\mathrm{i}\mathbf{k}\cdot\mathbf{x}-\mathrm{i}\omega t)$; this phase factor may or may not be explicitly written, but should always be understood to exist for a tilde-denoted quantity.

The three terms in Eq. (6.7) are respectively:

1. *The parallel quiver velocity*. This quiver velocity is the same as the quiver velocity of an unmagnetized particle, but is restricted to parallel motion. Because the magnetic force $q(\mathbf{v}\times\mathbf{B})$ vanishes for motion along the magnetic field, motion parallel to $\mathbf{B}$ in a magnetized plasma is identical to motion in an unmagnetized plasma.
2. *The generalized polarization drift*. This motion has a resonance at the cyclotron frequency but at low frequencies such that $\omega\ll\omega_{c\sigma}$, it reduces to the polarization drift $\mathbf{v}_{p\sigma}=m_\sigma\dot{\mathbf{E}}_\perp/q_\sigma B^2$ derived in Chapter 3.
3. *The generalized* $\mathbf{E}\times\mathbf{B}$ *drift*. This also has a resonance at the cyclotron frequency and for $\omega\ll\omega_{c\sigma}$ reduces to the drift $\mathbf{v}_E=\mathbf{E}\times\mathbf{B}/B^2$ derived in Chapter 3.

The particle velocities given by Eq. (6.7) produce a plasma current density

$$\begin{aligned}\tilde{\mathbf{J}}&=\sum_\sigma n_{0\sigma}q_\sigma\tilde{\mathbf{v}}_\sigma\\&=\mathrm{i}\varepsilon_0\sum_\sigma\frac{\omega_{p\sigma}^2}{\omega}\left[\tilde{E}_z\hat{z}+\frac{\tilde{\mathbf{E}}_\perp}{1-\omega_{c\sigma}^2/\omega^2}-\frac{\mathrm{i}\omega_{c\sigma}}{\omega}\frac{\hat{z}\times\tilde{\mathbf{E}}}{1-\omega_{c\sigma}^2/\omega^2}\right]\mathrm{e}^{\mathrm{i}\mathbf{k}\cdot\mathbf{x}-\mathrm{i}\omega t}.\end{aligned} \tag{6.8}$$

If these plasma currents are written out explicitly, then Ampère's law has the form

$$\nabla\times\tilde{\mathbf{B}} = \mu_0\tilde{\mathbf{J}} + \mu_0\varepsilon_0\frac{\partial\tilde{\mathbf{E}}}{\partial t}$$
$$= \mu_0\left(\mathrm{i}\varepsilon_0\sum_\sigma\frac{\omega_{p\sigma}^2}{\omega}\left[\tilde{E}_z\hat{z} + \frac{\tilde{\mathbf{E}}_\perp}{1-\omega_{c\sigma}^2/\omega^2} - \frac{\mathrm{i}\omega_{c\sigma}}{\omega}\frac{\hat{z}\times\tilde{\mathbf{E}}}{1-\omega_{c\sigma}^2/\omega^2}\right] - \mathrm{i}\omega\varepsilon_0\tilde{\mathbf{E}}\right), \tag{6.9}$$

where a factor $\exp(\mathrm{i}\mathbf{k}\cdot\mathbf{x}-\mathrm{i}\omega t)$ is implicit.

The cold plasma wave equation is established by combining Ampère's and Faraday's laws in a manner similar to the method used for vacuum electromagnetic waves. However, before doing so, it is useful to define the dielectric tensor $\overleftrightarrow{\mathbf{K}}$. This tensor contains the information in the right-hand side of Eq. (6.9) so that this equation is written as

$$\nabla\times\tilde{\mathbf{B}} = \mu_0\varepsilon_0\frac{\partial}{\partial t}\left(\overleftrightarrow{\mathbf{K}}\cdot\tilde{\mathbf{E}}\right), \tag{6.10}$$

where

$$\overleftrightarrow{\mathbf{K}}\cdot\tilde{\mathbf{E}} = \tilde{\mathbf{E}} - \sum_{\sigma=i,e}\frac{\omega_{p\sigma}^2}{\omega^2}\left[\tilde{E}_z\hat{z} + \frac{\tilde{\mathbf{E}}_\perp}{1-\omega_{c\sigma}^2/\omega^2} - \frac{\mathrm{i}\omega_{c\sigma}}{\omega}\frac{\hat{z}\times\tilde{\mathbf{E}}}{1-\omega_{c\sigma}^2/\omega^2}\right]$$
$$= \begin{bmatrix} S & -\mathrm{i}D & 0 \\ \mathrm{i}D & S & 0 \\ 0 & 0 & P \end{bmatrix}\cdot\tilde{\mathbf{E}} \tag{6.11}$$

and the elements of the dielectric tensor are

$$S = 1 - \sum_{\sigma=i,e}\frac{\omega_{p\sigma}^2}{\omega^2-\omega_{c\sigma}^2}, \qquad D = \sum_{\sigma=i,e}\frac{\omega_{c\sigma}}{\omega}\frac{\omega_{p\sigma}^2}{\omega^2-\omega_{c\sigma}^2}, \qquad P = 1 - \sum_{\sigma=i,e}\frac{\omega_{p\sigma}^2}{\omega^2}. \tag{6.12}$$

The nomenclature S, D, P for the matrix elements was introduced by Stix (1962) and is a mnemonic for "Sum," "Difference," and "Parallel." The reasoning behind "Sum" and "Difference" will become apparent later, but for now it is clear that the P element corresponds to the cold plasma limit of the parallel dielectric, i.e., $P = 1+\chi_i+\chi_e$, where $\chi_\sigma = -\omega_{p\sigma}^2/\omega^2$. This is the cold limit of the unmagnetized dielectric because behavior involving parallel motions in a magnetized plasma is identical to that in an unmagnetized plasma. In the limit of no plasma, $\overleftrightarrow{\mathbf{K}}$ becomes the unit tensor and describes the effect of the vacuum displacement current only.

This definition of the dielectric tensor means that Maxwell's equations, the Lorentz equation, and the plasma currents can now be summarized in just two coupled equations, namely

$$\nabla \times \mathbf{B} = \frac{1}{c^2}\frac{\partial}{\partial t}\left(\overleftrightarrow{\mathbf{K}} \cdot \mathbf{E}\right) \tag{6.13}$$

$$\nabla \times \mathbf{E} = -\frac{\partial \mathbf{B}}{\partial t}, \tag{6.14}$$

where, for clarity, the tildes have now been omitted and it is to be understood that **E** and **B** refer to the wave fields. The cold plasma wave equation is obtained by taking the curl of Eq. (6.14) and then substituting for $\nabla \times \mathbf{B}$ using (6.13) to obtain

$$\nabla \times (\nabla \times \mathbf{E}) = -\frac{1}{c^2}\frac{\partial^2}{\partial t^2}\left(\overleftrightarrow{\mathbf{K}} \cdot \mathbf{E}\right). \tag{6.15}$$

Since a phase dependence $\exp(\mathrm{i}\mathbf{k}\cdot\mathbf{x} - \mathrm{i}\omega t)$ is assumed, this can be written in algebraic form as

$$\mathbf{k} \times (\mathbf{k} \times \mathbf{E}) = -\frac{\omega^2}{c^2}\overleftrightarrow{\mathbf{K}} \cdot \mathbf{E}. \tag{6.16}$$

It is now convenient to define the refractive index $\mathbf{n} = c\mathbf{k}/\omega$, a renormalization of the wavevector **k** arranged so that light waves have a refractive index of unity. Using this definition Eq. (6.16) becomes

$$\mathbf{nn}\cdot\mathbf{E} - n^2\mathbf{E} + \overleftrightarrow{\mathbf{K}} \cdot \mathbf{E} = 0, \tag{6.17}$$

which is essentially a set of three homogeneous equations in the three components of **E**.

The refractive index $\mathbf{n} = c\mathbf{k}/\omega$ can be decomposed into parallel and perpendicular components relative to the equilibrium magnetic field $\mathbf{B} = B_0\hat{z}$. For convenience, the x axis of the coordinate system is defined to lie along the perpendicular component of **n** so that $n_y = 0$ by assumption. This simplification is possible for a spatially uniform equilibrium only; if the plasma is non-uniform in the $x-y$ plane, there can be a real distinction between x and y direction propagation and the refractive index in the y direction cannot be simply defined away by choice of coordinate system.

To set the stage for obtaining a dispersion relation, Eq. (6.17) is written in matrix form as

$$\begin{bmatrix} S - n_z^2 & -\mathrm{i}D & n_x n_z \\ \mathrm{i}D & S - n^2 & 0 \\ n_x n_z & 0 & P - n_x^2 \end{bmatrix} \cdot \begin{bmatrix} E_x \\ E_y \\ E_z \end{bmatrix} = 0. \tag{6.18}$$

It is now useful to introduce a spherical coordinate system in k-space (or equivalently refractive index space) with $\hat{z}$ defining the axis and θ the polar angle.

Thus, the Cartesian components of the refractive index are related to the spherical components by

$$\begin{aligned} n_x &= n\sin\theta \\ n_z &= n\cos\theta \\ n^2 &= n_x^2 + n_z^2 \end{aligned} \tag{6.19}$$

and so Eq. (6.18) becomes

$$\begin{bmatrix} S - n^2\cos^2\theta & -\mathrm{i}D & n^2\sin\theta\cos\theta \\ \mathrm{i}D & S - n^2 & 0 \\ n^2\sin\theta\cos\theta & 0 & P - n^2\sin^2\theta \end{bmatrix} \cdot \begin{bmatrix} E_x \\ E_y \\ E_z \end{bmatrix} = 0. \tag{6.20}$$

6.3.1 Mode behavior at $\theta = 0$

Non-trivial solutions to the set of three coupled equations for E_x, E_y, E_z prescribed by Eq. (6.20) exist only if the determinant of the matrix vanishes. For arbitrary values of θ, this determinant is complicated. Rather than examining the arbitrary-θ determinant immediately, two simpler limiting cases will first be considered, namely the situations where $\theta = 0$ (i.e., $\mathbf{k} \parallel \mathbf{B}_0$) and $\theta = \pi/2$ (i.e., $\mathbf{k} \perp \mathbf{B}_0$). These special cases are simpler than the general case because the off-diagonal matrix elements $n^2\sin\theta\cos\theta$ vanish for both $\theta = 0$ and $\theta = \pi/2$.

When $\theta = 0$ Eq. (6.20) becomes

$$\begin{bmatrix} S - n^2 & -\mathrm{i}D & 0 \\ \mathrm{i}D & S - n^2 & 0 \\ 0 & 0 & P \end{bmatrix} \cdot \begin{bmatrix} E_x \\ E_y \\ E_z \end{bmatrix} = 0. \tag{6.21}$$

The determinant of this system is

$$\left[\left(S - n^2\right)^2 - D^2\right] P = 0, \tag{6.22}$$

which has roots

$$P = 0 \tag{6.23}$$

and

$$n^2 - S = \pm D. \tag{6.24}$$

Equation (6.24) may be rearranged in the form

$$n^2 = R, \quad n^2 = L, \tag{6.25}$$

where

$$R = S + D, \quad L = S - D \tag{6.26}$$

have the mnemonics "right" and "left." The rationale behind the nomenclature "S(um)" and "D(ifference)" now becomes apparent since

$$S = \frac{R+L}{2}, \quad D = \frac{R-L}{2}. \tag{6.27}$$

What does all this algebra mean? Equation (6.25) states that for $\theta = 0$ the dispersion relation has two distinct roots, each corresponding to a natural mode (or characteristic wave) constituting a self-consistent solution to the Maxwell–Lorentz system. The definitions in Eqs. (6.12) and (6.26) show that

$$R = 1 - \sum_\sigma \frac{\omega_{p\sigma}^2}{\omega(\omega + \omega_{c\sigma})}, \quad L = 1 - \sum_\sigma \frac{\omega_{p\sigma}^2}{\omega(\omega - \omega_{c\sigma})} \tag{6.28}$$

so that R diverges when $\omega = -\omega_{c\sigma}$ whereas L diverges when $\omega = \omega_{c\sigma}$. Since $\omega_{c\sigma} = q_\sigma B/m_\sigma$, the ion cyclotron frequency is positive and the electron cyclotron frequency is negative. Hence, R diverges at the electron cyclotron frequency, whereas L diverges at the ion cyclotron frequency. When $\omega \to \infty$, both $R,\ L \to 1$. In the limit $\omega \to 0$, evaluation of $R,\ L$ must be done very carefully, since

$$\begin{aligned} \frac{\omega_{p\sigma}^2}{\omega_{c\sigma}} &= \frac{n_\sigma q_\sigma^2}{\varepsilon_0 m_\sigma} \frac{m_\sigma}{q_\sigma B} \\ &= \frac{n_\sigma q_\sigma}{\varepsilon_0 B} \end{aligned} \tag{6.29}$$

so

$$\frac{\omega_{pi}^2}{\omega_{ci}} = -\frac{\omega_{pe}^2}{\omega_{ce}}. \tag{6.30}$$

Thus

$$\begin{aligned} \lim_{\omega \to 0} R, L &= 1 - \frac{1}{\omega}\left[\frac{\omega_{pi}^2}{(\omega \pm \omega_{ci})} + \frac{\omega_{pe}^2}{(\omega \pm \omega_{ce})}\right] \\ &= 1 - \frac{\omega_{pi}^2 + \omega_{pe}^2}{\omega_{ci}\omega_{ce}} \\ &\simeq 1 - \frac{n_e q_e^2}{\varepsilon_0 m_e} \frac{m_i}{q_i B} \frac{m_e}{q_e B} \\ &= 1 + \frac{\omega_{pi}^2}{\omega_{ci}^2} \\ &= 1 + \frac{c^2}{v_A^2}, \end{aligned} \tag{6.31}$$

where $v_A^2 = B^2/\mu_0 \rho$ is the Alfvén velocity. Thus, at low frequency, both R and L are related to Alfvén modes. The $n^2 = L$ mode is the slow mode (larger k) and

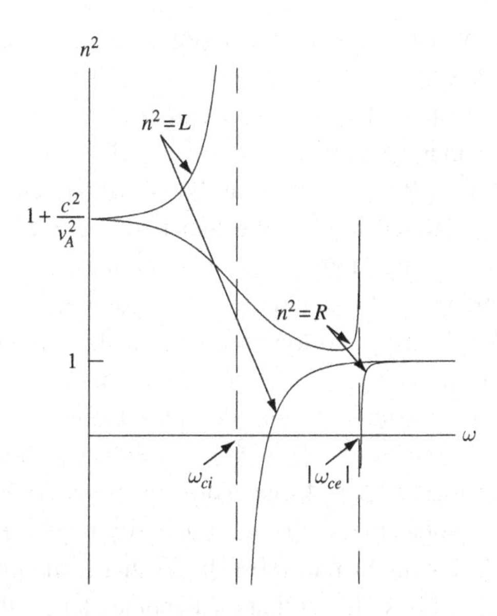

Fig. 6.1 Propagation parallel to the magnetic field.

the $n^2 = R$ mode is the fast mode (smaller k). Figure 6.1 shows the frequency dependence of the $n^2 = R, L$ modes.

Having determined the eigenvalues for $\theta = 0$, the associated eigenvectors can now be found. These are obtained by substituting the eigenvalue back into the original set of equations; for example, substitution of $n^2 = R$ into Eq. (6.21) gives

$$\begin{bmatrix} -D & -\mathrm{i}D & 0 \\ \mathrm{i}D & -D & 0 \\ 0 & 0 & P \end{bmatrix} \cdot \begin{bmatrix} E_x \\ E_y \\ E_z \end{bmatrix} = 0, \tag{6.32}$$

so that the eigenvector associated with $n^2 = R$ is

$$\frac{E_x}{E_y} = -\mathrm{i}, \quad \text{for eigenvalue } n^2 = R. \tag{6.33}$$

The implication of this eigenvector can be seen by considering the root $n = +\sqrt{R}$ so that the electric field in the plane orthogonal to $\hat{z}$ has the form

$$\begin{aligned} \mathbf{E}_\perp &= \mathrm{Re}\left\{E_\perp\left(\hat{x}+\mathrm{i}\hat{y}\right)\exp\left(\mathrm{i}k_z z - \mathrm{i}\omega t\right)\right\} \\ &= |E_\perp|\left\{\hat{x}\cos(k_z z - \omega t + \delta) - \hat{y}\sin(k_z z - \omega t + \delta)\right\}, \end{aligned} \tag{6.34}$$

where $E_\perp = |E_\perp|\mathrm{e}^{\mathrm{i}\delta}$. This is a *right-hand* circularly polarized wave propagating in the positive z direction; hence the nomenclature R. Similarly, the $n^2 = L$ root gives a left-hand circularly polarized wave. Linearly polarized waves may be constructed from appropriate sums and differences of these left- and right-hand circularly polarized waves.

In summary, two distinct modes exist when the wavevector happens to be exactly parallel to the magnetic field ($\theta = 0$): a right-hand circularly polarized wave with dispersion $n^2 = R$ with $n \to \infty$ at the electron cyclotron resonance and a left-hand circularly polarized mode with dispersion $n^2 = L$ with $n \to \infty$ at the ion cyclotron resonance. Since ion cyclotron motion is left-handed (mnemonic "Lion") it is reasonable that a left-hand circularly polarized wave resonates with ions, and vice versa for electrons. At low frequencies, these modes become Alfvén modes with dispersion $n_z^2 = 1 + c^2/v_A^2$ for $\theta = 0$. In the Chapter 4 discussion of Alfvén modes the dispersions of both compressional and shear modes were found to reduce to $c^2k_z^2/\omega^2 = n_z^2 = c^2/v_A^2$ for $\theta = 0$. One may ask why a "1" term did not appear in the Chapter 4 dispersion relations. The answer is that the "1" term comes from displacement current, a quantity neglected in the Chapter 4 derivations. The displacement current term shows that if the plasma density is so low (or the magnetic field is so high) that v_A becomes larger than c, then Alfvén modes become ordinary vacuum electromagnetic waves propagating at nearly the speed of light. In order for a plasma to demonstrate significant Alfvénic (i.e., MHD) behavior it must satisfy $B/\sqrt{\mu_0\rho} \ll c$ or equivalently have $\omega_{ci} \ll \omega_{pi}$.

6.3.2 Cutoffs and resonances

The general situation where $n^2 \to \infty$ is called a *resonance* and corresponds to the wavelength going to zero. Any slight dissipative effect in this situation will cause large wave damping. This is because if the wavelength becomes infinitesimal and the fractional attenuation per wavelength is constant, there will be a near-infinite number of wavelengths and the wave amplitude is reduced by the same fraction for each of these. Figure 6.1 also shows it is possible to have a situation where $n^2 = 0$. The general situation where $n^2 = 0$ is called a *cutoff* and corresponds to wave reflection, since n changes from being pure real to pure imaginary. If the plasma is non-uniform, it is possible for layers to exist in the plasma where either $n^2 \to \infty$ or $n^2 = 0$; these are called resonance or cutoff layers. Typically, if a wave intercepts a resonance layer it is absorbed, whereas if it intercepts a cutoff layer it is reflected.

6.3.3 Mode behavior at $\theta = \pi/2$

When $\theta = \pi/2$ Eq. (6.20) becomes

$$\begin{bmatrix} S & -\mathrm{i}D & 0 \\ \mathrm{i}D & S-n^2 & 0 \\ 0 & 0 & P-n^2 \end{bmatrix} \cdot \begin{bmatrix} E_x \\ E_y \\ E_z \end{bmatrix} = 0 \tag{6.35}$$

and again two distinct modes appear. The first mode has as its eigenvector the condition $E_z \neq 0$. The associated eigenvalue equation is $P - n^2 = 0$ or

$$\omega^2 = k^2c^2 + \sum_\sigma \omega_{p\sigma}^2, \tag{6.36}$$

which is just the dispersion for an electromagnetic plasma wave in an unmagnetized plasma. This is in accordance with the prediction that modes involving particle motion strictly parallel to the magnetic field are unaffected by the magnetic field. This mode is called the *ordinary* mode because it is unaffected by the magnetic field.

The second mode involves both E_x and E_y and has the eigenvalue equation $S(S - n^2) - D^2 = 0$, which gives the dispersion relation

$$n^2 = \frac{S^2 - D^2}{S} = 2\frac{RL}{R+L}. \tag{6.37}$$

Cutoffs occur here when either $R = 0$ or $L = 0$ and a resonance occurs when $S = 0$. Since this mode depends on the magnetic field, it is called the *extraordinary* mode. The $S = 0$ resonance is called a hybrid resonance because it depends on a hybrid of $\omega_{c\sigma}^2$ and $\omega_{p\sigma}^2$ terms (the $\omega_{c\sigma}^2$ terms depend on single-particle physics, whereas the $\omega_{p\sigma}^2$ terms depend on collective motion physics). Because S is quadratic in ω^2, the equation $S = 0$ has two distinct roots and these are found by explicitly writing

$$S = 1 - \frac{\omega_{pi}^2}{\omega^2 - \omega_{ci}^2} - \frac{\omega_{pe}^2}{\omega^2 - \omega_{ce}^2} = 0. \tag{6.38}$$

A plot of this expression shows the two roots are well separated. The large root may be found by assuming $\omega \sim \mathcal{O}(\omega_{ce})$, in which case the ion term becomes insignificant. Dropping the ion term shows the large root of S is simply

$$\omega_{uh}^2 = \omega_{pe}^2 + \omega_{ce}^2, \tag{6.39}$$

which is called the *upper hybrid frequency*. The small root may be found by assuming that $\omega^2 \ll \omega_{ce}^2$, which gives the *lower hybrid frequency*

$$\omega_{lh}^2 = \omega_{ci}^2 + \frac{\omega_{pi}^2}{1 + \dfrac{\omega_{pe}^2}{\omega_{ce}^2}}. \tag{6.40}$$

6.3.4 Very-low-frequency modes where θ is arbitrary

Equation (6.31) shows that for $\omega \ll \omega_{ci}$

$$\begin{aligned} S &\simeq R \simeq L \simeq 1 + c^2/v_A^2 \\ D &\simeq 0 \end{aligned} \tag{6.41}$$

so the cold plasma dispersion simplifies to

$$\begin{bmatrix} S-n_z^2 & 0 & n_x n_z \\ 0 & S-n^2 & 0 \\ n_x n_z & 0 & P-n_x^2 \end{bmatrix} \cdot \begin{bmatrix} E_x \\ E_y \\ E_z \end{bmatrix} = 0. \tag{6.42}$$

Because $D = 0$ the determinant factors into two modes, one where

$$\left(S-n^2\right) E_y = 0 \tag{6.43}$$

and the other where

$$\begin{bmatrix} S-n_z^2 & n_x n_z \\ n_x n_z & P-n_x^2 \end{bmatrix} \cdot \begin{bmatrix} E_x \\ E_z \end{bmatrix} = 0. \tag{6.44}$$

The former gives the dispersion relation

$$n^2 = S \tag{6.45}$$

with $E_y \neq 0$ as the eigenvector. This mode is the fast or compressional mode since, in the limit where the displacement current can be neglected, Eq. (6.45) becomes $\omega^2 = k^2 v_A^2$. The latter mode involves finite E_x and E_z and has the dispersion

$$n_x^2 = \frac{P}{S}\left(S - n_z^2\right), \tag{6.46}$$

which is the inertial Alfvén wave $\omega^2 = k_z^2 v_A^2 / \left(1 + k_x^2 c^2/\omega_{pe}^2\right)$ in the limit where the displacement current can be neglected.

6.3.5 Modes where ω and θ are arbitrary

The $\theta = 0,\ \pi/2$ limiting behaviors and the low-frequency Alfvén modes gave a useful introduction to the cold plasma modes and, in particular, showed how modes can be subject to cutoffs or resonances. We now evaluate the determinant of the matrix in Eq. (6.20) for arbitrary θ and arbitrary ω; after some algebra this determinant can be written as

$$An^4 - Bn^2 + C = 0, \tag{6.47}$$

where

$$\begin{aligned} A &= S\sin^2\theta + P\cos^2\theta \\ B &= (S^2 - D^2)\sin^2\theta + PS(1+\cos^2\theta) \\ C &= P(S^2 - D^2) = PRL. \end{aligned} \tag{6.48}$$

Equation (6.47) is quadratic in n^2 and has the two roots

$$n^2 = \frac{B \pm \sqrt{B^2 - 4AC}}{2A}. \tag{6.49}$$

Thus, the two distinct modes in the special cases of (i) $\theta = 0$, $\pi/2$ or (ii) $\omega \ll \omega_{ci}$ were just particular examples of the more general property that a cold plasma supports two distinct types of modes. Using a modest amount of algebraic manipulation (cf. assignments) it is straightforward to show that the quantity $B^2 - 4AC$ is positive definite for real θ, since

$$B^2 - 4AC = \left(S^2 - D^2 - SP\right)^2 \sin^4\theta + 4P^2D^2\cos^2\theta. \tag{6.50}$$

Thus, n is either pure real (corresponding to a propagating wave) or pure imaginary (corresponding to an evanescent wave).

From Eqs. (6.47) and (6.48) it is seen that cutoffs occur when $C = 0$, which happens if $P = 0$, $L = 0$, or $R = 0$. Also, resonances correspond to having $A \to 0$, in which case

$$S\sin^2\theta + P\cos^2\theta \simeq 0. \tag{6.51}$$

6.3.6 Wave normal surfaces

The information contained in a dispersion relation can be summarized in a qualitative, visual manner by a *wave normal surface*, which is a polar plot of the phase velocity of the wave normalized to c. Since $n = ck/\omega$, a wave normal surface is just a plot of $1/n(\theta)$ vs. θ. The most basic wave normal surface is obtained by considering the equation for a light wave in vacuum,

$$\left(\frac{\partial^2}{\partial t^2} - c^2\nabla^2\right)\mathbf{E} = 0, \tag{6.52}$$

which has the simple dispersion relation

$$\frac{1}{n^2} = \frac{\omega^2}{k^2c^2} = 1. \tag{6.53}$$

Thus, the wave normal surface of a light wave in vacuum is just a sphere of radius unity because $\omega/k = c/n$ is independent of direction. Wave normal surfaces of plasma waves are typically more complicated because n usually depends on θ.

The radius of the wave normal surface goes to zero at a resonance and goes to infinity at a cutoff (since $1/n \to 0$ at a resonance, $1/n \to \infty$ at a cutoff).

6.3.7 Taxonomy of modes – the CMA diagram

Equation (6.49) gives the general dispersion relation for arbitrary θ. While formally correct, this expression is of little practical value because of the complicated chain of dependence of n^2 on several variables. The CMA diagram (Clemmow and Mullaly (1955), Allis (1955)) provides an elegant method for revealing and classifying the large number of qualitatively different modes embedded in Eq. (6.49).

In principle, Eq. (6.49) gives the dependence of n^2 on the six parameters θ, ω, ω_{pe}, ω_{pi}, ω_{ce}, and ω_{ci}. However, ω_{pi} and ω_{pe} are not really independent parameters and neither are ω_{ci} and ω_{ce} because $\omega_{ce}^2/\omega_{ci}^2 = (m_i/m_e)^2$ and $\omega_{pe}^2/\omega_{pi}^2 = m_i/m_e$ for singly charged ions. Thus, once the ion species has been specified, the only free parameters are the density and the magnetic field. Once these have been specified, the plasma frequencies and the cyclotron frequencies are determined. It is reasonable to normalize these frequencies to the wave frequency in question since the quantities S, P, D depend only on the normalized frequencies. Thus, n^2 is effectively just a function of θ, ω_{pe}^2/ω^2, and ω_{ce}^2/ω^2. Pushing this simplification even further, we can say that for fixed ω_{pe}^2/ω^2 and ω_{ce}^2/ω^2, the refractive index n is just a function of θ. Then, once $n = n(\theta)$ is known, it can be used to plot a *wave normal surface*, i.e., ω/kc plotted vs. θ.

The CMA diagram is developed by first constructing a chart where the horizontal axis is $\ln\left(\omega_{pe}^2/\omega^2 + \omega_{pi}^2/\omega^2\right)$ and the vertical axis is $\ln\left(\omega_{ce}^2/\omega^2\right)$. For a given ω, any point on this chart corresponds to a unique density and a unique magnetic field. If we were ambitious, we could plot the wave normal surfaces $1/n$ vs. θ for a very large number of points on this chart, and so have plots of dispersions for a large set of cold plasmas. While conceivable, such a thorough examination of all possible combinations of density and magnetic field would require plotting an inconveniently large number of wave normal surfaces.

It is actually unnecessary to plot this very large set of wave normal surfaces because it turns out that the qualitative shape (i.e., topology) of the wave normal surfaces changes only at specific boundaries in parameter space. Away from these boundaries the wave normal surface deforms, but does not change its topology. The CMA diagram, shown in Fig. 6.2, charts these parameter space boundaries and so provides a powerful method for classifying cold plasma modes. Parameter space is divided up into a finite number of regions, called bounded volumes, separated by curves in parameter space, called bounding surfaces, across which the modes change *qualitatively*. Thus, within a bounded volume, modes change quantitatively but not qualitatively. For example, if Alfvén waves exist at one

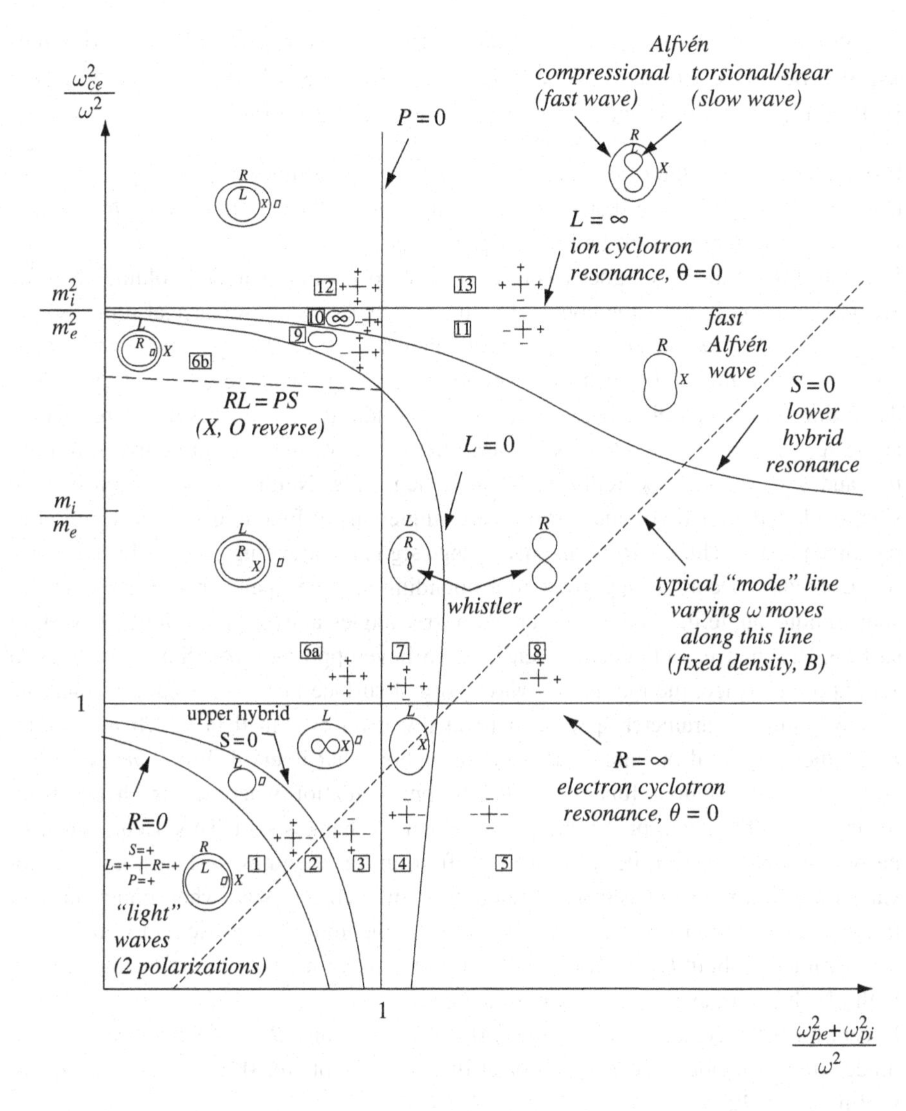

Fig. 6.2 CMA diagram.

point in a particular bounded volume, they must exist everywhere in that bounded volume, although the dispersion may not be quantitatively the same at different locations in the volume.

The appropriate choice of bounding surfaces consists of:

1. The *principle resonances*, which are the curves in parameter space where n^2 has a resonance at either $\theta = 0$ or $\theta = \pi/2$. Thus, the principal resonances are the curves $R = \infty$ (i.e., electron cyclotron resonance), $L = \infty$ (i.e., ion cyclotron resonance), and $S = 0$ (i.e., the upper and lower hybrid resonances).
2. The cutoffs $R = 0$, $L = 0$, and $P = 0$.

The behavior of wave normal surfaces inside a bounded volume and when crossing a bounded surface can be deduced using a set of five simple theorems (Stix 1962), each a consequence of the results derived so far:

1. Inside a bounded volume n cannot vanish. Proof: by setting $n=0$ in Eq. (6.47) and using Eq. (6.48) it is seen that n vanishes only when $PRL=0$, but $P=0$, $R=0$, and $L=0$ have been defined to be bounding surfaces.
2. If n^2 has a resonance (i.e., goes to infinity) at *any* point in a bounded volume, then for *every* other point in the same bounded volume, there exists a resonance at some unique angle θ_{res} and its associated mirror angles, namely $-\theta_{res}$, $\pi-\theta_{res}$, and $-(\pi-\theta_{res})$, but at *no* other angles. Proof: if $n^2\to\infty$ then $A\to 0$, in which case $\tan^2\theta_{res}=-P/S$ determines the unique θ_{res}. Now $\tan^2(\pi-\theta_{res})=\tan^2\theta_{res}$ so there is also a resonance at the supplement $\theta=\pi-\theta_{res}$. Also, since the square of the tangent is involved, both θ_{res} and $\pi-\theta_{res}$ may be replaced by their negatives. Neither P nor S can change sign inside a bounded volume and both are single valued functions of their location in parameter space. Thus, $-P/S$ can only change sign at a bounding surface. In summary, if a resonance occurs at any point in a bounding surface, then a resonance exists at some unique angle θ_{res} and its associated mirror angles at every point in the bounding surface. Resonances only occur when P and S have opposite signs. Since $1/n$ goes to zero at a resonance, the radius of a wave normal surface goes to zero at a resonance.
3. At any point in parameter space and for a given interval in θ in which n is finite, n is either pure real or pure imaginary throughout that interval. Proof: n^2 is always real and is a continuous function of θ. The only situation where n can change from being pure real to being pure imaginary is when n^2 changes sign. This occurs when n^2 passes through zero but, because of the definition for bounding surfaces, n^2 does not vanish inside a bounded volume. Although n^2 may change sign when going through infinity this situation is not relevant because the theorem was restricted to finite n.
4. n is symmetric about $\theta=0$ and $\theta=\pi/2$. Proof: n is a function of $\sin^2\theta$ and of $\cos^2\theta$, both of which are symmetric about $\theta=0$ and $\theta=\pi/2$.
5. Except for the special case where the surfaces $PD=0$ and $RL=PS$ intersect, the two modes may coincide only at $\theta=0$ or at $\theta=\pi/2$. Proof: for $0<\theta<\pi/2$ the square root in Eq. (6.49) is

$$\sqrt{B^2-4AC}=\sqrt{(RL-SP)^2\sin^4\theta+4P^2D^2\cos^2\theta} \tag{6.54}$$

and can only vanish if $PD=0$ and $RL=PS$ simultaneously.

These theorems provide sufficient information to characterize the morphology of wave normal surfaces throughout all of parameter space. In particular, the theorems show that only three types of wave normal surfaces exist. These are ellipsoid, dumbbell, and wheel as shown in Fig. 6.3(a)–(c) and each is a three-dimensional surface symmetric about the z axis.

We now discuss the features and interrelationships of these three types of wave normal surfaces. In this discussion, each of the two modes in Eq. (6.49)

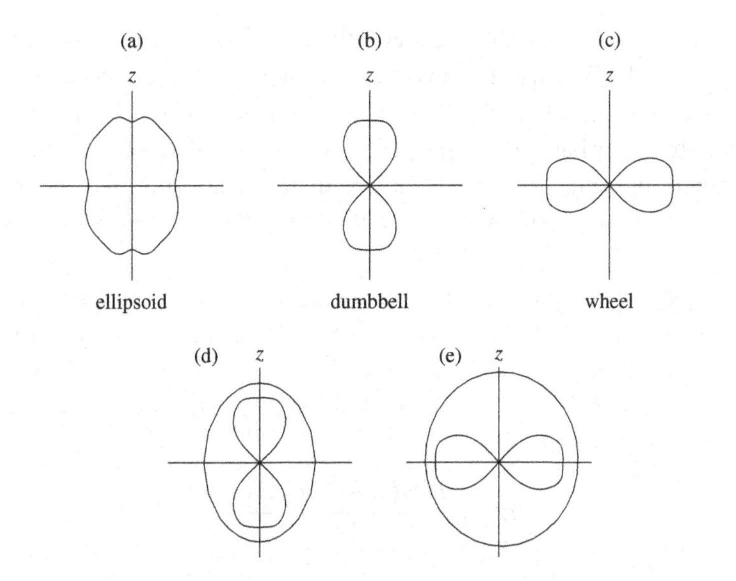

Fig. 6.3 (a), (b), (c) show types of wave normal surfaces; (d) and (e) show permissible overlays of wave normal surfaces.

is considered separately; i.e., either the plus or the minus sign is chosen. The convention is used that a mode is considered to exist (i.e., has a wave normal surface) only if $n^2 > 0$ for at least some range of θ; if $n^2 < 0$ for all angles, then the mode is evanescent (i.e., non-propagating) for all angles and is not plotted. The three types of wave normal surfaces are:

1. Bounded volume with no resonance and $n^2 > 0$ at some point in the bounded volume. Since $n^2 = 0$ occurs only at the bounding surfaces and $n^2 \to \infty$ only at resonances, n^2 must be positive and finite at every θ for each location in the bounded volume. The wave normal surface is thus ellipsoidal with symmetry about both $\theta = 0$ and $\theta = \pi/2$. The ellipse may deform as one moves inside the bounded volume, but will always have the morphology of an ellipse. This type of wave normal surface is shown in Fig. 6.3(a). The wave normal surface is three dimensional and is azimuthally symmetric about the z axis.
2. Bounded volume having a resonance at some angle θ_{res}, where $0 < \theta_{res} < \pi/2$ and $n^2(\theta)$ positive for $\theta < \theta_{res}$. At θ_{res}, $n \to \infty$ so the radius of the wave normal surface goes to zero. For $\theta < \theta_{res}$, the wave normal surface exists (n is pure real since $n^2 > 0$) and is plotted. At resonances $n^2(\theta)$ passes from $-\infty$ to $+\infty$ or vice versa. This type of wave normal surface is a dumbbell type as shown in Fig. 6.3(b).
3. Bounded volume having a resonance at some angle θ_{res}, where $0 < \theta_{res} < \pi/2$ and $n^2(\theta)$ positive for $\theta > \theta_{res}$. This is similar to case 2 above, except the wave normal surface now only exists for angles greater than θ_{res} and so has the wheel-type shape shown in Fig. 6.3(c).

Consider now the relationship between the two modes (plus and minus sign) given by Eq. (6.49). Because the two modes cannot intersect (cf. theorem 5 on p. 220) at angles other than $\theta = 0$, $\pi/2$, and mirror angles, if one mode is an ellipsoid and the other has a resonance (i.e., is a dumbbell or wheel), the ellipsoid must be outside the other dumbbell or wheel; for if not, the two modes would intersect at an angle other than $\theta = 0$ or $\theta = \pi/2$. This is shown in Figs. 6.3(d) and (e).

Also, only one of the modes can have a resonance, so at most one mode in a bounded volume can be a dumbbell or wheel. This can be seen by noting that a resonance occurs when $A \to 0$. In this case $B^2 \gg |4AC|$ in Eq. (6.49) and the two roots are *well separated*. This means the binomial expansion can be used on the square root in Eq. (6.49) to obtain

$$n^2 \simeq \frac{B \pm (B - 2AC/B)}{2A}$$

$$n_+^2 \simeq \frac{B}{A}, \quad n_-^2 \simeq \frac{C}{B}, \tag{6.55}$$

where $|n_+^2| \gg |n_-^2|$ since $B^2 \gg |4AC|$. The root n_+^2 has the resonance and the root n_-^2 has no resonance. Since the wave normal surface of the minus root has no resonance, its wave normal surface must be ellipsoidal (if it exists). Because the ellipsoidal surface must always lie outside the wheel or dumbbell surface, the ellipsoidal surface will have a larger value of ω/kc than the dumbbell or wheel at every θ and so the ellipsoidal mode will always be the *fast* mode. The mode with the resonance will be a dumbbell or wheel, will lie inside the ellipsoidal surface, and so will always be the *slow* mode. This concept of well-separated roots is quite useful and, if the roots are well separated, then Eq. (6.47) can be solved approximately for the large root (slow mode) by balancing the first two terms with each other, and for the small root (fast mode) by balancing the last two terms with each other.

Parameter space is subdivided into thirteen bounded volumes, each potentially containing two normal modes corresponding to two qualitatively distinct propagating waves. However, since two modes do not exist in all bounded volumes, the actual number of modes is smaller than 26. As an example of a bounded volume with two waves, the wave normal surfaces of the fast and slow Alfvén waves are in the upper right-hand corner of the CMA diagram, since this bounded volume corresponds to $\omega \ll \omega_{ci}$ and $\omega \ll \omega_{pe}$, i.e., above $L = 0$ and to the right of $P = 0$.

6.3.8 Use of the CMA diagram

The CMA diagram can be used in several ways. For example, it can be used to (i) identify all allowed cold plasma modes in a given plasma for various values of ω or (ii) investigate how a given mode evolves as it propagates through a spatially inhomogeneous plasma and possibly intersects resonances or cutoffs due to spatial variation of density or magnetic field.

Let us consider the first example. Suppose the plasma is uniform and has a prescribed density and magnetic field. Since $\ln\left[\left(\omega_{pe}^2+\omega_{pi}^2\right)/\omega^2\right]$ and $\ln\left[\omega_{ce}^2/\omega^2\right]$ are the coordinates of the CMA diagram, varying ω corresponds to tracing out a line having a slope of 45° and an offset determined by the prescribed density and magnetic field. High frequencies correspond to the lower left portion of this line and low frequencies to the upper right. Since the allowed modes lie along this line, if the line does not pass through a given bounded volume, then modes inside that bounded volume do not exist in the specified plasma.

Now consider the second example. Suppose the plasma is spatially non-uniform in such a way that both density and magnetic field are a function of position. To be specific, suppose density increases as one moves in the x direction while magnetic field increases as one moves in the y direction. Thus, the CMA diagram becomes a map of the actual plasma. A wave with prescribed frequency ω is launched at some position x, y and then propagates along some trajectory in parameter space as determined by its local dispersion relation. The wave will continuously change its character as determined by the local wave normal surface. Thus, a wave launched as a fast Alfvén mode from the upper-right bounded volume and propagating in the downward direction will only deform somewhat on traversing the $L=\infty$ bounding surface. In contrast, a wave launched as a slow Alfvén mode (dumbbell shape) from the same position will disappear when it reaches the $L=\infty$ bounding surface, because the slow mode does not exist on the lower side (high-frequency side) of the $L=\infty$ bounding surface. The slow Alfvén wave undergoes ion cyclotron resonance at the $L=\infty$ bounding surface and will be absorbed there.

6.4 Dispersion relation expressed as a relation between n_x^2 and n_z^2

The CMA diagram is very useful for classifying waves, but is often not so useful in practical situations where it is not obvious how to specify the angle θ. In a practical situation a wave is typically excited by an antenna that lies in a plane and the geometry of the antenna imposes the component of the wavevector in the antenna plane. The transmitter frequency determines ω.

For example, consider an antenna located in the $x=0$ plane and having some specified z dependence. When Fourier analyzed in z, such an antenna would excite a characteristic k_z spectrum. In the extreme situation of the antenna extending to infinity in the $x=0$ plane and having the periodic dependence $\exp(ik_z z)$, the antenna would excite just a single k_z. Thus, the antenna–transmitter combination in this situation would impose k_z and ω but leave k_x undetermined. The job of the dispersion relation would then be to determine k_x. It should be noted that antennas that are not both infinite and perfectly periodic will excite a spectrum of k_z modes rather than just a single k_z mode.

By writing $n_x^2 = n^2 \sin^2\theta$ and $n_z^2 = n^2\cos^2\theta$, Eqs. (6.47) and (6.48) can be expressed as a quadratic equation for n_x^2, namely

$$Sn_x^4 - [\bar{S}(S+P) - D^2]n_x^2 + P[\bar{S}^2 - D^2] = 0, \tag{6.56}$$

where

$$\bar{S} = S - n_z^2. \tag{6.57}$$

If the two roots of Eq. (6.56) are well separated, the large root is found by balancing the first two terms to obtain

$$n_x^2 \simeq \frac{\bar{S}(S+P) - D^2}{S}, \qquad \text{(large root)} \tag{6.58}$$

or in the limit of large P (i.e., low frequencies)

$$n_x^2 \simeq \frac{\bar{S}P}{S}, \qquad \text{(large root).} \tag{6.59}$$

The small root is found by balancing the last two terms of Eq. (6.56) to obtain

$$n_x^2 \simeq \frac{P[\bar{S}^2 - D^2]}{[\bar{S}(S+P) - D^2]}, \qquad \text{(small root)} \tag{6.60}$$

or in the limit of large P

$$n_x^2 \simeq \frac{(n_z^2 - R)(n_z^2 - L)}{S - n_z^2}, \qquad \text{(small root).} \tag{6.61}$$

Thus, any given n_z^2 always has an associated large n_x^2 mode and an associated small n_x^2 mode. Because the phase velocity is inversely proportional to the refractive index, the root with large n_x^2 is called the slow mode and the root with small n_x^2 is called the fast mode.

Using the quadratic formula it is seen that the exact form of these two roots of Eq. (6.56) is given by

$$n_x^2 = \frac{\bar{S}(S+P) - D^2 \pm \sqrt{\left[\bar{S}(S-P) - D^2\right]^2 + 4PD^2n_z^2}}{2S}. \tag{6.62}$$

It is clear that n_x^2 can become infinite only when $S = 0$. Situations where n_x^2 is complex (i.e., neither pure real or pure imaginary) can occur when P is large and negative, in which case the argument of the square root can become negative. In these cases, θ also becomes complex and is no longer a physical angle. This shows that considering real angles between $0 < \theta < 2\pi$ does not account for all possible types of wave behavior. The regions where n_x^2 becomes complex is called a region of inaccessibility and is a region where Eq. (6.56) does not have real roots. If a plasma is non-uniform in the x direction so that S, P, and D are functions of x and $\omega^2 < \omega_{pe}^2 + \omega_{pi}^2$ so that P is negative, the boundaries of a region of inaccessibility (if such a region exists) are the locations where the square root in Eq. (6.62) vanishes, i.e., where there is a solution for $\overline{S}(S-P) - D^2 = \pm\sqrt{-4PD^2 n_z^2}$.

6.5 A journey through parameter space

Imagine an enormous plasma where the density increases in the x direction and the magnetic field points in the z direction but increases in the y direction. Suppose further that a radio transmitter operating at a frequency ω is connected to a hypothetical antenna that emits *plane waves*, i.e., waves with spatial dependence $\exp(i\mathbf{k}\cdot\mathbf{x})$. These assumptions are somewhat self-contradictory because, in order to excite plane waves, an antenna must be infinitely long in the direction normal to $\mathbf{k}$ and if the antenna is infinitely long it cannot be localized. To circumvent this objection, it is assumed the plasma is so enormous that the antenna at any location is sufficiently large compared to the wavelength in question to emit waves that are nearly plane waves.

The antenna is located at some point x, y in the plasma and the emitted plane waves are detected by a phase-sensitive receiver. The position x, y corresponds to a point in CMA space. The antenna is rotated through a sequence of angles θ and, as the antenna is rotated, an observer walks in front of the antenna staying exactly one wavelength $\lambda = 2\pi/k$ from the face of the antenna. Since λ is proportional to $1/n = \omega/kc$ at fixed frequency, the locus of the observer's path will have the shape of a wave normal surface, i.e., a plot of $1/n$ versus θ.

Because of the way the CMA diagram was constructed, the topology of one of the two cold plasma modes always changes when a bounding surface is traversed. Which mode is affected and how its topology changes on crossing a bounding surface can be determined by monitoring the polarities of the four quantities S, P, R, L within each bounded volume. P changes polarity only at the $P = 0$ bounding surface, but R and L change polarity when they go through zero and also when they go through infinity. Furthermore, $S = (R+L)/2$ changes sign not only when $S = 0$ but also at $R = \infty$ and at $L = \infty$.

A straightforward way to establish how the polarities of S, P, R, L change as bounding surfaces are crossed is to start in the extreme lower left corner of parameter space, corresponding to $\omega^2 \gg \omega_{pe}^2, \omega_{ce}^2$. This is the limit of having no plasma and no magnetic field and so corresponds to unmagnetized vacuum. The cold plasma dispersion relation in this limit is simply $n^2 = 1$; i.e., vacuum electromagnetic waves such as ordinary light waves or radio waves. Here $S = P = R = L = 1$ because there are no plasma currents. Thus S, P, R, L are all *positive* in this bounded volume, denoted as Region 1 in Fig. 6.2 (regions are labeled by boxed numbers). To keep track of the respective polarities, a small cross is sketched in each of the 13 bounded volumes. The signs of L and R are noted on the left and right of the cross respectively, while the sign of S is shown at the top and the sign of P is shown at the bottom.

In traversing from Region 1 to Region 2, R passes through zero and so reverses polarity but the polarities of L, S, P are unaffected. Going from Region 2 to Region 3, S passes through zero so the sign of S reverses. By continuing from region to region in this manner, the plus or minus signs on the crosses in each bounded volume are established. It is important to remember that S changes sign at both $S = 0$ and the cyclotron resonances $L = \infty$ and $R = \infty$, but at all other bounding surfaces, only one quantity reverses sign.

Modes with resonances (i.e., dumbbells or wheels) only occur if S and P have opposite sign, which occurs in Regions 3, 7, 8, 10, and 13. The ordinary mode (i.e., $\theta = \pi/2$, $n^2 = P$) exists only if $P > 0$ and so exists only in regions to the left of the $P = 0$ bounding surface. Thus, to the right of the $P = 0$ bounding surface only extraordinary modes exist (i.e., only modes where $n^2 = RL/S$ at $\theta = \pi/2$). Extraordinary modes exist only if $RL/S > 0$, which cannot occur if an odd subset of the three quantities R, L, and S is negative. For example, in Region 5 all three quantities are negative so extraordinary modes do not exist in Region 5. The parallel modes $n^2 = R, L$ do not exist in Region 5 because R and L are negative there. Thus, no modes exist in Region 5 because if a mode were to exist there, it would need to have a limiting behavior of either ordinary or extraordinary at $\theta = \pi/2$ and of either right or left circularly polarized at $\theta = 0$.

When crossing a cutoff bounding surface ($R = 0, L = 0$, or $P = 0$), the outer (i.e., fast) mode has its wave normal surface become infinitely large, $\omega^2/k^2c^2 = 1/n^2 \to \infty$. Thus, immediately to the left of the $P = 0$ bounding surface, the fast mode (outer mode) is always the ordinary mode, because by definition this mode has the dispersion $n^2 = P$ at $\theta = \pi/2$ and so has a cutoff at $P = 0$. As one approaches the $P = 0$ bounding surface from the left, all the outer modes are ordinary modes and all disappear on crossing the $P = 0$ line so that to the right of the $P = 0$ line there are no ordinary modes.

In Region 13 where the modes are Alfvén waves, the slow mode is the $n^2 = L$ mode, since this is the mode that has the resonance at $L = \infty$. The slow Alfvén mode is the inertial Alfvén mode while the fast Alfvén mode is the compressional Alfvén mode. Going downwards from Region 13 to Region 11, the slow Alfvén wave undergoes ion cyclotron resonance and disappears, but the fast Alfvén wave remains. Similar arguments can be made to explain other boundary crossings in parameter space.

A subtle aspect of this taxonomy is the division of Region 6 into two sub-regions 6a, and 6b. This subtlety arises because the dispersion at $\theta = \pi/2$ has the form

$$n^2 = \frac{RL + PS \pm |RL - PS|}{2S} = P, RL/S. \tag{6.63}$$

In Region 6, both S and P are positive. If $RL - PS$ is also positive, then the plus sign gives the extraordinary mode, which is the slow mode (bigger n, inner of the two wave normal surfaces). On the other hand, if $RL - PS$ is negative, then the absolute value operator inverts the sign of $RL - PS$ and the minus sign now gives the extraordinary mode, which will be the fast mode (smaller n, outer of the two wave normal surfaces). Region 8 can also be divided into two regions (omitted here for clarity) separated by the $RL = PS$ line. In Region 8 the ordinary mode does not exist, but the extraordinary mode will be given by either the plus or minus sign in Eq. (6.63) depending on which side of the $RL = PS$ line one is considering.

For a given plasma density and magnetic field, varying the frequency corresponds to moving along a "mode" line that has a 45° slope on the log–log CMA diagram (see Fig. 6.2). If the plasma density is increased, the mode line moves to the right, whereas if the magnetic field is increased, the mode line moves up. Since any single mode line cannot pass through all 13 regions of parameter space, only a limited subset of the 13 regions of parameter space can be accessed for any given plasma density and magnetic field. Plasmas with $\omega_{pe}^2 > \omega_{ce}^2$ are often labeled "overdense" and plasmas with $\omega_{pe}^2 < \omega_{ce}^2$ are correspondingly labeled "underdense." For overdense plasmas, the mode line passes to the right of the intersection of the $P = 0$, $R = \infty$ bounding surfaces while for underdense plasmas the mode line passes to the left of this intersection. Two different plasmas will be self-similar if they have similar mode lines. For example, if a lab plasma has the same mode line as a space plasma it will support the same kind of modes, but do so in a scaled fashion. Because the CMA diagram is log–log the bounding surface curves extend infinitely to the left and right of the figure and also infinitely above and below it; however, no new regions exist outside of what is sketched in Fig. 6.2.

The weakly magnetized case corresponds to the lower parts of Regions 1–5, while the low-density case corresponds to the left parts of Regions 1, 2, 3, 6, 9, 10, and 12. The CMA diagram provides a visual way for categorizing a great deal of useful information. In particular, it allows identification of isomorphisms between modes in different regions of parameter space so that understanding developed about the behavior for one kind of mode can be exploited to explain the behavior of a different, but isomorphic, mode located in another region of parameter space.

6.6 High-frequency waves: Altar–Appleton–Hartree dispersion relation

Examination of the dielectric tensor elements S, P, and D shows that while both ion and electron terms are of importance for low-frequency waves, for high-frequency waves ($\omega \gg \omega_{ci}$, ω_{pi}) the ion terms are unimportant and may be dropped. Thus, for high-frequency waves the dielectric tensor elements simplify to

$$\begin{aligned} S &= 1 - \frac{\omega_{pe}^2}{\omega^2 - \omega_{ce}^2} \\ P &= 1 - \frac{\omega_{pe}^2}{\omega^2} \\ D &= \frac{\omega_{ce}}{\omega} \frac{\omega_{pe}^2}{\left(\omega^2 - \omega_{ce}^2\right)} \end{aligned} \tag{6.64}$$

and the corresponding R and L terms are

$$\begin{aligned} R &= 1 - \frac{\omega_{pe}^2}{\omega\left(\omega + \omega_{ce}\right)} \\ L &= 1 - \frac{\omega_{pe}^2}{\omega\left(\omega - \omega_{ce}\right)}. \end{aligned} \tag{6.65}$$

The development of long-distance short-wave radio communication in the 1930s motivated investigations into how radio waves bounce from the ionosphere. Because the bouncing involves a $P = 0$ cutoff and because the ionosphere has ω_{pe}^2 of order ω_{ce}^2 but usually larger, the relevant frequencies must be of the order of the electron plasma frequency and so are much higher than both the ion cyclotron and ion plasma frequencies. Thus, ion effects are unimportant and so all ion terms may be dropped in order to simplify the analysis.

Perhaps the most important result of this era was a peculiar, but useful, reformulation by Appleton, Hartree, and Altar[1] (Appleton 1932) of the $\omega \gg \omega_{pi}, \omega_{ci}$ limit of Eq. (6.49). The intent of this reformulation was to express n^2 in terms of its deviation from the vacuum limit, $n^2 = 1$. An obvious way to do this is to define $\xi = n^2 - 1$ and then rewrite Eq. (6.47) as an equation for ξ, namely

$$A(\xi^2 + 2\xi + 1) - B(\xi + 1) + C = 0 \tag{6.66}$$

or, after regrouping,

$$A\xi^2 + \xi(2A - B) + A - B + C = 0. \tag{6.67}$$

[1] See discussion by Swanson (1989) regarding the recent addition of Altar to this citation.

Unfortunately, when this expression is solved for ξ, the leading term is -1 and so this attempt to find the deviation of n^2 from its vacuum limit fails. However, a slight rewriting of Eq. (6.67) as

$$\frac{A-B+C}{\xi^2}+\frac{2A-B}{\xi}+A=0 \tag{6.68}$$

and then solving for $1/\xi$, gives

$$\xi=\frac{2(A-B+C)}{B-2A\pm\sqrt{B^2-4AC}}. \tag{6.69}$$

This expression does not have a leading term of -1 and so allows the solution of Eq. (6.49) to be expressed as

$$n^2=1+\frac{2(A-B+C)}{B-2A\pm\sqrt{B^2-4AC}}. \tag{6.70}$$

In the $\omega \gg \omega_{ci}, \omega_{pi}$ limit where S, P, D are given by Eq. (6.64), algebraic manipulation of Eq. (6.70) (cf. assignments) shows the numerator and denominator of the second term have a common factor. After canceling this common factor, Eq. (6.70) reduces to

$$n^2=1-\left[\frac{2\dfrac{\omega_{pe}^2}{\omega^2}\left(1-\dfrac{\omega_{pe}^2}{\omega^2}\right)}{2\left(1-\dfrac{\omega_{pe}^2}{\omega^2}\right)-\dfrac{\omega_{ce}^2}{\omega^2}\sin^2\theta\pm\Gamma}\right], \tag{6.71}$$

where

$$\Gamma=\sqrt{\frac{\omega_{ce}^4}{\omega^4}\sin^4\theta+4\frac{\omega_{ce}^2}{\omega^2}\left(1-\frac{\omega_{pe}^2}{\omega^2}\right)^2\cos^2\theta}. \tag{6.72}$$

Equation (6.71) is called the Altar–Appleton–Hartree dispersion relation (Appleton 1932) and has the desired property of showing the deviation of n^2 from the vacuum dispersion $n^2=1$.

We recall that the cold plasma dispersion relation simplified considerably when either $\theta=0$ or $\theta=\pi/2$. A glance at Eq. (6.71) shows that this expression reduces indeed to $n^2=R, L$ for $\theta=0$. Somewhat more involved manipulation shows that Eq. (6.71) also reduces to $n_+^2=P$ and $n_-^2=RL/S$ for $\theta=\pi/2$.

Equation (6.71) can be Taylor expanded in the vicinity of the principle angles $\theta=0$ and $\theta=\pi/2$ to give dispersion relations for quasi-parallel or quasi-perpendicular propagation. The terms quasi-longitudinal and quasi-transverse are commonly used to denote these situations. The nomenclature is somewhat unfortunate because of the possible confusion with the traditional convention that longitudinal and transverse refer to the orientation of **k** relative to the wave electric field. Here, longitudinal means **k** is nearly parallel to the static magnetic field while transverse means **k** is nearly perpendicular to the static magnetic field.

6.6.1 Quasi-transverse modes ($\theta \simeq \pi/2$)

For quasi-transverse propagation, the first term in Γ dominates, that is,

$$\frac{\omega_{ce}^4}{\omega^4}\sin^4\theta \gg 4\frac{\omega_{ce}^2}{\omega^2}\left(1-\frac{\omega_{pe}^2}{\omega^2}\right)^2\cos^2\theta. \tag{6.73}$$

In this case a binomial expansion of Γ gives

$$\begin{aligned}\Gamma &= \frac{\omega_{ce}^2}{\omega^2}\sin^2\theta\left[1+4\frac{\omega^2}{\omega_{ce}^2}\left(1-\frac{\omega_{pe}^2}{\omega^2}\right)^2\frac{\cos^2\theta}{\sin^4\theta}\right]^{1/2}\\ &\simeq \frac{\omega_{ce}^2}{\omega^2}\sin^2\theta+2\left(1-\frac{\omega_{pe}^2}{\omega^2}\right)^2\cot^2\theta.\end{aligned} \tag{6.74}$$

Substitution of Γ into Eq. (6.71) shows the generalization of the ordinary mode dispersion to angles in the vicinity of $\pi/2$ is

$$n_+^2 = \frac{1-\dfrac{\omega_{pe}^2}{\omega^2}}{1-\dfrac{\omega_{pe}^2}{\omega^2}\cos^2\theta}. \tag{6.75}$$

The subscript + here means that the positive sign has been used in Eq. (6.71). This mode is called the QTO mode as an acronym for "quasi-transverse-ordinary."

Choosing the − sign in Eq. (6.71) gives the quasi-transverse-extraordinary mode or QTX mode. After a modest amount of algebra (cf. assignments) the QTX dispersion is found to be

$$n_-^2 = \frac{\left(1-\dfrac{\omega_{pe}^2}{\omega^2}\right)^2-\dfrac{\omega_{ce}^2}{\omega^2}\sin^2\theta}{1-\dfrac{\omega_{pe}^2}{\omega^2}-\dfrac{\omega_{ce}^2}{\omega^2}\sin^2\theta}. \tag{6.76}$$

Note that the QTX mode has a resonance near the upper hybrid frequency.

6.6.2 Quasi-longitudinal dispersion ($\theta \simeq 0$)

Here, the term containing $\cos^2\theta$ dominates in Eq. (6.72). Because there are no cancelations of the leading terms in Γ with any remaining terms in the denominator of Eq. (6.71), it suffices to keep only the leading term of Γ. Thus, in this limit

$$\Gamma \simeq 2\left|\frac{\omega_{ce}}{\omega}\left(1-\frac{\omega_{pe}^2}{\omega^2}\right)\cos\theta\right| = -2\left(1-\frac{\omega_{pe}^2}{\omega^2}\right)\left|\frac{\omega_{ce}}{\omega}\cos\theta\right|, \qquad (6.77)$$

since $P = 1-\omega_{pe}^2/\omega^2$ is assumed to be negative. Upon substitution for Γ in Eq. (6.71) and then simplifying, one obtains

$$n_+^2 = 1 - \frac{\omega_{pe}^2/\omega^2}{1-\left|\dfrac{\omega_{ce}}{\omega}\cos\theta\right|}, \qquad \text{QLR mode} \qquad (6.78)$$

and

$$n_-^2 = 1 - \frac{\omega_{pe}^2/\omega^2}{1+\left|\dfrac{\omega_{ce}}{\omega}\cos\theta\right|}, \qquad \text{QLL mode.} \qquad (6.79)$$

These simplified dispersions are based on the implicit assumption that $|P|$ is large, because if $P \to 0$ the presumption that Eq. (6.77) gives the leading term in Γ would be inappropriate. When $\omega < |\omega_{ce}\cos\theta|$ the QLR mode (quasi-longitudinal, right-hand circularly polarized) is called the whistler or helicon wave. This wave is distinguished by having a descending whistling tone, which shows up at audio frequencies on sensitive amplifiers connected to long wire antennas. Whistlers may have been heard as early as the late nineteenth century by telephone linesmen installing long telephone lines. They became a subject of some interest in the trenches of the First World War when German scientist H. Barkhausen heard whistlers on a sensitive audio receiver while trying to eavesdrop on British military communications; the origin of these waves was a mystery at that time. After the war Barkhausen (1930) and Eckersley (1935) proposed that the descending tone was due to a dispersive propagation such that lower frequencies traveled more slowly, but did not explain the source location or propagation trajectory. The explanation had to wait over two more decades until Storey (1953) finally solved the mystery by showing that whistlers were caused by lightning bolts and identified two main types of propagation. The first type, called a short whistler, resulted from a lightning bolt in the opposite hemisphere exciting a wave that propagated dispersively along the Earth's magnetic field to the observer. The second type, called a long whistler, resulted from a lightning bolt in the vicinity of the observer exciting a wave that propagated dispersively along field lines to the opposite hemisphere, then reflected, and traveled back along the same path to the observer. The dispersion would be greater in this round-trip situation and also there would be a correlation with a click from the local lightning bolt. Whistlers are routinely observed by spacecraft flying through the Earth's magnetosphere and the magnetospheres of other planets.

The reason for the whistler's descending tone can be seen by representing each lightning bolt as a delta function in time

$$\delta(t) = \frac{1}{2\pi}\int \mathrm{e}^{-\mathrm{i}\omega t}\mathrm{d}\omega. \tag{6.80}$$

A lightning bolt therefore launches a very broad frequency spectrum. Because the ionospheric electron plasma frequency is in the range 10–30 MHz, audio frequencies are much lower than the electron plasma frequency, i.e., $\omega_{pe} \gg \omega$ and so $|P| \gg 1$. The electron cyclotron frequency in the ionosphere is of the order of 1 MHz so $\omega_{ce} \gg \omega$ also. Thus, the whistler dispersion for acoustic (a few kHz) waves in the ionosphere is

$$n_-^2 = \frac{\omega_{pe}^2}{\left|\dfrac{\omega_{ce}}{\omega}\cos\theta\right|} \tag{6.81}$$

or

$$k = \frac{\omega_{pe}}{c}\sqrt{\frac{\omega}{|\omega_{ce}\cos\theta|}}. \tag{6.82}$$

Each frequency ω in Eq. (6.80) has a corresponding k given by Eq. (6.82) so that the disturbance $g(x, t)$ excited by a lightning bolt has the form

$$g(x, t) = \frac{1}{2\pi}\int \mathrm{e}^{\mathrm{i}k(\omega)x-\mathrm{i}\omega t}\mathrm{d}\omega, \tag{6.83}$$

where $x = 0$ is the location of the lightning bolt. Because of the strong dependence of k on ω, contributions to the phase integral in Eq. (6.83) at adjacent frequencies will in general have substantially different phases. The integral can then be considered as the sum of contributions having all possible phases. Since there will be approximately equal amounts of positive and negative contributions, the contributions will cancel each other out when summed; this canceling is called phase mixing.

Suppose there exists some frequency ω at which the phase $k(\omega)x - \omega t$ has a local maximum or minimum with respect to variation of ω. In the vicinity of this extremum, the phase is independent of frequency and so the contributions from adjacent frequencies constructively interfere and produce a finite signal. Thus, an observer located at some position $x \neq 0$ will hear a signal only at the time when the phase in Eq. (6.83) is at an extremum. The phase extremum is found by setting to zero the derivative of the phase with respect to frequency, i.e., setting

$$\frac{\partial k}{\partial \omega}x - t = 0. \tag{6.84}$$

From Eq. (6.82) it is seen that

$$\frac{\partial k}{\partial \omega} = \frac{\omega_{pe}}{2c} \frac{1}{\sqrt{\omega |\omega_{ce} \cos \theta|}} \tag{6.85}$$

so the time at which a frequency ω is heard by an observer at location x is

$$t = \frac{\omega_{pe}}{2c} \frac{x}{\sqrt{\omega |\omega_{ce} \cos \theta|}}. \tag{6.86}$$

This shows that lower frequencies are heard at later times, resulting in the descending tone characteristic of whistlers.

6.7 Group velocity

Suppose that at time $t = 0$ the electric field of a particular fast or slow mode is decomposed into spatial Fourier modes, each varying as $\exp(\mathrm{i}\mathbf{k} \cdot x)$. The total wave field can then be written as

$$\mathbf{E}(\mathbf{x}) = \int \mathbf{dk}\tilde{\mathbf{E}}(\mathbf{k}) \exp(\mathbf{i}\mathbf{k} \cdot \mathbf{x}), \tag{6.87}$$

where $\tilde{\mathbf{E}}(\mathbf{k})$ is the amplitude of the mode with wavevector $\mathbf{k}$. The dispersion relation assigns an ω to each $\mathbf{k}$, so that at later times the field evolves as

$$\mathbf{E}(\mathbf{x}, t) = \int \mathbf{dk}\tilde{\mathbf{E}}(\mathbf{k}) \exp(\mathbf{i}\mathbf{k} \cdot \mathbf{x} - \mathrm{i}\omega(\mathbf{k})t), \tag{6.88}$$

where $\omega(\mathbf{k})$ is given by the dispersion relation. The integration over $\mathbf{k}$ may be viewed as a summation of rapidly oscillating waves, each having different rates of phase variation. In general, this sum vanishes because the waves add destructively or "phase mix." However, if the waves add constructively, a finite $\mathbf{E}(\mathbf{x}, t)$ will result. Denoting the phase by

$$\phi(\mathbf{k}) = \mathbf{k} \cdot \mathbf{x} - \omega(\mathbf{k})t \tag{6.89}$$

it is seen that the Fourier components add constructively at extrema (minima or maxima) of $\phi(\mathbf{k})$ because, in the vicinity of an extremum, the phase is stationary with respect to $\mathbf{k}$, that is, the phase does not vary with $\mathbf{k}$. Thus, the trajectory $\mathbf{x} = \mathbf{x}(t)$, along which $\mathbf{E}(\mathbf{x}, t)$ is finite, is the trajectory along which the phase is stationary. At time t the stationary phase is the place where $\partial\phi(\mathbf{k})/\partial\mathbf{k}$ vanishes, which is where

$$\frac{\partial \phi}{\partial \mathbf{k}} = \mathbf{x} - \frac{\partial \omega}{\partial \mathbf{k}} t = 0. \tag{6.90}$$

The trajectory of the points of stationary phase is therefore

$$\mathbf{x}(t) = \mathbf{v}_g t, \tag{6.91}$$

where $\mathbf{v}_g = \partial\omega/\partial\mathbf{k}$ is called the group velocity. The group velocity is the velocity at which a pulse propagates in a dispersive medium and is also the velocity at which energy propagates.

The phase velocity for a one-dimensional system is defined as $v_{ph} = \omega/k$. In three dimensions this definition can be extended to be $\mathbf{v}_{ph} = \hat{k}\omega/k$, i.e., a vector in the direction of $\mathbf{k}$ but with the magnitude ω/k.

Group and phase velocities are the same only for the special case where ω is linearly proportional to k, a situation that occurs only if there is no plasma. For example, the phase velocity of electromagnetic plasma waves (dispersion $\omega^2 = \omega_{pe}^2 + k^2c^2$) is

$$\mathbf{v}_{ph} = \hat{k}\frac{\sqrt{\omega_{pe}^2 + k^2c^2}}{k}, \tag{6.92}$$

which is faster than the speed of light. However, no paradox results because information and energy travel at the group velocity, not the phase velocity. The group velocity for this wave is evaluated by taking the derivative of the dispersion with respect to $\mathbf{k}$ giving

$$2\omega\frac{\partial\omega}{\partial\mathbf{k}} = 2\mathbf{k}c, \tag{6.93}$$

or

$$\frac{\partial\omega}{\partial\mathbf{k}} = \frac{\mathbf{k}c}{\sqrt{\omega_{pe}^2 + k^2c^2}}, \tag{6.94}$$

which is less than the speed of light.

This illustrates an important property of the wave normal surface concept – a wave normal surface is a polar plot of the *phase velocity* and should not be confused with the group velocity.

6.8 Quasi-electrostatic cold plasma waves

Another useful way of categorizing waves is according to whether the wave electric field is:

1. electrostatic so that $\nabla\times\mathbf{E} = 0$ and $\mathbf{E} = -\nabla\phi$

   or

2. inductive so that $\nabla\cdot\mathbf{E} = 0$ and in Coulomb gauge, $\mathbf{E} = -\partial\mathbf{A}/\partial t$, where $\mathbf{A}$ is the vector potential.

An electrostatic electric field is produced by net charge density whereas an inductive field is produced by time-dependent currents. Inductive electric fields are always associated with time-dependent magnetic fields via Faraday's law.

Waves involving purely electrostatic electric fields are called electrostatic waves, whereas waves involving inductive electric fields are called electromagnetic waves because these waves involve both electric and magnetic wave fields. In actuality, electrostatic waves must always have some slight inductive component, because there must always be a small current that establishes the net charge density. Thus, strictly speaking, the condition for electrostatic modes is $\nabla \times \mathbf{E} \simeq 0$ rather than $\nabla \times \mathbf{E} = 0$.

In terms of Fourier modes where ∇ is replaced by $i\mathbf{k}$, electrostatic modes are those for which $\mathbf{k} \times \mathbf{E} = 0$ so that $\mathbf{E}$ is parallel to $\mathbf{k}$; this means that electrostatic waves are longitudinal waves. Electromagnetic waves have $\mathbf{k} \cdot \mathbf{E} = 0$ and so are transverse waves. Here, we are using the usual wave terminology where longitudinal and transverse refer to whether $\mathbf{k}$ is parallel or perpendicular to $\mathbf{E}$.

The electron plasma waves and ion acoustic waves discussed in the previous chapter were electrostatic, the compressional Alfvén wave was inductive, and the inertial Alfvén wave was both electrostatic and inductive. We now wish to show that in a *magnetized* plasma, the wheel and dumbbell modes in the CMA diagram always have electrostatic behavior in the region where the wave normal surface comes close to the origin, i.e., near the cross-over in the figure-eight pattern of these wave normal surfaces. For these waves, when n becomes large (i.e., near the cross-over of the figure-eight pattern of the wheel or dumbbell), $\mathbf{n}$ becomes nearly parallel to $\mathbf{E}$ and the magnetic part of the wave becomes infinitesimal. We now prove this assertion and also take care to distinguish this situation from another situation where n becomes infinite, namely at cyclotron resonances.

When the two roots of the dispersion $An^4 - Bn^2 + C = 0$ are well separated (i.e., $B^2 \gg 4AC$) the slow mode is found by assuming that n^2 is large. In this case the dispersion can be approximated as $An^4 - Bn^2 \simeq 0$, which gives the slow mode as $n^2 \simeq B/A$. Resonance (i.e., $n^2 \to \infty$) can thus occur either from

1. $A = S\sin^2\theta + P\cos^2\theta$ vanishing, or
2. $B = RL\sin^2\theta + PS(1+\cos^2\theta)$ becoming infinite.

These two cases are different. In the first case S and P remain finite and the vanishing of A determines a critical angle $\theta_{res} = \tan^{-1}\sqrt{-P/S}$; this angle is the cross-over angle of the figure-eight pattern of the wheel or dumbbell. In the second case either R or L must become infinite, a situation occurring only at the R or L bounding surfaces, i.e., only at the electron or ion cyclotron resonances.

The first case results in quasi-electrostatic cold plasma waves, whereas the second case does not. To see this, the electric field is first decomposed into its longitudinal and transverse parts

$$\mathbf{E}^l = \hat{n}\hat{n}\cdot\mathbf{E}$$
$$\mathbf{E}^t = \mathbf{E}-\mathbf{E}^l, \tag{6.95}$$

where $\hat{n} = \hat{k} = \mathbf{n}/n$ is a unit vector in the direction of $\mathbf{n}$. The cold plasma wave equation, Eq. (6.17), can thus be written as

$$\mathbf{nn}\cdot\left(\mathbf{E}^t+\mathbf{E}^l\right)-n^2\left(\mathbf{E}^t+\mathbf{E}^l\right)+\overleftrightarrow{\mathbf{K}}\cdot\left(\mathbf{E}^t+\mathbf{E}^l\right)=0. \tag{6.96}$$

Since $\mathbf{n}\cdot\mathbf{E}^t=0$ and $\mathbf{nn}\cdot\mathbf{E}^l=n^2\mathbf{E}^l$, this expression can be recast as

$$\left(\overleftrightarrow{\mathbf{K}}-n^2\overleftrightarrow{\mathbf{I}}\right)\cdot\mathbf{E}^t+\overleftrightarrow{\mathbf{K}}\cdot\mathbf{E}^l=0, \tag{6.97}$$

where $\overleftrightarrow{\mathbf{I}}$ is the unit tensor. If the resonance is such that

$$n^2 \gg K_{ij}, \tag{6.98}$$

where K_{ij} are the elements of the dielectric tensor, then Eq. (6.97) may be approximated as

$$-n^2\mathbf{E}^t+\overleftrightarrow{\mathbf{K}}\cdot\mathbf{E}^l\simeq 0. \tag{6.99}$$

This shows that the transverse electric field is

$$\mathbf{E}^t=\frac{1}{n^2}\overleftrightarrow{\mathbf{K}}\cdot\mathbf{E}^l, \tag{6.100}$$

which is much smaller in magnitude than the longitudinal electric field by virtue of Eq. (6.98).

An easy way to obtain the dispersion relation (determinant of this system) is to dot Eq. (6.100) with $\mathbf{n}$ to obtain

$$\mathbf{n}\cdot\overleftrightarrow{\mathbf{K}}\cdot\mathbf{n}=n^2(S\sin^2\theta+P\cos^2\theta)\simeq 0, \tag{6.101}$$

which is just the first case discussed above. This argument is self-consistent because for the first case (i.e., $A\to 0$) the quantities S, P, D remain finite so the condition given by Eq. (6.98) is satisfied.

The second case, $B\to\infty$, occurs at the cyclotron resonances where S and D diverge so the condition given by Eq. (6.98) is not satisfied. Thus, for the second case the electric field is not quasi-electrostatic.

6.9 Resonance cones

The situation $A\to 0$ corresponds to Eq. (6.101) which is a dispersion relation having the surprising property of depending on θ, but *not* on the magnitude of n.

This limiting form of dispersion has some bizarre aspects, which will now be examined.

The group velocity in this situation can be evaluated by writing Eq. (6.101) as

$$k_x^2 S + k_z^2 P = 0 \tag{6.102}$$

and then taking the vector derivative with respect to $\mathbf{k}$ to obtain

$$2k_x \hat{x} S + 2k_z \hat{z} P + \left(k_x^2 \frac{\partial S}{\partial \omega} + k_z^2 \frac{\partial P}{\partial \omega} \right) \frac{\partial \omega}{\partial \mathbf{k}} = 0, \tag{6.103}$$

which may be solved to give

$$\frac{\partial \omega}{\partial \mathbf{k}} = -2 \left(\frac{k_x \hat{x} S + k_z \hat{z} P}{k_x^2 \dfrac{\partial S}{\partial \omega} + k_z^2 \dfrac{\partial P}{\partial \omega}} \right). \tag{6.104}$$

If Eq. (6.104) is dotted with $\mathbf{k}$ the surprising result

$$\mathbf{k} \cdot \frac{\partial \omega}{\partial \mathbf{k}} = 0 \tag{6.105}$$

is obtained, which means that the group velocity is *orthogonal* to the phase velocity. The same result may also be obtained in a quicker but more abstract way by using spherical coordinates in k-space, in which case the group velocity is just

$$\frac{\partial \omega}{\partial \mathbf{k}} = \hat{\mathbf{k}} \frac{\partial \omega}{\partial k} + \frac{\hat{\theta}}{k} \frac{\partial \omega}{\partial \theta}. \tag{6.106}$$

Applying this to Eq. (6.101), it is seen that the first term on the right-hand side vanishes because the dispersion relation is independent of the magnitude of $\mathbf{k}$. Thus, the group velocity is in the $\hat{\theta}$ direction, so the group and phase velocities are again *orthogonal*, since $\mathbf{k}$ is orthogonal to $\hat{\theta}$. Thus, energy and information propagate at right angles to the phase velocity.

A more physically intuitive interpretation of this phenomenon may be developed by "un-Fourier" analyzing the cold plasma wave equation, Eq. (6.17), giving

$$\nabla \times \nabla \times \mathbf{E} - \frac{\omega^2}{c^2} \overleftrightarrow{\mathbf{K}} \cdot \mathbf{E} = 0. \tag{6.107}$$

The modes corresponding to $A \to 0$ were obtained by dotting the dispersion relation with $\mathbf{n}$, an operation equivalent to taking the divergence in real space, and then arguing that the wave is mainly longitudinal. Let us therefore assume that $\mathbf{E} \simeq -\nabla \phi$ and take the divergence of Eq. (6.107) to obtain

$$\nabla \cdot \left(\overleftrightarrow{\mathbf{K}} \cdot \nabla \phi \right) = 0, \tag{6.108}$$

which is essentially Poisson's equation for a medium having dielectric tensor $\overleftrightarrow{\mathbf{K}}$. Equation (6.108) can be expanded to give

$$S\frac{\partial^2\phi}{\partial x^2}+P\frac{\partial^2\phi}{\partial z^2}=0. \tag{6.109}$$

If S and P have the same sign, Eq. (6.109) is an elliptic partial differential equation and so is just a distorted form of Poisson's equation. In fact, by defining the stretched coordinates $\xi=x/\sqrt{|S|}$ and $\eta=z/\sqrt{|P|}$, Eq. (6.109) becomes Poisson's equation in $\xi-\eta$ space.

Suppose now that waves are being excited by a line source $q\delta(x)\delta(z)\exp(-i\omega t)$, i.e., a wire antenna lying along the y axis oscillating at the frequency ω. In this case Eq. (6.109) becomes

$$\frac{\partial^2\phi}{\partial\xi^2}+\frac{\partial^2\phi}{\partial\eta^2}=\frac{q}{|SP|^{1/2}\varepsilon_0}\delta(\xi)\delta(\eta) \tag{6.110}$$

so that the equipotential contours excited by the line source are just static concentric circles in ξ, η or, equivalently, static concentric ellipses in x, z.

However, if S and P have *opposite signs*, the situation is entirely different, because now the equation is hyperbolic and has the form

$$\frac{\partial^2\phi}{\partial x^2}=\left|\frac{P}{S}\right|\frac{\partial^2\phi}{\partial z^2}. \tag{6.111}$$

Equation (6.111) is formally analogous to the standard hyperbolic wave equation

$$\frac{\partial^2\psi}{\partial t^2}=c^2\frac{\partial^2\psi}{\partial z^2}, \tag{6.112}$$

which has solutions propagating along the characteristics $\psi=\psi(z\pm ct)$. Thus, the solutions of Eq. (6.111) also propagate along characteristics, i.e.,

$$\phi=\phi\left(z\pm\sqrt{-P/S}x\right), \tag{6.113}$$

which are characteristics in the $x-z$ plane rather than the $x-t$ plane. For a line source, the potential is infinite at the line, and this infinite potential propagates from the source following the characteristics

$$z=\pm\sqrt{-\frac{P}{S}}x. \tag{6.114}$$

If the source is a point source, then the potential has the form

$$\phi(r,z)\sim\frac{q}{4\pi\varepsilon_0\left(\frac{r^2}{S}+\frac{z^2}{P}\right)^{1/2}}, \tag{6.115}$$

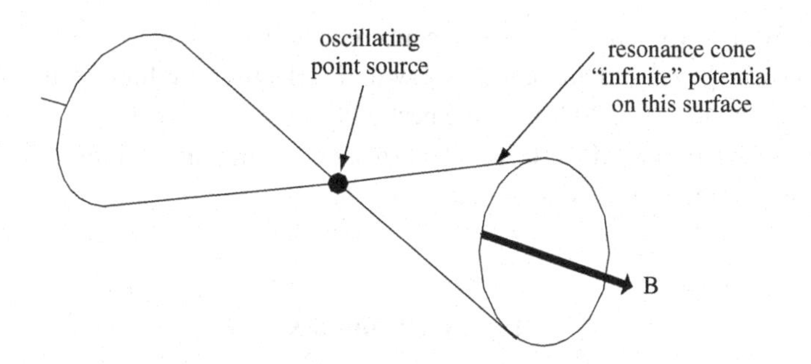

Fig. 6.4 Resonance cone excited by oscillating point source in magnetized plasma.

which diverges on the conical surface having cone angle $\tan\theta_{cone} = r/z = \pm\sqrt{-S/P}$, as shown in Fig. 6.4. These singular surfaces are called resonance cones and were first observed by Fisher and Gould (1969). The singularity results because the cold plasma approximation allows k to be arbitrarily large (i.e., allows infinitesimally short wavelengths). However, when k is made larger than ω/v_T, the cold plasma assumption $\omega/k \gg v_T$ becomes violated and warm plasma effects need to be taken into account. Thus, instead of becoming infinite on the resonance cone, the potential is large and finite and has a fine structure determined by thermal effects (Fisher and Gould 1971).

Resonance cones exist in the following regions of parameter space:

(i) Region 3, where they are called upper hybrid resonance cones; here $S < 0, P > 0$.

(ii) Regions 7 and 8, where they are a limiting form of the whistler wave and are also called lower hybrid resonance cones, since they are affected by the lower hybrid resonance (Briggs and Parker 1972, Bellan and Porkolab 1975). For $\omega_{ci} \ll \omega \ll \omega_{pe}, \omega_{ce}$ the P and S dielectric tensor elements become

$$P \simeq -\frac{\omega_{pe}^2}{\omega^2}, \quad S \simeq 1 - \frac{\omega_{pi}^2}{\omega^2} + \frac{\omega_{pe}^2}{\omega_{ce}^2}, \tag{6.116}$$

so the cone angle $\theta_{cone} = \tan^{-1} r/z$ is

$$\theta_{cone} = \tan^{-1}\sqrt{-\frac{S}{P}} = \tan^{-1}\sqrt{\left(\omega^2 - \omega_{lh}^2\right)\left(\omega_{pe}^{-2} + \omega_{ce}^{-2}\right)}. \tag{6.117}$$

If $\omega \gg \omega_{lh}$, the cone depends mainly on the smaller of ω_{pe}, ω_{ce}. For example, if $\omega_{ce} \ll \omega_{pe}$ then the cone angle is simply

$$\theta_{cone} \simeq \tan^{-1} \omega/\omega_{ce}, \tag{6.118}$$

whereas if $\omega_{ce} \gg \omega_{pe}$ then

$$\theta_{cone} \simeq \tan^{-1} \omega/\omega_{pe}. \tag{6.119}$$

For low-density plasmas this last expression can be used as the basis for a simple, accurate plasma density diagnostic.

(iii) Regions 10 and 13. The Alfvén resonance cones in Region 13 have a cone angle $\theta_{cone} = \omega/\sqrt{|\omega_{ce}\omega_{ci}|}$ and are associated with the electrostatic limit of inertial Alfvén waves (Stasiewicz *et al.* 2000). To the best of the author's knowledge, cones have not been investigated in Region 10, which corresponds to an unusual mix of parameters, namely ω_{pe} is the same order of magnitude as ω_{ci}.

6.10 Assignments

1. Prove that the cold plasma dispersion relation can be written as

$$An^4 - Bn^2 + C = 0,$$

where

$$A = S\sin^2\theta + P\cos^2\theta$$
$$B = (S^2 - D^2)\sin^2\theta + SP(1+\cos^2\theta)$$
$$C = P(S^2 - D^2)$$

so that the dispersion is

$$n^2 = \frac{B \pm \sqrt{B^2 - 4AC}}{2A}.$$

Prove that

$$RL = S^2 - D^2.$$

2. Prove that n^2 is always real if θ is real, by showing

$$B^2 - 4AC = \left[S^2 - D^2 - SP\right]^2 \sin^4\theta + 4P^2D^2\cos^2\theta.$$

3. Plot the bounding surfaces of the CMA diagram, by defining $m_e/m_i = \lambda$, $x = \left(\omega_{pe}^2 + \omega_{pi}^2\right)/\omega^2$, and $y = \omega_{ce}^2/\omega^2$. Show that

$$\frac{\omega_{pe}^2}{\omega^2} = \frac{x}{1+\lambda}$$

so

$$P = 1 - x$$

and S, R, L are functions of x and y with λ as a parameter. Hint: it is easier to plot x versus y for some of the functions.
4. Plot n^2 versus ω for $\theta = \pi/2$, showing the hybrid resonances.
5. Starting in Region 1 of the CMA diagram, establish the signs of S, P, R, L in all the regions.
6. Plot the CMA mode lines for plasmas having $\omega_{pe}^2 \gg \omega_{ce}^2$ and vice versa.

7. Consider a plasma with two ion species. By plotting S versus ω show there is an ion–ion hybrid resonance between the two ion cyclotron frequencies (Buchsbaum 1960, Ono 1979). Give an approximate expression for the frequency of this resonance in terms of the ratios of the densities of the two ion species. Hint: compare the magnitude of the electron term to that of the two ion terms. Using quasi-neutrality, obtain an expression depending only on the fractional density of each ion species.
8. Consider a two-dimensional plasma with an oscillatory delta function source at the origin. Suppose that slow waves are being excited, which satisfy the electrostatic dispersion

$$k_x^2 S + k_z^2 P = 0,$$

where $SP < 0$. By writing the source on the z axis as

$$f(z) = \frac{1}{2\pi} \int \mathrm{e}^{\mathrm{i}k_z z - \mathrm{i}\omega t} \mathrm{d}k_z$$

and by solving the dispersion to give $k_x = k_x(k_z)$ show the potential excited in the plasma is singular along the resonance cone surfaces. Explain why this happens. Draw the group and phase velocity directions.
9. What is the polarization (i.e., relative magnitude of E_x, E_y, E_z) of the QTO, QTX, and the two QL modes? How should a microwave horn be oriented (i.e., in which way should the **E** field of the horn point) when being used for (i) a QTO experiment, (ii) a QTX experiment? Which experiment would be best suited for heating the plasma and which best suited for measuring the density of the plasma?
10. Show there is a simple factoring of the cold plasma dispersion relation in the low frequency limit $\omega \ll \omega_{ci}$. Hint: first find approximate forms of S, P, and D in this limit and then show that the cold plasma dielectric tensor becomes diagonal. Consider a mode that has only E_y finite and a mode that only has E_x and E_z finite. What are the dispersion relations for these two modes, expressed in terms of ω as a function of k? Assume the Alfvén velocity is much smaller than the speed of light.

7

Waves in inhomogeneous plasmas and wave-energy relations

7.1 Wave propagation in inhomogeneous plasmas

Thus far in our discussion of wave propagation it has been assumed that the plasma is spatially uniform. While this assumption simplifies analysis, the real world is usually not so accommodating and it is plausible that spatial non-uniformity might modify wave propagation. The modification could be just a minor adjustment or it could be profound. Spatial non-uniformity might even produce entirely new kinds of waves. As will be seen, all these possibilities can occur.

To determine the effects of spatial non-uniformity, it is necessary to re-examine the original system of partial differential equations from which the wave dispersion relation was obtained. This is because the technique of substituting $i\mathbf{k}$ for ∇ is, in essence, a shortcut for spatial Fourier analysis, and so is mathematically valid *only* if the equilibrium is spatially uniform. The criteria for whether or not ∇ can be replaced by $i\mathbf{k}$ can be understood by considering the simple example of a high-frequency electromagnetic plasma wave propagating in an unmagnetized three-dimensional plasma having a gentle density gradient. The plasma frequency will be a function of position for this situation. To keep matters simple, the density non-uniformity is assumed to be in one direction only, which will be labeled the x direction. The plasma is thus uniform in the y and z directions, but non-uniform in the x direction. Because the frequency is high, ion motion may be neglected and the electron motion is just the quiver velocity

$$\mathbf{v}_1 = -\frac{q_e}{i\omega m_e}\mathbf{E}_1. \tag{7.1}$$

The current density associated with electron motion is therefore

$$\mathbf{J}_1 = -\frac{n_e(x)q_e^2}{i\omega m_e}\mathbf{E}_1 = -\varepsilon_0\frac{\omega_{pe}^2(x)}{i\omega}\mathbf{E}_1. \tag{7.2}$$

Inserting this current density into Ampère's law gives

$$\nabla\times\mathbf{B}_1 = -\frac{\omega_{pe}^2(x)}{\mathrm{i}\omega c^2}\mathbf{E}_1 - \frac{\mathrm{i}\omega}{c^2}\mathbf{E}_1 = -\frac{\mathrm{i}\omega}{c^2}\left(1-\frac{\omega_{pe}^2(x)}{\omega^2}\right)\mathbf{E}_1. \tag{7.3}$$

Substituting Ampère's law into the curl of Faraday's law gives

$$\nabla\times\nabla\times\mathbf{E}_1 = \frac{\omega^2}{c^2}\left(1-\frac{\omega_{pe}^2(x)}{\omega^2}\right)\mathbf{E}_1. \tag{7.4}$$

Attention is now restricted to waves for which $\nabla\cdot\mathbf{E}_1 = 0$; this is a generalization of the assumption that the waves are transverse (i.e., have $\mathrm{i}\mathbf{k}\cdot\mathbf{E} = 0$) or equivalently are electromagnetic, and so involve no density perturbation. In this case, expansion of the left-hand side of Eq. (7.4) yields

$$\left(\frac{\partial^2}{\partial x^2}+\frac{\partial^2}{\partial y^2}+\frac{\partial^2}{\partial z^2}\right)\mathbf{E}_1 + \frac{\omega^2}{c^2}\left(1-\frac{\omega_{pe}^2(x)}{\omega^2}\right)\mathbf{E}_1 = 0. \tag{7.5}$$

It should be recalled that Fourier analysis is restricted to equations with constant coefficients, so Eq. (7.5) can only be Fourier transformed in the y and z directions. It cannot be Fourier transformed in the x direction because the coefficient $\omega_{pe}^2(x)$ depends on x. Thus, after performing only the allowed Fourier transforms, the wave equation becomes

$$\left(\frac{\partial^2}{\partial x^2}-k_y^2-k_z^2\right)\tilde{\mathbf{E}}_1(x,k_y,k_z) + \frac{\omega^2}{c^2}\left(1-\frac{\omega_{pe}^2(x)}{\omega^2}\right)\tilde{\mathbf{E}}_1(x,k_y,k_z) = 0, \tag{7.6}$$

where $\tilde{\mathbf{E}}_1(x,k_y,k_z)$ is the Fourier transform in the y and z directions. This may be rewritten as

$$\left(\frac{\partial^2}{\partial x^2}+\kappa^2(x)\right)\tilde{\mathbf{E}}_1(x,k_y,k_z) = 0, \tag{7.7}$$

where

$$\kappa^2(x) = \frac{\omega^2}{c^2}\left(1-\frac{\omega_{pe}^2(x)}{\omega^2}\right) - k_y^2 - k_z^2. \tag{7.8}$$

We now realize that Eq. (7.7) is just the spatial analog of the WKB equation for a pendulum with slowly varying frequency, namely Eq. (3.17); the only difference is that the independent variable t has been replaced by the independent variable x. Since changing the name of the independent variable is of no consequence, the solution here is formally the same as the previously derived WKB solution, Eq. (3.24). Thus the approximate solution to Eq. (7.8) is

$$\tilde{\mathbf{E}}_1(x,k_y,k_z) \sim \frac{1}{\sqrt{\kappa(x)}}\exp\left(\mathrm{i}\int^x \kappa(x')\mathrm{d}x'\right). \tag{7.9}$$

Equation (7.9) shows that both the wave amplitude and effective wavenumber vary as the wave propagates in the x direction, i.e., in the direction of the inhomogeneity. It is clear that if the inhomogeneity is in the x direction, the wavenumbers k_y and k_z do not change as the wave propagates. This is because, unlike for the x direction, it was possible to Fourier transform in the y and z directions and so k_y and k_z are just coordinates in Fourier space. The effective wavenumber in the direction of the inhomogeneity, i.e., $\kappa(x)$, is not a coordinate in Fourier space because Fourier transformation was not allowed in the x direction. The spatial dependence of the effective wavenumber $\kappa(x)$ defined by Eq. (7.8) and the spatial dependence of the WKB amplitude together provide the means by which the system accommodates the spatial inhomogeneity. The invariance of the wavenumbers in the homogeneous directions is called Snell's law. An elementary example of Snell's law is the situation where light crosses an interface between two media having different dielectric constants and the refractive index parallel to the interface remains invariant.

One way of interpreting this result is to state that the WKB method gives qualified permission to Fourier analyze in the x direction. To the extent that such an x-direction Fourier analysis is allowed, $\kappa(x)$ can be considered as the effective wavenumber in the x direction, i.e., $\kappa(x) = k_x(x)$. The results in Chapter 3 imply that the WKB approximate solution, Eq. (7.9), is valid only when the criterion

$$\frac{1}{k_x}\frac{\mathrm{d}k_x}{\mathrm{d}x} \ll k_x \tag{7.10}$$

is satisfied. Inequality (7.10) is thus not satisfied when $k_x \to 0$, i.e., at a cutoff. At a resonance the situation is somewhat more complicated. According to cold plasma theory, k_x simply diverges at a resonance; however, when hot plasma effects are taken into account, it is found that instead of having k_x going to infinity, the resonant cold plasma mode coalesces with a hot plasma mode as shown in Fig. 7.1. At the point of coalescence $\mathrm{d}k_x/\mathrm{d}x \to \infty$ while all the other

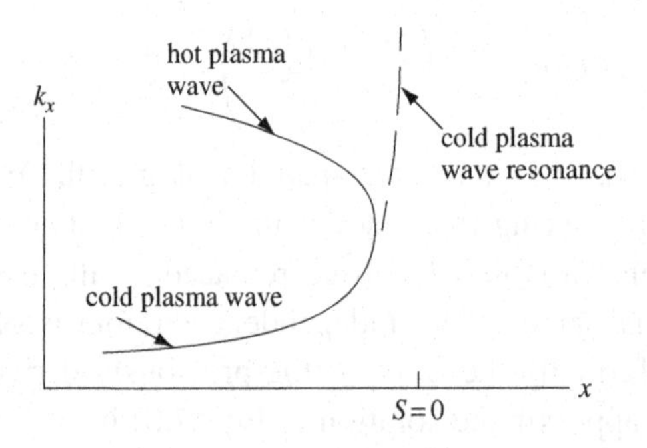

Fig. 7.1 Example of coalescence of a cold plasma wave and a hot plasma wave near the resonance of the cold plasma wave. Here a hybrid resonance causes the cold resonance.

terms in Eq. (7.9) remain finite, and so the WKB method also breaks down at a resonance.

An interesting and important consequence of this discussion is the very real possibility that inequality (7.10) could be violated in a plasma having only the mildest of inhomogeneities. This breakdown of WKB in an apparently benign situation occurs because the critical issue is how $k_x(x)$ changes and not how plasma parameters change. For example, k_x could go through zero at some critical plasma density and, no matter how gentle the density gradient is, there will invariably be a cutoff at the critical density.

7.2 Geometric optics

The WKB method can be generalized to a plasma that is inhomogeneous in more than one dimension. In the general case of inhomogeneity in all three dimensions, the three components of the wavenumber will be functions of position, i.e., $\mathbf{k} = \mathbf{k}(\mathbf{x})$. How is the functional dependence determined? The answer is to write the dispersion relation as

$$D(\mathbf{k}, \mathbf{x}) = 0. \tag{7.11}$$

The $\mathbf{x}$-dependence of D denotes an explicit spatial dependence of the dispersion relation due to density or magnetic field gradients. This dispersion relation is now presumed to be satisfied at some initial point $\mathbf{x}$ and then it is further assumed that all quantities evolve in such a way as to keep the dispersion relation satisfied at other positions. Thus, at some arbitrary nearby position $\mathbf{x} + \delta\mathbf{x}$, the dispersion relation is also satisfied so

$$D(\mathbf{k} + \delta\mathbf{k}, \mathbf{x} + \delta\mathbf{x}) = 0 \tag{7.12}$$

or, on Taylor expanding,

$$\delta\mathbf{k} \cdot \frac{\partial D}{\partial \mathbf{k}} + \delta\mathbf{x} \cdot \frac{\partial D}{\partial \mathbf{x}} = 0. \tag{7.13}$$

The general condition for satisfying Eq. (7.13) can be established by assuming that both $\mathbf{k}$ and $\mathbf{x}$ depend on some parameter that increases monotonically along the trajectory of the wave, for example the distance s along the wave trajectory. The wave trajectory itself can also be parametrized as a function of s. Then using both $\mathbf{k} = \mathbf{k}(s)$ and $\mathbf{x} = \mathbf{x}(s)$, it is seen that moving a distance δs corresponds to respective increments $\delta\mathbf{k} = \delta s \, \mathrm{d}\mathbf{k}/\mathrm{d}s$ and $\delta\mathbf{x} = \delta s \, \mathrm{d}\mathbf{x}/\mathrm{d}s$. This means that Eq. (7.13) can be expressed as

$$\left[\frac{\mathrm{d}\mathbf{k}}{\mathrm{d}s} \cdot \frac{\partial D}{\partial \mathbf{k}} + \frac{\mathrm{d}\mathbf{x}}{\mathrm{d}s} \cdot \frac{\partial D}{\partial \mathbf{x}}\right] \delta s = 0. \tag{7.14}$$

The general solutions to this equation are the two coupled equations

$$\frac{d\mathbf{k}}{ds} = -\frac{\partial D}{\partial \mathbf{x}}, \tag{7.15}$$

$$\frac{d\mathbf{x}}{ds} = \frac{\partial D}{\partial \mathbf{k}}. \tag{7.16}$$

These are just Hamilton's equations with the dispersion relation D acting as the Hamiltonian, the path length s acting like the time, $\mathbf{x}$ acting as the position, and $\mathbf{k}$ acting as the momentum. Thus, given the initial momentum at an initial position, the wavenumber evolution and wave trajectory can be calculated using Eqs. (7.15) and (7.16) respectively. The close relationship between wavenumber and momentum fundamental to quantum mechanics is plainly evident here. Snell's law states that the wavenumber in a particular direction remains invariant if the medium is uniform in that direction; this is clearly equivalent (cf. Eq. (7.15)) to the Hamilton–Lagrange result that the canonical momentum in a particular direction is invariant if the system is uniform in that direction.

This Hamiltonian point of view provides a useful way for interpreting cutoffs and resonances. Suppose that D is the dispersion relation for a particular mode and suppose that D can be written in the form

$$D(\mathbf{k}, \mathbf{x}) = \sum_{ij} \alpha_{ij} k_i k_j + g(\omega, n(\mathbf{x}), B(\mathbf{x})) = 0. \tag{7.17}$$

If D is construed to be the Hamiltonian, then the first term in Eq. (7.17) can be identified as the "kinetic energy" while the second term can be identified as the "potential energy." As an example, consider the simple case of an electromagnetic mode in an unmagnetized plasma that has non-uniform density, so that

$$D(\mathbf{k}, \mathbf{x}) = \frac{c^2 k^2}{\omega^2} - 1 + \frac{\omega_{pe}^2(x)}{\omega^2} = 0. \tag{7.18}$$

The wave propagation can be analyzed in analogy to the problem of a particle in a potential well. Here the kinetic energy is $c^2 k^2/\omega^2$, the potential energy is $-1 + \omega_{pe}^2(x)/\omega^2$, and the total energy is zero. This is a "wave–particle" duality formally like that of quantum mechanics, since there is a correspondence between wavenumber and momentum and between energy and frequency. Cutoffs give wave reflection in analogy to the reflection of a particle in a potential well at points where the potential energy equals the total energy. As shown in Fig. 7.2(a), when the potential energy has a local minimum, waves will be trapped in the potential well associated with this minimum. Electrostatic plasma waves can also exhibit wave trapping between two reflection points; these trapped waves are called cavitons (the analysis is essentially the same; one simply replaces c^2 by $3\omega_{pe}^2 \lambda_{De}^2$). The bouncing of short-wave radio waves from the ionospheric plasma

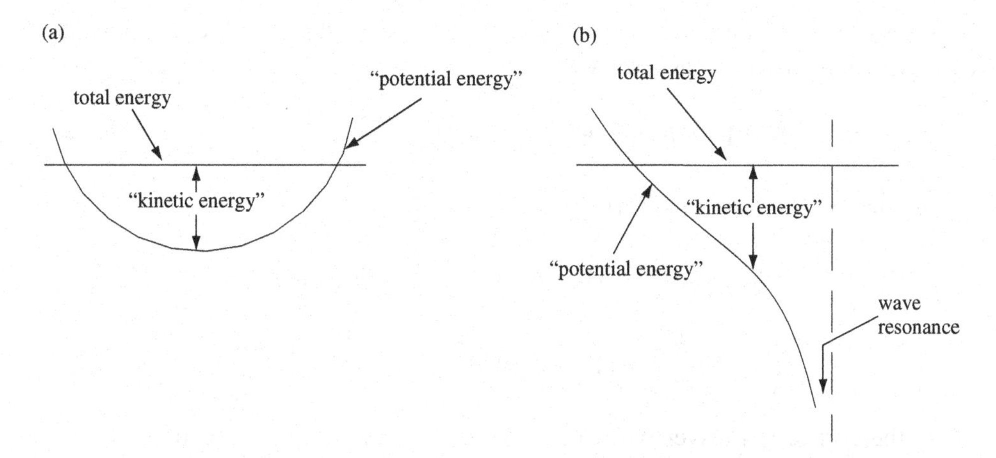

Fig. 7.2 (a) Effective potential energy for trapped wave; (b) for wave resonance.

can be analyzed using Eq. (7.18) together with Eqs. (7.15) and (7.16). As shown in Fig. 7.2(b) a wave resonance (i.e., $k^2 \to \infty$) would correspond to a deep crevice in the potential energy. One must be careful to use geometric optics only when the plasma is weakly inhomogeneous, so that the waves change sufficiently gradually to satisfy the WKB criterion.

7.3 Surface waves – the plasma-filled waveguide

An extreme form of plasma inhomogeneity occurs when there is an abrupt transition from plasma to vacuum – in other words, the plasma has an edge or surface. A qualitatively new mode, called a surface wave, appears in this circumstance. The physical basis of surface waves is closely related to the mechanism by which light waves propagate in an optical fiber.

Using the same analysis that led to Eq. (7.3), Maxwell's equations in an unmagnetized plasma may be expressed as

$$\nabla \times \mathbf{B} = -\frac{\mathrm{i}\omega}{c^2} P\mathbf{E}, \quad \nabla \times \mathbf{E} = \mathrm{i}\omega\mathbf{B}, \tag{7.19}$$

where P is the unmagnetized plasma dielectric function

$$P = 1 - \frac{\omega_{pe}^2}{\omega^2}. \tag{7.20}$$

We consider a plasma that is uniform in the z direction but non-uniform in the directions perpendicular to z. The electromagnetic fields and gradient operator can be separated into axial components (i.e., z direction) and transverse components (i.e., perpendicular to z) as follows:

$$\mathbf{B} = \mathbf{B}_t + B_z\hat{z}, \quad \mathbf{E} = \mathbf{E}_t + E_z\hat{z}, \quad \nabla = \nabla_t + \hat{z}\frac{\partial}{\partial z}. \tag{7.21}$$

Using these definitions Eqs. (7.19) become

$$\begin{aligned}\left(\nabla_t + \hat{z}\frac{\partial}{\partial z}\right) \times \left(\mathbf{B}_t + B_z\hat{z}\right) &= -\frac{\mathrm{i}\omega}{c^2}P\left(\mathbf{E}_t + E_z\hat{z}\right), \\ \left(\nabla_t + \hat{z}\frac{\partial}{\partial z}\right) \times \left(\mathbf{E}_t + E_z\hat{z}\right) &= \mathrm{i}\omega\left(\mathbf{B}_t + B_z\hat{z}\right).\end{aligned} \tag{7.22}$$

Since the curl of a transverse vector is in the z direction, these equations can be separated into axial and transverse components,

$$\hat{z}\cdot\nabla_t \times \mathbf{B}_t = -\frac{\mathrm{i}\omega}{c^2}PE_z, \tag{7.23}$$

$$\hat{z}\cdot\nabla_t \times \mathbf{E}_t = \mathrm{i}\omega B_z, \tag{7.24}$$

$$\hat{z}\times\frac{\partial \mathbf{B}_t}{\partial z} + \nabla_t B_z \times \hat{z} = -\frac{\mathrm{i}\omega}{c^2}P\mathbf{E}_t, \tag{7.25}$$

$$\hat{z}\times\frac{\partial \mathbf{E}_t}{\partial z} + \nabla_t E_z \times \hat{z} = \mathrm{i}\omega\mathbf{B}_t. \tag{7.26}$$

The transverse electric field on the left-hand side of Eq. (7.26) can be replaced using Eq. (7.25) to give

$$\mathrm{i}\omega\mathbf{B}_t = \hat{z}\times\frac{\partial}{\partial z}\left[\frac{\hat{z}\times\dfrac{\partial \mathbf{B}_t}{\partial z} + \nabla_t B_z\times\hat{z}}{-\dfrac{\mathrm{i}\omega}{c^2}P}\right] + \nabla_t E_z \times \hat{z}. \tag{7.27}$$

It is now assumed that all quantities have axial dependence $\sim \exp(\mathrm{i}kz)$ so Eq. (7.27) can be solved to give $\mathbf{B}_t$ solely in terms of E_z and B_z, i.e.,

$$\mathbf{B}_t = \left(\frac{\omega^2}{c^2}P - k^2\right)^{-1}\left[\nabla_t\frac{\partial B_z}{\partial z} - \frac{\mathrm{i}\omega}{c^2}P\nabla_t E_z \times \hat{z}\right]. \tag{7.28}$$

This result can also be used to solve for the transverse electric field by interchanging $-\mathrm{i}\omega P/c^2 \longleftrightarrow \mathrm{i}\omega$ and $\mathbf{E} \longleftrightarrow \mathbf{B}$ to obtain

$$\mathbf{E}_t = \left(\frac{\omega^2}{c^2}P - k^2\right)^{-1}\left[\nabla_t\frac{\partial E_z}{\partial z} + \mathrm{i}\omega\nabla_t B_z \times \hat{z}\right]. \tag{7.29}$$

Except for the plasma-dependent factor P, these are the standard waveguide equations. An important feature of these equations is that the transverse fields

$\mathbf{E}_t, \mathbf{B}_t$ are functions of the axial fields E_z, B_z only and so all that is required is to construct wave equations characterizing the axial fields. This is an enormous simplification because, instead of having to derive and solve six wave equations in the six components of $\mathbf{E}, \mathbf{B}$ as might be expected, it suffices to derive and solve wave equations for just E_z and B_z.

The sought-after wave equations are determined by eliminating $\mathbf{E}_t$ and $\mathbf{B}_t$ from Eqs. (7.23) and (7.24) to obtain

$$\hat{z}\cdot\nabla_t\times\left\{\left(\frac{\omega^2}{c^2}P-k^2\right)^{-1}\left[ik\nabla_t B_z-\frac{i\omega}{c^2}P\nabla_t E_z\times\hat{z}\right]\right\}=-\frac{i\omega}{c^2}PE_z, \tag{7.30}$$

$$\hat{z}\cdot\nabla_t\times\left\{\left(\frac{\omega^2}{c^2}P-k^2\right)^{-1}\left[ik\nabla_t E_z+i\omega\nabla_t B_z\times\hat{z}\right]\right\}=i\omega B_z. \tag{7.31}$$

In the special situation where $\nabla_t P\times\nabla_t B_z=\nabla_t P\times\nabla_t E_z=0$, the first terms in the square brackets of the above equations vanish. This simplification would occur, for example, in an azimuthally symmetric plasma having an azimuthally symmetric perturbation so that $\nabla_t P$, $\nabla_t E_z$, and $\nabla_t B_z$ are all in the r direction. It is now assumed that both the plasma and the mode have this azimuthal symmetry so Eqs. (7.30) and (7.31) reduce to

$$\hat{z}\cdot\nabla_t\times\left\{\left(\frac{\omega^2}{c^2}P-k^2\right)^{-1}\left(P\nabla_t E_z\times\hat{z}\right)\right\}=PE_z, \tag{7.32}$$

$$\hat{z}\cdot\nabla_t\times\left\{\left(\frac{\omega^2}{c^2}P-k^2\right)^{-1}\left(\nabla_t B_z\times\hat{z}\right)\right\}=B_z \tag{7.33}$$

or, equivalently,

$$\nabla_t\cdot\left\{\frac{P}{P-k^2c^2/\omega^2}\nabla_t E_z\right\}+\frac{\omega^2}{c^2}PE_z=0, \tag{7.34}$$

$$\nabla_t\cdot\left\{\frac{1}{P-k^2c^2/\omega^2}\nabla_t B_z\right\}+\frac{\omega^2}{c^2}B_z=0. \tag{7.35}$$

The assumption that both the plasma and the modes are azimuthally symmetric has the important consequence of decoupling the E_z and B_z modes so there are two distinct polarizations. These are (i) a mode where B_z is finite but $E_z=0$ and (ii) the reverse. Case (i) is called a transverse electric (TE) mode while case (ii) is called a transverse magnetic mode (TM), since in the first case the electric field is purely transverse while in the second case the magnetic field is purely transverse.

We now consider an azimuthally symmetric TM mode propagating in a uniform cylindrical plasma of radius a surrounded by vacuum. Since $B_z = 0$ for a TM mode, the transverse fields are the following functions of E_z:

$$\mathbf{B}_t = \left(\frac{\omega^2}{c^2}P - k^2\right)^{-1}\left[-\frac{i\omega}{c^2}P\nabla_t E_z \times \hat{z}\right], \tag{7.36}$$

$$\mathbf{E}_t = \left(\frac{\omega^2}{c^2}P - k^2\right)^{-1}\nabla_t \frac{\partial E_z}{\partial z}. \tag{7.37}$$

Additionally, because of the assumed symmetry, the TM mode Eq. (7.34) simplifies to

$$\frac{1}{r}\frac{\partial}{\partial r}\left(\frac{rP}{P - k^2c^2/\omega^2}\frac{\partial E_z}{\partial r}\right) + \frac{\omega^2}{c^2}PE_z = 0. \tag{7.38}$$

Since P is uniform within the plasma region and within the vacuum region, but has different values in these two regions, separate solutions to Eq. (7.38) must be obtained in the plasma and vacuum regions respectively and then matched at the interface. The jump in P is accommodated by defining distinct radial wave numbers

$$\kappa_p^2 = k^2 - \frac{\omega^2}{c^2}P, \tag{7.39}$$

$$\kappa_v^2 = k^2 - \frac{\omega^2}{c^2} \tag{7.40}$$

for the respective plasma and vacuum regions. The solutions to Eq. (7.38) in the respective plasma and vacuum regions are linear combinations of Bessel functions of order zero. If both of κ_p^2 and κ_v^2 are positive then the TM mode has the peculiar property of being radially evanescent in *both* the plasma and vacuum regions. In this case both the vacuum and plasma region solutions consist of modified Bessel functions I_0, K_0. These solutions are constrained by physical considerations as follows:

1. Because the parallel electric field is a physical quantity it must be finite. In particular, E_z must be finite at $r = 0$, in which case only the $I_0(\kappa_p r)$ solution is allowed in the plasma region (the K_0 solution diverges at $r = 0$). Similarly, because E_z must be finite as $r \to \infty$, only the $K_0(\kappa_v r)$ solution is allowed in the vacuum region (the $I_0(\kappa_v r)$ solution diverges at $r = \infty$).
2. The parallel electric field E_z must be continuous across the vacuum–plasma interface. This constraint is imposed by Faraday's law and can be seen by integrating Faraday's law over an area in the $r - z$ plane of axial length L and infinitesimal radial width.

The inner radius of this area is at r_- and the outer radius is at r_+, where $r_- < a < r_+$. Integrating Faraday's law over this area gives

$$\lim_{r_- \to r_+} \int d\mathbf{s} \cdot \nabla \times \mathbf{E} = \oint \mathbf{E} \cdot d\mathbf{l} = - \lim_{r_- \to r_+} \int d\mathbf{s} \cdot \frac{\partial \mathbf{B}}{\partial t}$$

or

$$E_z^{vac} L - E_z^{plasma} L = 0$$

showing E_z must be continuous at the plasma–vacuum interface.

3. Integration of Eq. (7.38) across the interface shows that the Quantity

$$P\left(P - k^2 c^2/\omega^2\right)^{-1} \partial E_z/\partial r$$

must be continuous across the interface.

In order to satisfy Constraint #1 the parallel electric field in the plasma must be

$$E_z(r) = E_z(a) \frac{I_0(\kappa_p r)}{I_0(\kappa_p a)} \tag{7.41}$$

and the parallel electric field in the vacuum must be

$$E_z(r) = E_z(a) \frac{K_0(\kappa_v r)}{K_0(\kappa_v a)}. \tag{7.42}$$

The normalization has been set so that E_z is continuous across the interface as required by Constraint #2.

Constraint #3 gives

$$\left[\left(\frac{\omega^2}{c^2} P - k^2\right)^{-1} P \frac{\partial E_z}{\partial r}\right]_{r=a_-} = \left[\left(\frac{\omega^2}{c^2} - k^2\right)^{-1} \frac{\partial E_z}{\partial r}\right]_{r=a_+}. \tag{7.43}$$

Inserting Eqs. (7.41) and (7.42) into the respective left- and right-hand sides of the above expression gives

$$\left(\frac{\omega^2}{c^2} P - k^2\right)^{-1} P \frac{\kappa_p I_0'(\kappa_p a)}{I_0(\kappa_p a)} = \left(\frac{\omega^2}{c^2} - k^2\right)^{-1} \frac{\kappa_v K_0'(\kappa_v a)}{K_0(\kappa_v a)}, \tag{7.44}$$

where a prime means a derivative with respect to the argument of the function. This expression is effectively a dispersion relation, since it prescribes a functional relationship between ω and k. It is qualitatively different from the previously discussed uniform plasma dispersion relations because it depends on a specific physical dimension, namely the plasma radius a. This dependence indicates that a plasma–vacuum interface is a necessary condition for the mode to exist. The mode amplitude is strongest in the vicinity of the interface because both the plasma and vacuum fields decay exponentially with increasing distance from the interface.

This surface wave dispersion depends on a combination of Bessel functions and the parallel dielectric P. However, a limit exists where the dispersion relation reduces to a simpler form, and this limit illustrates important features of these surface waves. Specifically, if the axial wavelength is sufficiently short for k^2 to be much larger than both $\omega^2 P/c^2$ and ω^2/c^2, one can then approximate $k^2 \simeq \kappa_v^2 \simeq \kappa_p^2$, in which case the dispersion simplifies to

$$P\frac{I_0'(ka)}{I_0(ka)} \simeq \frac{K_0'(ka)}{K_0(ka)}. \tag{7.45}$$

If, in addition, the axial wavelength is sufficiently long to satisfy $ka \ll 1$, then the small-argument limits of the modified Bessel functions can be used, namely

$$\lim_{\xi \ll 1} I_0(\xi) = 1 + \frac{\xi^2}{4}, \tag{7.46}$$

$$\lim_{\xi \ll 1} K_0(\xi) = -\ln \xi. \tag{7.47}$$

Thus, in the limit $\omega^2 P/c^2$, $\omega^2/c^2 \ll k^2 \ll 1/a^2$, Eq. (7.45) simplifies to

$$\left(1 - \frac{\omega_{pe}^2}{\omega^2}\right)\frac{ka}{2} \simeq -\frac{1}{ka \ln\left(\dfrac{1}{ka}\right)}. \tag{7.48}$$

Because $ka \ll 1$, the logarithmic term is negative. Hence, to satisfy Eq. (7.48) it is necessary to have $\omega \ll \omega_{pe}$ so that the dispersion further becomes

$$\frac{\omega}{\omega_{pe}} = ka\sqrt{\frac{1}{2}\ln\left(\frac{1}{ka}\right)}. \tag{7.49}$$

On the other hand, if $ka \gg 1$, then the large-argument limit of the Bessel functions can be used, namely

$$\lim_{\xi \gg 1} I_0(\xi) = \frac{1}{\sqrt{2\pi\xi}}\mathrm{e}^{\xi}, \quad \lim_{\xi \gg 1} K_0(\xi) = \sqrt{\frac{\pi}{2\xi}}\mathrm{e}^{-\xi} \tag{7.50}$$

so that the dispersion relation becomes

$$1 - \frac{\omega_{pe}^2}{\omega^2} = -1 \tag{7.51}$$

or

$$\omega = \frac{\omega_{pe}}{\sqrt{2}}. \tag{7.52}$$

This provides the curious result that a finite-radius plasma cylinder resonates at a lower frequency than a uniform plasma if the axial wavelength is much shorter than the cylinder radius.

These surface waves are slow waves since $\omega/k \ll c$ has been assumed. They were first studied by Trivelpiece and Gould (1959) and are seen in cylindrical plasmas surrounded by vacuum. For $ka \ll 1$ the phase velocity is $\omega/k \sim \mathcal{O}(\omega_{pe}a)$ and for $ka \gg 1$ the phase velocity goes to zero since ω is a constant at large ka. More complicated variations of the surface wave dispersion are obtained if the vacuum region is of finite extent and is surrounded by a conducting wall, i.e., if there is plasma for $r < a$, vacuum for $a < r < b$, and a conducting wall at $r = b$. In this case the vacuum region solution consists of a linear combination of $I_0(\kappa_v r)$ and $K_0(\kappa_v r)$ terms with coefficients chosen to satisfy Constraints #2 and #3 discussed earlier and also a new, additional constraint that E_z must vanish at the wall, i.e., at $r = b$.

7.4 Plasma wave-energy equation

The energy associated with a plasma wave is related in a subtle way to the dispersive properties of the wave. Quantifying this relation requires starting from first principles regarding the electromagnetic field energy density and taking into account specific features of dispersive waves. The basic equation characterizing electromagnetic energy density, called Poynting's theorem, is obtained by subtracting **B** dotted with Faraday's law from **E** dotted with Ampère's law,

$$\mathbf{E} \cdot \nabla \times \mathbf{B} - \mathbf{B} \cdot \nabla \times \mathbf{E} = \varepsilon_0 \mu_0 \mathbf{E} \cdot \frac{\partial \mathbf{E}}{\partial t} + \mathbf{B} \cdot \frac{\partial \mathbf{B}}{\partial t} + \mu_0 \mathbf{E} \cdot \mathbf{J},$$

and expressing this result as a conservation equation,

$$\frac{\partial w}{\partial t} + \nabla \cdot \mathbf{P} = 0. \tag{7.53}$$

The quantity

$$\mathbf{P} = \frac{\mathbf{E} \times \mathbf{B}}{\mu_0} \tag{7.54}$$

is called the Poynting flux and is the electromagnetic energy flux into the system while

$$\frac{\partial w}{\partial t} = \varepsilon_0 \mathbf{E} \cdot \frac{\partial \mathbf{E}}{\partial t} + \frac{1}{\mu_0} \mathbf{B} \cdot \frac{\partial \mathbf{B}}{\partial t} + \mathbf{E} \cdot \mathbf{J} \tag{7.55}$$

is the rate of change of energy density in the system. The energy density is obtained by time integration and is

$$\begin{aligned} w(t) &= w(t_0) + \int_{t_0}^{t} \mathrm{d}t \left\{ \varepsilon_0 \mathbf{E} \cdot \frac{\partial \mathbf{E}}{\partial t} + \frac{1}{\mu_0} \mathbf{B} \cdot \frac{\partial \mathbf{B}}{\partial t} + \mathbf{E} \cdot \mathbf{J} \right\} \\ &= w(t_0) + \left[\frac{\varepsilon_0}{2} E^2 + \frac{B^2}{2\mu_0} \right]_{t_0}^{t} + \int_{t_0}^{t} \mathrm{d}t \, \mathbf{E} \cdot \mathbf{J}, \end{aligned} \tag{7.56}$$

where $w(t_0)$ is the energy density at some reference time t_0.

The quantity $\mathbf{E}\cdot\mathbf{J}$ is the rate of change of kinetic energy density of the particles. This can be seen by first dotting the Lorentz equation with $\mathbf{v}$ to obtain

$$m\mathbf{v}\cdot\frac{\mathrm{d}\mathbf{v}}{\mathrm{d}t} = q\mathbf{v}\cdot(\mathbf{E}+\mathbf{v}\times\mathbf{B}) \tag{7.57}$$

or

$$\frac{\mathrm{d}}{\mathrm{d}t}\left(\frac{1}{2}mv^2\right) = q\mathbf{E}\cdot\mathbf{v}. \tag{7.58}$$

Since this is the rate of change of kinetic energy of a single particle, the rate of change of the kinetic energy density of all the particles, found by summing over all the particles, is

$$\begin{aligned}\frac{\mathrm{d}}{\mathrm{d}t}(\text{kinetic energy density}) &= \sum_\sigma \int \mathrm{d}\mathbf{v} f_\sigma q_\sigma \mathbf{E}\cdot\mathbf{v} \\ &= \sum_\sigma n_\sigma q_\sigma \mathbf{E}\cdot\mathbf{u}_\sigma \\ &= \mathbf{E}\cdot\mathbf{J}.\end{aligned} \tag{7.59}$$

This shows that positive $\mathbf{E}\cdot\mathbf{J}$ corresponds to work going into the particles (increase of particle kinetic energy) whereas negative $\mathbf{E}\cdot\mathbf{J}$ corresponds to work coming out of the particles (decrease of the particle kinetic energy). The latter situation is obviously possible only if the particles start with a finite initial kinetic energy. Since $\mathbf{E}\cdot\mathbf{J}$ accounts for changes in the particle kinetic energy density, w must be the sum of the electromagnetic field density and the particle kinetic energy density.

The time integration of Eq. (7.55) must be done with great care when $\mathbf{E}$ and $\mathbf{J}$ are wave fields. This is because it must always be kept in mind that writing a wave field as $\psi = \tilde{\psi}\exp(\mathrm{i}\mathbf{k}\cdot\mathbf{x} - \mathrm{i}\omega t)$ is simply a notational convenience and should never be taken to mean that the actual physical wave field is complex. The physically meaningful variable is, of course, $\psi = \mathrm{Re}\left[\tilde{\psi}\exp(\mathrm{i}\mathbf{k}\cdot\mathbf{x} - \mathrm{i}\omega t)\right]$. Explicitly taking the real part is critical when evaluating nonlinear relationships because, unlike for linear relationships, omission of this step would lead to a nonsensical result. When dealing with a product of two oscillating physical quantities, say $\psi(t) = \mathrm{Re}\left[\tilde{\psi}\mathrm{e}^{-\mathrm{i}\omega t}\right]$ and $\chi(t) = \mathrm{Re}\left[\tilde{\chi}\mathrm{e}^{-\mathrm{i}\omega t}\right]$, it is therefore essential to extract the real part *before* calculating the product and so the product must be written as

$$\psi(t)\chi(t) = \mathrm{Re}\left[\tilde{\psi}\mathrm{e}^{-\mathrm{i}\omega t}\right]\times\mathrm{Re}\left[\tilde{\chi}\mathrm{e}^{-\mathrm{i}\omega t}\right]. \tag{7.60}$$

In particular, if $\omega = \omega_r + \mathrm{i}\omega_i$ is a complex frequency, then the product in Eq. (7.60) assumes the form

$$\psi(t)\chi(t) = \left(\frac{\tilde{\psi}\mathrm{e}^{-\mathrm{i}\omega t} + \tilde{\psi}^*\mathrm{e}^{\mathrm{i}\omega^* t}}{2}\right)\left(\frac{\tilde{\chi}\mathrm{e}^{-\mathrm{i}\omega t} + \tilde{\chi}^*\mathrm{e}^{\mathrm{i}\omega^* t}}{2}\right)$$

$$= \frac{1}{4}\left[\begin{array}{l}\tilde{\psi}\tilde{\chi}\mathrm{e}^{-2\mathrm{i}\omega t} + \tilde{\psi}^*\tilde{\chi}^*\mathrm{e}^{2\mathrm{i}\omega^* t} \\ +\tilde{\psi}\tilde{\chi}^*\mathrm{e}^{-\mathrm{i}(\omega-\omega^*)t} + \tilde{\psi}^*\tilde{\chi}\mathrm{e}^{-\mathrm{i}(\omega-\omega^*)t}\end{array}\right]. \tag{7.61}$$

When considering the energy density of a wave, we are typically interested in time-averaged quantities, not rapidly fluctuating quantities. Thus the time average of the product $\psi(t)\chi(t)$ over one wave period will be considered. The $\tilde{\psi}\tilde{\chi}$ and $\tilde{\psi}^*\tilde{\chi}^*$ terms oscillate at the fast second harmonic of the frequency and vanish upon time-averaging. In contrast, the $\tilde{\psi}\tilde{\chi}^*$ and $\tilde{\psi}^*\tilde{\chi}$ terms survive time-averaging because these terms have no oscillatory factor since $\omega - \omega^* = 2\mathrm{i}\omega_i$. Thus, the time average, denoted by $\langle\rangle$, of the product is

$$\langle\psi(t)\chi(t)\rangle = \frac{1}{4}(\tilde{\psi}\tilde{\chi}^* + \tilde{\psi}^*\tilde{\chi})\mathrm{e}^{2\omega_i t};$$

$$= \frac{1}{2}\mathrm{Re}\left(\tilde{\psi}\,\tilde{\chi}^*\right)\mathrm{e}^{2\omega_i t}; \tag{7.62}$$

this is the correct rule for time-averaging products of oscillating physical quantities, which have been represented using complex notation.

7.5 Cold plasma wave-energy equation

The current density $\mathbf{J}$ in Ampère's law consists of the explicit plasma currents. The frequency dependence of these currents means that care is required when integrating $\mathbf{E}\cdot\mathbf{J}$. In order to prepare for this integration we express Eq. (6.9) and Eq. (6.10) as

$$\frac{1}{\mu_0}\nabla\times\tilde{\mathbf{B}} = \varepsilon_0\frac{\partial}{\partial t}\left(\overleftrightarrow{\mathbf{K}}(\omega)\cdot\tilde{\mathbf{E}}\mathrm{e}^{-\mathrm{i}\omega t}\right)$$

$$= \tilde{\mathbf{J}}\mathrm{e}^{-\mathrm{i}\omega t} + \varepsilon_0\frac{\partial}{\partial t}\left(\tilde{\mathbf{E}}\mathrm{e}^{-\mathrm{i}\omega t}\right), \tag{7.63}$$

where the time dependence is shown explicitly and $\tilde{\mathbf{J}}$ refers to the explicit plasma currents as distinct from the displacement current.

Integration of Eq. (7.55) taking into account the prescription given by Eq. (7.60) has the form

$$w(t) = w(-\infty) + \int_{-\infty}^{t} dt \left\langle \begin{matrix} \mathbf{E}(\mathbf{x},t) \cdot \left(\mathbf{J}(\mathbf{x},\mathbf{t}) + \varepsilon_0 \dfrac{\partial \mathbf{E}(\mathbf{x},t)}{\partial t} \right) \\ + \dfrac{1}{\mu_0} \mathbf{B}(\mathbf{x},t) \cdot \dfrac{\partial \mathbf{B}(\mathbf{x},t)}{\partial t} \end{matrix} \right\rangle$$

$$= w(-\infty) + \frac{1}{4} \int_{-\infty}^{t} dt \left\langle \begin{matrix} \tilde{\mathbf{E}} e^{-i\omega t} \cdot \varepsilon_0 \dfrac{\partial}{\partial t} \left(\overleftrightarrow{\mathbf{K}}(\omega) \cdot \tilde{\mathbf{E}} e^{-i\omega t} \right)^* \\ + \dfrac{1}{\mu_0} \tilde{\mathbf{B}} e^{-i\omega t} \cdot \dfrac{\partial}{\partial t} \left(\tilde{\mathbf{B}} e^{-i\omega t} \right)^* + \text{c.c.} \end{matrix} \right\rangle, \tag{7.64}$$

where c.c. means complex conjugate of all preceding terms in the expression. The term containing the rates of change of the electric field and the particle kinetic energy can be written using Eq. (7.63) as

$$\left\langle \mathbf{E} \cdot \left(\mathbf{J} + \varepsilon_0 \frac{\partial \mathbf{E}}{\partial t} \right) \right\rangle = \frac{\varepsilon_0}{4} \left\{ \tilde{\mathbf{E}} e^{-i\omega t} \cdot \left(i\omega^* \overleftrightarrow{\mathbf{K}}^* \cdot \tilde{\mathbf{E}}^* e^{i\omega^* t} \right) + \text{c.c.} \right\}$$

$$= \frac{\varepsilon_0}{4} \left\{ \tilde{\mathbf{E}} \cdot i\omega^* \overleftrightarrow{\mathbf{K}}^* \cdot \tilde{\mathbf{E}}^* - \tilde{\mathbf{E}}^* \cdot i\omega \overleftrightarrow{\mathbf{K}} \cdot \tilde{\mathbf{E}} \right\} e^{2\omega_i t}$$

$$= \frac{\varepsilon_0}{4} \left\{ \begin{matrix} i\omega_r \left[\tilde{\mathbf{E}} \cdot \overleftrightarrow{\mathbf{K}}^* \cdot \tilde{\mathbf{E}}^* - \tilde{\mathbf{E}}^* \cdot \overleftrightarrow{\mathbf{K}} \cdot \tilde{\mathbf{E}} \right] \\ + \omega_i \left[\tilde{\mathbf{E}} \cdot \overleftrightarrow{\mathbf{K}}^* \cdot \tilde{\mathbf{E}}^* + \tilde{\mathbf{E}}^* \cdot \overleftrightarrow{\mathbf{K}} \cdot \tilde{\mathbf{E}} \right] \end{matrix} \right\} e^{2\omega_i t}. \tag{7.65}$$

To proceed further, it is noted that

$$\tilde{\mathbf{E}} \cdot \overleftrightarrow{\mathbf{K}}^* \tilde{\mathbf{E}}^* = \sum_{pq} \tilde{E}_p K_{pq}^* \tilde{E}_q^* = \sum_{pq} \tilde{E}_p K_{qp}^{*t} \tilde{E}_q^* = \tilde{\mathbf{E}}^* \cdot \overleftrightarrow{\mathbf{K}}^\dagger \cdot \tilde{\mathbf{E}}, \tag{7.66}$$

where the superscript t means transpose and the superscript † means Hermitian conjugate, i.e., the complex conjugate of the transpose. Thus, Eq. (7.65) can be rewritten as

$$\left\langle \mathbf{E} \cdot \left(\mathbf{J} + \varepsilon_0 \frac{\partial \mathbf{E}}{\partial t} \right) \right\rangle = \frac{\varepsilon_0}{4} \left[i\omega_r \tilde{\mathbf{E}}^* \cdot \left(\overleftrightarrow{\mathbf{K}}^\dagger - \overleftrightarrow{\mathbf{K}} \right) \cdot \tilde{\mathbf{E}} + \omega_i \tilde{\mathbf{E}}^* \cdot \left(\overleftrightarrow{\mathbf{K}}^\dagger + \overleftrightarrow{\mathbf{K}} \right) \cdot \tilde{\mathbf{E}} \right] e^{2\omega_i t}. \tag{7.67}$$

Both the Hermitian part of the dielectric tensor,

$$\overleftrightarrow{\mathbf{K}}_h = \frac{1}{2} \left(\overleftrightarrow{\mathbf{K}} + \overleftrightarrow{\mathbf{K}}^\dagger \right), \tag{7.68}$$

and the anti-Hermitian part,

$$\overleftrightarrow{\mathbf{K}}_{ah} = \frac{1}{2} \left(\overleftrightarrow{\mathbf{K}} - \overleftrightarrow{\mathbf{K}}^\dagger \right), \tag{7.69}$$

occur in Eq. (7.67). The cold plasma dielectric tensor is a function of ω via the functions S, P, and D,

$$\overleftrightarrow{\mathbf{K}}(\omega) = \begin{bmatrix} S(\omega) & -\mathrm{i}D(\omega) & 0 \\ \mathrm{i}D(\omega) & S(\omega) & 0 \\ 0 & 0 & P(\omega) \end{bmatrix}. \tag{7.70}$$

If $\omega_i = 0$ then S, P, and D are all pure real. In this case $\overleftrightarrow{\mathbf{K}}(\omega)$ is Hermitian so that $\overleftrightarrow{\mathbf{K}}_h = \overleftrightarrow{\mathbf{K}}$ and $\overleftrightarrow{\mathbf{K}}_{ah} = 0$. However, if ω_i is finite but small, then $\overleftrightarrow{\mathbf{K}}(\omega)$ will have a small non-Hermitian part. This non-Hermitian part is extracted using a Taylor expansion in terms of ω_i, i.e.,

$$\overleftrightarrow{\mathbf{K}}(\omega_r + \mathrm{i}\omega_i) = \overleftrightarrow{\mathbf{K}}(\omega_r) + \mathrm{i}\omega_i \left[\frac{\partial}{\partial \omega} \overleftrightarrow{\mathbf{K}}(\omega) \right]_{\omega=\omega_r}. \tag{7.71}$$

The transpose of the complex conjugate of this expansion is

$$\left[\overleftrightarrow{\mathbf{K}}(\omega_r + \mathrm{i}\omega_i) \right]^{\dagger} = \overleftrightarrow{\mathbf{K}}(\omega_r) - \mathrm{i}\omega_i \left[\frac{\partial}{\partial \omega} \overleftrightarrow{\mathbf{K}}(\omega) \right]_{\omega=\omega_r}, \tag{7.72}$$

since $\overleftrightarrow{\mathbf{K}}$ is non-Hermitian only to the extent that ω_i is finite. Substituting Eqs. (7.71) and (7.72) into Eqs. (7.68) and (7.69) and assuming small ω_i gives

$$\overleftrightarrow{\mathbf{K}}_h = \overleftrightarrow{\mathbf{K}}(\omega_r) \tag{7.73}$$

and

$$\overleftrightarrow{\mathbf{K}}_{ah} = \mathrm{i}\omega_i \left[\frac{\partial}{\partial \omega} \overleftrightarrow{\mathbf{K}}(\omega) \right]_{\omega=\omega_r}. \tag{7.74}$$

Inserting Eqs. (7.73) and (7.74) in Eq. (7.67) yields

$$\begin{aligned} \left\langle \mathbf{E} \cdot \left(\mathbf{J} + \varepsilon_0 \frac{\partial \mathbf{E}}{\partial t} \right) \right\rangle &= \frac{2\varepsilon_0 \omega_i}{4} \left[\omega_r \tilde{\mathbf{E}}^* \cdot \left[\frac{\partial}{\partial \omega} \overleftrightarrow{\mathbf{K}}(\omega) \right]_{\omega=\omega_r} \cdot \tilde{\mathbf{E}} + \tilde{\mathbf{E}}^* \cdot \overleftrightarrow{\mathbf{K}}(\omega_r) \cdot \tilde{\mathbf{E}} \right] \mathrm{e}^{2\omega_i t} \\ &= \frac{2\varepsilon_0 \omega_i}{4} \tilde{\mathbf{E}}^* \cdot \left[\frac{\partial}{\partial \omega} \omega \overleftrightarrow{\mathbf{K}}(\omega) \right]_{\omega=\omega_r} \cdot \tilde{\mathbf{E}} \mathrm{e}^{2\omega_i t}. \end{aligned} \tag{7.75}$$

Similarly, the rate of change of the magnetic energy density is

$$\left\langle \frac{1}{\mu_0} \mathbf{B} \cdot \frac{\partial \mathbf{B}}{\partial t} \right\rangle = \frac{1}{4\mu_0} \left[2\omega_i \tilde{\mathbf{B}}^* \cdot \tilde{\mathbf{B}} \right] \mathrm{e}^{2\omega_i t}. \tag{7.76}$$

Using Eqs. (7.75) and (7.76) in Eq. (7.64) gives

$$w = w(-\infty) + \left\{ \frac{\varepsilon_0}{4} \tilde{\mathbf{E}}^* \cdot \left[\frac{\partial}{\partial \omega} \left(\omega \overleftrightarrow{\mathbf{K}}(\omega) \right) \right]_{\omega=\omega_r} \cdot \tilde{\mathbf{E}} + \frac{1}{4\mu_0} \left[\tilde{\mathbf{B}}^* \cdot \tilde{\mathbf{B}} \right] \right\} \int_{-\infty}^{t} \mathrm{d}t 2\omega_i \mathrm{e}^{2\omega_i t}, \tag{7.77}$$

which now may be integrated in time to give the total energy density *associated with bringing the wave into existence*

$$\begin{aligned} \bar{w} &= w - w(-\infty) \\ &= \left\{ \frac{\varepsilon_0}{4} \tilde{\mathbf{E}}^* \cdot \left[\frac{\partial}{\partial \omega} \left(\omega \overleftrightarrow{\mathbf{K}}(\omega) \right) \right]_{\omega=\omega_r} \cdot \tilde{\mathbf{E}} + \frac{1}{4\mu_0} \left[\tilde{\mathbf{B}}^* \cdot \tilde{\mathbf{B}} \right] \right\} \mathrm{e}^{2\omega_i t}. \end{aligned} \tag{7.78}$$

In the limit $\omega_i \to 0$ this reduces to

$$\bar{w} = \frac{\varepsilon_0}{4} \tilde{\mathbf{E}}^* \cdot \frac{\partial}{\partial \omega} \left[\omega \overleftrightarrow{\mathbf{K}}(\omega) \right] \cdot \tilde{\mathbf{E}} + \frac{|\tilde{B}|^2}{4\mu_0}. \tag{7.79}$$

Since the time-average of the energy density stored in the oscillating vacuum electric field is

$$w_E = \frac{\varepsilon_0 |\tilde{E}\,|^2}{4} \tag{7.80}$$

and the time-average of the energy density stored in the oscillating vacuum magnetic field is

$$w_B = \frac{|\tilde{B}|^2}{4\mu_0}, \tag{7.81}$$

the change in the time-average of the particle kinetic energy density associated with bringing the wave into existence is

$$\bar{w}_{part} = \frac{\varepsilon_0}{4} \tilde{\mathbf{E}}^* \cdot \left[\frac{\partial}{\partial \omega} \left(\omega \overleftrightarrow{\mathbf{K}}(\omega) \right) - \overleftrightarrow{\mathbf{I}} \right] \cdot \tilde{\mathbf{E}}. \tag{7.82}$$

This result has been established for the general case of the dielectric tensor $\overleftrightarrow{\mathbf{K}}(\omega)$ of a cold magnetized plasma. In order to illustrate the application of Eq. (7.79), we consider the simple example of high-frequency electrostatic oscillations in an unmagnetized plasma. In this situation $S = P = 1 - \omega_{pe}^2/\omega^2$ and $D = 0$

so $\overleftrightarrow{\mathbf{K}}(\omega) = \left(1 - \omega_{pe}^2/\omega^2\right)\overleftrightarrow{\mathbf{I}}$. Since the oscillations are electrostatic, $w_B = 0$. The energy density of the particles is therefore

$$\begin{aligned}\bar{w}_{part} &= \frac{\varepsilon_0|\tilde{E}|^2}{4}\left[\frac{\partial}{\partial\omega}\left\{\omega\left(1 - \frac{\omega_{pe}^2}{\omega^2}\right)\right\} - 1\right] \\ &= \frac{\varepsilon_0|\tilde{E}|^2}{4}\left[2\frac{\omega_{pe}^2}{\omega^2} - 1\right] \\ &= \frac{\varepsilon_0|\tilde{E}|^2}{4},\end{aligned} \tag{7.83}$$

where the dispersion relation $1 - \omega_{pe}^2/\omega^2 = 0$ has been used. Thus, in this simple example half of the average wave-energy density is contained in the electric field while the other half is contained in the coherent particle motion associated with the wave.

7.6 Finite-temperature plasma wave-energy equation

The dielectric tensor has no dependence on the wavevector $\mathbf{k}$ in a cold plasma, but does have such a dependence in a finite-temperature plasma. For example, the electrostatic unmagnetized cold plasma dielectric $P(\omega) = 1 - \omega_{pe}^2/\omega^2$ becomes $P(\omega, k) = 1 - (1 + 3k^2\lambda_{De}^2)\omega_{pe}^2/\omega^2$ in a warm plasma. The analysis of the previous section will now be extended to allow for the possibility that the dielectric tensor depends on $\mathbf{k}$ as well as on ω. In analogy to the method used in the previous section for treating complex ω, here $\mathbf{k}$ will also be assumed to have a small imaginary part. In this case, Taylor expansion of the dielectric tensor followed by extraction of the anti-Hermitian part shows that the anti-Hermitian part is

$$\overleftrightarrow{\mathbf{K}}_{ah} = \mathrm{i}\omega_i\left[\frac{\partial}{\partial\omega}\overleftrightarrow{\mathbf{K}}(\omega, \mathbf{k})\right]_{\omega=\omega_r, \mathbf{k}=\mathbf{k}_r} + \mathrm{i}\mathbf{k}_i\cdot\left[\frac{\partial}{\partial\mathbf{k}}\overleftrightarrow{\mathbf{K}}(\omega, \mathbf{k})\right]_{\omega=\omega_r, \mathbf{k}=\mathbf{k}_r} \tag{7.84}$$

while the Hermitian part remains as before. There is now a term involving $\mathbf{k}_i$. With the incorporation of this new term, Eq. (7.75) becomes

$$\left\langle\mathbf{E}\cdot\left(\mathbf{J} + \varepsilon_0\frac{\partial\mathbf{E}}{\partial t}\right)\right\rangle = \left[\begin{array}{l}\dfrac{2\varepsilon_0\omega_i}{4}\tilde{\mathbf{E}}^*\cdot\left[\dfrac{\partial}{\partial\omega}\omega\overleftrightarrow{\mathbf{K}}(\omega)\right]_{\omega=\omega_r}\cdot\tilde{\mathbf{E}} \\ +\dfrac{\varepsilon_0}{4}\tilde{\mathbf{E}}^*\cdot\left[2\omega\mathbf{k}_i\cdot\dfrac{\partial}{\partial\mathbf{k}}\overleftrightarrow{\mathbf{K}}(\omega, \mathbf{k})\right]_{\substack{\omega=\omega_r,\\ \mathbf{k}=\mathbf{k}_r}}\cdot\tilde{\mathbf{E}}\end{array}\right]\mathrm{e}^{-2\mathbf{k}_i\cdot\mathbf{x}+2\omega_i t}, \tag{7.85}$$

where we have explicitly written the exponential space-dependent factor $\exp(-2\mathbf{k}_i\cdot\mathbf{x})$.

What is the meaning of this new term involving $\mathbf{k}_i$? The answer to this question may be found by examining the Poynting flux for the situation where $\mathbf{k}_i$ is finite. Using the product rule and allowing for finite $\mathbf{k}_i$ shows

$$\begin{aligned}\langle\nabla\cdot\mathbf{P}\rangle &= \langle\nabla\cdot(\mathbf{E}\times\mathbf{H})\rangle \\ &= \frac{1}{4}\nabla\cdot\left[\left(\tilde{\mathbf{E}}^*\times\tilde{\mathbf{H}}+\tilde{\mathbf{E}}\times\tilde{\mathbf{H}}^*\right)\mathrm{e}^{-2\mathbf{k}_i\cdot\mathbf{x}+2\omega_i t}\right] \\ &= \frac{1}{4}\left(-2\mathbf{k}_i\cdot\left(\tilde{\mathbf{E}}^*\times\tilde{\mathbf{H}}+\tilde{\mathbf{E}}\times\tilde{\mathbf{H}}^*\right)\right)\mathrm{e}^{-2\mathbf{k}_i\cdot\mathbf{x}+2\omega_i t}.\end{aligned} \tag{7.86}$$

Comparison of Eqs. (7.85) and (7.86) shows that the factor $-2\mathbf{k}_i$ corresponds to the divergence operator acting on the product of two oscillating quantities and so the second term in Eq. (7.85) represents an *energy flux*. Since the Poynting vector $\mathbf{P}$ is the energy flux associated with the electromagnetic field, this additional energy flux must be identified as the energy flux associated with particle motion. Defining this flux as $\mathbf{T}$ it is seen that

$$T_j = -\frac{\omega\varepsilon_0}{4}\tilde{\mathbf{E}}^*\cdot\left[\frac{\partial}{\partial k_j}\overleftrightarrow{\mathbf{K}}(\omega,\mathbf{k})\right]\cdot\tilde{\mathbf{E}} \tag{7.87}$$

in the limit $\mathbf{k}_i\to 0$. For small but finite $\mathbf{k}_i$, ω_i the generalized Poynting theorem can be written as

$$-2\mathbf{k}_i\cdot(\mathbf{P}+\mathbf{T})+2\omega_i(w_E+w_B+\bar{w}_{part})=0. \tag{7.88}$$

We now define the generalized group velocity $\mathbf{v}_g$ to be the velocity with which wave energy is transported. This velocity is the total energy flux divided by the total energy density, i.e.,

$$\mathbf{v}_g=\frac{\mathbf{P}+\mathbf{T}}{w_E+w_B+\bar{w}_{part}}. \tag{7.89}$$

The bar in $\bar{w}_{part}$ indicates that this term is the *difference* between the particle energy with the wave and the particle energy without the wave and so could, in principle, be negative.

7.7 Negative energy waves

A curious consequence of this analysis is that a wave can have a *negative* energy density. The field energy densities w_E and w_B are positive-definite, but the particle energy density $\bar{w}_{part}$ can have either sign and in certain circumstances can be sufficiently negative to make the total wave-energy density negative. This surprising possibility can occur because the wave-energy density defined in Eq. (7.78) is actually the *change* in the total system energy density on going from a situation where there is no wave to a situation where there is a wave. Typically, negative energy waves occur when the equilibrium has a steady-state flow velocity and there exists a mode that reduces the average kinetic energy of the particles to a value below the initial equilibrium value. The growth of this type of mode taps the free energy in the flow.

As an example of a negative energy wave, we consider the situation where unmagnetized cold electrons stream with velocity v_0 through a background of infinitely massive ions. As shown in Section 5.2 the electrostatic dispersion for this simple 1-D situation with flow involves a parallel dielectric having a Doppler-shifted frequency, i.e., the dispersion relation is

$$P(\omega, k) = 1 - \frac{\omega_{pe}^2}{(\omega - kv_0)^2} = 0; \tag{7.90}$$

here it has been assumed that there exists a neutralizing background of infinitely massive ions. Since the plasma is unmagnetized, its dielectric tensor is simply $\overleftrightarrow{\mathbf{K}}(\omega, \mathbf{k}) = P(\omega, k)\overleftrightarrow{\mathbf{I}}$. Using Eq. (7.79), the wave-energy density is

$$w = \frac{\varepsilon_0 |\tilde{E}|^2}{4} \frac{\partial}{\partial \omega}(\omega P(\omega, k)) = \frac{\varepsilon_0 |\tilde{E}|^2}{2} \frac{\omega \omega_{pe}^2}{(\omega - kv_0)^3}. \tag{7.91}$$

However, the dispersion relation, Eq. (7.90), shows that

$$\omega = kv_0 \pm \omega_{pe} \tag{7.92}$$

so Eq. (7.91) can be recast as

$$w = \frac{\varepsilon_0 |\tilde{E}|^2}{2} \left(1 \pm \frac{kv_0}{\omega_{pe}} \right). \tag{7.93}$$

Thus, if $kv_0 > \omega_{pe}$ and the minus sign is selected, the wave has *negative* energy density.

This result can be verified by direct calculation of the change in system energy density due to growth of the wave. When there is no wave, the electric field is zero and the system energy density w^{sys} is simply the beam kinetic energy density

$$w_0^{sys} = \frac{1}{2} n_0 m_e v_0^2. \tag{7.94}$$

Now consider a one-dimensional electrostatic wave with electric field $E = \mathrm{Re}\left(\tilde{E} e^{ikx - i\omega t}\right)$. The average energy density of the system with this wave is

$$w_{wave}^{sys} = \frac{\varepsilon_0 |\tilde{E}|^2}{4} + \left\langle \frac{1}{2} n(x, t) m_e v(x, t)^2 \right\rangle \tag{7.95}$$

so the change in system energy density due to the wave is

$$\bar{w}^{sys} = w^{sys}_{wave} - w^{sys}_0 = \frac{\varepsilon_0|\tilde{E}\,|^2}{4} + \left\langle \frac{1}{2}\left[n_0 + n_1(x,t)\right] m_e \left[v_0 + v_1(x,t)\right]^2 \right\rangle - \frac{1}{2} n_0 m_e v_0^2, \tag{7.96}$$

where

$$n_1(x,t) = \mathrm{Re}\left(\tilde{n}\mathrm{e}^{ikx-\mathrm{i}\omega t}\right), \quad v_1(x,t) = \mathrm{Re}\left(\tilde{v}\mathrm{e}^{ikx-\mathrm{i}\omega t}\right). \tag{7.97}$$

Since odd powers of oscillating quantities vanish upon time averaging, Eq. (7.96) reduces to

$$\begin{aligned} \bar{w}^{sys} &= \frac{\varepsilon_0|\tilde{E}|^2}{4} + n_0 m_e \left\langle \frac{1}{2} v_1^2 + \frac{n_1}{n_0} v_1 v_0 \right\rangle \\ &= \frac{\varepsilon_0|\tilde{E}|^2}{4} + \frac{1}{2} n_0 m_e \left\langle v_1^2 \right\rangle + m_e v_0 \left\langle n_1 v_1 \right\rangle . \end{aligned} \tag{7.98}$$

The linearized continuity equation gives

$$-\mathrm{i}(\omega - kv_0)\tilde{n} + n_0 \mathrm{i}k\tilde{v} = 0 \tag{7.99}$$

or

$$\frac{\tilde{n}}{n_0} = \frac{k\tilde{v}}{\omega - kv_0}. \tag{7.100}$$

The electron quiver velocity in the wave is

$$\tilde{v} = \frac{q\tilde{E}}{-\mathrm{i}(\omega - kv_0)m_e} \tag{7.101}$$

so

$$\left\langle v_1^2 \right\rangle = \frac{1}{2} \frac{q^2|\tilde{E}|^2}{(\omega - kv_0)^2 m_e^2} \tag{7.102}$$

and

$$\left\langle n_1 v_1 \right\rangle = \frac{k}{2} \frac{n_0 q^2|\tilde{E}|^2}{(\omega - kv_0)^3 m_e^2}. \tag{7.103}$$

We may now evaluate Eq. (7.98) to obtain

$$\begin{aligned} \bar{w}^{sys} &= \frac{\varepsilon_0|\tilde{E}|^2}{4} + n_0 m_e \left[\frac{q^2|\tilde{E}|^2}{4(\omega - kv_0)^2 m_e^2} + \frac{kv_0}{2} \frac{q^2|\tilde{E}|^2}{(\omega - kv_0)^3 m_e^2} \right] \\ &= \frac{\varepsilon_0|\tilde{E}|^2}{4} \left[1 + \frac{\omega_{pe}^2}{(\omega - kv_0)^2} + \frac{2\omega_{pe}^2 kv_0}{(\omega - kv_0)^3} \right] \\ &= \frac{\varepsilon_0|\tilde{E}|^2}{2} \left[1 \pm \frac{kv_0}{\omega_{pe}} \right], \end{aligned} \tag{7.104}$$

where Eq. (7.92) has been repeatedly invoked. This is the same as Eq. (7.93). The energy flux associated with this wave is also negative (cf. assignments). However, the group velocity is positive (cf. assignments) because the group velocity is the ratio of a negative energy flux to a negative energy density.

Dissipation affects negative energy waves in a manner opposite to its effect on positive energy waves. This can be seen by Taylor expanding the dispersion relation $P(\omega, k) = 0$ as done in Eq. (5.87) to obtain

$$\omega_i = -\frac{P_i}{[\partial P_r/\partial\omega]_{\omega=\omega_r}}. \tag{7.105}$$

Expanding Eq. (7.91) gives

$$\bar{w} = \frac{\varepsilon_0 \omega |E|^2}{4}\frac{\partial P}{\partial \omega} \tag{7.106}$$

so a negative energy wave has $\omega \partial P/\partial\omega < 0$. If the dissipative term P_i is the same for both positive and negative waves, then, for a given sign of ω, the critical difference between positive and negative waves lies in the sign of $\partial P/\partial\omega$. Equation (7.105) shows that ω_i has opposite signs for positive and negative energy waves. This means that, in contrast to positive energy waves, dissipation *destabilizes* negative energy waves. This is because dissipation enables the tapping of available free energy by the negative energy wave. Since any real system generally has some dissipation, if a negative energy wave is an allowed mode in a given real system, the negative energy wave will spontaneously develop and grow at the expense of the free energy in the system (e.g., the free energy in the streaming particles).

7.8 Assignments

1. Consider the problem of short-wave radio transmission. Let x be the horizontal direction and z be the vertical direction. A short-wave radio antenna is designed to radiate most of the transmitter power into a specified k_x and k_z by careful control over the Fourier transform of the antenna geometry.
   (i) What is the frequency range of short-wave radio communications?
   (ii) For ionospheric parameters and the majority of the short-wave band, should ionospheric plasma be considered as magnetized or unmagnetized?
   (iii) What is the appropriate dispersion for short-wave radio waves (hint: it is very simple)?
   (iv) Using Snell's law and geometric optics, sketch the trajectory of a radio wave propagating from a terrestrial antenna (hint: consider k_x and k_z radiated by the antenna and consider the group velocity trajectory and also what happens when $\omega = \omega_{pe}(z)$).
2. Using geometric optics discuss qualitatively with sketches how a low-frequency wave could act as a lens for a high-frequency wave.

3. Using the example of an electrostatic electron plasma wave (dispersion $\omega^2 = \omega_{pe}^2(1+3k^2\lambda_{de}^2)$) show that the generalized group velocity as defined in Eq. (7.89) gives the same group velocity as found by direct evaluation of $\partial\omega/\partial\mathbf{k}$.
4. Calculate the wave-energy density, wave-energy flux, and group velocity for the electrostatic wave that can occur when a beam of cold electrons streams with velocity $\mathbf{v}_0$ through a neutralizing background of infinitely massive ions. Discuss the signs of these three quantities.

8

Vlasov theory of warm electrostatic waves in a magnetized plasma

8.1 Solving the Vlasov equation by tracking each particle's history

It has been tacitly assumed until now that the wave phase experienced by a particle is the phase the particle would have experienced if it had not deviated from its initial position $\mathbf{x}_0$. This means that the particle trajectory used when determining the wave phase experienced by the particle is effectively assumed to be $\mathbf{x} = \mathbf{x}_0$ instead of the actual trajectory $\mathbf{x} = \mathbf{x}(t)$. Thus, the wave phase seen by the particle was approximated as

$$\mathbf{k} \cdot \mathbf{x}(t) - \omega t \simeq \mathbf{k} \cdot \mathbf{x}_0 - \omega t. \tag{8.1}$$

This approximation is fine provided the deviation of the actual trajectory from the assumed trajectory satisfies the condition

$$|\mathbf{k} \cdot (\mathbf{x}(t) - \mathbf{x}_0)| \ll \pi/2, \tag{8.2}$$

in which case phase error due to the deviation would be insignificant. Two situations can occur where this condition is not satisfied, namely:

1. the wave amplitude is so large that the particle displacement due to the wave is significant compared to a wavelength,
2. the wave amplitude is small, but the particle has a large initial velocity so it moves substantially during a wave period.

The first case results in chaotic particle motion as discussed in Section 3.7.3 while the second case, the subject of this chapter, occurs when particles have significant thermal motion. If the motion is parallel to the magnetic field, significant thermal motion means that v_T is non-negligible compared to $\omega/k_\parallel$, a regime already discussed in Section 5.3 for unmagnetized plasmas. The subscript $\parallel$ is used here to denote the direction along the magnetic field. If the magnetic field is straight and given by $\mathbf{B} = B\hat{z}$, $\parallel$ would simply be the z direction, but in a more general situation the $\parallel$ component of a vector would be obtained by dotting the vector with $\hat{B}$. If the motion is perpendicular to the magnetic field (this direction is denoted as $\perp$), then significant thermal motion means that $k_\perp r_L$ becomes large, i.e., the Larmor orbit r_L becomes comparable to the wavelength. In the situation where $k_\perp r_L$ is not infinitesimal, the particle samples a continuum of wave phases as it traces out its Larmor orbit and these sampled wave phases are considerably different from the wave phase at the particle's gyrocenter.

Consider an electrostatic wave with potential

$$\phi_1(\mathbf{x}, t) = \tilde{\phi}_1 \exp(\mathrm{i}\mathbf{k}\cdot\mathbf{x} - \mathrm{i}\omega t). \tag{8.3}$$

As before, the convention will be used that a tilde refers to the amplitude of a perturbed quantity; if there is no tilde, then the exponential phase factor is understood to be included. Because the wave is electrostatic, Poisson's equation is the relevant equation relating particle motion to the electric field and is

$$k^2\phi_1 = \frac{1}{\varepsilon_0}\sum_\sigma n_{\sigma 1} q_\sigma, \tag{8.4}$$

where $n_{\sigma 1}$ is the density perturbation for each species σ. Since the density perturbation is just the zeroth moment of the perturbed distribution function,

$$n_{\sigma 1} = \int f_{\sigma 1} \mathrm{d}^3 v, \tag{8.5}$$

the problem reduces to determining the perturbed distribution function $f_{\sigma 1}$.

In the presence of a uniform magnetic field the linearized Vlasov equation is

$$\frac{\partial f_{\sigma 1}}{\partial t} + \mathbf{v}\cdot\frac{\partial f_{\sigma 1}}{\partial \mathbf{x}} + \frac{q_\sigma}{m_\sigma}(\mathbf{v}\times\mathbf{B})\cdot\frac{\partial f_{\sigma 1}}{\partial \mathbf{v}} = \frac{q_\sigma}{m_\sigma}\nabla\phi_1\cdot\frac{\partial f_{\sigma 0}}{\partial \mathbf{v}}, \tag{8.6}$$

where the subscript 0 refers to equilibrium quantities and the subscript 1 to first-order perturbations (no 0 subscript has been used to distinguish the equilibrium magnetic field **B** from a perturbed magnetic field because, being electrostatic, the assumed wave has no associated perturbed magnetic field).

Consider all particles of species σ located at an arbitrary point $\mathbf{x}, \mathbf{v}$ in phase-space at the present time t. These particles must all have identical phase-space trajectories in both the future and the past because they are all acted on by the same Lorentz force $q_\sigma(\mathbf{E}(\mathbf{x}, t) + \mathbf{v}\times\mathbf{B}(\mathbf{x}, t))$ and they all have the same initial condition at time t since, by assumption, they have the same $\mathbf{x}$ and $\mathbf{v}$ at time t. By integrating the equation of motion starting from this point in phase-space, the phase-space trajectory $\mathbf{x}(t')$, $\mathbf{v}(t')$ of this set of particles can be determined. The boundary conditions on such a phase-space trajectory are simply

$$\mathbf{x}(t) = \mathbf{x}, \quad \mathbf{v}(t) = \mathbf{v}. \tag{8.7}$$

Instead of treating $\mathbf{x}, \mathbf{v}$ as independent variables denoting a point in phase-space, let us think of these quantities as temporal boundary conditions for particles with phase-space trajectories $\mathbf{x}(t)$, $\mathbf{v}(t)$ that happen to be at location $\mathbf{x}$, $\mathbf{v}$ at time t. Thus, the velocity distribution function for all particles that happen to be at phase-space location $\mathbf{x}$, $\mathbf{v}$ at time t is $f_{\sigma 1} = f_{\sigma 1}(\mathbf{x}(t), \mathbf{v}(t), t)$ and since $\mathbf{x}$ and $\mathbf{v}$ were arbitrary, this expression is valid at any point in phase-space. The time derivative of this function is

$$\frac{\mathrm{d}}{\mathrm{d}t} f_{\sigma 1}(\mathbf{x}(t), \mathbf{v}(t), t) = \frac{\partial f_{\sigma 1}}{\partial t} + \frac{\partial f_{\sigma 1}}{\partial \mathbf{x}} \cdot \frac{\mathrm{d}\mathbf{x}}{\mathrm{d}t} + \frac{\partial f_{\sigma 1}}{\partial \mathbf{v}} \cdot \frac{\mathrm{d}\mathbf{v}}{\mathrm{d}t}. \tag{8.8}$$

In principle, the wave force acting on the particles ought to be taken into account when calculating their trajectories. However, if the wave amplitude is sufficiently small, the particle trajectory will not be significantly affected by the wave and so will be essentially the same as the unperturbed trajectory, namely the trajectory the particle would have had if there were no wave. Since the unperturbed particle trajectory equations are

$$\frac{\mathrm{d}\mathbf{x}}{\mathrm{d}t} = \mathbf{v}, \quad \frac{\mathrm{d}\mathbf{v}}{\mathrm{d}t} = \frac{q_\sigma}{m_\sigma} \mathbf{v} \times \mathbf{B}, \tag{8.9}$$

it is seen that Eq. (8.8) is identical to the left-hand side of Eq. (8.6). Equation (8.6) can thus be rewritten as

$$\left(\frac{\mathrm{d}}{\mathrm{d}t} f_{\sigma 1}(\mathbf{x}(t), \mathbf{v}(t), t) \right)_{\substack{\text{unperturbed} \\ \text{trajectory}}} = \frac{q_\sigma}{m_\sigma} \nabla \phi_1 \cdot \frac{\partial f_{\sigma 0}}{\partial \mathbf{v}}, \tag{8.10}$$

where the left-hand side is the derivative of the distribution function that would be measured by an observer sitting on a particle having the *unperturbed* phase-space trajectory $\mathbf{x}(t)$, $\mathbf{v}(t)$. Equation (8.10) may be integrated to give

$$f_{\sigma 1}(\mathbf{x}, \mathbf{v}, t) = \frac{q_\sigma}{m_\sigma} \int_{-\infty}^{t} \mathrm{d}t' \left[\nabla \phi_1 \cdot \frac{\partial f_{\sigma 0}}{\partial \mathbf{v}} \right]_{\mathbf{x}=\mathbf{x}(t'), \mathbf{v}=\mathbf{v}(t')}. \tag{8.11}$$

If the right-hand side of Eq. (8.10) is considered as a "force" acting to change the perturbed distribution function, then Eq. (8.11) is effectively a statement that the perturbed distribution function at $\mathbf{x}$, $\mathbf{v}$ for time t is a result of the sum of all the "forces" acting over times prior to t evaluated along the unperturbed trajectory of the particle. In effect, Eq. (8.11) states that, just as the present state of a person is in a sense the consequence of the cumulative effect of all the various influences experienced by the person at different times and places during his/her entire previous life, the present status of the particles, denoted by $f_{\sigma 1}(\mathbf{x}, \mathbf{v}, t)$, is the consequence of the cumulative effect of all the various influences (i.e., forces) experienced at different times and places by the particles during their entire past history. "Unperturbed trajectories" refers to the solution to Eqs. (8.9); these equations neglect any wave-induced changes to the particle trajectory and simply give the trajectory of a thermal particle. The "force" in Eq. (8.11) must be evaluated along the *past* phase-space trajectory because *that* is where the particles at $\mathbf{x}, \mathbf{v}$ were located at previous times and so that is where the particles "felt" the "force." This is called "integrating along the *unperturbed* orbits" and is valid only when the unperturbed orbits (trajectories) are a good approximation to the particles' actual orbits. Mathematically speaking, these unperturbed orbits are the characteristics of the left-hand side of Eq. (8.6), a homogeneous hyperbolic partial differential equation. The solutions of this homogeneous equation are constant along the characteristics. The right-hand side is the inhomogeneous or forcing term and acts to modify the homogeneous solution; the cumulative effect of this force is found by integrating along the characteristics of the homogeneous part.

The problem is now formally solved; all that is required is an explicit evaluation of the integrals. The functional form of the equilibrium distribution function is determined by the specific physical problem under consideration. Often the plasma has a uniform Maxwellian distribution

$$f_{\sigma 0}(\mathbf{v}) = \frac{n_{\sigma 0}}{\pi^{3/2} v_{T\sigma}^3} \mathrm{e}^{-v^2/v_{T\sigma}^2}, \tag{8.12}$$

where

$$v_{T\sigma} = \sqrt{2\kappa T_\sigma / m_\sigma}. \tag{8.13}$$

It must be understood that Eq. (8.12) represents one of an infinite number of possible choices for the equilibrium distribution function – *any* other function of the constants of the motion would also be valid. In fact, functions differing from Eq. (8.12) will later be used to model drifting plasmas and plasmas with density gradients.

Substitution of Eq. (8.12) into the orbit integral Eq. (8.11) gives

$$\begin{aligned} f_{\sigma 1}(\mathbf{x}, \mathbf{v}, t) = & -\frac{2 n_{\sigma 0} q_\sigma \tilde{\phi}_1}{\pi^{3/2} m_\sigma v_{T\sigma}^3} \exp\left[-v^2/v_{T\sigma}^2\right] \\ & \times \int_{-\infty}^{t} \mathrm{d}t' \left\{ \frac{\mathrm{i}\mathbf{k} \cdot \mathbf{v}(t')}{v_{T\sigma}^2} \exp\left[\mathrm{i}\mathbf{k} \cdot \mathbf{x}(t') - \mathrm{i}\omega t'\right] \right\}. \end{aligned} \tag{8.14}$$

Since the kinetic energy $mv^2/2$ of the unperturbed orbits is a constant of the motion, the quantity $\exp\left[-v^2/v_{T\sigma}^2\right]$ has been factored out of the time integral. This orbit integral may be simplified by noting

$$\frac{\mathrm{d}}{\mathrm{d}t'} \exp\left[\mathrm{i}\mathbf{k} \cdot \mathbf{x}(t')\right] = \mathrm{i}\mathbf{k} \cdot \mathbf{v}(t') \exp\left[\mathrm{i}\mathbf{k} \cdot \mathbf{x}(t')\right] \tag{8.15}$$

so Eq. (8.14) becomes

$$f_{\sigma 1}(\mathbf{x}, \mathbf{v}, t) = -\frac{q_\sigma}{\kappa T_\sigma}\tilde{\phi}_1 f_{\sigma 0}\int_{-\infty}^{t} dt' \left\{\exp\left[-i\omega t'\right]\frac{d}{dt'}\exp\left[i\mathbf{k}\cdot\mathbf{x}(t')\right]\right\}$$
$$= -\frac{q_\sigma}{\kappa T_\sigma}\tilde{\phi}_1 f_{\sigma 0}\left\{\left[\exp\left(i\mathbf{k}\cdot\mathbf{x}(t') - i\omega t'\right)\right]_{-\infty}^{t} + i\omega I_{phase}(\mathbf{x}, t)\right\}, \quad (8.16)$$

where the *phase-history integral* I_{phase} is defined as

$$I_{phase}(\mathbf{x}, t) = \int_{-\infty}^{t} dt' \exp\left(i\mathbf{k}\cdot\mathbf{x}(t') - i\omega t'\right).$$

Evaluation of I_{phase} requires knowledge of the unperturbed orbit trajectory $\mathbf{x}(t')$. This trajectory, determined by solving Eqs. (8.9) with boundary conditions specified by Eq. (8.7), has the velocity time-dependence

$$\mathbf{v}(t') = v_\parallel \hat{B} + \mathbf{v}_\perp \cos\left[\omega_{c\sigma}(t'-t)\right] - \hat{B}\times\mathbf{v}_\perp \sin\left[\omega_{c\sigma}(t'-t)\right]. \quad (8.17)$$

Equation (8.17) satisfies both the dynamics and the boundary condition $\mathbf{v}(t) = \mathbf{v}$ and so gives the correct helical "unwinding" into the past for a particle at its present position in phase-space. The position trajectory, found by integrating Eq. (8.17), is

$$\mathbf{x}(t') = \mathbf{x} + v_\parallel\left(t'-t\right)\hat{B}$$
$$+ \frac{1}{\omega_{c\sigma}}\left\{\mathbf{v}_\perp \sin\left[\omega_{c\sigma}(t'-t)\right] + \hat{B}\times\mathbf{v}_\perp\left(\cos\left[\omega_{c\sigma}(t'-t)\right] - 1\right)\right\}, \quad (8.18)$$

which satisfies the related boundary condition $\mathbf{x}(t) = \mathbf{x}$.

To proceed further, we define φ to be the velocity-space angle between the fixed quantity $\mathbf{k}_\perp$ and the dummy variable $\mathbf{v}_\perp$ so that $\mathbf{k}_\perp\cdot\mathbf{v}_\perp = k_\perp v_\perp \cos\varphi$, and $\mathbf{k}_\perp\cdot\hat{B}\times\mathrm{v}_\perp = k_\perp v_\perp \sin\varphi$. Using this definition, the time history of the spatial part of the phase can be written as

$$\mathbf{k}\cdot\mathbf{x}(t') = \mathbf{k}\cdot\mathbf{x} + k_\parallel v_\parallel\left(t'-t\right) + \frac{k_\perp v_\perp}{\omega_{c\sigma}}\left\{\sin\left[\omega_{c\sigma}(t'-t) + \varphi\right] - \sin\varphi\right\}. \quad (8.19)$$

The phase integral can now be expanded

$$I_{phase}(\mathbf{x},t) = e^{-i\omega t}\int_{-\infty}^{t} dt' e^{i\mathbf{k}\cdot\mathbf{x}(t') - i\omega(t'-t)}$$
$$= e^{i\mathbf{k}\cdot\mathbf{x} - i\omega t}\int_{-\infty}^{0} d\tau \exp\left\{i\left[\begin{array}{c}\left(k_\parallel v_\parallel - \omega\right)\tau + \\ \dfrac{k_\perp v_\perp}{\omega_{c\sigma}}\left\{\sin\left[\omega_{c\sigma}\tau + \varphi\right] - \sin\varphi\right\}\end{array}\right]\right\}, \quad (8.20)$$

where $\tau = t' - t$. Certain Bessel function relations are now of use. The first of these is the integral representation of the J_n Bessel function, namely

$$J_n(z) = \frac{1}{2\pi}\int_0^{2\pi} \mathrm{e}^{\mathrm{i}z\sin\theta - \mathrm{i}n\theta}\mathrm{d}\theta. \tag{8.21}$$

The inverse of this relation is

$$\mathrm{e}^{\mathrm{i}z\sin\theta} = \sum_{n=-\infty}^{\infty} J_n(z)\mathrm{e}^{\mathrm{i}n\theta}, \tag{8.22}$$

which may be validated by taking the θ Fourier transform of both sides over the interval from 0 to 2π.

The phase integral can be evaluated using Eq. (8.22) in Eq. (8.20) to obtain

$$\begin{aligned} I_{phase}(\mathbf{x}, t) &= \mathrm{e}^{\mathrm{i}\mathbf{k}\cdot\mathbf{x} - \mathrm{i}\omega t}\sum_{n=-\infty}^{\infty} J_n\left(\frac{k_\perp v_\perp}{\omega_{c\sigma}}\right)\mathrm{e}^{-\frac{\mathrm{i}k_\perp v_\perp \sin\varphi}{\omega_{c\sigma}}}\int_{-\infty}^{0}\mathrm{d}\tau \mathrm{e}^{\mathrm{i}\left(k_\parallel v_\parallel - \omega\right)\tau + \mathrm{i}n(\omega_{c\sigma}\tau + \varphi)} \\ &= \mathrm{e}^{\mathrm{i}\mathbf{k}\cdot\mathbf{x} - \mathrm{i}\omega t}\mathrm{e}^{-\frac{\mathrm{i}k_\perp v_\perp \sin\varphi}{\omega_{c\sigma}}}\sum_{n=-\infty}^{\infty} J_n\left(\frac{k_\perp v_\perp}{\omega_{c\sigma}}\right)\frac{\left[\mathrm{e}^{\mathrm{i}\left(k_\parallel v_\parallel - \omega\right)\tau + \mathrm{i}n(\omega_{c\sigma}\tau + \varphi)}\right]_{-\infty}^{0}}{-\mathrm{i}\left(\omega - k_\parallel v_\parallel - n\omega_{c\sigma}\right)}. \end{aligned} \tag{8.23}$$

In Eqs. (8.16) and (8.23) there is a lower limit at $t = -\infty$; this limit corresponds to the phase in the distant past and is essentially the information regarding the initial condition of the system in the distant past. We saw in our previous discussion of the Landau problem that initial value problems ought to be analyzed using Laplace transforms, not Fourier transforms. However, if we ignore the initial conditions and use the Plemelj formula with Fourier transforms, the same result as the Laplace method is obtained. This shortcut procedure will now be followed and so any reference to the initial conditions will be dropped and Fourier transforms will be used with invocation of the Plemelj formula whenever it is necessary to resolve any singularities in the integrations. Hence, terms evaluated at $t = -\infty$ in Eqs. (8.16) and (8.23) are dropped since these are initial values. After making these simplifications, the perturbed distribution function becomes

$$f_{\sigma 1}(\mathbf{x}, \mathbf{v}, t) = -\frac{q_\sigma \tilde{\phi}_1 f_{\sigma 0}\mathrm{e}^{\mathrm{i}\mathbf{k}\cdot\mathbf{x} - \mathrm{i}\omega t}}{\kappa T_\sigma}\left\{1 - \mathrm{e}^{-\frac{\mathrm{i}k_\perp v_\perp \sin\varphi}{\omega_{c\sigma}}}\sum_n \frac{J_n(\frac{k_\perp v_\perp}{\omega_{c\sigma}})\omega \mathrm{e}^{\mathrm{i}n\varphi}}{\left(\omega - k_\parallel v_\parallel - n\omega_{c\sigma}\right)}\right\}. \tag{8.24}$$

The next step is to evaluate the density perturbation Eq. (8.5), an operation that involves integrating the perturbed distribution function over velocity; the velocity integrals are evaluated using the Bessel identity (Watson 1922)

$$\int_0^\infty z J_n^2(\beta z) \mathrm{e}^{-\alpha^2 z^2} \mathrm{d}z = \frac{1}{2\alpha^2} \mathrm{e}^{-\beta^2/2\alpha^2} I_n\left(\frac{\beta^2}{2\alpha^2}\right). \tag{8.25}$$

Using this identity, substitution of Eq. (8.24) into Eq. (8.5) gives

$$\begin{aligned} n_{\sigma 1} &= -\frac{q_\sigma}{\kappa T_\sigma}\phi_1 n_{\sigma 0}\left[1 - \frac{\omega}{\pi^{3/2} v_{T\sigma}^3}\sum_n \int_{-\infty}^{\infty} \mathrm{d}v_\parallel \int_0^{2\pi} \mathrm{d}\varphi \int_0^\infty v_\perp \mathrm{d}v_\perp \frac{\mathrm{e}^{-\mathrm{i}\frac{k_\perp v_\perp}{\omega_{c\sigma}}\sin\varphi + \mathrm{i}n\varphi} J_n(\frac{k_\perp v_\perp}{\omega_{c\sigma}}) \mathrm{e}^{-\left(v_\parallel^2 + v_\perp^2\right)/v_{T\sigma}^2}}{\left(\omega - k_\parallel v_\parallel - n\omega_{c\sigma}\right)}\right] \\ &= -\frac{q_\sigma \phi_1 n_{\sigma 0}}{\kappa T_\sigma}\left[1 - \frac{2\omega}{\sqrt{\pi} v_{T\sigma}^3}\sum_n \int_{-\infty}^{\infty} \mathrm{d}v_\parallel \int_0^\infty v_\perp \mathrm{d}v_\perp \frac{J_n^2(\frac{k_\perp v_\perp}{\omega_{c\sigma}}) \mathrm{e}^{-\frac{v_\parallel^2 + v_\perp^2}{v_{T\sigma}^2}}}{\left(\omega - k_\parallel v_\parallel - n\omega_{c\sigma}\right)}\right] \\ &= -\frac{q_\sigma}{\kappa T_\sigma}\phi_1 n_{\sigma 0}\left[1 + \alpha_{0\sigma} \mathrm{e}^{-k_\perp^2 r_{L\sigma}^2} \sum_n \frac{I_n\left(k_\perp^2 r_{L\sigma}^2\right)}{\pi^{1/2}} \int_{-\infty}^{\infty} \mathrm{d}\xi \frac{\mathrm{e}^{-\xi^2}}{\xi - \alpha_{n\sigma}}\right] \\ &= -\frac{q_\sigma}{\kappa T_\sigma}\phi_1 n_{\sigma 0}\left[1 + \alpha_{0\sigma} \mathrm{e}^{-k_\perp^2 r_{L\sigma}^2} \sum_n I_n\left(k_\perp^2 r_{L\sigma}^2\right) Z(\alpha_{n\sigma})\right], \end{aligned} \tag{8.26}$$

where

$$\alpha_{n\sigma} = \frac{\omega - n\omega_{c\sigma}}{k_\parallel v_{T\sigma}}, \tag{8.27}$$

the Larmor radius is defined to be

$$r_{L\sigma} = \frac{\sqrt{\kappa T_\sigma / m_\sigma}}{\omega_{c\sigma}}, \tag{8.28}$$

and Z is the plasma dispersion function defined in Eq. (5.70). Finally, Eq. (8.26) is substituted into Eq. (8.4) to obtain the warm magnetized plasma electrostatic dispersion relation

$$\mathcal{D}(\omega, k) = 1 + \sum_\sigma \frac{1}{k^2 \lambda_{D\sigma}^2}\left[1 + \alpha_{0\sigma} \mathrm{e}^{-k_\perp^2 r_{L\sigma}^2} \sum_{n=-\infty}^{\infty} I_n\left(k_\perp^2 r_{L\sigma}^2\right) Z(\alpha_{n\sigma})\right] = 0. \tag{8.29}$$

Note that $\mathcal{D}(\omega, k)$ refers to the dispersion relation and should not be confused with D, the off-diagonal term of the cold plasma dielectric tensor, nor with $\mathbf{D}$, the displacement vector. A similar, but more involved calculation using unperturbed orbit phase integrals for the perturbed current density gives the hot plasma version of the full electromagnetic dispersion, i.e., the finite temperature generalization of the cold plasma dielectric tensor $\overleftrightarrow{\mathbf{K}}$. Although the calculation is essentially similar, it is considerably more tedious, and the interested reader is referred to specialized texts on plasma waves such as those by Stix (1992) or Swanson (2003).

8.2 Analysis of the warm plasma electrostatic dispersion relation

Equation (8.29) generalizes the unmagnetized warm plasma electrostatic dispersion relation Eq. (5.55) to the situation of a magnetized warm plasma. Thus, Eq. (8.29) ought to revert to Eq. (5.55) in the $B \to 0$ limit. The Bessel identity

$$\sum_{n=-\infty}^{\infty} I_n(\lambda) = \mathrm{e}^{\lambda} \tag{8.30}$$

together with the condition $\alpha_n \to \alpha_0$ if $B \to 0$ show that this is indeed the case.

Furthermore, Eq. (8.29) should also be the warm plasma generalization of the cold, magnetized plasma, electrostatic dispersion

$$k_x^2 S + k_z^2 P = 0; \tag{8.31}$$

demonstration of this correspondence will be presented later. Equation (8.30) can be used to recast Eq. (8.29) as

$$\mathcal{D}(\omega, k) = 1 + \sum_\sigma \frac{\mathrm{e}^{-k_\perp^2 r_{L\sigma}^2}}{k^2 \lambda_{D\sigma}^2} \sum_{n=-\infty}^{\infty} I_n\left(k_\perp^2 r_{L\sigma}^2\right)\left[1 + \alpha_{0\sigma} Z(\alpha_{n\sigma})\right] = 0. \tag{8.32}$$

Using $I_{-n}(z) = I_n(z)$, the summation over n can be rearranged to give

$$\mathcal{D}(\omega, k) = 1 + \sum_\sigma \frac{\mathrm{e}^{-k_\perp^2 r_{L\sigma}^2}}{k^2 \lambda_{D\sigma}^2} \left\{ \begin{array}{l} I_0\left(k_\perp^2 r_{L\sigma}^2\right)\left(1 + \alpha_{0\sigma} Z(\alpha_{0\sigma})\right) + \\ \sum\limits_{n=1}^{\infty} I_n\left(k_\perp^2 r_{L\sigma}^2\right)\left[2 + \alpha_{0\sigma}\left\{Z(\alpha_{n\sigma}) + Z(\alpha_{-n\sigma})\right\}\right] \end{array} \right\}. \tag{8.33}$$

In order to obtain the lowest order thermal correction, it is assumed $\alpha_0 = \omega/k_\parallel v_{T\sigma} \gg 1$, in which case the large argument expansion Eq. (5.77) can be used to evaluate $Z(\alpha_0)$. Invoking this expansion, and keeping only lowest order terms, shows that

$$\begin{aligned} 1 + \alpha_0 Z(\alpha_0) &= 1 + \alpha_0 \left\{ -\frac{1}{\alpha_0}\left[1 + \frac{1}{2\alpha_0^2} + \ldots\right] + \mathrm{i}\pi^{1/2}\exp\left(-\alpha_0^2\right) \right\} \\ &= -\frac{1}{2\alpha_0^2} + \mathrm{i}\alpha_0 \pi^{1/2} \exp\left(-\alpha_0^2\right). \end{aligned} \tag{8.34}$$

It is additionally assumed that $|\alpha_n| = |(\omega - n\omega_{c\sigma})/k_\parallel v_{T\sigma}| \gg 1$ for $n \neq 0$; this corresponds to assuming that the wave frequency is not too close to a cyclotron resonance. Because of this assumption, the large argument expansion of the plasma dispersion function is also appropriate for the $n \neq 0$ terms, and so we can write

$$\begin{aligned} 2+\alpha_0\left[Z(\alpha_n)+Z(\alpha_{-n})\right] &= 2+\alpha_0\left[-\frac{1}{\alpha_n}-\frac{1}{\alpha_{-n}}+\mathrm{i}\sqrt{\pi}\left(\mathrm{e}^{-\alpha_n^2}+\mathrm{e}^{-\alpha_{-n}^2}\right)\right] \\ &= 2-\frac{\omega}{(\omega-n\omega_{c\sigma})}-\frac{\omega}{(\omega+n\omega_{c\sigma})}+\mathrm{i}\alpha_0\sqrt{\pi}\left(\mathrm{e}^{-\alpha_n^2}+\mathrm{e}^{-\alpha_{-n}^2}\right) \\ &= -\frac{2n^2\omega_{c\sigma}^2}{(\omega^2-n^2\omega_{c\sigma}^2)}+\mathrm{i}\alpha_0\sqrt{\pi}\left(\mathrm{e}^{-\alpha_n^2}+\mathrm{e}^{-\alpha_{-n}^2}\right), \end{aligned} \tag{8.35}$$

where the subscript σ has been omitted from the α_n to keep the algebra uncluttered. Substitution of these expansions back into the dispersion relation gives

$$\mathcal{D}(\omega,k) = 1+\sum_\sigma \frac{\mathrm{e}^{-k_\perp^2 r_{L\sigma}^2}}{k^2\lambda_{D\sigma}^2}\left\{\begin{array}{l} I_0\left(k_\perp^2 r_{L\sigma}^2\right)\left(-\frac{1}{2\alpha_{0\sigma}^2}+\mathrm{i}\alpha_{0\sigma}\sqrt{\pi}\exp\left(-\alpha_{0\sigma}^2\right)\right)+ \\ \sum_{n=1}^{\infty} I_n\left(k_\perp^2 r_{L\sigma}^2\right)\left(\begin{array}{c} -\frac{2n^2\omega_{c\sigma}^2}{\omega^2-n^2\omega_{c\sigma}^2} \\ +\mathrm{i}\alpha_{0\sigma}\sqrt{\pi}\left(\begin{array}{c}\mathrm{e}^{-\alpha_n^2} \\ +\mathrm{e}^{-\alpha_{-n}^2}\end{array}\right)\end{array}\right)\end{array}\right\} = 0. \tag{8.36}$$

Equation (8.36) generalizes the magnetized cold plasma electrostatic dispersion to finite temperature and shows how Landau damping appears both at the wave frequency ω and also at cyclotron harmonics, i.e., in the vicinity of $n\omega_{c\sigma}$. Two important limits of Eq. (8.36) are discussed in the following sections.

8.3 Bernstein waves

Suppose the wave phase is uniform in the direction along the magnetic field. Such a situation would occur if the antenna exciting the wave were an infinitely long wire aligned parallel to a magnetic field line (in reality, the antenna would have to be sufficiently long to behave as if infinite). This situation corresponds to having $k_\parallel \to 0$, in which case the Landau damping terms and the $1/2\alpha_{0\sigma}^2$ term vanish. The dispersion relation Eq. (8.36) consequently reduces to

$$1 = \sum_\sigma \frac{e^{-\lambda_\sigma}}{\lambda_\sigma} \sum_{n=1}^{\infty} \frac{2n^2\omega_{p\sigma}^2}{\omega^2 - n^2\omega_{c\sigma}^2} I_n(\lambda_\sigma), \tag{8.37}$$

where

$$\lambda_\sigma = k_\perp^2 r_{L\sigma}^2 \tag{8.38}$$

and

$$r_{L\sigma}^2 = \frac{\omega_{p\sigma}^2 \lambda_{D\sigma}^2}{\omega_{c\sigma}^2} = \frac{\kappa T_\sigma}{m_\sigma \omega_{c\sigma}^2} \tag{8.39}$$

is the Larmor radius. The waves resulting from this dispersion were first derived by Bernstein (1958) and are called Bernstein or cyclotron harmonic waves; they are also sometimes referred to as hot plasma waves because their existence depends on the plasma having a finite temperature. Bernstein waves have been observed in both laboratory experiments and in spacecraft measurements of magnetospheric plasmas. Early measurements of electron Bernstein waves in laboratory experiments were reported by Crawford (1965) and by Leuterer (1969); measurements of ion Bernstein waves were reported by Schmitt (1973). An example of a spacecraft measurement was reported by Moncuquet, Meyervernet, and Hoang (1995) who used a fit of data to the Bernstein dispersion relation to infer the electron temperature in the magnetized plasma of Jupiter's moon, Io. Electron Bernstein waves involve ω being in the vicinity of an electron cyclotron harmonic, i.e., $\omega^2 \sim \mathcal{O}(n^2\omega_{ce}^2)$, whereas for ion Bernstein waves, ω is in the vicinity of an ion cyclotron harmonic.

The Bernstein wave dispersion relation, Eq. (8.37), provides a transcendental relation between ω and $k_\perp$. For any given $k_\perp$ the dispersion relation has an infinite number of roots ω, each associated with a different cyclotron harmonic. This is because as ω is increased from small values to infinity, the $\left(\omega^2 - n^2\omega_{c\sigma}^2\right)^{-1}$ term on the right-hand side of Eq. (8.37) will cause the entire right-hand side to oscillate between minus and plus infinity at each successive cyclotron harmonic. In particular, the right-hand side will assume a value of $+\infty$ when ω^2 is infinitesimally larger than $n^2\omega_{c\sigma}^2$ and a value of $-\infty$ when ω^2 is infinitesimally smaller than $n^2\omega_{c\sigma}^2$. Thus, as ω increases from the nth cyclotron harmonic to the $(n+1)$th cyclotron harmonic, the right-hand side of Eq. (8.37) will take on all values between minus and plus infinity and so will always equal the left-hand side at some value of ω in this range.

We now consider electron Bernstein waves and so drop the ion terms from the summation (the analysis for ion Bernstein waves is similar). The subscript σ

will be dropped and it will be understood that all quantities refer to electrons. On keeping the electron terms only, Eq. (8.37) reduces to

$$1 = 2\frac{e^{-\lambda}}{\lambda}\sum_{n=1}^{\infty}\frac{n^2\omega_p^2}{\omega^2 - n^2\omega_c^2}I_n(\lambda), \tag{8.40}$$

which has the following limiting forms depending on the ratio of ω_p^2 to ω_c^2:

1. $\omega_p^2 \ll \omega_c^2$ case:

   Here each $2n^2\omega_p^2/(\omega^2 - n^2\omega_c^2)$ factor is negligible compared to unity except when $\omega^2 \sim \mathcal{O}(n^2\omega_{ce}^2)$. Thus, for a given ω only one term in the summation is near resonance and the dispersion relation is satisfied by this one term on the right-hand side of Eq. (8.40) balancing the left-hand side. This results in an infinite set of modes, each slightly above a cyclotron harmonic, i.e.,

   $$\omega^2 = n^2\omega_c^2 + 2\frac{e^{-\lambda}}{\lambda}n^2\omega_p^2 I_n(\lambda), \quad \text{for } n = 1, 2, ..., \infty. \tag{8.41}$$

   The small and large λ limits of this expression are determined using the asymptotic values of the I_n Bessel function, namely for $n \geq 1$

   $$\lim_{\lambda \ll 1} I_n(\lambda) = \frac{1}{n!}\left(\frac{\lambda}{2}\right)^n$$
   $$\lim_{\lambda \gg 1} I_n(\lambda) = \frac{1}{\sqrt{2\pi\lambda}}e^{\lambda} \tag{8.42}$$

   to give

   $$\omega^2 = n^2\omega_c^2 + (1-\lambda)\,n^2\omega_p^2\frac{1}{n!}\left(\frac{\lambda}{2}\right)^{n-1} \quad \text{for } \lambda \ll 1,$$
   $$\omega^2 = n^2\omega_c^2 \text{ for } \lambda \gg 1. \tag{8.43}$$

   The $n = 1$ mode differs slightly from the $n \geq 2$ modes because the $n = 1$ mode has the small λ dispersion

   $$\omega^2 = \omega_{uh}^2 - \lambda\omega_p^2 \tag{8.44}$$

   showing this mode is the warm plasma generalization of the upper hybrid resonance. The frequency of the $n \geq 2$ modes starts at $\omega^2 = n^2\omega_c^2$ when $\lambda = 0$, takes on a maximum value at some finite λ, and then reverts to $\omega^2 = n^2\omega_c^2$ as $\lambda \to \infty$. To the left of the frequency maximum the group velocity $\partial\omega/\partial k_\perp$ is positive and to the right of the maximum the group velocity is negative. This system of modes is sketched in Fig. 8.1(a).

2. $\omega_p^2 \gg \omega_c^2$ case:

   For large λ, the product of the exponential factor and the modified Bessel function in Eq. (8.40) is unity, leaving a $\lambda^{3/2}$ factor in the denominator, so again the dispersion reduces to cyclotron harmonics, $\omega^2 = n^2\omega_c^2$ as $\lambda \to \infty$. For small λ, the situation is

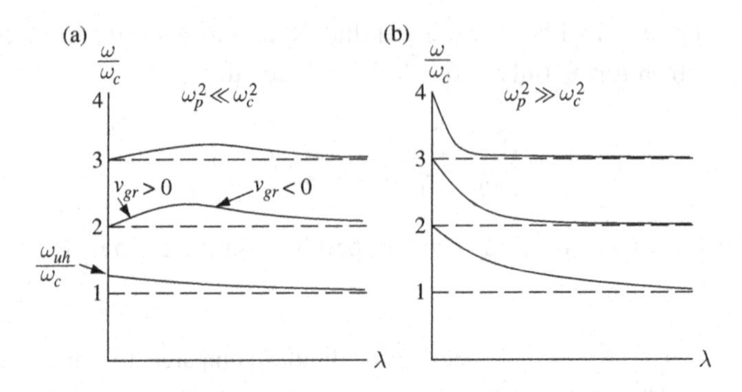

Fig. 8.1 Electron Bernstein wave dispersion relations.

more involved, because the $n = 1$ term is independent of λ to lowest order and so the $n = 1$ term must always be retained when approximating Eq. (8.40). Let us suppose some other term, say the $(n+1)$th term, is near resonance, i.e., $\omega^2 \sim (n+1)^2\omega_c^2$. Then, keeping the left-hand side, the $n = 1$ term, and the $(n+1)$th term, expansion of Eq. (8.40) results in

$$1 \simeq \frac{\omega_p^2}{\left[(n+1)^2 - 1\right]\omega_c^2} + \frac{(n+1)\,\omega_p^2}{\left[\omega^2 - (n+1)^2\,\omega_c^2\right]n!}\left(\frac{\lambda}{2}\right)^n, \tag{8.45}$$

which may be solved for ω to give

$$\omega = (n+1)\,\omega_c\left\{1 - \frac{\omega_p^2\left[(n+1)^2 - 1\right]}{\left(\omega_p^2 - \left[(n+1)^2 - 1\right]\omega_c^2\right)}\frac{\lambda^n}{2^{n+1}\,(n+1)!}\right\}. \tag{8.46}$$

Thus, the nth mode starts at $\omega \simeq (n+1)\,\omega_c$ for small λ and then the frequency decreases towards the asymptotic limit $n\omega_c$ as $\lambda \to \infty$. This dispersion is sketched in Fig. 8.1(b).

Excitation of Bernstein waves requires setting $k_\parallel = 0$, which turns out to be a quite stringent requirement. The antenna must be absolutely uniform in the direction along the magnetic field because otherwise a finite $k_\parallel$ will be excited, which would result in finite instead of infinite $\omega/k_\parallel$ and so cause the waves to be subject to strong Landau damping.

8.4 Finite $k_\parallel$ dispersion: linear mode conversion

An alternate limit for Eq. (8.36) would be to allow $k_\parallel$ to be finite but also have $k_\perp^2 r_L^2 \ll 1$ so only the lowest-order finite Larmor radius terms are retained. To keep matters simple, and also to relate to cold plasma theory, $k_\parallel$ will be assumed to be sufficiently small to make the wave-phase velocity much faster than the particle thermal velocities, i.e., $\omega/k_\parallel \gg v_{T_e}, v_{Ti}$. Since $r_{L\sigma}^2/\lambda_{D\sigma}^2 = \omega_{p\sigma}^2/\omega_{c\sigma}^2$, the lowest-order perpendicular thermal terms will be $\mathcal{O}(k_\perp^4)$ and so perpendicular quantities up to fourth order must be retained. This means that both the $n=1$ and the $n=2$ terms must be retained in the summation over n. With these approximations, and using $1/2\lambda_{D\sigma}^2\alpha_0^2 = k_\parallel^2\omega_{p\sigma}^2/\omega^2$, Eq. (8.36) becomes

$$k_\parallel^2 + k_\perp^2 + \sum_\sigma (1 - k_\perp^2 r_{L\sigma}^2) \left(\begin{array}{c} \left(1 + \dfrac{k_\perp^4 r_{L\sigma}^4}{4}\right)\left(-\dfrac{\omega_{p\sigma}^2}{\omega^2}k_\parallel^2 + \mathrm{i}\dfrac{\alpha_0\sqrt{\pi}}{\lambda_{D\sigma}^2}\mathrm{e}^{-\alpha_0^2}\right) \\ + \dfrac{k_\perp^2\omega_{p\sigma}^2}{2\omega_{c\sigma}^2}\left[\begin{array}{c} -\dfrac{2\omega_{c\sigma}^2}{(\omega^2-\omega_{c\sigma}^2)} \\ +\mathrm{i}\alpha_0\sqrt{\pi}\left(\begin{array}{c}\mathrm{e}^{-\alpha_1^2} \\ +\mathrm{e}^{-\alpha_{-1}^2}\end{array}\right)\end{array}\right] \\ + \dfrac{k_\perp^4 r_{L\sigma}^2\omega_{p\sigma}^2}{8\omega_{c\sigma}^2}\left[\begin{array}{c} -\dfrac{8\omega_{c\sigma}^2}{(\omega^2-4\omega_{c\sigma}^2)} \\ +\mathrm{i}\alpha_0\sqrt{\pi}\left(\begin{array}{c}\mathrm{e}^{-\alpha_2^2} \\ +\mathrm{e}^{-\alpha_{-2}^2}\end{array}\right)\end{array}\right] \end{array} \right) = 0. \tag{8.47}$$

The Landau damping terms will be assumed to be negligible to keep matters simple and the equation will now be grouped according to powers of $k_\perp^2$. Retaining only the lowest-order, finite-temperature perpendicular terms gives

$$k_\parallel^2\left(1 - \sum_\sigma \frac{\omega_{p\sigma}^2}{\omega^2}\right) + k_\perp^2\left(1 - \sum_\sigma \frac{\omega_{p\sigma}^2}{(\omega^2-\omega_{c\sigma}^2)}\right) - k_\perp^4 \sum_\sigma \left(\frac{3\omega_{p\sigma}^4\lambda_{D\sigma}^2}{(\omega^2-\omega_{c\sigma}^2)(\omega^2-4\omega_{c\sigma}^2)}\right) = 0. \tag{8.48}$$

This is of the form

$$-k_\perp^4\epsilon_{th} + k_\perp^2 S + k_\parallel^2 P = 0, \tag{8.49}$$

where the perpendicular fourth-order thermal coefficient is

$$\epsilon_{th} = \sum_\sigma \left(\frac{3\omega_{p\sigma}^4\lambda_{D\sigma}^2}{(\omega^2-\omega_{c\sigma}^2)(\omega^2-4\omega_{c\sigma}^2)}\right). \tag{8.50}$$

Equation (8.49) is a quadratic equation in $k_\perp^2$. The cold plasma model used earlier in effect set $\epsilon_{th} = 0$, in which case a wave propagating through an inhomogeneous plasma towards an $S=0$ hybrid resonance would have $k_\perp^2 \to \infty$. This non-physical prediction is resolved by the warm plasma theory because ϵ_{th}, while small, is finite and prevents $k_\perp^2 \to \infty$ from occurring. What happens instead is that Eq. (8.49) has two qualitatively distinct roots, namely

$$k_\perp^2 = \frac{S \pm \sqrt{S^2 + 4\epsilon_{th} k_\parallel^2 P}}{2\epsilon_{th}}. \tag{8.51}$$

At locations far from the hybrid resonance, S is large, in which case the two modes are well separated and given by

$$\begin{aligned} k_\perp^2 &= -\frac{P}{S} k_\parallel^2, \quad \text{cold plasma wave,} \\ k_\perp^2 &= \frac{S}{\epsilon_{th}}, \quad \text{hot plasma wave,} \end{aligned} \tag{8.52}$$

but at locations near the hybrid resonance S approaches zero and the two modes coalesce. The hot plasma wave results from balancing the first and middle terms in Eq. (8.49) while the cold plasma wave results from balancing the middle and last terms. The actual mode coalescence occurs where the square root term in Eq. (8.51) vanishes, i.e., where

$$S^2 = -4\epsilon_{th} k_\parallel^2 P. \tag{8.53}$$

At the location of mode coalescence $\mathrm{d}k_\perp/\mathrm{d}x$ is infinite, violating the $\mathrm{d}k_\perp/\mathrm{d}x \ll k_x^2$ requirement for the WKB approximation to be valid. Thus, a more sophisticated analysis than WKB is required. Such an analysis is presented in the next section where hot and cold waves are shown to become strongly coupled in the coalescence region. This strong coupling between the two modes should not be surprising because the two modes are essentially indistinguishable at this location. It will be demonstrated in particular that a cold wave propagating towards the hybrid resonance (e.g., see Fig. 8.2) can be linearly converted into a hot wave, which then propagates back outwards from the resonance.

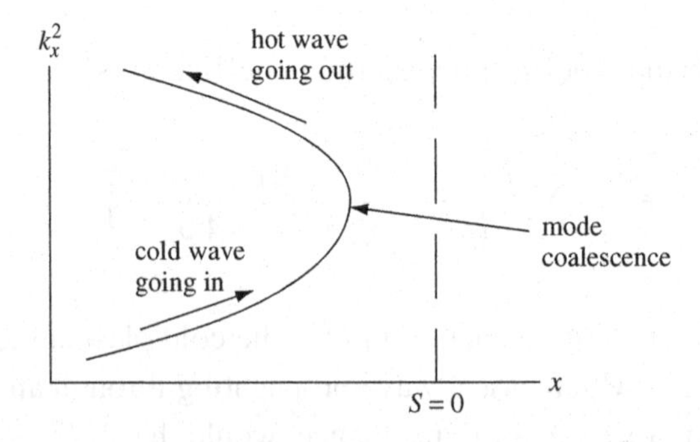

Fig. 8.2 Linear mode conversion of cold mode into hot mode near hybrid resonance.

8.5 Analysis of linear mode conversion

8.5.1 Airy equation

The procedure for analyzing linear mode conversion was developed by Stix (1965) and is an extension of the standard method for connecting solutions of the Airy equation

$$y'' + xy = 0 \tag{8.54}$$

from the $x < 0$ region to the $x > 0$ region. The Airy problem involves a second-order ordinary differential equation with a term containing a factor that vanishes at some location, namely the term xy vanishes at $x = 0$. The mode conversion problem involves a fourth-order equation, which also contains a term with a factor that vanishes at some location, namely S vanishes at a hybrid resonance. The Airy connection method will first be examined in order to introduce the relevant concepts and then these concepts will be applied to the actual mode conversion problem.

The sequence of steps for developing the solution to the Airy problem are as follows:

1. Laplace transform the equation. Equation (8.54) cannot be Fourier transformed because of the non-constant coefficient x in the second term; however, it can be Laplace transformed by taking advantage of the relation that the Laplace transform of $xy(x)$ is $-\mathrm{d}\tilde{y}(p)/\mathrm{d}p$. Thus, the Laplace transform of Eq. (8.54) is

$$p^2\tilde{y}(p) - \frac{\mathrm{d}\tilde{y}(p)}{\mathrm{d}p} = 0. \tag{8.55}$$

Strictly speaking, Laplace transforming y'' will also introduce terms involving initial conditions, but these additional terms will be ignored for now and this issue will be addressed later. Equation (8.55) is easily solved to give

$$\tilde{y}(p) = A\exp\left(p^3/3\right), \tag{8.56}$$

where A is a constant. The formal solution to Eq. (8.54) is the inverse transform

$$y(x) = (2\pi\mathrm{i})^{-1}\int_C \tilde{y}(p)\mathrm{e}^{px}\mathrm{d}p = a\int_C \mathrm{e}^{f(p)}\mathrm{d}p, \tag{8.57}$$

where $a = A/2\pi\mathrm{i}$, $f(p) = p^3/3 + px$, and C is some contour in the complex plane. By appropriate choice of this contour Eq. (8.57) is shown to be a valid solution to the original differential equation. In particular, if the endpoints p_1 and p_2 of the contour C are chosen to satisfy

$$\left[\mathrm{e}^{f(p)}\right]_{p_1}^{p_2} = 0, \tag{8.58}$$

then Eq. (8.57) turns out to be a valid solution to Eq. (8.54). The reason for this may be seen by explicitly calculating y'' and xy using Eq. (8.57) to obtain

$$y'' = \frac{\mathrm{d}^2}{\mathrm{d}x^2} a \int_C \mathrm{e}^{p^3/3+px} \mathrm{d}p = a \int_C p^2 \mathrm{e}^{p^3/3+px} \mathrm{d}p \tag{8.59}$$

and

$$\begin{aligned} xy &= a \int_C \mathrm{e}^{p^3/3} \left(\frac{\mathrm{d}}{\mathrm{d}p} \mathrm{e}^{px} \right) \mathrm{d}p \\ &= a \left[\mathrm{e}^{p^3/3+px} \right]_{p_1}^{p_2} - a \int_C \frac{\mathrm{d}}{\mathrm{d}p} \left(\mathrm{e}^{p^3/3} \right) \mathrm{e}^{px} \mathrm{d}p \\ &= -a \int_C p^2 \mathrm{e}^{p^3/3+px} \mathrm{d}p. \end{aligned} \tag{8.60}$$

Addition of these results corresponds to Eq. (8.54) if the end-point boundary conditions specified by Eq. (8.58) are satisfied.

2. Suppose the integrand is *analytic* in some region and consider two different contours in this region having the *same* pair of endpoints. We can take one of these contours and deform it into the other. Thus, the two contours give the same result and so correspond to the same solution.
3. Independence of contours. Since Eq. (8.54) is a second-order ordinary differential equation it must have two linearly independent solutions. This means there must be two linearly independent contours since a choice of contour corresponds to a solution. Linear independence of the contours means one contour cannot be deformed into the other. One possible way for the two contours to be linearly independent is for them to have different pairs of endpoints. In this case, one contour can only be deformed into the other if the endpoints can be moved without changing the integral. If moving the endpoints changes the integral, then the two contours are independent. Another possibility is the situation where the two contours have the same pairs of endpoints, but the integrand is *not analytic* in the region between the two contours. For example, there could be a pole or branch cut between the two contours. Then, one contour could not be deformed into the other because of the non-analytic region separating the two contours. The two contours would then be linearly independent.

8.5.2 Steepest descent contour

The solution to the Airy equation is thus

$$y(x) = \int_C \mathrm{e}^{f(p,x)} \mathrm{d}p, \tag{8.61}$$

where C satisfies the conditions listed above and the solution can be multiplied by an arbitrary constant. In order to evaluate the integral, it is useful to first separate the complex function $f(p, x)$ into its real and imaginary parts,

$$f(p, x) = f_r(p, x) + \mathrm{i} f_i(p, x). \tag{8.62}$$

Also, it should be remembered that, for the purposes of the p integration, x can be considered as a fixed parameter. Thus, when performing the p integration, x need not be written explicitly and so we may simply write $f(p) = f_r(p) + \mathrm{i}f_i(p)$, where $p = u + \mathrm{i}v$ and u, v are the coordinates in the complex p-plane. For an arbitrary contour in the p-plane, both f_r and f_i will vary. The variation of the phase factor f_i means there will be alternating positive and negative contributions to the integral in Eq. (8.61), making evaluation of this integral very delicate. However, evaluation of the integral can be made almost trivial if the contour is deformed to follow a certain optimum path.

To see this, it should first be recalled from the theory of complex variables that any function of a complex variable must satisfy the Cauchy–Riemann conditions; this means f_r and f_i must satisfy

$$\frac{\partial f_r}{\partial u} = \frac{\partial f_i}{\partial v}, \qquad \frac{\partial f_r}{\partial v} = -\frac{\partial f_i}{\partial u}. \tag{8.63}$$

Also, on defining the $u - v$ plane gradient operator $\nabla = \hat{u}\partial/\partial u + \hat{v}\partial/\partial v$ it is seen that

$$\nabla f_r \cdot \nabla f_i = \frac{\partial f_r}{\partial u}\frac{\partial f_i}{\partial u} + \frac{\partial f_r}{\partial v}\frac{\partial f_i}{\partial v} = 0. \tag{8.64}$$

This means that contours of constant f_r are everywhere orthogonal to contours of constant f_i.

Since f_i corresponds to the phase of $\exp(f_r(p) + \mathrm{i}f_i(p))$ and f_r corresponds to the amplitude, Eq. (8.64) shows that a path in the complex p-plane, which is arranged to follow the gradient of f_r, will automatically be a contour of constant f_i, i.e., a contour of constant phase. Thus, the path that is optimum for purposes of evaluation is the path obtained by following ∇f_r because on this path f_i will be constant and so there will not be any alternating positive and negative oscillating contributions to the integral resulting from variation in f_i. In fact, because f_i is constant on this special path, it can be factored from the integral, giving

$$y(x) = \mathrm{e}^{\mathrm{i}f_i} \int_{C\|\nabla f_r} \mathrm{e}^{f_r(p)} \mathrm{d}p. \tag{8.65}$$

Clearly, the maximum contribution to this integral comes from the vicinity of where f_r assumes its maximum value. Since this maximum occurs where $\nabla f_r = 0$, most of the contribution to the integral comes from the vicinity of where $\nabla f_r = 0$. The extrema of f are always saddle points because the Cauchy–Reimann conditions imply $\partial^2 f_{r,i}/\partial u^2 + \partial^2 f_{r,i}/\partial v^2 = 0$. Thus, the vicinity of $\nabla f_r = 0$ must be a saddle point.

The discussion in the previous paragraph implies that once the endpoints of the contour have been chosen, for purposes of evaluation it is advantageous to deform the contour to follow the gradient of f_r; this is called the steepest ascent/descent path (usually just called steepest descent). This optimum choice of path ensures that (i) there is a localized region where f_r assumes a maximum value and (ii) the phase does not vary along the path. The main contribution to the integral is concentrated into a small region of the complex p-plane in the vicinity of where f_r has its maximum value. Simple integration techniques may be used to evaluate the integral in this vicinity, and the contribution from this region dominates other contributions because $\exp(f_r(p))$ is exponentially larger at the maximum of f_r than at other locations. Furthermore, if the contour is chosen so that $f_r \to -\infty$ at the endpoints then Eq. (8.58) will be satisfied and the chosen contour will be a solution of the original differential equation.

The saddle points are located where $f'(p) = 0$. For the Airy function, the function $f(p)$ and its first and second derivatives are

$$\begin{aligned} f(p) &= p^3/3 + px \\ f'(p) &= p^2 + x \\ f''(p) &= 2p, \end{aligned} \tag{8.66}$$

so the saddle points, found by setting $f'(p) = 0$ and denoted by the subscript s, are located at $p_s = \pm(-x)^{1/2}$. In the vicinity of a saddle point, f can be Taylor expanded as

$$f(p) \simeq f(p_s) + \frac{(p - p_s)^2}{2} f''(p_s). \tag{8.67}$$

It is now convenient to define the origin of the coordinate system to be at the saddle point and also use cylindrical coordinates, r, θ, so $p - p_s = r\exp(\mathrm{i}\theta)$ and $\mathrm{d}p = \mathrm{d}r\mathrm{e}^{\mathrm{i}\theta}$ for fixed θ. Also, phasor notation is used for f'' so $f'' = |f''|\exp(\mathrm{i}\psi)$, where ψ is the phase of f''. Thus, in the vicinity of a saddle point we write

$$f(p) \simeq f(p_s) + r^2 \frac{|f''(p_s)|}{2} \exp(2\mathrm{i}\theta + \mathrm{i}\psi). \tag{8.68}$$

Choosing the contour to follow the path of steepest descent corresponds to choosing θ such that $2\theta + \psi = \pm\pi$, in which case

$$f(p) \simeq f(p_s) - r^2 \frac{|f''(p_s)|}{2} \tag{8.69}$$

and so Eq. (8.61) becomes

$$y(x) \simeq \int_{-\infty}^{\infty} \mathrm{e}^{f(p_s) - \frac{1}{2} r^2 |f''(p_s)|} \mathrm{d}r\mathrm{e}^{\mathrm{i}\theta}, \tag{8.70}$$

where the approximation has been made that nearly all the contribution to the integral comes from small r. Inaccuracy of the Taylor expansion at large r is of no consequence, since the contributions to the integral from large r are negligible because of the exponentially decaying behavior of the integrand at large r. Thus, f_r is maximum at the saddle point $r = 0$ and the main contribution to the integral comes from the contour going over the ridge of the saddle. The r integral is a Gaussian integral $\int \exp(-ar^2)\mathrm{d}r = \sqrt{\pi/a}$ so

$$\begin{aligned}
y(x) &\simeq \int_{-\infty}^{\infty} \mathrm{e}^{f(p_s)-r^2|f''(p_s)|/2}\mathrm{d}r\mathrm{e}^{\mathrm{i}\theta} \\
&= \mathrm{e}^{f(p_s)+\mathrm{i}\theta}\int_{-\infty}^{\infty} \mathrm{e}^{-r^2|f''(p_s)|/2}\mathrm{d}r \\
&= \mathrm{e}^{f(p_s)+\mathrm{i}\theta}\sqrt{\frac{2\pi}{|f''(p_s)|}} \\
&= \mathrm{e}^{f(p_s)}\sqrt{\frac{2\pi\mathrm{e}^{\mathrm{i}2\theta}}{|f''(p_s)|}} \\
&= \mathrm{e}^{f(p_s)}\sqrt{\frac{2\pi\mathrm{e}^{\mathrm{i}(\pm\pi-\psi)}}{|f''(p_s)|}} \\
&= \mathrm{e}^{f(p_s)}\sqrt{-\frac{2\pi}{f''(p_s)}}.
\end{aligned} \tag{8.71}$$

Since $f(p_s) = p\left(p^2/3+x\right) = \pm\sqrt{-x}\,(2x/3) = \mp 2(-x)^{3/2}/3$ and $f''(p_s) = \pm 2\sqrt{-x}$, it is seen that

$$\int_{\substack{vicinity\\ of\, saddle}} \mathrm{e}^{f(p)}\,dp \simeq \sqrt{\frac{\mp\pi}{(-x)^{1/2}}}\mathrm{e}^{\mp 2(-x)^{3/2}/3}. \tag{8.72}$$

Providing x is not too close to zero, the two saddle points determined by the $\pm$ signs are well separated and do not perturb each other; the integral will then be a summation over whichever saddle points the contour happens to pass over. The polarity of a saddle point is determined by the sense in which the contour passes through the saddle point.

8.5.3 Relationship between saddle-point solutions and WKB modes

When $|x|$ is large, Eq. (8.54) can be solved approximately using the WKB method, in which case it is assumed that $y(x) = A(x)\exp\left(\mathrm{i}\int^x k(x')\mathrm{d}x'\right)$. The wavenumber $k(x)$ is determined from the dispersion relation associated with the original differential equation and $A(x)$ is a function of $k(x)$. At large $|x|$ the dispersion relation associated with Eq. (8.54) is obtained by assuming $\mathrm{d}/\mathrm{d}x \to \mathrm{i}k$ and has the form

$$k^2 = x \tag{8.73}$$

so $k(x) = \pm x^{1/2}$ and $A(x) \sim (\pm x)^{-1/4}$. The WKB solution is thus

$$y_{WKB} \sim \frac{1}{(\pm x)^{1/4}} e^{\pm i \int^x x'^{1/2} dx'} = \frac{1}{(\pm x)^{1/4}} e^{\pm i2x^{3/2}/3}, \tag{8.74}$$

which is the same as the saddle-point solution. Thus, *each saddle-point solution corresponds to a WKB mode*, and in particular to a propagating wave if $x > 0$ and to an evanescent mode if $x < 0$.

8.5.4 Using boundary conditions to choose contours

The choice of contour is determined by the boundary conditions, which in turn are determined by the physical considerations. In this Airy equation problem, the boundary condition would typically be the physical constraint that no exponentially growing solutions are allowed for $x < 0$. This forces choice of a contour that does *not* pass through the saddle point having the plus sign in Eq. (8.74) when $x < 0$. This condition together with Eq. (8.58) uniquely determines the contour. Choosing a contour C is equivalent to specifying a particular solution to Eq. (8.57); this chosen solution can then be evaluated for $x > 0$. For $x > 0$ the saddle points are at different locations and so the steepest descent path (i.e., the path where evaluation is easiest) will be different, since it will have to pass through the saddle points as they exist for $x > 0$. The sum of the contributions of these saddle points gives the form of the solution for $x > 0$. The important point to realize here is that the *same* contour is used for both $x > 0$ and $x < 0$ because the linearly independent solution is determined by the contour. The contour is determined by its endpoints, but may be deformed provided analyticity is preserved. Thus, the endpoints are the same for both the $x > 0$ and $x < 0$ evaluations, but the deformations of the contour differ in order to pass through the respective $x > 0$ or $x < 0$ saddle points. The sign of x determines the character of the saddle points and so *for purposes of evaluation* the contour is deformed differently for $x > 0$ and $x < 0$.

This completes the discussion of the Airy problem and we now return to the linear mode conversion problem (Stix 1965). A comparison of Eqs. (6.108) and (8.49) shows that if Eq. (8.49) is "un-Fourier" analyzed in the x direction, it becomes

$$\frac{d}{dx}\left(\epsilon_{th}\frac{d^3}{dx^3} + S\frac{d}{dx}\right)\phi - k_\parallel^2 P\phi = 0, \tag{8.75}$$

where it is assumed that ϵ_{th} is approximately constant. To prove this is the correct form, consider the following two statements: (i) if $\epsilon_{th} \to 0$, Eq. (8.75) reverts to Eq. (6.108) (Fourier-analyzed in the z direction, but not the x direction), (ii) if

ϵ_{th} is finite and S and P are assumed uniform, then Fourier analysis in x restores Eq. (8.49).

We now suppose the plasma is non-uniform in a manner such that $S(x) = 0$ at some particular value of x. The x origin is defined to be at this location and so corresponds to the location of the hybrid resonance. In order to be specific, we assume that we are dealing with a mode where $S > 0$ and $P < 0$, since S and P must have opposite signs for the cold plasma wave to propagate. Then, in the vicinity of $x = 0$, Taylor expansion of S gives $S = S'x$ so that Eq. (8.75) becomes

$$\epsilon_{th}\frac{\mathrm{d}^4\phi}{\mathrm{d}x^4} + S'\frac{\mathrm{d}\phi}{\mathrm{d}x} + S'x\frac{\mathrm{d}^2\phi}{\mathrm{d}x^2} - k_\parallel^2 P\phi = 0, \tag{8.76}$$

which has a coefficient that vanishes at $x = 0$ just like the Airy equation. Although Eq. (8.76) could be analyzed as it stands, it is better to tidy its format by changing to a suitably chosen dimensionless coordinate. This is done by first defining

$$\xi = \frac{x}{\lambda}, \tag{8.77}$$

where λ is an as-yet undetermined characteristic length that will be chosen to provide maximum simplification of the coefficients. Replacing x by ξ in Eq. (8.76) gives

$$\frac{\epsilon_{th}}{\lambda^4}\frac{\mathrm{d}^4\phi}{\mathrm{d}\xi^4} + \frac{S'}{\lambda}\frac{\mathrm{d}\phi}{\mathrm{d}\xi} + \frac{S'}{\lambda}\xi\frac{\mathrm{d}^2\phi}{\mathrm{d}\xi^2} - k_\parallel^2 P\phi = 0. \tag{8.78}$$

This becomes

$$\frac{\mathrm{d}^4\phi}{\mathrm{d}\xi^4} + \frac{\mathrm{d}\phi}{\mathrm{d}\xi} + \xi\frac{\mathrm{d}^2\phi}{\mathrm{d}\xi^2} + \mu\phi = 0 \tag{8.79}$$

by choosing the scaling constants λ and μ to satisfy $\epsilon_{th}/S'\lambda^3 = 1$ and $\mu = -k_\parallel^2\lambda P/S'$. These choices imply

$$\lambda = \left(\frac{\epsilon_{th}}{S'}\right)^{1/3}, \quad \mu = -\frac{k_\parallel^2 P\epsilon_{th}^{1/3}}{(S')^{4/3}}. \tag{8.80}$$

Equation (8.79) describes a boundary layer problem having complicated behavior in the inner $|\xi| < 1$ region in the immediate vicinity of the hybrid resonance and, presumably, simple WKB-like behavior in the outer $|\xi| \gg 1$ region far from the hybrid resonance. To solve this problem the technique discussed for the Airy equation will now be applied and generalized:

1. Laplace transform: While Eq. (8.79) cannot be Fourier analyzed in the ξ direction, it can be Laplace transformed giving

$$p^4\tilde{\phi}(p) + p\tilde{\phi}(p) - \frac{\mathrm{d}}{\mathrm{d}p}\left[p^2\tilde{\phi}(p)\right] + \mu\tilde{\phi}(p) = 0 \tag{8.81}$$

or equivalently

$$\frac{1}{\left[p^2\tilde{\phi}(p)\right]}\frac{\mathrm{d}}{\mathrm{d}p}\left[p^2\tilde{\phi}(p)\right] = p^2 + \frac{1}{p} + \frac{\mu}{p^2}. \tag{8.82}$$

This has the solution

$$\tilde{\phi}(p) = A\exp\left(p^3/3 - \mu/p - \ln p\right), \tag{8.83}$$

where A is a constant. The inverse transform is

$$\phi(\xi) = A\int_C \mathrm{e}^{f(p)}\mathrm{d}p, \tag{8.84}$$

where

$$f(p) = \frac{p^3}{3} - \frac{\mu}{p} - \ln(p) + p\xi. \tag{8.85}$$

2. Boundary conditions: Since Eq. (8.79) is a fourth-order ordinary differential equation, *four* independent solutions must exist with four associated independent choices for the contour C in Eq. (8.84). The appropriate contour is determined by the imposed physical boundary conditions, which must be equal in number to the order of the equation. We consider a specific physical problem where an external antenna located at $\xi \gg 0$ generates a cold plasma wave propagating to the left. This physical prescription imposes the following four boundary conditions:
   (a) To the right of $\xi = 0$ there is a cold plasma wave with *energy* propagating to the left (i.e., towards the hybrid layer). It is important to recall that, as demonstrated in Eq. (6.105), the group velocity for these waves is orthogonal to the phase velocity so the wave is backwards in either the x or the z direction. The dispersion of these waves is given in Eq. (6.102) and the group velocity is given in Eq. (6.104). A plot of $S(\omega)$ and $P(\omega)$ shows that $\partial S/\partial\omega > 0$ and $\partial P/\partial\omega > 0$. Thus, for the situation where $S > 0$ and $P < 0$, the x components of the wave phase and group velocities have opposite signs. Hence, the cold wave phase velocity must propagate to the *right* to be consistent with the boundary condition that cold wave energy is propagating to the *left*.
   (b) To the right of $\xi = 0$ there is *no* warm plasma wave propagating to the left (for the hot wave, energy and phase propagate in the same direction).
   (c) To the left of $\xi = 0$ the cold plasma evanescent mode vanishes as $\xi \to -\infty$.
   (d) To the left of $\xi = 0$ the hot plasma evanescent mode vanishes as $\xi \to -\infty$.

   The question is: what happens at the hybrid layer? Possibilities include absorption of the incoming cold plasma wave (unlikely since there is no dissipation in this problem), reflection of the incoming cold plasma wave at the hybrid layer, or mode conversion.
3. Calculation of saddle points: The saddle points are the roots of

$$f'(p) = p^2 + \xi + \frac{\mu}{p^2} - \frac{1}{p} = 0. \tag{8.86}$$

   For large ξ, these roots separate into two large roots (hot mode), which are obtained when the first two terms in Eq. (8.86) balance each other and two small roots (cold mode), which are obtained when the second and third terms balance each other. The large roots satisfy $p^2 = -\xi$ and the small roots satisfy $p^2 = -\mu/\xi$. For large ξ, the fourth term is small compared to the dominant terms for both large and small roots. The quantity f can be approximated for large p as

$$\begin{aligned} f(p) &\simeq \frac{p^3}{3} + p\xi - \ln p \\ &= p\left(\frac{p^2}{3} + \xi\right) - \ln p \\ &= \frac{2}{3}p\xi - \ln p \end{aligned} \tag{8.87}$$

while for small p it can be approximated as

$$\begin{aligned} f(p) &\simeq p\xi - \frac{\mu}{p} - \ln p \\ &= p\left(\xi - \frac{\mu}{p^2}\right) - \ln p \\ &= 2p\xi - \ln p. \end{aligned} \tag{8.88}$$

For $\xi > 0$ and large roots, the quantities p_l, $f(p_l)$, and $f''(p_l)$ are

$$p_l = \pm(-\xi)^{1/2}, \quad f(p_l) = \mp\frac{2}{3}(-\xi)^{3/2} - \ln p_l, \quad f''(p_l) = \pm 2(-\xi)^{1/2}. \tag{8.89}$$

For $\xi > 0$ the corresponding quantities for the small roots are

$$p_s = \pm(-\mu/\xi)^{1/2}, \quad f(p_s) = \pm 2(-\xi\mu)^{1/2} - \ln p_s, \quad f''(p_s) = \mp 2\frac{(-\xi)^{3/2}}{\mu^{1/2}}. \tag{8.90}$$

The Gaussian integrals corresponding to steepest descent paths over these $\xi > 0$ saddle points are

$$\text{large roots:} \int_{\substack{vicinity \\ of\ saddle}} e^{f(p)} dp = \sqrt{\frac{\mp\pi}{(-\xi)^{3/2}}} e^{\mp\frac{2}{3}(-\xi)^{3/2}} \tag{8.91}$$

$$\text{small roots:} \int_{\substack{vicinity \\ of\ saddle}} e^{f(p)} dp = \sqrt{\frac{\pm\pi}{(-\mu\xi)^{1/2}}} e^{\pm 2(-\xi\mu)^{1/2}}, \tag{8.92}$$

where the logarithmic term in $f(p)$ has been taken into account. For $\xi < 0$, the large root quantities are

$$p_l = \pm|\xi|^{1/2}, \quad f(p_l) = \mp\frac{2}{3}|\xi|^{3/2} - \ln p_l, \quad f''(p_l) = \pm 2|\xi|^{1/2} \tag{8.93}$$

and the small root quantities are

$$p_s = \pm(\mu/|\xi|)^{1/2}, \quad f(p_s) = \mp 2(|\xi|\mu)^{1/2} - \ln p_s, \quad f''(p_s) = \mp 2\frac{|\xi|^{3/2}}{\mu^{1/2}}. \tag{8.94}$$

4. Choice of contour path: For $\xi < 0$ and assuming $\mu > 0$, the saddle points giving solutions that vanish when $\xi \to -\infty$ are the saddle points with the upper (i.e., minus) sign chosen in the argument of the exponential. Hence, for $\xi < 0$ a contour must be chosen that passes through one or both of these saddle points. For $\xi > 0$ the upper sign corresponds to waves propagating to the left (i.e., towards the hybrid layer). Hence, we allow the upper sign (plus) for the small root (i.e., cold mode), but not for the large root since one of the boundary conditions was that there is no inward propagating hot plasma wave. To proceed further, it is necessary to look at the topography of the real part of $f(p)$ for both signs of ξ; a method for plotting this topography is the subject of Assignment 2.

   From these plots it is seen that the correct joining is given by

$$\left\{ \begin{array}{c} \mathrm{i}\dfrac{\exp\left(-\frac{2}{3}|\xi|^{3/2}\right)}{|\xi|^{3/4}} + \\ \\ +\dfrac{\exp\left[-2\mu|\xi|^{1/2}\right]}{(|\xi|\mu)^{1/4}} \end{array} \right\} \Longleftrightarrow \left\{ \begin{array}{c} \dfrac{\exp\left[\frac{2}{3}\mathrm{i}(\xi)^{3/2}\right]}{\mathrm{i}^{1/2}\xi^{3/4}} + \\ \\ \mathrm{i}^{1/2}\dfrac{\exp\left[2\mathrm{i}(\mu\xi)^{1/2}\right]}{(\xi\mu)^{1/4}} \end{array} \right\}, \tag{8.95}$$

$$\text{evanescent side, } \xi < 0 \qquad \text{propagating side, } \xi > 0$$

   which shows that a cold wave with energy propagating into the $S = 0$ layer is converted into a hot wave that propagates back out.

5. WKB connection: The quantities in Eq. (8.95) can be expressed as integrals having the same form as WKB solutions. For example, the cold propagating term can be written as

$$\begin{aligned} \frac{\exp\left[2\mathrm{i}(\mu\xi)^{1/2}\right]}{(\xi\mu)^{1/4}} &= \sqrt{2}\frac{\exp\left[\mathrm{i}\int_0^{\xi}(\mu/\xi')^{1/2}\mathrm{d}\xi'\right]}{\sqrt{\int_0^{\xi}(\mu/\xi')^{1/2}\mathrm{d}\xi'}} \\ &= \sqrt{2}\frac{\exp\left[\mathrm{i}\int_0^{x}(-k_z^2 P/S)^{1/2}\mathrm{d}x'\right]}{\sqrt{\int_0^{x}(-k_z^2 P/S)^{1/2}\mathrm{d}x']}}, \end{aligned} \tag{8.96}$$

   where the last form is clearly the WKB solution. A similar identification exists for the hot plasma mode, so that Eq. (8.95) can also be written in terms of the WKB solutions.

This is just one example of mode conversion; other forms occur in different contexts but a similar analysis may be used and similar joining conditions are obtained. A curious feature of the mode conversion analysis is that the differential equation is never explicitly solved near $\xi = 0$; all that is done is match asymptotic solutions for the region where ξ is large and positive to the solutions for the region where ξ is large and negative.

8.6 Drift waves

Only textbook plasmas are uniform – real plasmas have both finite extent and gradients in various parameters such as pressure, magnetic field, etc. A finite-extent, magnetically confined warm plasma necessarily has a pressure gradient perpendicular to the magnetic field. For example, consider an azimuthally symmetric cylindrical plasma immersed in a strong axial magnetic field as sketched in Fig. 8.3; the pressure is assumed to be peaked on the z axis and to fall off radially. Particles stream freely in the axial direction but are constrained to make Larmor orbits in the perpendicular direction. Since the concept of pressure gradient has no meaning for an individual particle, consideration of the effect of pressure gradients requires a fluid or Vlasov point of view. This will be done first using a two-fluid analysis, then using a Vlasov analysis.

From the two-fluid point of view, the radial pressure gradient implies an equilibrium force balance

$$0 = q_\sigma \mathbf{u}_\sigma \times \mathbf{B} - n_\sigma^{-1} \nabla (n_\sigma \kappa T_\sigma) . \tag{8.97}$$

Solving Eq. (8.97) for $\mathbf{u}_\sigma$ shows that each species has a steady-state perpendicular motion at the *diamagnetic drift velocity*

$$\mathbf{u}_{d\sigma} = -\frac{\nabla (n_\sigma \kappa T_\sigma) \times \mathbf{B}}{q_\sigma n_\sigma B^2}, \tag{8.98}$$

which is in the azimuthal direction. The corresponding diamagnetic drift current is

$$\begin{aligned} \mathbf{J}_d &= \sum_\sigma n_\sigma q_\sigma \mathbf{u}_{d\sigma} \\ &= -\frac{1}{B^2} \sum_\sigma \nabla (n_\sigma \kappa T_\sigma) \times \mathbf{B} \\ &= -\frac{1}{B^2} \nabla P \times \mathbf{B}, \end{aligned} \tag{8.99}$$

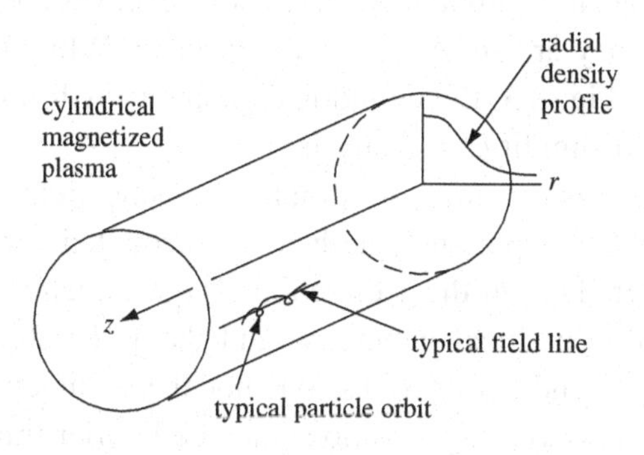

Fig. 8.3 Cylindrical magnetized plasma with radial density gradient.

which is just the azimuthal current associated with the MHD equilibrium equation $\mathbf{J} \times \mathbf{B} = \nabla P$, where $P = P_i + P_e$.

The electron and ion diamagnetic drift velocities thus provide the current necessary to establish the magnetic force that balances the MHD pressure gradient. It turns out that magnetized plasmas with pressure gradients are unstable to a class of electrostatic modes called "drift" waves. These modes exist in the same frequency regime as MHD but do not appear in standard MHD models, since standard MHD models do not provide a sufficiently detailed characterization of the differences between electron and ion dynamics.

From the particle point of view, the diamagnetic velocity is entirely fictitious because no particle actually moves at this velocity. The diamagnetic drift velocity is nevertheless quite genuine from the fluid point of view and so a fluid-based wave analysis must linearize about an equilibrium that includes this equilibrium drift. As will be seen later, the Vlasov point of view confirms and extends the conclusions of the fluid theory, providing care is taken to use an equilibrium velocity distribution function that is both valid and consistent with existence of a pressure gradient.

Drift waves will be examined using three progressively more realistic points of view. These are:

1. A collisionless two-fluid model where ions are assumed to be cold and electrons are assumed to be hot. This model establishes existence of the mode and provides a derivation for the intrinsic frequency, but provides no information regarding stability.
2. A collisional two-fluid model, which shows that *collisions* destabilize drift waves.
3. A Vlasov model including both finite ion temperature and net axial current. This model shows that both Landau damping and axial currents destabilize drift waves.

Drift waves involve physically distinct ion and electron motions, three-dimensional geometry, magnetized warm plasma effects, pressure gradients, collisionality, and Landau damping/instability. These waves are often of practical importance because they are so easily driven unstable. While real plasmas have geometry resembling Fig. 8.3, the cylindrical geometry will now be replaced by Cartesian geometry to simplify the analysis.

Using Cartesian geometry, the equilibrium magnetic field is assumed to be in the z direction and the pressure gradient is accounted for by assuming an exponential density gradient in the x direction. The x direction thus corresponds to the r direction of cylindrical geometry, and the y direction corresponds to the θ direction. The y and z directions, but not the x direction, are ignorable coordinates, and so the governing equations may be Fourier-transformed in the y and z directions, but not in the x direction.

8.6.1 Simple two-fluid model of drift waves

The plasma is assumed to have the exponential density gradient

$$n \sim \exp(-x/L), \tag{8.100}$$

where L is called the density-gradient scale length. We consider electrostatic, extremely low-frequency ($\omega \ll \omega_{ci}$) waves having the potential perturbation

$$\phi_1 = \tilde{\phi} \exp(\mathrm{i}k_y y + \mathrm{i}k_z z - \mathrm{i}\omega t). \tag{8.101}$$

The parallel phase velocity is assumed to lie in the range $v_{Ti} \ll \omega/k_z \ll v_{Te}$ so that ions are adiabatic and electrons are isothermal (this is the same regime as ion acoustic waves). The parallel and perpendicular wavelengths are both assumed to be much longer than the electron Debye length, so the plasma may be considered quasi-neutral and have $n_i \simeq n_e$. Since $\omega/k_z \ll v_{Te}$, the parallel component of the electron equation of motion is simply

$$0 \simeq -q_e \frac{\partial \phi_1}{\partial z} - \frac{1}{n_e} \frac{\partial}{\partial z}(n_e \kappa T_e), \tag{8.102}$$

which leads to a Boltzmann electron density,

$$n_e = n_{e0} \exp(-q_e \phi_1 / \kappa T_e). \tag{8.103}$$

Assuming ϕ_1 is small, linearization of this Boltzmann electron density gives the first-order electron density

$$\frac{n_{e1}}{n_{e0}} = -\frac{q_e \phi_1}{\kappa T_e}. \tag{8.104}$$

Since $\omega/k_z \gg v_{Ti}$, ions may be considered cold to first approximation and the zero-pressure limit of the ion equation of motion characterizes ion dynamics. Furthermore, since $\omega \ll \omega_{ci}$ the lowest-order perpendicular ion motion is just the $\mathbf{E} \times \mathbf{B}$ drift so

$$\mathbf{u}_{i1} = \frac{-\nabla \phi_1 \times \mathbf{B}}{B^2} = -\frac{\mathrm{i}k_y \phi_1}{B} \hat{x}. \tag{8.105}$$

Since the $\mathbf{E} \times \mathbf{B}$ drift is in the x direction, this drift is *in the direction of the density gradient* and leads to an ion density perturbation because of a convective interaction with the equilibrium density gradient. The ion density perturbation is found by linearizing the ion continuity equation

$$\frac{\partial n_{i1}}{\partial t} + \mathbf{u}_{i1} \cdot \nabla n_{i0} + n_{i1} \nabla \cdot \mathbf{u}_{i1} = 0. \tag{8.106}$$

After noting that $\nabla \cdot \mathbf{u}_{i1} = 0$, substitution for $\mathbf{u}_{i1}$ in the convective term gives

$$\frac{n_{i1}}{n_{i0}} = \frac{k_y \phi_1}{\omega L B}. \tag{8.107}$$

The density perturbation results from the ion $\mathbf{E}\times\mathbf{B}$ drift causing the ions with their density gradient to slosh back and forth in the x direction so that a stationary observer at a fixed point x sees oscillations of the ion density.

Quasi-neutrality means that the electron and ion density perturbations must be almost exactly equal although electrons and ions are governed by entirely different physics. Equating the electron and ion perturbation densities, respectively given by Eqs. (8.104) and (8.107), provides the most basic form of the drift wave dispersion relation,

$$\omega = -\frac{k_y\kappa T_e}{q_e LB} = k_y u_{de}, \tag{8.108}$$

where

$$u_{de} = -\kappa T_e/q_e LB \tag{8.109}$$

is the electron diamagnetic drift velocity given by Eq. (8.98). Equation (8.108) describes a normal mode where Boltzmann electron density perturbations caused by isothermal electron motion along field lines neutralize ion density perturbations caused by ion $\mathbf{E}\times\mathbf{B}$ motion sloshing the x-dependent equilibrium density profile in the x direction. This basic dispersion relation provides no information about the mode stability because collisions and Landau damping have not yet been considered. The wave phase velocity ω/k_y is equal to the electron diamagnetic drift velocity given by Eq. (8.98). The basic drift wave dispersion relation has the interesting feature that it does not depend on the mass of either species; this is because neither the electron Boltzmann dependence nor the ion $\mathbf{E}\times\mathbf{B}$ drift depend on mass. This basic dispersion relation provides a foundation for the more complicated models to be considered in the next two sections, and it is conventional to define the "drift frequency"

$$\omega^* = k_y u_{de}, \tag{8.110}$$

which will appear repeatedly in the more complicated models. The basic dispersion is therefore simply $\omega = \omega^*$ (the asterisk should not be confused with complex conjugate).

8.6.2 Two-fluid drift wave model with collisions

The next level of sophistication involves assuming the plasma is mildly collisional so that the electron equation of motion is now

$$m_e\frac{d\mathbf{u}_e}{dt} = q_e(-\nabla\phi + \mathbf{u}_e\times\mathbf{B}) - \frac{1}{n_e}\nabla(n_e\kappa T_e) - \nu_{ei}m_e(\mathbf{u}_e - \mathbf{u}_i). \tag{8.111}$$

The collision frequency ν_{ei} is assumed to be small compared to the electron cyclotron frequency $\omega_{ce} = q_e B/m_e$. Because $\omega/k_z \ll v_{Te}$, the electron inertia term on the left-hand side is negligible compared to the electron pressure gradient term on the right-hand side and so the parallel component of the electron equation reduces to

$$0 = -q_e \frac{\partial \phi}{\partial z} - \frac{\kappa T_e}{n_e} \frac{\partial n_e}{\partial z} - \nu_{ei} m_e u_{ez}; \tag{8.112}$$

the ion parallel velocity has been dropped since it is negligible compared to the electron parallel velocity. After linearization and assuming perturbed quantities have a space-time dependence given by Eq. (8.101), Eq. (8.112) can be solved to give the parallel electron velocity

$$u_{ez} = -\frac{\mathrm{i} k_z \kappa T_e}{\nu_{ei} m_e} \left(\frac{q_e \phi_1}{\kappa T_e} + \frac{n_{e1}}{n_{e0}} \right). \tag{8.113}$$

It is seen that in the limit of no collisions, this reverts to the Boltzmann result, Eq. (8.104).

As in the previous section, quasi-neutrality is invoked so the dispersion relation is obtained by equating the perturbed electron and ion densities. Equation (8.113) together with the continuity equation gives the perturbed electron density, provided the electron perpendicular velocity is also known. Since it has been assumed that $|\omega_{ce}| \gg \nu_{ei}$, the magnetic force term in Eq. (8.111) is much greater than the perpendicular component of the collision term. The perpendicular component of the electron equation of motion is therefore

$$0 = q_e \left(-\nabla_\perp \phi + \mathbf{u}_e \times \mathbf{B} \right) - \frac{1}{n_e} \nabla_\perp \left(n_e \kappa T_e \right). \tag{8.114}$$

In previous derivations the continuity equation was typically linearized right away, but here it turns out to be computationally advantageous to postpone linearization and instead solve Eq. (8.114) as it stands for the perpendicular electron flux to obtain

$$\Gamma_{e\perp} = n_e \mathbf{u}_{e\perp} = -\frac{n_e \nabla \phi \times \mathbf{B}}{B^2} - \frac{1}{q_e B^2} \nabla \left(n_e \kappa T_e \right) \times \mathbf{B}. \tag{8.115}$$

Using the vector identities $\nabla \cdot (\nabla \phi \times \mathbf{B}) = 0$ and $\nabla \cdot [\nabla (n_e \kappa T_e) \times \mathbf{B}] = 0$ it is seen that the divergence of the perpendicular electron flux is

$$\nabla \cdot \Gamma_{e\perp} = -\nabla n_e \cdot \frac{\nabla \phi \times \mathbf{B}}{B^2}. \tag{8.116}$$

The electron continuity equation can be expressed as

$$\frac{\partial n_e}{\partial t} + \frac{\partial}{\partial z} \left(n_e u_{ez} \right) + \nabla \cdot \Gamma_{e\perp} = 0, \tag{8.117}$$

and, after substitution for the perpendicular electron flux, this becomes

$$\frac{\partial n_e}{\partial t} + \frac{\partial}{\partial z}\left(n_e u_{ez}\right) - \nabla n_e \cdot \frac{\nabla\phi \times \mathbf{B}}{B^2} = 0. \tag{8.118}$$

At this stage of the derivation it is now appropriate to linearize the continuity equation, which becomes

$$\frac{\partial n_e}{\partial t} + n_{e0}\frac{\partial u_{ez1}}{\partial z} - \frac{\mathrm{d}n_{e0}}{\mathrm{d}x}\hat{x} \cdot \frac{\nabla\phi_1 \times \mathbf{B}}{B^2} = 0. \tag{8.119}$$

Substituting for ϕ_1 using Eq. (8.101), and using Eq. (8.113) for u_{z1} gives

$$\frac{n_{e1}}{n_{e0}}\left(-\mathrm{i}\omega + \frac{k_z^2 \kappa T_e}{\nu_{ei} m_e}\right) + \frac{q_e \phi_1}{\kappa T_e}\left(\frac{k_z^2 \kappa T_e}{\nu_{ei} m_e} + \frac{\mathrm{i}k_y \kappa T_e}{q_e B L}\right) = 0, \tag{8.120}$$

which may be solved to give the sought-after electron density perturbation

$$\frac{n_{e1}}{n_{e0}} = -\frac{q_e \phi_1}{\kappa T_e}\frac{\left(-\mathrm{i}\omega^* + \dfrac{1}{\tau_\|}\right)}{\left(-\mathrm{i}\omega + \dfrac{1}{\tau_\|}\right)}. \tag{8.121}$$

Here

$$\tau_\| = \nu_{ei} m_e / k_z^2 \kappa T_e \tag{8.122}$$

is the nominal time required for electrons to diffuse a distance of the order of a parallel wavelength. This is because the parallel collisional diffusion coefficient scales as $D_\| \sim (\text{random step})^2/(\text{collision period}) \sim \kappa T_e/m_e \nu_{ei}$ and, according to the diffusion equation, the distance diffused in time t is given by $z^2 \sim 4D_\| t$. Equation (8.121) shows that in the limit $\tau_\| \to 0$, the electron density perturbation reverts to being Boltzmann but, for finite $\tau_\|$, a phase lag occurs between n_{e1} and ϕ_1.

The next step is to calculate the ion density perturbation. The ion equation of motion is

$$m_i \frac{\mathrm{d}\mathbf{u}_i}{\mathrm{d}t} = q_i\left(-\nabla\phi + \mathbf{u}_i \times \mathbf{B}\right) - \nu_{ie} m_i (\mathbf{u}_i - \mathbf{u}_e); \tag{8.123}$$

because the ions are heavy, the left-hand side inertial term must now be retained. The ions are assumed to be cold and the ion inertial term is assumed to be much larger than the collisional term so the parallel component of the linearized Eq. (8.123) is

$$u_{iz1} = \frac{k_z q_i \phi_1}{\omega m_i} = \frac{k_z c_s^2}{\omega}\frac{q_i \phi_1}{\kappa T_e}, \tag{8.124}$$

where $c_s^2 = \kappa T_e/m_i$ is the ion acoustic velocity. The appearance of the ion acoustic velocity comes from normalizing the potential perturbation to κT_e; this normalization is used in order to be consistent with the normalization used for the electron dynamics. Noting again that the inertial term is assumed to be much larger than the collisional term, the linearized perpendicular ion equation becomes

$$m_i \frac{\mathrm{d}\mathbf{u}_{i1}}{\mathrm{d}t} = q_i \left(-\nabla\phi_1 + \mathbf{u}_{i_1} \times \mathbf{B} \right), \tag{8.125}$$

which can be solved to give the perpendicular velocity as a sum of an $\mathbf{E}\times\mathbf{B}$ drift and a polarization drift, i.e.,

$$\mathbf{u}_{i\perp} = \frac{-\nabla\phi \times \mathbf{B}}{B^2} + \frac{m_i}{q_i B^2} \frac{\mathrm{d}}{\mathrm{d}t} (-\nabla\phi). \tag{8.126}$$

The polarization drift is retained in the ion but not the electron equation, because polarization drift is proportional to mass. The perpendicular ion flux is

$$\Gamma_{i\perp} = \frac{-n_i \nabla\phi \times \mathbf{B}}{B^2} + \frac{n_i m_i}{q_i B^2} \frac{\mathrm{d}}{\mathrm{d}t} (-\nabla_\perp \phi), \tag{8.127}$$

which has a divergence

$$\nabla \cdot \Gamma_{i\perp} = -\nabla n_i \cdot \frac{\nabla\phi \times \mathbf{B}}{B^2} - \nabla \cdot \left(\frac{n_i m_i}{q_i B^2} \frac{\mathrm{d}}{\mathrm{d}t} (\nabla_\perp \phi) \right). \tag{8.128}$$

Substitution of Eqs. (8.128) and (8.124) into the ion continuity equation, linearizing, and solving for the ion density perturbation gives

$$\begin{aligned} \frac{n_{i1}}{n_{i_0}} &= \left[\frac{k_z^2 c_s^2}{\omega^2} + \frac{q_e}{q_i} \frac{k_y \kappa T_e}{\omega q_e L B} - \frac{m_i \kappa T_e}{q_i^2 B^2} k_y^2 \right] \frac{q_i \phi_1}{\kappa T_e} \\ &= \left[\frac{k_z^2 c_s^2}{\omega^2} - \frac{q_e}{q_i} \frac{\omega^*}{\omega} - k_y^2 \rho_s^2 \right] \frac{q_i \phi_1}{\kappa T_e}, \end{aligned} \tag{8.129}$$

where $\rho_s^2 \equiv \kappa T_e / m_i \omega_{ci}^2$ is a fictitious ion Larmor orbit defined using the electron temperature instead of the ion temperature. The fictitious length ρ_s^2 is analogous to the fictitious velocity $c_s^2 = \kappa T_e/m_i$, which appeared in the analysis of ion acoustic waves.

Since the plasma is quasi-neutral, the normalized electron and ion density perturbations must be the same. Equating the normalized density perturbations obtained from Eq. (8.121) and (8.129) gives

$$\frac{\left(-i\omega^* + \dfrac{1}{\tau_{\|}}\right)}{\left(-i\omega + \dfrac{1}{\tau_{\|}}\right)} = \frac{\omega^*}{\omega} + \frac{k_z^2 c_s^2}{\omega^2} - k_y^2 \rho_s^2. \tag{8.130}$$

For simplicity it is assumed here that $q_i/q_e = -1$, which is the situation for low-temperature plasmas where the ions are too cold to be multiply ionized ($q_i/q_e = -1$ is, of course, always true for hydrogen plasmas). It is also assumed that the collision frequency is sufficiently small to have $\omega\tau_{\|} = \omega\nu_{ei}m_e/k_z^2T_e \ll 1$ and that ω is of the order of ω^*; the self-consistency of these assumptions will be checked later. The left-hand side of Eq. (8.130) can now be expanded using the binomial theorem to obtain the collisional drift wave dispersion relation

$$D(\omega, k_y, k_z) = 1 - \frac{\omega^*}{\omega} + k_y^2\rho_s^2 - \frac{k_z^2 c_s^2}{\omega^2} + i(\omega - \omega^*)\tau_{\|} = 0. \tag{8.131}$$

This dispersion relation shows drift waves have an association with ion acoustic waves since in the limit $k_y^2\rho_s^2 \to 0$, the real part of the dispersion becomes

$$1 - \frac{\omega^*}{\omega} - \frac{k_z^2 c_s^2}{\omega^2} = 0; \tag{8.132}$$

this dispersion relation reduces to an ion acoustic wave dispersion relation in the limit of there being no equilibrium pressure gradient (i.e., when ω^* vanishes). Equation (8.132) has two roots

$$\omega = \frac{\omega^* \pm \sqrt{(\omega^*)^2 + 4k_z^2 c_s^2}}{2}, \tag{8.133}$$

which for small $k_z c_s$ are

1. $\omega = \omega^*$ and $\omega = -|k_z c_s|$ if $\omega^* > 0$
2. $\omega = \omega^*$ and $\omega = |k_z c_s|$ if $\omega^* < 0$.

Thus, the drift mode is a distinct mode compared to the ion acoustic wave and its parallel phase velocity is much faster than the ion acoustic wave since the drift wave occurs in the limit $\omega/k_z \gg c_s$.

We now address the important question of the stability properties of Eq. (8.131). To do this, k_z is assumed to be sufficiently small to have $k_z c_s \ll \omega$ so the dispersion describes the drift mode and not the ion acoustic mode. The real part of the dispersion is then set to zero to obtain the real part of the frequency

$$\omega_r = \frac{\omega^*}{1+k_y^2\rho_s^2} \tag{8.134}$$

showing that the actual frequency of the collisional drift wave is smaller than ω^*, but is of the same order, provided $k_y^2\rho_s^2$ is not larger than order unity. The $k_y^2\rho_s^2$ dependence results from ion polarization drift, an effect that was neglected in the initial simple model. The assumption $\omega\tau_\parallel = \omega\nu_{ei}m_e/k_z^2T_e \ll 1$ implies $k_y\nu_{ei}/k_z^2L\omega_{ce} \ll 1$, which is true provided the ratio ν_{ei}/ω_{ce} is sufficiently small for a given geometric factor k_y/k_z^2L. Using the Taylor expansion technique discussed in the treatment of Eqs. (5.85)–(5.87), the imaginary part of the frequency is found to be

$$\omega_i = -\frac{D_i}{(\partial D_r/\partial\omega)_{\omega=\omega_r}} = -\frac{(\omega-\omega^*)\tau_\parallel}{\omega^*/\omega_r^2} = \omega^*\frac{k_y^2\rho_s^2\omega^*\tau_\parallel}{\left(1+k_y^2\rho_s^2\right)^3}. \tag{8.135}$$

Equation (8.135) shows ω_i has several important features:

1. ω_i is positive so collisional drift waves are *always unstable.*
2. ω_i is proportional to $\tau_\parallel$ so modes with the smallest k_z and hence the fastest parallel phase velocities are the most unstable, subject to the proviso that $\omega\tau_\parallel \ll 1$ is maintained. This increase of growth rate with k_z^{-1} (i.e., increase with parallel wavelength) means drift waves typically have the longest possible parallel wavelength allowed by the boundary conditions. For example, a linear plasma of finite axial extent with grounded conducting end walls has the boundary condition $\phi_1 = 0$ at both end walls; the longest allowed parallel wavenumber is π/h, where h is the axial length of the plasma (i.e., half a wavelength is the minimum number of waves that can be fitted subject to the boundary condition).
3. Collisions make the wave unstable since $\tau_\parallel \sim \nu_{ei}$.
4. ω_i is proportional to the factor $k_y^2\rho_s^2/(1+k_y^2\rho_s^2)^3$, which has a maximum when $k_y\rho_s$ is of order unity.
5. ω_i is proportional to L^{-2}; for a realistic cylindrical plasma (cf. Fig. 8.3) the density is uniform near the axis and has a gradient near the edge (e.g., a Gaussian density profile where $n(r) \sim \exp(-r^2/L^2)$). The density gradient is localized near the edge of the plasma and so drift waves will have the largest growth rate near the edge. Drift waves in real plasmas are typically observed to have maximum amplitude in the region of maximum density gradient.

The free energy driving drift waves is the pressure gradient and so the drift waves might be expected to deplete their energy source eventually by flattening out the pressure gradient. This indeed happens and, in particular, it is the nonlinear behavior of drift waves that reduces the pressure gradient. This flattening is accomplished by the drift waves pumping plasma from regions of high pressure to regions of low pressure. The pumping can be calculated by considering the time-averaged, nonlinear, x-directed particle flux associated with drift waves, namely

$$\Gamma_x = \langle \mathrm{Re}\, n_1 \mathrm{Re}\, u_{1x} \rangle = \frac{1}{2}\mathrm{Re}\ (n_1 u_{1x}^*). \tag{8.136}$$

The species subscript σ has been omitted here, because to lowest order both species have the same u_{1x} given by the $\mathbf{E} \times \mathbf{B}$ drift, i.e.,

$$u_{1x} = \hat{x} \cdot \frac{-\nabla\phi_1 \times \mathbf{B}}{B^2} = -\mathrm{i}k_y \phi_1 / B. \tag{8.137}$$

Being careful to remember that ω^* is a real quantity (this is a conventional, but confusing, notation), Eqs. (8.121) and (8.137) are substituted into Eq. (8.136) to obtain the wave-induced particle flux

$$\begin{aligned}
\Gamma_x &= \frac{1}{2}\mathrm{Re}\left[-\frac{n_0 q_e}{\kappa T_e}\left(1 - \mathrm{i}(\omega^* - \omega)\tau_{\|}\right) \frac{\mathrm{i}k_y |\phi_1|^2}{B} \right] \\
&= -\frac{1}{2}\mathrm{Re}\left[\frac{n_0 q_e}{\kappa T_e}(\omega^* - \omega)\tau_{\|} \frac{k_y |\phi_1|^2}{B} \right] \\
&= \left(\frac{k_y^2 \rho_s^2}{1 + k_y^2 \rho_s^2} \right) \frac{k_y^2 |\phi_1|^2}{2LB^2} \frac{\nu_{ei} n_0 m_e}{k_z^2 \kappa T_e} \\
&= -D_{nl} \frac{\mathrm{d}n_0}{\mathrm{d}x}.
\end{aligned} \tag{8.138}$$

Thus, it is seen that collisional drift waves cause an outward diffusion of plasma characterized by the nonlinear wave-induced diffusion coefficient

$$D_{nl} = \left(\frac{k_y^2 \rho_s^2}{1 + k_y^2 \rho_s^2} \right) \frac{k_y^2 |\phi_1|^2}{2B^2} \frac{\nu_{ei} m_e}{k_z^2 \kappa T_e}. \tag{8.139}$$

This diffusion coefficient has a proportionality similar to ω_i. The drift wave thus has the following properties: (i) the pressure gradient provides free energy and also an equilibrium where drift waves are a normal mode, (ii) collisions allow the drift wave to feed on the available free energy and grow, (iii) nonlinear diffusion flattens the pressure gradient thereby depleting the free energy and also undermining the confining effect of the magnetic field. A nonlinear saturated amplitude can result if some external source continuously replenishes the pressure gradient.

It was shown earlier that the most unstable drift waves are those having $k_y \rho_s$ of order unity. This property can be adapted to the more physically realistic situation of a cylindrical plasma by realizing that in a cylindrical plasma k_y is replaced by m/r, where m is the azimuthal mode number. Periodicity in θ forces m to be an integer. If the most unstable m is a small integer, then the drift waves tend to be coherent, but if m is a large integer, many azimuthal wavelengths fit into the circumference of the cylinder. In this case, the periodicity condition is only a weak constraint and the waves typically become turbulent. Large m corresponds to small ρ_s, which occurs when the magnetic field is strong. Thus, plasmas with strong magnetic fields tend to have turbulent, short perpendicular wavelength drift waves, whereas plasmas with weak magnetic fields have coherent, long perpendicular wavelength drift waves.

Drift wave turbulence has been found to be the main reason why the radial diffusion in tokamak magnetic plasma confinement devices is much worse than can be explained by simple random walks due to particle collisions. The importance of drift wave turbulence can be estimated by assuming $k_y^2\rho_s^2 \sim 1$ and then taking the ratio of D_{nl} to the classical diffusion given by Eq. (2.114) to obtain

$$\frac{D_{nl}}{D_{classical}} \sim \frac{1}{k_z^2\rho_s^2}\left(\frac{e|\phi_1|}{\kappa T_e}\right)^2. \tag{8.140}$$

In the situation where the magnetic field is strong so ρ_s is a microscopic length (this length is $\sim r_{Li}$ if $T_e \sim T_i$) then $k_z\rho_s$ will be very small since k_z^{-1} is of the order of the axial extent of the configuration. Thus, if the turbulence level is sufficiently strong to satisfy the very modest requirement $e|\phi_1|/\kappa T_e > k_z\rho_s$, plasma transport across the magnetic field will be mainly from diffusion caused by drift wave turbulence, not classical diffusion.

8.6.3 Vlasov theory of drift waves: collisionless drift waves

Drift waves also exist when the plasma is so hot that the collision frequency becomes insignificant and the plasma can be considered collisionless. The previous section showed that a collision-induced phase lag between the density and potential fluctuations produces a destabilizing imaginary term in the dispersion relation. The analysis of collisionless plasma waves showed that Landau damping also causes a phase shift between the density and potential fluctuations resulting in an imaginary term in the dispersion relation. The Vlasov analysis in Assignment 8 of Chapter 5 showed that a relative motion between electrons and ions (i.e., a net current) destabilizes ion acoustic waves, and so it is reasonable to expect that a similar current-driven destabilization might also occur for drift waves.

The Vlasov analysis of drift waves is less intuitive than the fluid analysis, but the reward for abstraction is a more profound model. As in the fluid analysis, the plasma is assumed to have a uniform equilibrium magnetic field $\mathbf{B} = B\hat{z}$, an x-directed density gradient given by Eq. (8.100), and a perturbed potential given by Eq. (8.101). The Vlasov analysis is in a sense simpler than the fluid analysis, because the Vlasov analysis involves a modest extension of the warm plasma electrostatic Vlasov model discussed in Section 8.1. As was discussed in Section 8.1 the perturbed distribution function is evaluated by integrating along the unperturbed orbits, i.e.,

$$f_{\sigma 1}(\mathbf{x}, \mathbf{v}, t) = \int_{-\infty}^{t} \mathrm{d}t' \left[\frac{q_\sigma}{m_\sigma} \nabla \phi_1 \cdot \frac{\partial f_{\sigma 0}}{\partial \mathbf{v}} \right]_{\mathbf{x}=\mathbf{x}(t'),\mathbf{v}=\mathbf{v}(t')} . \tag{8.141}$$

The new feature here is that the equilibrium distribution function must incorporate the assumed density gradient.

A first instinct would be to accomplish this by simply multiplying the Maxwellian distribution function of Section 8.1 by the factor $\exp(-x/L)$ so that the assumed equilibrium distribution function would be $f_{\sigma 0}(x, \mathbf{v}) = (\pi v_{T\sigma})^{-3/2} \exp(-v^2/v_{T\sigma}^2 - x/L)$. This approach turns out to be wrong because $f_{\sigma 0}(x, \mathbf{v}) = (\pi v_{T\sigma})^{-3/2} \exp(-v^2/v_{T\sigma}^2 - x/L)$ is *not* a function of the constants of the motion and so is not a solution of the equilibrium Vlasov equation.

What is needed is some constant of the motion that includes the parameter x. The equilibrium distribution function could then be constructed from this constant of the motion and arranged to have the desired x-dependence. The appropriate constant of the motion is the canonical momentum in the y direction, namely

$$P_y = m_\sigma v_y + q_\sigma A_y = m_\sigma v_y + q_\sigma B_z x = \left(\frac{v_y}{\omega_{c\sigma}} + x \right) q_\sigma B_z \tag{8.142}$$

since $B_z = \partial A_y / \partial x$. Multiplying the original Maxwellian by the factor $\exp[-(v_y/\omega_{c\sigma} + x)/L]$ produces the desired spatial dependence while simultaneously satisfying the requirement that the distribution function depends only on constants of the motion. Note that P_y is a constant of the motion because y is an ignorable coordinate.

As suggested above, it turns out that z-directed currents can also destabilize collisionless drift waves, much like the situation where currents provide free energy that destabilizes ion acoustic waves. Since it takes little additional effort to include this possibility, a z-directed current will also be assumed so that electrons and ions are assumed to have unequal mean velocities $u_{\sigma z}$. In a frame moving with velocity $u_{\sigma z}$ the species σ is thus assumed to have a distribution function $\sim \exp[-mv'^2/2 - (v'_y/\omega_{c\sigma} + x)/L]$, where $\mathbf{v}' = \mathbf{v} - u_{\sigma z}\hat{z}$ is the velocity in the moving frame. This is a valid solution to the equilibrium Vlasov equation, since both the energy $mv'^2/2$ and the y-direction canonical momentum $(v'_y/\omega_{c\sigma} + x)/L$ are constants of the motion. The appropriately normalized lab-frame distribution function is

$$f_{\sigma 0}(x, \mathbf{v}) = \frac{n_{0\sigma}}{\pi^{3/2} v_{T\sigma}^3} \exp\left[- \left(\mathbf{v} - u_{\sigma z}\hat{z}\right)^2 / v_{T\sigma}^2 - (v_y/\omega_{c\sigma} + x)/L - \left(\frac{v_T}{2\omega_c L} \right)^2 \right], \tag{8.143}$$

where the normalization factor $\exp\left(-(v_T/2\omega_c L)^2\right)$ has been inserted so that the zeroth moment of $f_{\sigma 0}$ gives the density. The necessity of this factor is made evident by completing the squares for the velocities and writing Eq. (8.143) in the equivalent form

$$f(x,\mathbf{v}) = \frac{n_0}{\pi^{3/2} v_T^3} \exp\left(-\left[\frac{v_x^2 + \left(v_y + \dfrac{v_T^2}{2\omega_c L}\right)^2 + \left(v_z - u_{\sigma z}\right)^2}{v_T^2} \right] - \frac{x}{L} \right). \quad (8.144)$$

The term $v_y/\omega_{c\sigma}$, which crept into Eq. (8.143) because of the seemingly abstract requirement of having the distribution function depend on the constant of the motion P_y, corresponds to the fluid theory equilibrium diamagnetic drift. This correspondence is easily seen by calculating the mean y-direction velocity (i.e., first moment of Eq. (8.143)) and finding that the existence of the $v_y/\omega_{c\sigma}$ term results in a fluid velocity u_y, which is precisely the diamagnetic velocity given by Eq. (8.98).

We now insert Eq. (8.143) in Eq. (8.141) to calculate the perturbed distribution function. The unperturbed particle orbits are the same as in Section 8.1 because the concepts of both density gradient and axial current result from averaging over a distribution of many particles and so have no meaning for an individual particle. The integration over unperturbed orbits is thus identical to that of Section 8.1, except that a different equilibrium distribution function is used.

The essence of the calculation is in the term

$$\begin{aligned} \nabla\phi_1 \cdot \frac{\partial f_{\sigma 0}}{\partial \mathbf{v}} &= \mathrm{i}\phi_1 \mathbf{k} \cdot \frac{\partial f_{\sigma 0}}{\partial \mathbf{v}} \\ &= -\mathrm{i}\tilde{\phi} \mathrm{e}^{\mathrm{i}\mathbf{k}\cdot\mathbf{x}(t) - \mathrm{i}\omega t} \left[\frac{2\mathbf{k}\cdot\mathbf{v}}{v_{T\sigma}^2} - \frac{2k_z u_{\sigma z}}{v_{T\sigma}^2} + \frac{k_y}{\omega_{c\sigma} L} \right] f_{\sigma 0}, \end{aligned} \quad (8.145)$$

which causes Eq. (8.141) to become

$$\begin{aligned} f_{\sigma 1}(\mathbf{x},\mathbf{v},t) &= -\frac{q_\sigma \tilde{\phi} f_{\sigma 0}}{m_\sigma} \int_{-\infty}^{t} \mathrm{d}t' \mathrm{e}^{\mathrm{i}\mathbf{k}\cdot\mathbf{x}(t') - \mathrm{i}\omega t'} \left[\frac{2\mathrm{i}\mathbf{k}\cdot\mathbf{v}}{v_{T\sigma}^2} - \frac{2\mathrm{i}k_z u_{\sigma z}}{v_{T\sigma}^2} + \frac{\mathrm{i}k_y}{\omega_{c\sigma} L} \right] \\ &= -\frac{q_\sigma \tilde{\phi} f_{\sigma 0}}{m_\sigma} \int_{-\infty}^{t} \mathrm{d}t' \left\{ \begin{array}{c} \dfrac{2\mathrm{e}^{-\mathrm{i}\omega t'}}{v_{T\sigma}^2} \left[\dfrac{\mathrm{d}}{\mathrm{d}t'} \mathrm{e}^{\mathrm{i}\mathbf{k}\cdot\mathbf{x}(t')} \right] - \\ \left(\dfrac{2\mathrm{i}k_z u_{\sigma z}}{v_{T\sigma}^2} - \dfrac{\mathrm{i}k_y}{\omega_{c\sigma} L} \right) \mathrm{e}^{\mathrm{i}\mathbf{k}\cdot\mathbf{x}(t') - \mathrm{i}\omega t'} \end{array} \right\} \\ &= -\frac{2q_\sigma \tilde{\phi} f_{\sigma 0}}{m_\sigma v_{T\sigma}^2} \left[\mathrm{e}^{\mathrm{i}\mathbf{k}\cdot\mathbf{x}(t) - \mathrm{i}\omega t} + \mathrm{i}\left(\omega - k_z u_{\sigma z} - \omega_\sigma^*\right) \int_{-\infty}^{t} \mathrm{d}t' \mathrm{e}^{\mathrm{i}\mathbf{k}\cdot\mathbf{x}(t') - \mathrm{i}\omega t'} \right]; \end{aligned}$$

(8.146)

here

$$\frac{k_y v_{T\sigma}^2}{2\omega_{c\sigma} L} = \frac{k_y \kappa T}{qBL} = -\omega_\sigma^*, \tag{8.147}$$

has been used as a generalization of Eq. (8.110). The rest of the analysis is as before and gives the dispersion relation

$$1+\sum_\sigma \frac{1}{k^2\lambda_{D\sigma}^2}\left[1+\frac{(\omega - k_z u_{\sigma z} - \omega_\sigma^*)}{k_z v_{T\sigma}} \sum_{n=-\infty}^{\infty} I_n(\Lambda_\sigma)\mathrm{e}^{-\Lambda_\sigma} Z(\alpha_{n\sigma})\right] = 0, \tag{8.148}$$

where, as before, $\Lambda_\sigma = k_\perp^2 r_{L\sigma}^2$ and $\alpha_{n\sigma} = (\omega - n\omega_{c\sigma})/k_z v_{T\sigma}$. It is seen that the density gradient and the axial current both provide a Doppler shift in the $\alpha_{0\sigma}$ term. In both cases the Doppler shift is given by the wavenumber in the appropriate direction times the mean fluid velocity in that direction.

Because Eq. (8.148) contains so much detailed physics in addition to the sought-after collisionless drift wave,[1] some effort and guidance is required to flush out the drift wave information from all the other information. This is done by taking appropriate asymptotic limits of Eq. (8.148) and guidance is obtained using the results from the two-fluid analysis. These results showed that drift waves exist in the regime where:

1. $v_{Ti} \ll \omega/k_z \ll v_{Te}$ so that the ions are adiabatic and the electrons are isothermal,
2. $\omega \sim \omega_e^* \ll \omega_{ci}, \omega_{ce}$,
3. $k_\perp^2 r_{Le}^2 \sim 0$ because the extremely small electron mass means that the electron Larmor orbit radius is negligible compared to the perpendicular wavelength even though the electrons are warm,
4. $k_\perp^2 r_{Li}^2 < 1$, since the ions were assumed cold,
5. $k_\perp^2 \rho_s^2 \sim 1$,
6. $k^2\lambda_{De}^2 \ll 1$ so that the waves are quasi-neutral,
7. $|u_{zi}| \ll |u_{ze}|$, since the parallel equilibrium velocities satisfy $m_e u_{ez} + m_i u_{iz} = 0$; there is no parallel flow, just a parallel current.

Because $\Lambda_e \sim 0$, all electron terms in the summation over n vanish except for the $n = 0$ term. Since $\omega/k_z \ll v_{Te}$, the small-argument limit of the plasma dispersion function is used for electrons and it is seen that the imaginary term is the dominant term for the electrons. In contrast, since $v_{Ti} \ll \omega/k_z$, the large-argument limit of the plasma dispersion function is used for the ions and also equilibrium parallel motion is neglected for ions because of the large ion mass. For the $n \neq 0$ harmonics the terms $Z(\alpha_n) + Z(\alpha_{-n}) \to -k_z v_{Ti}[1/(\omega - n\omega_{ci}) + 1/(\omega + n\omega_{ci})]$

[1] Eq. (8.148) also describes Bernstein waves, the electrostatic limit of magnetized cold plasma waves, and mode conversion.

and hence cancel each other since $\omega \ll \omega_{ci}$. On making these approximations, Eq. (8.148) reduces to

$$1+\frac{T_e}{T_i}\left[1-\mathrm{e}^{-\Lambda_i}I_0(\Lambda_i)\right]-\frac{\omega_e^*}{\omega}\mathrm{e}^{-\Lambda_i}I_0(\Lambda_i)+\mathrm{i}\pi^{1/2}\frac{\left(\omega-k_z u_{ze}-\omega_e^*\right)}{k_z v_{Te}}=0. \quad (8.149)$$

The real part of this dispersion gives

$$\omega_r=\omega_e^*\left\{\frac{\mathrm{e}^{-\Lambda_i}I_0(\Lambda_i)}{1+\frac{T_e}{T_i}\left[1-\mathrm{e}^{-\Lambda_i}I_0(\Lambda_i)\right]}\right\} \quad (8.150)$$

or in the limit of small Λ_i

$$\omega_r=\omega_e^*\left\{\frac{1-\Lambda_i}{1+k_\perp^2\rho_s^2}\right\}; \quad (8.151)$$

this corresponds to the fluid dispersion and moreover shows how finite ion temperature affects the dispersion. Using the method of Eq. (8.135), the imaginary part of the frequency is obtained from Eq. (8.149) as

$$\begin{aligned}\omega_i &= -\frac{\omega^2\pi^{1/2}}{\omega_e^*\mathrm{e}^{-\Lambda_i}I_0(\Lambda_i)}\frac{\left(\omega_r-k_z u_{ze0}-\omega_e^*\right)}{k_z v_{Te}}\\ &= \frac{\pi^{1/2}\left(\omega_e^*\right)^2}{k_z v_{Te}}\left\{\frac{(1-\Upsilon)\left(1+T_e/T_i\right)+k_z u_{ze0}\Upsilon/\omega_r}{\left[1+(1-\Upsilon)\,T_e/T_i\right]^3}\right\},\end{aligned} \quad (8.152)$$

where $\Upsilon=\mathrm{e}^{-\Lambda_i}I_0(\Lambda_i)$. In the limit $\Lambda_i\to 0$, it is seen that $\Upsilon\to 1-\Lambda_i$ so the imaginary part of the frequency becomes

$$\omega_i=\frac{\pi^{1/2}\left(\omega_e^*\right)^2}{k_z v_{Te}\left[1+k_\perp^2\rho_s^2\right]^3}\left\{k_\perp^2\left(\rho_s^2+r_{Li}^2\right)+k_z u_{ze0}/\omega_r\right\}. \quad (8.153)$$

The imaginary part of the frequency is therefore always positive so there is always instability. Two collisionless destabilization mechanisms are seen to exist, *normal Landau damping*, represented by the term involving $k_\perp^2\left(\rho_s^2+r_{Li}^2\right)$ in the curly brackets, and *current*, represented by $k_z u_{z0}/\omega_r$. Since collisions can also destabilize drift waves, there are at least three mechanisms by which drift waves can be destabilized. Comparison of Eq. (8.153) with Eq. (8.135) shows which of these mechanisms will be dominant in a given plasma. Modes with long parallel wavelengths are the most unstable for drift waves driven unstable by Landau damping, just as for collisional drift waves.

There are many more varieties of drift waves besides the basic versions discussed here. The common feature is that the gradient of pressure perpendicular to the magnetic field provides a new mode and free energy from the pressure gradient or from parallel current can be tapped by this mode so the mode is spontaneously unstable.

8.7 Assignments

1. Electrostatic ion cyclotron waves. Using the electrostatic hot plasma dispersion relation show that there exists a mode in the vicinity of the ion cyclotron frequency having the dispersion

$$\omega^2 = \omega_{ci}^2 + k_\perp^2 c_s^2. \tag{8.154}$$

   Show this mode can be driven unstable by an axial current.
2. Mode conversion. Work through the algebra of the linear mode conversion problem and plot the contours of Re $f(p)$ for both positive and negative ξ. Hint: it is easier to plot the contours of $f(p)$ if one uses polar coordinates in the complex p-plane so $p = r\mathrm{e}^{\mathrm{i}\theta}$. Then the real part of the term p^3 is just Re $(r^3\mathrm{e}^{3\mathrm{i}\theta}) = r^3\cos(3\theta)$. Choose a set of contour paths for $\xi < 0$ that satisfy the boundary condition that there are no modes growing exponentially with increasing distance from $\xi = 0$. Then, determine which combination of these contours does not give an inward propagating hot plasma wave when $\xi > 0$ (this was another boundary condition).
3. Prove that the distribution function

$$f_{\sigma 0}(x, \mathbf{v}) = \frac{n_{0\sigma}}{\pi^{3/2} v_{T\sigma}^3} \exp\left[-(\mathbf{v} - u_{\sigma z}\hat{z})^2 / v_{T\sigma}^2 - (v_y/\omega_{c\sigma} + x)/L - \left(\frac{v_T}{2\omega_c L}\right)^2\right]$$

   provides a mean y-direction drift, which corresponds to the diamagnetic drift of fluid theory.

9
MHD equilibria

9.1 Why use MHD?

Of the three levels of plasma description – Vlasov, two-fluid, and MHD – Vlasov is the most accurate and MHD is the least accurate. So, why use MHD? The answer is that, because MHD is a more macroscopic point of view, it is more efficient to use MHD in situations where the greater detail and accuracy of the Vlasov or two-fluid models are unnecessary. MHD is particularly suitable for situations having complex geometry because it is very difficult to model such situations using the microscopically oriented Vlasov or two-fluid approaches and because geometrical complexities are often most important at the MHD level of description. The equilibrium and gross stability of three-dimensional, finite-extent plasma configurations are typically analyzed using MHD. Issues requiring a two-fluid or a Vlasov point of view can exist and be important, but these more subtle questions can be addressed after an approximate understanding has first been achieved using MHD. The MHD point of view is especially relevant to situations where magnetic forces are used to confine or accelerate plasmas or liquid conductors such as molten metals. Examples of such situations include magnetic fusion confinement plasmas, solar and astrophysical plasmas, planetary and stellar dynamos, arcs, and magnetoplasmadynamic thrusters. Although molten metals are not plasmas, they are described by MHD and, in fact, the MHD description is actually more appropriate and more accurate for molten metals than it is for plasmas.

We begin our discussion by examining certain general properties of magnetic fields in order to develop an intuitive understanding of the various stresses governing MHD equilibrium and stability. The MHD equation of motion,

$$\rho\left[\frac{\partial \mathbf{U}}{\partial t}+\mathbf{U}\cdot\nabla\mathbf{U}\right]=\mathbf{J}\times\mathbf{B}-\nabla P, \tag{9.1}$$

is a generalization of the equation of motion for an ordinary fluid because it includes the $\mathbf{J} \times \mathbf{B}$ magnetic force. Plasma viscosity is normally very small and is usually omitted from the MHD equation of motion. However, when torques exist, a viscous damping term needs to be included if one wishes to consider equilibria. This is because viscous damping is required to balance any torque in equilibrium; otherwise the plasma will spin up without limit. Such situations will be discussed in Section 9.9; until then, viscosity will be assumed to be negligible and will be omitted from Eq. (9.1).

9.2 Vacuum magnetic fields

The simplest non-trivial magnetic field results when the entire magnetic field is produced by electric currents located outside the volume of interest so there are no currents in the volume of interest. This type of magnetic field is called a vacuum field since it could exist in a vacuum. Because there are no local currents, a vacuum field satisfies

$$\nabla \times \mathbf{B}_{vac} = 0. \tag{9.2}$$

Since the curl of a gradient is always zero, a vacuum field must be the gradient of some scalar potential χ, i.e., the vacuum field can always be expressed as $\mathbf{B}_{vac} = \nabla\chi$. For this reason vacuum magnetic fields are also called potential magnetic fields. Because all magnetic fields must satisfy $\nabla \cdot \mathbf{B} = 0$, the potential χ satisfies Laplace's equation,

$$\nabla^2 \chi = 0. \tag{9.3}$$

Hence the entire mathematical theory of vacuum electrostatic fields can be brought into play when studying vacuum magnetic fields. Vacuum electrostatic theory shows that if either χ or its normal derivative is specified on the surface S bounding a volume V, then χ is uniquely determined in V. Also, if an equilibrium configuration has symmetry in some direction so that the coefficients of the relevant linearized partial differential equations do not depend on this direction, the linearized equations may be Fourier transformed in this "ignorable" direction. Vacuum is automatically symmetric in all directions and Poisson's equation reduces to Laplace's equation, which is intrinsically linear. The linearity of the equation and the symmetry of the physical medium cause Laplace's equation to reduce to one of the standard equations of mathematical physics. For example, consider a cylindrical configuration with coordinates r, θ, and z and suppose this configuration is axially and azimuthally uniform so that both θ and z are ignorable coordinates; i.e., the coefficients of the partial differential equation do not depend on θ or on z. Fourier analysis of Eq. (9.3) implies χ can be expressed as the linear superposition of modes varying as $\exp(im\theta + ikz)$. For each choice of m and k, Eq. (9.3) becomes

$$\frac{\partial^2 \chi}{\partial r^2} + \frac{1}{r}\frac{\partial \chi}{\partial r} - \left(\frac{m^2}{r^2} + k^2\right)\chi = 0. \tag{9.4}$$

Defining $s = kr$, this can be recast as

$$\frac{\partial^2 \chi}{\partial s^2} + \frac{1}{s}\frac{\partial \chi}{\partial s} - \left(1 + \frac{m^2}{s^2}\right)\chi = 0, \tag{9.5}$$

a modified Bessel's equation. The solutions of Eq. (9.5) are the modified Bessel functions $I_m(kr)$ and $K_m(kr)$. Thus, the general solution of Laplace's equation here is

$$\chi(r,\theta,z) = \sum_{m=-\infty}^{\infty} \int \left[a_m(k) I_m(kr) + b_m(k) K_m(kr)\right] \mathrm{e}^{im\theta + ikz} \mathrm{d}k, \tag{9.6}$$

where the coefficients $a_m(k)$, $b_m(k)$ are determined by specifying either χ or its normal derivative on the bounding surface. Analogous solutions can be found in geometries having other symmetries.

This behavior can be viewed in a more general way. Equation (9.3) states that the sum of partial second derivatives in two or three different directions is zero, so at least one of these terms must be negative and at least one term must be positive. Since negative χ''/χ corresponds to oscillatory (harmonic) behavior and positive χ''/χ corresponds to exponential (non-harmonic) behavior, any solution of Laplace's equation must be oscillatory in one or two directions (the θ and z directions for the cylindrical example here), and exponentially growing or decaying in the remaining direction or directions (the r direction for the cylindrical example here).

Non-vacuum magnetic fields are more complicated than vacuum fields and, unlike vacuum fields, are not uniquely determined by the surface boundary conditions. This is because non-vacuum fields are determined by both the current distribution within the volume and the surface boundary conditions. Vacuum fields are distinguished from non-vacuum fields because vacuum fields are the lowest energy fields satisfying given boundary conditions on the surface S of a volume V. Let us now prove this statement.

Consider a volume V bounded by a surface S over which boundary conditions are specified. Let $\mathbf{B}_{\min}(\mathbf{r})$ be the magnetic field having the *lowest* stored magnetic energy of all possible magnetic fields satisfying the prescribed boundary conditions. We use methods of variational calculus to prove this lowest energy field is the vacuum field.

Consider some slightly different field, denoted as $\mathbf{B}(\mathbf{r})$, that satisfies the same boundary conditions as $\mathbf{B}_{\min}$. This slightly different field can be expressed as

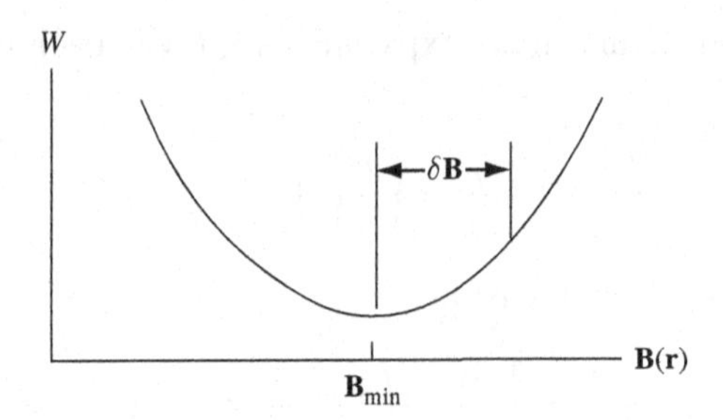

Fig. 9.1 Magnetic field energy for different configurations. $\mathbf{B}_{\text{min}}$ is the configuration with minimum magnetic energy for a given boundary condition.

$\mathbf{B}(\mathbf{r}) = \mathbf{B}_{\text{min}}(\mathbf{r}) + \delta\mathbf{B}(\mathbf{r})$, where $\delta\mathbf{B}(\mathbf{r})$ is a small arbitrary variation about $\mathbf{B}_{\text{min}}$. This situation is sketched in a qualitative fashion in Fig. 9.1, where the horizontal axis represents a continuum of different allowed choices for the vector function $\mathbf{B}(\mathbf{r})$. Since $\mathbf{B}(\mathbf{r})$ satisfies the same boundary conditions as $\mathbf{B}_{\text{min}}$, $\delta\mathbf{B}(\mathbf{r})$ must vanish on S. The magnetic energy W associated with $\mathbf{B}(\mathbf{r})$ can be evaluated as

$$\begin{aligned} 2\mu_0 W &= \int_V (\mathbf{B}_{\text{min}} + \delta\mathbf{B})^2 \, \mathrm{d}^3 r \\ &= \int_V \mathbf{B}_{\text{min}}^2 \mathrm{d}^3 r + 2\int_V \delta\mathbf{B} \cdot \mathbf{B}_{\text{min}} \mathrm{d}^3 r + \int_V (\delta\mathbf{B})^2 \, \mathrm{d}^3 r. \end{aligned} \tag{9.7}$$

If the middle term does not vanish, $\delta\mathbf{B}$ could always be chosen to be antiparallel to $\mathbf{B}_{\text{min}}$ inside V, in which case $\delta\mathbf{B} \cdot \mathbf{B}_{\text{min}}$ would be negative. Since $\delta\mathbf{B}$ is assumed small, $\delta\mathbf{B} \cdot \mathbf{B}_{\text{min}}$ would be larger in magnitude than $(\delta\mathbf{B})^2$. Such a choice for $\delta\mathbf{B}$ would make W lower than the energy of the assumed lowest energy field, thus contradicting the assumption that $\mathbf{B}_{\text{min}}$ is the lowest energy field. Thus, the only way to ensure that $\mathbf{B}_{\text{min}}$ is indeed the true minimum is to require

$$\int_v \delta\mathbf{B} \cdot \mathbf{B}_{\text{min}} \mathrm{d}^3 r = 0 \tag{9.8}$$

no matter how $\delta\mathbf{B}$ is chosen. Using $\delta\mathbf{B} = \nabla \times \delta\mathbf{A}$ and the vector identity $\nabla \cdot (\mathbf{A} \times \mathbf{B}) = \mathbf{B} \cdot \nabla \times \mathbf{A} - \mathbf{A} \cdot \nabla \times \mathbf{B}$, Eq. (9.8) can be integrated by parts to obtain

$$\int_V [\nabla \cdot (\delta\mathbf{A} \times \mathbf{B}_{\text{min}}) + \delta\mathbf{A} \cdot \nabla \times \mathbf{B}_{\text{min}}] \, \mathrm{d}^3 r = 0. \tag{9.9}$$

The first term can be transformed into a surface integral over S using Gauss' theorem. This surface integral vanishes because $\delta\mathbf{A}$ must vanish on the bounding surface (recall that the variation satisfies the same boundary condition as the minimum energy field). Because $\delta\mathbf{B}$ is arbitrary within V, $\delta\mathbf{A}$ must also be arbitrary within V and so the only way for the second term in Eq. (9.9) to vanish is to have $\nabla \times \mathbf{B}_{\text{min}} = 0$. Thus, $\mathbf{B}_{\text{min}}$ must be a vacuum field.

An important corollary is as follows: suppose boundary conditions are specified on the surface enclosing some volume. These boundary conditions can be considered as "rules" that must be satisfied by any solution to the equations. All configurations satisfying the imposed boundary condition and having finite current within the volume are *not* in the lowest energy state. Thus, non-vacuum fields can, in principle, have free energy available for driving boundary-condition-preserving instabilities.

9.3 Force-free fields

Although the vacuum field is the lowest energy configuration satisfying prescribed boundary conditions, non-vacuum configurations do not always "decay" to this lowest energy state. This is because there also exists a family of higher energy configurations to which the system may decay; these are the so-called force-free states. The current is not zero in a force-free state but the magnetic force is zero because the current density is everywhere parallel to the magnetic field. Thus $\mathbf{J} \times \mathbf{B}$ vanishes even though both $\mathbf{J}$ and $\mathbf{B}$ are finite. If a plasma not initially in a force-free state somehow evolves towards a force-free state, it will become "stuck" in the force-free state because, by definition, no forces can act on the plasma to move it out of the force-free state. The magnetic energy of a force-free field is not the absolute minimum energy for the specified boundary conditions, but it is a local minimum in configuration space as sketched in Fig. 9.2. This hierarchy of states is somewhat analogous to the states of a quantum system – the vacuum field is the analog of the ground state and the force-free states are the analogs of higher energy quantum states.

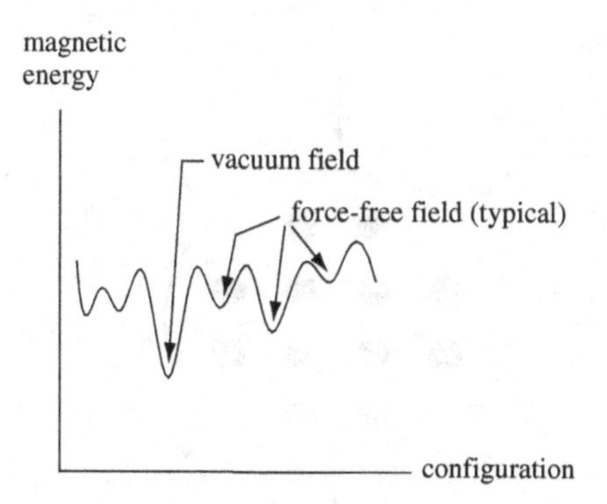

Fig. 9.2 Sketch of magnetic energy dependence on configuration of a system with fixed, specified boundary conditions. The variation of the configuration would correspond to different internal current profiles. The force-free configurations are local energy minima while the vacuum configuration has the absolute lowest minimum.

9.4 Magnetic pressure and tension

Much useful insight can be obtained by considering the force between two parallel current-carrying wires as shown in Fig. 9.3. Calculation of the field **B** observed at one wire due to the current in the other shows that the $\mathbf{J} \times \mathbf{B}$ force is such as to push the wires together; i.e., parallel currents attract each other. Conversely, antiparallel currents repel each other.

A bundle of parallel wires as shown in Fig. 9.4 will therefore mutually attract each other resulting in an effective net force that acts to reduce the diameter of the bundle. The bundle could be replaced by a distributed current such as the current carried by a finite-radius, cylindrical plasma. This contracting, inward-directed force is called the pinch force or the pinch effect.

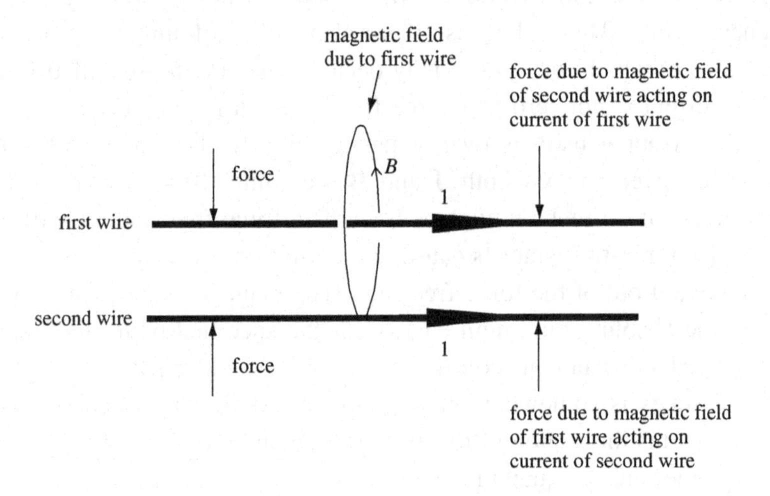

Fig. 9.3 Parallel currents attract each other.

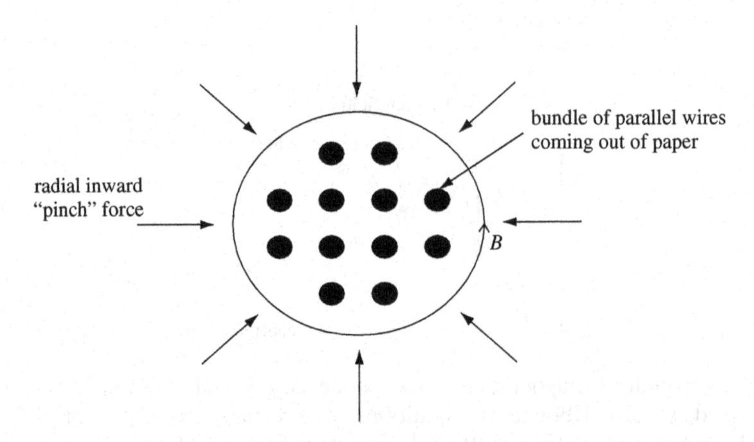

Fig. 9.4 Bundle of currents attract each other giving effective radial inward pinch force.

The pinch force may be imagined as being due to a "tension" in the azimuthal magnetic field that wraps around the distributed current. Extending this metaphor, the azimuthal magnetic field is visualized as acting like an "elastic band" which encircles the distributed current and squeezes or pinches the current to a smaller diameter. This concept is consistent with the situation of two permanent magnets attracting each other when the respective north and south poles face each other. The magnetic field lines go from the north pole of one magnet to the south pole of the other and so one can imagine that the attraction of the two magnets is due to tension in the field lines spanning the gap between the two magnets.

Now consider a current-carrying loop such as is shown in Fig. 9.5. Because the currents on opposite sides of the loop flow in opposing directions, there will be a repulsive force between the current element at each point and the current element on the opposite side of the loop. The net result is a force directed to expand the diameter of the loop. This force is called the hoop force or hoop stress. Hoop force may also be interpreted metaphorically by introducing the concept of magnetic pressure. As shown in Fig. 9.5, the magnetic field lines linking the current are more dense inside the loop than outside, a purely geometrical effect resulting from the curvature of the current path. Since magnetic field strength is proportional to field line density, the magnetic field is stronger on the inside of the loop than on the outside, i.e., B^2 is stronger on the inside than on the outside. The magnetic field in the plane of the loop is normal to the plane and so one can explain the hoop force by assigning a magnetic "pressure" proportional to B^2 acting in the direction perpendicular to $\mathbf{B}$. Because B^2 is stronger on the inside of the loop than on the outside, there is a larger magnetic pressure on the inside. The outward force due to this pressure imbalance is consistent with the hoop force.

The combined effects of magnetic pressure and tension can be visualized by imagining a current-carrying loop where the conductor has a finite diameter. Suppose this loop initially has the relative proportions of an automobile tire, as shown in Fig. 9.6. The hoop force will cause the diameter of the loop to increase, while the pinch force will cause the diameter of the conductor to decrease. The net

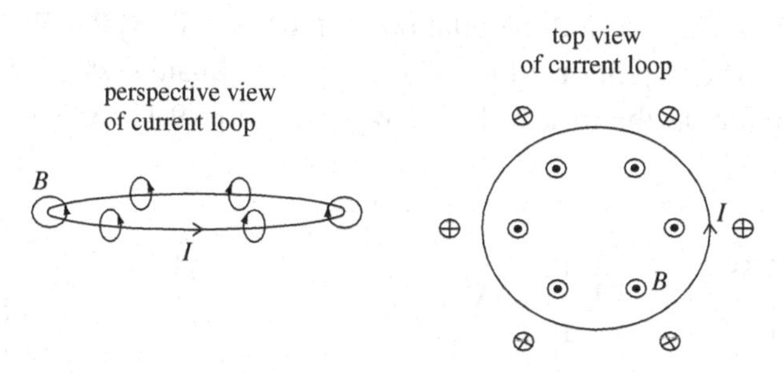

Fig. 9.5 Perspective and top view of magnetic fields generated by a current loop.

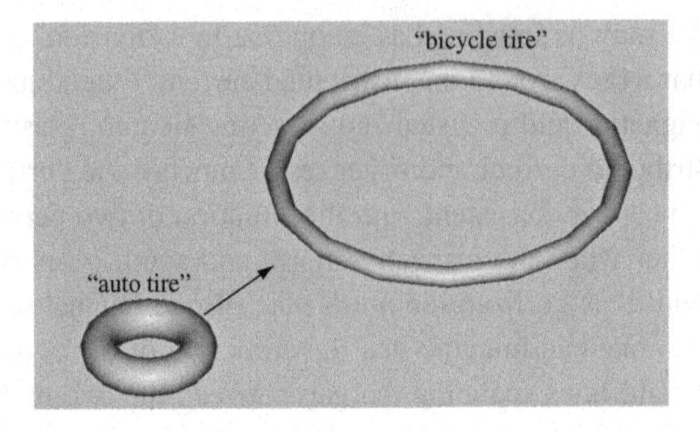

Fig. 9.6 Magnetic forces act to transform a fat, small current loop (auto tire) into a large, skinny loop (bicycle tire).

result is that these combined forces will cause the automobile tire to evolve towards the relative proportions of a bicycle tire (large major radius, small minor radius).

9.5 Magnetic stress tensor

The existence of magnetic pressure and tension shows that the magnetic force is different in different directions, and so the magnetic force ought to be characterized by an anisotropic stress tensor. To establish this mathematically, the vector identity $\nabla B^2/2 = \mathbf{B}\cdot\nabla\mathbf{B} + \mathbf{B}\times\nabla\times\mathbf{B}$ is invoked so that the magnetic force can be expressed as

$$\begin{aligned}\mathbf{J}\times\mathbf{B} &= \frac{1}{\mu_0}(\nabla\times\mathbf{B})\times\mathbf{B} \\ &= \frac{1}{\mu_0}\left[-\nabla\left(\frac{B^2}{2}\right) + \mathbf{B}\cdot\nabla\mathbf{B}\right] \\ &= -\frac{1}{\mu_0}\nabla\cdot\left[\frac{B^2}{2}\mathbf{I} - \mathbf{BB}\right],\end{aligned} \tag{9.10}$$

where $\mathbf{I}$ is the unit tensor and the relation $\nabla\cdot(\mathbf{BB}) = (\nabla\cdot\mathbf{B})\,\mathbf{B} + \mathbf{B}\cdot\nabla\mathbf{B} = \mathbf{B}\cdot\nabla\mathbf{B}$ has been used. At any point $\mathbf{r}$ a local Cartesian coordinate system can be defined with z axis parallel to the local value of $\mathbf{B}$ so that Eq. (9.1) can be written as

$$\rho\left[\frac{\partial\mathbf{U}}{\partial t} + \mathbf{U}\cdot\nabla\mathbf{U}\right] = -\nabla\cdot\begin{bmatrix} P + \dfrac{B^2}{2\mu_0} & & \\ & P + \dfrac{B^2}{2\mu_0} & \\ & & P - \dfrac{B^2}{2\mu_0}\end{bmatrix} \tag{9.11}$$

showing again that the magnetic field acts like a pressure in the directions transverse to **B** (i.e., x, y directions in the local Cartesian system) and like a tension in the direction parallel to **B**.

While the above interpretation is certainly useful, it can be somewhat misleading because it might be interpreted as implying the existence of a force in the direction of **B** when in fact no such force exists because $\mathbf{J}\times\mathbf{B}$ clearly does not have a component in the **B** direction. A more accurate way to visualize the relation between magnetic pressure and tension is to rearrange the second line of Eq. (9.10) as

$$\mathbf{J}\times\mathbf{B} = \frac{1}{\mu_0}\left[-\nabla\left(\frac{B^2}{2}\right) + B^2\hat{B}\cdot\nabla\hat{B} + \hat{B}\hat{B}\cdot\nabla\left(\frac{B^2}{2}\right)\right] = \frac{1}{\mu_0}\left[-\nabla_\perp\left(\frac{B^2}{2}\right) + B^2\boldsymbol{\kappa}\right] \tag{9.12}$$

or

$$\mathbf{J}\times\mathbf{B} = \frac{1}{\mu_0}\left[-\nabla_\perp\left(\frac{B^2}{2}\right) + B^2\boldsymbol{\kappa}\right]. \tag{9.13}$$

Here

$$\boldsymbol{\kappa} = \hat{B}\cdot\nabla\hat{B} = -\frac{\hat{R}}{R} \tag{9.14}$$

is a measure of the curvature of the magnetic field at a selected point on a field line and, in particular, **R** is the local radius of curvature vector. The vector **R** goes from the center of curvature to the selected point on the field line. The $\boldsymbol{\kappa}$ term in Eq. (9.13) describes a force that tends to straighten out magnetic curvature and is a more precise way for characterizing field line tension (recall that tension similarly acts to straighten out curvature). The term involving $\nabla_\perp B^2$ portrays a magnetic force due to pressure gradients perpendicular to the magnetic field and is a more precise expression of the hoop force.

In our earlier discussion it was shown that the vacuum magnetic field is the lowest energy state of all fields satisfying prescribed boundary conditions. A vacuum field might be curved for certain boundary conditions (e.g., a permanent magnet with finite dimensions) and, in such a case, the two terms in Eq. (9.13) are both finite but exactly cancel each other. We can think of the minimum-energy state for given boundary conditions as being analogous to the equilibrium state of a system of stiff rubber hoses that have been pre-formed into shapes having the morphology of the vacuum magnetic field and that have their ends fixed at the bounding surface. Currents will cause the morphology of this system to deviate from the equilibrium state. However, since any deformation requires work to be done on the system, currents invariably cause the system to be in a higher energy state.

9.6 Flux preservation, energy minimization, and inductance

Another useful way of understanding magnetic field behavior relates to the concept of electric circuit inductance. The self-inductance L of a circuit component is defined as the magnetic flux Φ linking the component divided by the current I flowing through the component, i.e.,

$$L = \frac{\Phi}{I}. \tag{9.15}$$

Consider an arbitrary short-circuited coil with current I located in an infinite volume V and let C denote the three-dimensional spatial contour traced out by the wire constituting the coil. The total magnetic flux linked by the turns of the coil can be expressed as

$$\Phi = \int \mathbf{B} \cdot \mathrm{d}\mathbf{s} = \int \nabla \times \mathbf{A} \cdot \mathrm{d}\mathbf{s} = \oint_C \mathbf{A} \cdot \mathrm{d}\mathbf{l}, \tag{9.16}$$

where the surface integral is over the area elements linked by the coil turns. The energy contained in the magnetic field produced by the coil is

$$\begin{aligned} W &= \int_V \frac{B^2}{2\mu_0} \mathrm{d}^3 r \\ &= \frac{1}{2\mu_0} \int_V \mathbf{B} \cdot \nabla \times \mathbf{A} \mathrm{d}^3 r. \end{aligned} \tag{9.17}$$

However, using the vector identity $\nabla \cdot (\mathbf{A} \times \mathbf{B}) = \mathbf{B} \cdot \nabla \times \mathbf{A} - \mathbf{A} \cdot \nabla \times \mathbf{B}$ this magnetic energy can be expressed as

$$W = \frac{1}{2\mu_0} \int_V \mathbf{A} \cdot \nabla \times \mathbf{B} \mathrm{d}^3 r + \frac{1}{2\mu_0} \int_{S_\infty} \mathrm{d}\mathbf{s} \cdot \mathbf{A} \times \mathbf{B}, \tag{9.18}$$

where Gauss' law has been invoked to obtain the second term, an integral over the surface at infinity, S_∞. This surface integral vanishes because (i) at infinity the magnetic field must fall off at least as fast as a dipole, i.e., $B \sim R^{-3}$, where R is the distance to the origin, (ii) the vector potential magnitude A scales as the integral of B so $A \sim R^{-2}$, and (iii) the surface at infinity scales as R^2.

Using Ampère's law, Eq. (9.18) can thus be rewritten as

$$W = \frac{1}{2} \int_V \mathbf{A} \cdot \mathbf{J} \mathrm{d}^3 r. \tag{9.19}$$

However, $\mathbf{J}$ is only finite in the coil wire and so the integral reduces to an integral over the volume of the wire. A volume element of wire can be expressed as

$d^3r = \mathbf{ds} \cdot \mathbf{dl}$, where $\mathbf{dl}$ is an element of length along the wire, and $\mathbf{ds}$ is the cross-sectional area of the wire. Since $\mathbf{J}$ and $\mathbf{dl}$ are parallel, they can be interchanged in Eq. (9.19) which becomes

$$\begin{aligned} W &= \frac{1}{2}\int_{V_{coil}} \mathbf{A} \cdot \mathbf{dl}\, \mathbf{J} \cdot \mathbf{ds} \\ &= \frac{I}{2}\oint \mathbf{A} \cdot \mathbf{dl} \\ &= \frac{I\Phi}{2} \\ &= \frac{\Phi^2}{2L} \\ &= \frac{1}{2}LI^2, \end{aligned} \tag{9.20}$$

where $\mathbf{J} \cdot \mathbf{ds} = I$ is the current flowing through the wire. Thus, the energy stored in the magnetic field produced by a coil is just the inductive energy of the coil.

If the coil is perfectly short-circuited, then it must be flux conserving, for if there were a change in flux, a voltage would appear across the ends of the coil. A closed current flowing in a perfectly conducting plasma is thus equivalent to a short-circuited current-carrying coil and so the perfectly conducting plasma can be considered as a flux-conserver. If flux is conserved, i.e., $\Phi = const.$, the second from last line in Eq. (9.20) shows that *the magnetic energy of the system will be lowered by any rearrangement of circuit topology that increases self-inductance.*

Hot plasmas are reasonably good flux conservers because of their high electrical conductivity. Thus, any inductance-increasing change in the topology of plasma currents will release free energy, which could be used to drive an instability. Since forces act so as to reduce the potential energy of a system, magnetic forces due to current flowing in a plasma will always act so as to increase the self-inductance of the configuration. One can therefore write the force $\mathbf{F}$ due to a flux-conserving change in inductance L as

$$\mathbf{F} = -\frac{\Phi^2}{2}\nabla\left(\frac{1}{L}\right). \tag{9.21}$$

The pinch force is consistent with this interpretation since the inductance of a conductor depends inversely on its radius. The hoop force is also consistent with this interpretation since inductance of a current loop increases with the major radius of the loop. More complicated behavior can also be explained, especially the kink instability to be discussed later. In the kink instability, current initially flowing in a straight line develops an instability that causes the current path to become helical. Since a coil (helix) has more inductance than a straight length of the same wire, the effect of the kink instability also acts so as to increase the circuit self-inductance.

9.7 Static versus dynamic equilibria

We define (i) a static equilibrium to be a time-independent solution to Eq. (9.1) having no flow velocity and (ii) a dynamic equilibrium as a solution with steady-state flow velocities. Thus, $\mathbf{U} = 0$ for a static equilibrium whereas $\mathbf{U}$ is finite and steady-state for a dynamic equilibrium. For a static equilibrium the MHD equation of motion reduces to

$$\nabla P = \mathbf{J} \times \mathbf{B}. \tag{9.22}$$

For a dynamic equilibrium the MHD equation of motion reduces to

$$\rho \mathbf{U} \cdot \nabla \mathbf{U} + \nabla P = \mathbf{J} \times \mathbf{B} + \nu \rho \nabla^2 \mathbf{U}, \tag{9.23}$$

where the last term represents a viscous damping and ν is the kinematic viscosity. By taking the curl of these last two equations we see that for static equilibria the magnetic force must be conservative, i.e., $\nabla \times (\mathbf{J} \times \mathbf{B}) = 0$, whereas for dynamic equilibria the magnetic force is typically not conservative since in general $\nabla \times (\mathbf{J} \times \mathbf{B}) \neq 0$. Thus, the character of the magnetic field is quite different for the two cases. Static equilibria are relevant to plasma confinement devices such as tokamaks, stellarators, reversed field pinches, and spheromaks, while dynamic equilibria are mainly relevant to arcs, jets, and magnetoplasmadynamic thrusters, but can also be relevant to tokamaks, etc., if there are flows. Both static and dynamic equilibria occur in space plasmas.

9.8 Static equilibria

9.8.1 *Static equilibria in two dimensions: the Bennett pinch*

The simplest static equilibrium was first investigated by Bennett (1934) and is called the Bennett pinch or z-pinch (here z refers to the direction of the current). This configuration, sketched in Fig. 9.7, consists of an infinitely long axisymmetric cylindrical plasma with axial current density $J_z = J_z(r)$ and no other currents.

The axial current flowing within a circle of radius r is

$$I(r) = \int_0^r 2\pi r' J_z(r') \mathrm{d}r' \tag{9.24}$$

and the axial current density is related to this integrated current by

$$J_z(r) = \frac{1}{2\pi r} \frac{\partial I}{\partial r}. \tag{9.25}$$

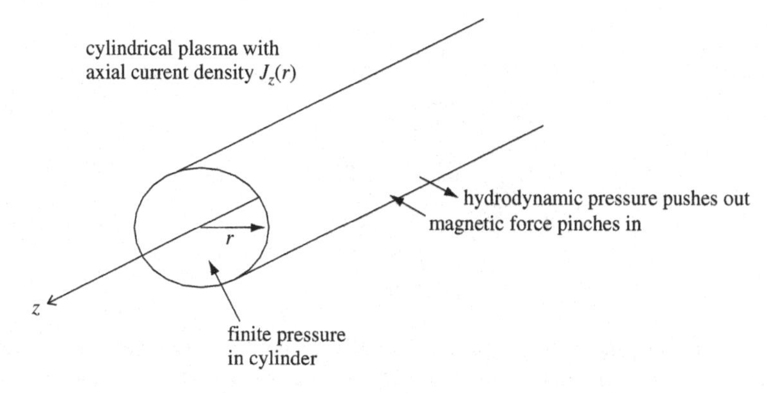

Fig. 9.7 Geometry of Bennett pinch.

Since Ampère's law gives

$$B_\theta(r) = \frac{\mu_0 I(r)}{2\pi r}, \tag{9.26}$$

Eq. (9.22) can be written

$$r^2 \frac{\partial P}{\partial r} = -\frac{\mu_0}{8\pi^2}\frac{\partial I^2}{\partial r}. \tag{9.27}$$

Integrating Eq. (9.27) from $r = 0$ to $r = a$, where a is the outer radius of the cylindrical plasma, gives the relation

$$\int_0^a r^2 \frac{\partial P}{\partial r} \mathrm{d}r = \left[r^2 P(r)\right]_0^a - 2\int_0^a rP(r)\mathrm{d}r = -\frac{\mu_0 I^2(a)}{8\pi^2}. \tag{9.28}$$

The integrated term vanishes at both $r = 0$ and $r = a$ since $P(a) = 0$ by definition. If the temperature is uniform, the pressure can be expressed as $P(r) = n(r)\kappa T$ and so Eq. (9.28) can be expressed as

$$I^2 = \frac{8\pi N\kappa T}{\mu_0}, \tag{9.29}$$

where $N = \int_0^a n(r)2\pi r\mathrm{d}r$ is the number of particles per axial length. Equation (9.29), called the Bennett relation, shows that the current required to confine a given N and T is independent of the details of the internal density profile. This relation describes the simplest non-trivial MHD equilibrium and suggests that quite modest currents could contain substantial plasma pressures. This relation motivated the design of early magnetic fusion confinement devices but, as will be seen, it is overly optimistic because simple z-pinch equilibria turn out to be highly unstable.

Confinement using currents flowing in the azimuthal direction is also possible, but this configuration, known as a θ-pinch, is fundamentally transient. In a θ-pinch, a rapidly changing azimuthal current in a coil surrounding a cylindrical

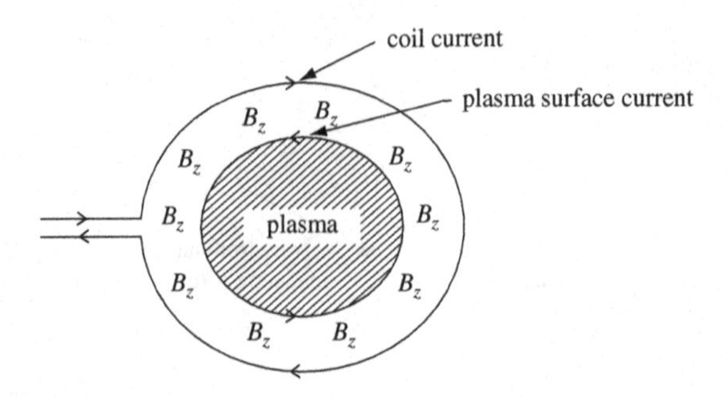

Fig. 9.8 Theta pinch configuration. Rapidly changing azimuthal coil current induces equal and opposite azimuthal current in plasma surface. Pressure of B_z between these two currents pushes in on plasma and provides confinement.

plasma creates a transient B_z field as shown in Fig. 9.8. Because the conducting plasma conserves magnetic flux, the transient B_z field cannot penetrate the plasma and so is confined to the vacuum region between the plasma and the coil. This exclusion of the B_z field by the plasma requires the existence in the plasma surface of an induced azimuthal current that creates in the plasma interior a magnetic field, which exactly cancels the coil-produced transient B_z field. The radial confining force results from a radially inward force $\mathbf{J}\times\mathbf{B} = J_\theta\hat{\theta}\times B_z\hat{z}$, where B_z is the magnetic field associated with the current in the coil and J_θ is the plasma surface current. An alternative but equivalent point of view is to invoke the concept that opposite currents repel and argue that the azimuthal current in the coil repels the oppositely directed azimuthal current in the plasma surface, thereby pushing the plasma inwards and so balancing the outwards force due to plasma pressure. The θ-pinch configuration is necessarily transient, because the induced azimuthally directed surface current cannot be sustained in steady state.

9.8.2 *Impossibility of self-confinement of current-carrying plasma in three dimensions: the virial theorem*

The Bennett analysis showed that axial currents flowing in an infinitely long cylindrical plasma generate a pinch force, which confines a finite pressure plasma. The inward pinch force balances the outward force associated with the pressure gradient. The question now is whether this two-dimensional result can be extended to three dimensions; i.e., is it possible to have a finite-pressure three-dimensional plasma, as shown in Fig. 9.9, that is confined entirely by currents circulating within the plasma? To be more specific, is it possible to have a finite-radius plasma sphere surrounded by vacuum where the confinement of the finite plasma pressure is entirely provided by the magnetic force of currents circulating in the plasma; i.e., can the plasma hold itself together by its own "bootstraps?" The

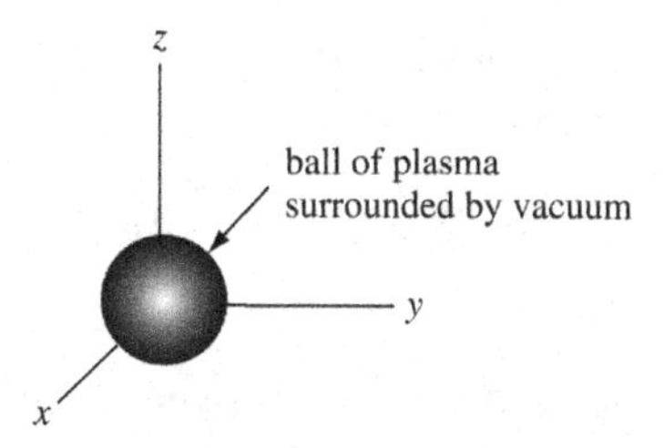

Fig. 9.9 Plasma sphere with finite pressure surrounded by vacuum.

answer, a resounding "no," is provided by a virial theorem due to Shafranov (1966). A virial is a suitably weighted integral over the entirety of a system and contains information about how an extensive property such as energy is partitioned in the system. For example, in mechanics the virial can be the time average of the potential or of the kinetic energy.

The MHD virial theorem is obtained by supposing a self-confining configuration exists and then showing this supposition leads to a contradiction.

We therefore postulate the existence of a spherical plasma with the following properties:

1. the plasma has finite radius a and is surrounded by vacuum,
2. the plasma is in static MHD equilibrium,
3. the plasma has finite internal pressure and the pressure gradient is entirely balanced by magnetic forces due to currents circulating in the plasma; i.e., there are no currents in the surrounding vacuum region.

Using Eqs. (9.10) and (9.11) the static MHD equilibrium can be expressed as

$$\nabla\cdot\mathbf{T}=0, \tag{9.30}$$

where the tensor $\mathbf{T}$ is defined as

$$\mathbf{T}=\left(P+\frac{B^2}{2\mu_0}\right)\mathbf{I}-\frac{1}{\mu_0}\mathbf{BB}. \tag{9.31}$$

Let $\mathbf{r}=x\hat{x}+y\hat{y}+z\hat{z}$ be the vector from the center of the plasma to the point of observation and consider the virial expression

$$\begin{aligned}\nabla\cdot(\mathbf{T}\cdot\mathbf{r}) &= \sum_{jk}\frac{\partial}{\partial x_j}\left(T_{jk}x_k\right)\\ &= (\nabla\cdot\mathbf{T})\cdot\mathbf{r}+\sum_{jk}T_{jk}\frac{\partial}{\partial x_j}x_k\\ &= \text{Trace }\mathbf{T},\end{aligned} \tag{9.32}$$

where Trace $\mathbf{T}=\sum_{jk}T_{jk}\delta_{jk}$. From Eq. (9.31) (or, equivalently from the matrix form in Eq. (9.11)) it is seen that

$$\text{Trace }\mathbf{T}=3P+B^2/2\mu_0$$

is *positive-definite.*

We now integrate both sides of Eq. (9.32) over all space. Since the right-hand side of Eq. (9.32) is positive definite, the integral of the right-hand side over all space is finite and positive. The integral of the left-hand side can be transformed to a surface integral at infinity using Gauss' theorem

$$\begin{aligned}\int \mathrm{d}^3r\nabla\cdot(\mathbf{T}\cdot\mathbf{r}) &= \int_{S_\infty}\mathrm{d}\mathbf{s}\cdot\mathbf{T}\cdot\mathbf{r}\\ &= \int_{S_\infty}\mathrm{d}\mathbf{s}\cdot\left[\left(P+\frac{B^2}{2\mu_0}\right)\mathbf{I}-\frac{1}{\mu_0}\mathbf{BB}\right]\cdot\mathbf{r}.\end{aligned} \tag{9.33}$$

Since the plasma is assumed to have finite extent, P is zero on the surface at infinity. The magnetic field can be expanded in multipoles with the lowest order multipole being a dipole. The magnetic field of a dipole scales as r^{-3} for large r while the surface area $\int \mathrm{d}\mathbf{s}$ scales as r^2. Thus the right hand side term scales as $\int \mathrm{d}sB^2r{\sim}r^{-3}$ and so vanishes as $r{\to}\infty$. This is in contradiction to the left hand side being positive definite and so the set of initial assumptions must be erroneous. Thus, any finite extent, three-dimensional, static plasma equilibrium must involve at least some currents external to the plasma. A finite extent, three-dimensional static plasma can therefore only be in equilibrium if at least some of the magnetic field is produced by currents in coils that are both external to the plasma and held in place by some mechanical structure. If the coils were not supported by a mechanical structure, then the current in the coils could be considered as part of the MHD plasma and the virial theorem would be violated. In summary, a finite extent three-dimensional plasma in static equilibrium with a finite internal hydrodynamic pressure P must ultimately have some tangible exterior object to "push against." The buttressing is provided by magnetic forces acting between currents in external coils and currents in the plasma.

9.8.3 *Three-dimensional static equilibria: the Grad–Shafranov equation*

Despite the simple appearance of Eq. (9.22), its three-dimensional solution is far from trivial. Before even attempting to find a solution, it is important to decide the appropriate way to pose the problem, i.e., it must be decided which quantities are prescribed and which are to be solved for. For example, one might imagine prescribing a pressure profile $P(\mathbf{r})$ and then using Eq. (9.22) to determine a corresponding $\mathbf{B}(\mathbf{r})$ with associated current $\mathbf{J}(\mathbf{r})$; alternatively one could imagine prescribing $\mathbf{B}(\mathbf{r})$ and then using Eq. (9.22) to determine $P(\mathbf{r})$. Unfortunately, neither of these approaches work in general because solutions to Eq. (9.22) exist for only a very limited set of functions.

The reason why an arbitrary magnetic field cannot be specified is that $\nabla \times \nabla P = 0$ is always true by virtue of a mathematical identity whereas $\nabla \times (\mathbf{J} \times \mathbf{B}) = 0$ is true only for certain types of $\mathbf{B}(\mathbf{r})$. It is also true that an arbitrary equilibrium pressure profile $P(\mathbf{r})$ cannot be prescribed (for example, see Assignment 2 where it is demonstrated that no $\mathbf{J} \times \mathbf{B}$ force exists that can confine a plasma with a spherically symmetric pressure profile). Equilibria thus exist only for certain specific situations, and these situations typically require symmetry in some direction. We shall now examine a very important example, namely static equilibria that are azimuthally symmetric about an axis (typically defined as the z axis). This symmetry applies to a wide variety of magnetic confinement devices used in magnetic fusion research, for example tokamaks, reversed field pinches, spheromaks, and field reversed theta pinches.

We start the analysis by assuming azimuthal symmetry about the z axis of a cylindrical coordinate system $\{r, \phi, z\}$ so any *physical* quantity f has the property $\partial f/\partial \phi = 0$. Because of the identity $\nabla \times \nabla \phi = 0$, algebraic manipulations become considerably simplified if vectors in the ϕ direction are expressed in terms of $\nabla \phi = \hat{\phi}/r$ rather than in terms of $\hat{\phi}$. As sketched in Fig. 9.10 the term toroidal denotes vectors in the ϕ direction (long way around a torus) and the term poloidal denotes vectors in the $r - z$ plane.

The most general form for an axisymmetric magnetic field is

$$\mathbf{B} = \frac{1}{2\pi} \left(\nabla \psi \times \nabla \phi + \mu_0 I \nabla \phi \right). \tag{9.34}$$

$\psi(r, z)$ is called the poloidal flux and $I(r, z)$ is the current linked by a circle of radius r with center on the axis at axial location z. The toroidal magnetic field then is

$$\mathbf{B}_{tor} = B_\phi \hat{\phi} = \frac{\mu_0 I}{2\pi} \nabla \phi = \frac{\mu_0 I}{2\pi r} \hat{\phi} \tag{9.35}$$

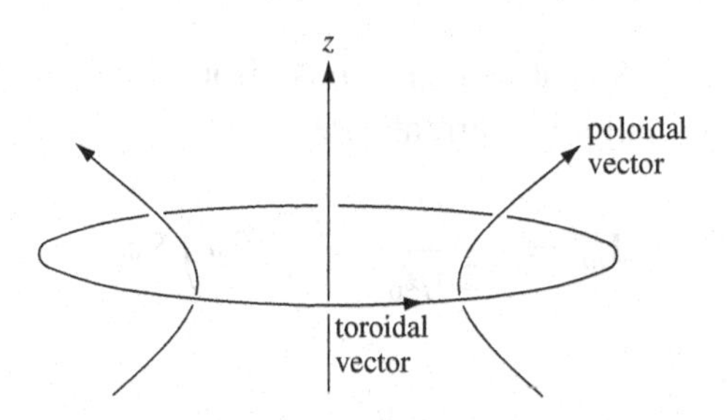

Fig. 9.10 Toroidal vectors and poloidal vectors.

showing that the functional form of Eq. (9.34) is consistent with Ampère's law $\oint \mathbf{B}\cdot d\mathbf{l} = \mu_0 I$. The poloidal magnetic field is

$$\mathbf{B}_{pol} = \frac{1}{2\pi}\left(\nabla\psi \times \nabla\phi\right). \tag{9.36}$$

Integration of the poloidal magnetic field over the area of a circle of radius r with center at axial location z gives

$$\int_0^r \mathbf{B}_{pol}\cdot d\mathbf{s} = \int_0^r \frac{1}{2\pi}\nabla\psi \times \nabla\phi \cdot \hat{z}2\pi r' dr' = \psi(r,z); \tag{9.37}$$

thus $\psi(r,z)$ is the poloidal flux at location r, z. The concept of poloidal flux depends on the existence of axisymmetry, which makes it always possible to associate any location r, z with a circular area of radius r centered at $z = 0$.

Axisymmetry also provides a useful relationship between toroidal and poloidal vectors. In particular, the curl of a toroidal vector is poloidal since

$$\nabla \times \mathbf{B}_{tor} = \frac{\mu_0}{2\pi}\nabla I \times \nabla\phi \tag{9.38}$$

and similarly the curl of a poloidal vector is toroidal since

$$\nabla \times \mathbf{B}_{pol} = \nabla \times \left(B_r\hat{r} + B_z\hat{z}\right) = \hat{\phi}\left(\frac{\partial B_r}{\partial z} - \frac{\partial B_z}{\partial r}\right). \tag{9.39}$$

The curl of the poloidal magnetic field is a Laplacian-like operator on ψ since

$$\nabla\phi\cdot\nabla\times\mathbf{B}_{pol} = \nabla\cdot\left(\mathbf{B}_{pol}\times\nabla\phi\right) = \nabla\cdot\left(\frac{1}{2\pi}\left[\nabla\psi\times\nabla\phi\right]\times\nabla\phi\right) = -\frac{1}{2\pi}\nabla\cdot\left(\frac{1}{r^2}\nabla\psi\right), \tag{9.40}$$

a relationship established using the vector identity $\nabla\cdot(\mathbf{F}\times\mathbf{G}) = \mathbf{G}\cdot\nabla\times\mathbf{F} - \mathbf{F}\cdot\nabla\times\mathbf{G}$. Because $\nabla\times\mathbf{B}_{pol}$ is purely toroidal and $\hat{\phi} = r\nabla\phi$, one can write

$$\nabla\times\mathbf{B}_{pol} = -\frac{r^2}{2\pi}\nabla\cdot\left(\frac{1}{r^2}\nabla\psi\right)\nabla\phi. \tag{9.41}$$

Ampère's law states $\nabla\times\mathbf{B} = \mu_0\mathbf{J}$. Thus, from Eqs. (9.38) and (9.41) the respective toroidal and poloidal currents are

$$\mathbf{J}_{tor} = -\frac{r^2}{2\pi\mu_0}\nabla\cdot\left(\frac{1}{r^2}\nabla\psi\right)\nabla\phi \tag{9.42}$$

and

$$\mathbf{J}_{pol} = \frac{1}{2\pi}\nabla I\times\nabla\phi. \tag{9.43}$$

We are now in a position to evaluate the magnetic force in Eq. (9.22). After decomposing the magnet field and current into toroidal and poloidal components, Eq. (9.22) becomes

$$\nabla P = \mathbf{J}_{pol} \times \mathbf{B}_{tor} + \mathbf{J}_{tor} \times \mathbf{B}_{pol} + \mathbf{J}_{pol} \times \mathbf{B}_{pol}. \tag{9.44}$$

The $\mathbf{J}_{pol} \times \mathbf{B}_{pol}$ term is in the toroidal direction and is the only toroidally directed term on the right-hand side of the equation. However, $\partial P/\partial \phi = 0$ because all physical quantities are independent of ϕ and so $\mathbf{J}_{pol} \times \mathbf{B}_{pol}$ must vanish. This implies

$$(\nabla I \times \nabla \phi) \times (\nabla \psi \times \nabla \phi) = 0, \tag{9.45}$$

which further implies that ∇I must be parallel to $\nabla \psi$. An arbitrary displacement $\mathrm{d}\mathbf{r}$ results in respective changes in current and poloidal flux $\mathrm{d}I = \mathrm{d}\mathbf{r} \cdot \nabla I$ and $\mathrm{d}\psi = \mathrm{d}\mathbf{r} \cdot \nabla \psi$ so

$$\frac{\mathrm{d}I}{\mathrm{d}\psi} = \frac{\mathrm{d}\mathbf{r} \cdot \nabla I}{\mathrm{d}\mathbf{r} \cdot \nabla \psi}. \tag{9.46}$$

Since ∇I is parallel to $\nabla \psi$, the derivative $\mathrm{d}I/\mathrm{d}\psi$ is always defined and so, in principle, may always be integrated. Thus, I must be a function of ψ and it is therefore always possible to write

$$\nabla I(\psi) = I'(\psi) \nabla \psi, \tag{9.47}$$

where prime means derivative with respect to the argument. The poloidal current can thus be expressed in terms of the poloidal flux function as

$$\mathbf{J}_{pol} = \frac{I'}{2\pi} \nabla \psi \times \nabla \phi. \tag{9.48}$$

Substitution for the currents and magnetic fields in Eq. (9.44) gives the expression

$$\begin{aligned} \nabla P &= \frac{I'}{2\pi} (\nabla \psi \times \nabla \phi) \times \frac{\mu_0 I}{2\pi} \nabla \phi - \nabla \phi \frac{r^2}{2\pi \mu_0} \nabla \cdot \left(\frac{1}{r^2} \nabla \psi \right) \times \frac{1}{2\pi} [\nabla \psi \times \nabla \phi] \\ &= - \left[\frac{\mu_0 I I'}{(2\pi r)^2} + \frac{1}{(2\pi)^2 \mu_0} \nabla \cdot \left(\frac{1}{r^2} \nabla \psi \right) \right] \nabla \psi. \end{aligned} \tag{9.49}$$

This gives the important result that ∇P must also be parallel to $\nabla \psi$, which in turn implies $P = P(\psi)$ so $\nabla P = P' \nabla \psi$. Equation (9.49) now has a common vector factor $\nabla \psi$, which may be divided out, in which case the original vector equation reduces to the *scalar* equation

$$\nabla \cdot \left(\frac{1}{r^2} \nabla \psi \right) + 4\pi^2 \mu_0 P' + \frac{\mu_0^2}{r^2} I I' = 0. \tag{9.50}$$

This equation, known as the Grad–Shafranov equation (Grad and Rubin 1958, Shafranov 1966), has the peculiarity that ψ shows up as both an independent variable and as a dependent variable, i.e., there are both derivatives of ψ and derivatives with respect to ψ. Axisymmetry has made it possible to transform a three-dimensional vector equation into a one-dimensional scalar equation. It is not surprising that axisymmetry would transform a three-dimensional system into a two-dimensional system, but the transformation of a three-dimensional system into a one-dimensional system suggests more profound physics is involved than just geometrical simplification.

The Grad–Shafranov equation can be substituted back into Eq. (9.42) to give

$$\mathbf{J}_{tor} = \left(2\pi r^2 P' + \frac{\mu_0}{2\pi} II'\right) \nabla\phi \tag{9.51}$$

so the total current can be expressed as

$$\mathbf{J} = \mathbf{J}_{pol} + \mathbf{J}_{tor} = \frac{I'}{2\pi} \nabla\psi \times \nabla\phi + \left(2\pi r^2 P' + \frac{\mu_0}{2\pi} II'\right) \nabla\phi = 2\pi r^2 P' \nabla\phi + I'\mathbf{B}. \tag{9.52}$$

The last term is called the "force-free" current because it is parallel to the magnetic field and so provides no force. The first term on the right-hand side is the diamagnetic current.

The Grad–Shafranov equation is a nonlinear equation in ψ and, in general, cannot be solved analytically. It does not in itself determine the equilibrium because it involves three independent quantities, ψ, $P(\psi)$, and $I(\psi)$. Thus, use of the Grad–Shafranov equation involves specifying two of these functions and then solving for the third. Typically $P(\psi)$ and $I(\psi)$ are specified (either determined from other equations or from experimental data) and then the Grad–Shafranov equation is used to determine ψ.

Although the Grad–Shafranov equation must in general be solved numerically, there exists a limited number of analytic solutions. These can be used as idealized examples that demonstrate typical properties of axisymmetric equilibria. We shall now examine one such analytic solution, the Solov'ev solution (Solov'ev 1976).

The Solov'ev solution is obtained by invoking two assumptions: first, the pressure is assumed to be a linear function of ψ,

$$P = P_0 + \lambda\psi \tag{9.53}$$

and second, I is assumed to be constant within the plasma so

$$I' = 0. \tag{9.54}$$

The second assumption corresponds to having all the z-directed current flowing on the z axis, so that away from the z axis, the toroidal field is a vacuum field (cf. Eq. (9.35)). This arrangement is equivalent to having a zero-radius current-carrying wire going up the z axis acting as the sole source for the toroidal magnetic field. The region over which the Solev'ev solution applies excludes the z axis and so the problem is analogous to a central force problem where the source of the central force is a singularity at the origin. The excluded region of the z axis corresponds to the "hole in the doughnut" of a tokamak. In the more general case where I' is finite, the plasma is either diamagnetic (toroidal field is weaker than the vacuum field) or paramagnetic (toroidal field is stronger than the vacuum field).

Using the two simplifying assumptions provided by Eqs. (9.53) and (9.54), the Grad–Shafranov equation reduces to

$$r\frac{\partial}{\partial r}\left(\frac{1}{r}\frac{\partial \psi}{\partial r}\right)+\frac{\partial^2 \psi}{\partial z^2}+4\pi^2 r^2 \mu_0 \lambda = 0, \tag{9.55}$$

which has the exact solution

$$\psi(r,z)=\psi_0\frac{r^2}{r_0^4}\left(2r_0^2-r^2-4\alpha^2 z^2\right), \tag{9.56}$$

where ψ_0, r_0, and α are constants; Eq. (9.56) is called the Solov'ev solution. Figure 9.11 shows a contour plot of ψ as a function of r/r_0 and z/z_0 for the case $\alpha = 1$. Note that there are three distinct types of curves in Fig. 9.11, namely (i) open curves going to $z = \pm\infty$, (ii) concentric closed curves, and (iii) a single curve, called the separatrix, that separates the first two types of curves.

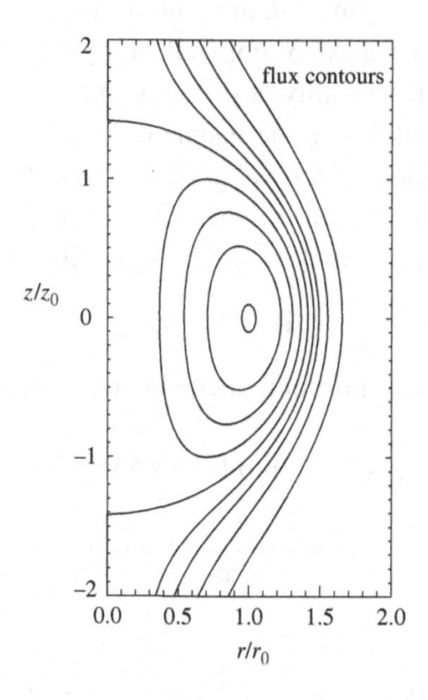

Fig. 9.11 Contours of constant flux of Solov'ev solution to Grad–Shafranov equation.

Even though it is an idealization, the Solov'ev solution illustrates many important features of three-dimensional, static MHD equilibria. Let us now examine some of these features. The constants in Eq. (9.56) have been chosen so that $\psi = \psi_0$ at $r = r_0, z = 0$. The location $r = r_0, z = 0$ is called the *magnetic axis* because it is the axis linked by the closed contours of poloidal flux.

Let us temporarily assume ψ_0 is positive. Examination of Eq. (9.56) shows that if ψ is positive, then the quantity $2r_0^2 - r^2 - 4\alpha^2 z^2$ is positive and vice versa. For any negative value of ψ, Eq. (9.56) can be satisfied by making r very small and z very large. In particular, if r is infinitesimal then z must become infinite. Thus, all contours for which ψ is negative go to $z = \pm\infty$; these contours are the open or type (i) contours.

On the other hand, if ψ is positive then it has a maximum value of ψ_0, which occurs on the magnetic axis and if $0 < \psi < \psi_0$ then ψ must be located at some point outside the magnetic axis, but inside the curve $2r_0^2 - r^2 - 4\alpha^2 z^2 = 0$. Hence all contours of positive ψ must lie inside the curve $2r_0^2 - r^2 - 4\alpha^2 z^2 = 0$ and correspond to field lines that do not go to infinity. These contours are the closed or type (ii) contours, since they form closed curves in the r, z plane.

The contour separating the closed contours from the contours going to infinity is the *separatrix* and is given by the ellipse

$$r^2 + 4\alpha^2 z^2 = 2r_0^2. \tag{9.57}$$

The magnetic axis is a local maximum of ψ if ψ_0 is positive or a local minimum if ψ_0 is negative (the sign depends on the sense of the toroidal current). Thus, one can imagine that a hill (or valley) of poloidal flux exists with apex (or bottom) at the magnetic axis and in the vicinity of the apex (bottom) the contours of constant ψ are circles or ellipses enclosing the magnetic axis. The degree of ellipticity of these surfaces is determined by the value of α. These closed flux surfaces are a set of nested toroidal surfaces sharing the same magnetic axis. The projection of the total magnetic field lies in a flux surface since Eq. (9.34) shows

$$\mathbf{B} \cdot \nabla\psi = 0,$$

i.e., the magnetic field has no component in the direction normal to the flux surface.

Direct substitution of Eq. (9.56) into Eq. (9.55) gives

$$\lambda = \frac{2\psi_0}{\pi^2 r_0^4 \mu_0}\left(1 + \alpha^2\right) \tag{9.58}$$

so the pressure is

$$P(\psi) = P_0 + \frac{2\psi_0}{\pi^2 r_0^4 \mu_0}\left(1 + \alpha^2\right)\psi. \tag{9.59}$$

Pressure vanishes at the plasma edge so if ψ_{edge} is the poloidal flux at the edge, then the constant P_0 is determined and Eq. (9.59) becomes

$$P(\psi) = \frac{2\psi_0}{\pi^2 r_0^4 \mu_0}\left(1+\alpha^2\right)\left[\psi - \psi_{edge}\right]. \tag{9.60}$$

In the earlier discussion of single particle motion it was shown that particles were attached to flux surfaces. Particles starting on a particular flux surface remain on that flux surface, although they may travel to any part of that flux surface. The equilibrium prescribed by Eq. (9.56) is essentially a three-dimensional vortex with pressure peaking on the magnetic axis and then falling monotonically to zero at the edge.

This is a particularly simple solution to the Grad–Shafranov equation, but nevertheless demonstrates several important features, namely:

1. Three-dimensional equilibria can incorporate closed nested poloidal flux surfaces that are concentric about a magnetic axis. The magnetic axis is the local maximum or minimum of the poloidal flux ψ.
2. There can also be open flux surfaces; i.e., flux surfaces that go to infinity.
3. The flux surface separating closed and open flux surfaces is called a separatrix.
4. The total magnetic field projects into the flux surfaces since $\mathbf{B}\cdot\nabla\psi = 0$.
5. Finite pressure corresponds to a depression (or elevation) of the poloidal flux. The pressure maximum and flux extrema are located at the magnetic axis.
6. The poloidal magnetic field and associated poloidal flux ψ are responsible for plasma confinement and result from the toroidal current. Thus *toroidal* current is essential for confinement in an axisymmetric geometry. For a given $P(\psi)$ the toroidal magnetic field does not contribute to confinement if $I' = 0$. If I' is finite, plasma diamagnetism or paramagnetism will affect $P(\psi)$. Although a vacuum toroidal field cannot directly provide confinement, it can affect the rate of cross-field particle and energy diffusion and hence the functional form of $P(\psi)$. For example, the frequency of drift waves will be a function of the toroidal field and drift waves can cause an outward diffusion of plasma across flux surfaces, thereby affecting $P(\psi)$.
7. The poloidal flux surfaces are related to the surfaces of constant canonical angular momentum since $\mathbf{B}_{pol} = \nabla\times(\psi\nabla\phi/2\pi) = \nabla\times\left(\psi\hat{\phi}/2\pi r\right) = \nabla\times A_\phi\hat{\phi}$ implies $A_\phi = \psi/2\pi r$. Thus, the canonical angular momentum can be expressed as

$$\begin{aligned} p_\phi &= mr^2\dot{\phi} + qrA_\phi \\ &= mrv_\phi + \frac{q\psi}{2\pi}. \end{aligned} \tag{9.61}$$

Surfaces of constant canonical angular momentum correspond to surfaces of constant poloidal flux in the limit $m \to 0$. Because the system is toroidally symmetric (axisymmetric) the canonical angular momentum of each particle is a constant of the motion and so, to the extent that the particles can be approximated as having zero mass, particle trajectories are constrained to lie on surfaces of constant ψ. Since p_ϕ is a conserved quantity, the maximum excursion a finite-mass particle can make from a poloidal flux surface can be estimated by writing

$$\delta p_\phi = \delta\left(mrv_\phi + qrA_\phi\right) = 0. \tag{9.62}$$

Thus,

$$\frac{v_\phi}{r}\delta r + \delta v_\phi + \frac{q}{m}\delta r \frac{1}{r}\frac{\partial}{\partial r}\left(rA_\phi\right) = 0 \tag{9.63}$$

or

$$\delta r = -\frac{r\delta v_\phi}{v_\phi + r\omega_{c,pol}}, \tag{9.64}$$

where $\omega_{c,pol} = qB_{pol}/m$ is the cyclotron frequency measured using the poloidal magnetic field. Since v_ϕ is of the order of the thermal velocity v_T or smaller, if the poloidal field is sufficiently strong to make the poloidal Larmor radius $r_{lpol} = v_T/\omega_{c,pol}$ much smaller than the nominal configuration radius r, then the first term in the denominator may be dropped. Since δv_ϕ is also of the order of the thermal velocity or smaller, this gives the important result that *a particle cannot deviate from its initial flux surface by more than a poloidal Larmor radius.* Since poloidal field is produced by toroidal current, it is clear that particle confinement to axisymmetric flux surfaces requires the existence of toroidal current.

Tokamaks, reverse field pinches, spheromaks, and field reversed configurations are all magnetic confinement configurations having three-dimensional axisymmetric equilibria similar to this Solov'ev solution and all have a toroidal current producing a set of nested, closed poloidal flux surfaces that link a magnetic axis. Stellarators are non-axisymmetric configurations that have poloidal flux surfaces without toroidal currents; the advantage of current-free operation is offset by the complexity of non-axisymmetry.

9.9 Dynamic equilibria: flows

MHD-driven flows are relevant to arcs, magnetoplasmadynamic thrusters, electric currents in molten metals, structures on the solar corona, and astrophysical jets. The basic mechanism driving MHD flows will be discussed in Section 9.9.1 using the simplified assumptions of incompressibility and self-field only. The more general situation where the plasma is compressible and where there are external, applied magnetic fields in addition to the self-field will then be addressed in Section 9.9.2.

9.9.1 Incompressible plasma with self-field only

MHD-driven flows involve situations where $\nabla\times(\mathbf{J}\times\mathbf{B})$ is finite. This means the MHD force $\mathbf{J}\times\mathbf{B}$ is non-conservative, in which case its finite curl acts as a torque or equivalently as a source for hydrodynamic vorticity. When the MHD force is non-conservative a closed line integral $\oint d\mathbf{l}\cdot\mathbf{J}\times\mathbf{B}$ will be finite, whereas in contrast a closed line integral of a pressure gradient $\oint d\mathbf{l}\cdot\nabla P$ is always zero. Since a pressure gradient cannot balance the torque produced by a non-conservative $\mathbf{J}\times\mathbf{B}$, it is necessary to include a vorticity-damping term (viscosity term) to allow for the possibility of balancing this torque. Plasma viscosity is mainly due to ion–ion collisions and is typically very small. With the addition of viscosity, the MHD equation of motion becomes

$$\rho\left(\frac{\partial\mathbf{U}}{\partial t}+\mathbf{U}\cdot\nabla\mathbf{U}\right)=\mathbf{J}\times\mathbf{B}-\nabla P+\rho\upsilon\nabla^2\mathbf{U}, \tag{9.65}$$

where υ is the kinematic viscosity. To focus attention on the generation, transport, and decay of vorticity, the following simplifying assumptions are made:

1. The motion is incompressible so that $\rho = const.$
   This is an excellent assumption for a molten metal, but questionable for a plasma. However, the assumption could be at least locally reasonable for a plasma if the region of vorticity generation is smaller in scale than regions of compression or rarefaction or is spatially distinct from these regions. The consequences of compressibility will be discussed later.
2. The system is cylindrically symmetric.
   A cylindrical coordinate system r, ϕ, z can then be used and ϕ is ignorable. This geometry corresponds to magnetohydrodynamic thrusters and arcs.
3. The flow velocities $\mathbf{U}$ and the current density $\mathbf{J}$ are in the poloidal direction (i.e., r and z directions). Restricting the current to be poloidal means that the magnetic field is purely toroidal (see Fig. 9.10).

The most general poloidal velocity for a constant-density, incompressible fluid has the form

$$\mathbf{U}=\frac{1}{2\pi}\nabla\mathcal{F}\times\nabla\phi, \tag{9.66}$$

where $\mathcal{F}(r,z)$ is the flux of fluid through a circle of radius r at location z. Note the analogy to the poloidal magnetic flux function $\psi(r,z)$ used in Eqs. (3.147) and (9.34). Since $\mathbf{U}$ lies entirely in the $r-z$ plane and since the system is axisymmetric, the curl of $\mathbf{U}$ will be in the ϕ direction. It is thus useful to define a scalar cylindrical "vorticity" $\chi = r\hat{\phi}\cdot\nabla\times\mathbf{U}$; this definition differs slightly from the conventional definition of vorticity (curl of velocity) because it includes an extra factor of r. This slight change of definition is essentially a matter of semantics, but makes the algebra more transparent.

Since $\nabla\phi = \hat{\phi}/r$ it is seen that

$$\nabla\times\mathbf{U}=\frac{\hat{\phi}}{r}r\hat{\phi}\cdot\nabla\times\mathbf{U}=\chi\nabla\phi \tag{9.67}$$

and so

$$\nabla \times \left(\frac{1}{2\pi} \nabla \mathcal{F} \times \nabla \phi \right) = \chi \nabla \phi. \tag{9.68}$$

Dotting Eq. (9.68) with $\nabla \phi$ and using the vector identity $\nabla \phi \cdot \nabla \times \mathbf{Q} = \nabla \cdot (\mathbf{Q} \times \nabla \phi)$ gives

$$\nabla \cdot \left(\frac{1}{2\pi} (\nabla \mathcal{F} \times \nabla \phi) \times \nabla \phi \right) = \frac{\chi}{r^2} \tag{9.69}$$

or

$$r^2 \nabla \cdot \left(\frac{1}{r^2} \nabla \mathcal{F} \right) = -2\pi \chi, \tag{9.70}$$

which is a Poisson-like relation between the vorticity and the stream function. The vorticity plays the role of the source (charge-density) and the stream function plays the role of the potential function (electrostatic potential). However, the elliptic operator is not exactly a Laplacian and when expanded has the form

$$r^2 \nabla \cdot \left(\frac{1}{r^2} \nabla \mathcal{F} \right) = r \frac{\partial}{\partial r} \left(\frac{1}{r} \frac{\partial \mathcal{F}}{\partial r} \right) + \frac{\partial^2 \mathcal{F}}{\partial z^2}, \tag{9.71}$$

which is just the operator in the Grad–Shafranov equation.

Since the current was assumed to be purely poloidal, the magnetic field must be purely toroidal and so can be written as

$$\mathbf{B} = \frac{\mu_0}{2\pi} I \nabla \phi, \tag{9.72}$$

where I is the current through a circle of radius r at location z. This is consistent with the integral form of Ampère's law, $\oint \mathbf{B} \cdot d\mathbf{l} = \mu_0 I$. The curl of Eq. (9.72) gives

$$\nabla \times \mathbf{B} = \frac{\mu_0}{2\pi} \nabla I \times \nabla \phi \tag{9.73}$$

showing that I acts like a stream-function for the current.

The magnetic force is therefore

$$\mathbf{J} \times \mathbf{B} = \frac{1}{2\pi} (\nabla I \times \nabla \phi) \times \frac{\mu_0}{2\pi} I \nabla \phi = -\frac{\mu_0}{(2\pi r)^2} \nabla \left(\frac{I^2}{2} \right). \tag{9.74}$$

It is seen that there is an axial force if I^2 depends on z; this is the essential condition that produces axial flows in MHD arcs (Maecker 1955) and plasma guns (Marshall 1960).

Using the identities $\nabla U^2/2 = \mathbf{U} \cdot \nabla \mathbf{U} + \mathbf{U} \times \nabla \times \mathbf{U}$ and $\nabla \times \nabla \times \mathbf{U} = -\nabla^2 \mathbf{U} = \nabla \chi \times \nabla \phi$ the equation of motion Eq. (9.65) becomes

$$\rho \left(\frac{\partial \mathbf{U}}{\partial t} + \nabla \frac{U^2}{2} - \mathbf{U} \times \chi \nabla \phi \right) = -\frac{\mu_0}{(2\pi r)^2} \nabla \left(\frac{I^2}{2} \right) - \nabla P - \rho \nu \nabla \chi \times \nabla \phi \tag{9.75}$$

or

$$\rho\left(\frac{\partial \mathbf{U}}{\partial t}+\nabla\frac{U^2}{2}-\mathbf{U}\times\chi\nabla\phi\right)=-\frac{\mu_0}{(2\pi)^2}\nabla\left(\frac{I^2}{2r^2}\right)+\frac{\mu_0 I^2}{(2\pi)^2}\nabla\left(\frac{1}{2r^2}\right)$$
$$-\nabla P-\rho\upsilon\nabla\chi\times\nabla\phi. \quad (9.76)$$

Every term in this equation is either a gradient of a scalar or else can be expressed as a cross-product involving $\nabla\phi$. The equation can thus be regrouped as

$$\nabla\left(\frac{\rho U^2}{2}+\frac{\mu_0}{(2\pi)^2}\frac{I^2}{2r^2}+P\right)+\left(\frac{\rho}{2\pi}\nabla\frac{\partial\mathcal{F}}{\partial t}-\rho\mathbf{U}\chi-\frac{\mu_0 I^2}{(2\pi r)^2}\hat{z}+\rho\upsilon\nabla\chi\right)\times\nabla\phi=0. \quad (9.77)$$

Similarly, the MHD Ohm's law $\mathbf{E}+\mathbf{U}\times\mathbf{B}=\eta\mathbf{J}$ can be written as

$$-\frac{\partial\mathbf{A}}{\partial t}-\nabla V+\mathbf{U}\times\frac{\mu_0}{2\pi}I\nabla\phi=\frac{\eta}{2\pi}\nabla I\times\nabla\phi, \quad (9.78)$$

where V is the electrostatic potential. Since $\mathbf{B}$ is purely toroidal, $\mathbf{A}$ is poloidal and so is orthogonal to $\nabla\phi$. Thus, both the equation of motion and Ohm's law are equations of the form

$$\nabla g+\mathbf{Q}\times\nabla\phi=0, \quad (9.79)$$

which is the most general form of an axisymmetric partial differential equation involving a potential. Two distinct scalar partial differential equations can be extracted from Eq. (9.79) by (i) operating with $\nabla\phi\cdot\nabla\times$ and (ii) taking the divergence. Doing the former it is seen that Eq. (9.79) becomes

$$\nabla\phi\cdot\nabla\times(\mathbf{Q}\times\nabla\phi)=0 \quad (9.80)$$

or

$$\nabla\cdot\left(\frac{1}{r^2}\mathbf{Q}_{pol}\right)=0. \quad (9.81)$$

Applying this procedure to Eq. (9.77) gives

$$\frac{\rho}{2\pi}\frac{\partial}{\partial t}\left(\nabla\cdot\frac{1}{r^2}\nabla\mathcal{F}\right)-\rho\nabla\cdot\left(\frac{1}{r^2}\mathbf{U}\chi\right)-\nabla\cdot\left(\frac{\mu_0 I^2}{(2\pi r^2)^2}\hat{z}\right)+\rho\upsilon\nabla\cdot\left(\frac{1}{r^2}\nabla\chi\right)=0 \quad (9.82)$$

or

$$\frac{\partial}{\partial t}\left(\frac{\chi}{r^2}\right)+\nabla\cdot\left(\mathbf{U}\frac{\chi}{r^2}\right)=-\frac{\mu_0}{4\pi^2\rho r^4}\frac{\partial I^2}{\partial z}+\upsilon\nabla\cdot\left(\frac{1}{r^2}\nabla\chi\right). \quad (9.83)$$

Since $\nabla \cdot \mathbf{U} = 0$, this can also be written as

$$\frac{\partial}{\partial t}\left(\frac{\chi}{r^2}\right) + \mathbf{U} \cdot \nabla \left(\frac{\chi}{r^2}\right) = -\frac{\mu_0}{4\pi^2 \rho r^4} \frac{\partial I^2}{\partial z} + \upsilon \nabla \cdot \frac{1}{r^2} \nabla \chi, \tag{9.84}$$

which shows that if there is no viscosity and if $\partial I^2/\partial z = 0$, then the scaled vorticity χ/r^2 convects with the fluid; i.e., is frozen into the fluid. The viscous term on the right-hand side describes a diffusive-like dissipation of vorticity. The remaining term $r^{-4}\partial I^2/\partial z$ acts as a *vorticity source* and is finite only if I^2 is non-uniform in the z direction. The vorticity source has a strong r^{-4} weighting factor so axial non-uniformities of I near $r = 0$ dominate. Positive χ corresponds to a clockwise rotation in the $r - z$ plane. If I^2 is an increasing function of z then the source term is negative, implying that a counterclockwise vortex is generated and vice versa. Suppose as shown in Fig. 9.12 that the current channel becomes wider with increasing z, corresponding to a fanning out of the current with increasing z. In this case $\partial I^2/\partial z$ will be negative and a clockwise vortex will be generated. Fluid will flow radially inwards at small z, then flow vertically upwards, and finally radially outwards at large z. The fluid flow produced by the vorticity source will convect the vorticity along with the flow until it is dissipated by viscosity.

Operating on Ohm's law, Eq. (9.78), with $\nabla\phi \cdot \nabla\times$, gives

$$\frac{\partial}{\partial t}\left(\frac{I}{r^2}\right) + \mathbf{U} \cdot \nabla \left(\frac{I}{r^2}\right) = \frac{\eta}{\mu_0} \nabla \cdot \left(\frac{1}{r^2} \nabla I\right) \tag{9.85}$$

showing that I/r^2 is similarly convected with the fluid and also has a diffusive-like term, this time with coefficient η/μ_0.

Thus, the system of equations can be summarized as (Bellan 1992)

$$\frac{\partial}{\partial t}\left(\frac{\chi}{r^2}\right) + \mathbf{U} \cdot \nabla \left(\frac{\chi}{r^2}\right) = \upsilon \nabla \cdot \frac{1}{r^2} \nabla \chi - \frac{\mu_0}{4\pi^2 \rho r^4} \frac{\partial I^2}{\partial z} \tag{9.86}$$

$$\frac{\partial}{\partial t}\left(\frac{I}{r^2}\right) + \mathbf{U} \cdot \nabla \left(\frac{I}{r^2}\right) = \frac{\eta}{\mu_0} \nabla \cdot \left(\frac{1}{r^2} \nabla I\right) \tag{9.87}$$

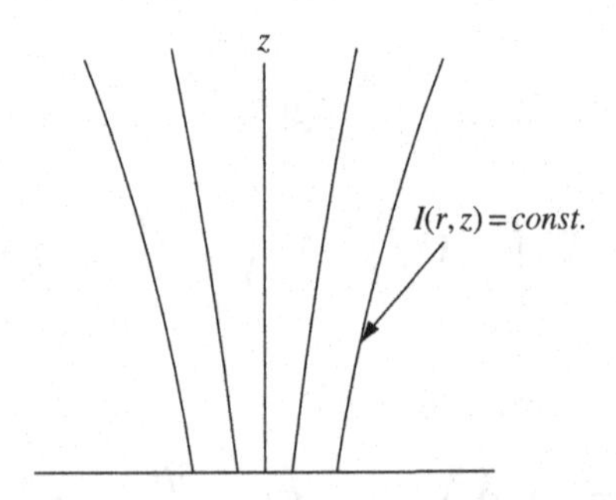

Fig. 9.12 Current channel $I(r, z)$.

$$r^2\nabla\cdot\left(\frac{1}{r^2}\nabla\mathcal{F}\right) = -2\pi\chi \tag{9.88}$$

$$\mathbf{U} = \frac{1}{2\pi}\nabla\mathcal{F}\times\nabla\phi. \tag{9.89}$$

This system cannot be solved analytically, but its general behavior can be described in a qualitative manner. The system involves three scalar variables, $\chi, \mathcal{F}$, and I, all of which are functions of r and z. Boundary conditions must be specified in order to have a well-posed problem. If a recirculating flow is driven, the fluid flux $\mathcal{F}$ will be zero on the boundary whereas I will be finite on electrodes at the boundary. In particular, the current I will typically flow from the anode into the plasma and then from the plasma into the cathode. The vorticity χ will usually have the boundary condition of vanishing on the bounding surface.

Initially, there is no flow so $\mathcal{F}$ is zero everywhere. An initial solution of Eq. (9.87) in this no-flow situation will establish a current profile between the two electrodes. For any reasonable situation, this $I(r,z)$ profile will have $\partial I^2/\partial z \neq 0$ so that a vorticity source will be created. The source will be quite localized because of the r^{-4} coefficient. Once some vorticity χ is created as determined by Eq. (9.86), this vorticity acts as a source term for the fluid flux $\mathcal{F}$ in Eq. (9.88), and so a finite $\mathcal{F}$ will be developed. Thus a flow $\mathbf{U}(r,z)$ will be created as specified by Eq. (9.89), and this flow will convect both χ/r^2 and I/r^2.

A good analogy is to think of the $r^{-4}\partial I^2/\partial z$ term as constituting a toroidally symmetric centrifugal pump, which accelerates fluid radially from large to small r. This radial acceleration takes place at z locations where the current channel radius is constricted. The pump then accelerates the ingested fluid up or down the z axis, in a direction away from the current constriction. This vortex generation can also be seen by drawing vectors showing the magnitude and orientation of the $\mathbf{J}\times\mathbf{B}$ force in the vicinity of a current constriction. It is seen that the $\mathbf{J}\times\mathbf{B}$ force is non-conservative and provides a centrifugal pumping as described above.

An arc or magnetoplasmadynamic thruster can thus be construed as a pump that draws fluid radially inward towards the smaller radius electrode, accelerates this fluid to high velocity, and then expels the accelerated fluid in the axial direction in a manner whereby the accelerated fluid shoots axially from the smaller electrode towards the larger radius electrode. The sign of the current does not matter, since the pumping action depends only on the z derivative of I^2. If the equations are put in dimensionless form, it is seen that the characteristic flow velocity is of the order of the Alfvén velocity.

Once χ, $\mathcal{F}$, and I have been determined, it is possible to determine the pressure and electrostatic potential profiles. This is done by taking the divergence of

Eq. (9.79) to obtain

$$\nabla^2 g = -\nabla\phi \cdot \nabla \times \mathbf{Q}. \tag{9.90}$$

The pressure and electrostatic potential are contained within the g term, whereas the $\mathbf{Q}$ term involves only χ, $\mathcal{F}$, and I or functions of χ, $\mathcal{F}$, and I (e.g., the velocity). Thus, the right-hand side of Eq. (9.90) is known and so can be considered as the source for a Poisson-like equation for the left-hand side. For the equation of motion

$$g_{motion} = \frac{\rho U^2}{2} + \frac{\mu_0}{(2\pi)^2}\frac{I^2}{2r^2} + P \tag{9.91}$$

so that

$$P = g_{motion} - \frac{\rho U^2}{2} - \frac{\mu_0}{(2\pi)^2}\frac{I^2}{2r^2}. \tag{9.92}$$

Regions where g_{motion} is constant satisfy an extended form of the Bernoulli theorem, namely

$$\frac{\rho U^2}{2} + \frac{B^2}{2\mu_0} + P = const. \tag{9.93}$$

Similarly, taking the divergence of Eq. (9.78) gives an equation for the electrostatic potential (using Coulomb gauge so that $\nabla \cdot \mathbf{A} = 0$). Thus, a complete solution for incompressible flow is obtained by first solving for χ, $\mathcal{F}$, and I using prescribed boundary conditions, and then solving for the pressure and electrostatic potential.

9.9.2 Compressible plasma and applied poloidal field

The previous discussion assumed that the plasma was incompressible and that the magnetic field was purely toroidal (i.e., generated by the prescribed poloidal current). The more general situation involves having a pre-existing poloidal magnetic field such as would be generated by external coils and also allowing for plasma compressibility. This situation will now be discussed qualitatively making reference to the sketch provided in Fig. 9.13 (a more thorough discussion is provided in Bellan (2003).

Consider the initial situation shown in Fig. 9.13(a) in which a plasma is immersed in an axisymmetric vacuum poloidal field $\psi(r, z)$ with a poloidal current $I(r, z)$. Because ψ is assumed to correspond to a vacuum field, it is produced by external coils with toroidal currents. There are thus no toroidal currents in the plasma volume under consideration (i.e., in the region described by Fig. 9.13(a))

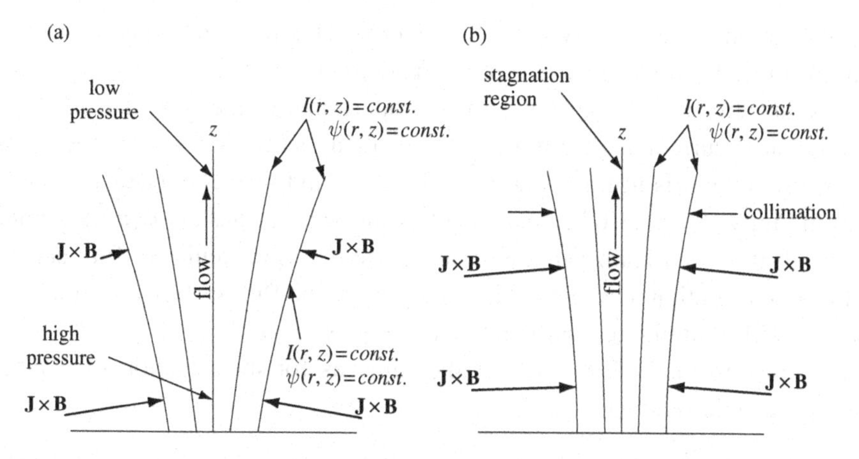

Fig. 9.13 (a) Current channel $I(r, z)$, which has same contours as poloidal flux surfaces $\psi(r, z)$ so that ∇I is parallel to $\nabla\psi$ in order to have no toroidal acceleration. The $\mathbf{J} \times \mathbf{B}$ force (shown by thick arrows) is larger at the bottom and canted giving both a higher on-axis pressure at the bottom and an axial upwards flow. (b) If the flow stagnates (slows down) for some reason, then mass accumulates at the top. The frozen-in convected toroidal flux also accumulates, which thus increases B_ϕ and so forces r to decrease since Ampère's law gives $2\pi B_\phi r = I(r, z)$, which is constant on a surface of constant $I(r, z)$. The upward-moving, stagnating plasma flow with embedded toroidal flux acts somewhat like a zipper for the surfaces of constant $I(r, z)$ and $\psi(r, z)$.

and so using Eq. (9.42) it is seen that $\psi(r, z)$ satisfies $\nabla \cdot \left(r^{-2}\nabla\psi\right) = 0$. The poloidal current $I(r, z)$ has an associated poloidal current density

$$\mathbf{J}_{pol} = \frac{1}{2\pi}\nabla I \times \nabla\phi \tag{9.94}$$

and an associated toroidal magnetic field

$$B_\phi = \frac{\mu_0 I(r, z)}{2\pi r}. \tag{9.95}$$

The poloidal flux $\psi(r, z)$ has an associated poloidal magnetic field

$$\mathbf{B}_{pol} = \frac{1}{2\pi}\nabla\psi \times \nabla\phi. \tag{9.96}$$

If $\mathbf{J}_{pol}$ is parallel to $\mathbf{B}_{pol}$ then ∇I would be parallel to $\nabla\psi$ so that constant $I(r, z)$ surfaces coincide with constant $\psi(r, z)$ surfaces as indicated in the figure. On the other hand, if $\mathbf{J}_{pol}$ were not parallel to $\mathbf{B}_{pol}$, these surfaces would not be coincident and there would be a force $\mathbf{J}_{pol} \times \mathbf{B}_{pol}$ in the toroidal direction, which would tend to cause a change in the toroidal velocity, i.e., an acceleration or deceleration of U_ϕ. We argue that $\mathbf{J}_{pol} \times \mathbf{B}_{pol}$ must be a transient force with zero time average because any current that flows perpendicular to the poloidal flux surfaces must be the polarization current as given by Eq. (3.145). This polarization current results from the time derivative of the electric field perpendicular to the poloidal flux surfaces and this electric field in turn is proportional to the time rate of change of the poloidal current as determined from Faraday's law. Thus, once the current is in steady state, it is necessary to have $I = I(\psi)$ and so the surfaces of constant $I(r, z)$ in Fig. 9.12 can also be considered as surfaces of constant $\psi(r, z)$. Another way to see this is to realize that a steady current perpendicular to flux surfaces would cause a continuous electrostatic charging of the flux surfaces, which would then violate the stipulation that the plasma is quasi-neutral.

The initial magnetic force is thus the same as for the situation of a purely toroidal magnetic field since

$$\mathbf{J}\times\mathbf{B} = \underbrace{\mathbf{J}_{tor}\times\mathbf{B}_{pol}}_{\text{zero}} + \mathbf{J}_{pol}\times\mathbf{B}_{tor}. \tag{9.97}$$

The remaining force $\mathbf{J}_{pol}\times\mathbf{B}_{tor} = \left(J_r\hat{r} + J_z\hat{z}\right)\times B_\phi\hat{\phi} = J_rB_\phi\hat{z} - J_zB_\phi\hat{r}$ has components in both the r and z directions. If the plasma axial length is much larger than its radius (i.e., it is long and skinny) then the plasma will develop a local radial pressure balance

$$\left(-\nabla P + \mathbf{J}_{pol}\times\mathbf{B}_{tor}\right)\cdot\hat{r} = 0. \tag{9.98}$$

However, this local radial pressure balance precludes the possibility of an axial pressure balance if $\partial\psi/\partial z \neq 0$ because, as discussed earlier, ∇P cannot balance $\mathbf{J}\times\mathbf{B}$ if the latter has a finite curl. Suppose that the radius of the current channel and the poloidal flux are both given by $a = a(z)$ and, furthermore, since $I = I(\psi)$ let us assume a simple linear dependence where $\mu_0 I(r, z) = \lambda\psi(r, z)$. Thus, ∇I is parallel to $\nabla\psi$ and I is just the current per flux. We assume a parabolic poloidal flux for $r \leq a(z)$, namely

$$\frac{\psi(r, z)}{\psi_0} = \left(\frac{r}{a(z)}\right)^2, \tag{9.99}$$

where ψ_0 is the flux at $r = a$; this is the simplest allowed form for the flux (if ψ depended linearly on r, then, as noted in the discussion of Eq.(3.155), there would be infinite fields at $r = 0$). The local radial pressure balance is essentially a local version of the Bennett pinch relation

$$\begin{aligned}\frac{\partial P}{\partial r} &= -J_zB_\phi \\ &= -\frac{\lambda^2}{8\pi^2\mu_0 r^2}\frac{\partial\psi^2}{\partial r}.\end{aligned} \tag{9.100}$$

On substitution of Eq. (9.99) this becomes

$$\frac{\partial P}{\partial r} = -\frac{\mu_0 \lambda^2 \psi_0^2}{8\pi^2 r^2}\frac{\partial}{\partial r}\left(\frac{r}{a(z)}\right)^4$$
$$= -\frac{\lambda^2 \psi_0^2 r}{2\pi^2 \mu_0 a^4}. \tag{9.101}$$

Integrating and using the boundary condition that $P = 0$ at $r = a$ gives

$$P(r) = \frac{\lambda^2 \psi_0^2}{4\pi^2 \mu_0 a^2}\left(1 - \frac{r^2}{a^2}\right), \tag{9.102}$$

which shows that the on-axis pressure is larger where $a(z)$ is smaller as indicated in Fig. 9.13(a). The axial component of the equation of motion is thus

$$\begin{aligned} F_z &= \hat{z}\cdot(-\nabla P + \mathbf{J}_{pol}\times\mathbf{B}_{tor}) \\ &= -\frac{\partial P}{\partial z} + J_r B_\phi \\ &= -\frac{\lambda^2\psi_0^2}{4\pi^2\mu_0}\frac{\partial}{\partial z}\left(\frac{1}{a^2} - \frac{r^2}{a^4}\right) - \frac{1}{\mu_0}B_\phi\frac{\partial B_\phi}{\partial z} \\ &= -\frac{\lambda^2\psi_0^2}{\pi^2\mu_0}\left(-\frac{1}{2a^3} + \frac{r^2}{a^5}\right)\frac{\partial a}{\partial z} - \frac{1}{2\mu_0}\frac{\partial}{\partial z}\left(\frac{\lambda\psi}{2\pi r}\right)^2 \\ &= \frac{\lambda^2\psi_0^2}{\pi^2 a^3\mu_0}\left(\frac{1}{2} - \frac{r^2}{a^2}\right)\frac{\partial a}{\partial z} - \frac{\lambda^2\psi_0^2}{8\pi^2 r^2\mu_0}\frac{\partial}{\partial z}\left(\frac{r}{a}\right)^4 \\ &= \frac{\mu_0 I_0^2}{2\pi^2 a^3}\left(1 - \frac{r^2}{a^2}\right)\frac{\partial a}{\partial z}, \end{aligned} \tag{9.103}$$

which shows there is an axial force that is maximum on the z axis. This force is proportional to the current flowing along the flux tube and to the axial non-uniformity of the flux tube, i.e., to $\partial a/\partial z$. The axial force accelerates plasma away from regions where a is small to regions where a is large.

If the flow stagnates, i.e., is such that the axial velocity is non-uniform so that $\nabla\cdot\mathbf{U}$ is negative, then there will be a net inflow of matter into the stagnation region and hence an increase of mass density at the locations where $\nabla\cdot\mathbf{U}$ is negative. Because magnetic flux is frozen to the plasma, there will be a corresponding accumulation of toroidal magnetic flux at the locations where $\nabla\cdot\mathbf{U}$ is negative. If U_ϕ is zero (as assumed on the basis of there being no steady toroidal acceleration and zero initial toroidal velocity), then the accumulation of toroidal flux will increase the toroidal field. This can be seen by considering the toroidal component of the induction equation

$$\frac{\partial B_\phi}{\partial t} = r\mathbf{B}_{pol} \cdot \nabla \left(\frac{U_\phi}{r} \right) - r\mathbf{U}_{pol} \cdot \nabla \left(\frac{B_\phi}{r} \right) - B_\phi \nabla \cdot \mathbf{U}_{pol} \tag{9.104}$$

or

$$\frac{DB_\phi}{Dt} = -B_\phi \nabla \cdot \mathbf{U}_{pol}, \tag{9.105}$$

where $D/DT = \partial/\partial t + \mathbf{U} \cdot \nabla$ and $U_\phi = 0$ has been assumed. Thus, it is seen that B_ϕ will increase in the frame of the plasma at locations where $\nabla \cdot \mathbf{U}_{pol}$ is negative. However, Ampère's law gives $\mu_0 I = 2\pi r B_\phi$ and since $I = I_0$ at the outer radius of the current channel, it is seen that if B_ϕ increases, then the current channel radius has to decrease in order to keep rB_ϕ fixed. The result is that stagnation of the flow tends to collimate the flux tube, i.e., make it axially uniform as shown in Fig. 9.13(b). The plasma accelerated upwards in Fig. 9.12 will then squeeze together the ψ and thus I surfaces so that these surfaces will become vertical lines; very roughly, stagnation of an upward moving plasma in Fig. 9.12 can be imagined as a sort of "zipper," which collimates the flux surfaces. Collimation is often seen in current-carrying magnetic flux tubes in laboratory experiments (e.g., see Hansen and Bellan (2001)) and is an important property of solar coronal loops (Klimchuk 2000) and of astrophysical jets (Livio 1999). The MHD pumping prescribed by Eq. (9.103) and its association with flux tube collimation has recently been observed in laboratory experiments (You, Yun, and Bellan 2005).

9.10 Assignments

1. Show that if a Bennett pinch equilibrium has J_z independent of r, then the azimuthal magnetic field is of the form $B_\theta(r) = B_\theta(a)r/a$. Then show that if the temperature is uniform, the pressure will be of the form $P = P_0(1 - r^2/a^2)$, where $P_0 = \mu_0 I^2/4\pi^2 a^2$. Show that this result is consistent with Eq. (9.29).
2. Show it is impossible to confine a spherically symmetric pressure profile using magnetic forces alone and therefore show that three-dimensional magnetostatic equilibria cannot be found for arbitrarily specified pressure profiles. To do this, suppose there exists a magnetic field satisfying $\mathbf{J} \times \mathbf{B} = \nabla P$, where $P = P(r)$ and r is the radius in spherical coordinates $\{r, \theta, \phi\}$. Then make the following analysis:
   (a) Show that neither $\mathbf{J}$ nor $\mathbf{B}$ can have a component in the $\hat{r}$ direction. Hint: assume B_r is finite and write out the three components of $\mathbf{J} \times \mathbf{B} = \nabla P$ in spherical coordinates. By eliminating either J_r or B_r from the equations resulting from the θ and ϕ components, obtain equations of the form $J_r \partial P/\partial r = 0$ and $B_r \partial P/\partial r = 0$. Then use the fact that $\partial P/\partial r$ is finite to develop a contradiction regarding the original assumption regarding J_r or B_r.

(b) By calculating J_θ and J_ϕ from Ampère's law (use the curl in spherical coordinates as given on p. 592), show that $\mathbf{J}\times\mathbf{B} = \nabla P$ implies $B_\theta^2 + B_\phi^2 = B^2(r)$. Argue that this means that the magnitude of $\mathbf{B} = B_\theta\hat{\theta} + B_\phi\hat{\phi}$ must be independent of θ and ϕ and therefore B_θ and B_ϕ must be of the form $B_\theta = B(r)\sin[\eta(\theta,\phi,r)]$ and $B_\phi = B(r)\cos[\eta(\theta,\phi,r)]$, where η is some arbitrary function.

(c) Using $J_r = 0$, $B_r = 0$, and $\nabla\cdot\mathbf{B} = 0$ (see p. 591 for the divergence in spherical polar coordinates) derive a pair of coupled equations for $\partial\eta/\partial\theta$ and $\partial\eta/\partial\phi$. Solve these coupled equations for $\partial\eta/\partial\theta$ and $\partial\eta/\partial\phi$ using the method of determinants (the solutions for $\partial\eta/\partial\theta$ and $\partial\eta/\partial\phi$ should be very simple). Integrate the $\partial\eta/\partial\phi$ equation to obtain η and show that this solution for η is inconsistent with the requirements of the solution for $\partial\eta/\partial\theta$, thereby demonstrating that the original assumption of existence of a spherically symmetric equilibrium must be incorrect.

3. Using numerical methods solve for the motion of charged particles in the Solov'ev field, $\mathbf{B} = \nabla\psi\times\nabla\phi$, where $\psi = B_0 r^2(2a^2 - r^2 - 4(\alpha z)^2)/a^2$ and B_0, a, and α are constants. Plot the surfaces of constant ψ and show there are open and closed surfaces. Select appropriate characteristic times, lengths, and velocities and choose appropriate time steps, initial conditions, and graph windows. Plot the x, y plane and the x, z plane. What interesting features are observed in the orbits? How do they relate to the flux surfaces $\psi =$ constant? What can be said about flux conservation?

4. Grad–Shafranov equation for an axisymmetric current-carrying magnetic flux tube. Because the Grad–Shafranov equation describes axisymmetric equilibria, in addition to describing axisymmetric configurations with closed flux surfaces (e.g., tokamaks), it can also be used to characterize an axisymmetric magnetic flux tube, i.e., a configuration with axisymmetric open flux surfaces that may or may not have an axially uniform cross-sectional area. Suppose that $\psi = \psi(r, z)$ and that ψ is a monotonically increasing function of r. The poloidal flux function thus defines the shape of an axisymmetric flux tube; in particular, the flux tube will be axially uniform only in the special case where ψ does not depend on z.

   Assume both the current I and pressure P are linear functions of ψ so

$$\mu_0 I = \lambda\psi$$

$$P = P_0(1 - \psi/\psi_0),$$

where ψ_0 is the flux surface on which P vanishes and P_0 is the pressure on the z axis. Show that the Grad–Shafranov equation can then be written in the form

$$r^2\nabla\cdot\left(\frac{1}{r^2}\nabla\bar{\psi}\right) + \lambda^2\bar{\psi} = 4\pi^2 r^2\mu_0\frac{P_0}{\psi_0^2},$$

where $\bar{\psi} = \psi(r, z)/\psi_0$. Show that if

$$P_0 = \frac{\lambda^2\psi_0^2}{4\pi^2\mu_0 a^2},$$

where a is the radius at which the pressure vanishes at $z = 0$, the flux tube must be axially uniform, i.e., ψ cannot have any z dependence. Express this condition in terms of the pitch angle of the magnetic field B_ϕ/B_z measured at $r = a$. Hint: note that the Grad–Shafranov equation has both a homogeneous part (left-hand side) and an inhomogeneous part (right-hand side). Show that $\bar{\psi} = r^2/a^2$ is a particular solution and then consider the form of the general solution (homogenous plus inhomogeneous solution).

5. Lawson Criterion. Of the many possible thermonuclear reactions, the deuterium-tritium (DT) reaction stands out as being the most feasible because it has the largest reaction cross-section at accessible temperatures. The DT reaction has the form

$$\mathrm{D}+\mathrm{T} \longrightarrow \mathrm{n}+\mathrm{He}^4+17.6\,\mathrm{MeV}.$$

The output energy consists of 14.1 MeV neutron kinetic energy and 3.5 MeV alpha particle kinetic energy. Fusion reaction cross-sections have an extreme dependence on the impact energy of reacting ions because Coulomb repulsion makes it energetically very difficult for two ions to approach one another. In order for a controlled fusion reaction to be economically useful, more energy must be generated than invested. The break-even condition is determined by equating the energy invested to the energy harvested from the fusion reaction. If the reaction dies prematurely then there is insufficient return on the investment. Thus, the lifetime of the fusion-producing plasma is a critical parameter and is characterized by the so-called *energy confinement* time τ_E of the configuration, a measure of the insulating capability of the plasma confinement device. Let all temperatures and energies be measured in kilovolts (keV) so the Boltzmann constant becomes $\kappa = 1.6 \times 10^{-16}$ and consider a volume V of reacting deuterium, tritium, and associated neutralizing electrons (the plasma needs to be overall neutral, otherwise there would be enormous electrostatic forces).

(a) Show that if the electrons and ions have the same temperature T, the energy required to heat this volume is $E_{in} \sim 3n_eV\kappa T$, where n_e is the electron density and equal densities of deuterium and tritium are assumed so that $n_D = n_T = n_e/2$. (Hint: the mean energy per degree of freedom is $\kappa T/2$ so the mean energy of a particle moving in three-dimensional space is $3\kappa T/2$).

(b) What is the fusion reaction rate R for a single incident ion of velocity v passing through stationary target ions where $\sigma(v)$ is the fusion reaction rate cross-section and n_t is the density of the target ions (assumed stationary). If $\langle\sigma(v)v\rangle$ denotes the velocity average of $\sigma(v)v$, what is the velocity-averaged reaction rate?

(c) Assume that the reaction of hot tritium with hot deuterium can be approximated as the sum of hot deuterium interacting with stationary tritium and hot tritium interacting with stationary deuterium. Show that the fusion output power in the volume is

$$P = n_D n_T \langle\sigma(v)v\rangle\, V\, E_{reaction},$$

where $E_{reaction} = 17.6 \times 10^3$ keV.

(d) Show that the ratio Q of the fusion output energy to the input energy satisfies the relation

$$n_e\tau_E \sim \frac{6.8 \times 10^{-4} T_{keV}}{\langle\sigma(v)v\rangle} Q.$$

At break-even $Q = 1$.

(e) For a thermal ion distribution and the experimentally measured reaction cross-section $\sigma(v)$, the velocity-averaged rate integral has the values tabulated below:

Temperature in keV	$\langle\sigma(v)v\rangle$ in m^3/s	$n_e\tau_E$
1	5.5×10^{-27}	
2	2.6×10^{-25}	
5	1.3×10^{-23}	
10	1.1×10^{-22}	
20	4.2×10^{-22}	
50	8.7×10^{-22}	
100	8.5×10^{-22}	

Calculate the $n_e\tau_E$ required for break-even at the listed temperatures (i.e., fill in the last column).

(f) The Joint European Tokamak (JET) located in Culham, UK, (in operation during the latter twentieth and early twenty first century) has a plasma volume of approximately 80 cubic meters. Suppose JET has an energy confinement time of $\tau_E \sim 0.3$ s at $n_e \sim 10^{20}$ m^{-3} and $T \sim 10$ keV. If the plasma is a 50%–50% mixture of D–T, how much fusion power would be produced in JET and what is Q? What is the weight of the plasma? How much does the confinement time have to be scaled up to achieve break-even? How much does the volume have to be scaled up to achieve 2000 MW thermal power output at break-even? In order to achieve this volume scale-up, how much does the linear dimension have to be scaled up?

10

Stability of static MHD equilibria

10.1 Introduction

Solutions to Eq. (9.50), the Grad–Shafranov equation, (or to some more complicated counterpart in the case of non-axisymmetric geometry) provide a static MHD equilibrium. The question now arises whether the equilibrium is stable. This issue was forced upon early magnetic fusion researchers who found that plasma that was expected to be well confined in a static MHD equilibrium configuration would instead became violently unstable and crash destructively into the wall in a few microseconds.

The difference between stable and unstable equilibria is shown schematically in Fig. 10.1. Here a ball, representing the plasma, is located at either the bottom of a valley or the top of a hill. If the ball is at the bottom of a valley, i.e., a minimum in the potential energy, then a slight lateral displacement results in a restoring force, which pushes the ball back. The ball then overshoots and oscillates about the minimum with a constant amplitude because energy is conserved. On the other hand, if the ball is initially located at the top of a hill, then a slight lateral displacement results in a force that pushes the ball further to the side so that there is an increase in the velocity. The perturbed force is not restoring, but rather the opposite. The velocity is always in the direction of the original displacement; i.e., there is no oscillation in velocity.

The equation of motion for this system is

$$m\frac{\mathrm{d}^2 x}{\mathrm{d}t^2} = \pm\kappa x, \tag{10.1}$$

where κ is assumed positive; the plus sign is chosen for the ball on hill case, and the minus sign is chosen for the ball in valley case. This equation has a solution $x \sim \exp(-\mathrm{i}\omega t)$, where $\omega = \pm\sqrt{\kappa/m}$ for the valley case and $\omega = \pm\mathrm{i}\sqrt{\kappa/m}$ for the hill case. The hill $\omega = +\mathrm{i}\sqrt{\kappa/m}$ solution is unstable and corresponds to the ball accelerating down the hill when perturbed from its initial equilibrium position.

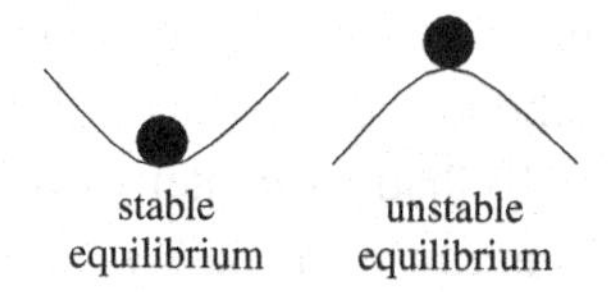

Fig. 10.1 Stable and unstable equilibria.

If this configuration is extended to two dimensions, then stability would require an absolute minimum in both directions. However, a saddle point potential would suffice for instability, since the ball could always roll down from the saddle point. Thus, a multi-dimensional system can only be stable if the equilibrium potential energy corresponds to an absolute minimum with respect to all possible displacements.

The problem of determining MHD stability is analogous to having a ball on a multi-dimensional hill. If the potential energy of the system increases for any allowed perturbation of the system, then the system is stable. However, if there exists even a single allowed perturbation that decreases the system's potential energy, then the system is unstable.

10.2 The Rayleigh–Taylor instability of hydrodynamics

An important subset of MHD instabilities is formally similar to the Rayleigh–Taylor instability of hydrodynamics; it is therefore useful to put aside MHD for the moment and examine this classical problem. As any toddler learns from stacking building blocks, it is possible to construct an equilibrium whereby a heavy object is supported by a light object, but such an equilibrium is unstable. The corresponding hydrodynamic situation has a heavy fluid supported by a light fluid as shown in Fig. 10.2(a); this situation is unstable with respect to the rippling shown in Fig. 10.2(b). The ripples are unstable because they effectively interchange volume elements of heavy fluid with equivalent volume elements of

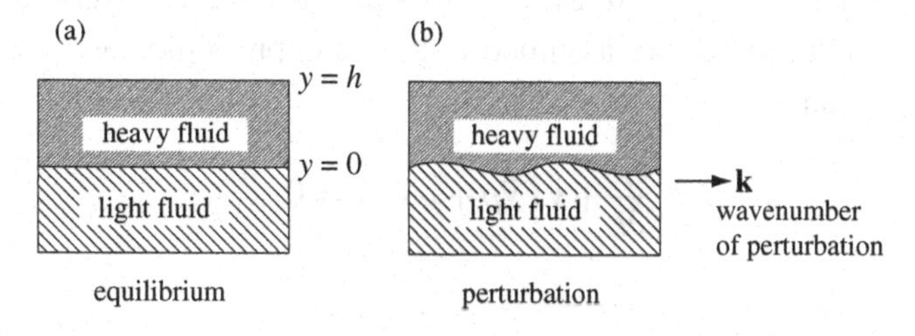

Fig. 10.2 (a) Top-heavy fluid equilibrium, (b) rippling instability.

light fluid. Each volume element of interchanged heavy fluid originally had its center of mass a distance Δ above the interface while each volume element of interchanged light fluid originally had its center of mass a distance Δ below the interface. Since the potential energy of a mass m at height h in a gravitational field g is mgh, the respective changes in potential energy of the heavy and light fluids are

$$\delta W_h = -2\rho_h V\Delta g, \quad \delta W_l = +2\rho_l V\Delta g, \tag{10.2}$$

where V is the volume of the interchanged fluid elements and ρ_h and ρ_l are the mass densities of the heavy and light fluids. The net change in the total system potential energy is

$$\delta W = -2\left(\rho_h - \rho_l\right) V\Delta g, \tag{10.3}$$

which is negative so the system *lowers* its potential energy by forming ripples. This is analogous to the ball falling off the top of the hill.

A well-known example of this instability is the situation of an inverted glass of water. The heavy fluid in this case is the water and the light fluid is the air. The system is stable when a piece of cardboard is located at the interface between the water and the air, but when the cardboard is removed the system becomes unstable and the water falls out. The function of the cardboard is to prevent ripple interchange from occurring. The system is in stable equilibrium when ripples are prevented because atmospheric pressure is adequate to support the inverted water. From a mathematical point of view, the cardboard places a constraint on the system by imposing a boundary condition that prevents ripple formation.

When the cardboard is removed so that there is no longer a constraint against ripple formation, the ripples grow from noise to large amplitude, and the water falls out. This is an example of an unstable equilibrium.

The geometry shown in Fig. 10.2 is now used to analyze the stability of a heavy fluid such as water supported by a light fluid such as air in the situation where no constraint exists at the interface. Here y corresponds to the vertical direction so that gravity is in the negative y direction. To simplify the analysis, it is assumed that $\rho_l \ll \rho_h$, in which case the mass of the light fluid can be ignored. The water and air are assumed to be incompressible, i.e., $\rho = const.$, so the continuity equation

$$\frac{\partial \rho}{\partial t} + \mathbf{v}\cdot\nabla\rho + \rho\nabla\cdot\mathbf{v} = 0 \tag{10.4}$$

reduces to

$$\nabla\cdot\mathbf{v} = 0. \tag{10.5}$$

The linearized continuity equation in the water therefore reduces to

$$\frac{\partial \rho_1}{\partial t} + \mathbf{v}_1 \cdot \nabla \rho_0 = 0 \tag{10.6}$$

and the linearized equation of motion in the water is

$$\rho_0 \frac{\partial \mathbf{v}_1}{\partial t} = -\nabla P_1 - \rho_1 g \hat{y}. \tag{10.7}$$

The location $y = 0$ is defined to be at the unperturbed air–water interface and the top of the glass is at $y = h$. The water fills the glass to the top and thus is constrained from moving at the top, giving the top boundary condition

$$v_y = 0 \text{ at } y = h. \tag{10.8}$$

The perturbation is assumed to have the form

$$\mathbf{v}_1 = \mathbf{v}_1(y) \mathrm{e}^{\gamma t + \mathrm{i}\mathbf{k}\cdot\mathbf{x}}, \tag{10.9}$$

where $\mathbf{k}$ lies in the $x - z$ plane and positive γ implies instability. The incompressibility condition, Eq. (10.5), can thus be written as

$$\frac{\partial v_{1y}}{\partial y} + \mathrm{i}\mathbf{k} \cdot \mathbf{v}_{1\perp} = 0, \tag{10.10}$$

where $\perp$ means perpendicular to the y direction. The y and $\perp$ components of Eq. (10.7) become respectively

$$\gamma \rho_0 v_{1y} = -\frac{\partial P_1}{\partial y} - \rho_1 g \tag{10.11}$$

$$\gamma \rho_0 \mathbf{v}_{1\perp} = -\mathrm{i}\mathbf{k} P_1. \tag{10.12}$$

The system is solved by dotting Eq. (10.12) with $\mathrm{i}\mathbf{k}$ and then using Eq. (10.10) to eliminate $\mathbf{k} \cdot \mathbf{v}_{1\perp}$ and so obtain

$$-\gamma \rho_0 \frac{\partial v_{1y}}{\partial y} = k^2 P_1. \tag{10.13}$$

The perturbed density, as given by Eq. (10.6), is

$$\gamma \rho_1 = -v_{1y} \frac{\partial \rho_0}{\partial y}. \tag{10.14}$$

Next, ρ_1 and P_1 are substituted for in Eq. (10.11) to obtain the eigenvalue equation

$$\frac{\partial}{\partial y}\left[\gamma^2 \rho_0 \frac{\partial v_{1y}}{\partial y}\right] = \left[\gamma^2 \rho_0 - g \frac{\partial \rho_0}{\partial y}\right] k^2 v_{1y}. \tag{10.15}$$

This equation is solved by considering the interior and the interface separately:

1. Interior: here $\partial \rho_0/\partial y = 0$ and $\rho_0 = const.$, in which case Eq. (10.15) becomes

$$\frac{\partial^2 v_{1y}}{\partial y^2} = k^2 v_{1y}. \tag{10.16}$$

The solution satisfying the boundary condition given by Eq. (10.8) is

$$v_{1y} = A \sinh(k(y-h)). \tag{10.17}$$

2. Interface: to find the properties of this region, Eq. (10.15) is integrated across the interface from $y = 0_-$ to $y = 0_+$ to obtain

$$\left[\gamma^2 \rho_0 \frac{\partial v_{1y}}{\partial y}\right]_{0_-}^{0_+} = -\left[g\rho_0 k^2 v_{1y}\right]_{0_-}^{0+} \tag{10.18}$$

or

$$\gamma^2 \frac{\partial v_{1y}}{\partial y} = -gk^2 v_{1y}, \tag{10.19}$$

where all quantities refer to the upper (water) side of the interface, since by assumption $\rho_0(y = 0_-) \simeq 0$.

Substitution of Eq. (10.17) into Eq. (10.19) gives the dispersion relation

$$\gamma^2 = kg \tanh(k_\perp h), \tag{10.20}$$

which shows that the configuration is always unstable since $\gamma^2 > 0$. Equation (10.20) furthermore shows that short wavelengths are most unstable, but a more detailed analysis taking into account surface tension (which is stronger for shorter wavelengths) would show γ^2 has a maximum at some wavelength. Above this most unstable wavelength, surface tension would decrease the growth rate.

10.3 MHD Rayleigh–Taylor instability

We define a "magnetofluid" as a fluid that satisfies the MHD equations. A given plasma may or may not behave as a magnetofluid, depending on the validity of the MHD approximation for the circumstances of the given plasma. The magnetofluid concept provides a legalism that allows consideration of the implications of MHD without necessarily accepting that these implications are relevant to a specific actual plasma. In effect, the magnetofluid concept can be considered as a tentative model of plasma.

Let us now replace the water in the Rayleigh–Taylor instability by magnetofluid. We further suppose that instead of atmospheric pressure supporting the magnetofluid, a vertical magnetic field gradient balances the downwards gravitational force, i.e., at each y the upward force of $-\nabla B^2/2\mu_0$ supports the downward force of the weight of the plasma above. Although gravity is normally unimportant in actual plasmas, the gravitational model is nevertheless quite useful for characterizing situations of practical interest, because gravity can be considered as a proxy for the actual forces that typically have a more complex structure. An important example is centrifugal force associated with thermal particle motion along curved magnetic field lines acting like a gravitational force in the direction of the radius of curvature of these field lines. The curved field with associated centrifugal force due to parallel thermal motion is replaced by a Cartesian geometry model having straight field lines and, perpendicular to the field lines, a gravitational force is invoked to represent the effect of the centrifugal force. The y direction corresponds to the direction of the radius of curvature.

In order for $-\nabla B^2$ to point upwards in the y direction, the magnetic field must depend on y such that its magnitude decreases with increasing y. Furthermore, it is required that $B_y = 0$ so ∇B^2 is perpendicular to the magnetic field and the field can be considered as locally straight (field line curvature has already been taken into account by introducing the fictitious gravity). Thus, the equilibrium magnetic field is assumed to be of the general form

$$\mathbf{B}_0 = B_{x0}(y)\hat{x} + B_{z0}(y)\hat{z}. \tag{10.21}$$

The unit vector associated with the equilibrium field is

$$\hat{B}_0 = \frac{B_{x0}(y)\hat{x} + B_{z0}(y)\hat{z}}{\sqrt{B_{x0}(y)^2 + B_{z0}(y)^2}}. \tag{10.22}$$

For the special case where $B_{x0}(y)$ and $B_{z0}(y)$ are proportional to each other, the field line direction is independent of y, but in the more general case where this is not so, $\hat{B}_0$ depends on y and thus rotates as a function of y. In this latter case, the magnetic field lines are said to be sheared, since adjacent y-layers of field lines are not parallel to each other.

The magnetofluid is assumed incompressible and, using Eq. (9.10), the linearized equation of motion becomes

$$\rho_0 \frac{\partial \mathbf{v}_1}{\partial t} = -\nabla \bar{P}_1 + \frac{\mathbf{B}_0 \cdot \nabla \mathbf{B}_1 + \mathbf{B}_1 \cdot \nabla \mathbf{B}_0}{\mu_0} - \rho_1 g\hat{y}, \tag{10.23}$$

where

$$\bar{P}_1 = P_1 + \frac{\mathbf{B}_0 \cdot \mathbf{B}_1}{\mu_0}$$

is the perturbation of the combined hydrodynamic and magnetic pressure, i.e., the perturbation of $P+B^2/2\mu_0$. It is again assumed that all quantities vary in the manner of Eq. (10.9) so Eq. (10.23) has the respective y and $\perp$ components

$$\gamma\rho_0 v_{1y} = -\frac{\partial \bar{P}_1}{\partial y} + \frac{\mathrm{i}\,(\mathbf{k}\cdot\mathbf{B}_0)\,B_{1y}}{\mu_0} - \rho_1 g \tag{10.24}$$

$$\gamma\rho_0 \mathbf{v}_{1\perp} = -\mathrm{i}\mathbf{k}\bar{P}_1 + \frac{1}{\mu_0}\left[\mathrm{i}\,(\mathbf{k}\cdot\mathbf{B}_0)\,\mathbf{B}_{1\perp} + B_{1y}\frac{\partial \mathbf{B}_0}{\partial y}\right]. \tag{10.25}$$

In analogy to the glass of water problem, Eq. (10.25) is dotted with $\mathrm{i}\mathbf{k}$ and Eq. (10.10) is invoked to obtain

$$-\gamma\rho_0\frac{\partial v_{1y}}{\partial y} = k^2\bar{P}_1 + \frac{1}{\mu_0}\left[-\,(\mathbf{k}\cdot\mathbf{B}_0)\,\mathbf{k}\cdot\mathbf{B}_{1\perp} + \mathrm{i}B_{1y}\frac{\partial\,(\mathbf{k}\cdot\mathbf{B}_0)}{\partial y}\right]. \tag{10.26}$$

Because $\nabla\cdot\mathbf{B}_1 = 0$, the perturbed perpendicular field is

$$\mathrm{i}\mathbf{k}\cdot\mathbf{B}_{1\perp} = -\frac{\partial B_{1y}}{\partial y} \tag{10.27}$$

and so Eq. (10.26) can be recast as

$$k^2\bar{P}_1 = -\gamma\rho_0\frac{\partial v_{1y}}{\partial y} - \frac{1}{\mu_0}\left[-\mathrm{i}\,(\mathbf{k}\cdot\mathbf{B}_0)\,\frac{\partial B_{1y}}{\partial y} + \mathrm{i}B_{1y}\frac{\partial\,(\mathbf{k}\cdot\mathbf{B}_0)}{\partial y}\right]. \tag{10.28}$$

Following a procedure analogous to that used in the inverted glass of water problem, $\bar{P}_1$ is eliminated in Eq. (10.24) by substitution of Eq. (10.28) to obtain

$$\begin{aligned}\gamma\rho_0 v_{1y} = &-\frac{1}{k^2}\frac{\partial}{\partial y}\left\{-\gamma\rho_0\frac{\partial v_{1y}}{\partial y} - \frac{1}{\mu_0}\left[-\,(\mathrm{i}\mathbf{k}\cdot\mathbf{B}_0)\,\frac{\partial B_{1y}}{\partial y} + B_{1y}\frac{\partial\,(\mathrm{i}\mathbf{k}\cdot\mathbf{B}_0)}{\partial y}\right]\right\}\\ &+\frac{\mathrm{i}\,(\mathbf{k}\cdot\mathbf{B}_0)\,B_{1y}}{\mu_0} - \rho_1 g.\end{aligned} \tag{10.29}$$

To proceed further, it is necessary to know B_{1y}. The complete vector $\mathbf{B}_1$ is found by first linearizing the MHD Ohm's law to obtain

$$\mathbf{E}_1 + \mathbf{v}_1\times\mathbf{B}_0 = 0, \tag{10.30}$$

then taking the curl, and finally using Faraday's law to obtain

$$\gamma\mathbf{B}_1 = \nabla\times[\mathbf{v}_1\times\mathbf{B}_0]. \tag{10.31}$$

The y component is found by dotting with $\hat{y}$ and then using the vector identity $\nabla\cdot(\mathbf{F}\times\mathbf{G}) = \mathbf{G}\cdot\nabla\times\mathbf{F} - \mathbf{F}\cdot\nabla\times\mathbf{G}$ to obtain

$$\gamma B_{1y} = \hat{y}\cdot\nabla\times[\mathbf{v}_1\times\mathbf{B}_0] = \nabla\cdot[(\mathbf{v}_1\times\mathbf{B}_0)\times\hat{y}] = \nabla\cdot\left[v_{1y}\mathbf{B}_0\right] = \mathrm{i}\mathbf{k}\cdot\mathbf{B}_0 v_{1y}. \tag{10.32}$$

Substituting into Eq. (10.29) using Eq. (10.32) and Eq. (10.14) and rearranging the order gives

$$\frac{\partial}{\partial y}\left\{\left[\gamma^2\rho_0+\frac{1}{\mu_0}(\mathbf{k}\cdot\mathbf{B}_0)^2\right]\frac{\partial v_{1y}}{\partial y}\right\}=k^2\left\{\gamma^2\rho_0-g\frac{\partial\rho_0}{\partial y}+\frac{(\mathbf{k}\cdot\mathbf{B}_0)^2}{\mu_0}\right\}v_{1y}, \tag{10.33}$$

which is identical to the inverted glass of water problem if $\mathbf{k}\cdot\mathbf{B}_0=0$.

Rather than have an abrupt interface between a heavy and light fluid as in the glass of water problem, it is assumed that the magnetofluid fills the container between $y=0$ and $y=h$ and that there is a density gradient in the y direction. This situation is more appropriate for a plasma, which would typically have a continuous density gradient. It is assumed that rigid boundaries exist at both $y=0$ and $y=h$ so that $v_{1y}=0$ at both $y=0$ and $y=h$. These boundary conditions mean that rippling is not allowed at $y=0,h$ but could occur in the interior, $0<y<h$. Equation (10.33) is now impossible to solve analytically because all the coefficients are functions of y. However, an approximate understanding for the behavior predicted by this equation can be found by multiplying the equation by v_{1y} and integrating from $y=0$ to $y=h$ to obtain

$$\left[\left\{\gamma^2\rho_0+\frac{1}{\mu_0}(\mathbf{k}\cdot\mathbf{B}_0)^2\right\}v_{1y}\frac{\partial v_{1y}}{\partial y}\right]_0^h-\int_0^h\left[\gamma^2\rho_0+\frac{1}{\mu_0}(\mathbf{k}\cdot\mathbf{B}_0)^2\right]\left(\frac{\partial v_{1y}}{\partial y}\right)^2\mathrm{d}y$$
$$=k^2\int_0^h\left\{\gamma^2\rho_0-g\frac{\partial\rho_0}{\partial y}+\frac{(\mathbf{k}\cdot\mathbf{B}_0)^2}{\mu_0}\right\}v_{1y}^2\mathrm{d}y. \tag{10.34}$$

The integrated term vanishes because of the boundary conditions (which could also have been $\partial v_{1y}/\partial y=0$). Solving for γ^2 gives

$$\gamma^2=\frac{\displaystyle\int_0^h\mathrm{d}y\left[k^2g\frac{\partial\rho_0}{\partial y}v_{1y}^2-\frac{(\mathbf{k}\cdot\mathbf{B}_0)^2}{\mu_0}\left(k^2v_{1y}^2+\left(\frac{\partial v_{1y}}{\partial y}\right)^2\right)\right]}{\displaystyle\int_0^h\mathrm{d}y\;\rho_0\left[k^2v_{1y}^2+\left(\frac{\partial v_{1y}}{\partial y}\right)^2\right]}. \tag{10.35}$$

If $\mathbf{k}\cdot\mathbf{B}_0=0$ and the density gradient is positive everywhere then $\gamma^2>0$ so there is instability. If the density gradient is negative everywhere except at one stratum with thickness Δy, then the system will be unstable with respect to an interchange at that one "top-heavy" stratum. The velocity will be concentrated at this unstable stratum and so the integrands will vanish everywhere except at the unstable stratum giving a growth rate $\gamma^2\sim g\rho_0^{-1}\partial\rho_0/\partial y$, where $\partial\rho_0/\partial y$ is the value in the unstable region. This MHD version of the Rayleigh–Taylor instability is called the Kruskal–Schwarzschild instability (Kruskal and Schwarzschild 1954).

Finite $\mathbf{k}\cdot\mathbf{B}_0$ opposes the effect of the destabilizing positive density gradient, reducing the growth rate to $\gamma^2 \sim g\rho_0^{-1}\,\partial\rho_0/\partial y - (\mathbf{k}\cdot\mathbf{B}_0)^2/\mu_0\rho_0$. A sufficiently strong field will stabilize the system. However, thermally excited noise excites modes with all possible values of $\mathbf{k}$ and those modes aligned such that $\mathbf{k}\cdot\mathbf{B}_0 = 0$ will not be stabilized. A shear in the magnetic field has the effect of making it possible to have $\mathbf{k}\cdot\mathbf{B}_0(y) = 0$ at only a single value of y. The lack of any spatial dependence of $\mathbf{k}$ results from the translational invariance of the plasma with respect to both x and z implying that $\mathbf{k} = k_x\hat{x} + k_z\hat{z}$ is independent of position. Thus, magnetic shear constrains the instability to a narrow y stratum.

The stabilizing effect of finite $\mathbf{k}\cdot\mathbf{B}_0$ can be understood by considering Eq. (10.32) which shows that the amount of B_{1y} associated with a given v_{1y} is proportional to $\mathbf{k}\cdot\mathbf{B}_0$. Since the original equilibrium field had no y component, introducing a finite B_{1y} corresponds to "plucking" the equilibrium field. The plucking varies sinusoidally along the equilibrium field and stretches the equilibrium field like a plucked violin string. The plucked field has a restoring force, which pulls the field back to its equilibrium position. If the energy associated with plucking the magnetic field exceeds the energy liberated by the interchange of heavy upper magnetofluid with light lower magnetofluid, the mode is stable.

As discussed earlier, the gravitational force in the magnetofluid model represents the centrifugal force resulting from guiding center motion along curved field lines. This leads to the concept of "good" and "bad" curvature illustrated in Fig. 10.3. A plasma has bad curvature if the field lines at the plasma–vacuum boundary have a convex shape as seen by an observer outside the plasma, since in this case the centrifugal force is outwards from the plasma. Bad curvature gives a centrifugal force that can drive interchange instabilities whereas good curvature corresponds to a concave shape so that the centrifugal force is always inwards. Mirror magnetic fields as sketched in Fig. 10.4 have good curvature in the vicinity of the mirrors and bad curvature in the vicinity of the mirror minimum so that a detailed analysis of interchange instabilities requires averaging the "goodness/badness" along the portion of the flux tube experienced by the particle. Cusp magnetic fields (cf. Fig. 10.4) have good curvature everywhere,

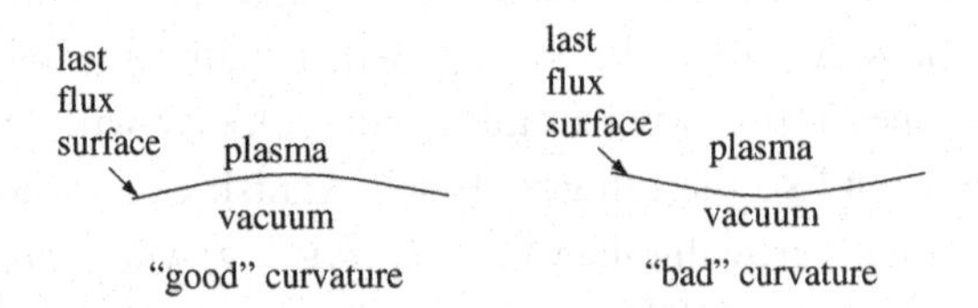

Fig. 10.3 Good and bad curvature of the plasma–vacuum interface.

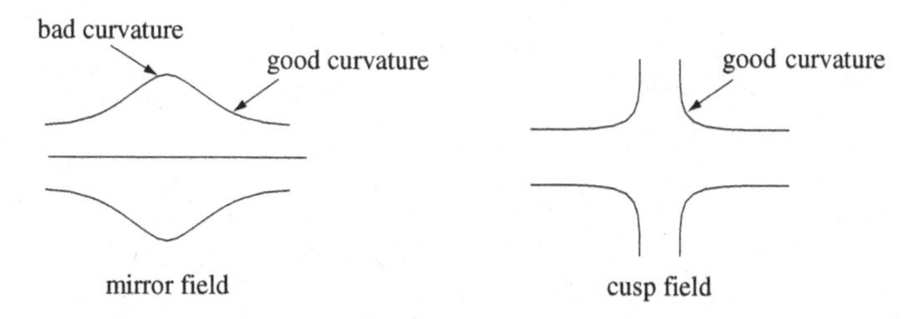

Fig. 10.4 Mirror fields have bad curvature at the mirror minimum, good curvature near the maximum; cusp fields have good curvature everywhere.

but have singular behavior at the cusps. Plasmas with internal currents such as tokamaks have significant shear everywhere, since $\partial B_x/\partial y \sim J_z$ and $\partial B_z/\partial y \sim J_x$; this shear inhibits interchange instabilities. However, as will be shown later these currents can be the source for another class of instability called a current-driven instability.

10.4 The MHD energy principle

Suppose a system initially in equilibrium is subject to a small perturbation instigated by random thermal noise. The perturbation will affect all the various dependent variables in a way that must be consistent with all the relevant equations and boundary conditions. Thus, the perturbation may be considered as a low-level excitation of some allowed normal mode of the system. The mode will be unstable if it reduces the potential energy of the system and the growth rate of the instability will be proportional to the amount by which the potential energy is reduced. In order to investigate whether a given system is stable, it suffices to show that no modes exist that reduce the system's potential energy. Demonstrating MHD stability this way was first done by Bernstein *et al.* (1958) and the method is called the MHD energy principle.

Each mode has its own specific pattern for displacing the magnetofluid volume elements from their equilibrium positions. The pattern of displacements can be represented by a vector function of position $\boldsymbol{\xi}(\mathbf{x})$, which prescribes how a fluid volume element originally at location $\mathbf{x}$ is displaced. Thus, the mode involves moving a fluid element initially at $\mathbf{x}$ to the position $\mathbf{x}+\boldsymbol{\xi}(\mathbf{x})$. The displacement of a magnetofluid element initially at the surface is shown in Fig. 10.5, a sketch of a two-dimensional cut of a three-dimensional magnetofluid (plasma) surrounded by a vacuum region in turn bounded by a conducting wall. The wall could be brought right up to the magnetofluid to eliminate the vacuum region or, alternatively, the wall could be placed at infinity to represent a system having no wall and surrounded by vacuum.

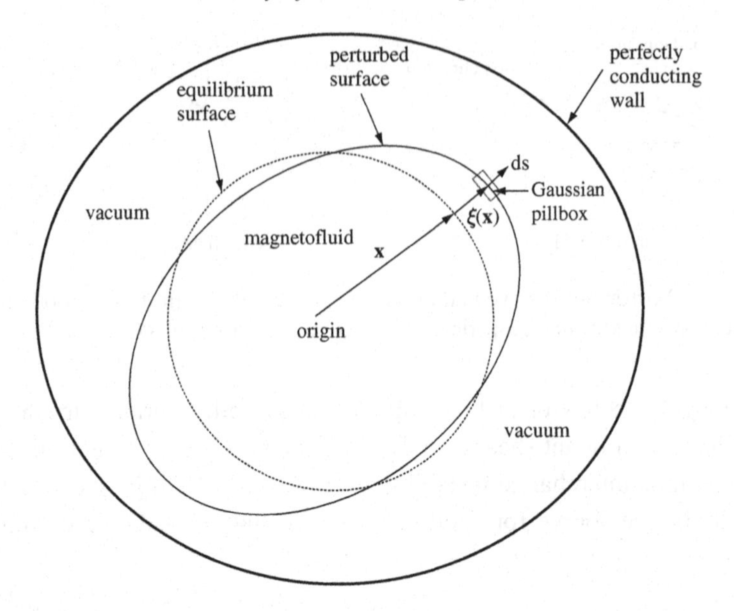

Fig. 10.5 Two-dimensional cut of three-dimensional plasma equilibrium and perturbation; Gaussian pillbox is used to relate quantities in vacuum to quantities in plasma. Each volume element at a position **x** is displaced by an amount $\boldsymbol{\xi}(\mathbf{x})$. The displacement of a volume element at the surface is illustrated.

10.4.1 Energy equation for a magnetofluid

The energy content of a magnetofluid can be obtained from the ideal MHD equations if it is assumed that all motions are sufficiently fast to be adiabatic but slow enough for collisions to keep the pressure isotropic. The ideal MHD equations are

$$\rho\left(\frac{\partial \mathbf{U}}{\partial t}+\mathbf{U}\cdot\nabla\mathbf{U}\right)=\mathbf{J}\times\mathbf{B}-\nabla P \tag{10.36}$$

$$\mathbf{E}+\mathbf{U}\times\mathbf{B}=0 \tag{10.37}$$

$$\nabla\times\mathbf{E}=-\frac{\partial \mathbf{B}}{\partial t} \tag{10.38}$$

$$\nabla\times\mathbf{B}=\mu_o\mathbf{J} \tag{10.39}$$

$$\frac{\partial \rho}{\partial t}+\nabla\cdot(\rho\mathbf{U})=0 \tag{10.40}$$

$$P\sim\rho^{\gamma}. \tag{10.41}$$

Ohm's law, Eq. (10.37), and Faraday's law, Eq. (10.38), are combined to give the *induction equation*

$$\frac{\partial \mathbf{B}}{\partial t} = \nabla \times (\mathbf{U} \times \mathbf{B}), \tag{10.42}$$

which prescribes the magnetic field evolution and, as discussed in the context of Eq. (2.83), shows that the magnetic flux is frozen into the magnetofluid. Equation (10.41), the adiabatic relation, implies the following relationships for spatial and temporal derivatives of density and pressure

$$\frac{\nabla P}{P} = \gamma \frac{\nabla \rho}{\rho}, \quad \frac{1}{P}\frac{\partial P}{\partial t} = \frac{\gamma}{\rho}\frac{\partial \rho}{\partial t}. \tag{10.43}$$

Combining these relationships with the continuity equation, Eq. (10.40), gives the pressure evolution equation

$$\frac{\partial P}{\partial t} + \mathbf{U} \cdot \nabla P + \gamma P \nabla \cdot \mathbf{U} = 0, \tag{10.44}$$

which can also be written as

$$\frac{\partial P}{\partial t} + \nabla \cdot (P\mathbf{U}) + (\gamma - 1) P \nabla \cdot \mathbf{U} = 0. \tag{10.45}$$

An expression for the overall energy can be derived by first writing the equation of motion, Eq. (10.36), as

$$\rho \frac{\partial \mathbf{U}}{\partial t} + \rho \nabla \left(\frac{U^2}{2} \right) - \rho \mathbf{U} \times \nabla \times \mathbf{U} = \mathbf{J} \times \mathbf{B} - \nabla P \tag{10.46}$$

and then dotting Eq. (10.46) with $\mathbf{U}$ to obtain

$$\rho \frac{\partial}{\partial t} \left(\frac{U^2}{2} \right) + \rho \mathbf{U} \cdot \nabla \left(\frac{U^2}{2} \right) = \mathbf{J} \times \mathbf{B} \cdot \mathbf{U} - \mathbf{U} \cdot \nabla P. \tag{10.47}$$

Multiplying the entire continuity equation by $U^2/2$ and adding the result gives

$$\frac{\partial}{\partial t} \left(\frac{\rho U^2}{2} \right) + \nabla \cdot \left(\frac{\rho U^2}{2} \mathbf{U} \right) = -\mathbf{J} \cdot \mathbf{U} \times \mathbf{B} - \nabla \cdot (P\mathbf{U}) + P \nabla \cdot \mathbf{U}. \tag{10.48}$$

Using Ampère's law, Eq. (10.39), to eliminate $\mathbf{J}$, then Ohm's law, Eq. (10.37), to eliminate $\mathbf{U} \times \mathbf{B}$, and the pressure evolution equation, Eq. (10.45), to eliminate $P \nabla \cdot \mathbf{U}$, this energy equation becomes

$$\frac{\partial}{\partial t} \left(\frac{\rho U^2}{2} \right) + \nabla \cdot \left(\frac{\rho U^2}{2} \mathbf{U} \right) = \frac{(\nabla \times \mathbf{B})}{\mu_0} \cdot \mathbf{E} - \nabla \cdot (P\mathbf{U}) - \frac{1}{(\gamma - 1)} \left(\frac{\partial P}{\partial t} + \nabla \cdot (P\mathbf{U}) \right). \tag{10.49}$$

Finally, using the vector identity

$$\begin{aligned}\nabla\cdot(\mathbf{E}\times\mathbf{B}) &= \mathbf{B}\cdot\nabla\times\mathbf{E}-\mathbf{E}\cdot\nabla\times\mathbf{B}\\ &= -\mathbf{B}\cdot\frac{\partial\mathbf{B}}{\partial t}-\mathbf{E}\cdot\nabla\times\mathbf{B}\end{aligned} \tag{10.50}$$

Eq. (10.49) can be written as

$$\frac{\partial}{\partial t}\left(\frac{\rho U^2}{2}+\frac{P}{\gamma-1}+\frac{B^2}{2\mu_0}\right)+\nabla\cdot\left(\rho\frac{U^2}{2}\mathbf{U}+\frac{\mathbf{E}\times\mathbf{B}}{\mu_0}+\frac{\gamma}{\gamma-1}P\mathbf{U}\right)=0. \tag{10.51}$$

This is a conservation equation relating energy density and energy flux. The time derivative operates on the magnetofluid energy density and the divergence operates on the energy flux. The energy density is comprised of kinetic energy density $\rho U^2/2$, thermal energy density $P/(\gamma-1)$, and magnetic energy $B^2/2\mu_0$. The energy flux is comprised of convection of kinetic energy $\rho U^2\mathbf{U}/2$, the Poynting vector $\mathbf{E}\times\mathbf{B}/\mu_0$, and convection of thermal energy density $\gamma P\mathbf{U}/(\gamma-1)$.

The pressure and density both vanish at the magnetofluid surface, but the Poynting flux $\mathbf{E}\times\mathbf{B}$ can be finite. However, if the tangential electric field vanishes at the surface, then the Poynting flux normal to the surface will be zero. This situation occurs if the magnetofluid is bounded by a perfectly conducting wall with no vacuum region between the magnetofluid and the wall. On the other hand, if a vacuum region bounds the magnetofluid, then a tangential electric field can exist at the vacuum–magnetofluid interface and allow a Poynting flux normal to the surface. Energy could then flow back and forth between the magnetofluid and the vacuum region. For example, if the magnetofluid were to move towards the wall thereby reducing the volume of the vacuum region, magnetic field in the vacuum region would be compressed and so the vacuum magnetic field energy would increase. This flow of energy into the vacuum region would require a Poynting flux from the magnetofluid into the vacuum region.

Let us now consider the energy properties of the vacuum region between the magnetofluid and a perfectly conducting wall. The equations characterizing the vacuum region are Faraday's law

$$\nabla\times\mathbf{E}=-\frac{\partial\mathbf{B}}{\partial t} \tag{10.52}$$

and Ampère's law

$$\nabla\times\mathbf{B}=0. \tag{10.53}$$

Dotting Faraday's law with $\mathbf{B}$, Ampère's law with $\mathbf{E}$, and subtracting gives

$$\frac{\partial}{\partial t}\left(\frac{B^2}{2\mu_0}\right)+\nabla\cdot\left(\frac{\mathbf{E}\times\mathbf{B}}{\mu_0}\right)=0, \tag{10.54}$$

which is just the limit of Eq. (10.51) for zero density and zero pressure. Thus, Eq. (10.51) characterizes not only the magnetofluid, but also the surrounding vacuum region. As mentioned earlier, the Poynting flux is the means by which electromagnetic energy flows between the magnetofluid and the vacuum region.

If the vacuum region is bounded by a conducting wall, then the tangential electric field must vanish on the wall. The integral of the energy equation over the volume of both the magnetofluid and the surrounding vacuum region then becomes

$$\frac{\partial}{\partial t}\int \mathrm{d}^3r\left(\frac{\rho U^2}{2}+\frac{P}{\gamma-1}+\frac{B^2}{2\mu_0}\right)=0, \tag{10.55}$$

since on the wall $\mathrm{d}\mathbf{s}\cdot\mathbf{E}\times\mathbf{B}=0$, where $\mathrm{d}\mathbf{s}$ is a surface element of the wall. If the wall is brought right up to the magnetofluid so there is no vacuum region, then Eq. (10.55) will also result, with the additional stipulation that the normal component of the velocity vanishes at the wall (i.e., the wall is impermeable).

A system consisting of a magnetofluid surrounded by a vacuum region enclosed by an impermeable perfectly conducting wall will therefore have its total internal energy conserved, that is

$$\int_{V_{mf}} \mathrm{d}^3r\left(\frac{\rho U^2}{2}+\frac{P}{\gamma-1}+\frac{B^2}{2\mu_0}\right)+\int_{V_{vac}} \mathrm{d}^3r\frac{B^2}{2\mu_0}=const., \tag{10.56}$$

where V_{mf} is the volume of the magnetofluid and V_{vac} is the volume of the vacuum region between the magnetofluid and the wall.

This total system energy can be split into a kinetic energy term

$$T=\int \mathrm{d}^3r\frac{\rho U^2}{2} \tag{10.57}$$

and a potential energy term

$$W=\int_V \mathrm{d}^3r\left(\frac{P}{\gamma-1}+\frac{B^2}{2\mu_0}\right) \tag{10.58}$$

so that

$$T+W=\mathcal{E}, \tag{10.59}$$

where the total energy $\mathcal{E}$ is a constant. Here V includes the volume of both the magnetofluid and any vacuum region between the magnetofluid and the wall.

We will now consider a static equilibrium (i.e., an equilibrium with $\mathbf{U}_0=0$) so that

$$0=\mathbf{J}_0\times\mathbf{B}_0-\nabla P_0. \tag{10.60}$$

Thus ∇P_0 is normal to the surface defined by the $\mathbf{J}_0$ and $\mathbf{B}_0$ vectors. Furthermore, dotting Eq. (10.60) with $\mathbf{B}_0$ and then with $\mathbf{J}_0$ gives the two relations

$$\mathbf{B}_0\cdot\nabla P_0=0,\quad \mathbf{J}_0\cdot\nabla P_0=0, \tag{10.61}$$

which indicate that the pressure is constant along both a magnetic field line and a line defined by the current density vector $\mathbf{J}_0$. Since the analysis of the Grad–Shafranov equation in Section 9.8.3 showed that $\mathbf{B}_0$ and $\mathbf{J}_0$ typically lie in a magnetic surface, the pressure is uniform on the magnetic surface, and the pressure gradient is normal to the magnetic surface.

10.4.2 Self-adjointness of potential energy as a consequence of energy integral

Static equilibrium means $\mathbf{U}_0 = 0$ and so Equation (10.57) indicates $T_0 = 0$ in static equilibrium. Thus, all internal energy in equilibrium must exist in the form of stored potential energy, i.e.,

$$W_0 = \mathcal{E}. \tag{10.62}$$

It is now supposed that random thermal noise causes small motions of order ϵ to develop at each point in the magnetofluid, i.e., at each point there exists a first-order velocity

$$\mathbf{U}_1 \sim \epsilon. \tag{10.63}$$

The displacement of a fluid element is obtained by time integration to be

$$\boldsymbol{\xi} = \int_0^t \mathbf{U}_1 \mathrm{d}t' \tag{10.64}$$

and so $\boldsymbol{\xi}$ is also of order ϵ. Because ϵ is small, any spatial dependence of $\mathbf{U}_1$ can be ignored when evaluating the time integral, i.e., terms of order $\boldsymbol{\xi} \cdot \nabla \mathbf{U}_1$ can be ignored since these are of order ϵ^2.

The kinetic energy associated with the mode is

$$\delta T = \int \mathrm{d}^3 r \frac{\rho_0 U_1^2}{2}, \tag{10.65}$$

which clearly is of order ϵ^2. Since total energy is conserved it is necessary to have

$$\delta T + \delta W = 0 \tag{10.66}$$

leading to the important conclusion that the perturbed potential energy δW must also be of order ϵ^2.

The perturbed magnetic field and pressure can be found to first-order in ϵ by integrating Eqs. (10.42) and (10.44) respectively to obtain

$$\mathbf{B}_1 = \nabla \times (\boldsymbol{\xi} \times \mathbf{B}_0) \tag{10.67}$$

and

$$P_1 = -\boldsymbol{\xi} \cdot \nabla P_0 - \gamma P_0 \nabla \cdot \boldsymbol{\xi}; \tag{10.68}$$

these relations show that both $\mathbf{B}_1$ and P_1 are linear functions of $\boldsymbol{\xi}$. However, since it was just shown that δW scales as ϵ^2, only terms that are second-order in $\boldsymbol{\xi}$ can contribute to δW, and so all first-order terms must average to zero when performing the volume integration required to evaluate δW. To specify that δW depends only to second-order on $\boldsymbol{\xi}$, we write

$$\delta W = \delta W(\boldsymbol{\xi}, \boldsymbol{\xi}), \tag{10.69}$$

where the double argument means that δW is a bilinear function, i.e., $\delta W(a\boldsymbol{\xi}, b\boldsymbol{\eta}) = ab\delta W(\boldsymbol{\xi}, \boldsymbol{\eta})$ for arbitrary $\boldsymbol{\xi}, \boldsymbol{\eta}$. The time derivative of δW is thus

$$\delta \dot{W} = \delta W(\dot{\boldsymbol{\xi}}, \boldsymbol{\xi}) + \delta W(\boldsymbol{\xi}, \dot{\boldsymbol{\xi}}). \tag{10.70}$$

Since $\dot{\boldsymbol{\xi}}$ is algebraically independent of $\boldsymbol{\xi}$, Eq. (10.70) implies $\delta \dot{W}$ is self-adjoint (i.e., $\delta \dot{W}$ is invariant when its two arguments are interchanged). Self-adjointness is a direct consequence of the existence of an energy integral.

10.4.3 Formal solution for perturbed potential energy

The self-adjointness property can be exploited by explicitly calculating the time derivative of the perturbed potential energy. This is done using the linearized equation of motion

$$\rho_0 \frac{\partial^2 \boldsymbol{\xi}}{\partial t^2} = \mathbf{F}_1, \tag{10.71}$$

where

$$\mathbf{F}_1 = \mathbf{J}_0 \times \mathbf{B}_1 + \mathbf{J}_1 \times \mathbf{B}_0 - \nabla P_1 \tag{10.72}$$

results from linear operations on $\boldsymbol{\xi}$, since $\mathbf{B}_1$, $\mu_0 \mathbf{J}_1 = \nabla \times \mathbf{B}_1$, and P_1 all result from linear operations on $\boldsymbol{\xi}$.

Multiplying Eq. (10.71) by the perturbed velocity $\dot{\boldsymbol{\xi}}$ and integrating over the volume of the magnetofluid and vacuum region gives

$$\int \mathrm{d}^3 r \rho_0 \frac{\partial}{\partial t}\left(\frac{\dot{\xi}^2}{2}\right) = \int \mathrm{d}^3 r \dot{\boldsymbol{\xi}} \cdot \mathbf{F}_1(\boldsymbol{\xi}). \tag{10.73}$$

Although the integration includes the vacuum region, both the right- and left-hand sides of this equation vanish in the vacuum region because no density, current, or pressure exist there.

The left-hand side of Eq. (10.73) is just the time derivative of the kinetic energy $\delta \dot{T}$. Since $\delta \dot{T} + \delta \dot{W} = 0$, it is seen that

$$\delta \dot{W} = -\int \mathrm{d}^3 r \dot{\boldsymbol{\xi}} \cdot \mathbf{F}_1(\boldsymbol{\xi}). \tag{10.74}$$

However, because $\delta\dot{W}$ was shown to be self-adjoint, Eq. (10.74) can also be written in an alternative form as $\delta\dot{W} = -\int \mathrm{d}^3r\boldsymbol{\xi}\cdot\mathbf{F}_1(\dot{\boldsymbol{\xi}})$. Combining Eq. (10.74) and its alternative form provides an integrable form for $\delta\dot{W}$, namely

$$\begin{aligned}\delta\dot{W} &= -\frac{1}{2}\left(\int \mathrm{d}^3r\boldsymbol{\xi}\cdot\mathbf{F}_1(\dot{\boldsymbol{\xi}}) + \int \mathrm{d}^3r\dot{\boldsymbol{\xi}}\cdot\mathbf{F}_1(\boldsymbol{\xi})\right)\\ &= -\frac{1}{2}\frac{\partial}{\partial t}\left(\int \mathrm{d}^3r\boldsymbol{\xi}\cdot\mathbf{F}_1(\boldsymbol{\xi})\right).\end{aligned} \tag{10.75}$$

Integrating the above equation with respect to time gives the change in system potential energy associated with an arbitrary fluid displacement, namely

$$\delta W = -\frac{1}{2}\int \mathrm{d}^3r\boldsymbol{\xi}\cdot\mathbf{F}_1(\boldsymbol{\xi}). \tag{10.76}$$

Standard techniques of normal mode analysis can now be invoked and used to provide some restrictions on the form of any dynamical behavior. In particular, by assuming that the displacement has the form

$$\boldsymbol{\xi} = \mathrm{Re}\left(\bar{\boldsymbol{\xi}}e^{-i\omega t}\right), \tag{10.77}$$

the equation of motion can be written as

$$-\omega^2\rho\bar{\boldsymbol{\xi}} = \mathbf{F}_1(\bar{\boldsymbol{\xi}}). \tag{10.78}$$

Then, multiplication by $\bar{\boldsymbol{\xi}}^*$ and integrating over volume gives

$$-\omega^2\int \mathrm{d}^3r\rho|\bar{\xi}|^2 = \int \mathrm{d}^3r\bar{\boldsymbol{\xi}}^*\cdot\mathbf{F}_1(\bar{\boldsymbol{\xi}}). \tag{10.79}$$

The discussion of Eq. (10.70) showed that $\int \mathrm{d}^3r\boldsymbol{\xi}\cdot\mathbf{F}_1(\boldsymbol{\eta})$ is self-adjoint when $\boldsymbol{\xi}$ and $\boldsymbol{\eta}$ are arbitrary *real* variables. However, the linearity of $\mathbf{F}_1$ means that $\int \mathrm{d}^3r\boldsymbol{\xi}\cdot\mathbf{F}_1(\boldsymbol{\eta})$ is self-adjoint even if $\boldsymbol{\xi}$ and $\boldsymbol{\eta}$ are complex since $\int d^3r\,(\boldsymbol{\xi}_r+i\boldsymbol{\xi}_i)\cdot \mathbf{F}_1(\boldsymbol{\eta}_r+i\boldsymbol{\eta}_i) = \int \mathrm{d}^3r\,(\boldsymbol{\xi}_r+i\boldsymbol{\xi}_i)\cdot(\mathbf{F}_1(\boldsymbol{\eta}_r)+i\mathbf{F}_1(\boldsymbol{\eta}_r)) = \int \mathrm{d}^3r\,(\boldsymbol{\eta}_r+i\boldsymbol{\eta}_i)\cdot(\mathbf{F}_1(\boldsymbol{\xi}_r+ i\boldsymbol{\xi}_i)))$. Equation (10.79) can therefore be recast as

$$-\omega^2\int \mathrm{d}^3r\rho|\bar{\xi}|^2 = \frac{1}{2}\left[\int \mathrm{d}^3r\bar{\boldsymbol{\xi}}^*\cdot\mathbf{F}_1(\bar{\boldsymbol{\xi}}) + \int \mathrm{d}^3r\bar{\boldsymbol{\xi}}\cdot\mathbf{F}_1(\bar{\boldsymbol{\xi}}^*)\right], \tag{10.80}$$

which shows that ω^2 must be pure real. Negative ω^2 corresponds to instability.

Equation (10.78) can be considered as an eigenvalue problem where ω^2 is the eigenvalue and $\boldsymbol{\xi}$ is the eigenvector. As in the usual linear algebra sense, eigenvectors having different eigenvalues are orthogonal and so, in principle, a basis set of normalized orthogonal eigenvectors $\{\boldsymbol{\zeta}_m\}$ can be constructed where

$$\mathbf{F}_1(\boldsymbol{\zeta}_m) = -\omega_m^2\rho\boldsymbol{\zeta}_m. \tag{10.81}$$

Thus, any arbitrary displacement can be expressed as a suitably weighted sum of eigenvectors,

$$\boldsymbol{\xi} = \sum_m \alpha_m \boldsymbol{\zeta}_m. \tag{10.82}$$

The orthogonality of the basis set gives the relation

$$\int \mathrm{d}^3 r \boldsymbol{\zeta}_m \cdot \mathbf{F}(\boldsymbol{\zeta}_n) = -\omega_m^2 \int \mathrm{d}^3 r \rho \boldsymbol{\zeta}_m \cdot \boldsymbol{\zeta}_n = 0 \text{ if } m \neq n. \tag{10.83}$$

Instability will result if there exists some perturbation that makes δW negative. Conversely, if all possible perturbations result in positive δW for a given equilibrium, then the equilibrium is MHD stable.

10.4.4 Evaluation of δW

Since $\mathbf{F}_1$ consists of three terms involving $\mathbf{J}_0, \mathbf{B}_1$, and P_1 respectively, δW can be decomposed into three terms,

$$\delta W = \delta W_{J_0} + \delta W_{B_1} + \delta W_{P_1}. \tag{10.84}$$

Using Eqs. (10.67), (10.68), and (10.72) in Eq. (10.76) shows that these terms are

$$\delta W_{J_0} = -\frac{1}{2} \int \mathrm{d}^3 r \boldsymbol{\xi} \cdot \mathbf{J}_0 \times \mathbf{B}_1, \tag{10.85}$$

$$\delta W_{B_1} = -\frac{1}{2\mu_0} \int \mathrm{d}^3 r \boldsymbol{\xi} \cdot (\nabla \times \mathbf{B}_1) \times \mathbf{B}_0, \tag{10.86}$$

$$\delta W_{P1} = \frac{1}{2} \int \mathrm{d}^3 r \boldsymbol{\xi} \cdot \nabla P_1. \tag{10.87}$$

Although the above three right-hand sides are in principle integrated over the volumes of both the magnetofluid and the vacuum regions, in fact, the integrands vanish in the vacuum region for all three cases and so the integration is effectively over the magnetofluid volume only. The second two integrals can be simplified using the vector identities $\nabla \cdot (\mathbf{C} \times \mathbf{D}) = \mathbf{D} \cdot \nabla \times \mathbf{C} - \mathbf{C} \cdot \nabla \times \mathbf{D}$ and $\nabla \cdot (f\mathbf{a}) = \mathbf{a} \cdot \nabla f + f \nabla \cdot \mathbf{a}$ to obtain

$$\begin{aligned} \delta W_{B_1} &= \frac{1}{2\mu_0} \int_{V_{mf}} \mathrm{d}^3 r (\boldsymbol{\xi} \times \mathbf{B}_0) \cdot \nabla \times \mathbf{B}_1 \\ &= \frac{1}{2\mu_0} \int_{V_{mf}} \mathrm{d}^3 r B_1^2 + \frac{1}{2\mu_0} \int_{S_{mf}} \mathrm{d}\mathbf{s} \cdot \mathbf{B}_1 \times (\boldsymbol{\xi} \times \mathbf{B}_0) \end{aligned} \tag{10.88}$$

and

$$\delta W_{P_1} = -\frac{1}{2}\int_{V_{mf}} d^3r\boldsymbol{\xi}\cdot\nabla\left[\boldsymbol{\xi}\cdot\nabla P_0 + \gamma P_0\nabla\cdot\boldsymbol{\xi}\right]$$
$$= \frac{1}{2}\int_{V_{mf}} d^3r\left[-\boldsymbol{\xi}\cdot\nabla\left(\boldsymbol{\xi}\cdot\nabla P_0\right) + \gamma P_0\left(\nabla\cdot\boldsymbol{\xi}\right)^2\right] - \frac{1}{2}\int_{S_{mf}} d\mathbf{s}\cdot\boldsymbol{\xi}\gamma P_0\nabla\cdot\boldsymbol{\xi}. \tag{10.89}$$

Since both $\mathbf{J}\times\mathbf{B}$ and ∇P are perpendicular to $\mathbf{B}$ at the surface of the magnetofluid, the force acting on the magnetofluid at the surface is perpendicular to $\mathbf{B}$ and so the displacement $\boldsymbol{\xi}$ of the magnetofluid at the surface is perpendicular to $\mathbf{B}$, i.e., at the surface $\boldsymbol{\xi} = \boldsymbol{\xi}_\perp$. The surface integral in Eq. (10.88) can be expanded

$$\frac{1}{2\mu_0}\int_{S_{mf}} d\mathbf{s}\cdot\mathbf{B}_1\times\left(\boldsymbol{\xi}\times\mathbf{B}_0\right) = \frac{1}{2\mu_0}\int_{S_{mf}} d\mathbf{s}\cdot\boldsymbol{\xi}_\perp\mathbf{B}_0\cdot\mathbf{B}_1, \tag{10.90}$$

since $d\mathbf{s}\cdot\mathbf{B}_0 = 0$ on the magnetofluid surface. This latter condition is true because the magnetic field was initially tangential to the magnetofluid surface (i.e., $\mathbf{J}_0\times\mathbf{B}_0 = \nabla P_0$ implies ∇P_0 is perpendicular to $\mathbf{B}_0$) and must remain so since the field is frozen into the magnetofluid.

Recombining and reordering these separate contributions gives

$$\delta W = \frac{1}{2}\int_{V_{mf}} d^3r\left\{\gamma P_0\left(\nabla\cdot\boldsymbol{\xi}\right)^2 + \frac{B_1^2}{\mu_0} - \boldsymbol{\xi}\cdot\left[\mathbf{J}_0\times\mathbf{B}_1 + \nabla\left(\boldsymbol{\xi}\cdot\nabla P_0\right)\right]\right\}$$
$$+\frac{1}{2\mu_0}\int_{S_{mf}} d\mathbf{s}\cdot\boldsymbol{\xi}_\perp\left[\mathbf{B}_0\cdot\mathbf{B}_1 - \mu_0\gamma P_0\nabla\cdot\boldsymbol{\xi}\right]. \tag{10.91}$$

The substitution $d\mathbf{s}\cdot\boldsymbol{\xi} = d\mathbf{s}\cdot\boldsymbol{\xi}_\perp$ has been made for the pressure contribution to the surface term on the grounds that $d\mathbf{s}$ must be perpendicular to $\mathbf{B}_0$. Further simplification is obtained by considering the dot product with $\mathbf{B}_0$ of the term in square brackets in Eq. (10.91), namely

$$\begin{aligned}
\mathbf{B}_0\cdot\left[\mathbf{J}_0\times\mathbf{B}_1 + \nabla\left(\boldsymbol{\xi}\cdot\nabla P_0\right)\right] &= -\nabla P_0\cdot\mathbf{B}_1 + \mathbf{B}_0\cdot\nabla\left(\boldsymbol{\xi}\cdot\nabla P_0\right)\\
&= -\nabla P_0\cdot\nabla\times\left(\boldsymbol{\xi}\times\mathbf{B}_0\right) + \mathbf{B}_0\cdot\nabla\left(\boldsymbol{\xi}\cdot\nabla P_0\right)\\
&= \nabla\cdot\left\{\nabla P_0\times\left(\boldsymbol{\xi}\times\mathbf{B}_0\right) + \left(\boldsymbol{\xi}\cdot\nabla P_0\right)\mathbf{B}_0\right\}\\
&= \nabla\cdot\left[\boldsymbol{\xi}\mathbf{B}_0\cdot\nabla P_0\right]\\
&= 0,
\end{aligned} \tag{10.92}$$

where Eq. (10.61) has been invoked to obtain the last line. Thus, $\boldsymbol{\xi}\cdot\left[\mathbf{J}_0\times\mathbf{B}_1 + \nabla\left(\boldsymbol{\xi}\cdot\nabla P_0\right)\right] = \boldsymbol{\xi}_\perp\cdot\left[\mathbf{J}_0\times\mathbf{B}_1 + \nabla\left(\boldsymbol{\xi}\cdot\nabla P_0\right)\right]$ since Eq. (10.92) shows that the factor in square brackets has no component parallel to the equilibrium magnetic field.

The potential energy variation δW is now decomposed into its magnetofluid volume and surface components,

$$\delta W_{F'} = \frac{1}{2}\int_{V_{mf}} d^3r \left\{ \begin{array}{l} \gamma P_0 (\nabla\cdot\boldsymbol{\xi})^2 + \dfrac{B_{1\perp}^2}{\mu_0} + \dfrac{B_{1\parallel}^2}{\mu_0} - \\ \boldsymbol{\xi}_\perp\cdot\mathbf{J}_0\times\mathbf{B}_1 - \boldsymbol{\xi}_\perp\cdot\nabla_\perp(\boldsymbol{\xi}_\perp\cdot\nabla P_0) \end{array} \right\} \tag{10.93}$$

and

$$\delta W_{S'} = \frac{1}{2\mu_0}\int_{S_{mf}} d\mathbf{s}\cdot\boldsymbol{\xi}_\perp(\mathbf{B}_0\cdot\mathbf{B}_1 - \mu_0\gamma P_0\nabla\cdot\boldsymbol{\xi}). \tag{10.94}$$

At this point it is useful to examine the parallel and perpendicular components of $\mathbf{B}_1$. Finite $\mathbf{B}_{1\perp}$ corresponds to changing the curvature of field lines (twanging or plucking) whereas finite $\mathbf{B}_{1\parallel}$ corresponds to compressing or rarefying the density of field lines. The equilibrium force balance can be written as

$$\mu_0\nabla P_0 = -\nabla_\perp\frac{B_0^2}{2} + \boldsymbol{\kappa}B_0^2, \tag{10.95}$$

where $\boldsymbol{\kappa} = \hat{B}_0\cdot\nabla\hat{B}_0$ is a measure of the local curvature of the equilibrium magnetic field (see discussion regarding Eq. (9.14)). Applying the vector identity $\nabla(\mathbf{C}\cdot\mathbf{D}) = \mathbf{C}\cdot\nabla\mathbf{D} + \mathbf{D}\cdot\nabla\mathbf{C} + \mathbf{D}\times\nabla\times\mathbf{C} + \mathbf{C}\times\nabla\times\mathbf{D}$ provides an alternate form for $\boldsymbol{\kappa}$, namely

$$\boldsymbol{\kappa} = -\hat{B}_0\times\nabla\times\hat{B}_0, \tag{10.96}$$

which will now be used in the evaluation of $B_{1\parallel}$, the parallel component of the perturbed magnetic field. From Eq. (10.67) it is seen that

$$\begin{aligned} B_{1\parallel} &= \hat{B}_0\cdot\nabla\times(\boldsymbol{\xi}_\perp\times\mathbf{B}_0) \\ &= (\boldsymbol{\xi}_\perp\times\mathbf{B}_0)\cdot\nabla\times\hat{B}_0 + \nabla\cdot\left[(\boldsymbol{\xi}_\perp\times\mathbf{B}_0)\times\hat{B}_0\right] \\ &= -B_0\boldsymbol{\xi}_\perp\cdot\boldsymbol{\kappa} - \nabla\cdot(B_0\boldsymbol{\xi}_\perp) \\ &= -B_0\boldsymbol{\xi}_\perp\cdot\boldsymbol{\kappa} - B_0(\nabla\cdot\boldsymbol{\xi}_\perp) - \frac{\boldsymbol{\xi}_\perp}{B_0}\cdot\nabla\frac{B_0^2}{2} \\ &= -B_0[2\boldsymbol{\xi}_\perp\cdot\boldsymbol{\kappa} + \nabla\cdot\boldsymbol{\xi}_\perp] + \frac{\mu_0}{B_0}\boldsymbol{\xi}_\perp\cdot\nabla P_0. \end{aligned} \tag{10.97}$$

The term involving $\mathbf{J}_0$ in Eq. (10.93) can be expanded to give

$$\begin{aligned} \boldsymbol{\xi}_\perp\cdot\mathbf{J}_0\times\mathbf{B}_1 &= \boldsymbol{\xi}_\perp\cdot\mathbf{J}_{0\perp}\times\hat{B}_0B_{1\parallel} + \boldsymbol{\xi}_\perp\cdot J_{0\parallel}\hat{B}_0\times\mathbf{B}_{1\perp} \\ &= (\boldsymbol{\xi}_\perp\cdot\nabla P_0)\frac{B_{1\parallel}}{B_0} + \boldsymbol{\xi}_\perp\cdot J_{0\parallel}\hat{B}_0\times\mathbf{B}_{1\perp}. \end{aligned} \tag{10.98}$$

Substituting for this term in Eq. (10.93), factoring out one power of $B_{1\|}$, and then substituting Eq. (10.97) gives

$$\begin{aligned}
\delta W_{F'} &= \frac{1}{2}\int_{V_{mf}} \mathrm{d}^3 r \left\{ \begin{array}{l} \gamma P_0 (\nabla\cdot\boldsymbol{\xi})^2 + \dfrac{B_{1\perp}^2}{\mu_0} + \\ \dfrac{B_{1\|}^2}{\mu_0} - (\boldsymbol{\xi}_\perp\cdot\nabla P_0)\dfrac{B_{1\|}}{B_0} - \\ \boldsymbol{\xi}_\perp\cdot J_{0\|}\hat{B}_0\times\mathbf{B}_{1\perp} - \boldsymbol{\xi}_\perp\cdot\nabla_\perp(\boldsymbol{\xi}_\perp\cdot\nabla P_0) \end{array} \right\} \\
&= \frac{1}{2}\int_{V_{mf}} \mathrm{d}^3 r \left\{ \begin{array}{l} \gamma P_0 (\nabla\cdot\boldsymbol{\xi})^2 + \dfrac{B_{1\perp}^2}{\mu_0} + \\ \dfrac{B_{1\|}}{\mu_0}\left(B_{1\|} - \dfrac{\mu_0\boldsymbol{\xi}_\perp\cdot\nabla P_0}{B_0}\right) - \\ \boldsymbol{\xi}_\perp\cdot J_{0\|}\hat{B}_0\times\mathbf{B}_{1\perp} - \boldsymbol{\xi}_\perp\cdot\nabla_\perp(\boldsymbol{\xi}_\perp\cdot\nabla P_0) \end{array} \right\} \\
&= \frac{1}{2}\int_{V_{mf}} \mathrm{d}^3 r \left\{ \begin{array}{l} \gamma P_0 (\nabla\cdot\boldsymbol{\xi})^2 + \dfrac{B_{1\perp}^2}{\mu_0} + \\ \left(\dfrac{B_0^2}{\mu_0}[2\boldsymbol{\xi}_\perp\cdot\boldsymbol{\kappa} + \nabla\cdot\boldsymbol{\xi}_\perp] - \boldsymbol{\xi}_\perp\cdot\nabla P_0\right)[2\boldsymbol{\xi}_\perp\cdot\boldsymbol{\kappa} + \nabla\cdot\boldsymbol{\xi}_\perp] - \\ \boldsymbol{\xi}_\perp\cdot J_{0\|}\hat{B}_0\times\mathbf{B}_{1\perp} + (\boldsymbol{\xi}_\perp\cdot\nabla P_0)\,\nabla\cdot\boldsymbol{\xi}_\perp - \\ \nabla\cdot[\boldsymbol{\xi}_\perp(\boldsymbol{\xi}_\perp\cdot\nabla P_0)] \end{array} \right\} \\
&= \frac{1}{2}\int_{V_{mf}} \mathrm{d}^3 r \left\{ \begin{array}{l} \gamma P_0 (\nabla\cdot\boldsymbol{\xi})^2 + \dfrac{B_{1\perp}^2}{\mu_0} + \dfrac{B_0^2}{\mu_0}[2\boldsymbol{\xi}_\perp\cdot\boldsymbol{\kappa} + \nabla\cdot\boldsymbol{\xi}_\perp]^2 - \\ (\boldsymbol{\xi}_\perp\cdot\nabla P_0)(2\boldsymbol{\xi}_\perp\cdot\boldsymbol{\kappa}) - \boldsymbol{\xi}_\perp\cdot J_{0\|}\hat{B}_0\times\mathbf{B}_{1\perp} - \\ \nabla\cdot[\boldsymbol{\xi}_\perp(\boldsymbol{\xi}_\perp\cdot\nabla P_0)] \end{array} \right\} \\
&= \frac{1}{2\mu_0}\int_{V_{mf}} \mathrm{d}^3 r \left\{ \begin{array}{l} \gamma\mu_0 P_0 (\nabla\cdot\boldsymbol{\xi})^2 + B_{1\perp}^2 + B_0^2[2\boldsymbol{\xi}_\perp\cdot\boldsymbol{\kappa} + \nabla\cdot\boldsymbol{\xi}_\perp]^2 - \\ 2\mu_0(\boldsymbol{\xi}_\perp\cdot\nabla P_0)(\boldsymbol{\xi}_\perp\cdot\boldsymbol{\kappa}) + \boldsymbol{\xi}_\perp\times\mathbf{B}_{1\perp}\cdot\left(\mu_0 J_{0\|}\hat{B}_0\right) \end{array} \right\} \\
&\quad -\frac{1}{2}\int_{S_{mf}} \mathrm{d}\mathbf{s}\cdot\boldsymbol{\xi}_\perp(\boldsymbol{\xi}_\perp\cdot\nabla P_0). \qquad (10.99)
\end{aligned}$$

A new surface term has appeared because of an integration by parts; this new term is absorbed into the previous surface term and the fluid and surface terms are redefined by removing the primes; thus,

$$\delta W_F = \frac{1}{2\mu_0}\int_{V_{mf}} \mathrm{d}^3 r \left\{ \begin{array}{l} \gamma\mu_0 P_0 (\nabla\cdot\boldsymbol{\xi})^2 + B_{1_\perp}^2 + B_0^2[2\boldsymbol{\xi}_\perp\cdot\boldsymbol{\kappa} + (\nabla\cdot\boldsymbol{\xi}_\perp)]^2 - \\ 2\mu_0(\boldsymbol{\xi}_\perp\cdot\nabla P_0)(\boldsymbol{\xi}_\perp\cdot\boldsymbol{\kappa}) + \boldsymbol{\xi}_\perp\times\mathbf{B}_{1\perp}\cdot\left(\mu_0 J_{0\|}\hat{B}_0\right) \end{array} \right\} \tag{10.100}$$

and

$$\delta W_S = \frac{1}{2\mu_0}\int_{S_{mf}} \mathrm{d}\mathbf{s}\cdot\boldsymbol{\xi}_\perp \left\{\mathbf{B}_0\cdot\mathbf{B}_1 - \mu_0\left(\gamma P_0\nabla\cdot\boldsymbol{\xi} + \boldsymbol{\xi}_\perp\cdot\nabla P_0\right)\right\}$$

$$= \frac{1}{2\mu_0}\int_{S_{mf}} \mathrm{d}\mathbf{s}\cdot\boldsymbol{\xi}_\perp \left\{\mathbf{B}_0\cdot\mathbf{B}_1 + \mu_0 P_1\right\}, \tag{10.101}$$

where $\mathbf{B}_0\cdot\mathrm{d}\mathbf{s} = 0$ and Eq. (10.68) have been used to simplify the surface integral. The surface integral can be further rearranged by considering the relationship between the vacuum and magnetofluid fields at the perturbed surface. If the equilibrium force balance is integrated over the volume of the small Gaussian pillbox located at the perturbed magnetofluid surface in Fig. 10.5, it is seen that

$$0 = \int_{V_{pillbox,ps}} \mathrm{d}^3r\nabla\cdot\left[\left(P + \frac{B^2}{2\mu_0}\right)\mathbf{I} - \frac{1}{\mu_0}\mathbf{B}\mathbf{B}\right] = \int_{S,ps} \mathrm{d}s\left[P + \frac{B^2}{2\mu_0}\right]_{magnetofluid}^{vac} \tag{10.102}$$

since $\mathrm{d}\mathbf{s}\cdot\mathbf{B} = 0$ at the perturbed surface. The subscripts *ps* indicate that the volume and surface integrals are at the perturbed surface.

Quantities evaluated at the perturbed surface are of the form

$$f_{pert\,sfc} = f_0 + f_1 + \boldsymbol{\xi}\cdot\nabla f_0; \tag{10.103}$$

i.e., both absolute and convective first-order terms must be included. Since the pillbox extent was arbitrary in Eq. (10.102), the integrand $\left[P + B^2/2\mu_0\right]_{magnetofluid}^{vac}$ must vanish at each point on the perturbed surface, giving the relation

$$\left(P_1 + \frac{\mathbf{B}_0\cdot\mathbf{B}_1}{\mu_0}\right)_{\substack{magneto\\fluid}} + \boldsymbol{\xi}\cdot\nabla\left(P_0 + \frac{B_0^2}{2\mu_0}\right)_{\substack{magneto\\fluid}} = \left(\frac{\mathbf{B}_0\cdot\mathbf{B}_1}{\mu_0}\right)_{vac} + \boldsymbol{\xi}\cdot\nabla\left(\frac{B_0^2}{2\mu_0}\right)_{vac}, \tag{10.104}$$

which can be rewritten as

$$\left(P_1 + \frac{\mathbf{B}_0\cdot\mathbf{B}_1}{\mu_0}\right)_{magnetofluid} = \boldsymbol{\xi}\cdot\nabla\left[P_0 + \frac{B_0^2}{2\mu_0}\right]_{magnetofluid}^{vac} + \left(\frac{\mathbf{B}_0\cdot\mathbf{B}_1}{\mu_0}\right)_{vac}. \tag{10.105}$$

Thus, the surface integral Eq. (10.101) can be rewritten as

$$\delta W_S = \frac{1}{2}\int_{S_{mf}} \mathrm{d}\mathbf{s}\cdot\boldsymbol{\xi}_\perp\boldsymbol{\xi}_\perp\cdot\nabla\left[P_0 + \frac{B_0^2}{2\mu_0}\right]_{magnetofluid}^{vac} + \frac{1}{2\mu_0}\int_{S_{mf}} \mathrm{d}\mathbf{s}\cdot\boldsymbol{\xi}_\perp\left(\mathbf{B}_0\cdot\mathbf{B}_1\right)_{vac}. \tag{10.106}$$

The volume integral in Eq. (10.100) is over the magnetofluid volume and the direction of $\mathrm{d}\mathbf{s}$ in Eq. (10.106) is outwards from the magnetofluid volume. The energy stored in the vacuum region between the magnetofluid and wall is

$$\begin{aligned}\delta W_{vac} &= \frac{1}{2\mu_0}\int_{V_{vac}} \mathrm{d}^3 r B_{1v}^2 \\ &= \frac{1}{2\mu_0}\int_{V_{vac}} \mathrm{d}^3 r \mathbf{B}_{1v}\cdot\nabla\times\mathbf{A}_{1v} \\ &= \frac{1}{2\mu_0}\int_{V_{vac}} \mathrm{d}^3 r \nabla\cdot(\mathbf{A}_{1v}\times\mathbf{B}_{1v}) \\ &= \frac{1}{2\mu_0}\int_{S_{vac}} \mathrm{d}\mathbf{s}\cdot(\mathbf{A}_{1v}\times\mathbf{B}_{1v}),\end{aligned} \tag{10.107}$$

where $\mathrm{d}\mathbf{s}$ points *out* from the vacuum region and $\nabla\times\mathbf{B}_{1v}=0$ has been used when integrating by parts. At the conducting wall (which might be at infinity) the integrand vanishes, since it contains the factor $\mathrm{d}\mathbf{s}\times\mathbf{A}_{1v}$, which is proportional to the tangential electric field, a quantity that vanishes at a conductor. Thus, only the surface integral over the magnetofluid–vacuum interface remains. Since an element of surface $\mathrm{d}\mathbf{s}$ pointing out of the vacuum points into the magnetofluid, on using $\mathrm{d}\mathbf{s}$ to mean out of the magnetofluid (as before), Eq. (10.107) becomes

$$\delta W_{vac} = -\frac{1}{2\mu_0}\int_{S_{mf}} \mathrm{d}\mathbf{s}\cdot(\mathbf{A}_{1v}\times\mathbf{B}_{1v}), \tag{10.108}$$

where S_{mf} has been used instead of S_{vac} to indicate that $\mathrm{d}\mathbf{s}$ points out of the magnetofluid.

The tangential component of the electric field must be continuous at the magnetofluid–vacuum interface. This field is not necessarily zero. The electric field inside the magnetofluid is determined by Ohm's law, Eq. (10.37). Thus, the electric field on the magnetofluid side of the magnetofluid–vacuum interface is

$$\mathbf{E}_{1p} = -\mathbf{v}_1\times\mathbf{B}_0 \tag{10.109}$$

while on the vacuum side of this interface the electric field is simply

$$\mathbf{E}_{1v} = -\frac{\partial\mathbf{A}_{1v}}{\partial t}, \tag{10.110}$$

where $\mathbf{A}_{1v}$ is the vacuum vector potential. Defining the surface normal unit vector $\hat{n}$ and integrating both electric fields with respect to time, the condition that the tangential electric field is continuous is seen to be

$$\hat{n}\times\mathbf{A}_{1v} = \hat{n}\times(\boldsymbol{\xi}\times\mathbf{B}_0) = -\hat{n}\cdot\boldsymbol{\xi}\mathbf{B}_0, \tag{10.111}$$

where $\hat{n}\cdot\mathbf{B}_0=0$ has been used. Thus, the second surface integral in Eq. (10.106) can be written as

$$\frac{1}{2\mu_0}\int_S \mathrm{d}s\hat{n}\cdot\boldsymbol{\xi}_\perp\mathbf{B}_0\cdot\mathbf{B}_{1v} = -\frac{1}{2\mu_0}\int_S \mathrm{d}s\hat{n}\times\mathbf{A}_{1v}\cdot\mathbf{B}_{1v} = \delta W_{vac}. \tag{10.112}$$

These results are now summarized. The perturbed potential energy can be expressed as

$$\delta W = \delta W_F + \delta W_{int} + \delta W_{vac}, \tag{10.113}$$

where the contribution from the fluid volume interior is

$$\delta W_F = \frac{1}{2\mu_0}\int_{V_{mf}} d^3r \left\{ \begin{array}{l} \gamma\mu_0 P_0 (\nabla\cdot\boldsymbol{\xi})^2 + B_{1\perp}^2 + B_0^2\left[2\boldsymbol{\xi}_\perp\cdot\boldsymbol{\kappa} + (\nabla\cdot\boldsymbol{\xi}_\perp)\right]^2 \\ +\boldsymbol{\xi}_\perp\times\mathbf{B}_{1\perp}\cdot\hat{B}_0\mu_0 J_{0\parallel} - 2\mu_0(\boldsymbol{\xi}_\perp\cdot\boldsymbol{\kappa})(\boldsymbol{\xi}_\perp\cdot\nabla P_0) \end{array} \right\}, \tag{10.114}$$

the contribution from the vacuum–magnetofluid interface is

$$\delta W_{int} = \frac{1}{2\mu_0}\int_{S_{mf}} d\mathbf{s}\cdot\boldsymbol{\xi}_\perp\boldsymbol{\xi}_\perp\cdot\nabla\left[\mu_0 P_0 + \frac{B_0^2}{2}\right]_{magnetofluid}^{vac}, \tag{10.115}$$

and the contribution from the vacuum region is

$$\delta W_{vac} = \frac{1}{2\mu_0}\int_{vac} d^3r\, B_{1v}^2. \tag{10.116}$$

10.5 Discussion of the energy principle

The terms in the integrands of δW_F, δW_{int}, and δW_{vac} are of two types: those that are positive-definite and those that are not. Positive-definite terms always increase δW and are therefore stabilizing whereas terms that could be negative are potentially destabilizing. Consideration of δW_F in particular shows that magnetic perturbations interior to the magnetofluid and perpendicular to the equilibrium field are always stabilizing, since these perturbations appear in the form $B_{1\perp}^2$. It is also seen that a positive-definite term $\sim(\nabla\cdot\boldsymbol{\xi})^2$ exists indicating that a compressible magnetofluid is always more stable than an otherwise identical incompressible magnetofluid. Hence, incompressible instabilities are more violent than compressible ones. It is also seen that two types of destabilizing terms exist. One gives instability if

$$(\boldsymbol{\kappa}\cdot\boldsymbol{\xi}_\perp)(\boldsymbol{\xi}_\perp\cdot\nabla P_0) > 0; \tag{10.117}$$

this is a generalization of the good curvature/bad curvature result obtained in the earlier Rayleigh–Taylor analysis. In the vicinity of the magnetofluid surface all quantities in Eq. (10.117) point in the same direction and so Eq. (10.117) can be written as

$$\boldsymbol{\kappa}\cdot\nabla P_0 > 0 \quad\Longrightarrow \text{instability}, \tag{10.118}$$

which gives instability for bad curvature ($\boldsymbol{\kappa}$ parallel to ∇P_0) and stability for good curvature ($\boldsymbol{\kappa}$ antiparallel to ∇P_0). A Bennett pinch has bad curvature and is therefore grossly unstable to Rayleigh–Taylor interchange modes. Instabilities associated with Eq. (10.117) are called pressure-driven instabilities, and are important in plasmas where there is significant energy stored in the pressure (high β where $\beta = 2\mu_0 P_0/B_0^2$).

The other type of destabilizing term depends on the existence of a force-free current (i.e., $J_{0\|} \neq 0$) and leads to internal kink instabilities. These sorts of instabilities are called current-driven instabilities (although strictly speaking, only the parallel component of the current is involved).

There also exist instabilities associated with the magnetofluid–vacuum interface. The energy δW_{int} can be thought of as the change in potential energy of a stretched membrane at the magnetofluid–vacuum interface where the stretching force is given by the difference between the vacuum and magnetofluid forces pushing on the membrane surface; instability will occur if $\delta W_{int} < 0$. When investigating these surface instabilities it is convenient to set δW_F to zero by idealizing the plasma to being incompressible, having uniform internal pressure, and no internal currents. Surface instabilities can exist only if the surface can move, and so require a vacuum region between the wall and the plasma. Thus, moving a conducting wall right up to the surface of a conducting plasma prevents surface instabilities. These surface instabilities will be investigated in detail later in this chapter.

10.6 Current-driven instabilities and helicity

We shall now discuss current driven instabilities and show that these are helical in nature and are driven by gradients in $J_{0\|}/B_0$. To simplify the analysis $P \to 0$ is assumed so that pressure-driven instabilities can be neglected, since they have already been discussed. On making this simplification, Eq. (10.97) shows that the parallel component of the perturbed magnetic field reduces to

$$B_{1\|} = -B_0 \left[2\boldsymbol{\xi}_\perp \cdot \boldsymbol{\kappa} + \nabla \cdot \boldsymbol{\xi}_\perp \right]. \tag{10.119}$$

Thus, we may identify

$$B_1^2 = B_{1\perp}^2 + B_{1\|}^2 = \mathbf{B}_{1\perp}^2 + B_0^2 \left[2\boldsymbol{\xi}_\perp \cdot \boldsymbol{\kappa} + \nabla \cdot \boldsymbol{\xi}_\perp \right]^2 \tag{10.120}$$

and so the perturbed potential energy of the magnetofluid volume reduces to

$$\delta W_F = \frac{1}{2\mu_0} \int d^3r \left\{ B_1^2 + \boldsymbol{\xi}_\perp \times \mathbf{B}_{1\perp} \cdot \mathbf{B}_0 \frac{\mu_0 J_{0\|}}{B_0} \right\}. \tag{10.121}$$

Equation (10.67) shows that the perturbed vector potential can be identified as

$$\mathbf{A}_1 = \boldsymbol{\xi} \times \mathbf{B}_0 \tag{10.122}$$

so Eq. (10.121) can be recast as

$$\delta W_F = \frac{1}{2\mu_0}\int \mathrm{d}^3 r \left\{ B_1^2 - \mathbf{A}_1 \cdot \mathbf{B}_1 \frac{\mu_0 J_{0\|}}{B_0} \right\}. \tag{10.123}$$

We now show that finite $\mathbf{A}_1 \cdot \mathbf{B}_1$ corresponds to a helical perturbation. Consider the simplest situation where $\mathbf{A}_1 \cdot \mathbf{B}_1$ is simply a constant and define a local Cartesian coordinate system with z axis parallel to the local $\mathbf{B}_0$. Equation (10.122) shows that $\mathbf{A}_1 = A_{1x}\hat{x} + A_{1y}\hat{y}$ so

$$\mathbf{A}_1 \cdot \mathbf{B}_1 = -A_{1x}\frac{\partial A_{1y}}{\partial z} + A_{1y}\frac{\partial A_{1x}}{\partial z}. \tag{10.124}$$

Suppose both components of $\mathbf{A}_1$ are non-trivial functions of z and, in particular, assume $A_{1x} = \mathrm{Re}\ \mathcal{A}_{1x}\exp(\mathrm{i}kz)$ and $A_{1y} = \mathrm{Re}\ \mathcal{A}_{1y}\exp(\mathrm{i}kz)$. In this case

$$\mathbf{A}_1 \cdot \mathbf{B}_1 = \frac{1}{2}\mathrm{Re}\left[-\mathcal{A}_{1x}^* \frac{\partial \mathcal{A}_{1y}}{\partial z} + \mathcal{A}_{1y}^* \frac{\partial \mathcal{A}_{1x}}{\partial z} \right] = -\frac{k}{2}\mathrm{Re}\left[\mathrm{i}\left(\mathcal{A}_{1x}^* \mathcal{A}_{1y} - \mathcal{A}_{1y}^* \mathcal{A}_{1x} \right) \right], \tag{10.125}$$

which can be finite only if $\mathcal{A}_{1x}^* \mathcal{A}_{1y}$ is not pure real. The simplest such case is where $\mathcal{A}_{1y} = \mathrm{i}\mathcal{A}_{1x}$ so

$$\mathbf{A}_1 \cdot \mathbf{B}_1 = k|\mathcal{A}_{1x}|^2 \tag{10.126}$$

and

$$\mathbf{A}_1 = \mathrm{Re}\left[\mathcal{A}_{1x}\left(\hat{x} + \mathrm{i}\hat{y} \right)\exp(\mathrm{i}kz) \right], \tag{10.127}$$

which is a helically polarized field since $A_{1x} \sim \cos kz$ and $A_{1y} \sim \sin kz$.

10.7 Magnetic helicity

Since finite $\mathbf{A}_1 \cdot \mathbf{B}_1$ corresponds to the local helical polarization of the perturbed fields, it is reasonable to define $\mathbf{A} \cdot \mathbf{B}$ as the density of magnetic helicity and to define the total magnetic helicity in a volume as

$$K = \int_V \mathrm{d}^3 r \mathbf{A} \cdot \mathbf{B}. \tag{10.128}$$

The question immediately arises whether this definition makes sense, i.e., is it reasonable to define $\mathbf{A} \cdot \mathbf{B}$ as an intensive property and K as an extensive property? An obvious problem is that $\mathbf{A}$ is undefined with respect to a gauge: $\mathbf{A}$ can be redefined to be $\mathbf{A}' = \mathbf{A} + \nabla f$, where f is an arbitrary scalar function since $\mathbf{B} = \nabla \times \mathbf{A} = \nabla \times (\mathbf{A} + \nabla f)$ is unaffected by the choice of gauge. The definition for magnetic helicity would be of little use if K were gauge-dependent because gauge has no physical significance. However, if no magnetic field penetrates the surface S enclosing the volume V, the proposed definition of K is gauge-independent. This is because if no magnetic field penetrates the surface then $\mathbf{B}\cdot d\mathbf{s}=0$ everywhere on the surface. If this is true, then

$$
\begin{aligned}
\int_V d^3r\,(\mathbf{A}+\nabla f)\cdot\mathbf{B} &= \int_V d^3r\mathbf{A}\cdot\mathbf{B}+\int_V d^3r\nabla\cdot(f\mathbf{B})\\
&= \int_V d^3r\mathbf{A}\cdot\mathbf{B}+\int_S d\mathbf{s}\cdot(f\mathbf{B})\\
&= \int_V d^3r\mathbf{A}\cdot\mathbf{B}
\end{aligned}
\tag{10.129}
$$

and so the total helicity defined by Eq. (10.128) is gauge-independent even though the helicity density $\mathbf{A}\cdot\mathbf{B}$ is gauge-dependent.

Let us consider the situation where there is no vacuum region between the plasma and an impermeable wall. Thus the normal fluid velocity $\mathbf{u}_{1\perp}$ must vanish at the wall. From Ohm's law the component of the perturbed electric field tangential to the wall $\mathbf{E}_{1t}$ must vanish since

$$
\mathbf{E}_{1t}=-\mathbf{u}_{1\perp}\times\mathbf{B}_0=0 \tag{10.130}
$$

and $\mathbf{B}_0$ lies in the plane of the wall. Thus, an impermeable wall is equivalent to a conducting wall if $\mathbf{B}_0$ lies in the plane of the wall. If the magnetic field initially does not penetrate the wall, i.e., $\mathbf{B}\cdot d\mathbf{s}=0$ initially, then the field will always remain tangential to the wall and the total helicity K in the volume bounded by the wall will always be a well-defined quantity (i.e., will always be gauge-invariant).

A conservation equation for helicity density can be obtained by combining the time derivative of the magnetic helicity density with Faraday's law. This is seen by direct calculation:

$$
\begin{aligned}
-\frac{\partial}{\partial t}(\mathbf{A}\cdot\mathbf{B}) &= -\frac{\partial\mathbf{A}}{\partial t}\cdot\mathbf{B}-\mathbf{A}\cdot\frac{\partial\mathbf{B}}{\partial t}\\
&= (\mathbf{E}+\nabla\varphi)\cdot\mathbf{B}+\mathbf{A}\cdot\nabla\times\mathbf{E}\\
&= 2\mathbf{E}\cdot\mathbf{B}+\nabla\cdot(\varphi\mathbf{B}+\mathbf{E}\times\mathbf{A}),
\end{aligned}
\tag{10.131}
$$

where φ is the electrostatic potential. Since the ideal MHD Ohm's law, Eq. (10.37), implies $\mathbf{E}\cdot\mathbf{B}=0$, Eq. (10.131) can be rearranged in the form of a conservation equation,

$$
\frac{\partial}{\partial t}(\mathbf{A}\cdot\mathbf{B})+\nabla\cdot(\varphi\mathbf{B}+\mathbf{E}\times\mathbf{A})=0. \tag{10.132}
$$

Integration of Eq. (10.132) over the entire magnetofluid volume gives the conservation relation for the total helicity to be

$$
\frac{\partial}{\partial t}\left[\int d^3r\mathbf{A}\cdot\mathbf{B}\right]+\int ds\hat{n}\cdot(\varphi\mathbf{B}+\mathbf{E}\times\mathbf{A})=0. \tag{10.133}
$$

Since both $\mathbf{B}\cdot\hat{n}=0$ and $\mathbf{E}_t=0$ at the wall, this reduces to

$$K=\int \mathrm{d}^3r\mathbf{A}\cdot\mathbf{B}=const.; \tag{10.134}$$

i.e., the total helicity is conserved for an ideal plasma surrounded by a rigid or conducting wall that is not penetrated by any magnetic field lines.

We now recall that $\mathbf{A}_1$ is a linear function of $\boldsymbol{\xi}$ prescribed by Eq. (10.122) and that $\boldsymbol{\xi}$ is assumed to be of order ϵ. Furthermore, we note that the energy principle derivation is unaffected by making the additional assumption that the variation $\delta\mathbf{A}=\mathbf{A}_1$ is exact to all orders in ϵ, i.e., the energy principle is unaffected if we assume no terms $\mathbf{A}_n\sim\epsilon^n$ exist for $n\geq 2$. Let us use these properties to consider the implications of Eq. (10.134) when the magnetic field is perturbed. On writing $\mathbf{B}=\mathbf{B}_0+\mathbf{B}_1$ and $\mathbf{A}=\mathbf{A}_0+\mathbf{A}_1$, helicity conservation as given by Eq. (10.134) implies

$$\begin{aligned}K&=\int \mathrm{d}^3r\mathbf{A}_0\cdot\nabla\times\mathbf{A}_0\\&=\int \mathrm{d}^3r\,(\mathbf{A}_0+\mathbf{A}_1)\cdot\nabla\times(\mathbf{A}_0+\mathbf{A}_1)\\&=\int \mathrm{d}^3r\mathbf{A}_0\cdot\nabla\times\mathbf{A}_0+\int \mathrm{d}^3r\mathbf{A}_1\cdot\nabla\times\mathbf{A}_0\\&\quad+\int \mathrm{d}^3r\mathbf{A}_0\cdot\nabla\times\mathbf{A}_1+\int \mathrm{d}^3r\mathbf{A}_1\cdot\nabla\times\mathbf{A}_1.\end{aligned} \tag{10.135}$$

Thus, to first-order in the perturbation (i.e., to order ϵ),

$$\int \mathrm{d}^3r\mathbf{A}_1\cdot\nabla\times\mathbf{A}_0+\int \mathrm{d}^3r\mathbf{A}_0\cdot\nabla\times\mathbf{A}_1=0 \tag{10.136}$$

and to second-order (i.e., to order ϵ^2, the order relevant to the energy principle),

$$\int \mathrm{d}^3r\mathbf{A}_1\cdot\nabla\times\mathbf{A}_1=\int \mathrm{d}^3r\mathbf{A}_1\cdot\mathbf{B}_1=0. \tag{10.137}$$

This can be compared to Eq. (10.123), the second term of which is the same as Eq. (10.137) except for a factor $-\mu_0J_{0\|}/B_0$. In general, this factor can be some complicated function of position. However, in the special case where $\mu_0J_{0\|}/B_0$ does not depend on position, $\mu_0J_{0\|}/B_0$ may be factored out of the second term in Eq. (10.123) so as to obtain, using Eq. (10.137),

$$\delta W_F=\frac{1}{2\mu_0}\int \mathrm{d}^3r\,B_1^2, \tag{10.138}$$

which is positive-definite and therefore gives absolute stability. Thus, equilibria having spatially uniform $\lambda=\mu_0J_{0\|}/B_0$ are stable against current-driven modes. Since these equilibria are helical or kinked, they may be considered as being the final relaxed state associated with a kink instability – once a system attains this final state no free energy remains to drive further instability. Since pressure has been assumed to be negligible, Eq. (10.60) implies that the equilibrium current must be parallel to the equilibrium magnetic field, but does not specify the proportionality factor. Thus, if a configuration satisfies

$$\mu_0 \mathbf{J}_0 = \lambda \mathbf{B}_0, \tag{10.139}$$

where λ is spatially uniform, the configuration is stable against any further helical perturbations. This gives rise to the relation

$$\nabla \times \mathbf{B}_0 = \lambda \mathbf{B}_0, \tag{10.140}$$

where λ is spatially uniform. This equation is called a force-free equilibrium and its solutions are helical vector fields, namely fields where the curl of the field is parallel to the field itself. If a field is confined to a plane, then its curl will be normal to the plane and so a field confined to a plane cannot be a solution to Eq. (10.140). The field must be three-dimensional.

To summarize, it has been shown that current-driven instabilities are helical and drive the plasma towards a force-free equilibrium as prescribed by Eq. (10.140). Current-driven instabilities are energized by gradients in $J_{0\|}/B_0$ and become stabilized when $J_{0\|}/B_0$ becomes spatially uniform. Gradients in $J_{0\|}/B_0$ can therefore be considered as free energy available for driving helical modes. When this free energy is depleted, the helical modes are stabilized and the plasma assumes a force-free equilibrium with spatially uniform $J_{0\|}/B_0$.

This tendency to coil up or kink is a means by which the plasma increases its inductance. However, when the plasma coils up into a state satisfying Eq. (10.140) it is in a stable equilibrium. This stable equilibrium represents a local minimum in potential energy. There might be several such local minima, each of which has a different energy level as sketched in Fig. 9.2. As mentioned earlier, this set of discrete energy levels is somewhat analogous to the ground and excited states of a quantum system. Here, the vacuum magnetic field is analogous to the ground state, while the various force-free equilibria (i.e., solutions of Eq. (10.140)) are analogous to the higher energy states.

10.8 Characterization of free-boundary instabilities

The previous section considered internal instabilities of a magnetofluid bounded by a rigid wall with no vacuum region between the magnetofluid and the wall. Let us now consider the other extreme, namely a situation where not only does a vacuum region exist and bound the magnetofluid but, in addition, the location of the vacuum–magnetofluid interface can move. To focus attention on the motion of the magnetofluid–vacuum boundary, the simplest non-trivial configuration will be considered, namely a configuration where the interior pressure is both uniform and finite. This means ∇P vanishes in the magnetofluid interior so the entire pressure gradient and therefore the entire $\mathbf{J} \times \mathbf{B}$ force is concentrated in an infinitesimally thin layer at the magnetofluid surface.

Compressibility, i.e., finite $\nabla \cdot \boldsymbol{\xi}$, was shown by the MHD energy principle to be stabilizing. Therefore, if a given system is stable with respect to incompressible modes, it will be even more stable with respect to compressible modes, or equivalently, with respect to modes having finite $\nabla \cdot \boldsymbol{\xi}$. By assuming $\nabla \cdot \boldsymbol{\xi} = 0$ the worst-case instability scenario is therefore considered and, furthermore, the analysis is simplified. Thus, in order to consider the simplest non-trivial instability resulting from motion of the magnetofluid–vacuum boundary, we assume $\nabla \cdot \boldsymbol{\xi} = 0$, cylindrical geometry, and axial uniformity. For example, the Bennett pinch would satisfy these assumptions if all the current were concentrated at the plasma surface. The physical basis of the two main types of current-driven instability, sausage and kink, will first be discussed qualitatively before proceeding with a detailed mathematical discussion.

10.8.1 Qualitative examination of the sausage instability

Consider a Bennett pinch (z-pinch) that is an axially uniform cylindrical plasma with an axial current and an associated azimuthal magnetic field. As discussed above, the pressure is assumed to be (i) uniform in the interior region $r < a$, where a is the plasma radius, and (ii) zero in the exterior region $r > a$. In equilibrium, the radially inward $\mathbf{J} \times \mathbf{B}$ pinch force balances the radially outward force associated with the pressure gradient. Thus $-J_{z0}B_{\theta 0} = \partial P_0/\partial r$ and both sides of this equation are finite only in an infinitesimal surface layer at $r = a$. We now suppose that random thermal noise causes the incompressible plasma to develop small axially periodic constrictions and bulges shown in an exaggerated fashion in Fig. 10.6. The azimuthal magnetic field $B_\theta = \mu_0 I/2\pi r$ at the constrictions is larger than its equilibrium value because $r < a$ at a constriction whereas I is fixed. Thus, at a constriction the pinch force $\sim B_\theta^2/r$ is greater than its equilibrium value and so is stronger than the outward force from the internally uniform pressure. The resulting net force is therefore inwards and so will cause an inwards radial motion, thereby enhancing the constriction. Because the configuration was assumed to be incompressible, the plasma squeezed inwards at constrictions must flow into the interspersed bulges shown in Fig. 10.6. The azimuthal magnetic field at a bulge is weaker than its equilibrium value because $r > a$ at a bulge. Thus, at a bulge the tables are now turned in the competition between outward pressure and inward pinch force. At a bulge the pressure exceeds the weakened pinch force and this force imbalance causes the bulge to increase. Hence, any combination of initial

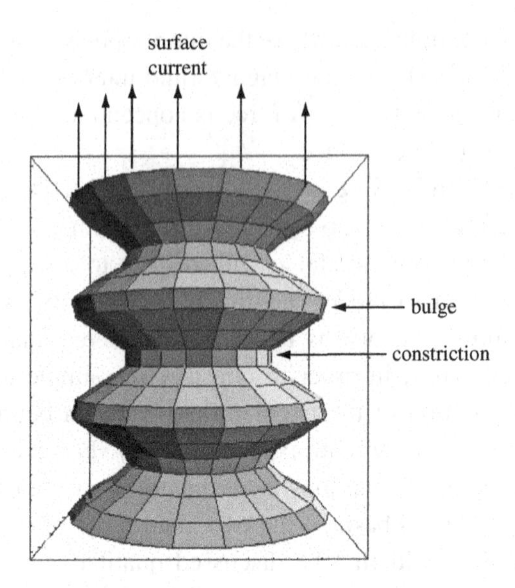

Fig. 10.6 Sausage instability: current is axial, magnetic field is azimuthal.

infinitesimal constrictions and bulges will spontaneously grow in amplitude and so the system is unstable. This behavior is called the "sausage" instability because the end result is a plasma resembling a string of sausages.

Sausage instability can be prevented by either surrounding the plasma with a perfectly conducting wall or by immersing the plasma in a strong axial vacuum magnetic field (i.e., an axial field generated by currents flowing in external solenoid coils that are coaxial with the plasma). If a plasma with embedded axial field attempts to sausage, the axial rippling of the plasma would have to bend the embedded axial field since this axial field is frozen into the plasma. As was shown in Section 9.2 any deformation of a vacuum magnetic field requires work and so the free energy available to drive the sausage instability would have to do work to bend the axial field. If bending the axial magnetic field absorbs more energy than is liberated by the sausaging motion, the plasma is stabilized. The vacuum axial field stabilizes the plasma in a manner analogous to a steel reinforcing rod embedded in concrete. The other method for stabilization, a close-fitting conducting wall, works because image currents induced in the wall by the sausage motion produce a magnetic field that interacts with the plasma current in such a way as to repel the plasma from the wall.

10.8.2 Qualitative examination of the kink instability

The combination of the axial magnetic field B_z proposed in the previous paragraph and the azimuthal magnetic field B_θ produced by the plasma current results in

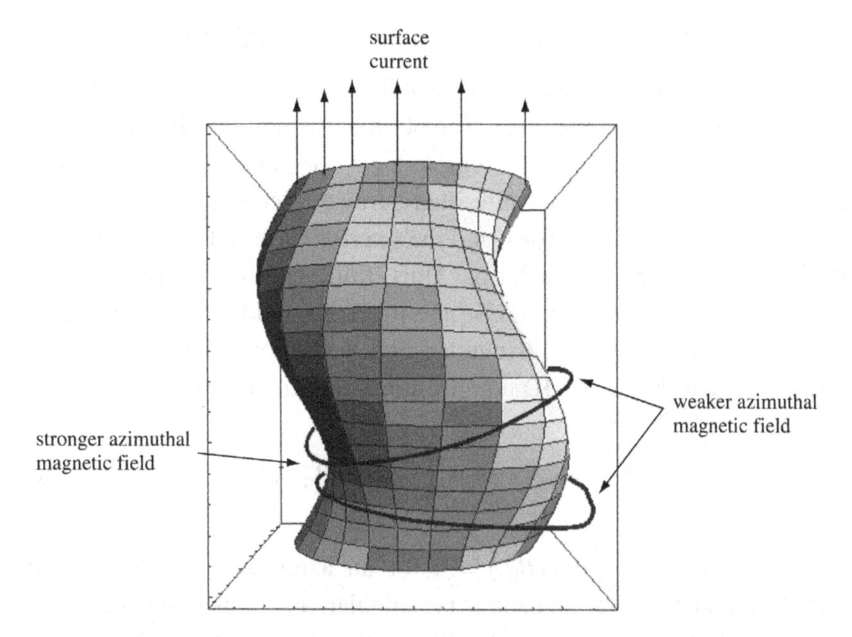

Fig. 10.7 Kink instability can occur when the equilibrium magnetic field is helical.

a helical magnetic field. This helical magnetic field is susceptible to the helical kink instability sketched in Fig. 10.7. At the concave parts of the plasma surface there is a concentration of the azimuthal field resulting in a magnetic pressure that increases the concavity (see lower left of Fig. 10.7). Similarly, at the convex portions of the surface, the azimuthal field is weaker so that the convex bulge will tend to increase (see lower right of Fig. 10.7). This tendency can also be viewed as an example of the hoop force trying to increase the major radius of a segment of curved current. If the externally imposed axial vacuum magnetic field is sufficiently strong, the energy required to deform this vacuum field will exceed the free energy available from the kinking and the instability will be prevented.

10.8.3 *Qualitative examination of wall stabilization*

Now consider a plasma with no externally produced B_z field but surrounded by a close-fitting perfectly conducting wall. The plasma axial current generates an azimuthal field in the vacuum region between the plasma and the wall. If an instability causes the plasma to move towards the wall, then the vacuum region becomes thinner and the magnetic field in the vacuum region becomes compressed since it cannot penetrate the wall or the plasma. The increased magnetic pressure acts as a restoring force, pushing the plasma back away from the wall. Thus, a close-fitting perfectly conducting wall stabilizes both kinks and sausages.

10.9 Analysis of free-boundary instabilities

Consider the azimuthally symmetric, axially uniform cylindrical plasma with nominal radius a shown in Fig. 10.8. The plasma is subdivided into two concentric regions consisting of (i) an interior region $0 < r < a_-$ and (ii) a thin surface layer $a_- < r < a_+$. In the limit of infinitesimal surface thickness, a_- approaches a_+. The interior pressure P_0 is uniform and the interior current density is zero so both $\mathbf{J} \times \mathbf{B}$ and ∇P are zero in the interior. Thus, finite current density exists *only* in the surface layer and everywhere else the magnetic field is a vacuum field. The plasma is also assumed to be surrounded by a perfectly conducting wall located at a radius b where $b > a$. MHD equilibrium, i.e., Eq. (10.60), can be written as

$$0 = -\mu_0 \nabla P_0 - \nabla \left(\frac{B_0^2}{2} \right) + \mathbf{B}_0 \cdot \nabla \mathbf{B}_0. \tag{10.141}$$

Because $\nabla \cdot \mathbf{B}_0 = r^{-1} \partial / \partial r \, (r B_{0r}) = 0$ for an azimuthally symmetric, axially uniform field and because B_r cannot be singular, B_{0r} must vanish everywhere. Since all equilibrium quantities depend on r only, the third term in Eq. (10.141) can be expanded

$$\mathbf{B}_0 \cdot \nabla \mathbf{B}_0 = \left(B_{0\theta}(r) \hat{\theta} + B_{0z}(r) \hat{z} \right) \cdot \nabla \left(B_{0\theta}(r) \hat{\theta} + B_{0z}(r) \hat{z} \right) = B_{0\theta}^2 \hat{\theta} \cdot \nabla \hat{\theta} = -\frac{B_{0\theta}^2}{r} \hat{r}. \tag{10.142}$$

Thus, Eq. (10.141) only has r components and so reduces to

$$0 = -\mu_0 \frac{\partial P_0}{\partial r} - \frac{\partial}{\partial r} \left(\frac{B_0^2}{2} \right) - \frac{B_{0\theta}^2}{r}. \tag{10.143}$$

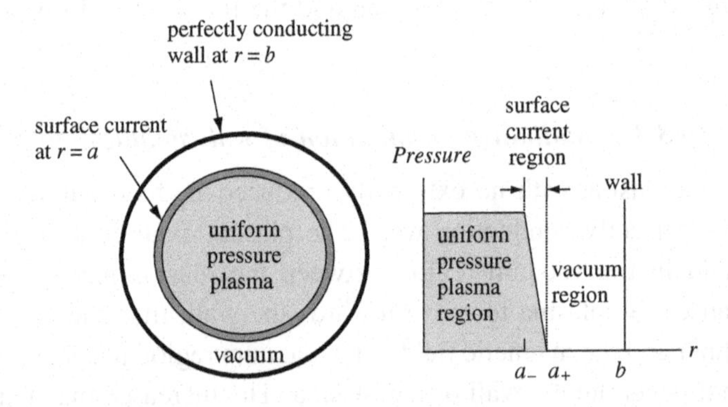

Fig. 10.8 Equilibrium for free-boundary stability analysis: assumed cross-section (left) and assumed pressure profile (right).

On integrating across the surface layer, this gives the pressure balance relation

$$P_0 + \frac{B_{0pz}^2}{2\mu_0} = \frac{B_{0v\theta}^2 + B_{0vz}^2}{2\mu_0}, \tag{10.144}$$

where the subscript p refers to the inner side (i.e., plasma side) of the surface layer and the subscript v refers to the outer side (i.e., vacuum side). The reason no $B_{0p\theta}^2$ term appears on the left-hand side of Eq. (10.144) is that there are no interior plasma currents (from Ampère's law $2\pi a B_{0p\theta} = \mu_0 I = 0$).

The location of the equilibrium surface can be described in a formal mathematical fashion by the function $r = a$ or, equivalently, by the surface-defining equation

$$S_0(r) = r - a = 0. \tag{10.145}$$

Using this formalism, an incompressible perturbation can be characterized as a harmonic deformation of the plasma surface such that $r = a + \xi_r e^{im\theta+ikz}$. The equation describing the perturbed surface is thus

$$S(r, \theta, z) = r - a - \xi_r e^{im\theta+ikz} = 0, \tag{10.146}$$

where, without loss of generality, $\xi_r > 0$ is assumed. The equilibrium magnetic field has $B_{0r} = 0$ so no equilibrium magnetic field lines penetrate the surface. This property can be stated equivalently as the equilibrium magnetic field being tangential to the surface or, in a more abstract mathematical way, as $\mathbf{B}_0 \cdot \nabla S_0 = 0$. Since the magnetic field is assumed frozen into the plasma, the magnetic field must *continue* to be tangential to the surface even when the surface becomes deformed from its equilibrium shape. Thus, the condition

$$\mathbf{B} \cdot \nabla S = 0 \tag{10.147}$$

must be satisfied at all times where ∇S is in the direction normal to the surface; this is essentially a statement that no magnetic field line penetrates the surface even when the surface becomes deformed by the instability.

A quantity f at the perturbed surface (denoted by the subscript ps) can be expressed as

$$f_{ps} = f_0 + f_1 + \boldsymbol{\xi} \cdot \nabla f_0, \tag{10.148}$$

where the the middle term is the absolute first-order change and the last term is the convective term due to the motion of the surface.

In order for the configuration to be stable, any perturbation must generate a restoring force, which pushes the perturbed surface back to its equilibrium location. The competition between destabilizing and restoring forces is found by integrating Eq. (10.141) across the perturbed surface. This integration gives Eq. (10.144) evaluated at the perturbed surface, i.e., using Eq. (10.148) to take into account the effect of the perturbation of the surface. Since ξ_r was assumed positive (outward bulge at $\theta = 0$, $z = 0$) the perturbed configuration will be stable if the effective pressure on the perturbed surface exterior exceeds the effective pressure on the perturbed surface interior, i.e., if

$$\left[\frac{B_{v\theta}^2 + B_{vz}^2}{2\mu_0}\right]_{\text{perturbed sfc}} > \left[P + \frac{B_{pz}^2}{2\mu_0}\right]_{\text{perturbed sfc}} \Longrightarrow \text{stable.} \tag{10.149}$$

In this case, the restoring force pushes back the bulge and makes the system revert to its equilibrium condition. Subtracting the equilibrium pressure balance relation Eq. (10.144) from Eq. (10.149) gives

$$\frac{B_{0v\theta}B_{1v\theta} + B_{0vz}B_{1vz}}{\mu_0} + \xi_r\frac{\partial}{\partial r}\left[\frac{B_{0v\theta}^2 + B_{0vz}^2}{2\mu_0}\right] > P_1 + \frac{B_{0pz}B_{1pz}}{\mu_0} + \xi_r\frac{\partial}{\partial r}\left[P_0 + \frac{B_{0pz}^2}{2\mu_0}\right], \tag{10.150}$$

where all quantities are evaluated at $r = a$.

Equation (10.150) can be simplified considerably because of the following relationships:

1. There are no currents in either the plasma interior or the external vacuum so

$$\frac{\partial B_{0pz}}{\partial r} = \frac{\partial B_{0vz}}{\partial r} = 0. \tag{10.151}$$

2. The adiabatic pressure equation gives

$$P_1 + \boldsymbol{\xi}\cdot\nabla P_0 + \gamma P_0 \nabla\cdot\boldsymbol{\xi} = 0. \tag{10.152}$$

Since $\nabla\cdot\boldsymbol{\xi} = 0$ has been assumed (i.e., incompressibility), the adiabatic relation reduces to

$$P_1 + \xi_r \partial P_0/\partial r = 0. \tag{10.153}$$

3. From Ampère's law it is seen that $B_{0v\theta}^2 \sim r^{-2}$ so $\left[\partial B_{0v\theta}^2/\partial r\right]_{r=a} = -2B_{0v\theta}^2/a$; thus Eq. (10.150) simplifies to

$$B_{0v\theta}B_{1v\theta} + B_{0vz}B_{1vz} - \frac{\xi_r}{a}B_{0v\theta}^2 > B_{0pz}B_{1pz} \Longrightarrow \text{stable,} \tag{10.154}$$

where again all fields are evaluated at $r = a$.

To simplify the algebra all fields are now normalized to $B_{0v\theta}(a)$. The normalized fields are denoted by a bar on top and are

$$\bar{B}_{0vz} = \frac{B_{0vz}}{B_{0v\theta}(a)}, \quad \bar{B}_{0pz} = \frac{B_{0pz}}{B_{0v\theta}(a)},$$

$$\bar{B}_{1vz} = \frac{B_{1vz}}{B_{0v\theta}(a)}, \quad \bar{B}_{1v\theta} = \frac{B_{1v\theta}}{B_{0v\theta}(a)}, \quad \bar{B}_{1pz} = \frac{B_{1pz}}{B_{0v\theta}(a)} \tag{10.155}$$

so, when expressed as a relationship between normalized fields, Eq. (10.154) becomes

$$\bar{B}_{1v\theta} + \bar{B}_{0vz}\bar{B}_{1vz} - \frac{\xi_r}{a} > \bar{B}_{0pz}\bar{B}_{1pz} \Longrightarrow \text{stable.} \tag{10.156}$$

What remains to be done is express all first-order magnetic fields in terms of ξ_r. This is relatively easy because the current is confined to the surface layer so the magnetic field is a vacuum magnetic field everywhere except exactly on the surface.

Since the magnetic field is a vacuum field for both $0 < r < a$ and $a < r < b$, the normalized fields under consideration here must be of the form $\bar{\mathbf{B}} = \nabla\chi$, where $\nabla^2\chi = 0$. This is true in both the plasma and vacuum regions but, due to the surface current at the vacuum–plasma interface, a jump discontinuity between the vacuum and plasma tangential magnetic fields occurs at $r = a$. Because the perturbed fields were assumed to have an $\exp(im\theta + ikz)$ dependence, the Laplacian becomes

$$\nabla^2\chi = \frac{\mathrm{d}^2\chi(r)}{\mathrm{d}r^2} + \frac{1}{r}\frac{\mathrm{d}\chi(r)}{\mathrm{d}r} - \left(\frac{m^2}{r^2} + k^2\right)\chi(r) = 0, \tag{10.157}$$

which has solutions consisting of the modified Bessel functions, $I_{|m|}(|k|r)$ and $K_{|m|}(|k|r)$. Absolute value signs have been used here to avoid possible confusion for situations to be encountered later where m or k could be negative. The $I_{|m|}(|k|r)$ function is finite at $r = 0$ but diverges at $r = \infty$, while the opposite is true for $K_{|m|}(|k|r)$.

The surface current causes a discontinuity in the tangential components of the perturbed magnetic field. Hence, different solutions must be used in the respective plasma and vacuum regions and then these solutions must be related to each other in such a way as to satisfy the requirements of this discontinuity. The forms of the plasma and vacuum solutions are also constrained by the respective boundary conditions at $r = 0$ and at $r = b$. In particular, the $K_{|m|}(|k|r)$ solution is not allowed in the plasma because of the constraint that the magnetic field must be finite at $r = 0$; inside the plasma the solution must therefore be of the form

$$\chi = \alpha I_{|m|}(|k|r) \quad \text{in plasma region } 0 \leq r \leq a_-. \tag{10.158}$$

On the other hand, both the $I_{|m|}(|k|r)$ and $K_{|m|}(|k|r)$ solutions are permissible in the vacuum region and so the vacuum region solution is of the general form

$$\chi = \beta_1 I_{|m|}(|k|r) + \beta_2 K_{|m|}(|k|r) \quad \text{in vacuum region } a_+ \leq r \leq b. \tag{10.159}$$

The objective now is to express the coefficients α, β_1, and β_2 in terms of ξ_r so all components of the perturbed magnetic field can be expressed as a function of ξ_r, both in the plasma and in the vacuum.

The functional dependence of these coefficients is determined by considering boundary conditions at the wall and then at the plasma–vacuum interface:

1. Wall: since the wall is perfectly conducting it is a flux conserver. There is no radial magnetic field initially and so there is no flux linking each small patch of the wall. Thus $\bar{B}_{1vr}(b)$ must vanish at the wall in order to maintain zero flux at each patch of the wall. Using Eq. (10.159) to calculate the perturbed radial magnetic field $\bar{B}_{1vr} = \partial\chi/\partial r$ at the wall, setting $\bar{B}_{1vr}(b) = 0$ implies

$$\beta_1 I'_{|m|}(|k|b) + \beta_2 K'_{|m|}(|k|b) = 0, \tag{10.160}$$

which may be solved for β_2 to obtain

$$\beta_2 = -\beta_1 \frac{\hat{I}'_{|m|}}{\hat{K}'_{|m|}}; \tag{10.161}$$

here the circumflex means the modified Bessel function is evaluated at the wall, i.e., with its argument set to $|k|b$. Equation (10.159) can then be recast as

$$\chi = \beta\left[I_{|m|}(|k|r)\hat{K}'_{|m|} - \hat{I}'_{|m|}K_{|m|}(|k|r)\right], \tag{10.162}$$

where $\beta = \beta_1/\hat{K}'_{|m|}$. The wall boundary condition has reduced the number of independent coefficients by one.

2. Vacuum side of plasma–vacuum interface: here Eq. (10.147) is linearized to obtain

$$\mathbf{B}_1 \cdot \nabla S_0 + \mathbf{B}_0 \cdot \nabla S_1 = 0 \tag{10.163}$$

or using Eqs. (10.145) and (10.146)

$$\bar{B}_{1vr} - \mathrm{i}\left(\frac{m}{a} + k\bar{B}_{0vz}\right)\xi_r = 0. \tag{10.164}$$

On the vacuum side Eq. (10.162) gives

$$\bar{B}_{1vr} = |k|\beta\left[I'_{|m|}\hat{K}'_{|m|} - \hat{I}'_{|m|}K'_{|m|}\right], \tag{10.165}$$

where omission of the argument means the modified Bessel function is evaluated at $r = a$. Substitution of Eq. (10.165) into Eq. (10.164) gives

$$\beta = \frac{\mathrm{i}\left(\dfrac{m}{a} + k\bar{B}_{0z}\right)}{|k|\left[I'_{|m|}\hat{K}'_{|m|} - \hat{I}'_{|m|}K'_{|m|}\right]}\xi_r \tag{10.166}$$

so the complete vacuum field can now be expressed in terms of ξ_r.

3. Plasma side of plasma–vacuum interface: the plasma-side version of Eq. (10.163) gives

$$\bar{B}_{1pr} - \mathrm{i}k\bar{B}_{0pz}\xi_r = 0 \tag{10.167}$$

since $B_{0p\theta}$ vanishes inside the plasma. From Eq. (10.158) the perturbed radial magnetic field on the plasma side of the interface is

$$\bar{B}_{1pr} = |k|\alpha I'_{|m|}. \tag{10.168}$$

The plasma-side version of Eq. (10.164) is thus

$$\alpha = \frac{\mathrm{i}k\bar{B}_{0pz}}{|k|I'_{|m|}}\xi_r \tag{10.169}$$

and so the plasma fields can now also be expressed in terms of ξ_r.

The stability condition, Eq. (10.156), can be written in terms of α and β to obtain

$$\beta\left(\frac{\mathrm{i}m}{a}+\mathrm{i}k\bar{B}_{0vz}\right)\left(I_{|m|}\hat{K}'_{|m|}-\hat{I}'_{|m|}K_{|m|}\right)-\frac{\xi_r}{a} > \alpha\bar{B}_{0pz}\mathrm{i}kI_{|m|}. \tag{10.170}$$

Substituting for α and β and rearranging the order gives

$$|k|a\bar{B}_{0pz}^2\left[\frac{I_{|m|}}{I'_{|m|}}\right]-\frac{(m+ka\bar{B}_{0vz})^2}{|k|a}\left[\frac{I_{|m|}\hat{K}'_{|m|}-\hat{I}'_{|m|}K_{|m|}}{I'_{|m|}\hat{K}'_{|m|}-\hat{I}'_{|m|}K'_{|m|}}\right] > 1 \Longrightarrow \text{stable}. \tag{10.171}$$

If we introduce the normalized pressure

$$\bar{P}_0 = \frac{2\mu_0 P_0}{B_{0v\theta}^2(a)}, \tag{10.172}$$

the MHD equilibrium, Eq. (10.144), can be written in normalized form as

$$\bar{P}_0+\bar{B}_{0pz}^2 = 1+\bar{B}_{0vz}^2. \tag{10.173}$$

Substitution for $\bar{B}_{0pz}^2$ into Eq. (10.171) and rearranging the order of the second term gives

$$|k|a\left[1+\bar{B}_{0vz}^2-\bar{P}_0\right]\left[\frac{I_{|m|}}{I'_{|m|}}\right]+\frac{(m+ka\bar{B}_{0vz})^2}{|k|a}\left[\frac{-I_{|m|}\hat{K}'_{|m|}+\hat{I}'_{|m|}K_{|m|}}{I'_{|m|}\hat{K}'_{|m|}-\hat{I}'_{|m|}K'_{|m|}}\right] > 1, \tag{10.174}$$

which is the general requirement for stability of a configuration having a specified internal pressure, vacuum axial field, plasma radius, and wall radius.

Conditions for sausage and kink instability

Let us now examine the effect of the various factors in Eq. (10.174). For large argument, the modified Bessel functions have the asymptotic form

$$\lim_{s\to\infty} I_{|m|}(s) \to \frac{1}{\sqrt{2\pi s}}\mathrm{e}^{s}; \quad \lim_{s\to\infty} K_{|m|}(s) \to \sqrt{\frac{\pi}{2s}}\mathrm{e}^{-s} \tag{10.175}$$

so if the wall radius goes to infinity, the factor

$$\left[\frac{-I_{|m|}\hat{K}'_{|m|}+\hat{I}'_{|m|}K_{|m|}}{I'_{|m|}\hat{K}'_{|m|}-\hat{I}'_{|m|}K'_{|m|}}\right] \to -\frac{K_{|m|}}{K'_{|m|}} = \text{positive-definite.} \tag{10.176}$$

Since bringing the wall closer is stabilizing, the factor

$$\frac{-I_{|m|}\hat{K}'_{|m|}+\hat{I}'_{|m|}K_{|m|}}{I'_{|m|}\hat{K}'_{|m|}-\hat{I}'_{|m|}K'_{|m|}}$$

will always be positive-definite. In particular, if $b \to a$ then $I'_{|m|}\hat{K}'_{|m|} \to \hat{I}'_{|m|}K'_{|m|}$, in which case the wall stabilization becomes arbitrarily large.

As $|k|a \to \infty$ the left-hand side of Eq. (10.174) becomes infinite. Thus, the configuration is stable with respect to modes having short axial wavelengths. This is because short wavelength perturbations cause more bending of the magnetic field than a long wavelength perturbation and so require more energy.

We therefore focus attention on modes with a *long* axial wavelength (i.e., modes with a small k) since these offer the only possibility of instability. The analysis can be subdivided into specific cases, such as $m = 0$, $|m| = 1$, $|m| > 1$, close-fitting wall, no wall, low pressure, high pressure, etc. Before getting into the details, let us take a broader look at the effect of the various terms in Eq. (10.174). Since $I_{|m|}/I'_{|m|} > 0$, we see that increasing $\bar{P}_0$ is destabilizing, whereas increasing $\bar{B}^2_{0vz}$ is stabilizing. Also, if $m + ka\bar{B}_{0vz} = 0$ the second term vanishes, leading to *reduced* stability; by defining the wavevector as $\mathbf{k} = (m/a)\hat{\theta} + k\hat{z}$ it is seen that $m + ka\bar{B}_{0vz} = 0$ corresponds to having $\mathbf{k}\cdot\mathbf{B}_0 = 0$ on the surface.

Sausage modes

The $m = 0$ modes are the sausage instabilities and here Eq. (10.174) reduces to

$$\left[1+\bar{B}^2_{0vz}-\bar{P}_0\right]\left[\frac{I_0}{I'_0}\right]+\bar{B}^2_{0vz}\left[\frac{-I_0\hat{K}'_0+\hat{I}'_0K_0}{I'_0\hat{K}'_0-\hat{I}'_0K'_0}\right] > \frac{1}{|k|a} \Longrightarrow \text{stable.} \tag{10.177}$$

For a given normalized plasma pressure and normalized wall radius b/a, this expression can be used to make a stability plot of $\bar{B}^2_{0vz}$ versus $|k|a$. Since the wall always provides stabilization if brought in close enough, let us consider situations where there is no wall (i.e., $b \to \infty$), in which case the stability condition reduces to

$$\left[1+\bar{B}_{0vz}^2-\bar{P}_0\right]\left[\frac{I_0}{I_0'}\right]+\bar{B}_{0vz}^2\left[\frac{-K_0}{K_0'}\right] > \frac{1}{|k|a} \Longrightarrow \text{stable.} \tag{10.178}$$

For small arguments, the modified Bessel functions of order zero have the asymptotic values

$$I_0(s) \simeq 1+\frac{s^2}{4}; \quad K_0(s) \simeq -\ln s \tag{10.179}$$

so the stability criterion becomes

$$\frac{1}{2}+\bar{B}_{0vz}^2\left[1-k^2a^2\ln\,(|k|a)\right] > \bar{P}_0 \Longrightarrow \text{stable.} \tag{10.180}$$

This gives a simple criterion for how much $\bar{B}_{0vz}^2$ is required to stabilize a given plasma pressure against sausage instabilities. The logarithmic term is stabilizing for $|k|a < 1$ but is destabilizing for $|k|a > 1$; however, this region of instability is limited because we showed that very large $|k|a$ is stable.

Kink modes

The finite m modes are the kink modes. It was shown earlier that large $|k|a$ is stable so again we confine attention to small $|k|a$. In addition, we again let the wall location go to infinity to simplify the analysis. The stability condition now reduces to

$$|k|a\left[1+\bar{B}_{0vz}^2-\bar{P}_0\right]\left[\frac{I_{|m|}}{I_{|m|}'}\right]+\frac{(m+ka\bar{B}_{0vz})^2}{|k|a}\left[\frac{-K_{|m|}}{K_{|m|}'}\right] > 1 \Longrightarrow \text{stable.} \tag{10.181}$$

For $m \neq 0$ the small argument asymptotic limits of the modified Bessel functions are

$$I_{|m|}(s) \simeq \frac{1}{|m|!}\left(\frac{s}{2}\right)^{|m|}; \quad K_{|m|}(s) \simeq \frac{|m-1|!}{2}\left(\frac{s}{2}\right)^{-|m|} \tag{10.182}$$

so the stability condition becomes

$$k^2a^2\left[1+\bar{B}_{0vz}^2-\bar{P}_0\right]+(m+ka\bar{B}_{0vz})^2 > |m| \Longrightarrow \text{stable,} \tag{10.183}$$

which is a quadratic equation in ka. Let us consider plasmas where $\bar{B}_{0vz}^2 \gg 1$ and $\bar{B}_{0vz}^2 \gg \bar{P}_0$; this corresponds to a low-beta plasma where the externally imposed axial field is much stronger than the field generated by the internal plasma currents (tokamaks are in this category). Let us define

$$x = ka\bar{B}_{0vz} \tag{10.184}$$

so the stability condition becomes

$$x^2+(m+x)^2 > |m| \Longrightarrow \text{stable.} \tag{10.185}$$

Without loss of generality x can be assumed to be positive, in which case instability occurs only when m is negative. Thus, Eq. (10.185) can be written as

$$2x^2 - 2|m|x + m^2 - |m| > 0 \Longrightarrow \text{stable}. \tag{10.186}$$

The threshold for instability occurs when the left-hand side of Eq. (10.186) vanishes, i.e., at the roots of the left-hand side. These roots are

$$x = \frac{|m| \pm \sqrt{2|m| - m^2}}{2}. \tag{10.187}$$

Since the left-hand side of Eq. (10.186) goes to positive infinity for $|x| \to \infty$, the left-hand side is negative only in the region between the two roots. Thus, the plasma is unstable only if $|x|$ lies between the two roots. The stability condition is that $|x|$ must lie outside the region between the two roots, i.e., for stability we must have

$$x > \frac{|m| + \sqrt{2|m| - m^2}}{2} \quad \text{or}$$
$$x < \frac{|m| - \sqrt{2|m| - m^2}}{2} \quad . \tag{10.188}$$

For $m = -1$ modes this gives the stability condition

$$x > 1. \tag{10.189}$$

In un-normalized quantities and using $k = 2\pi/L$, where L is the axial length of the system, the stability condition is

$$\frac{2\pi a}{L}\frac{B_{0z}}{B_{0\theta}} > 1; \tag{10.190}$$

this is known as the Kruskal–Shafranov kink stability criterion (Kruskal *et al.* 1958, Shafranov 1958).

For $m = -2$ modes, the two roots coalesce at $x = 1$ and so Eq. (10.189) also gives stability. For $m \geq 3$, the argument of the square root in Eq. (10.187) is negative so there is no region of instability.

In toroidal devices such as tokamaks, the axial wavenumber corresponds to the toroidal wavenumber since the dominant magnetic field is in the toroidal direction, i.e., $B_{0z} \to B_{0\phi}$, where ϕ is the toroidal angle. The axial wavenumber k becomes n/R, where n is the toroidal mode number. Since long axial wavelengths are the most unstable we assume $n = 1$ to have the worst-case scenario. The Kruskal–Shafranov kink stability condition then becomes

$$q = \frac{aB_{0\phi}}{RB_{0\theta}} > 1, \tag{10.191}$$

where q is called the safety factor. In order to avoid kink instability, tokamaks typically operate with $q \sim 3-4$ at the wall and q slightly above unity at the magnetic axis; this q condition is one of the most important design criteria for tokamaks since it dictates the size of the large and expensive toroidal field system.

10.10 Assignments

1. Interchange instabilities and volume per unit flux. Another way to consider pressure-driven instabilities is to calculate the consequence of interchanging two flux tubes having the same magnetic flux. The plasma moves across the magnetic field in such a way that the frozen-in flux condition is maintained. This interchange will not change the magnetic energy since the magnetic field is unaffected. However, if the flux tubes contain finite-pressure plasma and the volumes of the two flux tubes differ, then the interchange will result in compression of the plasma in the flux tube that initially had the larger volume and expansion of plasma in the flux tube that initially had the smaller volume. The former requires work on the plasma and the latter involves work by the plasma. If net work must be done on the plasma to effect the interchange, then the interchange is stable and vice versa. In a magnetic confinement configuration, the interior region has higher pressure than the exterior region. Thus, the question is whether interchanging a high pressure, interior region flux tube with a low pressure, exterior region flux tube requires positive or negative work.
   (a) Show that the volume per unit flux in a flux tube is given by
   $$V' = \oint \frac{\mathrm{d}l}{B},$$
   where the contour is over the length of the flux tube. Hint: use $BA = const.$ on a flux tube.
   (b) Show that instability corresponds to V' increasing on going from interior to exterior.
   (c) Consider the magnetic field external to a current-carrying straight wire. How does V' scale with distance from the wire? Would a plasma confined by such a magnetic field be stable or unstable to interchanges? Is this result consistent with the concepts of good and bad curvature?
2. Work through the algebra of the magnetic energy principle and verify Eqs. (10.113), (10.114), (10.115), and (10.116).
3. Show that the force on a plasma in an *arched magnetic field* with fixed ends tends to push the plasma towards the shape of a vacuum magnetic field arch having the same boundary conditions. Under what circumstances will the force (i) increase the major radius of an arched magnetic field, (ii) decrease the major radius, (iii) leave the major radius unchanged? Show that, in contrast, the force on a plasma containing an *arched current* always tends to increase the major radius of the arched current.

4. Show that if the wall radius $b \to \infty$, then Eq. (10.174) reduces to the condition

$$F(x, \bar{B}_{0vz}, \bar{P}_0) = x\bar{B}_{pvz}^2 \left[\frac{I_m(x)}{I'_m(x)}\right] - \frac{(m + x\bar{B}_{0vz})^2}{x}\frac{K_m(x)}{K'_m(x)} - 1 > 0 \text{ for stability,}$$

where $x = ka$. By assuming $\bar{B}_{0pz} = \bar{B}_{0vz}$ and evaluating the modified Bessel functions numerically, plot this expression in a parameter space where the vertical axis is x and the horizontal axis is $B_{0v\theta}/B_{0vz} = 1/\bar{B}_{0vz}$. In particular, shade the regions where $F < 0$ to indicate the regions of instability. Do this for both $m = 0$ and for $m = -1$ to show the regions of parameter space where kinks and sausages are unstable. For a given k show how raising the axial current would lead to kink or sausage instability. Which instability happens first (kink or sausage)?

5. Are kink instabilities diamagnetic or paramagnetic with respect to the axial field? To find this, consider the kinked current as a solenoid and determine whether the orientation of the solenoid is such as to increase or decrease the initial B_z field.

6. According to Eq. (10.139), current driven instabilities cause the plasma to relax to a situation where $\mu_0 \mathbf{J} = \lambda \mathbf{B}$.

   (a) Show that a class of solutions to $\mu_0 \mathbf{J} = \lambda \mathbf{B}$ can be found if

   $$\mathbf{B} = \lambda \nabla\psi \times \hat{z} + \nabla \times (\nabla\psi \times \hat{z}),$$

   where $\nabla^2\psi + \lambda^2\psi = 0$.

   (b) Suppose $\psi(x, z) = f(z)\sin(kx)$. What is the form of $f(z)$? Show that solutions that decay in z are only possible for a certain range of λ^2.

   (c) Consider a two-dimensional force-free solar coronal loop. Let the $z = 0$ plane denote the solar surface. Calculate the functional form of B_x and B_z using the solution in part (a) above.

   (d) Sketch the shape of the field line going from $x = \pi/2k - \varepsilon$ to $x = -\pi/2k + \varepsilon$ in the $z = 0$ plane where ε is a small displacement chosen so that $\cos(kx)$ is non-zero. Do this for a sequence of increasing values of λ^2. What happens to the shape of the field line as λ^2 is increased? How is current related to λ?

11

Magnetic helicity interpreted and Woltjer–Taylor relaxation

11.1 Introduction

The previous chapter introduced the concept of magnetic helicity via the energy principle and showed that total helicity $K = \int d^3r\mathbf{A}\cdot\mathbf{B}$ is a conserved quantity in an ideal plasma. This chapter shows that helicity can be interpreted in a topological sense as a count of the linkages of magnetic flux tubes with each other. Furthermore, it will be shown that when the plasma is not ideal so energy is not conserved, helicity conservation remains a rather good approximation.

The greater robustness of magnetic helicity compared to magnetic energy in the presence of dissipation leads to the Woltjer–Taylor relaxation theory, which shows that a dissipative plasma will spontaneously relax from an arbitrary initial state to a specific final state. Relaxation theory has two remarkable features, namely (i) it sidesteps describing the actual MHD dynamics and simply predicts the end state after all dynamics is over, and (ii) it thrives on complexity so the more complicated the dynamics, the more applicable is the theory. The second feature results because increased complication simply provides more channels whereby the plasma can relax to the specific final state. Relaxation theory has been very successful at predicting the approximate behavior of many laboratory, space, and astrophysical plasmas.

This chapter concludes by showing that magnetic helicity can be manifested in different forms. In particular, the kink instability will be shown to be a mechanism that converts helicity from one of these forms (twist) to another (writhe).

11.2 Topological interpretation of magnetic helicity

11.2.1 Linkage helicity

Consider the two thin linked untwisted flux tubes sketched in Fig. 11.1. The respective fluxes of these two tubes are Φ_1 and Φ_2, the flux tube axes follow the

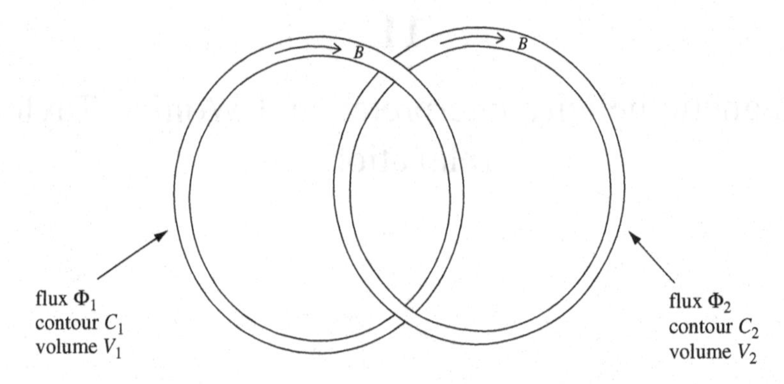

Fig. 11.1 Two linked thin untwisted flux tubes. Magnetic field is zero outside the flux tubes.

contours C_1 and C_2, and the flux tube volumes are V_1 and V_2. The magnetic field is assumed to vanish outside of the flux tubes.

The helicity of the volume V containing the two linked flux tubes is

$$K = \int_V \mathbf{A} \cdot \mathbf{B} d^3 r. \tag{11.1}$$

Because $\mathbf{B}$ vanishes outside of the two flux tubes, the helicity integral is finite only inside V_1 and V_2 so Eq. (11.1) reduces to (Moffatt 1978)

$$K = \int_{V_1} \mathbf{A} \cdot \mathbf{B} d^3 r + \int_{V_2} \mathbf{A} \cdot \mathbf{B} d^3 r. \tag{11.2}$$

The contribution to the helicity from integrating over the volume of flux tube #1 is

$$K_1 = \int_{V_1} \mathbf{A} \cdot \mathbf{B} d^3 r. \tag{11.3}$$

In order to evaluate this integral it is recalled that the magnetic flux through a surface S with perimeter C can be expressed as

$$\Phi = \int_S \mathbf{B} \cdot d\mathbf{s} = \oint_C \mathbf{A} \cdot d\mathbf{l}. \tag{11.4}$$

In flux tube #1, $d^3 r = d\mathbf{l} \cdot \hat{B} \Delta S$, where ΔS is the cross-sectional area of flux tube #1 and $d\mathbf{l}$ is an element of length along C_1. The integrand in Eq. (11.3) can thus be arranged as

$$\begin{aligned} \mathbf{A} \cdot \mathbf{B} d^3 r &= \mathbf{A} \cdot \mathbf{B} d\mathbf{l} \cdot \hat{B} \Delta S \\ &= \mathbf{A} \cdot d\mathbf{l} \Phi_1 \end{aligned} \tag{11.5}$$

since $\mathbf{dl}$ is parallel to $\mathbf{B}$ and $B\Delta S = \Phi_1$. Because Φ_1 is by definition constant along the length of flux tube #1, Φ_1 may be factored from the K_1 integral, giving

$$K_1 = \Phi_1 \int_{C_1} \mathbf{A} \cdot d\mathbf{l}. \tag{11.6}$$

However, the flux linked by contour C_1 is precisely the flux in tube #2, i.e., $\int_{C_1} \mathbf{A} \cdot d\mathbf{l} = \Phi_2$ and so

$$K_1 = \Phi_1 \Phi_2. \tag{11.7}$$

Application of the same analysis to flux tube #2 gives $K_2 = \Phi_1 \Phi_2$ and so the total helicity of the two linked flux tubes is

$$K = K_1 + K_2 = 2\Phi_1 \Phi_2. \tag{11.8}$$

The flux tubes did not actually have to be thin: a fat flux tube #2 could be decomposed into a number of adjacent thin flux tubes, in which case K_1 would be Φ_1 times the sum of the flux in all of the thin #2 flux tubes. The helicity is thus just the sum of flux tube linkages since if the two flux tubes each had unit flux, there would be one linkage of flux tube #1 with flux tube #2 and one linkage of flux tube #2 with flux tube #1. If flux tube #2 wrapped around flux tube #1 twice, then the contributions would be $K_1 = 2\Phi_1 \Phi_2$ and $K_2 = 2\Phi_1 \Phi_2$, in which case the helicity would be $K = 4\Phi_1 \Phi_2$. This would correspond to two linkages of flux tube #2 on flux tube #1 and two linkages of flux tube #1 on flux tube #2.

11.2.2 Twist helicity

Now suppose the major radius of flux tube #2 is reduced until flux tube #2 tightly encircles flux tube #1. Then, further suppose, as sketched in Fig. 11.2, that the cross-section of flux tube #2 is both squeezed and stretched until its field lines

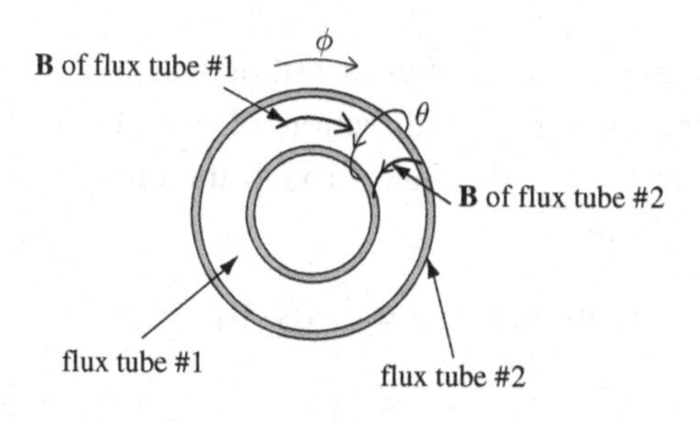

Fig. 11.2 Flux tube #2 deformed so that it tightly encircles flux tube #1 and its cross-section is squeezed and stretched so as to uniformly cover flux tube #1 like a coat of paint (shaded area). Flux in flux tube #1 is Φ, flux in flux tube #2 is $d\psi$.

are uniformly distributed over the length of flux tube #1; the field lines of flux tube #2 are normal to the shaded area in Fig. 11.2. Thus, the volume of flux tube #2 is like a thin coat of paint (shaded region in figure) applied to flux tube #1. Let Φ denote the flux in flux tube #1 and $\mathrm{d}\psi$ denote the flux in flux tube #2 so, using Eq. (11.8), the helicity of this configuration is seen to be

$$\mathrm{d}K = 2\Phi\mathrm{d}\psi. \tag{11.9}$$

We next add another layer of "paint" with more embedded ψ flux, and also with embedded Φ flux so both ψ and Φ increase. The value of Φ can be used to label the layers of "paint" so Φ is the amount of flux in flux tube #1 up to the layer of "paint" labeled by Φ. Furthermore, since ψ increases with added layers of "paint," ψ must be a function of the layer of "paint" and so $\psi = \psi(\Phi)$. It is therefore possible to write $\mathrm{d}\psi = \psi'\mathrm{d}\Phi$, where $\psi' = \mathrm{d}\psi/\mathrm{d}\Phi$. Thus, the amount of helicity added with each layer of "paint" is

$$\mathrm{d}K = 2\Phi\psi'\mathrm{d}\Phi \tag{11.10}$$

and so the sum of the helicity contributions from all the layers of "paint" is

$$K = 2\int_0^{\Phi} \Phi\psi'\mathrm{d}\Phi. \tag{11.11}$$

We now show that ψ' represents the twist of the embedded magnetic field. Let ϕ be the angle the long way around flux tube #1 and θ be the angle the long way around flux tube #2 as shown in Fig. 11.2. Thus, increasing ϕ is in the direction of contour C_1 and increasing θ is in the direction of contour C_2. The perimeter of a cross-section of flux tube #1 is in the θ direction and the perimeter of a cross-section of flux tube #2 is in the ϕ direction. The magnetic field in flux tube #1 can be written as

$$\mathbf{B}_1 = \frac{1}{2\pi}\nabla\Phi \times \nabla\theta = \frac{1}{2\pi}\nabla \times \Phi\nabla\theta, \tag{11.12}$$

which is in the ϕ direction since $\nabla\Phi$ is orthogonal to $\nabla\theta$. Here Φ is the flux linked by a contour going in the direction of $\nabla\theta$. To verify that this is the appropriate expression for $\mathbf{B}_1$, the flux through the cross-section S_1 of flux tube #1 is calculated as follows:

$$\begin{aligned}
\text{flux through } S_1 &= \int_{S_1} \mathrm{d}\mathbf{s}\cdot\mathbf{B}_1 \\
&= \frac{1}{2\pi}\int_{S_1} \mathrm{d}\mathbf{s}\cdot\nabla\times\Phi\nabla\theta \\
&= \frac{1}{2\pi}\int_{C_2} \mathrm{d}\mathbf{l}\cdot\Phi\nabla\theta
\end{aligned}$$

$$= \frac{\Phi}{2\pi} \int_{C_2} \mathrm{d}\mathbf{l} \cdot \nabla\theta$$

$$= \frac{\Phi}{2\pi} \oint \mathrm{d}\theta$$

$$= \Phi. \tag{11.13}$$

The flux Φ can be factored from the integral in the third line above, because Φ is the flux linked by contour C_2, which goes in the direction of $\nabla\theta$. Similarly, it is possible to write

$$\mathbf{B}_2 = \frac{1}{2\pi}\nabla\psi \times \nabla\phi = \frac{1}{2\pi}\nabla \times \psi\nabla\phi, \tag{11.14}$$

which is in the θ direction since $\nabla\psi$ is orthogonal to $\nabla\phi$. Since $\mathbf{B}_1$ is in the ϕ direction and $\mathbf{B}_2$ is in the θ direction, the total magnetic field can be written as $\mathbf{B} = B_1\hat{\phi} + B_2\hat{\theta}$, which is helical.

The twist of the magnetic field is defined as the number of times a field line goes around in the θ direction for one circuit in the ϕ direction. If $\mathrm{d}l_\phi$ is a displacement in the ϕ direction then

$$\mathrm{d}\phi = |\nabla\phi|\,\mathrm{d}l_\phi \tag{11.15}$$

and similarly

$$\mathrm{d}\theta = |\nabla\theta|\,\mathrm{d}l_\theta. \tag{11.16}$$

The trajectory of a magnetic field line is parallel to the magnetic field so if $\mathrm{d}\mathbf{l}$ is an increment along a magnetic field line then $\mathbf{B} \times \mathrm{d}\mathbf{l} = 0$ or

$$\frac{\mathrm{d}l_\theta}{B_\theta} = \frac{\mathrm{d}l_\phi}{B_\phi}. \tag{11.17}$$

Substituting for $\mathrm{d}l_\theta$ and $\mathrm{d}l_\phi$ using Eqs. (11.15) and (11.16) gives

$$\frac{\mathrm{d}\theta}{|\nabla\theta|\,B_\theta} = \frac{\mathrm{d}\phi}{|\nabla\phi|\,B_\phi}. \tag{11.18}$$

However, $B_\theta = |\nabla\psi|\,|\nabla\phi|\,/2\pi$ and $B_\phi = |\nabla\Phi|\,|\nabla\theta|\,/2\pi$ so

$$\frac{\mathrm{d}\theta}{\mathrm{d}\phi} = \frac{|\nabla\theta|\,B_\theta}{|\nabla\phi|\,B_\phi}$$

$$= \frac{|\nabla\theta|\,|\nabla\psi|\,|\nabla\phi|\,/2\pi}{|\nabla\phi|\,|\nabla\Phi|\,|\nabla\theta|\,/2\pi}$$

$$= \frac{|\nabla\psi|}{|\nabla\Phi|}. \tag{11.19}$$

Finally, if $\psi = \psi(\Phi)$ then $\nabla\psi = \psi'\nabla\Phi$, where $\psi' = \mathrm{d}\psi/\mathrm{d}\Phi$ and so

$$\frac{\mathrm{d}\theta}{\mathrm{d}\phi} = \psi'. \tag{11.20}$$

Thus, θ increases ψ' times faster than does ϕ and so if ϕ makes one complete circuit (i.e., goes from 0 to 2π), then θ makes ψ' circuits (i.e., θ goes from 0 to $2\pi\psi'$). The number of times the field line goes around in the θ direction for each time it goes around in the ϕ direction is called the twist

$$T(\Phi) = \psi'. \tag{11.21}$$

Hence Eq. (11.11), which gave the helicity when the ψ flux is embedded in the Φ flux, can be expressed in terms of the twist as

$$K = 2\int \Phi T(\Phi)\mathrm{d}\Phi \tag{11.22}$$

and if the twist is a constant (i.e., $T' = 0$) then

$$K = T\Phi^2. \tag{11.23}$$

11.2.3 Conservation of magnetic helicity during magnetic reconnection

Ideal MHD constrains magnetic flux to be frozen into the frame of the plasma. This means that the topology (connectedness) of magnetic field lines in a perfectly conducting plasma cannot change. As will be shown in Chapter 12, introduction of a small amount of resistivity allows the frozen-in condition to be violated at certain special locations where considerations of geometrical symmetry require the velocity to vanish. At these special locations, the approximation that the $\eta\mathbf{J}$ term in the resistive MHD Ohm's law $\mathbf{E}+\mathbf{U}\times\mathbf{B} = \eta\mathbf{J}$ can be neglected so as to obtain the ideal Ohm's law $\mathbf{E}+\mathbf{U}\times\mathbf{B} = 0$ necessarily fails because if $\mathbf{U}$ is zero, the left-hand side has only one term and there is no possibility of balancing this term unless finite resistivity is taken into account. Thus, at these special locations Ohm's law has the form $\mathbf{E} = \eta\mathbf{J}$, in which case magnetic field lines can diffuse across the plasma and reconnect with each other. Although localized to the immediate vicinity of where $\mathbf{U}$ vanishes, magnetic reconnection changes the overall topology of the magnetic field, much like switching railroad tracks at a single crossroads in the middle of a country alters the routing of trains across the entire country. Magnetic reconnection can both destroy and create linkages between flux tubes but, as will be shown, does so in a fashion whereby a replacement linkage is created for every destroyed linkage so that the total number of linkages, and hence the total helicity, is conserved. The ideal MHD constraint of perfect flux conservation is consequently replaced by the somewhat weaker constraint of having conservation of total magnetic helicity. Reconnection necessarily involves energy dissipation since reconnection requires finite resistivity. Thus, reconnection dissipates magnetic field energy while conserving total magnetic helicity.

Conservation of helicity during reconnection is demonstrated using the sequence shown in Fig. 11.3(a)–(f). An initial state consisting of two linked, untwisted ribbons of magnetic flux is shown in Fig. 11.3(a). If each ribbon is imagined to be a magnetic field line bundle having nominal flux Φ, then according to Eq. (11.8) this initial configuration has helicity $K = 2\Phi^2$. The ribbons are then cut at their line of overlap as in Fig. 11.3(b) and reconnected as in Fig. 11.3(c) to form one long ribbon. This long ribbon in Fig. 11.3(c) is then continuously deformed until in the shape shown in Fig. 11.3(f), a long ribbon having *two* twists. The twist parameter for Fig. 11.3(f) is therefore $T = 2$ and so according to Eq. (11.23) this two-twist ribbon has a helicity $K = T\Phi^2 = 2\Phi^2$. Since the initial and final helicities are both $K = 2\Phi^2$, this demonstrates that reconnection conserves magnetic helicity. Since the magnetic equivalent of cutting ribbons with scissors involves the dissipative process of resistive diffusion of magnetic field across the plasma, it can be concluded that magnetic reconnection conserves helicity, but dissipates magnetic energy.

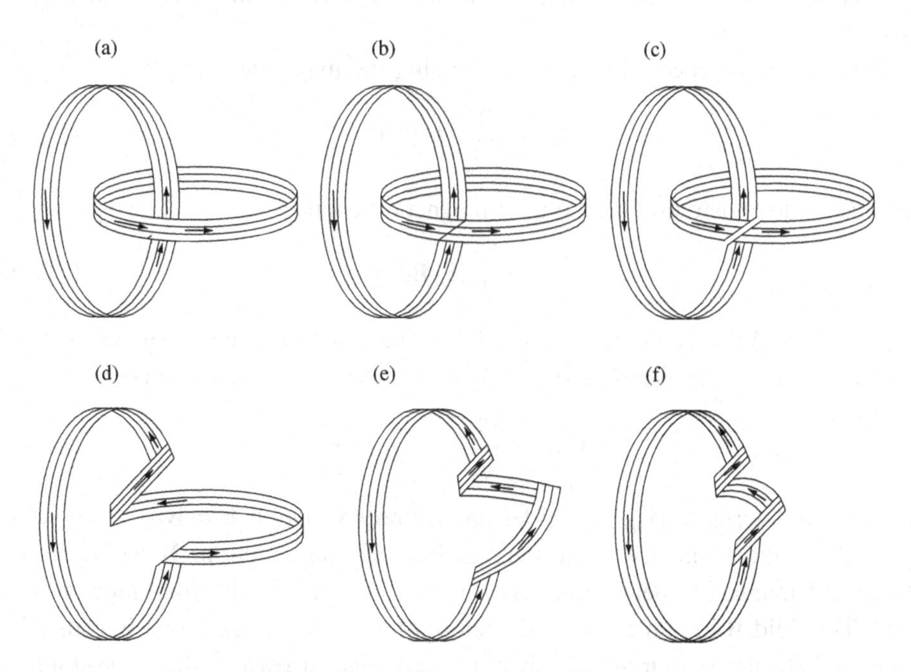

Fig. 11.3 (a) Two linked, untwisted flux tubes squeezed to be ribbons, magnetic field directions shown by arrows; (b) cut at overlay of two ribbons; (c) reconnection at cut; (d) deformation; (e) more deformation; (f) more deformation showing there are two complete twists of magnetic field.

11.3 Woltjer–Taylor relaxation

Woltjer (1958) presented a mathematical proof showing that the lowest energy state of an isolated zero-pressure (i.e., zero β) plasma with a fixed amount of magnetic helicity is a certain kind of force-free state, but did not provide any detailed explanation on how the system might attain this state. Taylor (1974) argued that because reconnection conserves helicity while dissipating magnetic energy, reconnection events ought to provide the operative mechanism whereby an isolated zero β plasma would relax towards a state having the lowest magnetic energy consistent with conservation of helicity. The discussion of Eq. (10.133), the helicity conservation equation, showed that the isolation requirement corresponds to having no field lines penetrate the surface bounding the plasma and also arranging for this surface to be an electrostatic equipotential. In order to investigate the relaxation process, we therefore consider an isolated zero β plasma bounded by a perfectly conducting wall where the normal component of the magnetic field is zero at the wall. Because the wall is a perfect conductor, the tangential electric field and hence the tangential component of the vector potential vanish at the wall. The zero β assumption is an extreme limit of the low-β situation, which corresponds to a plasma where magnetic forces dominate hydrodynamic forces, i.e., the $\mathbf{J}\times\mathbf{B}$ term is much larger than the ∇P term in the MHD equation of motion.

The relaxation process involves minimizing the magnetic energy

$$W = \frac{1}{2\mu_0}\int B^2 d^3r \tag{11.24}$$

subject to the constraint that the total magnetic helicity

$$K = \int \mathbf{A}\cdot\mathbf{B} d^3r \tag{11.25}$$

is conserved (Woltjer 1958, Taylor 1974). The minimum-energy magnetic field for the given helicity is called $\mathbf{B}_{ME}$ and its associated magnetic energy is

$$W_{ME} = \frac{1}{2\mu_0}\int B_{ME}^2 d^3r. \tag{11.26}$$

In order to determine $\mathbf{B}_{ME}$, consider an arbitrary variation $\mathbf{B} = \mathbf{B}_{ME} + \delta\mathbf{B}$ satisfying the same boundary conditions as $\mathbf{B}_{ME}$. By assumption, $\mathbf{B}$ has a higher associated magnetic energy than $\mathbf{B}_{ME}$ since the latter is the minimum energy field. The field $\mathbf{B}$ has an associated vector potential $\mathbf{A} = \mathbf{A}_{ME} + \delta\mathbf{A}$. Because the tangential electric field must vanish at the wall in accordance with the assumption of helicity conservation, the tangential component of $\delta\mathbf{A}$ must vanish at the wall. A naive attempt at minimizing W_{ME} would involve setting $\mathbf{B}_{ME}$ to zero, but such an approach is forbidden because it would make K vanish and violate the helicity conservation requirement. What must be done then is minimize W subject to the constraint that K remains constant. The variation of the magnetic energy relative to the minimum energy state is

$$\begin{aligned}
\delta W &= \frac{1}{2\mu_0}\int\left[(\mathbf{B}_{ME}+\delta\mathbf{B})^2 - B_{ME}^2\right]\mathrm{d}^3r \\
&= \frac{1}{\mu_0}\int \mathbf{B}_{ME}\cdot\delta\mathbf{B}\mathrm{d}^3r \\
&= \frac{1}{\mu_0}\int \mathbf{B}_{ME}\cdot\nabla\times\delta\mathbf{A}\mathrm{d}^3r \\
&= \frac{1}{\mu_0}\int [\nabla\cdot(\mathbf{B}_{ME}\times\delta\mathbf{A})+\delta\mathbf{A}\cdot\nabla\times\mathbf{B}_{ME}]\,\mathrm{d}^3r \\
&= \frac{1}{\mu_0}\int \mathrm{d}\mathbf{s}\cdot\mathbf{B}_{ME}\times\delta\mathbf{A}+\int\delta\mathbf{A}\cdot\nabla\times\mathbf{B}_{ME}\mathrm{d}^3r \\
&= \int \delta\mathbf{A}\cdot\nabla\times\mathbf{B}_{ME}\mathrm{d}^3r,
\end{aligned} \tag{11.27}$$

since $\mathrm{d}\mathbf{s}\cdot\mathbf{B}_{ME}\times\delta\mathbf{A}=0$ on the wall. On the other hand, the variation of the total helicity is

$$\begin{aligned}
\delta K &= \int(\delta\mathbf{A}\cdot\mathbf{B}_{ME}+\mathbf{A}_{ME}\cdot\delta\mathbf{B})\,\mathrm{d}^3r \\
&= \int(\delta\mathbf{A}\cdot\mathbf{B}_{ME}+\mathbf{A}_{ME}\cdot\nabla\times\delta\mathbf{A})\,\mathrm{d}^3r \\
&= \int(\delta\mathbf{A}\cdot\mathbf{B}_{ME}+\delta\mathbf{A}\cdot\nabla\times\mathbf{A}_{ME})+\nabla\cdot(\delta\mathbf{A}\times\mathbf{A}_{ME})\,\mathrm{d}^3r \\
&= 2\int\delta\mathbf{A}\cdot\mathbf{B}_{ME}\mathrm{d}^3r.
\end{aligned} \tag{11.28}$$

Minimization of δW subject to the constraint that K remains constant is characterized using a Lagrange multiplier λ so the constrained variational equation is

$$\delta W = \lambda\delta K. \tag{11.29}$$

Substitution for δW and δK gives

$$\int\delta\mathbf{A}\cdot\nabla\times\mathbf{B}_{ME}\mathrm{d}^3r = 2\lambda\int\delta\mathbf{A}\cdot\mathbf{B}_{ME}\;\mathrm{d}^3r \tag{11.30}$$

or, after redefining the arbitrary parameter λ,

$$\int\delta\mathbf{A}\cdot(\nabla\times\mathbf{B}_{ME}-\lambda\mathbf{B}_{ME})\,\mathrm{d}^3r = 0. \tag{11.31}$$

Since the variation $\delta\mathbf{A}$ is arbitrary within the volume, the quantity in parentheses must vanish and so

$$\nabla\times\mathbf{B}_{ME}=\lambda\mathbf{B}_{ME}. \tag{11.32}$$

The coefficient λ is necessarily constant (i.e., spatially uniform) because the Lagrange multiplier is a constant (see Assignment 2 of Chapter 2, p. 69). Thus, relaxation leads to the same force-free state predicted by Eq. (10.140). These states are a good approximation to many solar and astrophysical plasmas as well as certain laboratory plasmas such as spheromaks and reversed field pinches.

The energy per helicity of a minimum energy state can be written as

$$\begin{aligned}\frac{W}{K}&=\frac{\int B_{ME}^2\mathrm{d}^3r}{2\mu_0\int\mathbf{A}_{ME}\cdot\mathbf{B}_{ME}\mathrm{d}^3r}\\&=\frac{\int\mathbf{B}_{ME}\cdot\nabla\times\mathbf{A}_{ME}\mathrm{d}^3r}{2\mu_0\int\mathbf{A}_{ME}\cdot\mathbf{B}_{ME}\mathrm{d}^3r}\\&=\frac{\int[\mathbf{A}_{ME}\cdot\nabla\times\mathbf{B}_{ME}+\nabla\cdot(\mathbf{A}_{ME}\times\mathbf{B}_{ME})]\,\mathrm{d}^3r}{2\mu_0\int\mathbf{A}_{ME}\cdot\mathbf{B}_{ME}\mathrm{d}^3r}.\end{aligned} \tag{11.33}$$

However, Eq. (11.32) can be integrated to give

$$\mathbf{B}_{ME}=\lambda\mathbf{A}_{ME}+\nabla f, \tag{11.34}$$

where f is some arbitrary scalar function. This can be used to show

$$\nabla\cdot(\mathbf{A}_{ME}\times\mathbf{B}_{ME})=\nabla\cdot(\mathbf{A}_{ME}\times\nabla f)=\nabla f\cdot\nabla\times\mathbf{A}_{ME}=\mathbf{B}_{ME}\cdot\nabla f=\nabla\cdot(f\mathbf{B}_{ME}) \tag{11.35}$$

so

$$\int\nabla\cdot(\mathbf{A}_{ME}\times\mathbf{B}_{ME})\,\mathrm{d}^3r=\int\mathrm{d}\mathbf{s}\cdot(f\mathbf{B}_{ME})=0, \tag{11.36}$$

since $\mathbf{B}_{ME}\cdot\mathrm{d}\mathbf{s}=0$ by assumption. Thus, Eq. (11.33) reduces to

$$\frac{W}{K}=\frac{\lambda}{2\mu_0} \tag{11.37}$$

indicating that the minimum energy state must have the smallest λ consistent with the prescribed boundary conditions.

11.4 Kinking and magnetic helicity

Magnetic helicity can be manifested in various forms and plasma dynamics can cause transformation from one form to another. We now discuss an important example of helicity-conserving, morphology-altering dynamics, namely the situation where kink instability causes the twist of a straight-axis flux tube to be transformed into the writhe of the axis of an untwisted flux tube (Berger and Field 1984, Moffatt and Ricca 1992).

Consider an ideal plasma with volume V bounded by a surface S and suppose no magnetic field lines penetrate S so $\mathbf{B}\cdot\mathrm{d}\mathbf{s}=0$ over all of S. The volume V could extend to infinity or be finite; all that is required is $\mathbf{B}\cdot\mathrm{d}\mathbf{s}=0$ on the bounding surface. The helicity contained in V is

$$K_V=\int_V \mathbf{A}\cdot\mathbf{B}\mathrm{d}^3r. \tag{11.38}$$

Equation (10.129) showed K_V is gauge-independent because no field lines penetrate S and Eq. (10.134) showed K_V is a conserved quantity, i.e., the total helicity in the volume V is constant.

The volume V is now decomposed into two sub-volumes where $\mathbf{B}\cdot\mathrm{d}\mathbf{s}=0$ everywhere on the surface separating these sub-volumes. Thus, any specific field line in V is entirely in one or the other of the two sub-volumes because if a field line did travel from one sub-volume to the other it would have to penetrate the surface separating these sub-volumes and so violate the assumption $\mathbf{B}\cdot\mathrm{d}\mathbf{s}=0$ on this surface.

1 The first sub-volume, called V_{tube}, is a closed flux tube of minor radius a with a possibly helical axis as shown in Fig. 11.4. The axis of this flux tube is concealed in Fig. 11.4, but can be seen in Fig. 11.5 where it is labeled as "helical axis of flux tube." The length of this flux tube axis is denoted l_{axis} and the variable ξ denotes the distance along this axis from some fixed reference point $\mathbf{x}_0$ on the axis. Incrementing ξ from 0 to l_{axis} thus corresponds to going once around the flux tube axis. A pseudo-angular coordinate ϕ is defined as

$$\phi=2\pi\frac{\xi}{l_{axis}} \tag{11.39}$$

so that going once around the flux tube axis corresponds to incrementing ϕ from 0 to 2π. The unit vector $\hat{\phi}$ can then be used to define the local direction of the axis. Because the flux tube axis is helical, there exists at each point along the axis a radius of curvature vector $\boldsymbol{\kappa}=\hat{\phi}\cdot\nabla\hat{\phi}$, which is at right angles to $\hat{\phi}$. The local radius of curvature of the axis is $r_{curve}=1/\left|\hat{\phi}\cdot\nabla\hat{\phi}\right|$ (see discussion of Eq. (3.85)). From the point of view of an observer inside the flux tube, the flux tube appears as a long curved tunnel, which eventually closes upon itself. The flux tube minor radius a must always be smaller than r_{curve}; this restriction corresponds locally to the condition that the major radius of a torus must, by definition, exceed the minor radius. The flux tube interior volume is now imagined to be filled with fiduciary lines aligned parallel to the flux tube axis; the lengths of these fiduciary lines will vary according to their location relative to the axis. We define ξ' as the distance along a fiduciary line of length l' from a plane perpendicular to the flux tube axis at the reference point $\mathbf{x}_0$. The meaning of ϕ can then be extended to indicate the distance along any fiduciary line using the

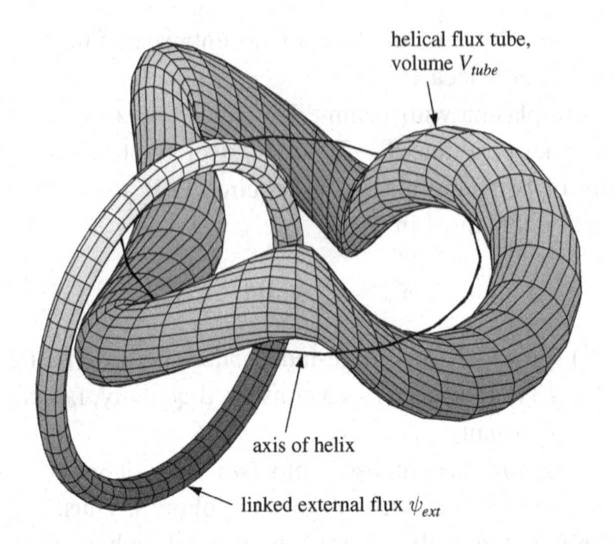

Fig. 11.4 Sub-volume #1 has a volume V_{tube} and is a flux tube with a possibly helical axis. Sub-volume #2 is all space external to sub-volume #1. All field lines in sub-volume #2 linking sub-volume #1 are represented by the thin vertical flux tube labeled ψ_{ext}.

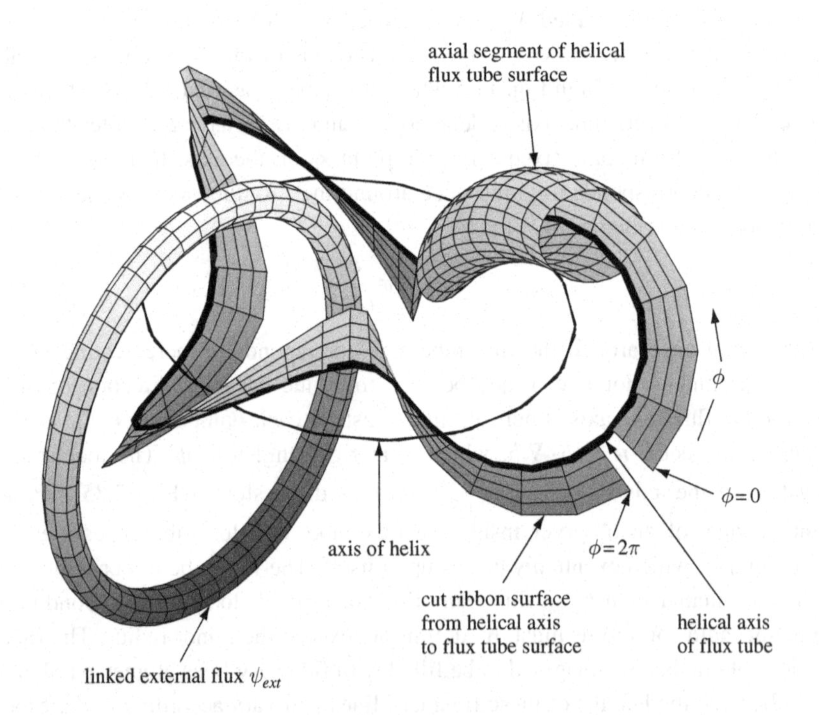

Fig. 11.5 Ribbon surface extending from helical axis of helical flux tube to surface of flux tube with ribbon surface normal to $\hat{\phi} \times \hat{\kappa}$.

relation $\xi' = \phi l'/2\pi$. Incrementing ϕ from 0 to 2π thus corresponds to going once around any or all of these fiduciary lines and so ϕ provides an unambiguous measure of fractional distance along the flux tube for any point within the flux tube even though the flux tube may be curved, twisted, or helical.

2. The second sub-volume in this decomposition is the remaining volume of V and is labeled as V_{ext}. This second sub-volume may have field lines linking flux tube V_{tube} and the flux due to these linkages is labeled ψ_{ext}. This external flux linkage is represented by the vertical flux tube labeled ψ_{ext} in Fig. 11.4.

If the dynamics within the volume V is governed by ideal MHD then, as mentioned above, the helicity K_V in V must be conserved. Since K_V is a volume integral, the volume of integration can be separated into the flux tube volume and the volume external to the flux tube, i.e., the total helicity can be expressed as the sum

$$K_V = K_{V_{tube}} + K_{V_{ext}}. \tag{11.40}$$

11.4.1 $K_{V_{tube}}$, helicity content of the flux tube

We now consider the helicity content of the possibly helical flux tube,

$$K_{V_{tube}} = \int_{V_{tube}} \mathbf{A} \cdot \mathbf{B} d^3 r. \tag{11.41}$$

The magnetic field in the flux tube is decomposed into a component $\mathbf{B}_{axis}$ parallel to the flux tube axis and an orthogonal component $\mathbf{B}_{azimuthal}$, which goes the short way around the axis; thus the magnetic field inside the flux tube is

$$\mathbf{B} = \mathbf{B}_{axis} + \mathbf{B}_{azimuthal}. \tag{11.42}$$

The field lines in the flux tube are assumed to lie in successive layers (magnetic surfaces) wrapped around the flux tube axis. The axial and azimuthal magnetic fields are derived from respective vector potentials $\mathbf{A}_{axis}$ and $\mathbf{A}_{azimuthal}$ so

$$\begin{aligned} \mathbf{B}_{axis} &= \nabla \times \mathbf{A}_{axis} \\ \mathbf{B}_{azimuthal} &= \nabla \times \mathbf{A}_{azimuthal}. \end{aligned} \tag{11.43}$$

These definitions say nothing about the direction of $\mathbf{A}_{axis}$ or $\mathbf{A}_{azimuthal}$ and so, unlike the axisymmetric situation considered when analyzing the Grad–Shafranov equation in Section 9.8.3, here neither $\mathbf{A}_{axis}$ nor $\mathbf{A}_{azimuthal}$ should be construed to be in any particular direction. All that can be said is that the curl of $\mathbf{A}_{axis}$ gives the flux tube axial magnetic field and the curl of $\mathbf{A}_{azimuthal}$ gives the azimuthal field.

The helicity content of the flux tube is thus

$$K_{V_{tube}} = \int_{V_{tube}} (\mathbf{A}_{axis} + \mathbf{A}_{azimuthal}) \cdot (\mathbf{B}_{axis} + \mathbf{B}_{azimuthal}) \, d^3r \tag{11.44}$$

and this is true even though the axis of the flux tube could be helical.

By definition, each layer of field lines constituting a magnetic surface encloses a flux Φ. An equivalent definition is to state that Φ is the magnetic flux linked by a contour in the magnetic surface going the short way around the flux tube axis. The magnetic surface is labeled by Φ and so Φ can be considered as a coordinate having its gradient always normal to the flux surface. In effect Φ is a rescaled minor radius, since Φ increases monotonically with minor radius.

These definitions are sufficiently general to allow for the axis of the flux tube to be a helix, a knot, or a simple closed curve lying in a plane. If the flux tube axis is just a simple closed curve lying in a plane, then $\nabla \times \hat{\phi}$ is normal to the plane and therefore normal to $\hat{\phi}$. In the slightly more general case where the axis is not in a plane but $\hat{\phi} \cdot \nabla \times \hat{\phi} = 0$, the path traced out by $\hat{\phi}$ can be considered to be the perimeter of some bumpy surface. However, in the most general case where $\hat{\phi} \cdot \nabla \times \hat{\phi}$ is finite, *no surface exists* for which the path traced out by $\hat{\phi}$ is the perimeter (Barnes 1977). There might also be situations where $\hat{\phi} \cdot \nabla \times \hat{\phi} \neq 0$ but the sign of $\hat{\phi} \cdot \nabla \times \hat{\phi}$ alternates. If the average of $\hat{\phi} \cdot \nabla \times \hat{\phi}$ over the length of the axis is zero, then the situation would be similar to the case where $\hat{\phi} \cdot \nabla \times \hat{\phi} = 0$ everywhere, because twists in the axis could be squeezed together axially until mutually canceling out.

Because any magnetic field line in the flux tube lies in some magnetic surface labeled by Φ, any field line in the flux tube has a component parallel to the flux tube axis and possibly also a component perpendicular to the flux tube axis, but never a component in the $\nabla\Phi$ direction. This suggests introduction of an azimuthal coordinate θ, which is defined as the angular distance around the flux tube axis in the $\nabla\phi \times \nabla\Phi$ direction, i.e., θ is defined such that

$$\frac{\nabla\theta}{|\nabla\theta|} = \frac{\nabla\phi \times \nabla\Phi}{|\nabla\phi \times \nabla\Phi|}. \tag{11.45}$$

The local direction of $\nabla\theta$ depends on both ϕ and θ; furthermore, for θ to denote some definite position, θ must be measured with respect to some unambiguous origin. An unambiguous θ origin can be established by using as a reference the direction of the local radius of curvature vector of the flux tube axis, i.e., the direction of $-\boldsymbol{\kappa} = -\hat{B} \cdot \nabla \hat{B}$ evaluated on the flux tube axis (see discussion of Eq. (9.14) on p. 313). The $\theta = 0$ surface is then a ribbon surface, as shown in Fig. 11.5, that extends from the flux tube axis to the flux tube outer surface and oriented so $\hat{\phi} \times \boldsymbol{\kappa}$ is normal to the ribbon surface. The ribbon is considered to

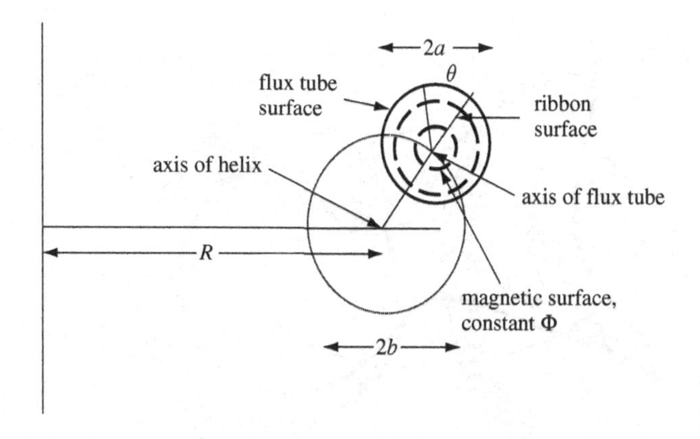

Fig. 11.6 Cross-section showing flux tube of minor radius a, with its axis tracing out a helix with radius b about the "axis of helix." Interior flux surfaces with $\Phi = const.$ are shown as dashed lines. The ribbon surface defining $\theta = 0$ is also indicated.

have a cut at $\phi = 0$ in order to distinguish the ribbon ends at $\phi = 0$ and $\phi = 2\pi$ from each other.

Figure 11.6 shows a cross-section of this system. The flux tube has minor radius a and the helical axis of the flux tube traces out a trajectory with minor radius b about an axis that has major radius R. Representative magnetic surfaces (surfaces of constant Φ), the ribbon surface, and the angle θ are also shown in this figure.

The flux Φ was a generalization of the toroidal flux of an axisymmetric system. We now define the corresponding generalization of the poloidal flux $\psi(\Phi)$ to be the magnetic flux penetrating a sub-ribbon extending inwards from the outer surface of the possibly helical flux tube to some given interior magnetic surface Φ. This sub-ribbon is shown in Fig. 11.7 and the definition is such that

$$\begin{aligned}\psi &= \int_{sub\text{-}ribbon} d\mathbf{s} \cdot \mathbf{B} \\ &= \int_{sub\text{-}ribbon} d\mathbf{s} \cdot \nabla \times \mathbf{A} \\ &= \oint_C d\mathbf{l} \cdot \mathbf{A},\end{aligned} \tag{11.46}$$

where the contour C follows the perimeter of the sub-ribbon and specifically follows the two ends at $\phi = 0$ and $\phi = 2\pi$. This definition implies $\psi = 0$ on the outer surface of the flux tube because the sub-ribbon area is zero at this location.

We now define the vector potential

$$\mathbf{A}_{azimuthal} = \frac{\psi(\Phi)}{2\pi} \nabla\phi \tag{11.47}$$

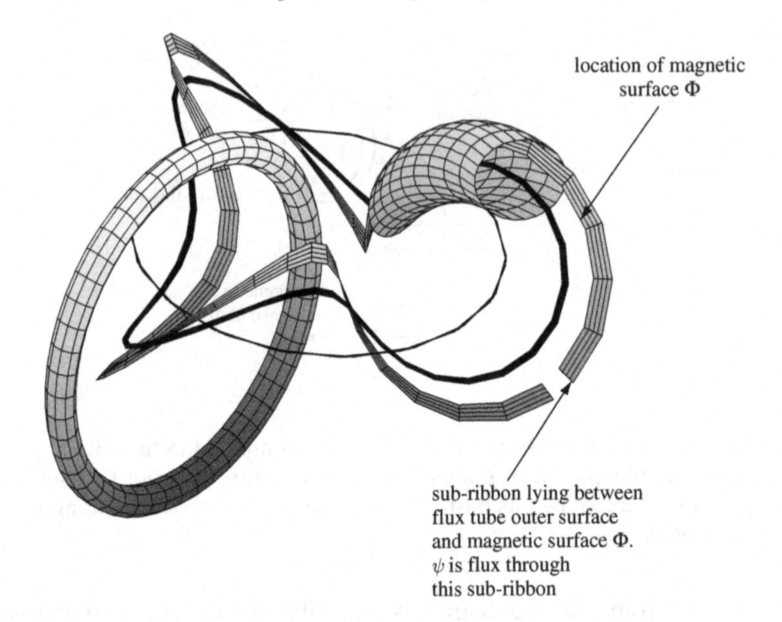

Fig. 11.7 ψ is flux through cut sub-ribbon located between magnetic surface Φ and outer surface of flux tube.

and note that this definition is valid only in the range $0 \leq \phi < 2\pi$. The validity of this definition is established by calculating the line integral of the presumed vector potential following the contour C around the perimeter of the sub-ribbon to obtain

$$\begin{aligned}
\oint_C \mathrm{d}\mathbf{l} \cdot \mathbf{A}_{azimuthal} &= \oint_C \mathrm{d}\mathbf{l} \cdot \frac{\psi(\Phi)}{2\pi} \nabla\phi \\
&= \frac{\psi(\Phi)}{2\pi} \int_{C_\Phi} \mathrm{d}\mathbf{l} \cdot \nabla\phi \\
&= \psi(\Phi),
\end{aligned} \tag{11.48}$$

where $\mathrm{d}\mathbf{l} \cdot \nabla\phi = \mathrm{d}\phi$. The justification for the steps in Eq. (11.48) is as follows: Figure 11.7 shows that the contour C around the perimeter of the sub-ribbon consists of four segments, namely a segment following the outside edge of the sub-ribbon (this segment lies in the outer surface of the flux tube where $\psi = 0$), the segment labeled C_Φ on the inside edge of the sub-ribbon, a segment at the $\phi = 0$ end, and an oppositely directed segment at the $\phi = 2\pi_-$ end. Only the C_Φ segment contributes to the integral because (i) $\mathrm{d}\mathbf{l} \cdot \nabla\phi = 0$ on the ends $\phi = 0$ and $\phi = 2\pi_-$ since ϕ is constant on both these ends and (ii) $\psi = 0$ on the outside edge of the sub-ribbon.

The magnetic field inside the flux tube can thus be written as

$$\mathbf{B} = \mathbf{B}_{axis} + \nabla \times \frac{\psi(\Phi)}{2\pi} \nabla\phi, \tag{11.49}$$

where $\mathbf{B}_{axis}$ is the magnetic field component parallel to the flux tube axis. This decomposition of the magnetic field is valid even for a twisted or knotted axis provided it is used only for situations where $0 \leq \phi < 2\pi$.

The vector potential for the entire field can be written as

$$\mathbf{A} = \mathbf{A}_{axis} + \frac{\psi(\Phi)}{2\pi} \nabla\phi. \tag{11.50}$$

These results can be used in Eq. (11.44) to give the helicity content of the flux tube as

$$\begin{aligned} K_{V_{tube}} &= \int_{V_{tube}} \mathrm{d}^3 r \left(\mathbf{A}_{axis} + \frac{\psi(\Phi)}{2\pi} \nabla\phi \right) \cdot \left(\mathbf{B}_{axis} + \frac{1}{2\pi} \nabla\psi \times \nabla\phi \right) \\ &= \int_{V_{tube}} \mathrm{d}^3 r \left(\mathbf{A}_{axis} \cdot \mathbf{B}_{axis} + \frac{\psi(\Phi)}{2\pi} \nabla\phi \cdot \mathbf{B}_{axis} + \mathbf{A}_{axis} \cdot \frac{1}{2\pi} \nabla\psi \times \nabla\phi \right). \end{aligned} \tag{11.51}$$

This expression may now be decomposed into writhe and twist terms,

$$K_{V_{tube}} = K_{writhe} + K_{twist}$$

respectively defined as

$$K_{writhe} = \int_{V_{tube}} \mathrm{d}^3 r \mathbf{A}_{axis} \cdot \mathbf{B}_{axis} \tag{11.52}$$

and

$$K_{twist} = \int_{V_{tube}} \mathrm{d}^3 r \left(\frac{\psi(\Phi)}{2\pi} \nabla\phi \cdot \mathbf{B}_{axis} + \mathbf{A}_{axis} \cdot \frac{1}{2\pi} \nabla\psi \times \nabla\phi \right). \tag{11.53}$$

As will be shown below, K_{twist} depends on ψ being finite whereas K_{writhe} depends on the extent to which the flux tube axis is helical. The flux tube helicity can thus be entirely due to K_{writhe}, entirely due to K_{twist}, or due to some combination of these two types of helicity.

11.4.2 Evaluation of K_{twist}

Since the K_{twist} integral is a *volume* integral, the flux tube may be cut at $\phi = 0$ without affecting this integral. Making such a cut means that ϕ is restricted to the range $0 \leq \phi < 2\pi$ and therefore does not make a complete circuit around the axis of the flux tube. The evaluation of this integral is insensitive to the connectivity of the axis because connectivity is a concept that makes sense only after making a complete circuit of the axis. The axial magnetic field within the cut volume may be expressed as

$$\mathbf{B}_{axis} = \frac{1}{2\pi}\nabla \times \Phi \nabla\theta, \tag{11.54}$$

where θ is the angular distance on a contour C_θ encircling the flux tube axis and lies in a magnetic surface. This representation for the axial magnetic field is appropriate here since at each ϕ for $0 \leq \phi < 2\pi$ we may write

$$\begin{aligned}
\Phi &= \int \mathrm{d}\mathbf{s} \cdot \mathbf{B}_{axis} \\
&= \frac{1}{2\pi}\int \mathrm{d}\mathbf{s} \cdot \nabla \times \Phi\nabla\theta \\
&= \frac{1}{2\pi}\oint_{C_\theta} \mathrm{d}\mathbf{l} \cdot \Phi\nabla\theta \\
&= \frac{\Phi}{2\pi}\oint_{C_\theta} \mathrm{d}\mathbf{l} \cdot \nabla\theta \\
&= \frac{\Phi}{2\pi}\oint_{C_\theta} \mathrm{d}\theta.
\end{aligned} \tag{11.55}$$

By uncurling Eq. (11.54) it is seen that vector potential associated with the axial magnetic field may be represented in the cut flux tube as

$$\mathbf{A}_{axis} = \frac{\Phi}{2\pi}\nabla\theta, \quad 0 \leq \phi < 2\pi \tag{11.56}$$

so

$$\begin{aligned}
K_{twist} &= \frac{1}{4\pi^2}\int_{V_{tube}} \mathrm{d}^3r\,(\psi\nabla\phi \cdot \nabla\Phi \times \nabla\theta + \Phi\nabla\theta \cdot \nabla\psi \times \nabla\phi) \\
&= \frac{1}{4\pi^2}\int_{V_{tube}} \mathrm{d}^3r\,(\psi\nabla\phi \cdot \nabla\Phi \times \nabla\theta + \Phi\psi'\nabla\theta \cdot \nabla\Phi \times \nabla\phi) \\
&= \frac{1}{4\pi^2}\int_{V_{tube}} \mathrm{d}^3r\,(-\psi + \Phi\psi')\,\nabla\theta \cdot \nabla\Phi \times \nabla\phi.
\end{aligned} \tag{11.57}$$

The three direction gradients $\nabla\theta$, $\nabla\Phi$, and $\nabla\phi$ form an orthonormal coordinate system and an element of volume in this system is given by

$$\mathrm{d}^3r = \mathrm{d}l_\Phi \mathrm{d}l_\theta \mathrm{d}l_\phi = \frac{\mathrm{d}\Phi\mathrm{d}\theta\mathrm{d}\phi}{\nabla\theta \cdot \nabla\Phi \times \nabla\phi} \tag{11.58}$$

since $\mathrm{d}\theta = \mathrm{d}l_\theta\,|\nabla\theta|$, $\mathrm{d}\phi = \mathrm{d}l_\phi\,|\nabla\phi|$, and $\mathrm{d}\Phi = \mathrm{d}l_\Phi\,|\nabla\Phi|$. Thus,

$$\begin{aligned} K_{twist} &= \frac{1}{4\pi^2}\int_0^\Phi \mathrm{d}\Phi \int_{\theta=0}^{2\pi} \mathrm{d}\theta \int_{\phi=0}^{(2\pi)_-} \mathrm{d}\phi\,(-\psi + \Phi\psi') \\ &= -\int_0^\Phi \psi \mathrm{d}\Phi + \int_0^\Phi \psi'\Phi \mathrm{d}\Phi \\ &= -\int_0^\Phi \psi \mathrm{d}\Phi + \int_0^\psi \Phi \mathrm{d}\psi \\ &= 2\int_0^\Phi \Phi \frac{\mathrm{d}\psi}{\mathrm{d}\Phi}\mathrm{d}\Phi \end{aligned} \tag{11.59}$$

where the integration limit $(2\pi)_-$ corresponds to being infinitesimally less than 2π. The last line has been obtained using the relationship $\int \mathrm{d}(\psi\Phi) = [\psi\Phi]_{axis}^{surface} = \int \Phi \mathrm{d}\psi + \int \psi \mathrm{d}\Phi$ and noting that the integrated term vanishes since $\psi = 0$ on the flux tube surface and $\Phi = 0$ on the flux tube axis. Since Eq. (11.59) is consistent with the definition of twist discussed in reference to Eqs. (11.21) and (11.22), the decision to use the name K_{twist} for the helicity term defined in Eq. (11.53) is validated.

11.4.3 Evaluation of K_{writhe}

Connectivity of the flux tube is the important issue here. In order to evaluate K_{writhe} the volume element is now expressed as

$$\mathrm{d}^3 r = \mathbf{dl}\cdot\mathbf{ds}, \tag{11.60}$$

where $\mathbf{dl}$ is an increment of length along the axis and $\mathbf{ds}$ is an element of surface in the plane perpendicular to the axis so $\mathbf{B}_{axis}\cdot\mathbf{ds} = \mathrm{d}\Phi$. Because the line integral involves a *complete circuit* of the flux tube axis, in contrast to the earlier evaluation of K_{twist}, we now avoid using the gradient of a scalar to denote distance along the axis. Using Eq. (11.60) in Eq. (11.52) the writhe helicity may be expressed as

$$\begin{aligned} K_{writhe} &= \int_{C_{axis}} \mathbf{A}_{axis}\cdot\mathbf{dl}_{axis} \int \mathrm{d}\Phi \\ &= \Phi \int_{C_{axis}} \mathbf{A}_{axis}\cdot\mathbf{dl}_{axis}. \end{aligned} \tag{11.61}$$

This integral differs topologically from the integrals of the previous section, because here the contour is a *complete circuit*, i.e., ϕ varies from 0 to 2π and not from 0 to $2\pi_-$.

A contour path C of an integral $\int_C \mathbf{A}\cdot\mathbf{dl}$ may be continuously deformed into a new contour path C' without changing the value of the integral, provided no magnetic flux is linked by the surface S bounded by C and C'. This property is a three-dimensional analog to the concept of analyticity for a contour integral in the complex plane and its validity is established as follows: since no magnetic flux is assumed to link the surface S bounded by C and C' we may write

$$\begin{aligned}0 &= \int_S \mathbf{B}\cdot d\mathbf{s} \\ &= \oint \mathbf{A}\cdot d\mathbf{l} \\ &= \int_C \mathbf{A}\cdot d\mathbf{l} - \int_{C'} \mathbf{A}\cdot d\mathbf{l}. \end{aligned} \tag{11.62}$$

The presumption that it is possible to continuously deform C into C' imposes the requirement that a surface S exists between C and C'; this is only true if C does not link C'.

We now consider some limiting cases for K_{writhe} and to aid in visualization it should be recalled that the flux tube axis is assumed to have a helical trajectory with helix major radius R and helix minor radius b as shown in Fig. 11.6.

Limiting case where flux tube axis is not helical

This situation corresponds to having $b = 0$, in which case the flux tube axis lies in a plane. The contour C_{axis} in Eq. (11.61) can thus be slipped from its original position through the $\mathbf{B}_{axis}$ field lines to the surface of the flux tube without crossing any $\mathbf{B}_{axis}$ field lines. This is topologically possible because, by definition, all $\mathbf{B}_{axis}$ field lines in the flux tube are parallel to the flux tube axis. Thus, here

$$K_{writhe} = \Phi \int_{C'} \mathbf{A}_{axis} \cdot d\mathbf{l}_{axis}, \tag{11.63}$$

where C' lies on the flux tube surface. The line integral C' encircles the external flux linked by the flux tube and so

$$K_{writhe} = \Phi \psi_{ext}, \tag{11.64}$$

when the flux tube axis is not helical.

When the flux tube axis is helical, two possibilities exist, namely $a < b$ and $a > b$. The distinction between the $a < b$ and $a > b$ cases is sketched in Fig. 11.8. The parameter b characterizes the amplitude of the kink so infinitesimal b corresponds to a very weak kinking of the flux tube.

Strong kink case: axis is helical and $b > a$

Because the helix minor radius b exceeds the flux tube minor radius a here, the entire flux tube revolves around the axis of the helix as shown in Fig. 11.9. Because the helical axis closes upon itself, it must have an integral number of periods and we let N be the number of helix periods. However, because $\mathbf{B}_{axis}$ is everywhere parallel to the flux tube axis, the flux tube axis may again be slipped

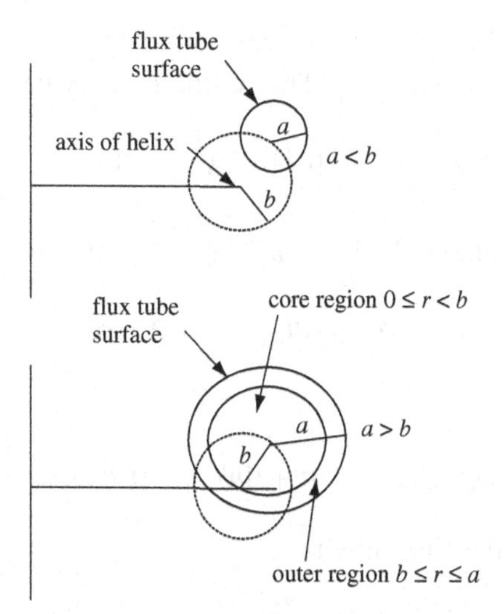

Fig. 11.8 Top: case where flux tube radius a is less than radius b of helix traced out by flux tube axis. Bottom: case where $a > b$; the core region $0 \leq r < b$ rotates around the axis of the helix whereas the outer region with $b < r < a$ wobbles about the axis of the helix.

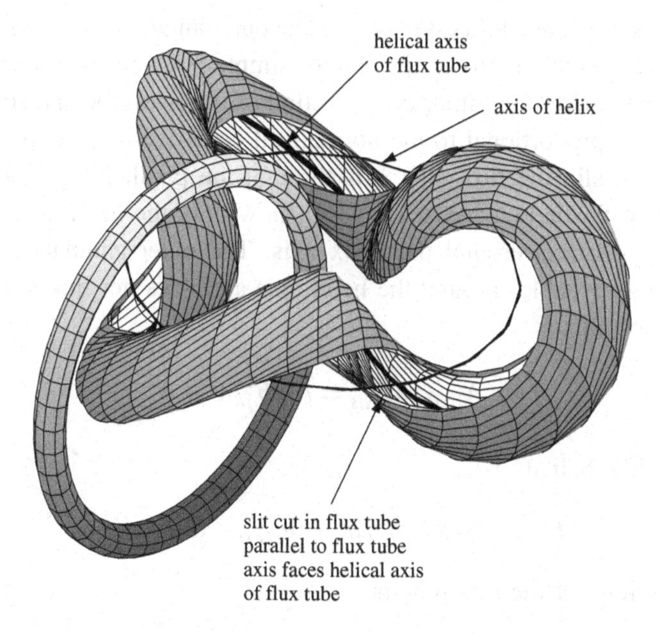

Fig. 11.9 Helical axis of flux tube may be slipped through $\mathbf{B}_{axis}$ field lines of flux tube to coincide with axis of helix (note that open slit in flux tube always faces axis of helix).

through the $\mathbf{B}_{axis}$ field lines and in this case moved towards the helix axis until it coincides with the helix axis C_{helix}. The writhe helicity can thus be expressed as

$$K_{writhe} = \Phi \int_{C_{helix}} \mathbf{A}_{axis} \cdot \mathbf{dl}_{axis}. \tag{11.65}$$

However, the flux tube links the helix axis N times and so

$$\int_{C_{helix}} \mathbf{A}_{axis} \cdot \mathbf{dl}_{axis} = N\Phi + \psi_{ext}. \tag{11.66}$$

Thus,

$$K_{writhe} = N\Phi^2 + \Phi\psi_{ext} \quad \text{if } b > a. \tag{11.67}$$

Hence, the helicity of the flux tube is

$$\begin{aligned} K_{V_{tube}} &= K_{twist} + K_{writhe} \\ &= 2\int \Phi \frac{\mathrm{d}\psi}{\mathrm{d}\Phi} \mathrm{d}\Phi + N\Phi^2 + \Phi\psi_{ext} \quad \text{if } b > a. \end{aligned} \tag{11.68}$$

Weak kink case: flux tube axis is helical but $a > b$

In this situation the flux tube may be subdivided into an inner core with minor radius $r < b$ and associated flux $\Phi b^2/a^2$ and an outer annular region with $b < r < a$ with the remainder of Φ (to keep matters simple, we are assuming that Φ is evenly distributed over the minor cross-section of the flux tube and so the flux in the inner core is proportional to the normalized cross-sectional area of the inner core). We again slip the flux tube axis through the parallel $\mathbf{B}_{axis}$ field lines to be coincident with the helix axis. However, now only the inner core of the flux tube ($r < b_{axis}$) rotates around the helix axis. The outer annular region of the flux tube merely wobbles around the helix axis and does not link the helix axis. Thus, the linked flux is

$$\int_{C_{helix}} \mathbf{A}_{axis} \cdot \mathbf{dl}_{axis} = N\Phi b^2/a^2 + \psi_{ext} \tag{11.69}$$

and so the writhe helicity is

$$K_{writhe} = N\Phi^2 b^2/a^2 + \Phi\psi_{ext} \quad \text{if } a > b. \tag{11.70}$$

Hence, the helicity of the flux tube is

$$\begin{aligned} K_{V_{tube}} &= K_{twist} + K_{writhe} \\ &= 2\int \Phi \frac{\mathrm{d}\psi}{\mathrm{d}\Phi} \mathrm{d}\Phi + N\Phi^2 b^2/a^2 + \Phi\psi_{ext} \quad \text{if } a > b. \end{aligned} \tag{11.71}$$

11.4.4 Twist and writhe exchange in a kink instability

A kink instability is governed by ideal MHD and so must be helicity-conserving. The number of turns of the kink is defined by the initial twist condition $\mathbf{k}\cdot\mathbf{B} = k_\phi B_\phi + k_\theta B_\theta = 0$ so if $k_\theta = 1/a$ and $k_\phi = n/R$ then the kink will result in a helix with $N = n$. Since helicity is conserved, it is seen that as the amplitude of the kink increases (i.e., b increases), the flux tube twist $\mathrm{d}\psi/\mathrm{d}\Phi$ will have to decrease. In particular if the twist is uniform so that $\psi = T\Phi$, where twist T is independent of Φ, then as in Section 11.2.2 we may write

$$2\int \Phi \frac{\mathrm{d}\psi}{\mathrm{d}\Phi}\mathrm{d}\Phi = T\Phi^2 \tag{11.72}$$

and so from conservation of helicity we conclude

$$T + Nb^2/a^2 = const. \text{ for } b < a; \tag{11.73}$$

this is an example of the Calugareanu theorem (Calugareanu 1959).

A kink instability that starts with $b = 0$ and an initial twist $T_{initial} = n$ will grow until at $b = a$ the twist vanishes and all helicity is contained by an N-turn writhe where $N = T_{initial}$. Thus, a kink will convert twist into writhe so in the strong kink case (i.e., $b > a$) the result is $T = 0$ and $N = T_{initial}$. This property can be confirmed by taking a length of garden hose with a stripe running along the length and connecting the two ends together to form a closed flux tube with a stripe running along the axis. Initially, $N + T = 0$ because the stripe is not twisted and the hose axis is not helical. If the hose is deformed into a right-handed helix, then the stripe will make a left-handed helix about the hose axis to keep $N + T = 0$. If the stripe is initially a right-handed helix when the hose is not a helix, then deformation of the hose into a right-handed helix will result in the stripe becoming parallel to the hose axis.

11.5 Assignments

1. Derive the helicity conservation equation for a resistive plasma

$$\frac{\mathrm{d}K}{\mathrm{d}t} + \int_S \mathrm{d}\mathbf{s}\cdot(\varphi\mathbf{B} + \mathbf{E}\times\mathbf{A}) = -2\int \mathrm{d}^3r\,\eta\mathbf{J}\cdot\mathbf{B} \tag{11.74}$$

and express it in the form

$$\frac{\mathrm{d}K}{\mathrm{d}t} + \int_S \mathrm{d}\mathbf{s}\cdot\left(2\varphi\mathbf{B} + \mathbf{A}\times\frac{\partial\mathbf{A}}{\partial\mathbf{t}}\right) = -2\int \mathrm{d}^3r\,\eta\mathbf{J}\cdot\mathbf{B}, \tag{11.75}$$

where φ is the electrostatic potential. What happens to the surface integral when the surface is a perfectly conducting wall? Hint: start by evaluating $-\partial\,(\mathbf{A}\cdot\mathbf{B})\,\partial t$ and make repeated use of Faraday's law $\nabla\times\mathbf{E} = -\partial\mathbf{B}/\partial t$. The electric field should be expressed as $\mathbf{E} = -\nabla\varphi - \partial\mathbf{A}/\partial t$ and use the resistive MHD Ohm's law $\mathbf{E}+\mathbf{U}\times\mathbf{B} = \eta\mathbf{J}$.

2. Alternative explanation for why helicity is conserved better than magnetic energy:

   (a) By subtracting **E** dotted with Ampère's law from **B** dotted with Faraday's law derive the MHD limit of Poynting's theorem:

$$\frac{\partial}{\partial t}\left(\frac{B^2}{2\mu_0}\right)+\nabla\cdot\left(\frac{\mathbf{E}\times\mathbf{B}}{\mu_0}\right)=-\mathbf{E}\cdot\mathbf{J}. \tag{11.76}$$

   (b) Consider a plasma in a perfectly conducting chamber with no vacuum gap between the plasma and the chamber wall. Show that integration of Eq. (11.74) and Eq. (11.76) over the entire volume respectively give

$$\frac{\mathrm{d}K}{\mathrm{d}t}=-2\int \mathrm{d}^3r\eta\mathbf{J}\cdot\mathbf{B} \tag{11.77}$$

   and

$$\frac{\mathrm{d}W}{\mathrm{d}t}=-\int \mathrm{d}^3r\eta J^2, \tag{11.78}$$

   where $W=\int \mathrm{d}^3rB^2/2\mu_0$ is the magnetic energy.

   (c) Show that the right-hand side of Eq. (11.78) is proportional to higher order spatial derivatives of **B** than is the right-hand side of Eq. (11.77). Use this property to argue that the rate of dissipation of magnetic energy greatly exceeds that of magnetic helicity when the dynamics is spatially complex so that most of the spectral power is in short characteristic scale lengths.

   (d) Explain why this difference in dissipation rates could be approximated by assuming that magnetic helicity remains constant during magnetic energy decay via dynamics having fine scales (e.g., localized reconnection). Argue that the decay of magnetic energy is thereby constrained by the requirement that helicity is conserved.

3. Show that the minimum-energy state given by Eq. (11.32) is a force-free configuration. Is λ spatially uniform? Is this the same result as given in Eqs. (10.139) and (10.140)? Take the curl again to obtain

$$\nabla^2\mathbf{B}+\lambda^2\mathbf{B}=0.$$

   What are the components of this equation in axisymmetric cylindrical geometry? Be careful when evaluating the components of $\nabla^2\mathbf{B}$ to take into account derivatives operating on unit vectors (e.g., $\nabla^2\left(B_\phi\hat{\phi}\right)\neq\hat{\phi}\nabla^2B_\phi$, see Appendix B).

   (a) Show that for axisymmetric cylindrical geometry the minimum-energy states have the magnetic field components

$$B_z(r)=\bar{B}J_0(\lambda r)$$
$$B_\phi(r)=\bar{B}J_1(\lambda r).$$

   Sketch $B_z(r)$ and $B_\phi(r)$. This is called the Bessel function model or Lundquist solution (Lundquist 1950) to the force-free equation and is often an excellent representation for nearly force-free equilibria such as spheromaks, reversed field pinches, and solar coronal loops.

4. For a cylindrical system with coordinates $\{r, \phi, z\}$ show that if χ satisfies the Helmholtz equation

$$\nabla^2\chi + \lambda^2\chi = 0$$

then

$$\mathbf{B} = \lambda\nabla\chi \times \nabla z + \nabla \times (\nabla\chi \times \nabla z)$$

is a solution of the force-free equation $\mu_0\mathbf{J} = \lambda\mathbf{B}$. By assuming χ is independent of ϕ and is of the form $\chi \sim f(r)\cos kz$, find $f(r)$ and then determine the components of $\mathbf{B}$. Show that if λ satisfies an eigenvalue condition, it is possible to have a finite force-free field having no normal component on the walls of a cylinder of length h and radius a. Give the magnetic field components for this situation (spheromak model). Calculate the poloidal flux $\psi(r, z)$ by direct integration of $B_z(r, z)$ and plot surfaces of constant poloidal flux. Hint: the answers will be in terms of Bessel functions J_0 and J_1.

5. Obtain a ribbon such as is used in gift-wrapping, make one complete twist in this ribbon and tape the ends together. By manipulating the twisted ribbon (Pfister and Gekelman 1991, Bellan 2000) show the following:
   (a) If the ribbon is manipulated to have no twists, then it has a figure-eight pattern with a cross-over. If the flux through the ribbon is Φ, show that the helicity of one full twist is Φ^2. Then argue that the helicity of one cross-over must also be Φ^2 and by drawing arrows on the ribbons show that there are two distinct types of cross-over corresponding to left- and right-handed helicity respectively.
   (b) Cut the ribbon all the way along its length so it becomes two interlinked ribbons of unequal width. Assuming the original ribbon had a flux Φ, calculate the helicity of the new configuration taking into account the helicity due to both twists and linkages and show that helicity is conserved.

6. Demonstrate that the sum of twist plus writhe is conserved using two ropes with radius of approximately 1 cm having distinct colors and a broomstick with a radius of approximately 2 cm. Wrap one rope around the other four times in a left-handed sense to form a simple braid of the two ropes. The wrapping should have a long pitch so that the distance between turns is 10–20 cm. Then take this braid and wrap it in a left-handed sense around the broomstick. What happens to the wrapping of the one rope around the other? Explain this in terms of twist and writhe helicity and conservation of helicity. Next start with the two ropes parallel to each other and wrap them four times in a left-handed sense around the broomstick. Check to see if the ropes become wrapped around each other and, if so, in what sense are they wrapped around each other and how many times? Explain your results in terms of twist, writhe, and cross-over helicity.

12

Magnetic reconnection

12.1 Introduction

Section 2.6.4 established the fundamental concept underlying ideal MHD, namely that magnetic flux is frozen into the plasma. This concept means that the magnetic topology of an ideal MHD plasma cannot change because a change in magnetic topology would require a change of magnetic flux within the frame of the plasma. Chapter 10 showed that ideal MHD plasmas are susceptible to two distinct types of instabilities, pressure-driven and current-driven. Pressure-driven modes draw on free energy associated with heavy fluids stacked on top of light fluids in an effective gravitational field whereas current-driven instabilities draw on free magnetic energy and involve the plasma attempting to increase its inductance in a flux-conserving manner. Both of these instabilities occur on the Alfvén time scale defined as some characteristic distance divided by v_A.

It is possible for an MHD equilibrium to be stable to all ideal MHD modes and yet not be in a lowest energy state. Because ideal MHD does not allow the topology to change, a plasma that is not initially in the lowest energy state will not be able to access this lowest energy state if the lowest energy state is topologically different from the initial state. However, the lowest energy state could be accessed by non-ideal modes, i.e., modes that violate the frozen-in flux condition, and so the available free energy could drive an instability involving these non-ideal modes. Magnetic reconnection is a non-ideal instability where the plasma is effectively ideal everywhere *except at a very thin boundary layer* where the ideal MHD frozen-in assumption fails so magnetic fields can leak across the plasma and change their topology. Even though this boundary layer is microscopically thin, the reconnection and associated change in magnetic topology at the boundary layer allow the configuration to relax to a lower energy state. Magnetic reconnection thus describes how a very slight departure from ideal MHD leads to important new behavior.

The simplest reconnection model is obtained by including finite resistivity in the MHD description, but more elaborate non-ideal physics is also possible due to two-fluid or Vlasov effects omitted from MHD (e.g., finite electron mass, Landau damping). Resistive reconnection is much slower than ideal MHD modes because it involves the diffusion (i.e., leakage) of magnetic field across the plasma in some very limited spatial region where ideal MHD breaks down. Although including finite resistivity in an MHD description is the simplest way to invoke non-ideal physics enabling magnetic diffusion, the actual resistivity of most relevant laboratory and space plasmas is inadequate by orders of magnitude for observed reconnection rates to be explained by resistive MHD. Observations and computer models show that the observed rapid reconnection rates are due to complicated two-fluid and Vlasov effects, typically involving wave excitation and acceleration of particles to very high energies.

Nevertheless, analysis of resistive MHD provides a good introductory overview of the main issues involved in magnetic reconnection. In fact, even though it is known that two-fluid and Vlasov models would be more appropriate, resistive MHD is often used as a convenient approximate description for actual situations by invoking a suitably large ad hoc "anomalous" resistivity. This is a hybrid approach that essentially separates the reconnection physics (e.g., location, rate, topology) from the question of what causes the dissipation enabling the reconnection. The anomalous resistivity approach simply assumes there exists some complicated collisionless mechanism (e.g., some form of wave turbulence) for slowing down electrons relative to ions and leaves the issue of determining this mechanism as a separate problem. Investigation of collisionless reconnection is an active research area at the time of writing and a single straightforward theory does not exist. Resistive MHD analysis should therefore be considered as a first step in attacking the problem and a means for identifying relevant issues, but not as a quantitatively accurate model for most real situations where complicated collisionless physics seems to be much more important than the effect of finite resistivity.

Magnetic reconnection (also known as tearing) is an intrinsically complicated boundary layer phenomenon because two very different length scales mutually interact. This imposes a strong burden on the mathematical description because the mathematics has to simultaneously characterize two different kinds of physics, namely (i) the diffusive physics of the non-ideal, microscopically thin reconnection layer and (ii) the ideal behavior of the macroscopic regions on either side of this thin layer. We shall begin the discussion with an analogy drawn from everyday experience and then use this analogy to argue why it is energetically favorable for a sheet current to break up into filaments. The time scale for the resistive MHD reconnection process will be estimated and shown to be much slower than ideal MHD physics. It will then be shown that the existence of current sheets corresponds to having β of order unity and so the current sheet model is not directly relevant to low-β plasmas. However, it will be shown that a modest generalization of the current sheet analysis to the situation of sheared magnetic fields provides a model for low-β situations.

12.2 Water-beading: an analogy to magnetic reconnection

Since magnetic reconnection is not a simple process, it is helpful to start by considering the dynamics of a somewhat analogous instability known from everyday experience, namely the process of water beading. The initial condition in water beading is shown in Fig. 12.1(a) and consists of a long, thin, two-dimensional, incompressible drop of water (approximate analog to the magnetic field) frictionally attached to a substrate (approximate analog to the plasma). The long, thin drop has surface tension "trying" to reduce the perimeter of the drop (this is analogous to the pinch force being interpreted as field line "tension" squeezing on a current channel). If the drop were not attached to the substrate then, as shown in Fig. 12.1(b), the surface tension would simply collapse the long, thin incompressible drop into a circular drop having an area equal to that of the initial long thin drop. Because of the frictional work involved in dragging the water over the substrate, this wholesale collapse is not energetically favorable and does not occur. On the other hand, if as shown in Fig. 12.1(c) the long, thin drop breaks up into a line of discrete segments (which does not involve significant dragging of

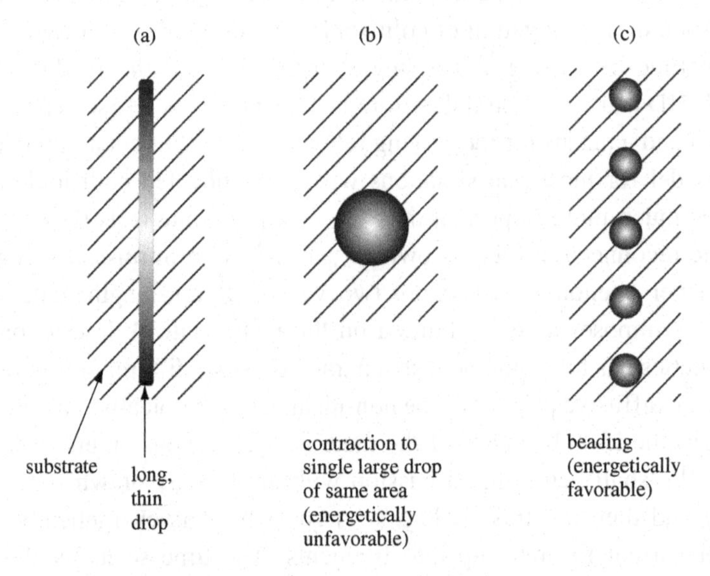

Fig. 12.1 (a) Initial long, thin drop, (b) contraction into big, round drop (too much scraping on substrate), (c) beading into chain of little drops.

water across the substrate), the surface tension of each line segment causes each discrete segment to contract in length and bulge in width until circular. Because only modest frictional dragging of water across the substrate is required to do this, the process can be energetically favorable, and so result in water beading.

12.3 Qualitative description of sheet current instability

Suppose an infinite-extent plasma has, as shown in Fig. 12.2(a), a thin sheet of current flowing in the z direction (out of the page). The sheet current is centered horizontally at $x = 0$ and extends vertically from $y = -\infty$ to $y = +\infty$. The current is uniform in both the y and the z directions and roughly corresponds to the initial long, thin water drop. This somewhat artificial situation can be considered as the Cartesian analog of a cylindrical shell current flowing in the z direction, localized at some radius $r = r_0$ and azimuthally symmetric. Thus, we identify r with x and y with θ.

Since all quantities depend on x only, Ampère's law reduces to

$$\mu_0 J_z(x) = \frac{\partial B_y}{\partial x}, \tag{12.1}$$

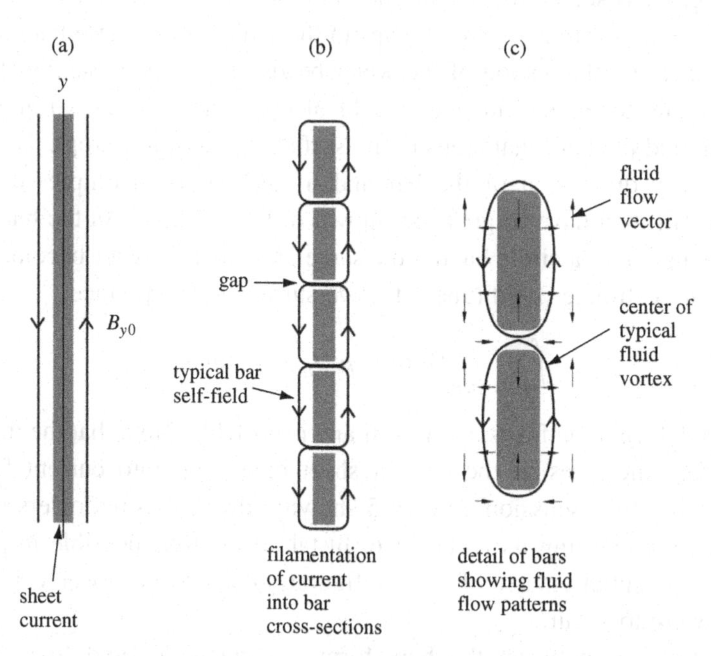

Fig. 12.2 (a) Initial sheet current (x axis is in horizontal direction, y axis is vertical, z axis is out of page); (b) breaking up into bar cross-section filaments; (c) detail showing fluid flow consists of set of small vortices that are antisymmetric in x.

which integrates to give

$$B_y(x) = B_y(0) + \int_0^x \mu_0 J_z(x')\mathrm{d}x'. \tag{12.2}$$

On the basis of symmetry the sheet current cannot generate a magnetic field at $x = 0$ so the field $B_y(0)$ must be entirely due to external currents. Let us assume for now that no external currents exist and set $B_y(0) = 0$; the situation of finite $B_y(0)$ will be treated later in Section 12.5. Thus $B_y(x)$ is a sheared magnetic field that is positive for $x > 0$ and negative for $x < 0$. The magnitude of $B_y(x)$ changes rapidly inside the current layer and becomes constant for $|x| \to \infty$.

This situation can be characterized analytically by using a magnetic field

$$B_y(x) = B\tanh(x/L), \tag{12.3}$$

where L is the scale width of the current layer. Substituting in Eq. (12.1) gives

$$J_z(x) = \frac{B}{\mu_0 L}\cosh^{-2}(x/L), \tag{12.4}$$

which is sharply peaked in the neighborhood of $x = 0$.

Suppose, as shown in Fig. 12.2(b), a perturbation is introduced whereby the current sheet cross-section breaks up into a number of bar-shaped structures each with current I_{bar} separated by small gaps in the y direction. These bars of current (i.e., filaments) are the analog of the water beads discussed in Section 12.2. The pinch force due to the self-magnetic field of each bar acts like an elastic band wrapped around the bar (analogous to the surface tension of a drop). This tension contracts the y dimension of the bar and, if the bar is incompressible, the x dimension will then have to grow, as shown in Fig. 12.2(c). As the bar deforms from a rectangle into a circle having the same area, its perimeter becomes shorter giving rise to a stronger field (and effective surface tension) since

$$\int_{perimeter} \mathbf{B}\cdot\mathrm{d}\mathbf{l} = \mu_0 I_{bar} = const. \tag{12.5}$$

Hence, this deformation feeds upon itself and is unstable. Note that the inductance of the system increases as the current sheet breaks up into current filaments, consistent with the discussion on p. 315 showing that a plasma lowers its potential energy by increasing its inductance. Instability is thus possible in principle, because the potential energy of the configuration would be lowered if the filamentation were to occur.

Before the gaps between the bars form, all magnetic field lines are open and straight, stretching from $y = -\infty$ to $y = +\infty$, whereas after the gaps form, some field lines circle the current filaments and so are closed. In order for the configuration to transform itself from the initial state to the filamented state, some magnetic field lines must move across plasma and thereby change the flux linked by plasma (this change in topology is approximately analogous to dragging the water drops across the substrate in the water-beading problem). This motion of field lines across the plasma is forbidden by ideal MHD because it would violate the frozen-in flux constraint of ideal MHD. Thus, even though this filamentation instability is energetically feasible, it is forbidden in an ideal plasma, i.e., in a plasma governed by ideal MHD.

The field lines have an "X" shape at the center of the gaps and the change in topology occurs at these X-points. The center of the bar is called an O-point since the field around this point has an O-shape. If some non-MHD mechanism exists that allows these X-points to develop, an instability involving this mechanism becomes possible because of the available free energy. Since finite resistivity allows magnetic field to diffuse across plasma, a resistive plasma ought to be susceptible to this instability and the instability should occur at locations where the magnetic field attachment to the plasma is weakest. The location of the weak spot can be deduced by examining Ohm's law for a nearly ideal plasma (i.e., small but finite resistivity),

$$\mathbf{E}+\mathbf{U}\times\mathbf{B}=\eta\mathbf{J}. \tag{12.6}$$

At most locations in the plasma, the two left-hand side terms are both much larger than the right-hand side resistive term and so plasma behavior is determined by these two left-hand terms balancing each other. This balancing causes the electric field in the plasma frame to be zero, which corresponds to the magnetic field being frozen into the plasma. However, if there exists a point, line, or plane of symmetry where either **U** or **B** vanishes, then in the vicinity of this special region, Eq. (12.6) reduces to

$$\mathbf{E}\approx\eta\mathbf{J}. \tag{12.7}$$

The curl of Eq. (12.7) gives

$$\frac{\partial\mathbf{B}}{\partial t}=\frac{\eta}{\mu_0}\nabla^2\mathbf{B}, \tag{12.8}$$

a *diffusion equation* for the magnetic field. Thus, in this *special region* the magnetic field is not frozen to the plasma and diffuses across the plasma.

Let us now examine the fluid flow pattern associated with the bars sketched in Fig. 12.2(b) as these bars contract in the y direction and expand in the x direction. As shown in Fig. 12.2(c), each bar has y-directed velocities pointing from the gap towards the bar center and x-directed velocities pointing out from the center of the bar. To complete the incompressible flow there must also be oppositely directed x and y velocities just outside the bar with the net result that there is a set of small fluid vortices that are antisymmetric in x and, in addition, have a

y-dependence 90° out of phase with respect to the y-direction periodicity of the bars. In particular, there is an outward x-directed velocity at the y location of the O-points and an inward x-directed velocity at the y location of the X-points. The fluid motion thus consists of a spatially periodic set of vortices that are antisymmetric with respect to $x = 0$. Each bar has a pair of opposite vortices for positive x and a mirror image pair of vortices for negative x, so there are four vortices for each bar.

12.4 Semi-quantitative estimate of the tearing process

An exact, self-consistent description of tearing and reconnection is beyond the capability of standard analytic methods because of the multi-scale nature of this process. However, the essential features (geometry, critical parameters, growth rate) and a reasonable physical understanding can be deduced using a semi-quantitative analysis, which outlines the basic physics and determines the relevant orders of magnitude. The starting point for this analysis involves solving for the vector potential associated with the magnetic field in Eq. (12.3), obtaining

$$A_z(x) = -\int^x B_y(x')dx' = -BL\ln\left[\cosh(x/L)\right] + const. \tag{12.9}$$

The constant is chosen to give $A_z = 0$ at $x = 0$ so

$$A_z(x) = -BL\ln\left[\cosh(x/L)\right]. \tag{12.10}$$

For $|x| \ll L$, $\cosh(x/L) \simeq 1 + x^2/L^2$ while for $|x| \gg L$, $\cosh(x/L) \simeq \exp(|x|/L)/2$. Thus, the limiting forms of the vector potential are

$$\lim_{|x| \ll L} A_z(x) = -\frac{Bx^2}{L} \tag{12.11}$$

and

$$\lim_{|x| \gg L} A_z(x) = -BL\left(\frac{|x|}{L} - \ln 2\right). \tag{12.12}$$

Near $x = 0$, A_z is an inverted parabola with a maximum value of zero, while far from $x = 0$, A_z is linear and becomes more negative with increasing displacement from $x = 0$. This behavior of the vector potential is consistent with the field being uniform far from $x = 0$, but reversing sign on going across $x = 0$. The behavior is also consistent with the relationship between the current density and the second derivative of the vector potential,

$$\mu_0 J_z = \frac{\partial B_y}{\partial x} = -\frac{\partial^2 A_z}{\partial x^2}. \tag{12.13}$$

Current density is therefore associated with curvature in $A_z(x)$ or, in a more extreme form, with a discontinuity in the first derivative of $A_z(x)$. Specifying the vector potential is sufficient to characterize the problem since the magnetic field and currents are the first and second derivatives of $A_z(x)$ respectively. This general idea can be extended to more complicated geometries if there is sufficient symmetry so that specification of an equilibrium flux profile uniquely gives both the equilibrium field and the current distribution.

The reconnection process is characterized by the MHD equation of motion

$$\rho \frac{\mathrm{d}\mathbf{U}}{\mathrm{d}t} = \mathbf{J} \times \mathbf{B} - \nabla P, \tag{12.14}$$

Faraday's law expressed as

$$\mathbf{E} = -\frac{\partial \mathbf{A}}{\partial t}, \tag{12.15}$$

Ampère's law

$$\nabla \times \mathbf{B} = \mu_0 \mathbf{J}, \tag{12.16}$$

and the resistive Ohm's law

$$\mathbf{E} + \mathbf{U} \times \mathbf{B} = \eta \mathbf{J}. \tag{12.17}$$

The analysis involves relating the velocity vortices to the linearized Ohm's law, and in particular to its z component

$$E_{z1} + U_{x1} B_{y0} = \eta J_{z1}. \tag{12.18}$$

The sense of the vortices sketched in Fig. 12.2(c) indicate that the velocity perturbation is uniform in the z direction and U_{x1} is *antisymmetric* with respect to x. Also, since the motion consists of vortices, there is no net divergence of the fluid velocity and so it is reasonable and appropriate to stipulate that the flow is *incompressible* with $\nabla \cdot \mathbf{U} = 0$. Since the perturbed current density is in the z direction, and since for straight geometries the vector potential is parallel to the current density, the perturbed vector potential may also be assumed to be in the z direction. Hence, both equilibrium and perturbed vector potentials are in the z direction and so the total magnetic field is related to the total vector potential by

$$\mathbf{B} = \nabla \times A_z \hat{z} = \nabla A_z \times \hat{z}. \tag{12.19}$$

Equation (12.18) can be recast using Eqs. (12.15) and (12.19) as an *induction equation*,

$$-\frac{\partial A_{z1}}{\partial t} - U_{x1} \frac{\partial A_{z0}}{\partial x} = \eta J_{z1}. \tag{12.20}$$

The current has the form

$$\mu_0 J_z = \hat{z}\cdot\nabla\times\mathbf{B} = \hat{z}\cdot\nabla\times\left(\nabla\times A_z\hat{z}\right) = \hat{z}\cdot\nabla\times\left(\nabla A_z\times\hat{z}\right) = -\nabla_\perp^2 A_z, \tag{12.21}$$

where the subscript $\perp$ means perpendicular to $\hat{z}$. Thus, the induction equation becomes

$$\frac{\partial A_{z1}}{\partial t} + U_{x1}\frac{\partial A_{z0}}{\partial x} = \frac{\eta}{\mu_0}\nabla_\perp^2 A_{z1}. \tag{12.22}$$

To proceed, it is necessary to express the perturbed velocity U_{x1} in terms of A_{z1}; this relation is obtained from the equation of motion.

While we could just plow ahead and manipulate the equation of motion to obtain U_{x1} in terms of A_{z1}, it is more efficient to exploit the incompressibility relation. In two-dimensional hydrodynamics, incompressibility simplifies flow dynamics so that flow is described by two related scalars, the stream-function f and the vorticity Ω. For two-dimensional motion in the $x-y$ plane of interest here, the general incompressible velocity can be expressed as

$$\mathbf{U} = \nabla f\times\hat{z}, \tag{12.23}$$

since

$$\nabla\cdot\mathbf{U} = \nabla\cdot(\nabla f\times\hat{z}) = \hat{z}\cdot\nabla\times\nabla f = 0. \tag{12.24}$$

The vorticity is the curl of the velocity and because the velocity lies in the $x-y$ plane, the vorticity vector is in the z direction. The vorticity magnitude Ω is given by

$$\Omega = \hat{z}\cdot\nabla\times\mathbf{U} = \hat{z}\cdot\nabla\times(\nabla f\times\hat{z}) = \nabla\cdot[(\nabla f\times\hat{z})\times\hat{z}] = -\nabla_\perp^2 f, \tag{12.25}$$

where $\perp$ means perpendicular with respect to z. Given the vorticity, f can be found by solving the Poisson-like Eq. (12.25), and then, knowing f, the velocity can be evaluated using Eq. (12.23). Appropriate boundary conditions must be specified for both f and Ω; these boundary conditions are that the vorticity is antisymmetric in x and is only large in the vicinity of $x = 0$ as indicated in Fig. 12.2(c).

The curl of the equation of motion provides the vorticity evolution and also annihilates ∇P; this elimination of P from consideration is why the vorticity/stream-line method is a more efficient approach than direct solution of the equation of motion.

Let us now solve for U_{1x} following this procedure. The linearized equation of motion is

$$\rho_0\frac{\partial \mathbf{U}_1}{\partial t} = (\mathbf{J}\times\mathbf{B})_1 - \nabla P_1; \tag{12.26}$$

taking the curl and dotting with $\hat{z}$ gives

$$\begin{aligned}\rho_0 \frac{\partial \Omega_1}{\partial t} &= \hat{z}\cdot\nabla\times(\mathbf{J}\times\mathbf{B})_1 \\ &= \hat{z}\cdot\nabla\times\left[J_z\hat{z}\times\left(\nabla A_z\times\hat{z}\right)\right]_1 \\ &= \hat{z}\cdot\nabla\times\left(J_z\nabla A_z\right)_1 \\ &= \hat{z}\cdot\left(\nabla J_z\times\nabla A_z\right)_1 . \end{aligned} \tag{12.27}$$

Using Eq. (12.21) this can be written as

$$\frac{\partial \Omega_1}{\partial t} = \frac{1}{\mu_0\rho_0}\hat{z}\cdot\left[\nabla A_z\times\nabla\left(\nabla_\perp^2 A_z\right)\right]_1 . \tag{12.28}$$

From Fig. 12.2(c) it is expected that the vortices have significant amplitude only in the vicinity of where the current bars are deforming and that at large $|x|$ there will be negligible vorticity. Thus, it is assumed that the vorticity evolution equation has the following behavior:

1. Inner (tearing/reconnection) region: here it is assumed that the perturbation has much steeper gradients than the equilibrium so

$$\frac{|\nabla\left(\nabla_\perp^2 A_{z0}\right)|}{|\nabla A_{z0}|} \ll \frac{|\nabla\left(\nabla_\perp^2 A_{z1}\right)|}{|\nabla A_{z1}|}. \tag{12.29}$$

This allows Eq. (12.28) to be approximated as

$$\begin{aligned}\frac{\partial \Omega_1}{\partial t} &\simeq \frac{1}{\mu_0\rho_0}\hat{z}\cdot\left[\nabla A_{z0}\times\nabla\left(\nabla_\perp^2 A_{z1}\right)\right] \\ &= \frac{1}{\mu_0\rho_0}\frac{\mathrm{d}A_{z0}}{\mathrm{d}x}\frac{\partial}{\partial y}\left(\nabla_\perp^2 A_{z1}\right), \end{aligned} \tag{12.30}$$

which shows that J_{z1} crossed with B_{y0} generates vorticity. Since B_{y0} is antisymmetric with respect to x, the vortices have the assumed antisymmetry. Furthermore, because J_{z1} is symmetric with respect to x and localized in the vicinity of $x=0$ the vortices are localized to the vicinity of $x=0$.
2. Outer (ideal) region: here $\Omega_1 \simeq 0$ is assumed so Eq. (12.28) becomes

$$\frac{\mathrm{d}A_{z0}}{\mathrm{d}x}\frac{\partial}{\partial y}\left(\nabla_\perp^2 A_{z1}\right) - \frac{\partial A_{z1}}{\partial y}\frac{\mathrm{d}^3 A_{z0}}{\mathrm{d}x^3} = 0, \tag{12.31}$$

which is a specification for A_{z1} in the outer region for a given A_{z0}. Thus, it is effectively assumed that the outer perturbed field is force-free, i.e., $(\mathbf{J}\times\mathbf{B})_1 = 0$. This is consistent with there being no generation of vorticity in the outer region.

The perturbed quantities will now be assumed to have the space-time dependence

$$\begin{aligned} A_{z1} &= A_{z1}(x)\mathrm{e}^{iky+\gamma t} \\ \Omega_1 &= \Omega_1(x)\mathrm{e}^{iky+\gamma t} \end{aligned} \tag{12.32}$$

so Eq. (12.30) gives the inner region vorticity as

$$\Omega_1 = \frac{1}{\mu_0 \gamma \rho_0}\frac{\mathrm{d}A_{z0}}{\mathrm{d}x}ik\left(\nabla_\perp^2 A_{z1}\right) = -\frac{1}{\gamma\rho_0}\frac{\mathrm{d}A_{z0}}{\mathrm{d}x}ikJ_{z1}. \tag{12.33}$$

This satisfies all the geometric conditions noted earlier, namely the antisymmetric dependence on x, the localization near $x = 0$, and, consistent with Fig. 12.2(c), a y-periodicity 90° out of phase with the periodicity of J_{z1}.

Using Eq. (12.23), it is seen that

$$U_{x1} = \frac{\partial f_1}{\partial y} = ikf_1. \tag{12.34}$$

The stream-function f_1 is a solution of the Poisson-like system Eq. (12.25) and Eq. (12.33),

$$\frac{\partial^2 f_1}{\partial x^2} - k^2 f_1 = \frac{1}{\gamma\rho_0}\frac{\mathrm{d}A_{z0}}{\mathrm{d}x}ikJ_{z1}. \tag{12.35}$$

Since the perturbed current peaks at $x = 0$ and has a width of the order of ϵ, it may be characterized by the Gaussian profile

$$J_{z1} \simeq \frac{\lambda}{\epsilon\sqrt{\pi}}\mathrm{e}^{-x^2/\epsilon^2}, \tag{12.36}$$

where

$$\lambda = \int_{layer} J_{z1}\mathrm{d}x \tag{12.37}$$

is the total perturbed current in the tearing layer. The gradient of the vector potential can be written as

$$\frac{\mathrm{d}A_{z0}}{\mathrm{d}x} = -B_{y0}(x) \simeq -\frac{x}{L}B'_{y0}. \tag{12.38}$$

Assuming that the tearing layer is very narrow gives

$$\frac{\partial^2 f_1}{\partial x^2} \gg k^2 f_1 \tag{12.39}$$

so Eq. (12.35) becomes

$$\frac{\partial^2 f_1}{\partial x^2} = -\frac{B'_{y0}}{\gamma L\rho_0}ik\frac{\lambda}{\epsilon\sqrt{\pi}}x\mathrm{e}^{-x^2/\epsilon^2} = \frac{ikB'_{y0}}{2\gamma L\rho_0}\frac{\lambda\epsilon}{\sqrt{\pi}}\frac{\mathrm{d}}{\mathrm{d}x}\mathrm{e}^{-x^2/\epsilon^2}. \tag{12.40}$$

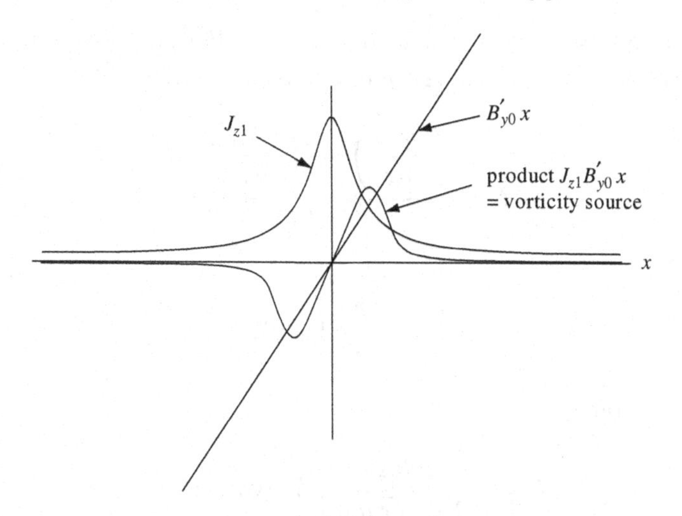

Fig. 12.3 Product of symmetric perturbed current with antisymmetric equilibrium field results in antisymmetric vorticity source localized near $x = 0$.

The profiles of J_{z1}, $B_{y0}(x)$, and their product (right-hand side of Eq. (12.40)) are shown in Fig. 12.3.

Integrating Eq. (12.40) with respect to x gives

$$\frac{\partial f_1}{\partial x} = \frac{\mathrm{i}kB'_{y0}}{2\gamma L\rho_0}\frac{\lambda\epsilon}{\sqrt{\pi}}\mathrm{e}^{-x^2/\epsilon^2}, \tag{12.41}$$

which incidentally gives $U_{1y} = -\partial f_1/\partial x$. Since it is desired to find the magnitude of U_{1x} in the region $x \sim \epsilon$, a rough "order of magnitude" integration of Eq. (12.41) in this region gives

$$f_1 \sim \frac{\mathrm{i}kB'_{yo}\lambda\epsilon^2}{2\gamma L\sqrt{\pi}\rho_0} sign(x) \text{ for } x \sim \epsilon \tag{12.42}$$

and so

$$U_{x1} \sim -\frac{k^2 B\lambda\epsilon^2}{2\gamma L\sqrt{\pi}\rho_0} sign(x) \text{ for } x \sim \epsilon. \tag{12.43}$$

The amplitude factor λ can be expressed using Eq. (12.21) as

$$\begin{aligned}\lambda &= -\frac{1}{\mu_0}\int_{layer}\nabla_\perp^2 A_{z1}\mathrm{d}x \\ &\simeq -\frac{1}{\mu_0}\int_{layer}\frac{\partial^2 A_{z1}}{\partial x^2}\mathrm{d}x \\ &= -\frac{1}{\mu_0}\left[\left(\frac{\partial A_{z1}}{\partial x}\right)_+ - \left(\frac{\partial A_{z1}}{\partial x}\right)_-\right],\end{aligned} \tag{12.44}$$

where the subscripts $\pm$ mean evaluated at $x = \pm\epsilon$. For purposes of joining to the outer ideal solution, the normalized jump derivative is defined as

$$\Delta' = \frac{\left(\frac{\partial A_{z1}}{\partial x}\right)_+ - \left(\frac{\partial A_{z1}}{\partial x}\right)_-}{A_{z1}(0)} \tag{12.45}$$

so

$$\lambda = -\frac{\Delta'}{\mu_0} A_{z1}(0). \tag{12.46}$$

The velocity becomes

$$U_{x1} \sim \frac{k^2 B \Delta' \epsilon^2}{2\gamma L \sqrt{\pi} \mu_0 \rho_0} A_{z1}(0) sign(x) \tag{12.47}$$

and using Eq. (12.36), the current density becomes

$$J_{z1} \sim -\frac{\Delta'}{\epsilon \mu_0 \sqrt{\pi}} A_{z1}(0). \tag{12.48}$$

We now repeat the induction equation, Eq. (12.20),

$$-\frac{\partial A_{z1}}{\partial t} - U_{x1} \frac{\partial A_{z0}}{\partial x} = \eta J_{z1} \tag{12.49}$$

and substitute for U_{x1}, J_{z1} and assume that $\partial A_{z0}/\partial x = -B_{y0} \simeq -B\epsilon/L$ in the tearing layer. This gives

$$\underset{\#1}{\gamma} - \underset{\#2}{\frac{k^2 B^2 \Delta' \epsilon^3}{2\gamma L^2 \sqrt{\pi} \mu_0 \rho_0}} = \underset{\#3}{\frac{\eta \Delta'}{\epsilon \mu_0 \sqrt{\pi}}}, \tag{12.50}$$

where the terms have been numbered for reference in the following discussion.

In the ideal plasma limit, terms #1 and #2 balance each other while term #3 is small; this gives the frozen-in condition. At exactly $x = 0$, term #2 vanishes and so terms #1 and #3 must balance each other, resulting in diffusion of the magnetic field. At the edge of the tearing layer, i.e., at the transition from the ideal limit to the diffusive limit, *all three terms are of the same size.* Thus, the three terms may be equated; this gives two equations that may be solved for γ and ϵ with Δ' as a parameter (Furth, Killeen, and Rosenbluth 1963). Equating terms #1 and #3 gives

$$\gamma = \frac{\eta \Delta'}{\epsilon \mu_0 \sqrt{\pi}} \tag{12.51}$$

while equating terms #2 and #3 gives

$$\gamma = \frac{(kB')^2 \epsilon^4}{2\eta\rho_0}, \tag{12.52}$$

where $B' = B/L$ is the derivative of the equilibrium field at $x = 0$. Equating these last two equations to eliminate γ gives the width of the tearing layer to be

$$\epsilon \simeq \left[\frac{2\eta^2 \rho_0 \Delta'}{\mu_0 \sqrt{\pi}\, (kB')^2} \right]^{1/5}. \tag{12.53}$$

Substituting ϵ back into Eq. (12.51) gives

$$\gamma = 0.55 \left(\Delta'\right)^{4/5} \left[\frac{\eta}{\mu_0} \right]^{3/5} \left[\frac{(kB')^2}{\rho_0 \mu_0} \right]^{1/5}. \tag{12.54}$$

This result can be put in a more physically intuitive form by defining characteristic times for ideal processes and for resistive processes. The characteristic time for ideal processes is the Alfvén time τ_A, defined as the time to move the characteristic length L when traveling at the Alfvén velocity, i.e.,

$$\tau_A^{-1} = \frac{v_A}{L} = \frac{\sqrt{B^2/\rho_0 \mu_0}}{L} = \sqrt{\frac{(B')^2}{\rho_0 \mu_0}}. \tag{12.55}$$

The Alfvén time is the characteristic time of ideal MHD and is typically a very fast time. The characteristic time for resistive processes τ_R is defined as the time to diffuse resistively a distance L, so

$$\tau_R^{-1} = \frac{\eta}{L^2 \mu_0}. \tag{12.56}$$

For nearly ideal plasmas the resistive time scale is very slow. Using these definitions, Eq. (12.54) can be written as (Furth, Killeen and Rosenbluth 1963)

$$\gamma = 0.55 \left(\Delta' L\right)^{4/5} (kL)^{2/5}\, \tau_R^{-3/5} \tau_A^{-2/5}. \tag{12.57}$$

All that is needed now is Δ'. This jump condition is found from Eq. (12.31) which gives the form of A_{z1} in the ideal region outside the tearing layer. This can be expressed as

$$\nabla_\perp^2 A_{z1} + \left[B_{y0}^{-1} \frac{\mathrm{d}^2 B_{y0}}{\mathrm{d}x^2} \right] A_{z1} = 0, \tag{12.58}$$

which shows that the equilibrium magnetic field acts like a "potential" for the perturbed vector potential "wavefunction." If boundary conditions are specified at large $|x|$ for the perturbed vector potential, then there will typically be a discontinuity in the first derivative of A_{z1} at $x = 0$; this discontinuity gives Δ'. The jump depends on the existence of a localized equilibrium current since

$$\frac{\mathrm{d}^2 B_{y0}}{\mathrm{d}x^2} = \mu_0 \frac{\mathrm{d}J_{z0}}{\mathrm{d}x}. \tag{12.59}$$

Equation (12.58) must in general be solved numerically.

The main result, as given by Eq. (12.57), is that if $\Delta' > 0$ an instability develops having a growth rate *intermediate* between the fast Alfvén time scale and the slow resistive time scale. Since a nearly ideal plasma is being considered, η is extremely small. The width of the tearing layer is therefore very narrow, since, as shown by Eq. (12.53), this width is proportional to $\eta^{2/5}$.

12.5 Generalization of tearing to sheared magnetic fields

The sheet current discussed above can occur in real situations but is a special case of the more general situation where the equilibrium magnetic field does not have a null, but instead is simply sheared. This means that the equilibrium magnetic field is straight, has components in both the y and z directions, and has a direction that is a varying function of x. The sheared situation thus has a uniform magnetic field in the z direction and, instead of the current being concentrated in a sheet, there is simply a non-uniform $B_{y0}(x)$.

There is a non-trivial difference between the special field-null equilibria and the more general types of equilibria where there is no field null. This difference occurs because equilibria involve balancing the magnetic and hydrodynamic pressures in such a way that $P + B^2/2\mu_0$ is continuous within the plasma and also across the plasma boundary. If a field null exists, then equilibrium consists of magnetic pressure $B^2/2\mu_0$ exterior to the null balancing hydrodynamic pressure P at the null and so a reconnection region involving a field null must have $\beta \simeq 1$. This situation occurs in the reconnection region associated with the Earth's magnetotail and has also been studied in certain laboratory plasma experiments (Trintchouk *et al.* 2003). However, in the more general case where there is no field null, B^2 is nearly continuous across the reconnection layer. This means that P will be small, in which case $\beta \ll 1$. Thus, the assumption of low β precludes the possibility of a field null and so precludes the possibility of the simple current sheet discussed in the previous section.

The more general situation where the equilibrium magnetic field has the form

$$\mathbf{B}_0 = B_{y0}(x)\hat{y} + B_{z0}\hat{z} \tag{12.60}$$

would thus be appropriate for low-β plasmas such as tokamaks, spheromaks, and the solar corona. A non-trivial feature of this situation is that, unlike the previously considered sheet current equilibrium, here $B_{y0}(x)$ does not vanish at any particular x. Instead, as will be seen later, what matters is the vanishing of $\mathbf{k}\cdot\mathbf{B}_0$. Equation (12.60) can be used as a slab representation of the straight cylindrical geometry equilibrium field

$$\mathbf{B}_0 = \nabla\psi_0(r)\times\nabla z + B_{z0}\hat{z}. \tag{12.61}$$

Equation (12.61) in turn can be thought of as the straight cylindrical approximation of a toroid with z corresponding to the toroidal angle (see Eq. (9.34)).

Figure 12.4 shows the magnetic field given by Eq. (12.60) as viewed in the $x=0$ plane. In the sheet current analysis discussed in the previous section, the perturbed vector potential pointed in the z direction and was periodic in the y direction. This corresponded to having $\mathbf{k}\cdot\mathbf{A}_1 = 0$ so that the wavevector was orthogonal to the vector potential and both were orthogonal to x. Other important properties were that $\mathbf{k}\cdot\mathbf{B}$ vanished at the reconnection layer, the fluid vorticity vector was pointed in the z direction, the perturbed currents and perturbed magnetic fields were such that $J_1/J_0 \gg B_1/B_0$ in the reconnection layer, and $J_1/J_0 \sim B_1/B_0$ in the exterior region.

These relationships and approximations are generalized here and, in particular, it is assumed all perturbed quantities have functional dependence $\sim g(x)\exp(ik_y y + ik_z z + \gamma t)$. As before, the vorticity equation is the curl of the linearized equation of motion, i.e.,

$$\rho_0\frac{\partial\Omega_1}{\partial t} = \nabla\times(\mathbf{J}_1\times\mathbf{B}_0 + \mathbf{J}_0\times\mathbf{B}_1) \tag{12.62}$$

and in the reconnection layer where $J_1/J_0 \gg B_1/B_0$ this becomes

$$\begin{aligned}\rho_0\frac{\partial\Omega_1}{\partial t} &= \nabla\times(\mathbf{J}_1\times\mathbf{B}_0)\\ &= \mathbf{B}_0\cdot\nabla\mathbf{J}_1 - \mathbf{J}_1\cdot\nabla\mathbf{B}_0\end{aligned} \tag{12.63}$$

since $\nabla\cdot\mathbf{B}_0 = 0$ and $\nabla\cdot\mathbf{J}_1 = 0$.

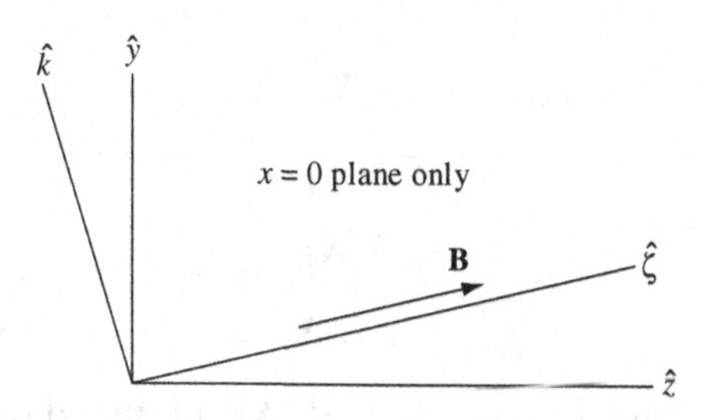

Fig. 12.4 Tilted coordinate system for general sheared field.

An essential feature of the reconnection topology is that the vorticity must be antisymmetric about the reconnection layer, as can be seen from examination of the fluid flow vectors in Fig. 12.2(c). Since the vorticity is created by the torque (i.e., the curl of the force), it is clear that the torque must be antisymmetric about the reconnection layer. As will be seen in the next paragraphs, the condition that the torque is antisymmetric does not imply that either $\mathbf{B}_0$ or $\mathbf{J}_1$ are antisymmetric, but rather implies some more subtle conditions.

Thus, if $x = 0$ is defined to be the location of the reconnection layer, then Ω_1 must be an odd function of x and therefore must vanish at $x = 0$. Hence, the right-hand side of Eq. (12.63) must vanish at $x = 0$ and, since there need be no particular functional relationship between $\mathbf{B}_0$ and its gradient, the two terms on the right-hand side of Eq. (12.63) must separately vanish. Therefore, one of the requirements is to choose the origin of the x axis such that $\mathbf{B}_0 \cdot \nabla \mathbf{J}_1 = 0$ at $x = 0$ or, equivalently,

$$\mathbf{k} \cdot \mathbf{B}_0 = 0 \text{ at } x = 0 \tag{12.64}$$

so that

$$k_y B_{y0}(0) + k_z B_{z0} = 0. \tag{12.65}$$

Having $\mathbf{k} \cdot \mathbf{B}$ vanish at the reconnection layer is physically reasonable, since finite $\mathbf{k} \cdot \mathbf{B}$ implies a periodic bending of the equilibrium field; such a bending absorbs energy and so is stabilizing. Having $\mathbf{k} \cdot \mathbf{B}$ vanish is like letting the instability cleave the system at a weak point so the instability can develop without requiring much free energy.

The second requirement for the right-hand side of Eq. (12.63) to vanish is to have $\mathbf{J}_1 \cdot \nabla \mathbf{B}_0 = 0$. If the flow is incompressible, then the perturbed magnetic field must be orthogonal to the equilibrium magnetic field and since the perturbed current is the curl of the perturbed magnetic field, the perturbed current must be parallel to the equilibrium magnetic field (this is essentially shear Alfvén wave physics). Since $\mathbf{B}_0$ is assumed straight, $\mathbf{B}_0 \cdot \nabla \mathbf{B}_0 = 0$ and, since $\mathbf{J}_1$ is parallel to $\mathbf{B}_0$, it is seen that $\mathbf{J}_1 \cdot \nabla \mathbf{B}_0 = (J_1/B_0)\,\mathbf{B}_0 \cdot \nabla \mathbf{B}_0 = 0$. Because $\mathbf{J}_1$ is parallel to $\mathbf{B}_0$, $\mathbf{J}_1$ must also be straight and so a Coulomb gauge vector potential will be parallel to $\mathbf{J}_1$, since

$$\begin{aligned} \mu_0 \mathbf{J}_1 &= \nabla \times \nabla \times \mathbf{A}_1 \\ &= \nabla \nabla \cdot \mathbf{A}_1 - \nabla^2 \mathbf{A}_1 \\ &= -\nabla^2 \mathbf{A}_1 \\ &= -\hat{\zeta} \nabla^2 A_1 \end{aligned} \tag{12.66}$$

can be satisfied by having both $\mathbf{J}_1$ and $\mathbf{A}_1$ in the direction of $\hat{\zeta}$, where $\hat{\zeta} = \mathbf{B}_0(0)/B_0(0)$ does not depend on position.

It is therefore assumed that $\mathbf{A}_1$ is parallel to $\mathbf{B}_0(0)$ and so

$$\mathbf{A}_1(x,y,z,t) = A_{1\zeta}(x)\hat{\zeta}\mathrm{e}^{\mathrm{i}k_y y + \mathrm{i}k_z z + \gamma t}. \tag{12.67}$$

Thus, $\mathbf{k}\cdot\mathbf{A}_1 = 0$ in this tilted coordinate system since $\mathbf{k}\cdot\hat{\zeta} = \mathbf{k}\cdot\mathbf{B}_0(0)/|\mathbf{B}_0(0)|$. Also $\nabla\cdot\mathbf{A}_1 = 0$ so the Coulomb gauge assumption is satisfied.

Reconsideration of the right-hand side of Eq. (12.63) shows that although the two terms both vanish at $x=0$ there is a difference in the geometrical dependence of these terms. In particular, since $\mathbf{J}_1$ is in the $\hat{\zeta}$ direction, it has no x component, whereas $\mathbf{B}_0$ depends only on x so

$$\mathbf{J}_1\cdot\nabla\mathbf{B}_0 = \left(J_{y1}\frac{\partial}{\partial y} + J_{z1}\frac{\partial}{\partial z}\right)\mathbf{B}_0(x) = 0. \tag{12.68}$$

Thus, Eq. (12.63) reduces to

$$\begin{aligned}\rho\frac{\partial\Omega_1}{\partial t} &= \mathbf{B}_0\cdot\nabla\mathbf{J}_1 \\ &= \mathrm{i}\left(\mathbf{k}\cdot\mathbf{B}_0\right)\mathbf{J}_1,\end{aligned} \tag{12.69}$$

where, by assumption, $\mathbf{k}\cdot\mathbf{B}_0$ vanishes at $x=0$.

Continuing this discussion of the ramifications of the antisymmetry of $\mathbf{k}\cdot\mathbf{B}_0$ about $x=0$, we now define an artificial reference magnetic field $\bar{\mathbf{B}}$ that is parallel to the real field at $x=0$, but has no shear (i.e., has no x dependence). The reference field therefore has the form

$$\bar{\mathbf{B}} = B_{y0}(0)\hat{y} + B_{z0}\hat{z} = B_0(0)\hat{\zeta} \text{ for all values of } x. \tag{12.70}$$

We now define $\mathbf{b}_0(x)$ as the difference between the real field and $\bar{\mathbf{B}}$ so that

$$\mathbf{B}_0(x) = \bar{\mathbf{B}} + \mathbf{b}_0(x). \tag{12.71}$$

Thus $\mathbf{b}_0(x)$ has the same x dependence as the B_{y0} field used for the sheet current instability in the previous section in that (i) it is antisymmetric about x and (ii) it is in the y direction. One way of thinking about this is to realize that $\mathbf{b}_0(x)$ is the component of $\mathbf{B}_0(x)$ that is antisymmetric about $x=0$. With this definition

$$\mathbf{k}\cdot\mathbf{B}_0 = \mathbf{k}\cdot\mathbf{b}_0(x) = k_y b_0 \tag{12.72}$$

and so Eq. (12.69) becomes

$$\gamma\rho_0\Omega_{\zeta 1} = \mathrm{i}k_y b_0 J_{\zeta 1}. \tag{12.73}$$

This equation provides the essence of the dynamics. It shows how the component of the magnetic field that is antisymmetric about $x=0$ creates fluid vortices that are antisymmetric with respect to x. In particular, since $\mathbf{b}_0(x)$ is an odd function of x and $J_{\zeta 1}$ is an even function of x, $\Omega_{\zeta 1}$ is an odd function of x.

The complete self-consistent description is obtained by, in addition, taking into account the induction equation, which shows how fluid motion acts to create perturbations of the electromagnetic field.

In the sheet current case the component of Ohm's law in the direction of symmetry of the perturbation was considered, i.e., the component in the z direction. Here, the corresponding symmetry direction for the perturbation is the ζ direction and so the relevant component of Ohm's law is the ζ component,

$$\hat{\zeta}\cdot[\mathbf{E}_1+\mathbf{U}_1\times\mathbf{B}_0(x)]=\hat{\zeta}\cdot\eta\mathbf{J}_1, \tag{12.74}$$

which becomes

$$-\frac{\partial A_{\zeta 1}}{\partial t}+\left(\hat{y}\times\hat{\zeta}\cdot\mathbf{U}_1\right)b_0(x)=\eta J_{\zeta 1}. \tag{12.75}$$

Since the vorticity vector lies along $\hat{\zeta}$, the incompressible velocity must be orthogonal to $\hat{\zeta}$ and so has the form

$$\mathbf{U}_1=\nabla f_1\times\hat{\zeta}. \tag{12.76}$$

Thus,

$$\hat{y}\times\hat{\zeta}\cdot\mathbf{U}_1=\left(\hat{y}\times\hat{\zeta}\right)\cdot\left(\nabla f_1\times\hat{\zeta}\right)=ik_y f_1. \tag{12.77}$$

Taking into account the ζ direction of the vorticity once again, it is seen that $\Omega_1=\Omega_{\zeta 1}\hat{\zeta}$, where

$$\Omega_{\zeta 1}=\hat{\zeta}\cdot\nabla\times\mathbf{U}_1=-\nabla_\perp^2 f_1 \tag{12.78}$$

and now $\perp$ means perpendicular to $\hat{\zeta}$.

Substituting for $\Omega_{\zeta 1}$ in Eq. (12.73) using Eq. (12.78) gives

$$\nabla_\perp^2 f_1=-\frac{ik_y b_0(x)J_{\zeta 1}}{\gamma\rho_0}; \tag{12.79}$$

this equation provides the essential dynamics of fluid vortices that are antisymmetric with respect to x and driven by the torque associated with the non-conservative $\mathbf{J}\times\mathbf{B}$ force. This is essentially the same as Eq. (12.35) and the rest of the analysis is the same as for the sheet current problem except that now k_y is used instead of k and b_0' is used instead of B_{y0}' since $b_0(x)$ only differs from $B_{y0}(x)$ by a constant, $b_0'=B_{y0}'$. Thus, using Eq. (12.54) the growth rate will be

$$\gamma=0.55\left(\Delta'\right)^{4/5}\left[\frac{\eta}{\mu_0}\right]^{3/5}\left[\frac{\left(k_y B_{y0}'\right)^2}{\rho_0\mu_0}\right]^{1/5}. \tag{12.80}$$

The global system has to be periodic in both the y and z direction in order for well-defined k_y and k_z to exist. In particular, the physical arrangement and dimensions of the global system determine the quantized spectra of k_y and k_z and so determine the allowed planes where $\mathbf{k}\cdot\mathbf{B}_0$ can vanish. As suggested earlier, the allowed planes can be considered as "cleavage" planes where the magnetic field can most easily become unglued from the plasma.

Let us express this result in the context of toroidal geometry such as that of a tokamak. This is done by letting B_{z0} correspond to the toroidal field B_ϕ and B_y correspond to B_θ the poloidal field. The Alfvén time is now defined in terms of B_ϕ as

$$\tau_A^{-1} = \frac{B_\phi}{a\sqrt{\rho_0\mu_0}}, \tag{12.81}$$

where a is the minor radius. The safety factor, a measure of the twist, is defined as

$$q = \frac{aB_\phi}{RB_\theta} \tag{12.82}$$

so

$$B_\theta = \frac{aB_\phi}{Rq} \tag{12.83}$$

and

$$B'_{y0} \to -\frac{aB_\phi}{Rq^2}q'. \tag{12.84}$$

Thus, it is possible to replace $k_y \to m/a$ and

$$\frac{\left(k_y B'_{y0}\right)^2}{\rho_0\mu_0} \to \left(\frac{ma}{Rq^2}q'\right)^2 \frac{1}{\tau_A}. \tag{12.85}$$

At the tearing layer $\mathbf{k}\cdot\mathbf{B} = 0$ or

$$\frac{m}{a}B_\theta + \frac{n}{R}B_\phi = 0 \tag{12.86}$$

so

$$q = -\frac{m}{n} \tag{12.87}$$

and Eq. (12.85) becomes

$$\frac{\left(k_y B'_{y0}\right)^2}{\rho_0\mu_0} \to \left(\frac{na}{R}\frac{q'}{q}\right)^2 \frac{1}{\tau_A}. \tag{12.88}$$

Thus, Eq. (12.80) becomes

$$\gamma = 0.55\left(\Delta' a\right)^{4/5} \tau_R^{-3/5}\tau_A^{-2/5}\left(\frac{na^2}{R}\frac{q'}{q}\right)^{2/5}, \tag{12.89}$$

where

$$\tau_R^{-1} = \frac{\eta}{a^2 \mu_0}. \tag{12.90}$$

Equation (12.89) shows that the essential source for the tearing instability is q'. From this point of view, the "free energy" is in the gradient of q and so, as the tearing mode uses up this free energy, the q profile will be flattened. The concept that gradients in q drive reconnection is closely related to the concept that relaxation is driven by gradients in the force-free parameter λ discussed with regards to Eq. (10.140). This is because λ is essentially the axial current I flowing through a flux tube with flux Φ. This can be seen by integrating Eq. (10.140) over the cross-section of a current-carrying flux tube to obtain

$$\frac{\mu_0 I}{\Phi} = \frac{\int \nabla \times \mathbf{B} \cdot d\mathbf{s}}{\int \mathbf{B} \cdot d\mathbf{s}} = \lambda. \tag{12.91}$$

If the torus is approximated as a straight cylinder so B_ϕ is the axial field and $B_\theta = \mu_0 I / 2\pi a$ is the azimuthal field then the axial flux is $\Phi = \pi a^2 B_\phi$ and

$$q = \frac{aB_\phi}{RB_\theta} = \frac{2\pi a^2 B_\phi}{R\mu_0 I} = \frac{2\Phi}{R\mu_0 I}. \tag{12.92}$$

We see that

$$q = \frac{2}{\lambda R} \tag{12.93}$$

so

$$q' = -\frac{2}{\lambda^2 R}\lambda'. \tag{12.94}$$

Thus, q and λ are inversely related and the concept that gradients in λ drive instability is essentially equivalent to the concept that gradients in q drive instability.

12.6 Magnetic islands

The tearing instability changes the topology of the magnetic field and causes the formation of magnetic "islands" (Rutherford 1973, Bateman 1978). The equilibrium magnetic field has the form

$$\mathbf{B} = \bar{\mathbf{B}} + b_0(x)\hat{y} = \bar{\mathbf{B}} + xB'_{y0}\hat{y} = \bar{\mathbf{B}} + \hat{z} \times \nabla A_{z0}(x), \tag{12.95}$$

where

$$A_{z0}(x) = \frac{x^2 B'_{y0}}{2}. \tag{12.96}$$

It is seen that

$$\mathbf{B}_\perp \cdot \nabla A_{z0} = 0, \tag{12.97}$$

where $\mathbf{B}_\perp$ is the component perpendicular to z. Thus, the surfaces $A_z = const.$ give the projection of the field lines in the plane perpendicular to z; these projections are the Cartesian geometry equivalents to the poloidal flux surfaces of toroidal geometry.

If it is assumed that $k_z \ll k_y$, so ζ is very nearly parallel to z, then it is seen that the tearing instability adds a perturbation to A_z giving

$$A_z(x, y) = \frac{x^2 B'_{y0}}{2} + A_{z1} \cos k_y y. \tag{12.98}$$

A sketch of a set of surfaces of constant $A_z(x, y)$ is shown in Fig. 12.5. These surfaces consist of (i) closed curves called islands, (ii) a separatrix that passes through the X-point, and (iii) open outer surfaces. On the $x = 0$ line an O-point corresponds to a maximum in $\cos k_y y$ (e.g., $k_y y = 0$) and an X-point corresponds to a minimum (e.g., $k_y y = \pm\pi$). The maximum width w of the separatrix can be calculated by noting that at the X-point (i.e., where $\cos(k_y y) = 1$ and $x = 0$)

$$A_z = 0 + A_{z1} \tag{12.99}$$

while at the point of maximum width on the separatrix (i.e., where $\cos(k_y y) = -1$ and $x = w/2$)

$$A_z = \frac{(w/2)^2 B'_{y0}}{2} - A_{z1}. \tag{12.100}$$

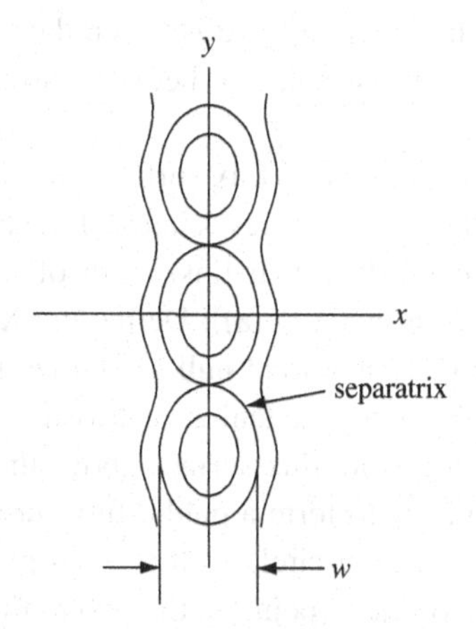

Fig. 12.5 Surfaces of constant A_z showing magnetic islands and separatrix with width w.

Since both these points are on the separatrix, they must have the same value of A_z, and so equating the last two equations gives the island width

$$w = 4\sqrt{\frac{A_{z1}}{B'_{y0}}} = 4\sqrt{\left|\frac{B_{x1}}{k_y B'_{y0}}\right|} \qquad (12.101)$$

using $B_{x1} = ik_y A_z$. In tokamak terminology we identify $B_{x1} \rightarrow B_{r1}$, where r is the minor radius, and using Eq. (12.84), Eq. (12.87), and $k_y \rightarrow m/a$ this becomes

$$w = 4\sqrt{\left|\frac{Rq}{nq'}\frac{B_{r1}}{B_\phi}\right|}. \qquad (12.102)$$

It is important to realize that the width of the island is much larger than the width of the tearing layer. Since particles tend to be attached to magnetic flux surfaces, the formation of islands means that particles can circulate around the island, thereby causing a flattening of the pressure gradient because the pressure is constant along a magnetic field line.

12.7 Assignments

1. Sweet–Parker type reconnection (Sweet 1958, Parker 1957, Trintchouk, Yamada, Ji, Kulsrud and Carter 2003). Consider the two identical flux-conserving current loops shown in Fig. 12.6(a). Because the system is axisymmetric, the magnetic field can be expressed as

$$\mathbf{B} = \frac{1}{2\pi}\nabla\psi \times \nabla\phi,$$

where ψ is the poloidal flux.

(a) Explain why there is an attractive force between the current loops (hint: consider the force between parallel currents or between north and south poles of two magnets).

(b) Define private flux to be a poloidal flux surface that links only one of the current loops (examples are the flux surfaces labeled 1 and 2 in Fig. 12.6(a)). Define public flux to be a flux surface that links both of the current loops (examples are flux surfaces 3, 4, 5 in Fig. 12.6(a)). Define the X-point to be the location in the $z = 0$ plane where there is a field null as shown in Fig. 12.6(a); let r_0 be the radius of the X-point. Show by sketching that as the two current loops approach each other in *vacuum*, a private flux surface above the midplane will merge with a private flux surface below to form a public flux surface.

(c) Show that the flux linked by a circle in the $z = 0$ plane with radius r_0 (i.e., the circle follows the locus of the X-point) is the public flux. Argue that if the current loops approach each other, this public flux will increase at the rate that private flux is converted into public flux. By integrating Faraday's law over the surface of

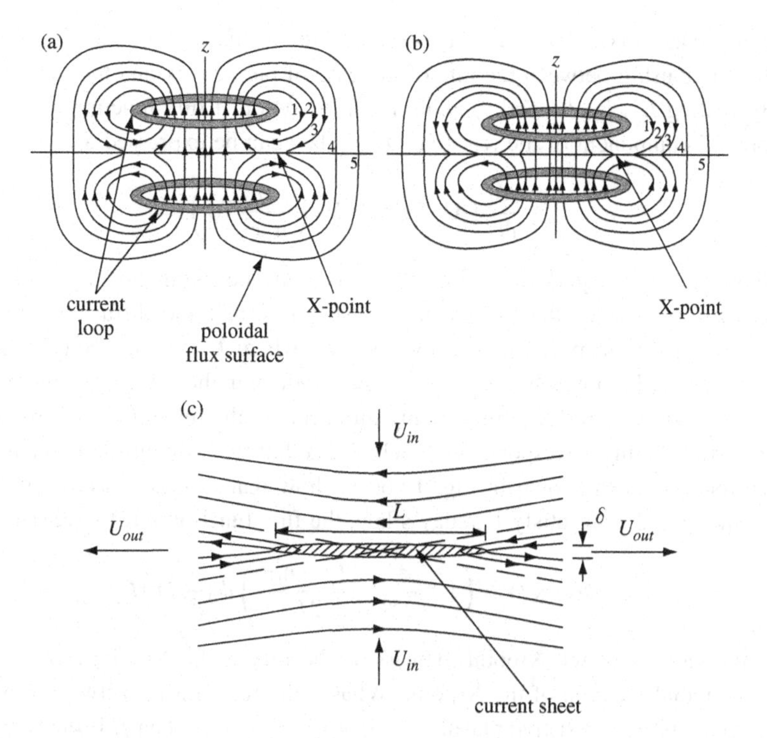

Fig. 12.6 (a) Two identical coaxial current loops with associated poloidal magnetic flux surfaces; (b) current loops approaching each other; (c) detail of X-point region of (b) showing inflows/outflows and current sheet (shaded region) having nominal length L and thickness δ.

this circle in the z-plane with radius r_0, show there exists a toroidal electric field at the X-point given by

$$E_\phi = -\frac{1}{2\pi r_0}\frac{\partial \psi_{public}}{\partial t},$$

where $\partial \psi_{public}/\partial t$ is the rate of increase of public flux.

(d) Now suppose that the vacuum is replaced by *perfectly conducting plasma* so that plasma is frozen to flux surfaces. Show from symmetry that the plasma velocity at the X-point must vanish. What is E_ϕ at the X-point if the two current loops are immersed in a plasma that satisfies the ideal Ohm's law,

$$\mathbf{E}+\mathbf{U}\times\mathbf{B}=0?$$

Can there be any conversion of private flux into public flux in an ideal plasma? Sketch contours of private and public flux when two current loops approach each other in an ideal plasma.

(e) Now suppose the two current loops are immersed in a non-ideal plasma that satisfies the resistive Ohm's law

$$\mathbf{E}+\mathbf{U}\times\mathbf{B}=\eta\mathbf{J}.$$

Can private flux be converted into public flux in this situation, and if so, what is the relationship between the rate of increase of public flux and the resistivity?

(f) By writing E_ϕ, B_r, B_z, and J_ϕ in terms of the poloidal flux function show that the toroidal component of the resistive Ohm's law can be expressed as

$$\frac{\partial\psi}{\partial t}+\mathbf{U}\cdot\nabla\psi = r^2\nabla\cdot\left(\frac{\eta}{\mu_0 r^2}\nabla\psi\right).$$

Show that this equation implies that flux convects with the plasma if $\eta = 0$, but diffuses across the plasma if $\mathbf{U} = 0$. Show that the diffusion increases if ψ develops a steep gradient at the location where $\mathbf{U} = 0$. By considering your response to (d) above, discuss how the flux gradient in the z direction might steepen in the vicinity of the X-point. Taking into account the relationship between $\mathbf{J}$ and ψ, show that this corresponds to developing a thin sheet of current in the midplane (current sheet) as shown in Fig. 12.6(c). Show that in order to have an X-point geometry and also satisfy Faraday's law, the flux function must have the form

$$\psi(r,z,t) = \left(1+\frac{z^2}{2\delta^2}-\frac{(r-r_0)^2}{2L^2}\right)\psi_0 - 2\pi r E_\phi t$$

in the vicinity of the X-point; here ψ_0 is the flux at the X-point at $t = 0$ and r_0 is the radial location of the X-point. What is the relationship between δ and L in vacuum and in a resistive plasma? Is δ larger or smaller than L if the two current loops are approaching each other? If $\delta < L$ is there a current in the vicinity of the X-point? What is the direction of this current with respect to the currents in the two current loops? Will the force on the current loops from the current in the current sheet accelerate or retard the motion of the two current loops towards each other? What is the direction of E_ϕ? Is ψ increasing or decreasing at the X-point?

(g) Suppose that the current sheet has a nominal thickness of δ in the z direction and a nominal width of L in the r direction as shown in Fig. 12.6(c). Let U_{in} be the nominal vertical velocity with which the plasma approaches the X-point and let U_{out} be the nominal horizontal velocity with which the plasma leaves the X-point region as shown in Fig. 12.6(c). If the plasma motion is incompressible show that

$$U_{in}L \simeq U_{out}\delta. \tag{12.103}$$

(h) Consider the transition from the ideal MHD form of Ohm's law to the form at the X-point. Argue that in the region of this transition the terms $\mathbf{U}\times\mathbf{B}$ and $\eta\mathbf{J}$ should have the same order of magnitude (this is essentially the same argument used to analyze Eq. (12.49)). Use this result and Ampère's law to find a relationship between ηJ, B_{in}, and δ. Use this to show

$$U_{in} \sim \eta/\mu_0\delta \tag{12.104}$$

and explain this result in terms of the convective velocity for motion of flux surfaces outside the current sheet and the "diffusive velocity" for motion of flux surfaces inside the current sheet.

(i) Using Eq. (9.42) show that the MHD force acting on the plasma is

$$\mathbf{F}_{MHD} = -\frac{\nabla\psi}{(2\pi)^2 \mu_0}\nabla\cdot\left(\frac{1}{r^2}\nabla\psi\right).$$

Sketch the direction of this force and indicate where this force is finite (hint: consider where $\mathbf{J}$ is finite). Estimate the work $\int \mathbf{F}\cdot d\mathbf{l}$ done on plasma accelerated through the current sheet by using relationships involving B_{in}, B_{out}, δ, and L. Use this estimate to show that

$$U_{out} \simeq v_{A,in}, \tag{12.105}$$

where $v_{A,in} = B_{in}/\sqrt{\mu_0\rho}$ is the Alfvén velocity at the input. (Hint: let $\Delta\psi$ be the jump in flux experienced by a fluid element as it moves across the current sheet. Note that $B_{out} \sim \Delta\psi/2\pi rL$ and $B_{in} \sim \Delta\psi/2\pi r\delta$, also note that $\nabla\cdot\left(r^{-2}\nabla\psi\right)$ can be expressed in terms of B_{in} and δ).

(j) By eliminating U_{in} and U_{out} between Eqs. (12.103), (12.104), and (12.105) show that

$$\frac{\delta}{L} = \frac{U_{in}}{v_{a,in}} = \frac{1}{\sqrt{S_{in}}},$$

where the inflow region Lundquist number S_{in} is defined as

$$S_{in} = \frac{\mu_0 v_{a,in} L}{\eta}.$$

(k) If the initial separation between the two current loops is also of the order of L, how much time is required for the two loops to merge, thus completing the reconnection process? Structures in the solar corona have scale lengths $L \sim 10^6$–10^8 m and it is observed that Alfvén wave disturbances propagate with velocities $v_A \sim 10^6$ ms^{-1}. The temperature of these structures is 10–100 eV. Assuming a nominal temperature of 50 eV, use Spitzer resistivity to calculate the nominal value of S_{in}. How long would it take for two solar current loops to merge if they are governed by the mechanism discussed here (what is the appropriate time unit to use: seconds, days, weeks, years, ...?). Actual current loops change topology on time scales ranging from minutes to hours – how does the observed time scale compare with the predictions of the resistive reconnection model?

13

Fokker–Planck theory of collisions

13.1 Introduction

Logically, this chapter ought to be located at the beginning of Chapter 2, just after the discussion of phase-space concepts. This chapter is not located there because the theory in this chapter is too advanced to be so close to the beginning of the book and its location near the beginning would have delayed the introduction of other important topics that do not need the detail of this chapter.

The discussion of collisions in Chapter 1 was very approximate. Collisions were shown to scale as an inverse power of temperature, but this was based on a "one size fits all" analysis since it was assumed that collision frequencies of slow and fast particles were nominally the same as that of a particle moving at the thermal velocity. Because the collision frequency scales as v^{-3}, it is quite dubious to assume that the collision rates of both super-thermal and sub-thermal particles can be well represented by a single collision frequency and a more careful averaging over velocities is clearly warranted. This careful averaging is provided by a Fokker–Planck analysis due to Rosenbluth, Macdonald, and Judd (1957). If this much more detailed analysis simply provided more accuracy, it would not be worth the considerable effort it requires except for occasional situations where high accuracy is important. However, the Fokker–Planck theory not only provides more accuracy, but also reveals new and important phenomena and, in particular, indicates when resistive MHD fails.

We begin by reviewing collisions between two particles. Suppose a test particle having mass m_T and charge q_T collides with a field particle having mass m_F and charge q_F. Since the electric field associated with a charge q is $\mathbf{E} = \hat{r}q/4\pi\varepsilon_0 r^2$, where r is the distance from the charge, the respective test and field particle equations of motion are

$$m_T \ddot{\mathbf{r}}_T = \frac{q_T q_F}{4\pi\varepsilon_0 |\mathbf{r}_T - \mathbf{r}_F|^3} (\mathbf{r}_T - \mathbf{r}_F) \tag{13.1a}$$

$$m_F \ddot{\mathbf{r}}_F = \frac{q_T q_F}{4\pi\varepsilon_0 |\mathbf{r}_T - \mathbf{r}_F|^3} (\mathbf{r}_F - \mathbf{r}_T). \tag{13.1b}$$

The center-of-mass vector is defined to be

$$\mathbf{R} = \frac{m_T \mathbf{r}_T + m_F \mathbf{r}_F}{m_T + m_F}. \tag{13.2}$$

Adding Eqs. (13.1a) and (13.1b) gives

$$(m_T + m_F)\ddot{\mathbf{R}} = m_T \ddot{\mathbf{r}}_T + m_F \ddot{\mathbf{r}}_F = 0 \tag{13.3}$$

showing that the center-of-mass velocity $\dot{\mathbf{R}}$ does not change as a result of the collision. If $\Delta\dot{\mathbf{r}}_T$ and $\Delta\dot{\mathbf{r}}_F$ are defined as the change in respective velocities of the two particles as result of a collision, then integration of Eq. (13.3) over the duration of the collision shows $\Delta\dot{\mathbf{r}}_T$ and $\Delta\dot{\mathbf{r}}_F$ are related by

$$(m_T + m_F)\Delta\dot{\mathbf{R}} = 0 = m_T \Delta\dot{\mathbf{r}}_T + m_F \Delta\dot{\mathbf{r}}_F. \tag{13.4}$$

It is useful to define the relative position vector

$$\mathbf{r} = \mathbf{r}_T - \mathbf{r}_F \tag{13.5}$$

and the reduced mass

$$\frac{1}{\mu} = \frac{1}{m_T} + \frac{1}{m_F}. \tag{13.6}$$

Then, dividing Eqs. (13.1a) and (13.1b) by their respective masses and taking the difference between the resulting equations gives an equation of motion for the relative velocity

$$\mu\ddot{\mathbf{r}} = \frac{q_T q_F}{4\pi\varepsilon_0 r^2}\hat{r}. \tag{13.7}$$

Solving Eqs. (13.2) and (13.5) for $\mathbf{r}_T$ and $\mathbf{r}_F$ gives

$$\mathbf{r}_F = \mathbf{R} - \frac{\mu}{m_F}\mathbf{r} \tag{13.8}$$

$$\mathbf{r}_T = \mathbf{R} + \frac{\mu}{m_T}\mathbf{r}$$

so the respective test and field velocities are

$$\dot{\mathbf{r}}_F = \dot{\mathbf{R}} - \frac{\mu}{m_F}\dot{\mathbf{r}} \tag{13.9}$$

$$\dot{\mathbf{r}}_T = \dot{\mathbf{R}} + \frac{\mu}{m_T}\dot{\mathbf{r}}. \tag{13.10}$$

Since $\Delta\dot{\mathbf{R}} = 0$, the change of the test and field particle velocities as measured in the lab frame can be related to the change in the relative velocity by

$$\Delta\dot{\mathbf{r}}_F = -\frac{\mu}{m_F}\Delta\dot{\mathbf{r}} \tag{13.11}$$

$$\Delta\dot{\mathbf{r}}_T = +\frac{\mu}{m_T}\Delta\dot{\mathbf{r}}. \tag{13.12}$$

The collision problem is first solved in the center-of-mass frame to find the change in the relative velocity and then the center-of-mass result is transformed to the lab frame to determine the change in the lab-frame velocity of the particles.

Let us now consider a many-particle point of view. Suppose a mono-energetic beam of particles impinges upon a background plasma. Several effects are expected to occur due to collisions between this beam and the background plasma. First, there will be a slowing down of the beam as it loses momentum due to collisions with the particles in the background plasma. Second, there should be a broadening of the beam particle velocity distribution since the collisions will also tend to randomize the velocity of the beam particles. Meanwhile, the background plasma should be heated and also should gain momentum due to the collisions. Eventually, the beam should be so slowed down and so spread out that it becomes indistinguishable from the background plasma, which will be warmer because of the energy transferred from the beam.

13.2 Statistical argument for the development of the Fokker–Planck equation

The Fokker–Planck theory (Rosenbluth, Macdonald and Judd 1957) is based on a logical argument describing how a distribution function attains its present form as a result of collisions that occurred at some earlier time. Consider a particle subject to random collisions that change its velocity. We define $F(\mathbf{v}, \Delta\mathbf{v})$ as the conditional probability that if this particle has a velocity $\mathbf{v}$ at time t, then at some later time $t + \Delta t$, collisions will have caused its velocity to become $\mathbf{v} + \Delta\mathbf{v}$.

Clearly at time $t + \Delta t$ the particle must have some velocity, so the sum of all the conditional probabilities must be unity, i.e., F must have the normalization

$$\int F(\mathbf{v}, \Delta\mathbf{v})\mathrm{d}\Delta\mathbf{v} = 1. \tag{13.13}$$

This definition of conditional probability can be used to show how a present distribution function $f(\mathbf{v}, t)$ comes to be the way it is because of the way it was at time $t - \Delta t$. To see how this works, consider all particles that have velocity $\mathbf{v}$ at the present time t. At some previous time $t - \Delta t$ these particles would have had velocities $\mathbf{v} - \Delta\mathbf{v}$, where $\Delta\mathbf{v}$ would have had a range of values. A particle with some specific $\mathbf{v}-\Delta\mathbf{v}$ at the previous time would need to have undergone a collision that increased its velocity by $\Delta\mathbf{v}$ in order to get to its present state. In order to get to the present *number* of particles having velocity $\mathbf{v}$ one has to take into account how many particles had the previous velocity $\mathbf{v}-\Delta\mathbf{v}$ and then multiply this number by the probability that a $\mathbf{v}-\Delta\mathbf{v}$ particle undergoes a collision that transforms it into a particle with velocity $\mathbf{v}$. This can be expressed mathematically as

$$f(\mathbf{v}, t) = \int f(\mathbf{v}-\Delta\mathbf{v}, t-\Delta t)F(\mathbf{v}-\Delta\mathbf{v}, \Delta\mathbf{v})\mathrm{d}\Delta\mathbf{v}, \tag{13.14}$$

which sums up all the possible ways for obtaining a given present velocity weighted by the probability of each of these ways occurring. This analysis presumes the present status depends only on what happened during the previous collision and so is independent of all events prior to the previous collision. A partial differential equation for f can be constructed by Taylor expanding the integrand as follows:

$$\begin{aligned} f(\mathbf{v}-\Delta\mathbf{v}, t-\Delta t)F(\mathbf{v}-\Delta\mathbf{v}, \Delta\mathbf{v}) = {} & f(\mathbf{v}, t)F(\mathbf{v}, \Delta\mathbf{v}) - \Delta t\frac{\partial f}{\partial t}F(\mathbf{v}, \Delta\mathbf{v}) \\ & -\Delta\mathbf{v}\cdot\frac{\partial}{\partial\mathbf{v}}(f(\mathbf{v}, t)F(\mathbf{v}, \Delta\mathbf{v})) \\ & +\frac{1}{2}\Delta\mathbf{v}\Delta\mathbf{v} : \frac{\partial}{\partial\mathbf{v}}\frac{\partial}{\partial\mathbf{v}}(f(\mathbf{v}, t)F(\mathbf{v}, \Delta\mathbf{v})). \end{aligned} \tag{13.15}$$

Substitution of Eq. (13.15) into Eq. (13.14) gives

$$\begin{aligned} f(\mathbf{v}, t) = {} & f(\mathbf{v}, t)\int F(\mathbf{v}, \Delta\mathbf{v})\mathrm{d}\Delta\mathbf{v} - \Delta t\frac{\partial f}{\partial t}\int F(\mathbf{v}, \Delta\mathbf{v})\mathrm{d}\Delta\mathbf{v} \\ & -\int \Delta\mathbf{v}\cdot\frac{\partial}{\partial\mathbf{v}}(f(\mathbf{v}, t)F(\mathbf{v}, \Delta\mathbf{v}))\,\mathrm{d}\Delta\mathbf{v} \\ & +\frac{1}{2}\int \Delta\mathbf{v}\Delta\mathbf{v} : \frac{\partial}{\partial\mathbf{v}}\frac{\partial}{\partial\mathbf{v}}(f(\mathbf{v}, t)F(\mathbf{v}, \Delta\mathbf{v}))\,\mathrm{d}\Delta\mathbf{v}, \end{aligned} \tag{13.16}$$

where in the top line advantage has been taken of $f(\mathbf{v}, t)$ not depending on $\Delta\mathbf{v}$. Upon invoking Eq. (13.13) this can be recast as

$$\frac{\partial f}{\partial t} = \left[\begin{array}{l} -\dfrac{\partial}{\partial\mathbf{v}}\cdot\left(\dfrac{\int \Delta\mathbf{v}F(\mathbf{v}, \Delta\mathbf{v})\mathrm{d}\Delta\mathbf{v}}{\Delta t}f(\mathbf{v}, t)\right) + \\ \dfrac{1}{2}\dfrac{\partial}{\partial\mathbf{v}}\dfrac{\partial}{\partial\mathbf{v}} : \left(\dfrac{\int \Delta\mathbf{v}\Delta\mathbf{v}F(\mathbf{v}, \Delta\mathbf{v})\mathrm{d}\Delta\mathbf{v}}{\Delta t}f(\mathbf{v}, t)\right) \end{array} \right]. \tag{13.17}$$

By defining

$$\left\langle \frac{\Delta \mathbf{v}}{\Delta t} \right\rangle = \frac{\int \Delta \mathbf{v} F(\mathbf{v}, \Delta \mathbf{v}) \mathrm{d}\Delta \mathbf{v}}{\Delta t} \tag{13.18}$$

and

$$\left\langle \frac{\Delta \mathbf{v} \Delta \mathbf{v}}{\Delta t} \right\rangle = \frac{\int \Delta \mathbf{v} \Delta \mathbf{v} F(\mathbf{v}, \Delta \mathbf{v}) \mathrm{d}\Delta \mathbf{v}}{\Delta t} \tag{13.19}$$

the standard form of the Fokker–Planck equation is obtained,

$$\frac{\partial f}{\partial t} = -\frac{\partial}{\partial \mathbf{v}} \cdot \left(\left\langle \frac{\Delta \mathbf{v}}{\Delta t} \right\rangle f(\mathbf{v}, t) \right) + \frac{1}{2} \frac{\partial}{\partial \mathbf{v}} \frac{\partial}{\partial \mathbf{v}} : \left(\left\langle \frac{\Delta \mathbf{v} \Delta \mathbf{v}}{\Delta t} \right\rangle f(\mathbf{v}, t) \right). \tag{13.20}$$

The first term gives the slowing down of a beam and is called the frictional term while the second term gives the spreading out of a beam and is called the diffusive term.

The goal now is to compute $\langle \Delta \mathbf{v}/\Delta t \rangle$ and $\langle \Delta \mathbf{v} \Delta \mathbf{v}/\Delta t \rangle$.To do this, it is necessary to consider all the ways collisions can cause the velocity of a specific test particle to change in a time Δt and then average all these possible values of $\Delta \mathbf{v}$ and $\Delta \mathbf{v} \Delta \mathbf{v}$ weighted according to their respective probability of occurrence. We will first evaluate $\langle \Delta \mathbf{v}/\Delta t \rangle$ and then do $\langle \Delta \mathbf{v} \Delta \mathbf{v}/\Delta t \rangle$. The problem will first be solved in the center-of-mass frame and then transformed back to the lab frame later. The center-of-mass frame analysis assumes that a particle with reduced mass μ, speed $v_{rel} = |\mathbf{v}_T - \mathbf{v}_F|$, and impact parameter b collides with a stationary scattering center located at the origin. This was discussed in Chapter 1 and the geometry was sketched in Fig. (1.3). The deflection angle θ for small-angle scattering was shown to be

$$\theta = \frac{q_T q_F}{2\pi \varepsilon_0 b \mu v_{rel}^2}. \tag{13.21}$$

The averaging procedure is done in two stages:

1. The effect of collisions on a test particle in time Δt is calculated for a specific field particle velocity. The functional dependence of $\Delta \mathbf{v}$ on the scattering angle θ is determined and then used to calculate the weighted average $\Delta \mathbf{v}$ for all possible impact parameters and all possible azimuthal angles of incidence. This result is then used to calculate the weighted average change in test particle velocity in the lab frame.
2. An averaging is then performed over all possible field particle velocities weighted according to their probability, i.e., weighted according to the field particle distribution function $f_F(\mathbf{v}_F)$.

Energy is conserved in the center-of-mass frame and since the energy before and after the collision is composed entirely of kinetic energy, the magnitude of v_{rel}^2 must be the same before and after the collision. The collision therefore simply rotates the relative velocity vector by the angle θ. Let z be the direction of the relative velocity before the collision and let $\hat{b}$ be the direction of the impact parameter. If $\mathbf{v}_{rel1}$, $\mathbf{v}_{rel2}$ are the respective relative velocities before and after the collision, then

$$\begin{aligned}\mathbf{v}_{rel1} &= v_{rel}\hat{z} \\ \mathbf{v}_{rel2} &= v_{rel}\hat{z}\cos\theta + v_{rel}\hat{b}\sin\theta.\end{aligned} \tag{13.22}$$

Thus,

$$\begin{aligned}\Delta\mathbf{v}_{rel} &= \mathbf{v}_{rel2} - \mathbf{v}_{rel1} \\ &= v_{rel}\hat{z}\,(\cos\theta - 1) + v_{rel}\hat{b}\sin\theta \\ &\simeq -\frac{v_{rel}\theta^2}{2}\hat{z} + v_{rel}\theta(\hat{x}\cos\phi + \hat{y}\sin\phi),\end{aligned} \tag{13.23}$$

where the scattering angle θ is assumed to be small and ϕ is the angle between $\hat{b}$ and the x axis. The negative sign for the z component of $\Delta\mathbf{v}_{rel}$ is the fundamental reason for the drag-producing nature of collisions and is simply a consequence of the relative velocity vector rotating during the course of the collision.

The cross-section associated with impact parameter b and the range of azimuthal impact angle $\mathrm{d}\phi$ is $b\mathrm{d}\phi\mathrm{d}b$. In time Δt the incident particle moves a distance $v\Delta t$ and so the volume swept out for this cross-section is $b\mathrm{d}\phi\mathrm{d}bv\Delta t$. The number of field particles with velocity $\mathbf{v}_F$ encountered in time Δt will be the density $f_F(\mathbf{v}_F)\mathrm{d}\mathbf{v}_F$ of these field particles multiplied by this volume, i.e., $f_F(\mathbf{v}_F)\mathrm{d}\mathbf{v}_F b\mathrm{d}\phi\mathrm{d}bv\Delta t$, and so the change in relative velocity for all possible impact parameters and all possible azimuthal angles for a given $\mathbf{v}_F$ will be

$$\Delta\mathbf{v}_{|\text{all } b, \text{ all } \phi} = v\Delta t f_F(\mathbf{v}_F)\mathrm{d}\mathbf{v}_F \int \mathrm{d}\phi \int \Delta\mathbf{v} b\mathrm{d}b. \tag{13.24}$$

The limits of integration for the azimuthal angle are from 0 to 2π and the limits of integration of the impact parameter are from the 90° impact parameter to the Debye length. On writing $\Delta\mathbf{v}$ in component form, this becomes

$$\Delta\mathbf{v}_{rel|\text{all } b, \text{ all } \phi} = v_{rel}\Delta t f_F(\mathbf{v}_F)\mathrm{d}\mathbf{v}_F \int_0^{2\pi} \mathrm{d}\phi \int_{b_{\pi/2}}^{\lambda_D} v_{rel}\left(\theta\cos\phi, \theta\sin\phi, -\theta^2/2\right) b\mathrm{d}b. \tag{13.25}$$

The x and y integrals vanish upon integration over ϕ so

$$\begin{aligned}
\Delta \mathbf{v}_{rel|\text{all } b, \text{ all } \phi} &= -\hat{z} v_{rel}^2 \Delta t f_F(\mathbf{v}_F) \mathrm{d}\mathbf{v}_F \pi \int_{b_{\pi/2}}^{\lambda_D} \theta^2 b db \\
&= -\hat{z} v_{rel}^2 \Delta t f_F(\mathbf{v}_F) \mathrm{d}\mathbf{v}_F \pi \int_{b_{\pi/2}}^{\lambda_D} \left(\frac{q_T q_F}{2\pi\varepsilon_0 b \mu v_{rel}^2} \right)^2 b db \\
&= -\hat{z} \Delta t f_F(\mathbf{v}_F) \mathrm{d}\mathbf{v}_F \frac{q_T^2 q_F^2}{4\pi\varepsilon_0^2 \mu^2 v_{rel}^2} \ln \Lambda,
\end{aligned} \tag{13.26}$$

where $\Lambda = \lambda_D / b_{\pi/2}$.

The above expression can be transformed using Eq. (13.12) to give the lab-frame change in test particle velocity averaged over all impact parameters and resulting from collisions with field particles having velocity $\mathbf{v}_F$,

$$\begin{aligned}
\Delta \mathbf{v}_{T|\text{all } b, \text{ all } \phi} &= -\hat{z} \Delta t f_F(\mathbf{v}_F) \mathrm{d}\mathbf{v}_F \frac{q_T^2 q_F^2}{4\pi\varepsilon_0^2 \mu m_T v_{rel}^2} \ln \Lambda \\
&= -\Delta t \frac{q_T^2 q_F^2 \ln \Lambda}{4\pi\varepsilon_0^2 \mu m_T} \frac{(\mathbf{v}_T - \mathbf{v}_F)}{|\mathbf{v}_T - \mathbf{v}_F|^3} f_F(\mathbf{v}_F) \mathrm{d}\mathbf{v}_F.
\end{aligned} \tag{13.27}$$

Next, summing over all field particle velocities gives the fully averaged change in lab-frame test particle velocity

$$\Delta \mathbf{v}_{T|\text{all } b, \text{ all } \phi, \text{ all } \mathbf{v}_F} = -\Delta t \frac{q_T^2 q_F^2 \ln \Lambda}{4\pi\varepsilon_0^2 \mu m_T} \int \frac{\mathbf{v}_{rel}}{v_{rel}^3} f_F(\mathbf{v}_F) \mathrm{d}\mathbf{v}_F. \tag{13.28}$$

The integrand on the right-hand side of Eq. (13.28) can be simplified using the relations

$$\frac{\partial v}{\partial \mathbf{v}} = \left(\hat{x} \frac{\partial}{\partial v_x} + \hat{y} \frac{\partial}{\partial v_y} + \hat{z} \frac{\partial}{\partial v_z} \right) \sqrt{v_x^2 + v_y^2 + v_z^2} = \frac{\mathbf{v}}{v} \tag{13.29}$$

and $\partial / \partial \mathbf{v}_T = \partial / \partial \mathbf{v}_{rel}$ to write

$$\frac{\partial v_{rel}}{\partial \mathbf{v}_T} = \frac{\mathbf{v}_{rel}}{v_{rel}} \tag{13.30}$$

and

$$\frac{\partial}{\partial \mathbf{v}_T} \frac{1}{v_{rel}} = -\frac{1}{v_{rel}^2} \frac{\mathbf{v}_{rel}}{v_{rel}}. \tag{13.31}$$

Equation (13.28) can therefore be written as

$$\Delta \mathbf{v}_{T|\text{all } b, \text{ all } \phi, \text{ all } \mathbf{v}_F} = \Delta t \frac{q_T^2 q_F^2 \ln \Lambda}{4\pi\varepsilon_0^2 \mu m_T} \frac{\partial}{\partial \mathbf{v}_T} \int \frac{1}{v_{rel}} f_F(\mathbf{v}_F) \mathrm{d}\mathbf{v}_F. \tag{13.32}$$

The $\partial/\partial \mathbf{v}_T$ has been factored out of the integral because $\mathbf{v}_T$ is independent of $\mathbf{v}_F$, the variable of integration. Since $\mathbf{v}_F$ is a dummy variable, it can be renamed $\mathbf{v}'$ and, since $\mathbf{v}_T$ is the velocity of the given incident particle, the subscript T can be dropped. After making these rearrangements we obtain the desired result,

$$\left\langle \frac{\Delta \mathbf{v}}{\Delta t} \right\rangle = \frac{q_T^2 q_F^2 \ln \Lambda}{4\pi\varepsilon_0^2 \mu m_T} \frac{\partial}{\partial \mathbf{v}} \int \frac{f_F(\mathbf{v}')}{|\mathbf{v}-\mathbf{v}'|} \mathrm{d}\mathbf{v}'. \tag{13.33}$$

The averaging procedure for $\Delta\mathbf{v}\Delta\mathbf{v}$ is performed in a similar manner, but instead starting with

$$\Delta\mathbf{v}\Delta\mathbf{v} = \left(\theta\cos\phi,\, \theta\sin\phi,\, -\theta^2/2\right)\left(\theta\cos\phi,\, \theta\sin\phi,\, -\theta^2/2\right), \tag{13.34}$$

in which case Eq. (13.25) is replaced by

$$\begin{aligned}\Delta\mathbf{v}_{rel}\Delta\mathbf{v}_{rel|\text{all } b, \text{ all } \phi} &= v_{rel}\Delta t f_F(\mathbf{v}_F)\mathrm{d}\mathbf{v}_F \int_0^{2\pi} \mathrm{d}\phi \int_{b_{\pi/2}}^{\lambda_D} b\mathrm{d}b \\ &\times v_{rel}^2 \left(\theta\cos\phi,\, \theta\sin\phi,\, -\theta^2/2\right)\left(\theta\cos\phi,\, \theta\sin\phi,\, -\theta^2/2\right).\end{aligned} \tag{13.35}$$

The terms that are linear in $\sin\phi$ or $\cos\phi$ vanish upon performing the ϕ integration. Also, the $\hat{z}\hat{z}$ term scales as θ^4 and so may be neglected relative to the $\hat{x}\hat{x}$ and $\hat{y}\hat{y}$ terms, which scale as θ^2. Thus, we obtain

$$\begin{aligned}\Delta\mathbf{v}_{rel}\Delta\mathbf{v}_{rel|\text{all } b,\phi} &= v_{rel}^3\Delta t f_F(\mathbf{v}_F)\mathrm{d}\mathbf{v}_F \int_0^{2\pi} \mathrm{d}\phi \int_{b_{\pi/2}}^{\lambda_D} \theta^2 \left(\hat{x}\hat{x}\cos^2\phi + \hat{y}\hat{y}\sin^2\phi\right) b\mathrm{d}b \\ &= v_{rel}^3\Delta t f_F(\mathbf{v}_F)\mathrm{d}\mathbf{v}_F \pi \left(\hat{x}\hat{x}+\hat{y}\hat{y}\right) \int_{b_{\pi/2}}^{\lambda_D} \theta^2 b\mathrm{d}b \\ &= \Delta t f_F(\mathbf{v}_F)\mathrm{d}\mathbf{v}_F \left(\hat{x}\hat{x}+\hat{y}\hat{y}\right) \frac{q_T^2 q_F^2 \ln\Lambda}{4\pi\varepsilon_0^2\mu^2 v_{rel}}.\end{aligned} \tag{13.36}$$

Since $\mathbf{v}_{rel}$ defines the z direction, we may write

$$\hat{x}\hat{x}+\hat{y}\hat{y} = \mathbf{I} - \frac{\mathbf{v}_{rel}\mathbf{v}_{rel}}{v_{rel}^2} \tag{13.37}$$

where $\mathbf{I}$ is the unit tensor. Thus,

$$\Delta\mathbf{v}_{rel}\Delta\mathbf{v}_{rel|\text{all } b, \text{ all } \phi} = \Delta t \frac{q_T^2 q_F^2 \ln\Lambda}{4\pi\varepsilon_0^2\mu^2} f_F(\mathbf{v}_F)\mathrm{d}\mathbf{v}_F \left(\frac{v_{rel}^2\mathbf{I} - \mathbf{v}_{rel}\mathbf{v}_{rel}}{v_{rel}^3}\right). \tag{13.38}$$

This can be transformed using Eq. (13.12) to give

$$\Delta\mathbf{v}_T\Delta\mathbf{v}_{T|\text{all } b, \text{ all } \phi} = \Delta t \frac{q_T^2 q_F^2 \ln\Lambda}{4\pi\varepsilon_0^2 m_T^2} f_F(\mathbf{v}_F)\mathrm{d}\mathbf{v}_F \left(\frac{v_{rel}^2\mathbf{I} - \mathbf{v}_{rel}\mathbf{v}_{rel}}{v_{rel}^3}\right). \tag{13.39}$$

The tensor may be expressed in a simpler form by noting

$$\begin{aligned}\frac{\partial}{\partial \mathbf{v}_T}\frac{\partial v_{rel}}{\partial \mathbf{v}_T} &= \frac{\partial}{\partial \mathbf{v}_T}\frac{\mathbf{v}_{rel}}{v_{rel}} \\ &= \frac{\mathbf{I}}{v_{rel}} - \frac{\mathbf{v}_{rel}}{v_{rel}^2}\frac{\mathbf{v}_{rel}}{v_{rel}} \\ &= \frac{v_{rel}^2\mathbf{I} - \mathbf{v}_{rel}\mathbf{v}_{rel}}{v_{rel}^3}. \end{aligned} \tag{13.40}$$

Inserting Eq. (13.40) in Eq. (13.39) and then integrating over all the field particles gives

$$\left\langle \frac{\Delta \mathbf{v}_T \Delta \mathbf{v}_T}{\Delta t} \right\rangle = \frac{q_T^2 q_F^2 \ln \Lambda}{4\pi\varepsilon_0^2 m_T^2} \int f_F(\mathbf{v}_F) \mathrm{d}\mathbf{v}_F \frac{\partial}{\partial \mathbf{v}_T}\frac{\partial}{\partial \mathbf{v}_T} |\mathbf{v}_T - \mathbf{v}_F|. \tag{13.41}$$

However, as before, $\mathbf{v}_T$ is constant in the integrand and so $\partial/\partial\mathbf{v}_T$ may be factored from the integral. Dropping the subscript T from the velocity and renaming $\mathbf{v}_F$ to be the integration variable $\mathbf{v}'$, this can be rewritten as

$$\left\langle \frac{\Delta \mathbf{v} \Delta \mathbf{v}}{\Delta t} \right\rangle = \frac{q_T^2 q_F^2 \ln \Lambda}{4\pi\varepsilon_0^2 m_T^2} \frac{\partial}{\partial \mathbf{v}}\frac{\partial}{\partial \mathbf{v}} \int |\mathbf{v} - \mathbf{v}'| f_F(\mathbf{v}') \mathrm{d}\mathbf{v}'. \tag{13.42}$$

It is convenient to define the Rosenbluth potentials

$$g_F(\mathbf{v}) = \int |\mathbf{v} - \mathbf{v}'| f_F(\mathbf{v}') \mathrm{d}\mathbf{v}' \tag{13.43}$$

$$h_F(\mathbf{v}) = \frac{m_T}{\mu} \int \frac{f_F(\mathbf{v}')}{|\mathbf{v} - \mathbf{v}'|} \mathrm{d}\mathbf{v}', \tag{13.44}$$

in which case

$$\begin{aligned}\left\langle \frac{\Delta \mathbf{v}}{\Delta t} \right\rangle &= \frac{q_T^2 q_F^2 \ln \Lambda}{4\pi\varepsilon_0^2 m_T^2} \frac{\partial h_F}{\partial \mathbf{v}} \\ \left\langle \frac{\Delta \mathbf{v} \Delta \mathbf{v}}{\Delta t} \right\rangle &= \frac{q_T^2 q_F^2 \ln \Lambda}{4\pi\varepsilon_0^2 m_T^2} \frac{\partial^2 g_F}{\partial \mathbf{v}\partial \mathbf{v}}. \end{aligned} \tag{13.45}$$

The Fokker–Planck equation, Eq. (13.20), thus becomes

$$\frac{\partial f_T}{\partial t} = \sum_{F=i,e} \frac{q_T^2 q_F^2 \ln \Lambda}{4\pi\varepsilon_0^2 m_T^2} \left[-\frac{\partial}{\partial \mathbf{v}} \cdot \left(f_T \frac{\partial h_F}{\partial \mathbf{v}} \right) + \frac{1}{2}\frac{\partial}{\partial \mathbf{v}}\frac{\partial}{\partial \mathbf{v}} : \left(f_T \frac{\partial^2 g_F}{\partial \mathbf{v}\partial \mathbf{v}} \right) \right]. \tag{13.46}$$

The first term on the right-hand side is a friction term, which describes the slowing down of the mean velocity associated with f_T while the second term is an isotropization term, which describes the spreading out (i.e., diffusive broadening) of the velocity distribution described by f_T. The right-hand side of Eq. (13.46) thus gives the rate of change of the distribution function due to collisions and so is the correct quantity to put on the right-hand side of Eq. (2.12).

13.2.1 Slowing down

Mean velocity is defined by

$$\mathbf{u} = \frac{\int \mathbf{v} f \mathrm{d}\mathbf{v}}{n}, \tag{13.47}$$

where $n = \int f \mathrm{d}\mathbf{v}$. The rate of change of the mean velocity of species T is thus found by taking the first velocity moment of Eq. (13.46). Integration by parts on the right hand side terms respectively gives

$$-\int \mathbf{v}\frac{\partial}{\partial \mathbf{v}} \cdot \left(f_T \frac{\partial h_F}{\partial \mathbf{v}} \right) \mathrm{d}\mathbf{v} = \int f_T \frac{\partial h_F}{\partial \mathbf{v}} \mathrm{d}\mathbf{v} \tag{13.48}$$

and

$$\int \mathbf{v}\frac{\partial}{\partial \mathbf{v}}\frac{\partial}{\partial \mathbf{v}} : \left(f_T \frac{\partial^2 g_F}{\partial \mathbf{v} \partial \mathbf{v}} \right) \mathrm{d}\mathbf{v} = -\int \frac{\partial}{\partial \mathbf{v}} \cdot \left(f_T \frac{\partial^2 g_F}{\partial \mathbf{v} \partial \mathbf{v}} \right) \mathrm{d}\mathbf{v} = 0, \tag{13.49}$$

where Gauss' theorem has been used in Eq. (13.49). The first velocity moment of Eq. (13.46) is therefore

$$\frac{\partial \mathbf{u}_T}{\partial t} = \sum_{F=i,e} \frac{q_T^2 q_F^2 \ln \Lambda}{4\pi\varepsilon_0^2 n_T m_T^2} \int f_T \frac{\partial h_F}{\partial \mathbf{v}} \mathrm{d}\mathbf{v}. \tag{13.50}$$

Let us now suppose that the test particles consist of a mono-energetic beam so that

$$f_T(\mathbf{v}) = n_T \delta(\mathbf{v} - \mathbf{u}_0). \tag{13.51}$$

In this case Eq. (13.50) becomes

$$\frac{\partial \mathbf{u}_T}{\partial t} = \sum_{F=i,e} \frac{q_T^2 q_F^2 \ln \Lambda}{4\pi\varepsilon_0^2 m_T^2} \left(\frac{\partial h_F}{\partial \mathbf{v}} \right)_{\mathbf{v}=\mathbf{u}_0}. \tag{13.52}$$

Let us further suppose that the field particles have a Maxwellian distribution so that

$$f_F(\mathbf{v}) = n_F \left(\frac{m_F}{2\pi\kappa T_F} \right)^{3/2} \exp\left(-m_F v^2/2\kappa T_F\right) \tag{13.53}$$

and so, using Eq. (13.44),

$$h_F(\mathbf{v}) = \frac{n_F m_T}{\mu} \left(\frac{m_F}{2\pi\kappa T_F} \right)^{3/2} \int \frac{\exp\left(-m_F v'^2/2\kappa T_F\right)}{|\mathbf{v} - \mathbf{v}'|} \mathrm{d}\mathbf{v}'. \tag{13.54}$$

The velocity integral in Eq. (13.54) can be evaluated using standard means (see assignments) to obtain

$$h_F(\mathbf{v}) = \frac{n_F m_T}{\mu v} \operatorname{erf}\left(\sqrt{\frac{m_F}{2\kappa T_F}} v\right), \tag{13.55}$$

where

$$\operatorname{erf}(x) = \frac{2}{\sqrt{\pi}} \int_0^x \exp(-w^2) \mathrm{d}w \tag{13.56}$$

is the error function.

Thus, Eq. (13.52) becomes

$$\begin{aligned} \frac{\partial \mathbf{u}_T}{\partial t} = & \frac{n_i q_T^2 q_i^2 \ln \Lambda}{4\pi\varepsilon_0^2 m_T^2} \frac{m_T}{\mu_i} \left\{ \frac{\partial}{\partial \mathbf{v}} \left[v^{-1} \operatorname{erf}\left(\sqrt{\frac{m_i}{2\kappa T_i}} v\right) \right] \right\}_{\mathbf{v}=\mathbf{u}_0} \\ & + \frac{n_e q_T^2 q_e^2 \ln \Lambda}{4\pi\varepsilon_0^2 m_T^2} \frac{m_T}{\mu_e} \left\{ \frac{\partial}{\partial \mathbf{v}} \left[v^{-1} \operatorname{erf}\left(\sqrt{\frac{m_e}{2\kappa T_e}} v\right) \right] \right\}_{\mathbf{v}=\mathbf{u}_0}, \end{aligned} \tag{13.57}$$

where $\mu_{i,e}^{-1} = m_{i,e}^{-1} + m_T^{-1}$.

This can be further simplified by noting (i) quasi-neutrality implies

$$n_i Z = n_e, \tag{13.58}$$

where Z is the charge of the ions, (ii) the masses are related by

$$\frac{m_T}{\mu_{i,e}} = 1 + \frac{m_T}{m_{i,e}}, \tag{13.59}$$

and (iii) the velocity gradient of the error function must be in the direction of $\mathbf{u}_0$ using Eq. (13.29). Using these relationships and realizing that both the left and right sides are in the direction of $\mathbf{u}$, Eq. (13.57) becomes

$$\begin{aligned} \frac{\partial u}{\partial t} = & \frac{n_e e^2 \ln \Lambda}{4\pi\varepsilon_0^2} \frac{q_T^2}{m_T^2} \left[Z\left(1 + \frac{m_T}{m_i}\right) \frac{\mathrm{d}}{\mathrm{d}u} \left(\frac{\operatorname{erf}\left(\sqrt{\frac{m_i}{2\kappa T_i}} u\right)}{u} \right) \right. \\ & \left. + \left(1 + \frac{m_T}{m_e}\right) \frac{\mathrm{d}}{\mathrm{d}u} \left(\frac{\operatorname{erf}\left(\sqrt{\frac{m_e}{2\kappa T_e}} u\right)}{u} \right) \right]. \end{aligned} \tag{13.60}$$

Let us define

$$\xi_{i,e} = \sqrt{\frac{m_{i,e}}{2\kappa T_{i,e}}} u, \tag{13.61}$$

which is the ratio of the beam velocity to the thermal velocity of the ions or electrons. The slowing down relationship can then be expressed as

$$\frac{\partial u}{\partial t} = \frac{n_e e^2 \ln \Lambda}{4\pi\varepsilon_0^2} \frac{q_T^2}{m_T^2} \left[Z \left(1 + \frac{m_T}{m_i}\right) \frac{m_i}{2\kappa T_i} \frac{\mathrm{d}}{\mathrm{d}\xi_i} \left(\frac{\mathrm{erf}\,(\xi_i)}{\xi_i} \right) + \left(1 + \frac{m_T}{m_e}\right) \frac{m_e}{2\kappa T_e} \frac{\mathrm{d}}{\mathrm{d}\xi_e} \left(\frac{\mathrm{erf}\,(\xi_e)}{\xi_e} \right) \right]; \tag{13.62}$$

the above expression describes the slowing down of a beam of mono-energetic test particles due to collisions with the electrons and ions in a background plasma.

The derivative $(\mathrm{erf}(x)/x)'$ in Eq. (13.62) is always negative so the right-hand side of Eq. (13.62) describes a frictional drag on the test particle beam. Figure 13.1 plots $-(\mathrm{erf}(x)/x)'$ as a function of x; this quantity has a maximum at $x = 0.9$ indicating that frictional drag is an increasing function of x when $x < 0.9$ but when $x > 0.9$ the drag is a decreasing function of x. The $x < 0.9$ situation is consistent with ordinary notions of friction, but the $x > 0.9$ situation is like being on a "slippery slope" since the faster the particle goes, the less friction it encounters. Examination of this figure suggests $-(\mathrm{erf}(x)/x)'$ depends linearly on x well to the left of the maximum but varies as an inverse power of x well to the right of the maximum.

This conjecture is verified by noting that Eq. (13.56) has the limiting values

$$\lim_{x \ll 1} \mathrm{erf}(x) \simeq \frac{2}{\sqrt{\pi}} \int_0^x (1 - w^2) \mathrm{d}w = \frac{2}{\sqrt{\pi}} \left(x - \frac{x^3}{3} \right)$$
$$\lim_{x \gg 1} \mathrm{erf}(x) \simeq 1 \tag{13.63}$$

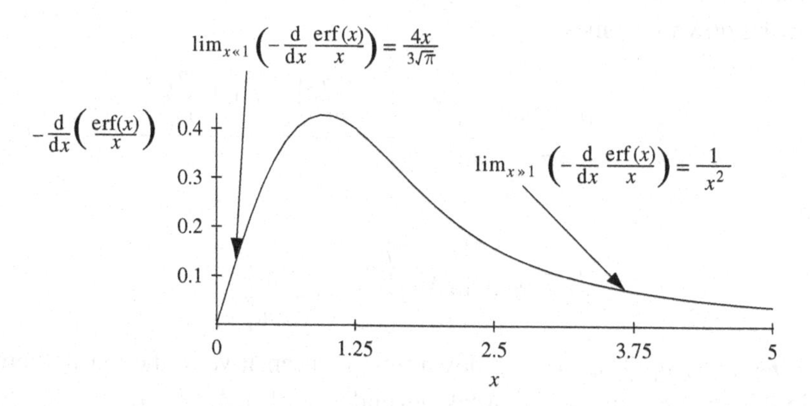

Fig. 13.1 Plot of $-(x^{-1}\,\mathrm{erf}(x))'$ vs. x.

and so, as indicated in the figure,

$$\lim_{x\ll 1}\left(-\frac{\mathrm{d}}{\mathrm{d}x}\frac{\mathrm{erf}(x)}{x}\right)=\frac{4x}{3\sqrt{\pi}} \tag{13.64a}$$

$$\lim_{x\gg 1}\left(-\frac{\mathrm{d}}{\mathrm{d}x}\frac{\mathrm{erf}(x)}{x}\right)=\frac{1}{x^2}. \tag{13.64b}$$

The existence of these two different asymptotic limits and their dependence on the ratio of the test particle velocity (i.e., beam velocity) to the thermal velocity of the particles in the background plasma indicates the existence of three distinct regimes, namely:

1. Test particle is much faster than both ion and electron thermal velocities, i.e., $\xi_i, \xi_e \gg 1$.
2. Test particle is much slower than both ion and electron thermal velocities, i.e., $\xi_i, \xi_e \ll 1$.
3. Test particle is much faster than the ion thermal velocity, but much slower than the electron thermal velocity, i.e., $\xi_i \gg 1$ and $\xi_e \ll 1$.

A nominal slowing-down time can be defined by writing the generic slowing down equation

$$\frac{\partial u}{\partial t}=-\frac{u}{\tau_s} \tag{13.65}$$

so

$$\tau_s=-\frac{u}{\partial u/\partial t}. \tag{13.66}$$

This can be used to compare slowing down of test particles in the three situations listed above.

Test particle faster than both electrons, ions

Here the limit given by Eq. (13.64b) is used for both electrons and ions so that the slowing down becomes

$$\frac{\partial u}{\partial t}=-\frac{n_e e^2\ln\Lambda}{4\pi\varepsilon_0^2}\frac{q_T^2}{m_T^2}\frac{Z\left(1+\dfrac{m_T}{m_i}\right)+\left(1+\dfrac{m_T}{m_e}\right)}{u^2} \tag{13.67}$$

and so

$$\tau_s\simeq\frac{4\pi\varepsilon_0^2}{n_e e^2\ln\Lambda}\frac{m_T^2}{q_T^2}\frac{u^3}{Z+1+\dfrac{m_T}{m_e}} \tag{13.68}$$

since $1/m_e \gg 1/m_i$. The slowing-down time is insensitive to the plasma temperature in this case except for the weak dependence of $\ln\Lambda$ on temperature. Equation (13.67) shows that if the test particle beam is composed of electrons, the ion friction scales as $Z/(Z+2)$ of the total friction while the electron friction scales as $2/(2+Z)$ of the total friction. In contrast, if the test particle is an ion, the friction is almost entirely from collisions with electrons. The slowing down time for ions is of the order of m_i/m_e times longer than the slowing down time for electrons having the same velocity.

Test particle slower than both electrons, ions

Here the limit given by Eq. (13.64a) is used for both electrons and ions so that the slowing-down equation becomes

$$\frac{\partial u}{\partial t} = -\frac{u}{3\sqrt{\pi}} \frac{n_e e^2 \ln \Lambda}{\pi \varepsilon_0^2} \frac{q_T^2}{m_T^2} \left(Z \left(1+\frac{m_T}{m_i}\right) \left(\frac{m_i}{2\kappa T_i}\right)^{3/2} + \left(1+\frac{m_T}{m_e}\right) \left(\frac{m_e}{2\kappa T_e}\right)^{3/2} \right) \tag{13.69}$$

and the slowing-down time becomes

$$\tau_s = \frac{3\sqrt{\pi}\pi\varepsilon_0^2 \, m_T^2}{n_e e^2 \ln \Lambda \; q_T^2} \frac{1}{\left(Z \left(1+\frac{m_T}{m_i}\right) \left(\frac{m_i}{2\kappa T_i}\right)^{3/2} + \left(1+\frac{m_T}{m_e}\right) \left(\frac{m_e}{2\kappa T_e}\right)^{3/2} \right)}. \tag{13.70}$$

The temperature terms scale as the inverse cube of the thermal velocity and so if the ion and electron temperatures are the same order of magnitude, then the ion contribution dominates. Thus, the slowing down is mainly done by collisions with ions and the slowing-down time is

$$\tau_s \simeq \frac{3\sqrt{\pi}\pi\varepsilon_0^2 \, m_T^2}{n_e e^2 \ln \Lambda \; q_T^2} \frac{(2\kappa T_i/m_i)^{3/2}}{Z\left(1+\frac{m_T}{m_i}\right)}. \tag{13.71}$$

The slowing down of a very slow beam is thus temperature-sensitive and, in particular, this sort of beam takes a long time to slow down in hot plasmas.

Intermediate case: beam faster than ions, slower than electrons

In this case, the slowing-down equation becomes

$$\frac{\partial u}{\partial t} = -\frac{n_e e^2 \ln \Lambda}{4\pi\varepsilon_0^2} \frac{q_T^2}{m_T^2} \left(Z\left(1+\frac{m_T}{m_i}\right)\frac{1}{u^2} + \left(1+\frac{m_T}{m_e}\right)\left(\frac{m_e}{2\kappa T_e}\right)^{3/2} \frac{4u}{3\sqrt{\pi}} \right) \tag{13.72}$$

and the slowing-down time is

$$\tau_s = \frac{4\pi\varepsilon_0^2 m_T^2}{n_e q_T^2 e^2 \ln\Lambda \left(\left(1+\frac{m_T}{m_i}\right)\frac{Z}{u^3} + \frac{4}{3\sqrt{\pi}}\left(1+\frac{m_T}{m_e}\right)\left(\frac{m_e}{2\kappa T_e}\right)^{3/2}\right)}. \tag{13.73}$$

13.3 Electrical resistivity

A uniform steady electric field imposed on a plasma accelerates ions and electrons in opposite directions. The accelerated particles will collide with other particles and this frictional drag will oppose the acceleration. A steady state might be achieved where the accelerating force due to the electric field balances the drag force due to collisions. Because the electric field causes equal and opposite momentum gains by the electrons and ions it does not alter the center-of-mass momentum of the entire plasma. Furthermore, electron–electron collisions cannot change the average momentum of the electrons nor can ion–ion collisions change the average momentum of the ions. The only way for the average electron momentum to change is by collisions with ions and vice versa. Thus, electrical resistivity must only depend on collisions between electrons and ions because electrical resistivity is a relation involving current, which is a function that depends on average electron momentum and average ion momentum.

We postulate that this balance between acceleration and drag results in an equilibrium where the electrons and ions have shifted Maxwellian distribution functions

$$f_i(\mathbf{v}) = \frac{n_i}{\pi^{3/2}(2\kappa T_i/m_i)^{3/2}} \exp\left(-m_i\left(\mathbf{v}-\mathbf{u}_i\right)^2/2\kappa T_i\right)$$
$$f_e(\mathbf{v}) = \frac{n_e}{\pi^{3/2}(2\kappa T_e/m_e)^{3/2}} \exp\left(-m_e\left(\mathbf{v}-\mathbf{u}_e\right)^2/2\kappa T_e\right). \tag{13.74}$$

Since $\mathbf{u}_i$ and $\mathbf{u}_e$ are the respective average ion and electron velocities, the current density is

$$\mathbf{J} = n_i q_i \mathbf{u}_i + n_e q_e \mathbf{u}_e. \tag{13.75}$$

It is convenient to transform to a frame moving with the electrons, i.e., with the velocity $\mathbf{u}_e$. In this frame the electron mean velocity will be zero and the ion mean velocity will be $\mathbf{u}_{rel} = \mathbf{u}_i - \mathbf{u}_e$. The current density remains the same because it is proportional to $\mathbf{u}_{rel}$, which is frame-independent. The distribution functions in this frame will be

$$f_i(\mathbf{v}) = \frac{n_i}{\pi^{3/2}(2\kappa T_i/m_i)^{3/2}} \exp\left(-m_i\left(\mathbf{v}-\mathbf{u}_{rel}\right)^2/2\kappa T_i\right)$$

$$f_e(\mathbf{v}) = \frac{n_e}{\pi^{3/2}(2\kappa T_e/m_e)^{3/2}} \exp\left(-m_e v^2/2\kappa T_e\right). \tag{13.76}$$

Since the ion thermal velocity is much smaller than the electron thermal velocity, the ion distribution function is much narrower than the electron distribution function and the ions can be considered as constituting a mono-energetic beam impinging upon the electrons. Using Eq. (13.57), the net force on this ion beam due to the combination of frictional drag on electrons and acceleration due to the electric field is

$$m_i \frac{\partial \mathbf{u}_{rel}}{\partial t} = \frac{n_e q_i^2 q_e^2 \ln \Lambda}{4\pi\varepsilon_0^2 \mu_e} \left\{ \frac{\partial}{\partial \mathbf{v}} \left[v^{-1} \operatorname{erf}\left(\sqrt{\frac{m_e}{2\kappa T_e}} v \right) \right] \right\}_{\mathbf{v}=\mathbf{u}_{rel}} + q_i \mathbf{E}. \tag{13.77}$$

In steady state the two terms on the right-hand side balance each other, in which case

$$\mathbf{E} = -\frac{n_e q_i e^2 \ln \Lambda}{4\pi\varepsilon_0^2 \mu_e} \left\{ \frac{\partial}{\partial \mathbf{v}} \left[v^{-1} \operatorname{erf}\left(\sqrt{\frac{m_e}{2\kappa T_e}} v \right) \right] \right\}_{\mathbf{v}=\mathbf{u}_{rel}}. \tag{13.78}$$

If u_{rel} is much smaller than the electron thermal velocity, then Eq. (13.64a) can be used to obtain

$$\mathbf{E} = \frac{n_e q_i e^2 \ln \Lambda}{3\sqrt{\pi}\pi\varepsilon_0^2 m_e} \frac{\mathbf{u}_{rel}}{(2\kappa T_e/m_e)^{3/2}} = \frac{Z e^2 m_e^{1/2} \ln \Lambda}{3\sqrt{\pi}\pi\varepsilon_0^2} \frac{\mathbf{J}}{(2\kappa T_e)^{3/2}}, \tag{13.79}$$

where $\mathbf{u}_{rel} = \mathbf{J}/n_e e$ and $\mu_e^{-1} = 1/m_i + 1/m_e \simeq 1/m_e$ have been used. Equation (13.79) can be used to define the electrical resistivity

$$\eta = \frac{Z e^2 m_e^{1/2}}{3\sqrt{\pi}\pi\varepsilon_0^2} \frac{\ln \Lambda}{(2\kappa T_e)^{3/2}} \tag{13.80}$$

so that

$$\mathbf{E} = \eta \mathbf{J}. \tag{13.81}$$

If the temperature is expressed in terms of electron volts then $\kappa = e$ and the resistivity is

$$\eta = 1.03 \times 10^{-4} \frac{Z \ln \Lambda}{T_e^{3/2}} \text{ Ohm m}. \tag{13.82}$$

This resistivity is more accurate than the rough calculation given by Eq. (1.25); the numerical coefficient here is smaller by about a factor of three.

13.4 Runaway electric field

The evaluation of the error function in Eq. (13.78) involved the assumption that u_{rel}, the relative drift between the electron and ion mean velocities, is much smaller than the electron thermal velocity. This assumption implies $-(x^{-1}\,\mathrm{erf}(x))' \simeq 4x/3\sqrt{\pi}$, so x must lie well to the left of 0.9 in the curve plotted in Fig. 13.1. Consideration of Fig. 13.1 shows that $-(x^{-1}\,\mathrm{erf}(x))'$ has a maximum value of 0.43, which occurs when $x = 0.9$ so that Eq. (13.78) cannot be satisfied if

$$E > \max\left\{-\frac{n_e q_i e^2 \ln\Lambda}{4\pi\varepsilon_0^2 m_e}\frac{\partial}{\partial v}\left[v^{-1}\,\mathrm{erf}\left(\sqrt{\frac{m_e}{2\kappa T_e}}v\right)\right]\right\} = 0.43\frac{n_e Z e^3 \ln\Lambda}{8\pi\varepsilon_0^2 \kappa T_e}. \tag{13.83}$$

When the electric field is large enough to satisfy this inequality, the entire concept of electrical resistivity fails (Dreicer 1959) because the underlying presumption of the resistivity calculation, namely a steady-state balance between drag and electric field acceleration, becomes false. If the electron temperature is expressed in terms of electron volts, attaining a balance between the acceleration due to the electric field and the deceleration due to frictional drag thus becomes impossible if

$$E > E_{Dreicer}, \tag{13.84}$$

where the Dreicer electric field is defined by

$$\begin{aligned} E_{Dreicer} &= 0.43\frac{n_e Z e^3 \ln\Lambda}{8\pi\varepsilon_0^2 \kappa T_e} \\ &= 5.6\times 10^{-18} n_e Z \frac{\ln\Lambda}{T_e}\ \mathrm{V/m}\,. \end{aligned} \tag{13.85}$$

If the electric field exceeds $E_{Dreicer}$ then the frictional drag lies to the right of the maximum in Fig. 13.1. No equilibrium is possible in this case as can be seen by considering the sequence of collisions of a nominal particle. The acceleration due to E between collisions causes the particle to go faster, but since it is to the right of the maximum, the particle has less drag when it goes faster. If the particle has less drag, then it will have a longer mean free path between collisions and so be accelerated to an even higher velocity. The particle velocity will therefore increase without bound if the system is infinite and uniform. In reality, the particle might exit the system if the system is finite or it might radiate energy. Very high, even relativistic velocities can easily develop in these runaway situations.

This runaway analysis is important to the characterization of magnetic reconnection. In particular, if the electric field in a reconnecting plasma satisfies Eq. (13.84), then the equation $\mathbf{E} = \eta\mathbf{J}$ cannot be used, in which case resistive models of reconnection become inappropriate and collisionless models must be used.

13.5 Assignments

1. Evaluate the integral in Eq. (13.54) and show that it leads to Eq. (13.55). Hint: since $\mathbf{v}$ is a fixed parameter in the integral, let the direction of $\mathbf{v}$ define the axis of a spherical polar coordinate system. Let $\boldsymbol{\xi} = \mathbf{v}' - \mathbf{v}$ and let θ be the angle between $\boldsymbol{\xi}$ and $\mathbf{v}$. Then note that $(v')^2 = \xi^2 + 2\boldsymbol{\xi}\cdot\mathbf{v} + v^2$ and $\mathrm{d}\mathbf{v}' = \mathrm{d}\boldsymbol{\xi}$. Express $\mathrm{d}\boldsymbol{\xi}$ in spherical polar coordinates and let $x = \cos\theta$.
2. Interaction of a low-density, fast electron beam with a cold background plasma. Assume that the velocity of a beam of electrons impinging on a plasma is much faster than the velocities of the background electrons and ions. The ions have charge Z so that, ignoring the beam density, the quasi-neutrality condition is $Zn_i = n_e$.

   (a) Show that the Rosenbluth h potential can be approximated as

$$h_F(\mathbf{v}) = \frac{m_T}{\mu}\int \frac{f_F(\mathbf{v}')}{|\mathbf{v}-\mathbf{v}'|}\mathrm{d}\mathbf{v}' \simeq \frac{m_e n_F}{v\mu},$$

   where

$$\frac{1}{\mu} = \frac{1}{m_e} + \frac{1}{m_F}.$$

   What is the form of $h_F(\mathbf{v})$ for beam electrons interacting with background plasma (i) electrons and (ii) ions?

   (b) By approximating

$$\left(v^2 - 2\mathbf{v}\cdot\mathbf{v}' + v'^2\right)^{1/2} = v\left(1 - \frac{\mathbf{v}\cdot\mathbf{v}'}{v^2} + \frac{v'^2}{2v^2}\right),$$

   where the beam velocity v is much larger than the background species velocity v', show that the Rosenbluth g potential can be approximated as

$$g_F(\mathbf{v}) = \int |\mathbf{v}-\mathbf{v}'| f_F(\mathbf{v}')\mathrm{d}\mathbf{v}' \simeq \left(v + \frac{1}{v}\frac{3\kappa T_F}{2m_F}\right)n_F.$$

   Hint: use symmetry arguments when considering the term involving $\mathbf{v}\cdot\mathbf{v}'$.

   (c) Assume the beam velocity is z-directed and let $F_T = \int \mathrm{d}^2 v_\perp f_T$ so F_T describes the projection of the 3-D beam distribution onto the z axis of velocity space. Since $v = v_z$ and $v_y = v_x \approx 0$ show it is possible to write

$$h_F(\mathbf{v}) \simeq \frac{m_e n_F}{v_z \mu}$$

$$g_F(\mathbf{v}) \simeq \left(v_z + \frac{1}{v_z}\frac{3\kappa T_F}{2m_F}\right)n_F.$$

Show that the Fokker–Planck equation can be written as

$$\frac{\partial F_T}{\partial t} \simeq \sum_{F=i,e} \frac{n_F q_T^2 q_F^2 \ln\Lambda}{4\pi\varepsilon_0^2 m_T^2} \left[\begin{array}{l} \dfrac{\partial}{\partial v_z}\left(F_T \dfrac{m_T}{\mu v_z^2}\right) + \\ \dfrac{1}{2}\dfrac{\partial^2}{\partial v_z^2}\left(F_T \dfrac{\partial^2}{\partial v_z^2}\left(v_z + \dfrac{3\kappa T_F}{2m_F v_z}\right)\right) \end{array} \right]$$

$$\simeq \sum_{F=i,e} \frac{n_F q_T^2 q_F^2 \ln\Lambda}{4\pi\varepsilon_0^2 m_T^2} \frac{\partial}{\partial v_z}\left[F_T \frac{m_T}{\mu v_z^2} + \frac{3}{2v_z^3}\frac{\kappa T_F}{m_F}\frac{\partial F_T}{\partial v_z}\right].$$

(d) Taking into account charge neutrality and $v_z \gg v_{Ti}$ show this can be recast as

$$\frac{\partial F_T}{\partial t} \simeq \frac{n_e e^4 \ln\Lambda}{4\pi\varepsilon_0^2 m_e^2} \frac{\partial}{\partial v_z}\left[F_T \frac{2+Z}{v_z^2} + \frac{3}{2v_z^3}\frac{\kappa T_e}{m_e}\frac{\partial F_T}{\partial v_z}\right].$$

(e) Show that a steady-state equilibrium can develop where, because of collisions with background electrons and ions, the fast beam distribution has the form

$$F_T \sim \exp\left(-\frac{(2+Z)\, m_e v_z^2}{3\kappa T_e}\right).$$

Is this consistent with the original assumption that the beam is fast compared to the background plasma?

3. An axisymmetric plasma has a magnetic field that can be expressed as

$$\mathbf{B} = \frac{1}{2\pi}\left(\nabla\psi \times \nabla\phi + \mu_0 I \nabla\phi\right),$$

where ϕ is the toroidal angle, $\psi(r, z)$ is the poloidal flux, and I is the current flowing through a circle of radius r at axial position z.

(a) Show that the toroidal component of the vector potential is

$$A_\phi(r, z, t) = \frac{1}{2\pi r}\psi(r, z, t).$$

(b) Assume that the plasma obeys the resistive Ohm's law

$$\mathbf{E} + \mathbf{U} \times \mathbf{B} = \eta \mathbf{J}$$

and assume that the plasma is stationary so that the toroidal component is simply

$$E_\phi = \eta J_\phi$$

and use Eq. (9.42) to show that

$$J_\phi = -\frac{r}{2\pi\mu_0}\nabla\cdot\left(\frac{1}{r^2}\nabla\psi\right)$$

so that the toroidal component of Ohm's law is

$$E_\phi = -\frac{r\eta}{2\pi\mu_0}\nabla\cdot\left(\frac{1}{r^2}\nabla\psi\right).$$

(c) Assuming classical resistivity $\eta \sim T_e^{-3/2}$ sketch the temperature dependence of $|E_\phi|$ as given above and also sketch the temperature dependence of $E_{Dreicer}$ as given by Eq. (13.85). For a plasma with given ψ and physical dimensions, at what electron temperature limit do runaway electrons develop (high or low temperature)? For a given temperature, do runaways develop with high ψ or low ψ? For a given temperature and flux, do runaways develop in a large device or in a small device? If the plasma density decays, will runaways develop?

4. Discuss qualitatively (no mathematics) what would happen if the electric field at the X-point of the Sweet–Parker reconnection configuration discussed in Assignment 1 of Chapter 12 exceeded $E_{Dreicer}$. By taking into account azimuthal symmetry discuss what constraints exist for the canonical angular momentum of a particle in the vicinity of the X-point. Discuss the particle orbits taking into account the results of Assignment 2 of Chapter 3. How might the particle motion relate to the excitation of waves and what sort of waves might be excited? How might this constitute an anomalous resistivity?

14
Wave–particle nonlinearities

14.1 Introduction

As seen in Chapters 4 through 7, linear models of plasma wave dynamics are straightforward and rich in descriptive power. The principle of superposition constitutes the very essence of linearity and underlies all concepts and methods inherent to linear models. In particular, the concepts of eigenmodes, eigenvalues, eigenvectors, orthogonality, and the method of integral transforms all ultimately depend on the validity of the principle of superposition.

While many important phenomena are adequately characterized by linear models, there nevertheless exist many other important phenomena where this is not so because the principle of superposition breaks down either partially or completely. Typically the phenomenon in question becomes amplitude-dependent above some critical amplitude threshold and then nonlinearity becomes important. Breakdown of the superposition principle means that modes with different eigenvalues (to the extent eigenvalues still exist) are no longer independent and start to interact with each other. These nonlinear interactions result when products of dependent variables become important in the system of equations. As an example, consider the product of two modes having respective frequencies ω_1 and ω_2. This product can be decomposed into sum and difference frequencies according to standard trigonometric identities, e.g.,

$$\cos(\omega_1 t)\cos(\omega_2 t) = \frac{\cos\left[(\omega_1+\omega_2)t\right]+\cos\left[(\omega_1-\omega_2)t\right]}{2}. \tag{14.1}$$

The beat waves at frequencies $\omega_1 \pm \omega_2$ can act as source terms driving oscillations at $\omega_1 \pm \omega_2$. In the special case where $\omega_1 = \omega_2$, the difference frequency is zero and the nonlinear product can act as a source term that modulates equilibrium parameters and thereby changes the mode dynamics. In particular, feedback loops can develop where modes affect their own stability properties. Another possibility occurs when the nonlinear product is very small, but happens to resonantly drive a linear mode. In this case even a weak nonlinear coupling between two modes can resonantly drive another linear mode to large amplitude. Similar beating can occur with spatial factors $\sim \exp(i\mathbf{k}\cdot\mathbf{x})$ so that the nonlinear product of modes with wavevectors $\mathbf{k}_1$ and $\mathbf{k}_2$ drives spatial oscillations with wavevectors $\mathbf{k}_1 \pm \mathbf{k}_2$. In analogy to quantum mechanics the momentum associated with a wave is found to be proportional to $\mathbf{k}$ and the energy proportional to ω. The beating together of two waves with respective space-time dependence $\exp(i\mathbf{k}_1\cdot\mathbf{x} - i\omega_1 t)$ and $\exp(i\mathbf{k}_2\cdot\mathbf{x} - i\omega_2 t)$ can then be interpreted in terms of conservation of wave energy and momentum,

$$\begin{aligned} \omega_3 &= \omega_1 + \omega_2 \\ \mathbf{k}_3 &= \mathbf{k}_1 + \mathbf{k}_2, \end{aligned} \tag{14.2}$$

where wave 3 is what results from beating together waves 1 and 2.

Because of the large variety of possible nonlinear effects, there are many ways to categorize nonlinear behavior. For example, one categorization is according to whether the nonlinearity involves velocity space (Vlasov nonlinearity) or position space (fluid nonlinearity). Vlasov nonlinearities are characterized by energy exchange between wave electric fields, resonant particles, and non-resonant particles. Fluid instabilities involve nonlinear mixing of two or more fluid modes and can be interpreted as one wave modulating the equilibrium seen by another wave. Another categorization is according to whether the nonlinearity is weak or strong. In situations where the nonlinear coupling is weak, linear theory may be invoked as a reasonable first approximation and then used as a basis for developing the nonlinear model. In situations where the nonlinear coupling is strong, linear assumptions fail completely and the nonlinear behavior must be addressed directly without any help from linear theory. Weak nonlinear theories can be further categorized into (i) mode-coupling models, where a small number of modes mutually interact in a coherent manner and (ii) weak turbulence models, where a statistically large number of modes mutually interact with random phase so some sort of averaging is required.

An important feature of nonlinear theory concerns energy. Because energy is a quadratic function of amplitude, energy does not satisfy the principle of superposition and so energy cannot be properly accounted for in linear models. Thus, nonlinear models are essential for tracking the flow of energy between modes and also between modes and the equilibrium.

This chapter will consider nonlinearities involving velocity space so a Vlasov description is required and wave–particle interactions are important. The following chapter will consider wave–wave nonlinearities and so will involve a fluid description of the plasma.

14.2 Vlasov nonlinearity and quasi-linear velocity space diffusion

14.2.1 Derivation of the quasi-linear diffusion equation

Quasi-linear theory (Vedenov, Velikhov, and Sagdeev 1962, Drummond and Pines 1962, Bernstein and Engelmann 1966), a surprisingly complete extension to the Landau model of plasma waves, shows how plasma waves can alter the equilibrium velocity distribution. In order to focus attention on the most essential features of this theory, a simplified situation will be considered where the plasma is assumed to be one dimensional, uniform, and unmagnetized. Furthermore, only electrostatic modes will be considered and, by assuming the ions are infinitely massive, ion motion will be neglected. Thus, the plasma is characterized by the coupled Vlasov and Poisson equations for electrons and the ions simply provide a static, uniform neutralizing background. It is assumed that the electron velocity distribution function can be decomposed into (i) a spatially independent equilibrium term, which is allowed to have a slow temporal variation, and (ii) a small, high-frequency perturbation having a space-time dependence resulting from a spectrum of linear plasma waves of the sort discussed in Section 5.3. Thus, it is assumed that the electron velocity distribution has the form

$$f(x, v, t) = f_0(v, t) + f_1(x, v, t) + f_2(x, v, t) + \ldots, \tag{14.3}$$

where it is implicit that the magnitude of terms with subscript n is of order ϵ^n, where $\epsilon \ll 1$. At $t = 0$, the terms f_n, where $n \geq 2$, all vanish because the perturbation was prescribed to be f_1 at $t = 0$. Other variables such as the electric field will have some kind of nonlinear dependence on the distribution function and so, for example, the electric field will have the form

$$E = E_0 + E_1 + E_2 + E_3 + \ldots, \tag{14.4}$$

where it is again implicit that the nth term is of order ϵ^n.

Since by assumption, $f_0(v, t)$ does not depend on position, it is convenient to define a velocity-normalized order zero distribution function

$$f_0(v, t) = n_0 \bar{f}_0(v, t) \tag{14.5}$$

so

$$\int \bar{f}_0(v, t) \mathrm{d}v = 1. \tag{14.6}$$

This definition causes n_0 to show up explicitly so terms such as $(e^2/m\varepsilon_0)\, \partial f_0/\partial v$ can be written as $\omega_p^2 \partial \bar{f}_0/\partial v$. A possible initial condition for $f_0(v, t)$, namely a monotonically decreasing velocity distribution, is shown in Fig. 14.1(a).

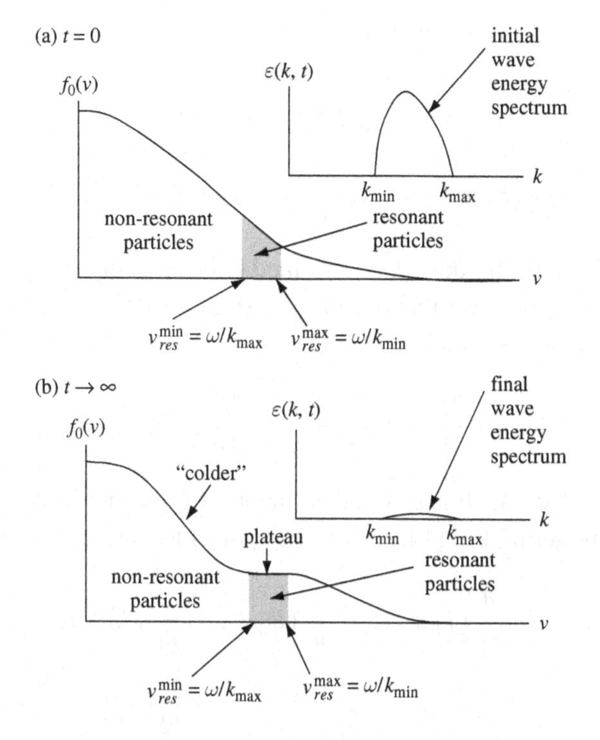

Fig. 14.1 (a) At $t = 0$ the equilibrium distribution function $f_0(v)$ is monotonically decreasing resulting in Landau damping of any waves and there is a wave spectrum (insert) with wave energy in the spectral range $k_{\min} < k < k_{\max}$. Resonant particles are shown as shaded in distribution function and lie in velocity range $v_{res}^{\min} < v < v_{res}^{\max}$. (b) As $t \to \infty$ the resonant particles develop a plateau (corresponding to absorbing energy from the wave), the wave spectrum goes to zero, and the non-resonant particles appear to become colder.

Spatial averaging will be used in the mathematical procedure to filter out some types of terms while retaining others. This averaging will be denoted by $\langle\rangle$ so, for example, the average of f_1 is

$$\langle f_1(x, v, t)\rangle = \frac{1}{L}\int \mathrm{d}x\, f_1(x, t), \tag{14.7}$$

where L is the length of the one-dimensional system and the integration is over this length. The spatial average of a quantity is independent of x and so, since f_0 is assumed to be independent of position, $\langle f_0\rangle = f_0$. (In an alternate version of the theory, the averaging is instead understood to be over a statistically large ensemble of systems containing turbulent waves and, in this version, the averaged quantity can be position-dependent.)

The 1-D electron Vlasov equation is

$$\frac{\partial}{\partial t}(f_0+f_1+f_2+\ldots)+v\frac{\partial}{\partial x}(f_1+f_2+\ldots)$$
$$-\frac{e}{m}(E_1+E_2+\ldots)\frac{\partial}{\partial v}(f_0+f_1+f_2+\ldots)=0; \tag{14.8}$$

there is no $\partial f_0/\partial x$ term because f_0 is assumed to be spatially independent and also there is no E_0 term because the system is assumed to be neutral in equilibrium. The linear portion of this equation,

$$\frac{\partial f_1}{\partial t}+v\frac{\partial f_1}{\partial x}-\frac{e}{m}E_1\frac{\partial f_0}{\partial v}=0, \tag{14.9}$$

forms the basis for the linear Landau theory of plasma waves discussed in Section 5.3. Subtracting Eq. (14.9) from Eq. (14.8) leaves the remainder equation

$$\frac{\partial}{\partial t}(f_0+f_2+\ldots)+v\frac{\partial}{\partial x}(f_2+\ldots)-\frac{e}{m}(E_2+\ldots)\frac{\partial}{\partial v}(f_0+f_1+f_2+\ldots)$$
$$-\frac{e}{m}E_1\frac{\partial}{\partial v}(f_1+f_2+\ldots)=0. \tag{14.10}$$

We assume quantities with subscripts $n\geq 1$ are waves and therefore have spatial averages that vanish, i.e., $\langle f_1\rangle=0$, $\langle E_1\rangle=0$, etc. Also, since f_0 is independent of position, $\langle E_2 f_0\rangle=f_0\langle E_2\rangle=0$, etc. Spatial averaging of Eq. (14.10) thus annihilates many terms, leaving

$$\frac{\partial f_0}{\partial t}-\frac{e}{m}\frac{\partial}{\partial v}[\langle E_1 f_1\rangle+\langle E_2 f_1\rangle+\ldots], \tag{14.11}$$

where $\partial/\partial v$ has been factored out of the spatial averaging because v is an independent variable in phase-space. The term $\langle E_1 f_1\rangle$ is of order ϵ^2 whereas $\langle E_2 f_1\rangle$ is of order ϵ^3 and the terms represented by the $+\ldots$ are of still higher order. The essential postulate of quasi-linear theory is that all terms of order ϵ^3 and higher may be neglected because ϵ is small. On applying this postulate, Eq. (14.11) reduces to the *quasi-linear velocity-space diffusion equation*

$$\frac{\partial f_0}{\partial t}=\frac{e}{m}\frac{\partial}{\partial v}\langle E_1 f_1\rangle\,. \tag{14.12}$$

This implies that even though f_0 is of order ϵ^0, the time derivative of f_0 is of order ϵ^2 so f_0 is a very slowly changing equilibrium.

The nonlinear term $\langle E_1 f_1 \rangle$ can be explicitly calculated by first expressing the perturbations as a sum (i.e., integral) over spatial Fourier modes, for example the first-order electric field can be expressed as

$$E_1(x,t) = \frac{1}{2\pi}\int \mathrm{d}k\, \tilde{E}_1(k,t)\mathrm{e}^{\mathrm{i}kx}. \tag{14.13}$$

This allows the product on the right-hand side of Eq. (14.12) to be evaluated as

$$\begin{aligned}\langle E_1 f_1 \rangle &= \frac{1}{L}\int \mathrm{d}x\left[\left(\frac{1}{2\pi}\int \mathrm{d}k\, \tilde{E}_1(k,t)\mathrm{e}^{\mathrm{i}kx}\right)\left(\frac{1}{2\pi}\int \mathrm{d}k'\, \tilde{f}_1(k',v,t)\mathrm{e}^{\mathrm{i}k'x}\right)\right] \\ &= \frac{1}{2\pi L}\int \mathrm{d}k \int \mathrm{d}k' \tilde{E}_1(k,t)\,\tilde{f}_1(k',v,t)\frac{1}{2\pi}\int \mathrm{d}x \mathrm{e}^{\mathrm{i}(k+k')x},\end{aligned} \tag{14.14}$$

where the order of integration has been changed in the second line. Then, invoking the representation of the Dirac delta function

$$\delta(k) = \frac{1}{2\pi}\int \mathrm{d}x\, \mathrm{e}^{\mathrm{i}kx}, \tag{14.15}$$

Eq. (14.14) reduces to

$$\langle E_1 f_1 \rangle = \frac{1}{2\pi L}\int \mathrm{d}k\, \tilde{E}_1(-k,t)\,\tilde{f}_1(k,v,t). \tag{14.16}$$

The linear perturbations E_1 and f_1 are governed by the system of linear equations discussed in Section 5.3. This means that associated with each wavevector k there is a complex frequency $\omega(k)$, which is determined by the linear dispersion relation $\mathcal{D}(\omega(k),k) = 0$. This gives the explicit time dependence of the modes,

$$\tilde{E}_1(k,t) = \tilde{E}_1(k)\mathrm{e}^{-\mathrm{i}\omega(k)t} \tag{14.17a}$$

$$\tilde{f}_1(k,v,t) = \tilde{f}_1(k,v)\mathrm{e}^{-\mathrm{i}\omega(k)t}. \tag{14.17b}$$

Furthermore, Eq. (14.9) provides a relationship between $\tilde{E}_1(k)$ and $\tilde{f}_1(k,v)$ since the spatial Fourier transform of this equation is

$$-\mathrm{i}\omega f_1 + \mathrm{i}kv f_1 - \frac{e}{m}E_1\frac{\partial f_0}{\partial v} = 0, \tag{14.18}$$

which leads to the familiar linear relationship

$$\tilde{f}_1(k,v) = \mathrm{i}\frac{e}{m}\tilde{E}_1(k)\frac{\partial f_0/\partial v}{\omega - kv}. \tag{14.19}$$

Inserting Eqs. (14.19) and (14.17a) into Eq. (14.16) gives

$$\langle E_1 f_1 \rangle = \frac{\mathrm{i}}{2\pi L}\frac{e}{m}\int \mathrm{d}k\, \tilde{E}_1(-k)\tilde{E}_1(k)\frac{\partial f_0/\partial v}{\omega - kv}\mathrm{e}^{-\mathrm{i}[\omega(-k)+\omega(k)]t}. \tag{14.20}$$

This expression can be further evaluated by invoking the parity properties of $\tilde{E}_1(k)$ and $\omega(k)$. These parity properties are established by writing

$$E_1(x,t) = \frac{1}{2\pi}\int \mathrm{d}k\,\tilde{E}_1(k)\mathrm{e}^{\mathrm{i}kx-\mathrm{i}\omega(k)t} \tag{14.21}$$

and noting that the left-hand side is real because it describes a physical quantity. Taking the complex conjugate of Eq. (14.21) gives

$$E_1(x,t) = \frac{1}{2\pi}\int \mathrm{d}k\,\tilde{E}_1^*(k)\mathrm{e}^{-\mathrm{i}kx+\mathrm{i}\omega^*(k)t}. \tag{14.22}$$

The parity properties are determined by defining a temporary new integration variable $k' = -k$ and noting that $\int_{-\infty}^{\infty}\mathrm{d}k$ corresponds to $\int_{-\infty}^{\infty}\mathrm{d}k'$ since $\mathrm{d}k' = -\mathrm{d}k$ and the k limits $(-\infty,\infty)$ correspond to the k' limits $(\infty,-\infty)$. Thus Eq. (14.22) can be recast as

$$\begin{aligned} E_1(x,t) &= \frac{1}{2\pi}\int \mathrm{d}k'\,\tilde{E}_1^*(-k')\mathrm{e}^{\mathrm{i}k'x+\mathrm{i}\omega^*(-k')t} \\ &= \frac{1}{2\pi}\int \mathrm{d}k\,\tilde{E}_1^*(-k)\mathrm{e}^{\mathrm{i}kx+\mathrm{i}\omega^*(-k)t}, \end{aligned} \tag{14.23}$$

where the primes have now been removed in the second line. Since Eq. (14.21) and (14.23) have the same left-hand sides, the right-hand sides must also be the same. Because $\tilde{E}_1(k)$ and $\omega(k)$ are arbitrary, they must individually satisfy the respective parity conditions

$$\begin{aligned} \tilde{E}_1(k) &= \tilde{E}_1^*(-k) \\ \omega(k) &= -\omega^*(-k) \end{aligned} \tag{14.24}$$

or equivalently

$$\begin{aligned} \tilde{E}_1(-k) &= \tilde{E}_1^*(k) \\ \omega(-k) &= -\omega^*(k). \end{aligned} \tag{14.25}$$

If the complex frequency is written in terms of explicit real and imaginary parts $\omega(k) = \omega_r(k) + \mathrm{i}\omega_i(k)$ then the frequency parity condition becomes

$$\begin{aligned} \omega_r(-k) + \mathrm{i}\omega_i(-k) &= -\left[\omega_r(k) + \mathrm{i}\omega_i(k)\right]^* \\ &= -\omega_r(k) + \mathrm{i}\omega_i(k) \end{aligned} \tag{14.26}$$

from which it can be concluded

$$\begin{aligned} \omega_r(-k) &= -\omega_r(k) \\ \omega_i(-k) &= \omega_i(k). \end{aligned} \tag{14.27}$$

The real part of ω is therefore an odd function of k whereas the imaginary part is an even function of k.

Application of the parity conditions gives

$$\tilde{E}_1(-k)\tilde{E}_1(k) = \tilde{E}_1^*(k)\tilde{E}_1(k) = \left|\tilde{E}_1(k)\right|^2 \tag{14.28}$$

and

$$\omega(-k) + \omega(k) = 2\mathrm{i}\omega_i(k), \tag{14.29}$$

in which case Eq. (14.20) reduces to

$$\langle E_1 f_1 \rangle = \frac{\mathrm{i}}{2\pi L}\frac{e}{m}\int \mathrm{d}k \frac{\left|\tilde{E}_1(k)\right|^2 \mathrm{e}^{2\omega_i(k)t}}{\omega - kv}\frac{\partial f_0}{\partial v}. \tag{14.30}$$

Substitution of $\langle E_1 f_1 \rangle$ into Eq. (14.11) gives the time evolution of the equilibrium distribution function,

$$\frac{\partial f_0}{\partial t} = \frac{\mathrm{i}}{2\pi L}\frac{e^2}{m^2}\frac{\partial}{\partial v}\int \mathrm{d}k \frac{\left|\tilde{E}_1(k)\right|^2 \mathrm{e}^{2\omega_i(k)t}}{\omega - kv}\frac{\partial f_0}{\partial v}. \tag{14.31}$$

The physical meaning of the quantity $\left|\tilde{E}_1(k)\right|^2$ can be understood by considering the volume average of the electric field energy

$$\begin{aligned}
\langle W_E \rangle &= \left\langle \frac{\varepsilon_0 E_1^2}{2} \right\rangle \\
&= \frac{\varepsilon_0}{2L}\int \mathrm{d}x \left(\frac{1}{2\pi}\int \mathrm{d}k\, \tilde{E}_1(k)\mathrm{e}^{\mathrm{i}kx - \mathrm{i}\omega(k)t}\right)\left(\frac{1}{2\pi}\int \mathrm{d}k'\, \tilde{E}_1(k')\mathrm{e}^{\mathrm{i}k'x - \mathrm{i}\omega(k')t}\right) \\
&= \frac{\varepsilon_0}{4\pi L}\int \mathrm{d}k \int \mathrm{d}k'\, \tilde{E}_1(k)\tilde{E}_1(k')\mathrm{e}^{-\mathrm{i}[\omega(k)+\omega(k')]t}\frac{1}{2\pi}\int \mathrm{d}x\, \mathrm{e}^{\mathrm{i}(k+k')x}.
\end{aligned} \tag{14.32}$$

However, using Eq. (14.15) this can be written as

$$\begin{aligned}
\langle W_E \rangle &= \frac{\varepsilon_0}{4\pi L}\int \mathrm{d}k\, \tilde{E}_1(k)\tilde{E}_1(-k)\mathrm{e}^{-\mathrm{i}[\omega(k)+\omega(-k)]t} \\
&= \int \mathrm{d}k\, \mathcal{E}(k,t),
\end{aligned} \tag{14.33}$$

where

$$\mathcal{E}(k,t) = \frac{\varepsilon_0}{4\pi L}\left|\tilde{E}_1(k)\right|^2 \mathrm{e}^{2\omega_i(k)t} \tag{14.34}$$

is the time-dependent electric field energy density associated with the wavevector k. A possible initial condition for the distribution function and wave-energy spectrum is shown in Fig. 14.1(a); the wave-energy spectrum is shown in the insert and is finite in the range $k_{\min} < k < k_{\max}$. The distribution function is monotonically decreasing so as to cause Landau damping of the waves.

Combination of Eqs. (14.31) and (14.34) shows the evolution of the equilibrium distribution function is given by

$$\frac{\partial f_0}{\partial t} = \frac{2\mathrm{i}}{\varepsilon_0}\frac{e^2}{m^2}\frac{\partial}{\partial v}\int \mathrm{d}k \frac{\mathcal{E}(k,t)}{\omega - kv}\frac{\partial f_0}{\partial v}. \tag{14.35}$$

This can be summarized as a velocity-space diffusion equation

$$\frac{\partial f_0}{\partial t} = \frac{\partial}{\partial v}\left(D_{QL}\frac{\partial f_0}{\partial v}\right), \tag{14.36}$$

where the *quasi-linear velocity-space diffusion coefficient* is

$$D_{QL} = \frac{2\mathrm{i}e^2}{\varepsilon_0 m^2}\int \mathrm{d}k \frac{\mathcal{E}(k,t)}{\omega - kv}. \tag{14.37}$$

The factor i in the velocity diffusion tensor seems surprising because f_0 is a real quantity. However, the i factor is entirely appropriate and is intimately related to the parity properties of $\omega(k)$ since

$$\frac{\mathrm{i}}{\omega(k) - kv} = \frac{\mathrm{i}\left[\omega_r(k) - kv\right] + \omega_i(k)}{\left[\omega_r(k) - kv\right]^2 + \omega_i^2(k)}. \tag{14.38}$$

The denominator in this expression is an even function of k, the term $[\omega_r(k) - kv]$ in the numerator is an odd function of k, and $\omega_i(k)$ is an even function of k. Since $\mathcal{E}(k,t)$ is an even function of k, integration over k in Eq. (14.37) annihilates the imaginary component (this component is an odd function of k) and so

$$D_{QL} = \frac{e^2}{\varepsilon_0 m^2}\int \mathrm{d}k \frac{2\omega_i(k)\mathcal{E}(k,t)}{\left[\omega_r(k) - kv\right]^2 + \omega_i^2(k)}. \tag{14.39}$$

In summary, the self-consistent coupled system of equations for the nonlinear evolution of the equilibrium consists of Eq. (14.36), (14.39), and

$$\frac{\partial}{\partial t}\mathcal{E}(k,t) = 2\omega_i(k)\mathcal{E}(k,t), \tag{14.40}$$

which is obtained from Eq. (14.34). The real and imaginary parts of the frequency $\omega_r(k)$, $\omega_i(k)$ appear as parameters in these equations and are determined from the linear wave dispersion relation, which in turn depends on f_0. This linear wave dispersion relation is obtained as in Section 5.3 by writing $E_1 = -\partial\phi_1/\partial x$ and then combining Eq. (14.19) with Poisson's equation

$$k^2\tilde{\phi}_1(k) = -\frac{e}{\varepsilon_0}\int \mathrm{d}v\tilde{f}_1 \tag{14.41}$$

to obtain

$$k^2\tilde{\phi}_1(k) = -\frac{e^2}{\varepsilon_0 m}\int \mathrm{d}v\frac{\partial f_0}{\partial v}\frac{k\tilde{\phi}_1(k)}{\omega - kv}, \tag{14.42}$$

which can be expressed, using Eq. (14.5), as

$$1+\frac{\omega_p^2}{k}\int \mathrm{d}v \frac{\partial \bar{f}_0/\partial v}{\omega - kv} = 0. \tag{14.43}$$

Thus, given the instantaneous value of $f_0(v,t) = n_0\bar{f}_0(v,t)$, the complex frequency is determined from Eq. (14.43) and then, given the instantaneous value of the wave spectral energy, both the evolution of f_0 and the wave spectral energy are determined from Eqs. (14.36) and (14.40).

14.2.2 Conservation properties of the quasi-linear diffusion equation

Conservation of particles

Conservation of particles occurs automatically because Eq. (14.36) has the form of a derivative in velocity space. Thus, the zeroth moment of Eq. (14.36) is simply

$$\frac{\partial}{\partial t}\int \mathrm{d}v f_0 = \int \mathrm{d}v \frac{\partial}{\partial v}\left(D_{QL}\frac{\partial f_0}{\partial v}\right) = \left[D_{QL}\frac{\partial f_0}{\partial v}\right]_{v=-\infty}^{v=\infty} = 0 \tag{14.44}$$

and so the quasi-linear diffusion equation conserves the density $n = \int \mathrm{d}v f_0$.

Conservation of momentum

Examination of momentum conservation requires taking the first moment of Eq. (14.36),

$$\frac{\partial}{\partial t}(nmu) = m\int \mathrm{d}v\, v\frac{\partial}{\partial v}\left(D_{QL}\frac{\partial f_0}{\partial v}\right) = -m\int \mathrm{d}v\, D_{QL}\frac{\partial f_0}{\partial v}. \tag{14.45}$$

Using Eq. (14.37) this becomes

$$\begin{aligned}\frac{\partial}{\partial t}(nmu) &= -m\int \mathrm{d}v\, \frac{2\mathrm{i}e^2}{\varepsilon_0 m^2}\int \mathrm{d}k\, \frac{\mathcal{E}(k,t)}{\omega - kv}\frac{\partial f_0}{\partial v} \\ &= -2\mathrm{i}\omega_p^2\int \mathrm{d}k\mathcal{E}(k,t)\int \mathrm{d}v \frac{1}{\omega - kv}\frac{\partial \bar{f}_0}{\partial v}.\end{aligned} \tag{14.46}$$

However, the linear dispersion relation Eq. (14.42) implies

$$\omega_p^2\int \mathrm{d}v \frac{1}{\omega - kv}\frac{\partial \bar{f}_0}{\partial v} = -k \tag{14.47}$$

and so

$$\frac{\partial}{\partial t}(nmu) = 2\mathrm{i}\int \mathrm{d}k\mathcal{E}(k,t)k = 0, \tag{14.48}$$

which vanishes because the integrand is an odd function of k. Thus, the constraint provided by the linear dispersion relation shows that the quasi-linear velocity diffusion equation also conserves momentum.

Conservation of energy

Consideration of energy conservation starts out in a similar manner but leads to some interesting, non-trivial results. The mean particle kinetic energy is defined to be

$$W_P = \int \mathrm{d}v \frac{mv^2}{2} f_0. \tag{14.49}$$

The time evolution of W_P is obtained by taking the second moment of the quasi-linear diffusion equation, Eq. (14.36),

$$\begin{aligned}\frac{\partial W_P}{\partial t} &= \int \mathrm{d}v \frac{mv^2}{2} \frac{\partial}{\partial v} D_{QL} \frac{\partial f_0}{\partial v} \\ &= -\int \mathrm{d}v\, mv D_{QL} \frac{\partial f_0}{\partial v}.\end{aligned} \tag{14.50}$$

Using Eq. (14.37) gives

$$\begin{aligned}\frac{\partial W_P}{\partial t} &= -\int \mathrm{d}v\, mv \frac{2\mathrm{i}e^2}{\varepsilon_0 m^2} \int \mathrm{d}k \frac{\mathcal{E}(k,t)}{\omega - kv} \frac{\partial f_0}{\partial v} \\ &= 2\mathrm{i}\omega_p^2 \int \mathrm{d}k \frac{\mathcal{E}(k,t)}{k} \int \mathrm{d}v \frac{\omega - kv - \omega}{\omega - kv} \frac{\partial \bar{f}_0}{\partial v} \\ &= -\omega_p^2 \int \mathrm{d}k \frac{2\mathrm{i}\omega(k)\mathcal{E}(k,t)}{k} \int \mathrm{d}v \frac{1}{\omega - kv} \frac{\partial \bar{f}_0}{\partial v}.\end{aligned} \tag{14.51}$$

Invoking Eq. (14.43) to substitute for the velocity integral results in

$$\frac{\partial W_P}{\partial t} = \int \mathrm{d}k\, 2\mathrm{i} \left[\omega_r(k) + \mathrm{i}\omega_i(k)\right] \mathcal{E}(k,t). \tag{14.52}$$

Since $\omega_r(k)$ is an odd function of k and $\omega_i(k)$ is an even function of k, only the term involving ω_i survives the k integration and so

$$\frac{\partial W_P}{\partial t} + \int \mathrm{d}k\, 2\omega_i(k)\mathcal{E}(k,t) = 0. \tag{14.53}$$

Using Eq. (14.40) this becomes

$$\frac{\partial}{\partial t}\left[W_P + \int \mathrm{d}k\, \mathcal{E}(k,t)\right] = 0, \tag{14.54}$$

which can now be integrated to give

$$W_P + \int \mathrm{d}k\, \mathcal{E}(k,t) = W_P + W_E = const.$$

showing that the sum of the particle and electric field energies is conserved. The particle energy and the electric field energy therefore need not be individually conserved – only the sum of these two types of energy is conserved. This result allows for energy exchange between the particles and the electric field.

14.2.3 Energy exchange with resonant particles

More detailed insight is obtained by considering the role of resonant particles, i.e., those particles having velocity $v \approx \omega/k$ as indicated by the shaded region in Fig. 14.1. This is done by using Eq. (14.38) to rewrite the top line of Eq. (14.51) as

$$\begin{aligned}\frac{\partial W_P}{\partial t} &= -\frac{2e^2}{\varepsilon_0 m}\int \mathrm{d}k\, \mathcal{E}(k,t)\int \mathrm{d}v\, v\frac{\mathrm{i}\,[\omega_r(k)-kv]+\omega_i(k)}{[\omega_r(k)-kv]^2+\omega_i^2(k)}\frac{\partial f_0}{\partial v}\\ &= -2\omega_p^2\int \mathrm{d}k\, \mathcal{E}(k,t)\int \mathrm{d}v\frac{v\omega_i(k)}{[\omega_r(k)-kv]^2+\omega_i^2(k)}\frac{\partial \bar{f}_0}{\partial v},\end{aligned} \tag{14.55}$$

where, because $\omega_r(k)-kv$ is an odd function of k, only the $\omega_i(k)$ numerator term survives the k integration. The velocity integral can be decomposed into a resonant portion, which is the velocity range where $\omega_r \simeq kv$, and the remaining or non-resonant portion. In the resonant portion, it is possible to approximate

$$\frac{\omega_i}{(\omega_r - kv)^2+\omega_i^2} \simeq \pi\delta(\omega_r - kv) = \frac{\pi}{k}\delta(v-\frac{\omega_r}{k}) \tag{14.56}$$

while the non-resonant portion can be written as a principle-part integral. Thus, Eq. (14.55) becomes

$$\frac{\partial W_P}{\partial t} = -\omega_p^2\int \mathrm{d}k\, 2\mathcal{E}(k,t)\left[P\int \mathrm{d}v\,\frac{v\omega_i(k)}{(\omega_r-kv)^2}\frac{\partial \bar{f}_0}{\partial v}+\pi\frac{\omega_r}{k^2}\left(\frac{\partial \bar{f}_0}{\partial v}\right)_{v=\omega_r/k}\right]. \tag{14.57}$$

A relationship between the principle part and resonant terms in this expression can be constructed by similarly decomposing Eq. (14.43) into a resonant portion and a non-resonant or principle part. The principle part is Taylor expanded as a function of $\omega_r + \mathrm{i}\omega_i$, where ω_i is assumed to be much smaller than ω_r. This procedure is essentially the scheme discussed in the development of Eq. (5.85), except that here the expansion of the principle part is written explicitly in order to emphasize certain details. Thus, the linear dispersion relation Eq. (14.43) can be expanded as

$$
\begin{aligned}
0 &= 1-\frac{\omega_p^2}{k^2}\int \mathrm{d}v \frac{\partial \bar{f}_0/\partial v}{v-\dfrac{\omega_r}{k}-\dfrac{\mathrm{i}\omega_i}{k}} \\
&= 1-\frac{\omega_p^2}{k^2}\left[P\int \mathrm{d}v \frac{\partial \bar{f}_0/\partial v}{v-\dfrac{\omega_r}{k}-\dfrac{\mathrm{i}\omega_i}{k}} + \mathrm{i}\pi\left(\frac{\partial \bar{f}_0}{\partial v}\right)_{v=\omega_r/k}\right] \\
&= 1-\frac{\omega_p^2}{k^2}\left[P\int \mathrm{d}v \frac{\partial \bar{f}_0/\partial v}{v-\dfrac{\omega_r}{k}} + \frac{\mathrm{i}\omega_i}{k}\frac{\partial}{\partial(\frac{\omega}{k})}P\int \mathrm{d}v \frac{\partial \bar{f}_0/\partial v}{v-\dfrac{\omega}{k}} + \mathrm{i}\pi\left(\frac{\partial \bar{f}_0}{\partial v}\right)_{v=\omega_r/k}\right] \\
&= 1-\frac{\omega_p^2}{k^2}\left[P\int \mathrm{d}v \frac{\partial \bar{f}_0/\partial v}{v-\dfrac{\omega_r}{k}} + \frac{\mathrm{i}\omega_i}{k}P\int \mathrm{d}v \frac{\partial \bar{f}_0/\partial v}{\left(v-\dfrac{\omega}{k}\right)^2} + \mathrm{i}\pi\left(\frac{\partial \bar{f}_0}{\partial v}\right)_{v=\omega_r/k}\right].
\end{aligned}
\tag{14.58}
$$

The imaginary part of the last line must vanish and so

$$
\frac{\omega_i}{k}P\int \mathrm{d}v \frac{\partial \bar{f}_0/\partial v}{(v-\omega_r/k)^2} + \pi\left(\frac{\partial \bar{f}_0}{\partial v}\right)_{v=\omega_r/k} = 0, \tag{14.59}
$$

which leads to the usual expression for Landau damping.

If it is assumed that $\omega_r/k \gg v_T$ then the principle-part integral in Eq. (14.57) can be approximated as

$$
P\int \mathrm{d}v \frac{v}{(\omega_r-kv)^2}\frac{\partial \bar{f}_0}{\partial v} \simeq \frac{1}{\omega_r^2}\int \mathrm{d}v\, v\frac{\partial \bar{f}_0}{\partial v} = -\frac{1}{\omega_r^2}\int \mathrm{d}v \bar{f}_0 = -\frac{1}{\omega_r^2} \tag{14.60}
$$

and the principle-part integral in Eq. (14.59) can similarly be approximated as

$$
\begin{aligned}
P\int \mathrm{d}v \frac{\partial \bar{f}_0/\partial v}{(v-\omega_r/k)^2} &= -\int \mathrm{d}v \bar{f}_0 \frac{\partial}{\partial v}\frac{1}{(v-\omega_r/k)^2} \\
&= 2\int \mathrm{d}v \frac{\bar{f}_0}{(v-\omega_r/k)^3} \\
&\simeq -2\frac{k^3}{\omega_r^3}.
\end{aligned}
\tag{14.61}
$$

Using Eq. (14.60), Eq. (14.57) becomes

$$\frac{\partial W_P}{\partial t} = -\omega_p^2 \int \mathrm{d}k\, 2\mathcal{E}(k,t) \left[-\frac{\omega_i(k)}{\omega_r^2} + \frac{\omega_r}{k^2}\pi \left(\frac{\partial \bar{f}_0}{\partial v} \right)_{v=\omega_r/k} \right] \tag{14.62}$$

and using Eq. (14.61), Eq. (14.59) becomes

$$-2\omega_i \frac{k^2}{\omega_r^3} + \pi \left(\frac{\partial \bar{f}_0}{\partial v} \right)_{v=\omega_r/k} = 0. \tag{14.63}$$

Comparison of the above two expressions shows that the second term in the square brackets of Eq. (14.62) has twice the magnitude of the first term and is of the opposite sign. Since $\omega_r^2 \simeq \omega_p^2$, this means Eq. (14.62) is of the form

$$\frac{\partial W_P}{\partial t} = \frac{\partial}{\partial t} W_{P,non-resonant} + \frac{\partial}{\partial t} W_{P,resonant} = \int \mathrm{d}k \mathcal{E}(k,t)\left[2\omega_i - 4\omega_i\right], \tag{14.64}$$

where the $2\omega_i$ term prescribes the rate of change of the kinetic energy of the non-resonant particles and the $-4\omega_i$ term prescribes the rate of change of the kinetic energy of the resonant particles. On the other hand, Eq. (14.53) showed that the rate of change of the total particle kinetic energy was equal and opposite to the total rate of change of the electric field energy. These two statements can be reconciled by asserting that the wave energy consists of equal parts of non-resonant particle kinetic energy and electric field energy and that the resonant particles act as a source or sink for this wave energy. This energy budgeting is shown schematically as

$$\underbrace{\underbrace{\int \mathrm{d}k 2\omega_i(k)\mathcal{E}(k,t)}_{\substack{\text{kinetic energy} \\ \text{of non-resonant particles}}} + \underbrace{\int \mathrm{d}k 2\omega_i(k)\mathcal{E}(k,t)}_{\substack{\text{energy stored in} \\ \text{electric field}}}}_{\substack{\text{wave} \\ \text{energy}}} \Longleftrightarrow \underbrace{\int \mathrm{d}k 4\omega_i(k)\mathcal{E}(k,t)}_{\substack{\text{kinetic energy} \\ \text{of resonant particles}}}. \tag{14.65}$$

Behavior of the resonant particles

We define f_{res} as the velocity distribution of the resonant particles, i.e., the particles with velocities $v \simeq \omega/k$ for which $\mathcal{E}(k,t)$ is finite. Since the velocity range $v = \omega(k)/k$ of the resonant particles maps to the spectrum $\mathcal{E}(k,\, t)$, the upper and lower bounds of the resonant particle velocity range respectively map to the lower and upper bounds of the values of k for which $\mathcal{E}(k,\, t)$ is finite. For these particles, the delta function approximation, Eq. (14.56), can be used to evaluate the quasi-linear diffusion coefficient given by Eq. (14.39) and obtain

$$\begin{aligned} D_{QL,res}(v) &\simeq \frac{2\pi e^2}{\varepsilon_0 m^2}\int \mathrm{d}k\,\delta(\omega_r - kv)\mathcal{E}(k,t) \\ &= \frac{2\pi e^2}{\varepsilon_0 m^2}\int \mathrm{d}k\,\delta(\omega_r/v - k)\frac{\mathcal{E}(k,t)}{v} \\ &= \frac{2\pi e^2}{\varepsilon_0 m^2}\frac{\mathcal{E}(\omega_r/v,t)}{v}. \end{aligned} \tag{14.66}$$

Using this coefficient the quasi-linear velocity diffusion for the resonant particles is

$$\frac{\partial f_{0,res}}{\partial t} = \frac{2\pi e^2}{\varepsilon_0 m^2}\frac{\partial}{\partial v}\left(\frac{\mathcal{E}(\omega_r/v,\,t)}{v}\frac{\partial f_{0,res}}{\partial v}\right). \tag{14.67}$$

It is seen from Eq. (14.63), the generalized formula for Landau damping, that

$$\pi\left(\frac{\partial f_0}{\partial v}\right)_{v=\omega_r/k} = 2\omega_i\frac{k^2}{\omega_r^3}n_0 \tag{14.68}$$

and so Eq. (14.67) becomes

$$\begin{aligned} \frac{\partial f_{0,res}}{\partial t} &= \frac{2e^2}{\varepsilon_0 m^2}\frac{\partial}{\partial v}\left(\frac{\mathcal{E}(\omega_r/v,t)}{v}2\omega_i\frac{k^2}{\omega_r^3}n_0\right) \\ &\simeq \frac{2\omega_p}{m}\frac{\partial}{\partial v}\left(\frac{\mathcal{E}(\omega_r/v,t)}{v^3}2\omega_i\right) \\ &= \frac{2\omega_p}{m}\frac{\partial}{\partial t}\frac{\partial}{\partial v}\left(\frac{\mathcal{E}(\omega_r/v,t)}{v^3}\right), \end{aligned} \tag{14.69}$$

where $\omega_r \simeq \omega_p = \sqrt{n_0 e^2/\varepsilon_0 m}$ has been used and also, because the particles are resonant, $v \simeq \omega_r/k$. This can be integrated with respect to time to obtain

$$f_{0,res}(v,\,t) - f_{0,res}(v,\,0) = \frac{2\omega_p}{m}\frac{\partial}{\partial v}\left(\frac{\mathcal{E}(\omega_r/v,t) - \mathcal{E}(\omega_r/v,0)}{v^3}\right). \tag{14.70}$$

By definition $\mathcal{E}(\omega_r/v,t)$ vanishes for v lying outside the resonant particle velocity range $v_{res}^{\min} < v < v_{res}^{\max}$ because outside this range there is no wave energy with which the particles can resonate (see Fig. 14.1). Hence, integration of Eq. (14.70) over the velocity range $v_{res}^{\min} < v < v_{res}^{\max}$ of the resonant particles gives

$$\int_{v_{res}^{\min}}^{v_{res}^{\max}} \mathrm{d}v\, f_{0,res}(v,\,t) = \int_{v_{res}^{\min}}^{v_{res}^{\max}} \mathrm{d}v\, f_{0,res}(v,\,0), \tag{14.71}$$

which demonstrates that the number of resonant particles is conserved.

Equation (14.65) showed that the resonant particle energy is not conserved and can be exchanged with the wave energy. Thus, the zeroth moment of the resonant particles is conserved, but the second moment is not. Change of the resonant particle energy while conserving the total number of resonant particles is achieved by simultaneously adjusting the number of resonant particles that are slightly slower than ω/k and the number of resonant particles that are slightly faster. For example, if there is a decrease in the number of resonant particles having $v \simeq (\omega/k)_-$ there must be a corresponding increase in the number of resonant particles having $v \simeq (\omega/k)_+$. This process provides a net transfer of energy from the wave to the resonant particles, involves wave Landau damping, and requires having $\partial f_{0,res}/\partial v < 0$. The result, increasing the number of resonant particles having $v \simeq (\omega/k)_+$ while decreasing the number having $v \simeq (\omega/k)_-$, flattens $f_{0,res}(v)$ and so makes it plateau-like as indicated in Fig. 14.1(b).

Behavior of the non-resonant particles

The quasi-linear diffusion coefficient for the non-resonant particles comes from the principle part of the integral in Eq. (14.39), i.e.,

$$D_{QL,non\text{-}res} = \frac{e^2}{\varepsilon_0 m^2} P \int \mathrm{d}k \frac{2\omega_i(k)\mathcal{E}(k,t)}{[\omega_r(k) - kv]^2 + \omega_i^2(k)}. \tag{14.72}$$

The vast majority of the non-resonant particles have velocities much slower than the wave, so for the non-resonant particles it can be assumed $v \ll \omega_r/k$, in which case

$$\begin{aligned} D_{QL,non\text{-}res} &\simeq \frac{e^2}{\varepsilon_0 m^2} \int \mathrm{d}k \frac{2\omega_i(k)\mathcal{E}(k,t)}{\omega_r^2(k)} \\ &\simeq \frac{1}{mn_0} \int \mathrm{d}k\, 2\omega_i(k)\mathcal{E}(k,t). \end{aligned} \tag{14.73}$$

Since this non-resonant particle velocity-space diffusion coefficient is velocity-independent, it can be factored from velocity integrals or derivatives. Equation (14.12) showed that the change in f_0 is order ϵ^2, where $\epsilon \ll 1$. This means that changes in f_0 are small compared to f_0. On the other hand, there is no zero-order wave energy since the wave energy scales as E_1^2 and so is entirely constituted of terms that are order ϵ^2. Thus, the wave-energy spectrum can change substantially (for example, disappear altogether) whereas there is a only slight corresponding change to f_0. Thus, Eq. (14.36), becomes

$$\begin{aligned} \frac{\partial f_{0,non\text{-}res}}{\partial t} &= \frac{1}{mn_0} \frac{\partial}{\partial v} \int \mathrm{d}k\, 2\omega_i(k)\mathcal{E}(k,t) \frac{\partial f_{0,non-res}}{\partial v} \\ &\simeq \frac{1}{mn_0} \left(\frac{\mathrm{d}}{\mathrm{d}t} \int \mathrm{d}k\, \mathcal{E}(k,t) \right) \frac{\partial^2 f_{0,non-res}}{\partial v^2}, \end{aligned} \tag{14.74}$$

which can be integrated to give

$$f_{0,non\text{-}res}(v,t) - f_{0,non\text{-}res}(v,0) = \frac{1}{mn_0}\left(\int \mathrm{d}k\,[\mathcal{E}(k,t) - \mathcal{E}(k,0)]\right)\frac{\partial^2 f_{0,non\text{-}res}}{\partial v^2}. \tag{14.75}$$

Since the number of resonant particles is conserved, the number of non-resonant particles must also be conserved.

If an initial wave spectrum becomes damped at $t = \infty$ then it is possible to write

$$f_{0,non\text{-}res}(v,\infty) = f_{0,non\text{-}res}(v,0) - \frac{1}{mn_0}\left[\int \mathrm{d}k\,\mathcal{E}(k,0)\right]\frac{\partial^2 f_{0,non\text{-}res}(v,0)}{\partial v^2}. \tag{14.76}$$

In the range of velocities where $\partial^2 f_{0,non\text{-}res}/\partial v^2 > 0$ there will be a decrease in the number of non-resonant particles and vice versa in the range where $\partial^2 f_{0,non\text{-}res}/\partial v^2 < 0$. An initially Maxwellian distribution (which has $\partial^2 f_{0,non\text{-}res}/\partial v^2 < 0$ for very small velocities and vice versa for very large velocities) will develop a double-plateau shape (high plateau at low velocities and low plateau at high velocities) with a sharp gradient between the two plateaus. The lower plateau will merge smoothly with the plateau of the resonant particles. The non-resonant portion of the velocity distribution will appear to become colder as the waves damp as indicated in Fig. 14.1(b).

This apparent cooling can be quantified by introducing an effective temperature, T_{eff}, which is defined to have the time derivative

$$\frac{\mathrm{d}}{\mathrm{d}t}\left(\kappa T_{eff}\right) = \frac{2}{n_0}\frac{\mathrm{d}}{\mathrm{d}t}\int \mathrm{d}k\,\mathcal{E}(k,t). \tag{14.77}$$

This expression can be integrated to obtain

$$\kappa T_{eff}(t) = \kappa T_0 + \frac{2}{n_0}\int \mathrm{d}k\mathcal{E}(k,t), \tag{14.78}$$

where T_0 is the temperature for the situation where there are no waves.

Using Eq. (14.77), Eq. (14.74) can be written as

$$\frac{\partial}{\partial t}f_{0,non\text{-}res} = \frac{\mathrm{d}}{\mathrm{d}t}\left(\frac{\kappa T_{eff}}{2m}\right)\frac{\partial^2 f_{0,non\text{-}res}}{\partial v^2}. \tag{14.79}$$

If $f_{0,non\text{-}res}$ is considered a function of κT_{eff} instead of t, Eq. (14.79) can be written as

$$\frac{\partial}{\partial(\kappa T_{eff})} f_{0,non\text{-}res} = \frac{1}{2m}\frac{\partial^2 f_{0,non\text{-}res}}{\partial v^2}, \tag{14.80}$$

which has the appropriately normalized solution

$$\begin{aligned} f_{0,non\text{-}res} &= n_0\sqrt{\frac{m}{2\pi\kappa T_{eff}(t)}}\exp\left(-\frac{mv^2}{2\kappa T_{eff}(t)}\right) \\ &= n_0\sqrt{\frac{m/2\pi}{\kappa T_0 + \dfrac{2}{n_0}\int \mathrm{d}k\,\mathcal{E}(k,t)}}\exp\left(-\frac{mv^2/2}{\kappa T_0 + \dfrac{2}{n_0}\int \mathrm{d}k\,\mathcal{E}(k,t)}\right). \end{aligned} \tag{14.81}$$

Thus, wave damping (i.e., the reduction of $\mathcal{E}(k,t)$) corresponds to an effective cooling of the non-resonant particles. As shown in Eq. (14.65) this kinetic energy reduction is accompanied by an equal reduction in the electric field energy and all this energy is transferred to the resonant particles.

14.3 Echoes

Plasma wave echoes (Gould, O'Neil, and Malmberg 1967, Malmberg *et al.* 1968) are a nonlinear effect whereby (i) a first wave is excited and then dies out from Landau damping, (ii) a second wave is excited and similarly dies out, and then finally (iii) at a much later time (or much more distant location), long after the first two waves have disappeared, a ghost-like third wave appears, seemingly from nowhere but with properties depending on the first two waves. Echoes provide useful insights into the Landau damping mechanism and also raise interesting questions about how Landau damping relates to entropy. These questions arise from consideration of the following apparently conflicting observations:

1. Wave damping should destroy the information content of the wave by converting ordered motion into heat. Thus, wave damping should increase the entropy of the system.
2. The collisionless Vlasov equation conserves entropy. This is because collisions are the agent that increases randomness and hence entropy. A collisionless system is in principle completely deterministic, so the future of the system can be predicted with complete precision simply by integrating the system of equations forward in time. The

entropy-conserving property of the collisionless Vlasov equation can be verified by direct calculation of the rate of change of the entropy,

$$\begin{aligned}
\frac{\mathrm{d}S}{\mathrm{d}t} &= \frac{\mathrm{d}}{\mathrm{d}t}\int \mathrm{d}x \int \mathrm{d}v\, f(x,v,t) \ln f(x,v,t) \\
&= \int \mathrm{d}x \int \mathrm{d}v\, (\ln f + 1)\, \frac{\partial f}{\partial t} \\
&= -\int \mathrm{d}x \int \mathrm{d}v\, (\ln f + 1) \left(v\frac{\partial f}{\partial x} + \frac{q}{m}E\frac{\partial f}{\partial v} \right) \\
&= -\int \mathrm{d}x \int \mathrm{d}v\, \ln f \left(v\frac{\partial f}{\partial x} + \frac{q}{m}E\frac{\partial f}{\partial v} \right) \\
&= -\int \mathrm{d}x \int \mathrm{d}v \left(v\frac{\partial}{\partial x} + \frac{q}{m}E\frac{\partial}{\partial v} \right) (f \ln f - f) \\
&= 0
\end{aligned} \tag{14.82}$$

since

$$\begin{gathered}
\int \mathrm{d}v \frac{\partial f}{\partial v} = 0, \quad \int \mathrm{d}x \frac{\partial f}{\partial x} = 0, \\
\int \mathrm{d}v \frac{\partial}{\partial v}(f \ln f - f) = 0, \quad \int \mathrm{d}x \frac{\partial}{\partial x}(f \ln f - f) = 0.
\end{gathered} \tag{14.83}$$

So, what really happens – does Landau damping increase entropy or not? The answer goes right to the heart of what is meant by entropy. In particular, it should be recalled that entropy is defined as the natural logarithm of the number of microscopic states corresponding to a given macroscopic state (see Section. 2.5.1). The concept "macroscopic state" implies the existence of a statistically large number of microscopic states that are indistinguishable from each other for all intents and purposes using all available means of observation. Collisions would cause the system to continuously evolve through all the various microscopic states and an observer of the system would not be able to distinguish one microscopic state from another, even if equipped with the best available measuring equipment. The system could then evolve through the various microscopic states that map to a specific macroscopic state and the observer would not notice.

The paradox of whether Landau damping increases entropy or not is resolved because the concept of many microscopic states mapping to a single macroscopic state is invalid for Landau damping. In actual fact, the physical system for the Landau damping problem is in just one well-defined, calculable state. The macroscopic state therefore maps to just one microscopic state and so the system does not continuously and randomly evolve through a sequence of microscopic states.

Landau damping does not involve turning ordered information into heat. Instead, macroscopically ordered information is turned into microscopically ordered information. The information still exists, but is encoded in a macroscopically invisible form. A good analogy is the process of making a holographic image of an object. The hologram looks blank unless illuminated in a special way that provides the proper decoding. If an observer does not know how to do the decoding, the information is lost to the observer, and the entropy content of the hologram appears high compared to a regular photograph. However, if the observer knows how to decode the microscopically stored information, then no information is lost.

The Landau damping process scrambles the phase of the macroscopic information and encodes this as microscopic information, which is normally irretrievable. If one does not know the appropriate trick for unscrambling the microscopically encoded information, it would be tempting to state that entropy has increased. However, the appropriate trick exists and has been demonstrated in laboratory experiments. Information that appeared to have been lost is retrieved and so entropy is not increased.

To get an idea for the issues involved, we first consider the simple problem of a one-dimensional beam of electrons having initial velocity v_0. The beam is transiently accelerated by an externally generated, spatially periodic electric field pulse $E = \bar{E}\cos(kx)\delta(t)$, where $\bar{E}$ is considered to be infinitesimal. The equation of motion for the electron beam is

$$m\left(\frac{\partial v}{\partial t} + v\frac{\partial v}{\partial x}\right) = -e\bar{E}\cos(kx)\delta(t) \tag{14.84}$$

and linearization of this equation gives

$$m\left(\frac{\partial v_1}{\partial t} + v_0\frac{\partial v_1}{\partial x}\right) = -e\bar{E}\cos(kx)\delta(t). \tag{14.85}$$

It is convenient to use complex notation so that $\cos kx \to \mathrm{e}^{ikx}$ and it is understood that, eventually, the real part of a complex solution will be used to give the physical solution. It is therefore assumed that $v_1 \sim \mathrm{e}^{ikx}$ and so the linearized equation becomes

$$m\left(\frac{\partial v_1}{\partial t} + \mathrm{i}kv_0v_1\right) = -e\bar{E}\delta(t). \tag{14.86}$$

Next it is argued that since the delta function can be considered as the superposition of an infinite spectrum of harmonic oscillations, i.e.,

$$\delta(t) = \frac{1}{2\pi}\int \mathrm{e}^{-\mathrm{i}\omega t}\mathrm{d}\omega, \tag{14.87}$$

the response of the beam to each frequency component can be considered. Thus, we consider the equation

$$\frac{\partial \tilde{v}_1}{\partial t} + \mathrm{i}kv_0\tilde{v}_1 = -\frac{e\bar{E}}{m}\mathrm{e}^{-\mathrm{i}\omega t}, \tag{14.88}$$

so the net velocity is

$$v_1(t) = \frac{1}{2\pi}\int \tilde{v}_1(\omega, t)\mathrm{d}\omega. \tag{14.89}$$

Note that $\tilde{v}_1(\omega, t)$ is *not* a Fourier transform because it contains the explicit time dependence.

Since v_0 is the initial beam velocity, $v_1(t)$ must vanish at $t = 0$ providing a boundary condition for Eq. (14.88) at $t = 0$. One way to solve this equation is first to assume that $\tilde{v}_1(\omega, t)$ consists of a particular solution satisfying the inhomogeneous part of the equation (i.e., balances the driving term on the right-hand side) and a homogeneous solution (i.e., a solution of the homogeneous equation [left-hand side of Eq. (14.88)]). The coefficient of the homogeneous solution is chosen to satisfy the boundary condition at $t = 0$. The particular solution is assumed to vary as $\mathrm{e}^{ikx-\mathrm{i}\omega t}$ and so is the solution of the equation

$$(-\mathrm{i}\omega + \mathrm{i}kv_0)\,\tilde{v}_1 = -\frac{e\bar{E}}{m}\mathrm{e}^{-\mathrm{i}\omega t}. \tag{14.90}$$

The homogenous solution is the solution of

$$\frac{\partial \tilde{v}_1}{\partial t} + \mathrm{i}kv_0\tilde{v}_1 = 0 \tag{14.91}$$

and has the form $\tilde{v}_1^h = \lambda \exp(-\mathrm{i}kv_0 t)$, where λ is a constant to be determined. Adding the particular and homogeneous solutions together gives the general solution

$$\tilde{v}_1 = -\frac{e\bar{E}}{m}\frac{\mathrm{i}\mathrm{e}^{-\mathrm{i}\omega t}}{(\omega - kv_0)} + \lambda \mathrm{e}^{-\mathrm{i}kv_0 t}, \tag{14.92}$$

where λ is chosen to satisfy the initial condition. The initial condition $v_1 = 0$ at $t = 0$ determines λ and gives

$$\tilde{v}_1(\omega, t) = -\frac{\mathrm{i}e\bar{E}}{m}\frac{\left(\mathrm{e}^{-\mathrm{i}\omega t} - \mathrm{e}^{-\mathrm{i}kv_0 t}\right)}{(\omega - kv_0)} \tag{14.93}$$

as the solution that satisfies both Eq. (14.88) and the initial condition. The term involving $\mathrm{e}^{-\mathrm{i}kv_0 t}$ is called the ballistic term. This term contains information about the initial conditions, is a solution of the homogeneous equation, is missed by Fourier treatments, is incorporated by Laplace transform treatments, and keeps v_1 from diverging when $\omega - kv_0 \to 0$.

If we wished to revert to the time domain, then the contributions of all the harmonics would have to be summed, giving

$$v_1(t) = -\frac{\mathrm{i}e\bar{E}}{2\pi m}\int \mathrm{d}\omega \frac{\left(\mathrm{e}^{-\mathrm{i}\omega t} - \mathrm{e}^{-\mathrm{i}kv_0 t}\right)}{(\omega - kv_0)}. \tag{14.94}$$

14.3.1 Ballistic terms and Laplace transforms

The discussion above used an approach related to Fourier transforms, but added additional structure to account for the initial condition $v_1 = 0$ at $t = 0$. This suggests Laplace transforms ought to be used, since Laplace transforms automatically take into account initial conditions. Let us therefore Laplace transform Eq. (14.88) to see if indeed the particular and ballistic terms are appropriately characterized. The Laplace transform of Eq. (14.88) gives

$$\begin{aligned} p\tilde{v}_1 + \mathrm{i}kv_0\tilde{v}_1 &= -\frac{e\bar{E}}{m}\int_0^\infty \mathrm{d}t\, \mathrm{e}^{-\mathrm{i}\omega t - pt} \\ &= -\frac{e\bar{E}}{m}\frac{1}{\mathrm{i}\omega + p} \end{aligned} \tag{14.95}$$

so

$$\tilde{v}_1 = -\frac{e\bar{E}/m}{(p+\mathrm{i}\omega)(p+\mathrm{i}kv_0)}. \tag{14.96}$$

The inverse Laplace transform gives

$$\tilde{v}_1 = -\frac{1}{2\pi\mathrm{i}}\frac{e\bar{E}}{m}\int_{b-\mathrm{i}\infty}^{b+\mathrm{i}\infty} \frac{\mathrm{e}^{pt}}{(p+\mathrm{i}\omega)(p+\mathrm{i}kv_0)}\mathrm{d}p. \tag{14.97}$$

Analytic continuation allows the contour to be completed on the left-hand side and it is seen that there are two poles, one at $p = -\mathrm{i}\omega$ and the other at $p = -\mathrm{i}kv$. Evaluation of the residues for the two poles gives

$$\begin{aligned} \tilde{v}_1 &= -\frac{1}{2\pi\mathrm{i}}\frac{e\bar{E}}{m}\left\{ \begin{array}{l} 2\pi\mathrm{i} \lim\limits_{p\to -\mathrm{i}\omega}(p+\mathrm{i}\omega)\left[\dfrac{\mathrm{e}^{pt}}{(p+\mathrm{i}\omega)(p+\mathrm{i}kv_0)}\right] + \\ 2\pi\mathrm{i} \lim\limits_{p\to -\mathrm{i}kv}(p+\mathrm{i}kv)\left[\dfrac{\mathrm{e}^{pt}}{(p+\mathrm{i}\omega)(p+\mathrm{i}kv_0)}\right] \end{array} \right\} \\ &= -\frac{e\bar{E}}{m}\left\{\frac{\mathrm{e}^{-\mathrm{i}\omega t}}{(-\mathrm{i}\omega + \mathrm{i}kv_0)} + \frac{\mathrm{e}^{-\mathrm{i}kv_0 t}}{(-\mathrm{i}kv_0 + \mathrm{i}\omega)}\right\}, \end{aligned} \tag{14.98}$$

which is the same as Eq. (14.93). The ballistic term $\sim \mathrm{e}^{-\mathrm{i}kv_0 t}$ thus results from a pole originating from the Laplace transform of the source term, whereas the homogeneous term $\sim \mathrm{e}^{-\mathrm{i}\omega t}$ results from the pole originating from the factor $(p+\mathrm{i}kv_0)$ appearing on the left-hand side of Eq. (14.95).

14.3.2 Phase mixing of the ballistic term for multiple beams

Now suppose that instead of just one electron beam, there are multiple beams where the density of each beam is proportional to $\exp(-v_0^2/v_T^2)$; this would be one way of characterizing a Maxwellian velocity distribution. Superposition of the ballistic terms from all these beams leads to a vanishing sum, because the ballistic terms each have a velocity-dependent phase and so the superposition would give destructive interference due to phase mixing. In particular, the superposition would involve integrals of the form

$$\int \mathrm{d}v_0 \mathrm{e}^{-v_0^2/v_T^2+\mathrm{i}kv_0t} = \mathrm{e}^{-k^2v_T^2t^2/4}\int \mathrm{d}v_0 \mathrm{e}^{-(v_0/v_T+\mathrm{i}kv_Tt/2)^2} = \sqrt{\pi}v_T\mathrm{e}^{-k^2v_T^2t^2/4}, \tag{14.99}$$

which would quickly become extremely small. This suggests the ballistic term has no enduring macroscopic physical importance and, in fact, this is true for linear problems where the ballistic term is typically ignored. However, the ballistic term can assume importance when nonlinearities are considered.

14.3.3 Beam echoes

The linearized continuity equation corresponding to Eq. (14.85) is

$$\frac{\partial n_1}{\partial t} + v_0\frac{\partial n_1}{\partial x} + n_0\frac{\partial v_1}{\partial x} = 0 \tag{14.100}$$

and so, invoking the assumed e^{ikx} dependence, this becomes

$$\frac{\partial n_1}{\partial t} + \mathrm{i}kv_0n_1 + \mathrm{i}kn_0v_1 = 0. \tag{14.101}$$

In analogy to Eq. (14.89) the linearized density can be expressed as

$$n_1(t) = \frac{1}{2\pi}\int \tilde{n}_1(\omega, t)\mathrm{d}\omega, \tag{14.102}$$

where again $\tilde{n}_1(\omega, t)$ is not a Fourier transform because $\tilde{n}_1(\omega, t)$ contains a ballistic term incorporating information about initial conditions. The equation for each frequency component thus is

$$\frac{\partial \tilde{n}_1}{\partial t} + \mathrm{i}kv_0\tilde{n}_1 + \mathrm{i}kn_0\tilde{v}_1 = 0, \tag{14.103}$$

which may be Laplace transformed to give

$$(p + \mathrm{i}kv_0)\,\tilde{n}_1 + \mathrm{i}kn_0\tilde{v}_1 = 0 \tag{14.104}$$

or, using Eq. (14.96),

$$\tilde{n}_1(p, \omega) = \frac{\mathrm{i}kn_0e\bar{E}/m}{(p+\mathrm{i}\omega)\,(p+\mathrm{i}kv_0)^2}. \tag{14.105}$$

There is now a second-order pole at $p = ikv_0$ and so there will be a ballistic term $\sim \exp(ikv_0 t)$ associated with the density perturbation. This ballistic term will also phase mix away if there is a Gaussian velocity distribution of beams.

In order to consider nonlinear consequences, we consider the second-order continuity equation

$$\frac{\partial n_2}{\partial t} + v_0 \frac{\partial n_2}{\partial t} + v_1 \frac{\partial n_1}{\partial x} + \frac{\partial}{\partial x}(n_1 v_1) = 0, \tag{14.106}$$

which has inhomogeneous (forcing) terms such as $n_1 v_1$ involving products of linear quantities. Suppose that in addition to the original pulse with spatial wavenumber k at time $t = 0$ an additional pulse is also imposed with wavenumber k' at time $t = \tau$. This additional pulse would introduce ballistic terms having a time dependence $\sim \exp(ik' v_0 (t - \tau))$. Thus, the nonlinear product $n_1 v_1$ includes a dependence

$$n_1 v_1 \sim \mathrm{Re}\left[\exp(ikv_0 t)\right] \times \mathrm{Re}\left[\exp(ik' v_0 (t - \tau))\right], \tag{14.107}$$

which contains terms of the form $\exp(ikv_0 t - k' v_0 (t - \tau))$. In general, this product of ballistic terms would phase mix away if there were a Gaussian distribution of beams, just as for the linear ballistic term. However, at the special time given by

$$kt - k'(t - \tau) = 0, \tag{14.108}$$

the phase of the nonlinear ballistic term vanishes for *all* velocities, and so no phase mixing occurs when the velocity contributions are summed. Thus, at the special time

$$t = \frac{k' \tau}{k' - k}, \tag{14.109}$$

the nonlinear product is immune to phase mixing and the superposition of the nonlinear ballistic terms of a Gaussian distribution of beams gives a macroscopic signal. The time at which the phase of the nonlinear ballistic term becomes stationary can greatly exceed τ and so a macroscopic nonlinear signal appears long after both initial pulses have gone. This ghost-like nonlinear signal is called the fluid echo.

14.3.4 The self-consistent nonlinear Vlasov–Poisson problem

These ideas carry over into the Vlasov–Poisson analysis of plasma waves, but several additional issues occur that complicate and obscure matters. First, the problem takes place in phase-space so, instead of having a pair of coupled equations for velocity and density, there is only the Vlasov equation. The Vlasov equation not only contains a convective term analogous to the fluid convective terms but also incorporates an acceleration term, which describes how the velocity distribution function is modified when particles undergo acceleration and change their velocity. As in the fluid equations, there is a coupling with Poisson's equation. This coupling not only provides the self-consistent interaction giving plasma waves, but also provides Landau damping. Landau damping is essentially a phase mixing of the Fourier-like driven terms, each of which scales as $\int dv(\omega - kv)^{-1} \exp(-i\omega t)\partial f_0/\partial v$. However, linear ballistic terms must also be excited in order to satisfy initial conditions. These ballistic terms scale as $\int dv(\omega - kv)^{-1} \exp(-ikvt)\partial f_0/\partial v$ and also phase mix away because of the $\exp(-ikvt)$ factor.

If there are two successive pulses, then the nonlinear ballistic term will again have a stationary phase (i.e., phase independent of velocity) at the special time given by Eq. (14.109). This time could be arranged to be long after the linear plasma responses to the two pulses have Landau-damped away so the echo would seem to appear from nowhere. Thus, the Vlasov–Poisson analysis contains essentially similar echo physics, but in addition has a self-consistent treatment of the plasma waves and their associated Landau damping.

To proceed with the Vlasov–Poisson analysis, we begin by considering Eq. (14.8) again, which is rewritten below for convenience

$$\begin{aligned}&\frac{\partial}{\partial t}(f_0+f_1+f_2+\ldots)+v\frac{\partial}{\partial x}(f_0+f_1+f_2+\ldots)\\&-\frac{e}{m}(E_0+E_1+E_2+\ldots)\frac{\partial}{\partial v}(f_0+f_1+f_2+\ldots)=0.\end{aligned} \tag{14.110}$$

Equilibrium is defined as being the solution obtained by balancing all the zeroth order terms, i.e.,

$$\frac{\partial f_0}{\partial t}+v\frac{\partial f_0}{\partial x}-\frac{e}{m}E_0\frac{\partial f_0}{\partial v}=0. \tag{14.111}$$

This is trivially satisfied by having $E_0 = 0, \partial f_0/\partial t = 0, \partial f_0/\partial x = 0,$ and $f_0(v)$ arbitrary; we will make these assumptions here. This equilibrium solution is then subtracted from Eq. (14.110) leaving

$$\begin{aligned}&\frac{\partial}{\partial t}(f_1+f_2+\ldots)+v\frac{\partial}{\partial x}(f_1+f_2+\ldots)-\frac{e}{m}E_1\frac{\partial}{\partial v}(f_0+f_1+f_2+\ldots)\\&\qquad-\frac{e}{m}(E_2+\ldots)\frac{\partial}{\partial v}(f_0+f_1+\ldots)=0.\end{aligned} \tag{14.112}$$

The first-order solution is defined as the solution to

$$\frac{\partial}{\partial t}f_1+v\frac{\partial f_1}{\partial x}-\frac{e}{m}E_1\frac{\partial f_0}{\partial v}=0, \tag{14.113}$$

the equation obtained by retaining all the first-order terms. The first-order solution is then subtracted from Eq. (14.112) and what remains are second and higher order terms. Dropping terms higher than second-order gives the second-order equation

$$\frac{\partial f_2}{\partial t}+v\frac{\partial f_2}{\partial x}-\frac{eE_2}{m}\frac{\partial f_0}{\partial v}-\frac{e}{m}E_1\frac{\partial f_1}{\partial v}=0. \tag{14.114}$$

The special case where $E_1 f_1$ has a zero frequency component was discussed in the previous section on quasi-linear diffusion and so here only the situation where $E_1 f_1$ has a finite frequency will be considered. Since the first-order system has been solved, the last term in Eq. (14.114) can be considered as a source term and so we rearrange the equation to be

$$\frac{\partial f_2}{\partial t}+v\frac{\partial f_2}{\partial x}-\frac{eE_2}{m}\frac{\partial f_0}{\partial v}=\frac{e}{m}\frac{\partial}{\partial v}\left(E_1 f_1\right), \tag{14.115}$$

where E_1 and f_1 are known and f_2 is to be determined. The specification that $f=f_0+f_1$ at $t=0$ provides the boundary condition $f_2=0$ at time $t=0$. Because Eq. (14.115) has both a self-consistent electric field contribution (E_2 term on the left-hand side) and a prescribed electric field term (E_1 term on the right-hand side), it has aspects of both the self-consistent problem and of the problem where the electric field is prescribed.

The linear problem

Suppose periodic grids are inserted into a plasma and the grids are transiently pulsed thereby creating the potential $\phi_{ext}(x,t)$. The charged particles will move in response to this applied potential and the resulting displacement produces a perturbation of the plasma charge density. Poisson's equation must therefore take into account both the charge density on the grid and the resulting plasma charge density and so has the form

$$\nabla^2\phi_1=-\frac{1}{\varepsilon_0}\left(\text{grid charge density}+\text{plasma charge density}\right). \tag{14.116}$$

Since the grid charge density is related to the grid potential by

$$\nabla^2\phi_{ext}=-\frac{1}{\varepsilon_0}\left(\text{grid charge density}\right) \tag{14.117}$$

Poisson's equation can be recast as

$$\nabla^2\phi_1=\nabla^2\phi_{ext}+\frac{e}{\varepsilon_0}\int f_1 \mathrm{d}v. \tag{14.118}$$

A Fourier–Laplace transform operation is then applied to Eq. (14.118) to obtain

$$-k^2\tilde{\phi}_1=-k^2\tilde{\phi}_{ext}+\frac{e}{\varepsilon_0}\int \tilde{f}_1 \mathrm{d}v. \tag{14.119}$$

However, using $E_1 = -\nabla\phi_1$, the Fourier–Laplace transform of Eq. (14.113) gives

$$\tilde{f}_1 = -\mathrm{i}\frac{e}{m}\frac{k\tilde{\phi}_1}{(p+\mathrm{i}kv)}\frac{\partial f_0}{\partial v} \tag{14.120}$$

so Eq. (14.119) becomes

$$\tilde{\phi}_1(p,k) = \frac{\tilde{\phi}_{ext}(p,k)}{\mathcal{D}(p,k)}, \tag{14.121}$$

where

$$\mathcal{D}(p,k) = 1 - \mathrm{i}\frac{e^2}{k^2 m\varepsilon_0}\int \frac{k}{p+\mathrm{i}kv}\frac{\partial f_0}{\partial v}\mathrm{d}v \tag{14.122}$$

is the usual self-consistent linear dielectric response function. Thus, the driven first-order velocity distribution function can be written as

$$\tilde{f}_1 = -\mathrm{i}\frac{e}{m}\frac{k}{(p+\mathrm{i}kv)}\frac{\tilde{\phi}_{ext}(p,k)}{\mathcal{D}(p,k)}\frac{\partial f_0}{\partial v}. \tag{14.123}$$

If an inverse transform were to be performed on $\tilde{f}_1$, then three types of contribution to the final result need to be taken into account. These contributions are (i) a ballistic term $\sim \exp(\mathrm{i}kvt)$ associated with the pole due to $p+\mathrm{i}kv$ in the denominator, (ii) a Landau-damped plasma oscillation due to the pole resulting from the root of $\mathcal{D}(p,k)$ in the denominator, and (iii) the direct inverse transform of the numerator $\tilde{\phi}_{ext}(p,k)$ modified by the presence of the other factors.

Fourier–Laplace transform of the non-linear equation

As discussed in Eq. (14.1), nonlinear quantities provide sum and difference frequencies. For example, an oscillation at frequency ω would result from the nonlinear product of a term oscillating at $\omega-\omega'$ and a term oscillating at ω' since $(\omega-\omega')+\omega' = \omega$. This can be written in a more formal and more general way using convolution integrals for both Fourier and Laplace transforms.

Since Laplace transforms will be used for the temporal dependence and since the right-hand side of Eq. (14.115) involves a product term, it is necessary to consider the Laplace transform of a product,

$$\begin{aligned}
\mathcal{L}(g(t)h(t)) &= \int_0^\infty g(t)h(t)\mathrm{e}^{-pt}\mathrm{d}t \\
&= \int_0^\infty \left[\frac{1}{2\pi\mathrm{i}}\int_{b-\mathrm{i}\infty}^{b+\mathrm{i}\infty} \tilde{g}(p')\mathrm{e}^{p't}\mathrm{d}p'\right] h(t)\mathrm{e}^{-pt}\mathrm{d}t \\
&= \frac{1}{2\pi\mathrm{i}}\int_{b-\mathrm{i}\infty}^{b+\mathrm{i}\infty} \tilde{g}(p')\left(\int_0^\infty h(t)\mathrm{e}^{(p'-p)t}\mathrm{d}t\right)\mathrm{d}p' \\
&= \frac{1}{2\pi\mathrm{i}}\int_{b-\mathrm{i}\infty}^{b+\mathrm{i}\infty} \tilde{g}(p')\tilde{h}(p-p')\mathrm{d}p'.
\end{aligned} \tag{14.124}$$

This means that the Laplace transform with argument p results from all possible products of $\tilde{g}(p')$ with $\tilde{h}(p-p')$. If $\exp(\alpha_1 t)$ is the fastest-growing term in $g(t)$ and $\exp(\alpha_2 t)$ is the fastest growing term in $h(t)$, then $b > \alpha_1$ is required in order for the Laplace transform of $g(t)$ to be defined. Furthermore, in order for the Laplace transform of $h(t)$ to be defined, it is necessary to have $\mathrm{Re}\,[p] - \mathrm{Re}\,[p'] > \alpha_2$ or $\mathrm{Re}\,[p] > \alpha_2 + \mathrm{Re}\,[p']$, which implies $\mathrm{Re}\,[p] > b + \alpha_2$. These requirements can be summarized as $\mathrm{Re}\,[p] - \alpha_2 > b > \alpha_1$.

Using

$$\tilde{g}(k) = \int_{-\infty}^{\infty} g(x)\mathrm{e}^{-\mathrm{i}kx}\mathrm{d}x \tag{14.125a}$$

$$g(x) = \frac{1}{2\pi}\int_{-\infty}^{\infty} \tilde{g}(k)\mathrm{e}^{+\mathrm{i}kx}\mathrm{d}k, \tag{14.125b}$$

a similar procedure for Fourier transforms gives

$$\begin{aligned}
\mathcal{F}\,(g(x)h(x)) &= \int_{-\infty}^{\infty} g(x)h(x)\mathrm{e}^{\mathrm{i}kx}\mathrm{d}x \\
&= \int_{-\infty}^{\infty}\left[\frac{1}{2\pi}\int_{-\infty}^{\infty}\tilde{g}(k')\mathrm{e}^{-\mathrm{i}k'x}\,\mathrm{d}k'\right]h(x)\mathrm{e}^{\mathrm{i}kx}\mathrm{d}x \\
&= \frac{1}{2\pi}\int_{-\infty}^{\infty}\mathrm{d}k'\tilde{g}(k')\int_{-\infty}^{\infty}h(x)\mathrm{e}^{\mathrm{i}(k-k')x}\mathrm{d}x \\
&= \frac{1}{2\pi}\int_{-\infty}^{\infty}\mathrm{d}k'\tilde{g}(k')\tilde{h}(k-k').
\end{aligned} \tag{14.126}$$

Thus, the Fourier–Laplace transform of Eq. (14.115) gives

$$\begin{aligned}
&(p+\mathrm{i}kv)\,\tilde{f}_2(p,k) + \mathrm{i}k\frac{e}{m}\frac{\partial f_0}{\partial v}\tilde{\phi}_2(p,k) \\
&= -\frac{e}{m}\frac{\partial}{\partial v}\left[\int_{-\infty}^{\infty}\frac{\mathrm{d}k'}{2\pi}\int_{b-\mathrm{i}\infty}^{b+\mathrm{i}\infty}\frac{\mathrm{d}p'}{2\pi}k'\tilde{\phi}_1(p',k')\tilde{f}_1(p-p',k-k')\right]
\end{aligned} \tag{14.127}$$

The convolution integrals are notationally unwieldy and so to simplify the notation we define the new dummy variables

$$\begin{aligned}
\bar{k}' &= k - k' \\
\bar{p}' &= p - p',
\end{aligned} \tag{14.128}$$

in which case Eq. (14.127) becomes

$$\begin{aligned}
&(p+\mathrm{i}kv)\,\tilde{f}_2(p,k) + \mathrm{i}k\frac{e}{m}\frac{\partial f_0}{\partial v}\tilde{\phi}_2(p,k) \\
&= -\frac{e}{m}\frac{\partial}{\partial v}\left[\int_{-\infty}^{\infty}\frac{\mathrm{d}k'}{2\pi}\int_{b-\mathrm{i}\infty}^{b+\mathrm{i}\infty}\frac{\mathrm{d}p'}{2\pi}k'\tilde{\phi}_1(p',k')\tilde{f}_1(\bar{p}',\bar{k}')\right].
\end{aligned} \tag{14.129}$$

The factors in the convolution integral can be expressed in terms of the original driving potential using Eqs. (14.121) and (14.123) to obtain

$$(p+\mathrm{i}kv)\tilde{f}_2(p,k)+\mathrm{i}k\frac{e}{m}\frac{\partial f_0}{\partial v}\tilde{\phi}_2(p,k)=\frac{\partial}{\partial v}\chi(p,k,v), \tag{14.130}$$

where the nonlinear convolution term is

$$\chi(p,k,v)=\left(\frac{e}{m}\right)^2\int_{-\infty}^{\infty}\frac{\mathrm{d}k'}{2\pi}\int_{b-\mathrm{i}\infty}^{b+\mathrm{i}\infty}\frac{\mathrm{d}p'}{2\pi}\left\{k'\frac{\tilde{\phi}_{ext}(p',k')}{\mathcal{D}(p',k')}\times\frac{\mathrm{i}\bar{k}'}{\bar{p}'+\mathrm{i}\bar{k}'v}\frac{\tilde{\phi}_{ext}(\bar{p}',\bar{k}')}{\mathcal{D}(\bar{p}',\bar{k}')}\frac{\partial f_0}{\partial v}\right\}. \tag{14.131}$$

Equation (14.130) may be solved for $\tilde{f}_2(p,k)$ to give

$$\tilde{f}_2(p,k)=-\mathrm{i}\frac{e}{m}\frac{k}{(p+\mathrm{i}kv)}\frac{\partial f_0}{\partial v}\tilde{\phi}_2(p,k)+\frac{1}{(p+\mathrm{i}kv)}\frac{\partial}{\partial v}\chi(p,k,v). \tag{14.132}$$

However, the Fourier–Laplace transform of the second-order Poisson's equation gives

$$-k^2\tilde{\phi}_2(p,k)=\frac{e}{\varepsilon_0}\int\tilde{f}_2(p,k)\mathrm{d}v \tag{14.133}$$

and so substituting for $\tilde{f}_2(p,k)$ gives

$$-k^2\tilde{\phi}_2(p,k)=\frac{e}{\varepsilon_0}\int\left[-\mathrm{i}\frac{e}{m}\frac{k}{(p+\mathrm{i}kv)}\frac{\partial f_0}{\partial v}\tilde{\phi}_2(p,k)+\frac{1}{(p+\mathrm{i}kv)}\frac{\partial\chi}{\partial v}\right]\mathrm{d}v \tag{14.134}$$

or

$$\begin{aligned}\tilde{\phi}_2(p,k)&=-\frac{e}{k^2\varepsilon_0\mathcal{D}(p,k)}\int\frac{1}{(p+\mathrm{i}kv)}\frac{\partial\chi}{\partial v}\mathrm{d}v\\&=\frac{e}{k^2\varepsilon_0\mathcal{D}(p,k)}\int\frac{\mathrm{i}k\chi}{(p+\mathrm{i}kv)^2}\mathrm{d}v.\end{aligned} \tag{14.135}$$

Substitution for χ and using the velocity-normalized distribution function $\bar{f}_0=f_0/n_0$ discussed in Eq. (14.5) gives

$$\tilde{\phi}_2(p,k)=\frac{\omega_p^2}{k^2\mathcal{D}(p,k)}\frac{e}{m}\int\mathrm{d}v\int_{-\infty}^{\infty}\frac{\mathrm{d}k'}{2\pi}\int_{b-\mathrm{i}\infty}^{b+\mathrm{i}\infty}\frac{\mathrm{d}p'}{2\pi}\left\{\frac{\mathrm{i}k}{(p+\mathrm{i}kv)^2}\frac{k'\tilde{\phi}_{ext}(p',k')}{\mathcal{D}(p',k')}\frac{\mathrm{i}\bar{k}'}{\bar{p}'+\mathrm{i}\bar{k}'v}\frac{\tilde{\phi}_{ext}(\bar{p}',\bar{k}')}{\mathcal{D}(\bar{p}',\bar{k}')}\frac{\partial\bar{f}_0}{\partial v}\right\}. \tag{14.136}$$

Double-impulse source function and its transform

In order to proceed further, the form of the external source must be specified. We assume the external source consists of two sets of periodic grids, which are pulsed sequentially. The first set of grids has wavenumber k_a and is pulsed at $t = 0$ whereas the second set of grids has wavenumber k_b and is pulsed after a delay τ. Thus, the external source has the form

$$\phi_{ext}(x,t) = \phi_a \cos(k_a x)\,\delta(\omega_p t) + \phi_b \cos(k_b x)\,\delta(\omega_p(t-\tau)). \tag{14.137}$$

The Fourier–Laplace transform of this source function gives

$$\begin{aligned}\tilde{\phi}_{ext}(p,k) &= \int_{-\infty}^{\infty}\int_0^{\infty} \phi_{ext}(x,t)\, \mathrm{e}^{-\mathrm{i}kx-pt}\mathrm{d}x\mathrm{d}t \\ &= \frac{\pi}{\omega_p}\sum_{\pm}\{\phi_a\delta(k\pm k_a) + \phi_b\delta(k\pm k_b)\,\mathrm{e}^{-p\tau}\}.\end{aligned} \tag{14.138}$$

Since we are interested in the nonlinear interaction between the a and b pulses, only the contribution from the a pulse in the first $\tilde{\phi}_{ext}$ factor in Eq. (14.136) and the contribution from the b pulse in the second $\tilde{\phi}_{ext}$ factor in Eq. (14.136) will be considered. We therefore substitute the a contribution in Eq. (14.138) for the first $\tilde{\phi}_{ext}$ factor in Eq. (14.136) to obtain

$$\begin{aligned}\tilde{\phi}_2(p,k) = &\frac{\omega_p \pi \phi_a}{k^2 \mathcal{D}(p,k)}\frac{e}{m}\int \mathrm{d}v \int_{-\infty}^{\infty}\frac{\mathrm{d}k'}{2\pi}\int_{b-\mathrm{i}\infty}^{b+\mathrm{i}\infty}\frac{\mathrm{d}p'}{2\pi}\frac{\mathrm{i}kk'\delta(k'\pm k_a)}{(p+\mathrm{i}kv)^2\,\mathcal{D}(p',k')} \\ &\times\frac{\mathrm{i}\bar{k}'}{(p-p')+\mathrm{i}\bar{k}'v}\frac{\tilde{\phi}_{ext}((p-p'),\bar{k}')}{\mathcal{D}(p-p',\bar{k}')}\frac{\partial \bar{f}_0}{\partial v}.\end{aligned} \tag{14.139}$$

The effect of the $\delta(k'\pm k_a)$ factor when evaluating the k' integral is to force $k' \to \pm k_a$ and also $\bar{k}' \to k \mp k_a$ so

$$\begin{aligned}\tilde{\phi}_2(p,k) = &\frac{\omega_p}{k^2 \mathcal{D}(p,k)}\frac{e\phi_a}{2m}\sum_{\pm}\int \mathrm{d}v \int_{b-\mathrm{i}\infty}^{b+\mathrm{i}\infty}\frac{\mathrm{d}p'}{2\pi}\frac{(\mp \mathrm{i}kk_a)}{(p+\mathrm{i}kv)^2\,\mathcal{D}(p',\mp k_a)} \\ &\times\frac{\mathrm{i}\,(k\mp k_a)}{\bar{p}'+\mathrm{i}\,(k\mp k_a)\,v}\frac{\tilde{\phi}_{ext}(\bar{p}',(k\mp k_a))}{\mathcal{D}(\bar{p}',k\mp k_a)}\frac{\partial \bar{f}_0}{\partial v}.\end{aligned} \tag{14.140}$$

Now let us substitute for the second $\tilde{\phi}_{ext}$ factor using the contribution from the b pulse to obtain

$$\begin{aligned}\tilde{\phi}_2(p,k) = &\frac{\pi}{2k^2 \mathcal{D}(p,k)}\frac{e}{m}\phi_a\phi_b\sum_{\pm}\sum_{\pm}\int \mathrm{d}v \int_{b-\mathrm{i}\infty}^{b+\mathrm{i}\infty}\frac{\mathrm{d}p'}{2\pi}\frac{(\mp \mathrm{i}kk_a)}{(p+\mathrm{i}kv)^2\,\mathcal{D}(p',\mp k_a)} \\ &\times\frac{\mathrm{i}\,(k\mp k_a)}{\bar{p}'+\mathrm{i}\,(k\mp k_a)\,v}\frac{\delta(k\mp k_a\pm k_b)\mathrm{e}^{-\bar{p}'\tau}}{\mathcal{D}(\bar{p}',k\mp k_a)}\frac{\partial \bar{f}_0}{\partial v}.\end{aligned} \tag{14.141}$$

The upper choice of the $\pm$ and $\mp$ signs is selected and the inverse Fourier transform performed to obtain

$$\tilde{\phi}_2^{upper}(p,x) = \frac{\phi_a\phi_b}{4k^2\mathcal{D}(p,k)}\frac{e}{m}\int_{-\infty}^{\infty}\mathrm{d}k\int \mathrm{d}v\int_{b-\mathrm{i}\infty}^{b+\mathrm{i}\infty}\frac{\mathrm{d}p'}{2\pi}\frac{(-\mathrm{i}kk_a)}{(p+\mathrm{i}kv)^2\,\mathcal{D}(p',-k_a)}$$
$$\times\frac{\mathrm{i}\,(k-k_a)}{\bar{p}'+\mathrm{i}\,(k-k_a)\,v}\frac{\delta(k-k_a+k_b)\mathrm{e}^{-\bar{p}'\tau+\mathrm{i}kx}}{\mathcal{D}(\bar{p}',k-k_a)}\frac{\partial\bar{f}_0}{\partial v}$$
$$= \frac{\phi_a\phi_b}{4\,(k_a-k_b)^2\,\mathcal{D}(p,k_a-k_b)}\frac{e}{m}\int \mathrm{d}v\int_{b-\mathrm{i}\infty}^{b+\mathrm{i}\infty}\frac{\mathrm{d}p'}{2\pi}$$
$$\times\frac{\mathrm{i}\,(k_a-k_b)}{(p+\mathrm{i}\,(k_a-k_b)\,v)^2}\frac{k_a}{\mathcal{D}(p',-k_a)}\frac{\mathrm{i}k_b}{(\bar{p}'-\mathrm{i}k_bv)}\frac{\mathrm{e}^{-\bar{p}'\tau+\mathrm{i}(k_a-k_b)x}}{\mathcal{D}(\bar{p}',-k_b)}\frac{\partial\bar{f}_0}{\partial v}. \tag{14.142}$$

The lower choice of the $\pm$ and $\mp$ signs means that $k_a \to -k_a$ and $k_b \to -k_b$ and so at the end of the calculation $\tilde{\phi}_2^{lower}(p,x)$ can also be determined by simply letting $k_a \to -k_a$ and $k_b \to -k_b$. The inverse Laplace transform gives

$$\tilde{\phi}_2^{upper}(t,x) = \frac{\phi_a\phi_b}{4\,(k_a-k_b)^2}\frac{e}{m}\int_{b-\mathrm{i}\infty}^{b+\mathrm{i}\infty}\frac{\mathrm{d}p}{2\pi\mathrm{i}}\frac{1}{\mathcal{D}(p,k_a-k_b)}\int \mathrm{d}v\int_{b-\mathrm{i}\infty}^{b+\mathrm{i}\infty}\frac{\mathrm{d}p'}{2\pi}$$
$$\times\frac{k_a}{(p+\mathrm{i}\,(k_a-k_b)\,v)^2\,\mathcal{D}(p',-k_a)}\frac{\mathrm{i}k_b}{(p-p'-\mathrm{i}k_bv)}$$
$$\times\frac{\mathrm{e}^{(pt-(p-p')\tau+\mathrm{i}(k_a-k_b)x)}}{\mathcal{D}(p-p',-k_b)}\frac{\partial\bar{f}_0}{\partial v}. \tag{14.143}$$

We first consider the p' integral and only retain the ballistic term due to the pole $(p-p')-\mathrm{i}k_bv$. Evaluating the residue associated with this pole gives $p' = p-\mathrm{i}k_bv$ so

$$\tilde{\phi}_2^{upper}(t,x) = -\frac{\phi_a\phi_b}{4\,(k_a-k_b)^2}\frac{e}{m}\int_{b-\mathrm{i}\infty}^{b+\mathrm{i}\infty}\mathrm{d}p\frac{1}{\mathcal{D}(p,k_a-k_b)}\int \mathrm{d}v$$
$$\times\frac{\mathrm{i}\,(k_a-k_b)}{(p+\mathrm{i}\,(k_a-k_b)\,v)^2}\frac{\mathrm{i}k_ak_b}{\mathcal{D}(p-\mathrm{i}k_bv,-k_a)}\frac{\mathrm{e}^{pt-\mathrm{i}k_bv\tau+\mathrm{i}(k_a-k_b)x}}{\mathcal{D}(\mathrm{i}k_bv,-k_b)}\frac{\partial\bar{f}_0}{\partial v}. \tag{14.144}$$

The pole at $(p+\mathrm{i}\,(k_a-k_b)\,v)^2$ is second order, in which case the rule for a second-order residue must be used, i.e.,

$$\oint \mathrm{d}p\frac{1}{(p-p_0)^2}g(p) = \pi\mathrm{i}\lim_{p\to p_0}\frac{\mathrm{d}}{\mathrm{d}p}g(p). \tag{14.145}$$

Thus, we obtain

$$\phi_2(t,x) = -\frac{\phi_a \phi_b \pi \mathrm{i}}{4\,(k_a-k_b)^2}\frac{e}{m}\int \mathrm{d}v \lim_{p\to \mathrm{i}(k_b-k_a)v}$$

$$\frac{\mathrm{d}}{\mathrm{d}p}\left(\frac{\mathrm{i}\,(k_a-k_b)}{\mathcal{D}(p,k_a-k_b)}\frac{\mathrm{i}k_a k_b}{\mathcal{D}(p-\mathrm{i}k_b v,-k_a)}\frac{\mathrm{e}^{pt-\mathrm{i}k_b v\tau+\mathrm{i}(k_a-k_b)x}}{\mathcal{D}(\mathrm{i}k_b v,-k_b)}\frac{\partial \bar{f}_0}{\partial v}\right). \tag{14.146}$$

The strongest dependence on p is in the exponential e^{pt}, and so retaining only the contribution of this term to the $\mathrm{d}/\mathrm{d}p$ operator gives

$$\tilde{\phi}_2^{upper}(t,x) = -\frac{\pi \mathrm{i}\phi_a\phi_b}{4\,(k_a-k_b)^2}\frac{e}{m}\int \mathrm{d}v \lim_{p\to \mathrm{i}(k_b-k_a)v}$$

$$\left(\frac{t\mathrm{i}\,(k_a-k_b)}{\mathcal{D}(p,k_a-k_b)}\frac{\mathrm{i}k_a k_b}{\mathcal{D}(p-\mathrm{i}k_b v,-k_a)}\frac{\mathrm{e}^{pt-\mathrm{i}k_b v\tau+\mathrm{i}(k_a-k_b)x}}{\mathcal{D}(\mathrm{i}k_b v,-k_b)}\frac{\partial \bar{f}_0}{\partial v}\right)$$

$$= \frac{\pi}{4}\frac{k_a k_b}{(k_a-k_b)^2}\frac{e\phi_a\phi_b}{m}\mathrm{e}^{\mathrm{i}(k_a-k_b)x}\int \mathrm{d}v$$

$$\frac{\mathrm{i}\,(k_a-k_b)\,t}{\mathcal{D}(\mathrm{i}(k_b-k_a)v,k_a-k_b)\,\mathcal{D}(-\mathrm{i}k_a v,-k_a)}$$

$$\times\frac{\mathrm{e}^{\mathrm{i}(k_b-k_a)vt-\mathrm{i}k_b v\tau}}{\mathcal{D}(\mathrm{i}k_b v,-k_b)}\frac{\partial \bar{f}_0}{\partial v}. \tag{14.147}$$

The $\tilde{\phi}_2^{lower}(t,x)$ term is obtained by letting $k_a-k_b \to -(k_a-k_b)$ and so

$$\tilde{\phi}_2(t,x) = \tilde{\phi}_2^{upper}(t,x)+\tilde{\phi}_2^{lower}(t,x). \tag{14.148}$$

If it is assumed that $\bar{f}_0 = (\pi v_T)^{-1/2}\exp(-v^2/v_T^2)$ then the velocity integrals in the upper and lower terms will be

$$\frac{1}{\sqrt{\pi}}\int \mathrm{d}v\mathrm{e}^{-v^2/v_T^2\pm \mathrm{i}[(k_b-k_a)t-k_b\tau]v} = \exp\left(-\,[(k_b-k_a)t-k_b\tau]^2\,v_T^2/4\right)$$

$$= \exp\left(-\left[t-\frac{k_b}{k_b-k_a}\tau\right]^2\frac{(k_b-k_a)^2 v_T^2}{4}\right). \tag{14.149}$$

Phase mixing due to the extreme velocity dependence of the $\exp(\mathrm{i}(k_b-k_a)vt-\mathrm{i}k_b v\tau)$ factor will cause the velocity integral to vanish except at the special time

$$t_{echo} = \frac{k_b}{k_b-k_a}\tau \tag{14.150}$$

when there is no phase mixing and so a finite $\tilde{\phi}_2(t, x)$ signal results. This is the echo. The half-width of the echo can be determined by writing Eq. (14.149) as

$$\frac{1}{\sqrt{\pi}} \int dv e^{-v^2/v_T^2 \pm i[(k_b - k_a)t - k_b\tau]v} = e^{-(t-t_{echo})^2/(\Delta t)^2}, \tag{14.151}$$

where

$$\Delta t = \frac{2}{|k_b - k_a| v_T} \tag{14.152}$$

is the width of the echo.

14.3.5 Spatial echoes

Creating spatially periodic sources with temporal delta functions is experimentally more difficult than creating temporally periodic sources with spatial delta functions. In the latter arrangement, two spatially separated grids are placed in a plasma and each grid is excited at a different frequency. This system is then characterized by a Fourier transform in time and a Laplace transform in space so the convective derivative in the linearized Vlasov equation has the form $-i\omega f_1 + v\partial f_1 \partial x$ giving ballistic terms proportional to $\exp(i\omega x/v)$ instead of proportional to $\exp(-ikvt)$. Thus, if two grids separated by a distance L are excited at respective frequencies ω_1 and ω_2, one grid will excite a ballistic term $\sim \exp(i\omega_1(x-L)/v)$ while the other will excite a ballistic term $\sim \exp(i\omega_2 x/v)$, and so the nonlinear product of these two ballistic terms will include a factor $\sim \exp(i\omega_1(x-L)/v - i\omega_2 x/v)$, which will phase mix to zero except at the spatial location

$$x_{echo} = \frac{\omega_1 L}{\omega_1 - \omega_2}. \tag{14.153}$$

If the spatial Landau damping length of the two linear modes is much less than x_{echo}, then the linear modes will appear to damp away spatially and then the echo will appear at $x = x_{echo}$, which will be much further away. This sort of arrangement was used to demonstrate echoes in lab experiments by Malmberg, Wharton, Gould and O'Neil (1968)

14.3.6 Higher order echoes

If one expands the Vlasov equation to higher than second order, then higher order products of ballistic terms can result. For spatial echoes, high order products of the form

$$\left[e^{i\omega_1(x-L)/v}\right]^N \times \left[e^{-i\omega_2 x/v}\right]^M \tag{14.154}$$

will appear and will phase mix away except at locations where

$$N\omega_1(x-L) = M\omega_2 x \tag{14.155}$$

so there will be a higher order echo at the location

$$x_{echo} = \frac{M\omega_2 L}{N\omega_1 - M\omega_2}. \tag{14.156}$$

14.4 Assignments

1. Competition between collisions and quasi-linear diffusion.
   (a) Let $v_{Te} = \sqrt{2\kappa T_e/m_e}$ and $w = v_z/v_{Te}$ and show that the Fokker–Planck equation in Assignment 2 of Chapter 13 can be written in dimensionless form as

   $$\frac{\partial F_T}{\partial \tau} \simeq \frac{\partial}{\partial w}\left(F_T \frac{2+Z}{w^2} + \frac{3}{4w^3}\frac{\partial F_T}{\partial w}\right),$$

   where $\tau = \nu t$ and

   $$\nu = \frac{n_e e^4 \ln \Lambda}{4\pi\varepsilon_0^2 m_e^2 v_{Te}^3}$$

   is an effective collision frequency.
   Now show that the quasi-linear diffusion equation can similarly be written in dimensionless form as

   $$\frac{\partial F_T}{\partial \tau} = \frac{\partial}{\partial w}\left(\bar{D}_{QL}\frac{\partial F_T}{\partial w}\right),$$

   where

   $$\bar{D}_{QL}(w) = \frac{D_{QL}}{D_{\text{col}}}$$

   and

   $$D_{\text{col}} = \frac{2\nu\kappa T_e}{m_e}$$

   is an effective collisional velocity-space diffusion coefficient.
   (b) Suppose both collisions and quasi-linear diffusion are operative so

   $$\frac{\partial F_T}{\partial \tau} = \frac{\partial}{\partial w}\left(\bar{D}_{QL}\frac{\partial F_T}{\partial w} + F_T\frac{2+Z}{w^2} + \frac{3}{4w^3}\frac{\partial F_T}{\partial w}\right).$$

   Show that the steady-state solution to this equation has the form (Fisch 1978)

   $$F_T \sim \exp\left(-\int^w dw \frac{(2+Z)\,w}{\left(w^3\bar{D}_{QL} + \dfrac{3}{4}\right)}\right).$$

   Sketch F_T for the cases where (i) $\bar{D}_{QL} \gg 1$ over a certain range of velocities (i.e., where wave spectrum is resonant with particles) and (ii) $\bar{D}_{QL} \ll 1$. How could this be used to drive a toroidal current in a tokamak? Using the appropriate form of D_{QL} show that the normalized quasi-linear diffusion for resonant particles is

$$\bar{D}_{QL}(w) = \frac{8\pi^2 \mathcal{E}(\omega_r/v,t)}{w m_e \omega_{p_e}^2 \ln \Lambda}.$$

If the wave is electrostatic then show

$$\mathcal{E}(\omega_r/v,t) \sim k^2|\phi|^2,$$

where k is the wavenumber of the waves resonantly interacting with the particles so

$$\bar{D}_{QL}(w) \sim \frac{8\pi^2 k^2|\phi|^2}{m_e w \omega_{p_e}^2 \ln \Lambda}.$$

What sort of waves would be most efficient at driving current, large or small k? Would this current drive work best in high- or low-density plasmas?

15
Wave–wave nonlinearities

15.1 Introduction

Wave nonlinearity is a vast subject and is not specific to plasma physics because nonlinear wave behavior occurs in virtually any physical medium where waves can propagate. However, because of the enormous variety of plasma waves, there is usually at least one plasma context where any given type of wave nonlinearity is an important issue. Three general types of nonlinear wave behavior will be discussed in this chapter: mode–mode coupling instabilities, self-modulation, and solitons. Before discussing these phenomena in detail, we first present a qualitative overview showing how certain basic wave nonlinearities are manifested.

Mode–mode coupling instabilities

Suppose a linear wave is excited in a plasma by an antenna driven by an appropriately tuned sine wave generator. In particular, suppose the plasma frequency is $\omega_{pe}/2\pi = 100\,\text{MHz}$ and the sine wave generator is tuned to a frequency well above the plasma frequency, say $\omega/2\pi = 500\,\text{MHz}$, so as to cause the antenna to radiate an electromagnetic plasma wave with dispersion relation $\omega^2 = \omega_{pe}^2 + k^2c^2$. This wave propagates through the plasma and is picked up by a distant receiving probe connected to a spectrum analyzer, a device that provides a graphic display of signal amplitude versus frequency. The received signal shows up on the spectrum analyzer display as a sharp peak at 500 MHz, as shown in Fig. 15.1(a). If the amplitude of the sine wave generator is increased or decreased, the peak on the spectrum analyzer display moves up or down proportionately, indicating the strength of the received signal.

An odd behavior is observed when the generator amplitude is increased above some critical threshold. At generator amplitudes below this threshold, the spectrum analyzer displays just the single peak at 500 MHz, but above this threshold, two additional peaks abruptly appear at two different frequencies and these new peaks have amplitudes considerably lower than the 500 MHz peak. These two additional

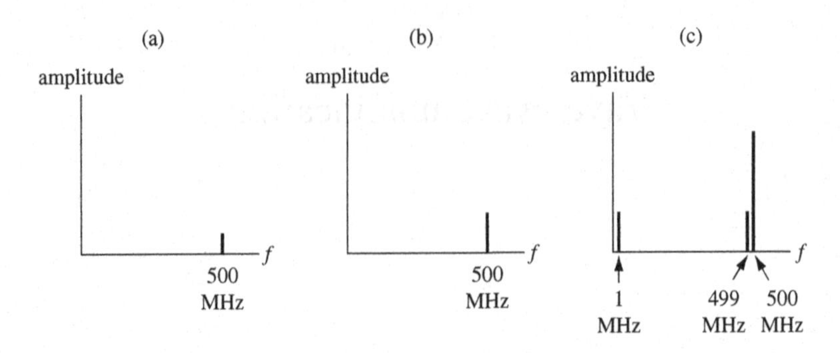

Fig. 15.1 Frequency spectra showing onset of decay instability at high power when a 500 MHz sine wave generator excites a wave in plasma. Low and intermediate generator power as in (a), (b) provides linear response, but high generator power as in (c) results in decay instability with daughter modes appearing at 1 MHz and 499 MHz.

peaks are typically at frequencies that sum to 500 MHz and the frequencies of the two additional peaks are typically very different from each other. For example, one of the two additional peaks might be at 1 MHz and the other at 499 MHz. These two peaks are called daughter waves and their spectrum is usually broader and their amplitude "springier" than the 500 MHz "pump" wave. If the generator amplitude is increased further, the pump wave amplitude no longer increases in proportion to the generator amplitude, and instead the daughter wave amplitudes increase. One gets the impression that the plasma does not like to have too much power in the pump wave and when the above-mentioned threshold is exceeded, excess power spills into the daughter waves at the expense of the pump wave. If the power is increased further, the daughters become quite large, and then at another threshold, the upper daughter might suddenly spawn its own pair of low- and high-frequency daughters. The spectrum becomes quite complicated and the amplitudes of the various spectral components depend sensitively on both the generator amplitude and on the plasma parameters. One gets the impression that energy is sloshing around between the pump wave and the daughter waves.

This behavior can occur at surprisingly modest generator power levels where it might have reasonably been expected that nonlinear behavior would be negligible. The springiness of the daughter wave amplitudes suggests some kind of resonant effect makes the nonlinearities more important than would be expected. The behavior sketched here is sometimes called a parametric decay instability (Silin 1965) because the pump wave is considered as decaying into the daughter waves. It has analogies to photon decay and in a sense can be considered as the classical limit of photon decay. Whether this sort of instability is good or bad depends on the context. If the goal is to propagate a large amplitude wave through a plasma and the wave decays into daughters, then the instability would be bad and efforts would be required to avoid it. On the other hand, if one of the daughter waves is normally difficult to excite and has some beneficial aspects, then the decay instability provides a means to access the desired daughter wave. Another possibility is that the onset of the decay could be used as a diagnostic to provide information about the plasma.

Self-modulation

A closely related situation is where the low-frequency daughter wave is at zero frequency. In this case the pump wave beats with itself and so modulates the equilibrium via a so-called "ponderomotive" nonlinear pressure, which acts to expel plasma. If this happens, increasing the generator amplitude above the nonlinear threshold causes the pump wave to dig a hole in the plasma density. The pump wave can dig itself a channel in the plasma and then this channel can guide, focus, and even trap the pump wave.

Solitons

Plasma waves are typically dispersive so, after propagating some distance, an initially sharp pulse will become less sharp because the various components of the Fourier spectrum constituting the pulse propagate at different phase velocities. In certain situations, however, pulses with an amplitude exceeding some critical value will propagate indefinitely without broadening, even though the medium is nominally dispersive. This high-amplitude non-dispersive pulse is called a soliton and its existence results from nonlinearities competing against dispersion in a way such that the two effects cancel each other.

15.2 Manley–Rowe relations

Before investigating wave–wave nonlinearities, it is instructive to examine a closely related, but simpler, system consisting of three nonlinearly coupled harmonic oscillators (Manley and Rowe 1956). This simple system consists of a particle of mass m moving in a three-dimensional space with dynamics governed by the Hamiltonian

$$H = \frac{1}{2m}\left(P_1^2 + P_2^2 + P_3^2\right) + \frac{1}{2}\left(\kappa_1 Q_1^2 + \kappa_2 Q_2^2 + \kappa_3 Q_3^2\right) + \lambda Q_1 Q_2 Q_3. \tag{15.1}$$

In the $\lambda \to 0$ limit this Hamiltonian describes three independent harmonic oscillators having respective frequencies $\omega_1 = \sqrt{\kappa_1/m}$, $\omega_2 = \sqrt{\kappa_2/m}$, and $\omega_3 = \sqrt{\kappa_3/m}$.

In the more general case of finite λ, Hamilton's equations for the P_1, Q_1 conjugate coordinates are

$$\dot{P}_1 = -\frac{\partial H}{\partial Q_1} = -\kappa_1 Q_1 - \lambda Q_2 Q_3, \tag{15.2a}$$

$$\dot{Q}_1 = \frac{\partial H}{\partial P_1} = P_1/m \tag{15.2b}$$

with similar equations for the P_2, Q_2 and P_3, Q_3 conjugates. Using relationships such as $\ddot{Q}_1 = \dot{P}_1/m$, three coupled oscillator equations result, namely

$$\begin{aligned}
\ddot{Q}_1 + \omega_1^2 Q_1 &= -\frac{\lambda}{m} Q_2 Q_3 \\
\ddot{Q}_2 + \omega_2^2 Q_2 &= -\frac{\lambda}{m} Q_1 Q_3 \\
\ddot{Q}_3 + \omega_3^2 Q_3 &= -\frac{\lambda}{m} Q_1 Q_2.
\end{aligned} \tag{15.3}$$

For small λ, each oscillator may be assumed to oscillate nearly independently so that during each cycle of a given oscillator, the oscillator experiences only slight amplitude and phase changes due to the nonlinear coupling with the other two oscillators. Thus, approximate solutions for the oscillators can be written as

$$\begin{aligned}
Q_1(t) &= A_1(t) \cos\left[\omega_1 t + \delta_1(t)\right] \\
Q_2(t) &= A_2(t) \cos\left[\omega_2 t + \delta_2(t)\right] \\
Q_3(t) &= A_3(t) \cos\left[\omega_3 t + \delta_3(t)\right],
\end{aligned} \tag{15.4}$$

where it is assumed

$$\frac{\dot{A}_j}{A_j} \ll \omega_j \text{ and } \dot{\delta}_j \ll \omega_j. \tag{15.5}$$

Each oscillator is considered as a linear mode of the system and, from now on, the term mode will be used interchangeably with oscillator. The energy associated with each mode is

$$W_j = \frac{1}{2m} P_j^2 + \frac{1}{2}\kappa_j Q_j^2 = \frac{m}{2}\dot{Q}_j^2 + \frac{1}{2}\kappa_j Q_j^2 = \frac{m}{2}\omega_j^2 A_j^2 \tag{15.6}$$

and the associated action is

$$\begin{aligned}
S = \oint P_j \mathrm{d}Q_j &= m \oint \dot{Q}_j \mathrm{d}Q_j = -m\omega_j A_j^2 \int_0^{2\pi} \sin\psi_j \mathrm{d}\cos\psi_j \\
&= m\omega_j A_j^2 \int_0^{2\pi} \sin^2\psi_j \mathrm{d}\psi_j = \frac{m}{2}\omega_j A_j^2,
\end{aligned} \tag{15.7}$$

where

$$\psi_j = \omega_j t + \delta_j(t) \tag{15.8}$$

is the phase of the mode.

Let us now use Eq. (15.8) to calculate the first and second derivatives of $Q_j = A_j \cos\psi_j$. Temporarily omitting the j subscript for clarity, the first derivative is seen to be $\dot{Q} = \dot{A}\cos\psi - A\dot{\psi}\sin\psi$ and the second derivative is

$$\begin{aligned}\ddot{Q} =& \ddot{A}\cos\psi - 2\dot{A}\dot{\psi}\sin\psi - A\ddot{\psi}\sin\psi - A\dot{\psi}^2\cos\psi \\ =& \ddot{A}\cos\psi - 2\dot{A}\left(\omega + \dot{\delta}\right)\sin\psi - A\ddot{\delta}\sin\psi - A\left(\omega + \dot{\delta}\right)^2\cos\psi. \end{aligned} \tag{15.9}$$

Since the amplitude A and phase δ are assumed to change only slightly during a wave period, ω is very large compared to time derivatives of both the amplitude and the phase, i.e.,

$$\begin{aligned}\frac{d\dot{A}}{dt} &\ll \omega\dot{A} \\ \frac{d\dot{\delta}}{dt} &\ll \omega\dot{\delta}. \end{aligned} \tag{15.10}$$

Thus, any term in Eq. (15.9) involving *two* derivatives of slowly varying quantities will be negligible compared to the corresponding term containing ω and *one* derivative of the slowly varying quantity. In particular, we may drop $\ddot{A}$ compared to $\omega\dot{A}$, $\ddot{\delta}$ compared to $\omega\dot{\delta}$, and $\dot{\delta}^2$ compared to $\omega\dot{\delta}$. On making these WKB-like approximations, Eq. (15.9) becomes

$$\ddot{Q}_j = -\omega_j^2 A_j\cos(\omega_j t + \delta_j) - 2\omega_j\dot{A}_j\sin(\omega_j t + \delta_j) - 2\omega_j A_j\dot{\delta}_j\cos(\omega_j t + \delta_j), \tag{15.11}$$

where the j subscript has now been restored. When the above expression is inserted into Eqs. (15.3), the terms involving ω_j^2 cancel and what remains is

$$\begin{aligned}\begin{bmatrix} 2\omega_j\dot{A}_j\sin(\omega_j t + \delta_j) \\ +2\omega_j A_j\dot{\delta}_j\cos(\omega_j t + \delta_j)\end{bmatrix} &= \frac{\lambda}{m}A_k A_l\cos(\omega_k t + \delta_k)\cos(\omega_l t + \delta_l) \\ &= \frac{\lambda}{2m}A_k A_l\begin{bmatrix}\cos((\omega_k + \omega_l)\,t + \delta_k + \delta_l) \\ +\cos((\omega_k - \omega_l)\,t + \delta_k - \delta_l)\end{bmatrix}. \end{aligned} \tag{15.12}$$

It is necessary to write out the three coupled equations explicitly because the coupling terms on the right-hand side are not fully symmetric. To identify resonant interactions it is assumed

$$\omega_3 = \omega_1 + \omega_2 \tag{15.13}$$

$$\theta(t) = \delta_1 + \delta_2 - \delta_3, \tag{15.14}$$

in which case the coupled mode equations can be written

$$\begin{bmatrix} \dot{A}_1 \sin(\omega_1 t + \delta_1) \\ + A_1 \dot{\delta}_1 \cos(\omega_1 t + \delta_1) \end{bmatrix} = \frac{\lambda A_2 A_3}{4m\omega_1} \begin{bmatrix} \cos((\omega_2 + \omega_3)t + \delta_2 + \delta_3) \\ \underbrace{+\cos((\omega_2 - \omega_3)t + \delta_2 - \delta_3)}_{\text{resonant at } -\omega_1} \end{bmatrix}$$

$$\begin{bmatrix} \dot{A}_2 \sin(\omega_2 t + \delta_2) \\ + A_2 \dot{\delta}_2 \cos(\omega_2 t + \delta_2) \end{bmatrix} = \frac{\lambda A_1 A_3}{4m\omega_2} \begin{bmatrix} \cos((\omega_1 + \omega_3)t + \delta_1 + \delta_3) \\ \underbrace{+\cos((\omega_1 - \omega_3)t + \delta_1 - \delta_3)}_{\text{resonant at } -\omega_2} \end{bmatrix}$$

$$\begin{bmatrix} \dot{A}_3 \sin(\omega_3 t + \delta_3) \\ + A_3 \dot{\delta}_3 \cos(\omega_3 t + \delta_3) \end{bmatrix} = \frac{\lambda A_1 A_2}{4m\omega_3} \begin{bmatrix} \overbrace{\cos((\omega_1 + \omega_2)t + \delta_1 + \delta_2)}^{\text{resonant at } +\omega_3} \\ +\cos((\omega_1 - \omega_2)t + \delta_1 - \delta_2) \end{bmatrix}. \tag{15.15}$$

Using $\omega_2 - \omega_3 = -\omega_1$, $\delta_2 - \delta_3 = \theta - \delta_1$, $\omega_1 - \omega_3 = -\omega_2$, $\delta_1 - \delta_3 = \theta - \delta_2$, and discarding non-resonant terms, Eq. (15.15) becomes

$$\begin{aligned} \dot{A}_1 \sin(\omega_1 t + \delta_1) + A_1 \dot{\delta}_1 \cos(\omega_1 t + \delta_1) &= \frac{\lambda A_2 A_3}{4m\omega_1} \cos(\theta - \omega_1 t - \delta_1) \\ \dot{A}_2 \sin(\omega_2 t + \delta_2) + A_2 \dot{\delta}_2 \cos(\omega_2 t + \delta_2) &= \frac{\lambda A_1 A_3}{4m\omega_2} \cos(\theta - \omega_2 t - \delta_2) \\ \dot{A}_3 \sin(\omega_3 t + \delta_3) + A_3 \dot{\delta}_3 \cos(\omega_3 t + \delta_3) &= \frac{\lambda A_1 A_2}{4m\omega_3} \cos(\theta + \omega_3 t + \delta_3) \end{aligned} \tag{15.16}$$

or

$$\begin{aligned} \begin{bmatrix} \dot{A}_1 \sin(\omega_1 t + \delta_1) \\ + A_1 \dot{\delta}_1 \cos(\omega_1 t + \delta_1) \end{bmatrix} &= \frac{\lambda A_2 A_3}{4m\omega_1} \begin{bmatrix} \cos\theta \cos(\omega_1 t + \delta_1) \\ + \sin\theta \sin(\omega_1 t + \delta_1) \end{bmatrix} \\ \begin{bmatrix} \dot{A}_2 \sin(\omega_2 t + \delta_2) \\ + A_2 \dot{\delta}_2 \cos(\omega_2 t + \delta_2) \end{bmatrix} &= \frac{\lambda A_1 A_3}{4m\omega_2} \begin{bmatrix} \cos\theta \cos(\omega_2 t + \delta_2) \\ + \sin\theta \sin(\omega_2 t + \delta_2) \end{bmatrix} \\ \begin{bmatrix} \dot{A}_3 \sin(\omega_3 t + \delta_3) \\ + A_3 \dot{\delta}_3 \cos(\omega_3 t + \delta_3) \end{bmatrix} &= \frac{\lambda A_1 A_2}{4m\omega_3} \begin{bmatrix} \cos\theta \cos(\omega_3 t + \delta_3) \\ - \sin\theta \sin(\omega_3 t + \delta_3) \end{bmatrix}. \end{aligned} \tag{15.17}$$

Matching the time-dependent sine and cosine terms on both sides gives

$$\dot{A}_1 = \frac{\lambda A_2 A_3}{4m\omega_1} \sin\theta \tag{15.18a}$$

$$\dot{A}_2 = \frac{\lambda A_1 A_3}{4m\omega_2} \sin\theta \tag{15.18b}$$

$$\dot{A}_3 = -\frac{\lambda A_1 A_2}{4m\omega_3} \sin\theta \tag{15.18c}$$

and

$$A_1\dot{\delta}_1 = \frac{\lambda A_2 A_3}{4m\omega_1}\cos\theta$$

$$A_2\dot{\delta}_2 = \frac{\lambda A_1 A_3}{4m\omega_2}\cos\theta$$

$$A_3\dot{\delta}_3 = \frac{\lambda A_1 A_2}{4m\omega_3}\cos\theta. \quad (15.19)$$

These equations can be used to establish various conservation relations. Multiplying Eqs. (15.18) by $\omega_j^2 A_j$ and summing gives

$$\frac{1}{2}\frac{\mathrm{d}}{\mathrm{d}t}\left(\omega_1^2 A_1^2 + \omega_2^2 A_2^2 + \omega_3^2 A_3^2\right) = \frac{(\omega_1 + \omega_2 - \omega_3)\,\lambda A_1 A_2 A_3}{4m}\sin\theta = 0$$

so

$$\omega_1^2 A_1^2 + \omega_2^2 A_2^2 + \omega_3^2 A_3^2 = const. \quad (15.20)$$

This shows that the total energy in the three modes is constant, yet allows portions of this energy to slosh back and forth between modes.

A corresponding set of relations for the action can be obtained by multiplying pairs of equations in Eq. (15.18) by $\omega_j A_j$ and then either adding or subtracting. For example, multiplying the first pair by $\omega_j A_j$ and subtracting gives

$$\frac{1}{2}\frac{\mathrm{d}}{\mathrm{d}t}\left(\omega_1 A_1^2 - \omega_2 A_2^2\right) = 0. \quad (15.21)$$

Appropriate adding and subtracting in this manner gives

$$\omega_1 A_1^2 - \omega_2 A_2^2 = const.$$

$$\omega_1 A_1^2 + \omega_3 A_3^2 = const.$$

$$\omega_2 A_2^2 + \omega_3 A_3^2 = const. \quad (15.22)$$

These relationships can be expressed in a manner analogous to quantum principles by defining the effective "quantum number" of a mode as its ratio of energy to frequency,

$$N_j = \frac{W_j}{\omega_j} = \frac{m}{2}\omega_j A_j^2. \quad (15.23)$$

It is clear that N_j is the same as the action except for an unimportant constant factor. Thus, the action conservation rules can be recast as

$$N_1 - N_2 = const.$$

$$N_1 + N_3 = const.$$

$$N_2 + N_3 = const. \quad (15.24)$$

or if changes in action are considered

$$\begin{aligned}\Delta N_1 &= +\Delta N_2\\ \Delta N_1 &= -\Delta N_3\\ \Delta N_2 &= -\Delta N_3.\end{aligned} \tag{15.25}$$

This provides an action accounting scheme such that a change $\Delta N_3 = -1$ can be considered as a mode 3 "photon" decaying (or equivalently disintegrating) into a mode 1 photon ($\Delta N_1 = +1$) and a mode 2 photon ($\Delta N_2 = +1$), all the while satisfying conservation of energy; this is sketched in Fig. 15.2.

An additional conservation equation can be obtained by subtracting the last of Eqs. (15.19) from the sum of the first two to obtain

$$\begin{aligned}\dot{\theta} &= \dot{\delta}_1 + \dot{\delta}_2 - \dot{\delta}_3\\ &= \left(\frac{\lambda A_2 A_3}{4mA_1\omega_1} + \frac{\lambda A_1 A_3}{4mA_2\omega_2} - \frac{\lambda A_1 A_2}{4mA_3\omega_3}\right)\cos\theta\\ &= \left(\frac{\dot{A}_1}{A_1} + \frac{\dot{A}_2}{A_2} + \frac{\dot{A}_3}{A_3}\right)\frac{\cos\theta}{\sin\theta}\end{aligned} \tag{15.26}$$

and then integrating to find

$$A_1 A_2 A_3 \cos\theta = const. \tag{15.27}$$

We now consider some solutions to the system of equations given by Eqs. (15.18a)–(15.18c). Suppose that initially $A_3 \gg A_2, A_1$. In this case Eq.(15.18c) gives $A_3 \simeq const$. Solving Eq. (15.18b) for A_1 and substituting the

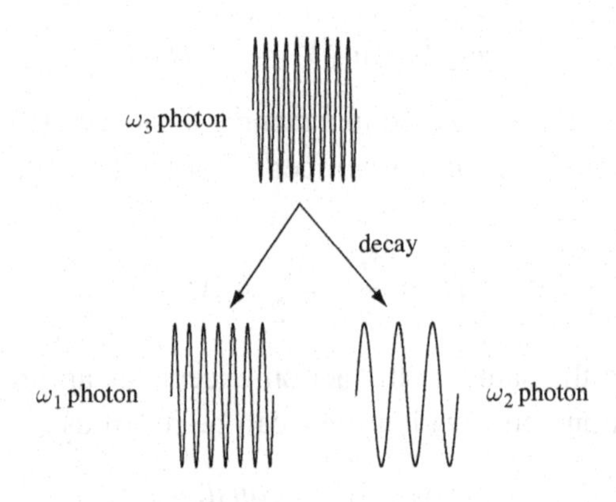

Fig. 15.2 Photon at frequency ω_3 decaying into a photon at frequency ω_1 and a photon at frequency ω_2.

result in Eq. (15.18a) gives

$$\frac{1}{\sin\theta}\frac{\mathrm{d}}{\mathrm{d}t}\left(\frac{1}{\sin\theta}\frac{\mathrm{d}A_2}{\mathrm{d}t}\right)=\frac{\lambda^2A_3^2}{16m^2\omega_1\omega_2}A_2, \tag{15.28}$$

which has exponentially growing solutions if $\omega_1\omega_2>0$. By defining

$$\tau=\int_0^t \mathrm{d}t'\sin\theta(t') \tag{15.29}$$

it is seen that

$$\frac{\mathrm{d}A_2}{\mathrm{d}t}=\frac{\mathrm{d}\tau}{\mathrm{d}t}\frac{\mathrm{d}A_2}{\mathrm{d}\tau}=\sin\theta\frac{\mathrm{d}A_2}{\mathrm{d}\tau} \tag{15.30}$$

and so

$$\frac{\mathrm{d}^2A_2}{\mathrm{d}\tau^2}=\frac{\lambda^2A_3^2}{16m^2\omega_1\omega_2}A_2. \tag{15.31}$$

Since $\omega_1\omega_2>0$ and $\omega_1+\omega_2=\omega_3$, if ω_3 is chosen to be positive then both ω_1 and ω_2 must both be positive and so $\omega_3>\omega_1,\omega_2$. If A_2 grows, then, because of the action rules, A_1 must grow in the same proportion. Furthermore, as A_1 and A_2 grow, A_3 must decrease until the approximation $A_3\gg A_1, A_3$ fails. This process can be considered as a high-energy "photon" with frequency ω_3 decaying into an ω_1 photon and an ω_2 photon where the last two photons have lower energy than ω_1. Equation (15.31) shows that the decay of the mode 3 photon into mode 1,2 photons can be characterized by an exponential growth of the mode 2 amplitude,

$$A_2\sim \mathrm{e}^{\gamma\tau}, \tag{15.32}$$

where

$$\gamma=\frac{\lambda A_3}{4m\sqrt{\omega_1\omega_2}} \tag{15.33}$$

is the instability growth rate. Mode 3 is called the pump mode since it supplies the energy for the instability. Modes 1 and 2 are called the daughter modes.

15.3 Application to waves

Nonlinearities in wave equations give rise to coupled systems of equations similar to Eqs. (15.3). Suppose Q now refers to a plasma parameter, say density, and that the plasma can support three distinct waves (modes) having respective linear density fluctuations

$$\begin{aligned}Q_1(\mathbf{x},t)&=A_1(t)\cos(\mathbf{k}_1\cdot\mathbf{x}-\omega_1t-\delta_1(t))\\Q_2(\mathbf{x},t)&=A_2(t)\cos(\mathbf{k}_2\cdot\mathbf{x}-\omega_2t-\delta_2(t))\\Q_3(\mathbf{x},t)&=A_3(t)\cos(\mathbf{k}_3\cdot\mathbf{x}-\omega_3t-\delta_3(t));\end{aligned} \tag{15.34}$$

the density Q therefore has the functional form

$$Q(\mathbf{x}, t) = Q_1(\mathbf{x}, t) + Q_2(\mathbf{x}, t) + Q_3(\mathbf{x}, t). \tag{15.35}$$

This situation is mathematically analogous to the coupled oscillator system discussed in Section 15.2 if the amplitude Q_j of each of the three waves is considered to be an effective canonical coordinate. Each wave satisfies a linear dispersion relation $\omega_j = \omega_j(\mathbf{k})$ and nonlinearities in the wave equations provide a mutual coupling between the modes analogous to the coupling between the different directions of motion for the oscillating mass discussed in Section 15.2. The $\exp(i\mathbf{k}\cdot\mathbf{x})$ dependence of the waves suggests the need for a wavenumber selection rule

$$\mathbf{k}_3 = \mathbf{k}_1 + \mathbf{k}_2 \tag{15.36}$$

analogous to the frequency selection rule Eq. (15.13). Using the wavenumber selection rule and generalizing the phase offset definition to be $\delta'_j(t) = \delta_j(t) - \mathbf{k}_j\cdot\mathbf{x}$, it is seen that the coupled wave equations become identical to the coupled oscillator equations with δ'_j replacing δ_j. Hence, so long as the selection rules $\omega_3 = \omega_1 + \omega_2$ and $\mathbf{k}_3 = \mathbf{k}_1 + \mathbf{k}_2$ are satisfied, a high-frequency wave ω_3 should decay spontaneously into two low-frequency waves ω_1, ω_2, providing a suitable coupling coefficient exists.

15.3.1 Examples of nonlinearities

Nonlinearities can be divided into two general types: (1) two high-frequency waves beating together to give a low-frequency driving term and (2) a high-frequency wave beating with a low-frequency wave to give a high-frequency driving term. As shown below, the first type of nonlinearity has the form of a radiation pressure (also known as a ponderomotive force) while the second type of nonlinearity has the form of a density modulation.

1. *Beat of two high-frequency waves driving a low-frequency wave (ponderomotive force).* Since ion motion is negligible in a high-frequency wave, all that is required when considering nonlinearities is the electron equation of motion,

$$\frac{\partial \mathbf{u}_e}{\partial t} = -\mathbf{u}_e\cdot\nabla\mathbf{u}_e + \frac{q_e}{m_e}(\mathbf{E} + \mathbf{u}_e\times\mathbf{B}) - \frac{1}{m_e n_e}\nabla P_e. \tag{15.37}$$

For simplicity, it is assumed that no equilibrium magnetic field exists, so the only magnetic field is the wave magnetic field. Tildes are used to denote linear quantities to avoid confusion with the subscripts 1, 2, 3, which denote the low-frequency daughter, high-frequency daughter, and pump wave respectively. If the high-frequency wave is electromagnetic, then the mode is transverse, i.e., the wavevector $\mathbf{k}$ is perpendicular to the mode electric field $\tilde{\mathbf{E}}$ and so there is no coupling to pressure oscillations because

$\nabla\tilde{P}_e$ is in the direction of $\mathbf{k}$. If the high-frequency mode is electrostatic, i.e., is a Langmuir wave, then there is a finite $\nabla\tilde{P}_e$ contribution, but this term is small because of the adiabatic assumption $\omega/k \gg v_{Te}$ for a Langmuir wave (see Eq. (4.28)) and so can be ignored to first approximation. Thus, when calculating the linear motion for purposes of determining nonlinear coupling coefficients, the $\nabla\tilde{P}_e$ term resulting from linearizing Eq. (15.37) may be dropped and so we simply have

$$\frac{\partial\tilde{\mathbf{u}}_e}{\partial t} = \frac{q_e}{m_e}\tilde{\mathbf{E}}. \tag{15.38}$$

This is exactly true for an electromagnetic wave and true to lowest order in the adiabatic assumption for an electrostatic wave. The electron quiver velocity is defined to be

$$\tilde{\mathbf{u}}_e^h = \frac{q_e}{m_e}\int^t \tilde{\mathbf{E}}dt', \tag{15.39}$$

which is the solution to Eq. (15.38) for both electron plasma waves and electromagnetic waves. The main nonlinear terms in Eq. (15.37) are $-\tilde{\mathbf{u}}_e\cdot\nabla\tilde{\mathbf{u}}_e + q_e\left(\tilde{\mathbf{u}}_e\times\tilde{\mathbf{B}}\right)/m_e$; nonlinearity in the pressure gradient is ignored because, by assumption, this term is already small. In order to obtain $\tilde{\mathbf{B}}$, Faraday's law is integrated with respect to time giving

$$\tilde{\mathbf{B}} = -\nabla\times\int^t \tilde{\mathbf{E}}dt' = -\frac{m_e}{q_e}\nabla\times\tilde{\mathbf{u}}_e^h. \tag{15.40}$$

The two nonlinear terms can be combined using Eq. (15.40) to form what is called the ponderomotive force or radiation pressure,

$$\begin{aligned} -\tilde{\mathbf{u}}_e\cdot\nabla\tilde{\mathbf{u}}_e + \frac{q_e}{m_e}\left(\tilde{\mathbf{u}}_e\times\tilde{\mathbf{B}}\right) &= -\tilde{\mathbf{u}}_e^h\cdot\nabla\tilde{\mathbf{u}}_e^h - \left(\tilde{\mathbf{u}}_e^h\times\nabla\times\tilde{\mathbf{u}}_e^h\right) \\ &= -\frac{1}{2}\nabla\left(\tilde{u}_e^h\right)^2. \end{aligned} \tag{15.41}$$

If only one high-frequency mode exists and beats with itself, then the ponderomotive force $-\nabla\left(\tilde{u}_e^h\right)^2/2$ is at zero frequency but if the beating is between two distinct high-frequency modes, then the ponderomotive force contains a term at the difference (i.e., beat) frequency between the frequencies of the two modes. Thus, at low frequencies the electron equation of motion, Eq. (15.37) becomes

$$\frac{\partial\tilde{\mathbf{u}}_e}{\partial t} = \frac{q_e}{m_e}\tilde{\mathbf{E}} - \frac{1}{m_e n}\nabla\tilde{P}_e - \frac{1}{2}\nabla\left(\tilde{u}_e^h\right)^2, \tag{15.42}$$

where only the beat frequency component in $\left(\tilde{u}_e^h\right)^2$ is used. The ponderomotive force provides a mechanism for high-frequency waves to couple to low-frequency waves. It acts as an effective pressure scaling as $m_e n\left(\tilde{u}_e^h\right)^2/2$ and so, in a sense, the quiver velocity $\tilde{u}_e^h$ acts as a thermal velocity. The ratio of ion radiation pressure to the ion pressure is smaller than the corresponding ratio for electrons by a factor of m_e/m_i because the ion quiver velocity is smaller by this factor. Thus, ion ponderomotive force is ignored since it is so small.

2. *Beating of a low-frequency wave with a high-frequency wave to drive another high-frequency wave (modulation).* By writing Ampère's law as

$$\nabla\times\tilde{\mathbf{B}}=\mu_0\sum_\sigma\left(\tilde{n}_\sigma q_\sigma\tilde{\mathbf{u}}_\sigma+n_0q_\sigma\tilde{\mathbf{u}}_\sigma\right)+\mu_0\varepsilon_0\frac{\partial\tilde{\mathbf{E}}}{\partial t}\tag{15.43}$$

it is seen that density fluctuations provide a nonlinear component to the current density. The nonlinear term can be put on the right-hand side to emphasize its role as a nonlinear driving term so that Ampère's law becomes

$$\nabla\times\tilde{\mathbf{B}}-\mu_0\varepsilon_0\frac{\partial\tilde{\mathbf{E}}}{\partial t}-\mu_0\sum_\sigma n_0q_\sigma\tilde{\mathbf{u}}_\sigma=+\mu_0\sum_\sigma\tilde{n}_\sigma q_\sigma\tilde{\mathbf{u}}_\sigma.\tag{15.44}$$

The nonlinear term is assumed to be a product of a high-frequency wave and a low-frequency wave. The linearized continuity equation gives

$$\frac{\partial\tilde{n}_\sigma}{\partial t}=-n_\sigma\nabla\cdot\tilde{\mathbf{u}}_\sigma\tag{15.45}$$

showing that $\tilde{n}_\sigma/n_\sigma=\mathbf{k}\cdot\tilde{\mathbf{u}}_\sigma/\omega$. The low-frequency wave typically has a smaller phase velocity than the high-frequency wave so, in terms of magnitudes,

$$\frac{\tilde{n}_\sigma^l}{\tilde{n}_\sigma^h}\sim\frac{\tilde{u}_\sigma^l}{\tilde{u}_\sigma^h}\frac{(\omega/k)_h}{(\omega/k)_l}\gg\frac{\tilde{u}_\sigma^l}{\tilde{u}_\sigma^h}.\tag{15.46}$$

Thus, the product $\tilde{n}_\sigma^l\tilde{\mathbf{u}}_\sigma^h$ is much larger than the product $\tilde{n}_\sigma^h\tilde{\mathbf{u}}_\sigma^l$, where l and h refer to low- and high-frequency waves. Thus, the dominant effect of a low-frequency wave is to modulate the density profile seen by a high-frequency wave.

15.3.2 Possible types of wave interaction

As discussed in Section 4.2, three distinct types of waves can propagate in an unmagnetized uniform plasma and these waves have the dispersion relations:

$$\begin{aligned}\omega^2&=\omega_{pe}^2+k^2c^2,\ \text{electromagnetic wave}\\ \omega^2&=\omega_{pe}^2\left(1+3k^2\lambda_{de}^2\right),\ \text{electron plasma wave}\\ \omega^2&=\frac{k^2c_s^2}{1+k^2\lambda_{De}^2},\ \text{ion acoustic wave.}\end{aligned}\tag{15.47}$$

If three waves satisfy the selection rules

$$\omega_3=\omega_1+\omega_2\tag{15.48}$$

$$\mathbf{k}_3=\mathbf{k}_1+\mathbf{k}_2,\tag{15.49}$$

they can only have the *same* dispersion relation if ω^2 depends linearly on k^2 so that the magnitude of $\mathbf{k}$ is linearly proportional to the magnitude of ω. This is not so for the modes listed in Eq. (15.47) except for the $k^2\lambda_{De}^2\ll1$ limit of the ion

Table 15.1

Pump	Hf daughter	Lf daughter	Common name
ω_3	ω_2	ω_1	
em	em	Langmuir	stimulated Raman backscatter
em	em	acoustic	stimulated Brillouin backscatter
em	em	zero frequency	self-focusing
em	Langmuir	Langmuir	two-plasmon decay
em	Langmuir	acoustic	parametric decay instability
Langmuir	Langmuir	acoustic	electron decay instability
Langmuir	Langmuir	zero frequency	caviton

acoustic wave; applying the selection rules to this $\omega^2 = k^2 c_s^2$ limit of ion acoustic waves gives nonlinear interactions between various sound wave harmonics, e.g., the first and second harmonic can interact with the third harmonic. Since this limit corresponds to the well-known nonlinear steepening of ordinary sound waves it will not be discussed further.

The situations that are of particular relevance to plasmas are where modes satisfying different types of dispersion relations interact with each other, for example an electromagnetic wave interacting with an electron plasma wave and an ion acoustic wave. The various possibilities for nonlinear interactions of the three types of plasma waves are shown in Table 15.1.

Here the abbreviations em, Langmuir, and acoustic stand for electromagnetic wave, electron plasma wave, and ion acoustic wave respectively and the ordering is by progressively lower frequency, column by column, taking into account (i) $\omega_3 > \omega_2 > \omega_1$ and (ii) electromagnetic waves have higher frequencies than Langmuir waves, which in turn have higher frequencies than ion acoustic waves.

In each case the low-frequency daughter modulates the density and beats with either the pump or the high-frequency daughter to provide a high-frequency nonlinear current as given by Eq. (15.44) while the pump and the high-frequency daughter beat together to provide a ponderomotive force that couples to the low-frequency wave. Consider a coordinate system where the x axis is in the direction of $\mathbf{k}_3$ and $\perp$ refers to the direction perpendicular to $\mathbf{k}_3$. Thus, $(\mathbf{k}_1 + \mathbf{k}_2)_\perp = 0$ so any daughter wavevector components in the direction perpendicular to the pump wave must be equal in magnitude and opposite in direction.

The high-frequency wave has a faster phase velocity than the daughter waves so

$$\frac{\omega_3^2}{k_3^2} > \frac{\omega_2^2}{k_2^2} \text{ and } \frac{\omega_3^2}{k_3^2} > \frac{\omega_1^2}{k_1^2}. \tag{15.50}$$

Let us suppose that all waves are in the same direction so we may assume k_1, k_2 and k_3, are all positive. Then taking the square root of all quantities in Eq. (15.50) and using $\omega_3 = \omega_1 + \omega_2$, $k_3 = k_1 + k_2$ gives

$$\frac{\omega_1 + \omega_2}{k_1 + k_2} > \frac{\omega_2}{k_2} \text{ and } \frac{\omega_1 + \omega_2}{k_1 + k_2} > \frac{\omega_1}{k_1}. \tag{15.51}$$

However, rearranging these two expressions leads to the contradictory relationships

$$\frac{\omega_1}{\omega_2} > \frac{k_1}{k_2} \text{ and } \frac{\omega_1}{\omega_2} < \frac{k_1}{k_2}. \tag{15.52}$$

This means it is impossible to have all waves in the same direction and satisfy the selection rules when the pump wave is faster then the daughter waves. Thus, the frequency and wavenumber selection rules can only be satisfied if $\mathbf{k}_1$, $\mathbf{k}_2$, and $\mathbf{k}_3$ are not all in the same direction.

If the daughter waves have no wavevector components perpendicular to the pump wave, then one of the daughter waves must have a wavevector antiparallel to the pump while the other is parallel to the pump. This is shown schematically in Fig. 15.3 for the parametric decay instability where it is seen that the ion acoustic daughter wave propagates backwards relative to the em and Langmuir waves. Since $v_{Te} \ll c$, the phase velocity of the electromagnetic wave is actually much faster than suggested by the figure, i.e., the speed of light is shown artificially slowed down in the figure to make the vector relationships more obvious. In actual parametric decay situations the approximation $k_1 + k_2 \simeq 0$ may be used since $|k_3| \ll |k_1|, |k_2|$.

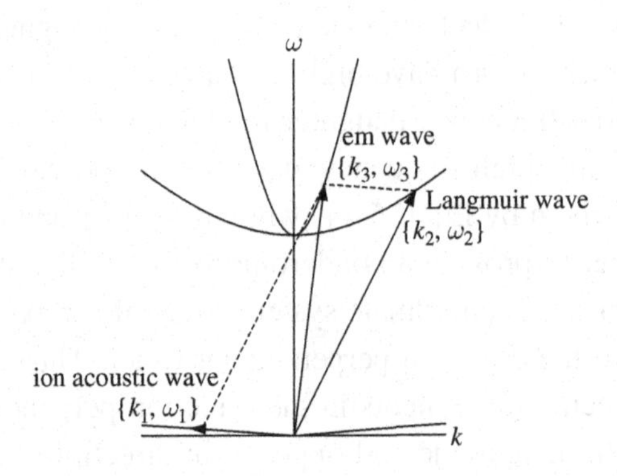

Fig. 15.3 Electromagnetic wave $\omega = \sqrt{\omega_{pe}^2 + k^2c^2}$ decaying into a Langmuir wave $\omega = \omega_{pe}\sqrt{1 + 3k^2\lambda_{de}^2}$ and an ion acoustic wave $\omega = kc_s/\sqrt{1 + k^2\lambda_{De}^2}$. The ion acoustic wave propagates backwards in order to have $\omega_3(k_1 + k_2) = \omega_1(k_1) + \omega_2(k_2)$ as can be seen from the addition of the $\{k, \omega\}$ vectors in the diagram.

The electromagnetic wave in a given plasma always has a higher frequency than the electron plasma or ion acoustic waves. As a consequence, the low-frequency daughter cannot be an electromagnetic wave because then all three waves would have to be electromagnetic waves, which would then violate the requirement that the dispersion relations cannot all be the same. The low-frequency daughter must therefore be either an ion acoustic wave or an electron plasma wave. Similarly, the high-frequency wave (pump or high-frequency daughter) cannot be an ion acoustic wave and so must be either an electromagnetic wave or an electron plasma wave. Thus, the various interactions tabulated above can be accounted for by establishing the appropriate coupling equations for the following four possibilities: high-frequency wave is an electromagnetic wave or a Langmuir wave, low-frequency wave is a Langmuir wave or an ion acoustic wave.

Low-frequency wave is an ion acoustic wave

On assuming quasi-neutrality, which corresponds to assuming $k^2\lambda_{De}^2 \ll 1$, the low frequency electron and ion equations of motion may be approximated as

$$\frac{\partial \tilde{\mathbf{u}}_e}{\partial t} = \frac{q_e}{m_e}\tilde{\mathbf{E}} - \frac{\kappa T_e}{m_e n}\nabla\tilde{n} - \frac{1}{2}\nabla\left(\tilde{u}_e^h\right)^2 \tag{15.53}$$

$$\frac{\partial \tilde{\mathbf{u}}_i}{\partial t} = \frac{q_i}{m_i}\tilde{\mathbf{E}}; \tag{15.54}$$

here only the low-frequency beat component of $\left(\tilde{u}_e^h\right)^2$ is used. Also, as shown earlier, the ion ponderomotive force is negligible and therefore ignored. Because the electron mass is very small, the left-hand side of the electron equation of motion is dropped, in which case this equation reduces to the simple force balance relation

$$\frac{q_e}{m_e}\tilde{\mathbf{E}} - \frac{\kappa T_e}{m_e n}\nabla\tilde{n} - \frac{1}{2}\nabla\left(\tilde{u}_e^h\right)^2 \simeq 0. \tag{15.55}$$

Using Eq. (15.55) to eliminate $\tilde{\mathbf{E}}$ from the ion equation gives

$$\frac{\partial \tilde{\mathbf{u}}_i}{\partial t} = -\frac{\kappa T_e}{m_i n}\nabla\tilde{n} - \frac{1}{2}\frac{m_e}{m_i}\nabla\left(\tilde{u}_e^h\right)^2. \tag{15.56}$$

Because quasi-neutrality is assumed, $\tilde{n}_i = \tilde{n}_e = n$, and so the time derivative of the ion continuity equation can be written as

$$\frac{\partial^2 \tilde{n}}{\partial t^2} + n\nabla\cdot\frac{\partial \tilde{\mathbf{u}}_i}{\partial t} = 0. \tag{15.57}$$

On substituting for $\partial\tilde{\mathbf{u}}_i/\partial t$, the above equation becomes

$$\frac{\partial^2 \tilde{n}}{\partial t^2} - n\nabla\cdot\left(\frac{\kappa T_e}{m_i n}\nabla\tilde{n} + \frac{1}{2}\frac{m_e}{m_i}\nabla\left(\tilde{u}_e^h\right)^2\right) = 0, \tag{15.58}$$

which can be written as an ion acoustic wave equation with a nonlinear coupling term due to the electron ponderomotive force,

$$\frac{\partial^2 \tilde{n}}{\partial t^2} - c_s^2 \nabla^2 \tilde{n} = \frac{n}{2} \frac{m_e}{m_i} \nabla^2 \left(\tilde{u}_e^h\right)^2. \tag{15.59}$$

If the electron quiver velocity is considered to behave as an effective thermal velocity, then the low-frequency beat component of $\left(\tilde{u}_e^h\right)^2$ can be considered as modulating the effective electron temperature. Specifically,

$$\tilde{n}_e \kappa T_e + n \kappa \tilde{T}_e = \tilde{n}_e \kappa T_e + n \frac{m_e \left(\tilde{u}_e^h\right)^2}{2} \tag{15.60}$$

and so

$$\tilde{n} c_s^2 \to \tilde{n} c_s^2 + n \widetilde{\left(c_s^2\right)} = \tilde{n} c_s^2 + n \frac{\kappa \tilde{T}_e}{m_i} = \tilde{n} c_s^2 + n \frac{m_e \left(\tilde{u}_e^h\right)^2}{2 m_i}, \tag{15.61}$$

which is consistent with Eq. (15.59).

Low-frequency wave is an electron plasma wave

In this case the low-frequency wave frequency is above ω_{pe} so the ions cannot respond and therefore can be considered to be a stationary, neutralizing background. Now only the electron dynamic response matters and the density perturbation is not quasi-neutral. The electron equation of motion for the low-frequency wave is

$$\frac{\partial \tilde{\mathbf{u}}_e}{\partial t} = \frac{q_e}{m_e} \tilde{\mathbf{E}} - \frac{3 \kappa T_e}{m_e n} \nabla \tilde{n}_e - \frac{1}{2} \nabla \left(\tilde{u}_e^h\right)^2, \tag{15.62}$$

where the 3 comes from the adiabatic pressure perturbation. Combining this with the time derivative of the electron continuity equation,

$$\frac{\partial^2 \tilde{n}_e}{\partial t^2} + n \nabla \cdot \left(\frac{\partial \tilde{\mathbf{u}}_e}{\partial t} \right) = 0, \tag{15.63}$$

gives

$$\frac{\partial^2 \tilde{n}_e}{\partial t^2} + n \nabla \cdot \left(\frac{q_e}{m_e} \tilde{\mathbf{E}} - \frac{3 \kappa T_e}{m_e n} \nabla \tilde{n}_e - \frac{1}{2} \nabla \left(\tilde{u}_e^h\right)^2 \right) = 0. \tag{15.64}$$

This may be simplified by invoking Poisson's equation

$$\nabla \cdot \tilde{\mathbf{E}} = \frac{1}{\varepsilon_0} \tilde{n}_e q_e \tag{15.65}$$

to obtain

$$\frac{\partial^2 \tilde{n}_e}{\partial t^2} + \omega_{pe}^2 \tilde{n}_e - \frac{3\kappa T_e}{m_e} \nabla^2 \tilde{n}_e = \frac{n}{2} \nabla^2 \left(\tilde{u}_e^h\right)^2 . \tag{15.66}$$

The left-hand side is the electron plasma wave equation (Langmuir wave) and the right-hand side provides the nonlinear drive (or coupling) due to ponderomotive force. Again, using Eq. (15.60) it is seen that the ponderomotive force term acts like a modulation of the electron temperature such that $\widetilde{n\kappa T_e} \rightarrow \tilde{n}\kappa T_e + n\kappa \tilde{T}_e$, where $\tilde{T}_e$ is due to the high-frequency electron quiver velocity.

High-frequency wave is an electron plasma wave

In this case the wave is electrostatic and so there is no high-frequency oscillating magnetic field. The nonlinear continuity equation is

$$\frac{\partial \tilde{n}_e}{\partial t} + n\nabla \cdot \tilde{\mathbf{u}}_e = -\nabla \cdot (\tilde{n}_e \tilde{\mathbf{u}}_e) \tag{15.67}$$

and taking a time derivative this becomes

$$\frac{\partial^2 \tilde{n}_e}{\partial t^2} + n\nabla \cdot \frac{\partial \tilde{\mathbf{u}}_e}{\partial t} = -\frac{\partial}{\partial t} \nabla \cdot (\tilde{n}_e \tilde{\mathbf{u}}_e) . \tag{15.68}$$

However, the high-frequency linear electron equation of motion is

$$\frac{\partial \tilde{\mathbf{u}}_e}{\partial t} = \frac{q_e}{m_e} \tilde{\mathbf{E}} - \frac{3\kappa T_e}{m_e n} \nabla \tilde{n}_e \tag{15.69}$$

so Eq. (15.68) becomes

$$\frac{\partial^2 \tilde{n}_e}{\partial t^2} + \omega_{pe}^2 \tilde{n}_e - \frac{3\kappa T_e}{m_e} \nabla^2 \tilde{n}_e = -\frac{\partial}{\partial t} \nabla \cdot (\tilde{n}_e \tilde{\mathbf{u}}_e) . \tag{15.70}$$

In principle, there are two possible components in the right-hand side nonlinear term since $\tilde{n}_e \tilde{\mathbf{u}}_e = \tilde{n}_e^l \tilde{\mathbf{u}}_e^h + \tilde{n}_e^h \tilde{\mathbf{u}}_e^l$, where l and h refer to low- and high-frequency waves. However, as was shown by Eq. (15.46), the $\tilde{n}_e^l \tilde{\mathbf{u}}_e^h$ term is the dominant term and so the $\tilde{n}_e^h \tilde{\mathbf{u}}_e^l$ term can be discarded. If the high-frequency wave associated with $\tilde{\mathbf{u}}_e^h$ is an electromagnetic wave, then the electric field is transverse so $\nabla \cdot \tilde{\mathbf{u}}_e^h \sim \nabla \cdot \tilde{\mathbf{E}}^h = 0$ and, in this case, $\tilde{n}_e^h \tilde{\mathbf{u}}_e^l$ is not just small, but in fact is exactly zero since $\tilde{n}_e^h = 0$. Hence, the high-frequency wave equation becomes

$$\frac{\partial^2 \tilde{n}_e}{\partial t^2} + \omega_{pe}^2 \tilde{n}_e - \frac{3\kappa T_e}{m_e} \nabla^2 \tilde{n}_e = -\frac{\partial}{\partial t} \nabla \cdot \left(\tilde{n}_e^l \tilde{\mathbf{u}}_e^h\right) . \tag{15.71}$$

If the high-frequency wave is electromagnetic, then $\nabla \cdot \left(\tilde{n}_e^l \tilde{\mathbf{u}}_e^h\right) = \tilde{\mathbf{u}}_e^h \cdot \nabla \tilde{n}_e^l$ showing that the nonlinearity consists of a density modulation due to the high-frequency motion across the density ripples of the low-frequency mode. For purposes of consistency, it is worthwhile to express Eq. (15.71) in terms of electric field using

$$\nabla\cdot\tilde{\mathbf{E}} = \frac{1}{\varepsilon_0}\tilde{n}_e q_e \tag{15.72}$$

so

$$\nabla\cdot\left(\frac{\partial^2\tilde{\mathbf{E}}}{\partial t^2} + \omega_{pe}^2\tilde{\mathbf{E}} - \frac{3\kappa T_e}{m_e}\nabla^2\tilde{\mathbf{E}}\right) = -\frac{\partial}{\partial t}\frac{1}{\varepsilon_0}\nabla\cdot\left(\tilde{n}_e^l q_e \tilde{\mathbf{u}}_e^h\right), \tag{15.73}$$

which can be integrated in space to give the general expression for a high-frequency electron plasma wave with nonlinear coupling term,

$$\frac{\partial^2\tilde{\mathbf{E}}}{\partial t^2} + \omega_{pe}^2\tilde{\mathbf{E}} - \frac{3\kappa T_e}{m_e}\nabla^2\tilde{\mathbf{E}} = -\frac{1}{\varepsilon_0}\frac{\partial}{\partial t}\left(\tilde{n}_e^l q_e \tilde{\mathbf{u}}_e^h\right). \tag{15.74}$$

High-frequency wave is an electromagnetic wave

Here the wave is transverse so $\nabla\cdot\tilde{\mathbf{E}} = 0$ and the curl of Faraday's law becomes

$$\nabla^2\tilde{\mathbf{E}} = \frac{\partial\nabla\times\tilde{\mathbf{B}}}{\partial t}. \tag{15.75}$$

Substituting Eq. (15.44) and using the linear equation of motion gives the general expression for a high-frequency electromagnetic wave with nonlinear coupling term,

$$\frac{\partial^2\tilde{\mathbf{E}}}{\partial t^2} + \omega_{pe}^2\tilde{\mathbf{E}} - c^2\nabla^2\tilde{\mathbf{E}} = -\frac{1}{\varepsilon_0}\frac{\partial}{\partial t}\left(\tilde{n}_e^l q_e \tilde{\mathbf{u}}_e^h\right); \tag{15.76}$$

the term $\tilde{n}_e^h q_e \tilde{\mathbf{u}}_e^l$ has been dropped for the reasons given in the previous paragraph.

Summary of mode interactions

Examination of the various combinations considered above shows that the nonlinear coupling acting on high-frequency modes (either pump or high-frequency daughter) is an effective modulation of the density experienced by the high-frequency wave, i.e., $\omega_{pe}^2 \to \omega_{pe}^2 + \omega_{pe}^2\tilde{n}^l/n$. On the other hand, the nonlinear coupling acting on the low-frequency daughter comes from the electron ponderomotive force, which effectively modulates the electron temperature experienced by the low-frequency mode, i.e., $\tilde{n}\kappa T_e \to \tilde{n}\kappa T_e + n\kappa m\left(\tilde{u}_e^h\right)^2/2$.

Coupled oscillator formulation

To see how the coupled wave equations can be expressed in terms of coupled oscillators, the specific situation of an electromagnetic wave interacting with a

Langmuir wave and an ion acoustic wave is now considered. In this case, the three coupled equations are

$$\frac{\partial^2 \tilde{\mathbf{E}}_3}{\partial t^2} + \omega_{pe}^2 \tilde{\mathbf{E}}_3 - c^2 \nabla^2 \tilde{\mathbf{E}}_3 = -\frac{q_e}{\varepsilon_0} \frac{\partial}{\partial t} (\tilde{n}_1 \tilde{\mathbf{u}}_2)$$

$$\frac{\partial^2 \tilde{\mathbf{E}}_2}{\partial t^2} + \omega_{pe}^2 \tilde{\mathbf{E}}_2 - \frac{3\kappa T_e}{m_e} \nabla^2 \tilde{\mathbf{E}}_2 = -\frac{q_e}{\varepsilon_0} \frac{\partial}{\partial t} (\tilde{n}_1 \tilde{\mathbf{u}}_3)$$

$$\frac{\partial^2 \tilde{n}_1}{\partial t^2} - c_s^2 \nabla^2 \tilde{n}_1 = n \frac{m_e}{m_i} \nabla^2 \langle \tilde{\mathbf{u}}_2 \cdot \tilde{\mathbf{u}}_3 \rangle, \tag{15.77}$$

where the relation

$$\left(\frac{1}{2} \left(\tilde{u}_e^h \right)^2 \right)_{\omega_1, \mathbf{k}_1} = \frac{1}{2} \langle (\tilde{\mathbf{u}}_2 + \tilde{\mathbf{u}}_3) \cdot (\tilde{\mathbf{u}}_2 + \tilde{\mathbf{u}}_3) \rangle = \langle \tilde{\mathbf{u}}_2 \cdot \tilde{\mathbf{u}}_3 \rangle \tag{15.78}$$

has been used and the angle brackets refer to the component oscillating at the beat frequency $\omega_1 = \omega_3 - \omega_2$ and having the beat wavevector $\mathbf{k}_1 = \mathbf{k}_3 - \mathbf{k}_2$.

The subscript e has been dropped from all dependent variables because they all refer to electrons. Since the acoustic wave frequency is much smaller than ω_{pe}, it is possible to approximate

$$\frac{\partial}{\partial t} (\tilde{n}_1 \tilde{\mathbf{u}}_2) \simeq \tilde{n}_1 \frac{\partial \tilde{\mathbf{u}}_2}{\partial t} = \tilde{n}_1 \frac{q_e}{m_e} \tilde{\mathbf{E}}_2. \tag{15.79}$$

Using the quiver relation Eq. (15.39), the product of the high-frequency velocities can be expressed as

$$\langle \tilde{\mathbf{u}}_2 \cdot \tilde{\mathbf{u}}_3 \rangle = \frac{q_e^2}{m_e^2 \omega_2 \omega_3} \langle \tilde{\mathbf{E}}_2 \cdot \tilde{\mathbf{E}}_3 \rangle. \tag{15.80}$$

Although the two velocities are oscillating 90° out of phase relative to their respective electric fields, the nonlinear product of the velocities has the same phase behavior as the corresponding electric fields. This is because the nonlinear product of the velocities scales as $\langle \cos \omega_2 t \cos \omega_3 t \rangle$ and the nonlinear product of the electric fields scales as $\langle \sin \omega_2 t \sin \omega_3 t \rangle$ so that the difference between these scalings is $\langle \cos \omega_2 t \cos \omega_3 t - \sin \omega_2 t \sin \omega_3 t \rangle = \langle \cos(\omega_2 + \omega_3) t \rangle$, which vanishes because $\omega_2 + \omega_3$ is non-resonant. The system of equations thus becomes

$$\left(\frac{\partial^2}{\partial t^2} + \omega_{pe}^2 - c^2 \nabla^2 \right) \tilde{\mathbf{E}}_3 = -\omega_{pe}^2 \frac{\tilde{n}_1}{n} \tilde{\mathbf{E}}_2$$

$$\left(\frac{\partial^2}{\partial t^2} + \omega_{pe}^2 - \frac{3\kappa T_e}{m_e} \nabla^2 \right) \tilde{\mathbf{E}}_2 = -\omega_{pe}^2 \frac{\tilde{n}_1}{n} \tilde{\mathbf{E}}_3$$

$$\left(\frac{\partial^2}{\partial t^2} - c_s^2 \nabla^2 \right) \frac{\tilde{n}_1}{n} = \frac{q_e^2}{m_i m_e \omega_2 \omega_3} \nabla^2 \langle \tilde{\mathbf{E}}_2 \cdot \tilde{\mathbf{E}}_3 \rangle. \tag{15.81}$$

If different modes are used, e.g., a Langmuir mode decaying into another Langmuir mode, or if the low-frequency mode is a Langmuir wave, the left-hand side wave terms will be changed accordingly, but the right-hand coupling terms will stay the same except if the low-frequency wave is a Langmuir wave, in which case the product $m_i m_e$ must be replaced by m_e^2 in the denominator of the right-hand side of the third equation.

The right-hand coupling terms can be made identical by defining a renormalized density perturbation

$$\tilde{\psi} = \alpha \frac{\tilde{n}_1}{n} \tag{15.82}$$

so the equations become

$$\begin{aligned}
\left(\frac{\partial^2}{\partial t^2} + \omega_{pe}^2 - c^2\nabla^2\right)\tilde{\mathbf{E}}_3 &= -\omega_{pe}^2 \frac{\tilde{\psi}}{\alpha}\tilde{\mathbf{E}}_2 \\
\left(\frac{\partial^2}{\partial t^2} + \omega_{pe}^2 - \frac{3\kappa T_e}{m_e}\nabla^2\right)\tilde{\mathbf{E}}_2 &= -\omega_{pe}^2 \frac{\tilde{\psi}}{\alpha}\tilde{\mathbf{E}}_3 \\
\left(\frac{\partial^2}{\partial t^2} - c_s^2\nabla^2\right)\tilde{\psi} &= -\frac{\alpha q_e^2 k_1^2}{m_i m_e \omega_2 \omega_3}\left\langle \tilde{\mathbf{E}}_2 \cdot \tilde{\mathbf{E}}_3 \right\rangle .
\end{aligned} \tag{15.83}$$

The replacement $\nabla^2 \langle \tilde{\mathbf{E}}_2 \cdot \tilde{\mathbf{E}}_3 \rangle \to -\, k_1^2 \langle \tilde{\mathbf{E}}_2 \cdot \tilde{\mathbf{E}}_3 \rangle$ has been made in the third line because, as stipulated in the discussion of Eq. (15.78), $\langle \tilde{\mathbf{E}}_2 \cdot \tilde{\mathbf{E}}_3 \rangle$ refers to the component of the product oscillating at the beat frequency $\omega_1 = \omega_3 - \omega_2$ and having the beat wavevector $\mathbf{k}_1 = \mathbf{k}_3 - \mathbf{k}_2$. Equating the coefficients of the nonlinear coupling terms in the second and third lines of Eq. (15.83) gives

$$\alpha = \omega_{pe} \frac{\sqrt{m_i m_e \omega_2 \omega_3}}{q_e k_1} \tag{15.84}$$

so the equations become

$$\begin{aligned}
\left(\frac{\partial^2}{\partial t^2} + \omega_{pe}^2 - c^2\nabla^2\right)\tilde{\mathbf{E}}_3 &= -\lambda \tilde{\psi}\,\tilde{\mathbf{E}}_2 \\
\left(\frac{\partial^2}{\partial t^2} + \omega_{pe}^2 - \frac{3\kappa T_e}{m_e}\nabla^2\right)\tilde{\mathbf{E}}_2 &= -\lambda \tilde{\psi}\,\tilde{\mathbf{E}}_3 \\
\left(\frac{\partial^2}{\partial t^2} - c_s^2\nabla^2\right)\tilde{\psi} &= -\lambda \tilde{\mathbf{E}}_2 \cdot \tilde{\mathbf{E}}_3,
\end{aligned} \tag{15.85}$$

where

$$\lambda = \frac{\omega_{pe} q_e k_1}{\sqrt{m_i m_e \omega_2 \omega_3}} \tag{15.86}$$

and the angle brackets have now been omitted for clarity.

Defining the mode frequencies as

$$\begin{aligned}
\omega_3^2 &= \omega_{pe}^2 + k_3^2 c^2 \\
\omega_2^2 &= \omega_{pe}^2 + 3k_2^2 \kappa T_e / m_e \\
\omega_1^2 &= k_1^2 c_s^2
\end{aligned} \tag{15.87}$$

the coupled equations become

$$\begin{aligned}
\left(\frac{\partial^2}{\partial t^2} + \omega_3^2 \right) \tilde{\mathbf{E}}_3 &= -\lambda \tilde{\psi} \tilde{\mathbf{E}}_2 \\
\left(\frac{\partial^2}{\partial t^2} + \omega_2^2 \right) \tilde{\mathbf{E}}_2 &= -\lambda \tilde{\psi} \tilde{\mathbf{E}}_3 \\
\left(\frac{\partial^2}{\partial t^2} + \omega_1^2 \right) \tilde{\psi} &= -\lambda \tilde{\mathbf{E}}_2 \cdot \tilde{\mathbf{E}}_3,
\end{aligned} \tag{15.88}$$

which is identical to the coupled oscillator system described by Eq. (15.3) if m is set to unity in Eq. (15.3) and $\tilde{\mathbf{E}}_2$ is parallel to $\tilde{\mathbf{E}}_3$. Using Eq. (15.33), the nonlinear growth rate is found to be

$$\begin{aligned}
\gamma &= \frac{\lambda E_3}{4\sqrt{\omega_1 \omega_2}} \\
&= \frac{1}{4} \frac{\omega_{pi}}{\omega_2} \sqrt{\frac{\omega_3}{\omega_1}} k_1 \frac{q_e E_3}{m_e \omega_3}.
\end{aligned} \tag{15.89}$$

15.4 Instability onset via nonlinear dispersion method

An equivalent way of considering the effect of nonlinearity is to derive a so-called nonlinear dispersion relation (Nishikawa 1968a, Nishikawa 1968b). This method has the virtue that both wave damping and frequency mismatches can easily be incorporated. To see how damping can be introduced, consider an electrical circuit consisting of an inductor, capacitor, and resistor all in series. The circuit equation is

$$L \frac{\mathrm{d}^2 Q}{\mathrm{d}t^2} + R \frac{\mathrm{d}Q}{\mathrm{d}t} + \frac{Q}{C} = 0, \tag{15.90}$$

where Q is the charge stored in the capacitor and the current is $I = \mathrm{d}Q/\mathrm{d}t$. The general solution is $Q \sim \mathrm{e}^{-\mathrm{i}\omega t}$, where ω satisfies

$$\omega^2 + \mathrm{i}\omega \frac{R}{L} - \frac{1}{LC} = 0. \tag{15.91}$$

Solving for ω gives the usual damped harmonic oscillator solution

$$\omega = -\frac{\mathrm{i}R}{2L} \pm \sqrt{\frac{1}{LC} - \frac{R^2}{4L^2}}. \tag{15.92}$$

Thus, if $1/LC \gg R/L$ the imaginary part of the frequency is $\omega_i = -R/2L$ and the real part of the frequency is $\omega_r = \pm 1/\sqrt{LC}$. This means that the resistor can be identified with damping by the relation $R \to -2\omega_i L$. Interpreting Q_j now as the dependent variable for oscillation mode j, the appropriate differential equation for the damped mode can be written as

$$\frac{\mathrm{d}^2 Q_j}{\mathrm{d}t^2} + 2\Gamma_j \frac{\mathrm{d}Q_j}{\mathrm{d}t} + \omega_j^2 Q_j = 0, \tag{15.93}$$

where $\Gamma_j = \left|\mathrm{Im}\,\omega_j\right|$ is the linear damping rate for mode j. Mode j therefore has the time dependence $\exp(\pm \mathrm{i}\omega_j t - \Gamma_j t)$, where both ω_j and Γ_j are positive quantities.

Thus, when dissipation is included, the typical system given by Eq. (15.88) generalizes to

$$\left(\frac{\partial^2}{\partial t^2} + 2\Gamma_3 \frac{\partial}{\partial t} + \omega_3^2\right) \tilde{\mathbf{E}}_3 = -\lambda \tilde{\psi} \tilde{\mathbf{E}}_2$$

$$\left(\frac{\partial^2}{\partial t^2} + 2\Gamma_2 \frac{\partial}{\partial t} + \omega_2^2\right) \tilde{\mathbf{E}}_2 = -\lambda \tilde{\psi} \tilde{\mathbf{E}}_3$$

$$\left(\frac{\partial^2}{\partial t^2} + 2\Gamma_1 \frac{\partial}{\partial t} + \omega_1^2\right) \tilde{\psi} = -\lambda \tilde{\mathbf{E}}_2 \cdot \tilde{\mathbf{E}}_3. \tag{15.94}$$

Let us consider once again the situation where the pump wave $\tilde{\mathbf{E}}_3$ is very large and so may be considered as having approximately constant amplitude. The decay waves then do not grow to sufficient amplitude to deplete the pump energy and so only the last two of the three coupled equations have to be considered. We define $\hat{s}_2$ as a unit vector in the direction of $\tilde{\mathbf{E}}_2$ so that

$$\tilde{E}_2 = \hat{s}_2 \cdot \tilde{\mathbf{E}}_2 \tag{15.95}$$

and the last two of the three coupled equations become

$$\left(\frac{\partial^2}{\partial t^2} + 2\Gamma_2 \frac{\partial}{\partial t} + \omega_2^2\right) \tilde{E}_2 = -\lambda \tilde{\psi} \hat{s}_2 \cdot \tilde{\mathbf{E}}_3$$

$$\left(\frac{\partial^2}{\partial t^2} + 2\Gamma_1 \frac{\partial}{\partial t} + \omega_1^2\right) \tilde{\psi} = -\lambda \tilde{E}_2 \hat{s}_2 \cdot \tilde{\mathbf{E}}_3. \tag{15.96}$$

The wavevector selection rules are assumed to be satisfied but a slight mismatching in the frequency selection rules will be allowed. The pump wave is now written as

$$\hat{s}_2 \cdot \tilde{\mathbf{E}}_3 = Z_3 \cos \omega_3 t, \tag{15.97}$$

where Z_3 is the effective pump wave amplitude. The coupled daughter equations then become

$$\begin{aligned} \left(\frac{\partial^2}{\partial t^2} + 2\Gamma_2 \frac{\partial}{\partial t} + \omega_2^2\right) \tilde{E}_2 &= -\frac{\lambda Z_3}{2}\left(\mathrm{e}^{\mathrm{i}\omega_3 t} + \mathrm{e}^{-\mathrm{i}\omega_3 t}\right) \tilde{\psi} \\ \left(\frac{\partial^2}{\partial t^2} + 2\Gamma_1 \frac{\partial}{\partial t} + \omega_1^2\right) \tilde{\psi} &= -\frac{\lambda Z_3}{2}\left(\mathrm{e}^{\mathrm{i}\omega_3 t} + \mathrm{e}^{-\mathrm{i}\omega_3 t}\right) \tilde{E}_2. \end{aligned} \tag{15.98}$$

The Fourier transform of a quantity $f(t)$ is defined as

$$f(\omega) = \int \mathrm{d}t f(t) \mathrm{e}^{\mathrm{i}\omega t}, \tag{15.99}$$

where the sign of the exponent is chosen to be consistent with the convention that a single Fourier mode has the form $\exp(-\mathrm{i}\omega t)$. The Fourier transform of $f(t)\mathrm{e}^{\pm \mathrm{i}\omega_3 t}$ will therefore be

$$\int \mathrm{d}t \left(f(t)\mathrm{e}^{\pm \mathrm{i}\omega_3 t}\right) \mathrm{e}^{\mathrm{i}\omega t} = f(\omega \pm \omega_3) \tag{15.100}$$

and the Fourier transform of $\partial f/\partial t$ will be

$$\int \mathrm{d}t \left(\frac{\partial}{\partial t} f(t)\right) \mathrm{e}^{\mathrm{i}\omega t} = -\mathrm{i}\omega \int \mathrm{d}t f(t) \mathrm{e}^{\mathrm{i}\omega t} = -\mathrm{i}\omega f(\omega). \tag{15.101}$$

Application of Eqs. (15.100) and (15.101) to Eqs. (15.98) gives

$$\begin{aligned} \left(-\omega^2 - 2\mathrm{i}\omega\Gamma_2 + \omega_2^2\right) \tilde{E}_2(\omega) &= -\frac{\lambda Z_3}{2}\left[\tilde{\psi}(\omega + \omega_3) + \tilde{\psi}(\omega - \omega_3)\right] \\ \left(-\omega^2 - 2\mathrm{i}\omega\Gamma_1 + \omega_1^2\right) \tilde{\psi}(\omega) &= \frac{\lambda Z_3}{2}\left[\tilde{E}_2(\omega + \omega_3) + \tilde{E}_2(\omega - \omega_3)\right]. \end{aligned} \tag{15.102}$$

By defining a generic linear dispersion relation

$$\varepsilon_j(\omega) = -\omega^2 - 2\mathrm{i}\omega\Gamma_j + \omega_j^2, \tag{15.103}$$

these equations can be written as

$$\varepsilon_2(\omega)\tilde{E}_2(\omega) = -\frac{\lambda Z_3}{2}\left[\tilde{\psi}(\omega + \omega_3) + \tilde{\psi}(\omega - \omega_3)\right] \tag{15.104}$$

$$\varepsilon_1(\omega)\tilde{\psi}(\omega) = -\frac{\lambda Z_3}{2}\left[\tilde{E}_2(\omega + \omega_3) + \tilde{E}_2(\omega - \omega_3)\right]. \tag{15.105}$$

The frequency ω in Eq. (15.105) can be replaced by $\omega \pm \omega_3$ to give

$$\tilde{\psi}(\omega \pm \omega_3) = -\frac{\lambda Z_3}{2\varepsilon_1(\omega \pm \omega_3)} \left[\tilde{E}_2(\omega) + \tilde{E}_2(\omega \pm 2\omega_3)\right], \tag{15.106}$$

which is then substituted into Eq. (15.104) to obtain

$$\varepsilon_2(\omega)\tilde{E}_2(\omega) = \left(\frac{\lambda Z_3}{2}\right)^2 \left[\frac{\tilde{E}_2(\omega) + \tilde{E}_2(\omega + 2\omega_3)}{\varepsilon_1(\omega + \omega_3)} + \frac{\tilde{E}_2(\omega) + \tilde{E}_2(\omega - 2\omega_3)}{\varepsilon_1(\omega - \omega_3)}\right]. \tag{15.107}$$

The terms $\tilde{E}_2(\omega \pm 2\omega_3)$ can be discarded because, being non-resonant, they have insignificant amplitude. What remains is

$$\varepsilon_2(\omega) = \left(\frac{\lambda Z_3}{2}\right)^2 \left[\frac{1}{\varepsilon_1(\omega + \omega_3)} + \frac{1}{\varepsilon_1(\omega - \omega_3)}\right], \tag{15.108}$$

which is called the *nonlinear dispersion relation* (Nishikawa 1968a, Nishikawa 1968b).

The nonlinear dispersion relation is investigated by first writing

$$\omega = x + \mathrm{i}y, \tag{15.109}$$

where it is assumed

$$x \simeq \omega_2 \tag{15.110}$$

and also

$$\omega - \omega_3 = x + \mathrm{i}y - \omega_3 \simeq -\omega_1. \tag{15.111}$$

These assumptions and definitions have been made so that $\varepsilon_1(\omega - \omega_3)$ is close to zero, $\varepsilon_1(\omega + \omega_3)$ is not close to zero, and positive y corresponds to instability. The term $1/\varepsilon_1(\omega + \omega_3)$ can therefore be discarded as being non-resonant and the nonlinear dispersion relation simplifies to

$$\varepsilon_2(\omega)\varepsilon_1(\omega - \omega_3) = \left(\frac{\lambda Z_3}{2}\right)^2. \tag{15.112}$$

The linear dispersion relations ε_2 and ε_1 on the left-hand side are each Taylor-expanded to give

$$\begin{aligned}
\varepsilon_2(\omega) &= \varepsilon_2(\omega_2 + x - \omega_2 + \mathrm{i}y) \\
&\simeq \varepsilon_2(\omega_2) + (x - \omega_2 + \mathrm{i}y) \left.\frac{\mathrm{d}\varepsilon_2}{\mathrm{d}\omega}\right|_{\omega=\omega_2} \\
&= -2\mathrm{i}\omega_2\Gamma_2 - 2\omega_2(x - \omega_2 + \mathrm{i}y)
\end{aligned} \tag{15.113}$$

and

$$\begin{aligned}\varepsilon_1(\omega-\omega_3) &= \varepsilon_1(-\omega_1+x+\mathrm{i}y+\omega_1-\omega_3)\\ &\simeq \varepsilon_2(-\omega_1)+(x+\mathrm{i}y+\omega_1-\omega_3)\left.\frac{\mathrm{d}\varepsilon_1}{\mathrm{d}\omega}\right|_{\omega=-\omega_1}\\ &= 2\mathrm{i}\omega_1\Gamma_1+2\omega_1\left(x+\mathrm{i}y+\omega_1-\omega_3\right). \end{aligned} \tag{15.114}$$

Using these expansions, the nonlinear dispersion relation becomes

$$\{-\mathrm{i}\left(\Gamma_2+y\right)-(x-\omega_2)\}\left\{\mathrm{i}\left(\Gamma_1+y\right)+(x+\omega_1-\omega_3)\right\}=\frac{1}{\omega_1\omega_2}\left(\frac{\lambda Z_3}{4}\right)^2 \tag{15.115}$$

or, in more compact form,

$$(\bar{x}+\mathrm{i}\left(\Gamma_2+y\right))\left(\bar{x}-\Delta+\mathrm{i}\left(\Gamma_1+y\right)\right)=-\frac{1}{\omega_1\omega_2}\left(\frac{\lambda Z_3}{4}\right)^2, \tag{15.116}$$

where

$$\bar{x}=x-\omega_2 \tag{15.117}$$

and

$$\Delta=\omega_3-(\omega_1+\omega_2) \tag{15.118}$$

is the frequency mismatch.

The real and imaginary parts of Eq. (15.116) are

$$(\bar{x}-\Delta)\,\bar{x}-\left(\Gamma_2+y\right)\left(\Gamma_1+y\right)=-\frac{1}{\omega_1\omega_2}\left(\frac{\lambda Z_3}{4}\right)^2 \tag{15.119}$$

$$\bar{x}\left(\Gamma_1+y\right)+(\bar{x}-\Delta)\left(\Gamma_2+y\right)=0. \tag{15.120}$$

Solving Eq. (15.120) gives

$$\bar{x}=\frac{(\Gamma_2+y)}{(2y+\Gamma_1+\Gamma_2)}\Delta, \quad \bar{x}-\Delta=-\frac{(\Gamma_1+y)}{(2y+\Gamma_1+\Gamma_2)}\Delta \tag{15.121}$$

and substitution for $\tilde{x}$ and $\tilde{x}-\Delta$ in Eq. (15.119) results in the expression for the growth rate

$$(\Gamma_2+y)(\Gamma_1+y)\left(1+\frac{\Delta^2}{(2y+\Gamma_1+\Gamma_2)^2}\right)=\frac{1}{\omega_1\omega_2}\left(\frac{\lambda Z_3}{4}\right)^2. \tag{15.122}$$

Onset of instability corresponds to y being close to zero and at marginal stability Eq. (15.122) becomes

$$(\Gamma_2+\Gamma_1)\,y+\Gamma_1\Gamma_2=\frac{\lambda^2 Z_3^2}{16\omega_1\omega_2\left(1+\dfrac{\Delta^2}{(\Gamma_2+\Gamma_1)^2}\right)}. \tag{15.123}$$

The growth rate near threshold thus is

$$y=\frac{1}{(\Gamma_2+\Gamma_1)}\left(\frac{\lambda^2 Z_3^2}{16\omega_1\omega_2\left(1+\dfrac{\Delta^2}{(\Gamma_2+\Gamma_1)^2}\right)}-\Gamma_1\Gamma_2\right). \tag{15.124}$$

This shows that the pump amplitude threshold for instability is

$$Z_3=\frac{4\sqrt{\omega_1\omega_2\Gamma_1\Gamma_2}}{\lambda}\sqrt{1+\frac{\Delta^2}{(\Gamma_2+\Gamma_1)^2}}; \tag{15.125}$$

this threshold is proportional to the geometric mean of the linear damping rates of the two modes. If the pump amplitude is below threshold then the nonlinear instability does not occur. The lowest threshold occurs when $\Delta=0$, i.e., when the frequency selection rule is exactly satisfied. Instability when the pump amplitude exceeds a threshold is routinely observed in actual experimental situations as was discussed earlier, i.e., decay instability of a pump wave is observed to begin only when the pump wave amplitude exceeds threshold (for example, see Stenzel and Wong (1972)) . For a pump amplitude well above threshold, Eq. (15.122) becomes

$$y=\sqrt{\frac{1}{\omega_1\omega_2}\left(\frac{\lambda Z_3}{4}\right)^2-\frac{\Delta^2}{4}}, \tag{15.126}$$

which shows that frequency mismatch reduces the growth rate. The situation reduces to Eqs. (15.33) and (15.89) when $\Delta=0$. Thus the nonlinear dispersion relation formalism extends the Manley–Rowe coupled oscillator model to include the effects of both dissipation and frequency mismatch.

15.5 Digging a density hole via ponderomotive force

If the low-frequency daughter mode is at zero frequency, then the ion acoustic mode ceases to be a wave. Instead, it becomes a density depletion caused by the ponderomotive force and so Eq. (15.56) reduces to

$$\frac{1}{n}\nabla\tilde{n} = -\frac{1}{2}\frac{m_e}{\kappa T_e}\nabla\left(\tilde{u}_e^h\right)^2, \tag{15.127}$$

which can be integrated to give a Boltzmann-like relation

$$\frac{\tilde{n}}{n} = -\frac{1}{2}\frac{m_e\left\langle\left(\tilde{u}_e^h\right)^2\right\rangle}{\kappa T_e}. \tag{15.128}$$

Since the high-frequency daughter wave is the same as the pump wave there is now only one high-frequency wave and so there is no need to have subscripts distinguishing the modes. The consequence is that the high-frequency wave propagates in a plasma having a density depletion dug out by the ponderomotive force associated with the wave. For example, the Langmuir wave equation in this case would become

$$\left(\frac{\partial^2}{\partial t^2} + 2\Gamma\frac{\partial}{\partial t} + \omega_{pe}^2\left(1 - \frac{1}{2}\frac{m_e\left\langle\left(\tilde{u}_e^h\right)^2\right\rangle}{\kappa T_e}\right) - \frac{3\kappa T_e}{m_e}\nabla^2\right)\tilde{\mathbf{u}}_e^h = 0, \tag{15.129}$$

where $\tilde{\mathbf{u}}_e^h$ has been used as the linear variable instead of $\tilde{\mathbf{E}}$ and a linear damping term involving the linear damping rate Γ has been introduced.

An undamped linear Langmuir wave in a uniform plasma satisfies the dispersion relation $\omega^2 = \omega_{pe}^2(1+3k^2\lambda_{De}^2)$, where $k\lambda_{De} \ll 1$ so that the wave frequency is very close to ω_{pe}. It is reasonable to presume that the nonlinear wave also oscillates at a frequency very close to ω_{pe} and so what is important is the deviation of the frequency from ω_{pe}. To investigate this, the electron fluid velocity is assumed to be of the form

$$\begin{aligned}\tilde{\mathbf{u}}_e^h(\mathbf{x}, t) &= \mathrm{Re}\left[\mathbf{A}(x, t)\mathrm{e}^{-\mathrm{i}\omega_{pe}t}\right] \\ &= \frac{1}{2}\left\{\mathbf{A}(x, t)\mathrm{e}^{-\mathrm{i}\omega_{pe}t} + \mathbf{A}^*(x, t)\mathrm{e}^{\mathrm{i}\omega_{pe}t}\right\},\end{aligned} \tag{15.130}$$

in which case

$$\left\langle\left(\tilde{u}_e^h\right)^2\right\rangle = \frac{1}{2}|\mathbf{A}|^2 \tag{15.131}$$

and the time dependence of $\mathbf{A}(x, t)$ characterizes the extent to which the wave frequency deviates from ω_{pe}. Because this deviation is small, $\mathbf{A}$ changes slowly compared to ω_{pe}, and so in analogy to Eq. (3.22) it is possible to approximate

$$\frac{\partial^2}{\partial t^2}\left[\mathbf{A}(x, t)\mathrm{e}^{-\mathrm{i}\omega_{pe}t}\right] \simeq -\omega_{pe}^2\mathbf{A}(x, t)\mathrm{e}^{-\mathrm{i}\omega_{pe}t} - 2\mathrm{i}\omega_{pe}\frac{\partial \mathbf{A}}{\partial t}\mathrm{e}^{-\mathrm{i}\omega_{pe}t} \tag{15.132}$$

and

$$2\Gamma\frac{\partial}{\partial t}\left[\mathbf{A}(x, t)\mathrm{e}^{-\mathrm{i}\omega_{pe}t}\right] \approx -\mathrm{i}2\omega_{pe}\Gamma\mathbf{A}\mathrm{e}^{-\mathrm{i}\omega_{pe}t}.$$

Substitution of Eq. (15.132) into Eq. (15.129) gives

$$2\mathrm{i}\omega_{pe}\frac{\partial \mathbf{A}}{\partial t} + \mathrm{i}2\omega_{pe}\Gamma\mathbf{A} + \frac{\omega_{pe}^2}{4}\frac{m_e}{\kappa T_e}\frac{|\mathbf{A}|^2}{} A + c^2\nabla^2\mathbf{A} = 0. \tag{15.133}$$

By defining the normalized variables

$$\begin{aligned}
\tau &= \omega_{pe}t/2 \\
\chi &= \frac{\mathbf{A}}{2\sqrt{\kappa T_e/m_e}} \\
\xi &= \mathbf{x}\omega_{pe}/c \\
\eta &= 2\Gamma/\omega_{pe}
\end{aligned} \tag{15.134}$$

Eq. (15.133) can be put in the standardized form

$$\mathrm{i}\frac{\partial \chi}{\partial \tau} + \mathrm{i}\eta\chi + |\chi|^2\chi + \nabla_\xi^2\chi = 0; \tag{15.135}$$

this is called a nonlinear Schrödinger equation since, if $\eta = 0$, this equation resembles a Schrödinger equation where $|\chi|^2$ plays the role of a potential energy.

In order to exploit this analogy, we recall the relationship between the Schrödinger equation and the classical conservation of energy relation for a particle in a potential well $V(x)$. According to classical mechanics, the sum of the kinetic and potential energies gives the total energy, i.e.,

$$\frac{p^2}{2m} + V(\mathbf{x}) = E. \tag{15.136}$$

However, in quantum mechanics, the momentum and the energy are expressed as spatial and temporal operators, $\mathbf{p} = -\mathrm{i}\hbar\nabla$ and $E = \mathrm{i}\hbar\partial/\partial t$, which act on a wave function ψ so that Eq. (15.136) becomes

$$-\frac{\hbar^2}{2m}\nabla^2\psi + V\psi = \mathrm{i}\hbar\frac{\partial\psi}{\partial t} \tag{15.137}$$

or, after rearrangement,

$$\mathrm{i}\hbar\frac{\partial\psi}{\partial t} - V\psi + \frac{\hbar^2}{2m}\nabla^2\psi = 0. \tag{15.138}$$

Equation (15.136) shows that a particle will be trapped in a potential well if $V(\pm\infty) > E > V_{\min}$, where $V_{\min}$ is the minimum value of V. From the quantum mechanical point of view, $|\psi|^2$ is the probability of finding the particle at position $\mathbf{x}$. Thus, existence of solutions to Eq. (15.138) localized to the vicinity of $V_{\min}$ is the quantum mechanical way of stating that a particle can be trapped in a potential well. Comparison of Eqs. (15.135) and (15.138) shows that $-\left|\bar{\mathbf{A}}\right|^2$ plays the role of V and so a local maximum of $\left|\bar{\mathbf{A}}\right|^2$ should act as an effective potential well. This makes physical sense because Langmuir waves reflect from regions of high density and the amplitude-dependent ponderomotive force digs a hole in the plasma. Thus, regions of high wave amplitude create a density depression and the Langmuir wave reflects from the high-density regions surrounding this density depression. The Langmuir wave then becomes trapped in a depression of its own making. Formation of this depression can be an unstable process because if a wave is initially trapped in a shallow well, its energy $|\chi|^2$ will concentrate at the bottom of this well, but this concentration of $|\chi|^2$ will make the well deeper and so concentrate the wave energy into a smaller region, making $|\chi|^2$ even larger, and so on.

Caviton instability

The instability outlined above can be described in a quantitative manner using the 1-D version of Eq. (15.135), namely

$$\mathrm{i}\frac{\partial\chi}{\partial\tau} + \mathrm{i}\eta\chi + |\chi|^2\chi + \frac{\partial^2\chi}{\partial\xi^2} = 0. \tag{15.139}$$

It is assumed that a stable solution $\chi_0(x, t)$ exists initially and satisfies

$$\mathrm{i}\frac{\partial\chi_0}{\partial\tau} + \mathrm{i}\eta\chi_0 + |\chi_0|^2\chi_0 + \frac{\partial^2\chi_0}{\partial\xi^2} = 0, \tag{15.140}$$

where $|\chi_0(x, t)|$ is bounded in both time and space. Next, a slightly different solution is considered,

$$\chi(x, t) = \chi_0(x, t) + \tilde{\chi}(x, t), \tag{15.141}$$

where the perturbation $\tilde{\chi}(x, t)$ is assumed to be small compared to $\chi_0(x, t)$. The equation for $\chi(x, t)$ is thus

$$\mathrm{i}\frac{\partial}{\partial\tau}(\chi_0 + \tilde{\chi}) + \mathrm{i}\eta(\chi_0 + \tilde{\chi}) + |\chi_0 + \tilde{\chi}|^2(\chi_0 + \tilde{\chi}) + \frac{\partial^2}{\partial\xi^2}(\chi_0 + \tilde{\chi}) = 0. \tag{15.142}$$

Subtracting Eq. (15.140) from (15.142) yields

$$\mathrm{i}\frac{\partial\tilde{\chi}}{\partial\tau}+\mathrm{i}\eta\tilde{\chi}+|\chi_0+\tilde{\chi}|^2(\chi_0+\tilde{\chi})-|\chi_0|^2\chi_0+\frac{\partial^2\tilde{\chi}}{\partial\xi^2}=0. \tag{15.143}$$

Expansion of the potential-energy-like terms while keeping only terms linear in the perturbation gives

$$|\chi_0+\tilde{\chi}|^2(\chi_0+\tilde{\chi})-|\chi_0|^2\chi_0\approx\chi_0^2\tilde{\chi}^*+2|\chi_0|^2\tilde{\chi} \tag{15.144}$$

so Eq. (15.143) becomes

$$\mathrm{i}\frac{\partial\tilde{\chi}}{\partial\tau}+\mathrm{i}\eta\tilde{\chi}+\chi_0^2\tilde{\chi}^*+2|\chi_0|^2\tilde{\chi}+\frac{\partial^2\tilde{\chi}}{\partial\xi^2}=0. \tag{15.145}$$

It is now assumed that the perturbation is unstable and has the space-time dependence

$$\tilde{\chi}\sim\mathrm{e}^{\mathrm{i}k\xi+\gamma t}, \tag{15.146}$$

in which case Eq. (15.145) becomes

$$\left(\mathrm{i}\gamma+\mathrm{i}\eta+2|\chi_0|^2-k^2\right)\tilde{\chi}=-\chi_0^2\tilde{\chi}^*, \tag{15.147}$$

which has the complex conjugate

$$\left(-\mathrm{i}\gamma-\mathrm{i}\eta+2|\chi_0|^2-k^2\right)\tilde{\chi}^*=-\chi_0^{*2}\tilde{\chi}. \tag{15.148}$$

Combining the above two equations gives a dispersion relation for the growth rate

$$(\gamma+\eta)^2=-k^4+4k^2|\chi_0|^2-3|\chi_0|^4. \tag{15.149}$$

The maximum γ is found by taking the derivative of both sides with respect to k^2 and setting this derivative to zero, obtaining

$$k_{\mathrm{max}}^2=2|\chi_0|^2 \tag{15.150}$$

as the value of k^2 giving the maximum for γ. Substitution of k_{max}^2 into Eq. (15.149) gives

$$\gamma_{\mathrm{max}}=-\eta+|\chi_0|^2. \tag{15.151}$$

Thus, the configuration is unstable when $|\chi_0|^2>\eta$ or, in terms of the original variables, when

$$\frac{m_e\left\langle\left(\tilde{u}_e^h\right)^2\right\rangle}{4\kappa T_e}>\frac{\Gamma}{\omega_{pe}}. \tag{15.152}$$

This is called a caviton instability and it tends to dig a sharp hole in the plasma density. This is because each successive stage of growth can be considered a quasi-equilibrium, which is destabilized, and from Eq. (15.150) it is seen that the most unstable k becomes progressively larger as the amplitude increases. Another way of seeing this hole-digging tendency is to note that Eqs. (15.147) and (15.148) are coupled by their respective right-hand terms and these terms are proportional to the amplitude of the original wave. It is this coupling that leads to instability of the perturbation and so the perturbation is most unstable where the amplitude of the original wave was largest. The digging of a density cavity in a plasma by a Langmuir wave was observed experimentally by Kim, Stenzel and Wong (1974); the density cavity was called a caviton.

Stationary Envelope Soliton

A special, fully nonlinear solution of Eq. (15.139) can be found in the limit where damping is sufficiently small to be neglected so that the nonlinear Schrödinger equation reduces to

$$\mathrm{i}\frac{\partial \chi}{\partial \tau}+|\chi|^2\chi+\frac{\partial^2 \chi}{\partial \xi^2}=0. \tag{15.153}$$

We now search for a solution that vanishes at infinity, propagates at some fixed velocity, and oscillates so that

$$\chi = g(\xi)\mathrm{e}^{\mathrm{i}\Omega\tau}. \tag{15.154}$$

With this assumption, Eq. (15.153) becomes

$$-\Omega g+g^3+g''=0. \tag{15.155}$$

After multiplying by the integrating factor g', this becomes

$$\frac{\mathrm{d}}{\mathrm{d}\xi}\left(-\frac{\Omega}{2}g^2+\frac{1}{4}g^4+\frac{1}{2}\left(g'\right)^2\right)=0. \tag{15.156}$$

Since the solution is assumed to vanish at infinity, integration with respect to ξ from $\xi=-\infty$ gives

$$\left(g'\right)^2=\Omega g^2-\frac{1}{2}g^4 \tag{15.157}$$

or

$$\frac{\mathrm{d}g}{g\sqrt{\Omega-\frac{1}{2}g^2}}=\mathrm{d}\xi. \tag{15.158}$$

By letting

$$g=\frac{\sqrt{2\Omega}}{\cosh\theta} \tag{15.159}$$

it is seen that

$$dg = -\frac{\sqrt{2\Omega}}{\cosh^2\theta}\sinh\theta d\theta, \tag{15.160}$$

in which case

$$\theta = -\sqrt{\Omega}\xi + \delta, \tag{15.161}$$

where δ is an arbitrary constant. Thus, the solution is

$$\chi(\xi, \tau) = \frac{\sqrt{2\Omega}e^{i\Omega\tau}}{\cosh\left(\sqrt{\Omega}\xi - \delta\right)}, \tag{15.162}$$

which is localized in space. For large Ω, the oscillation frequency and amplitude both increase, and the localization is more pronounced.

Propagating envelope soliton

One may ask whether the above solution can be generalized to a propagating solution, i.e., can ξ be replaced by $\xi - vt$, where v is a velocity? Making only this replacement is clearly inadequate and so a solution of the form

$$\chi = g(\xi - vt)e^{i\Omega\tau + ih(\xi,\tau)} \tag{15.163}$$

is assumed, where $h(\xi, \tau)$ is an unknown function to be determined. Substitution of this assumed solution into Eq. (15.153) gives

$$-ivg' - \Omega g - g\frac{\partial h}{\partial \tau} + g^3 + g'' + 2i\frac{\partial h}{\partial \xi}g' - \left(\frac{\partial h}{\partial \xi}\right)^2 g = 0. \tag{15.164}$$

Setting the imaginary part to zero gives

$$v = 2\frac{\partial h}{\partial \xi}, \tag{15.165}$$

which can be integrated to give

$$h = \frac{v\xi}{2} + f(\tau), \tag{15.166}$$

where $f(\tau)$ is to be determined. The real part of the equation becomes

$$-\Omega g - g\frac{\partial h}{\partial \tau} + g^3 + g'' - \frac{v^2}{4}g = 0. \tag{15.167}$$

If we set

$$\frac{\partial h}{\partial \tau} = -\frac{v^2}{4} \tag{15.168}$$

then the second and last terms in Eq. (15.167) cancel each other. Equation (15.168) implies $f(\tau) = -v^2\tau/4$ so

$$h(\xi, \tau) = \frac{v\xi}{2} - \frac{v^2}{4}\tau. \tag{15.169}$$

Thus, Eq. (15.167) reverts to Eq. (15.155) which is solved as in Eqs. (15.156)–(15.162) to give the propagating envelope soliton

$$\chi(\xi, \tau) = \frac{\sqrt{2\Omega}\exp\left(\mathrm{i}\Omega\tau + \mathrm{i}v\xi/2 - \mathrm{i}v^2\tau/4\right)}{\cosh\left(\sqrt{\Omega}(\xi - vt) - \delta\right)}. \tag{15.170}$$

15.6 Ion acoustic wave soliton

The ion acoustic wave dispersion relation is

$$\omega^2 = \frac{k^2 c_s^2}{1 + k^2\lambda_D^2}, \tag{15.171}$$

which has the forward propagating solution

$$\omega = kc_s\left(1 + k^2\lambda_D^2\right)^{-1/2} \tag{15.172}$$

or, for small $k\lambda_D$,

$$\omega = kc_s - k^3\lambda_D^2 c_s/2, \tag{15.173}$$

where the last term is small. Since $\partial/\partial x \to \mathrm{i}k$ and $\partial/\partial t \to -\mathrm{i}\omega$ the inverse substitution is $k \to -\mathrm{i}\partial/\partial x$ and $\omega \to \mathrm{i}\partial/\partial t$. Applying this inverse substitution to Eq. (15.173) shows that the forward propagating ion acoustic wave implies that the partial differential equation for, say, ion velocity would be

$$\mathrm{i}\frac{\partial u_i}{\partial t} = -\mathrm{i}\frac{\partial u_i}{\partial x}c_s - (-\mathrm{i})^3\frac{\partial^3 u_i}{\partial x^3}\frac{\lambda_D^2 c_s}{2}. \tag{15.174}$$

After multiplying by $-\mathrm{i}$, this gives the dispersive forward propagating wave equation

$$\frac{\partial u_i}{\partial t} = -c_s\frac{\partial u_i}{\partial x} - \frac{\lambda_D^2 c_s}{2}\frac{\partial^3 u_i}{\partial x^3}. \tag{15.175}$$

This wave equation was derived using a linearized version of the ion fluid equation of motion, namely

$$m_i\frac{\partial u_i}{\partial t} = q_i E. \tag{15.176}$$

If, instead, the complete nonlinear ion equation had been used, the ion fluid equation of motion would contain a convective nonlinear term and be

$$m_i\left(\frac{\partial u_i}{\partial t}+u_i\frac{\partial u_i}{\partial x}\right)=q_iE. \tag{15.177}$$

This suggests that inclusion of convective ion nonlinearity corresponds to making the generalization

$$\frac{\partial u_i}{\partial t}\to\frac{\partial u_i}{\partial t}+u_i\frac{\partial u_i}{\partial x} \tag{15.178}$$

and so the forward propagating ion acoustic wave equation with inclusion of ion convective nonlinearity is

$$\frac{\partial u_i}{\partial t}+(u_i+c_s)\frac{\partial u_i}{\partial x}+\frac{\lambda_D^2 c_s}{2}\frac{\partial^3 u_i}{\partial x^3}=0. \tag{15.179}$$

This suggests defining a new variable

$$U=u_i+c_s, \tag{15.180}$$

which differs only by a constant from the ion fluid velocity. The forward propagating nonlinear ion acoustic wave equation can thus be rewritten as

$$\frac{\partial U}{\partial t}+U\frac{\partial U}{\partial x}+\frac{\lambda_D^2 c_s}{2}\frac{\partial^3 U}{\partial x^3}=0. \tag{15.181}$$

We now introduce dimensionless variables by normalizing lengths to the Debye length, velocities to c_s, and time to $\omega_{pi}=c_s/\lambda_{de}$. The dimensionless variables are thus

$$\chi=\frac{U}{c_s},\ \xi=\frac{x}{\lambda_D},\ \tau=\omega_{pi}t, \tag{15.182}$$

and the dimensionless wave equation becomes

$$\frac{\partial\chi}{\partial\tau}+\chi\frac{\partial\chi}{\partial\xi}+\frac{1}{2}\frac{\partial^3\chi}{\partial\xi^3}=0; \tag{15.183}$$

this is called the Korteweg–de Vries (KdV) equation (Korteweg and de Vries 1895). Modern interest in this equation was stimulated with the discovery of a general solution to the soliton problem by Gardner *et al.* (1967).

Because the linear wave is forward traveling with unity velocity in these dimensionless variables, it is reasonable to postulate that the nonlinear solution has the forward propagating form

$$\chi=\chi(\xi-V\tau), \tag{15.184}$$

where V is of order unity. A special solution can be found by introducing the wave-frame position variable

$$\eta = \xi - V\tau \tag{15.185}$$

so the lab-frame space and time derivatives can be written as

$$\begin{aligned} \frac{\partial}{\partial \xi} &= \frac{\partial}{\partial \eta} \\ \frac{\partial}{\partial \tau} &= -V\frac{\partial}{\partial \eta}. \end{aligned} \tag{15.186}$$

Substitution of these into the wave equation gives an ordinary differential equation in the wave-frame,

$$-V\frac{\mathrm{d}\chi}{\mathrm{d}\eta} + \frac{1}{2}\frac{\mathrm{d}\chi^2}{\mathrm{d}\eta} + \frac{1}{2}\frac{\mathrm{d}^3\chi}{\mathrm{d}\eta^3} = 0, \tag{15.187}$$

where d has been used instead of ∂ because the equation is now an ordinary differential equation. A solution is sought that vanishes at both plus and minus infinity; such a solution is called a solitary wave. To find this solution, Eq. (15.187) is integrated using the boundary condition that χ vanishes at infinity to obtain

$$-V\chi + \frac{\chi^2}{2} + \frac{1}{2}\frac{\mathrm{d}^2\chi}{\mathrm{d}\eta^2} = 0. \tag{15.188}$$

Multiplying by the integration factor $\mathrm{d}\chi/\mathrm{d}\eta$ allows this to be recast as

$$\frac{\mathrm{d}}{\mathrm{d}\eta}\left(-\frac{V\chi^2}{2} + \frac{\chi^3}{6} + \frac{1}{4}\left(\frac{\mathrm{d}\chi}{\mathrm{d}\eta}\right)^2\right) = 0 \tag{15.189}$$

and then integrating gives

$$\frac{\mathrm{d}\chi}{\mathrm{d}\eta} = \chi\sqrt{2V - \frac{2}{3}\chi}, \tag{15.190}$$

where the boundary condition at infinity has been used again. This can be written as

$$\frac{\mathrm{d}\,(\chi/3)}{2^{1/2}\frac{\chi}{3}\sqrt{V - \frac{\chi}{3}}} = \mathrm{d}\eta. \tag{15.191}$$

The substitution

$$\frac{\chi}{3} = \frac{V}{\cosh^2\theta} \tag{15.192}$$

allows simplification of Eq. (15.191) to

$$\mathrm{d}\left(\frac{\chi}{3}\right) = -\frac{2V\sinh\theta}{\cosh^3\theta}\mathrm{d}\theta. \tag{15.193}$$

Thus, Eq. (15.191) becomes

$$\frac{-\dfrac{2V\sinh\theta}{\cosh^3\theta}\mathrm{d}\theta}{2^{1/2}\dfrac{V}{\cosh^2\theta}\sqrt{V-\dfrac{V}{\cosh^2\theta}}} = \mathrm{d}\eta \tag{15.194}$$

or

$$-\sqrt{\frac{2}{V}}\mathrm{d}\theta = \mathrm{d}\eta,$$

which can be integrated to give

$$\theta = \sqrt{\frac{V}{2}}\left(\xi_0 - \eta\right), \tag{15.195}$$

where ξ_0 is a constant. The propagating solitary wave solution to Eq. (15.183) is therefore

$$\chi(\eta) = \frac{3V}{\cosh^2\left(\sqrt{\dfrac{V}{2}}\left(\xi_0 - \eta\right)\right)} \tag{15.196}$$

or

$$\chi(\xi,\tau) = \frac{3V}{\cosh^2\left(\sqrt{\dfrac{V}{2}}\left(\xi_0 - (\xi - V\tau)\right)\right)}. \tag{15.197}$$

This solution, called a soliton, has the following properties:

1. As required, the solution vanishes when $|\xi| \to \infty$.
2. The spatial profile consists of a solitary pulse centered around the position $\xi = \xi_0 + V\tau$.
3. The pulse width scales as $V^{-1/2}$ and the pulse height scales as V, so larger amplitude solitons are sharper and propagate faster.

An important property of solitons is that they obey a form of superposition principle even though they are essentially nonlinear. In particular, when a fast soliton overtakes a slow soliton, it is seen that after the collision or interaction, both the fast and slow solitons retain their identity. The underlying mathematical theory explaining this surprising behavior is complex and beyond the scope of this text (the interested reader should consult Drazin and Johns (1989)). This theory is called inverse scattering (Gardner *et al.* 1967) and involves mapping the nonlinear equation to a special linear equation, which is then solved and then mapped back again to give the time evolution of the nonlinear solution. The special linear equation describes a quantum mechanical particle trapped in a potential well, where $\chi(\xi, t)$ plays the role of the potential. In particular, the linear equation is of the form $\partial^2\psi/\partial\xi^2 + (\lambda - \chi(\xi, t))\,\psi = 0$, where λ is an eigenvalue and time t is treated as a parameter. When this linear equation is solved for $\chi(\xi, t)$ and the solution is inserted in the KdV equation it is found that $\partial\lambda/\partial t = 0$. This invariance of the eigenvalues is a key feature, which makes possible the {nonlinear →linear} and then {linear→nonlinear} mappings required to construct the solution.

15.7 Assignments

1. Pump depletion for the system of three coupled oscillators discussed in Section 15.2.
   (a) Suppose $A_1 = 0$ at $t = 0$ but A_2 and A_3 are finite. What is the value of the constant in Eq. (15.27)?
   (b) Suppose that $\sin\theta$ is not zero. Will A_1 become finite at times $t > 0$? If A_1 becomes finite, what constraint does Eq. (15.27) put on the value of $\cos\theta$ and hence on $\sin\theta$?
   (c) Use Eqs. (15.24) to write $A_1^2(t)$ and $A_2^2(t)$ in terms of $A_3^2(t)$ and the initial conditions $A_1^2(0)$, $A_2^2(0)$, and $A_3^2(0)$.
   (d) Square both sides of Eq. (15.18c) and use the results in (b) and (c) above to obtain an equation of the form

$$\frac{1}{2}\left(\frac{dA_3}{dt}\right)^2 + U(A_3) = E, \tag{15.198}$$

   where E is a constant. Sketch the dependence of $U(A_3)$ on A_3 indicating the locations of maxima and minima.
   (e) Consider Eq. (15.198) as the energy equation for a pseudo-particle with velocity dA_3/dt in a potential well $U(A_3)$. What is the total energy of this pseudo-particle in terms of $A_1^2(0)$, $A_2^2(0)$, and $A_3^2(0)$? To what does the position of the pseudo-particle in the potential well correspond?
   (f) Considering the initial conditions given in (a), where is the pseudo-particle initially located in its potential well? Sketch the qualitative time dependence of A_3 and indicate how features of this time dependence correspond to the location of the pseudo-particle in its potential well.
   (g) Give an integral expression for the time required for A_3 to go to its first zero.
   (h) Solve Eqs. (15.18a)–(15.18c) numerically for the initial conditions given in (a) and use the numerical results to show that the time dependence of A_3 can be interpreted as the position of a pseudo-particle in a potential well prescribed by Eq. (15.198). Note: the reduction of the pump wave amplitude as it transfers energy to the daughter waves is called pump depletion; the mathematical behavior discussed in

this problem can be expressed in terms of Jacobi elliptic functions (Sagdeev and Galeev 1969).

2. Stimulated Raman scattering in laser fusion. Laser fusion is a proposed method for attaining a controlled fusion reaction. This method involves illuminating a millimeter radius pellet with intense laser light so as to ablate the outer layer of the pellet. The radial outflow of the ablating material constitutes a radial outflow of momentum and so, in order to conserve radial momentum, the rest of the pellet accelerates inwards in rocket-like fashion. An important issue is the decay of the incident laser light into other waves because such a decay would reduce the power available to drive the ablation process. For example, there could be stimulated Raman scattering where the incident laser light decays into an outward propagating electromagnetic wave (often called backscattered light) and an inward propagating Langmuir wave.

   (a) Let the incident laser beam be denoted by subscript 3, the outward propagating electromagnetic wave be denoted by subscript 2, and the inward propagating Langmuir wave be denoted by subscript 1. Assume the laser wavelength is much shorter than the characteristic density scale length. Show that if the backward scattered electromagnetic wave has a frequency only slightly above ω_{pe} then it is possible to satisfy the frequency and wavenumber matching conditions at a location where the density is approximately 1/4 of the density at which the incident wave would reflect.

   (b) Draw a sketch of ω versus k for the incident em wave, the backscattered em wave, and the Langmuir wave. Note that $v_{Te} \ll c$ so if this plot is scaled to show the dispersion of the electromagnetic waves, the Langmuir wave dispersion is almost a horizontal line.

   (c) Draw vectors on the sketch in (a) with coordinates $\{k, \omega\}$ so the incident electromagnetic wave is a vector $\mathbf{v}_3 = \{k_3, \omega_3\}$, the backscattered wave is a vector $\mathbf{v}_2 = \{k_2, \omega_2\}$, and the Langmuir wave is a vector $\mathbf{v}_1 = \{k_1, \omega_1\}$. Show on the sketch how the vectors can add up in a manner consistent with the selection rules $\mathbf{v}_3 = \mathbf{v}_1 + \mathbf{v}_2$.

3. Parametric decay instability. The minimum pump amplitude for the parametric decay instability (electromagnetic wave decays into a Langmuir wave and an ion acoustic wave) is given by Eq. (15.125) to be

$$E_3 = 4\frac{\sqrt{\omega_1\omega_2\Gamma_1\Gamma_2}}{\lambda},$$

where the coupling parameter was defined in Eq. (15.86) to be

$$\lambda = \frac{\omega_{pe} q k_1}{\sqrt{m_i m_e \omega_2 \omega_3}}.$$

   (a) How do the Langmuir wave frequency ω_2 and the ion acoustic wave frequency ω_1 compare to ω_{pe} (nearly same, much larger, or much smaller)?

(b) Taking into account the selection rules, how does the pump frequency ω_3 compare to ω_{pe}? What does this imply for k_3 and hence k_1 and k_2?

(c) Using the results from (a) express the instability threshold in a form where $\varepsilon_0 E^2/n\kappa T_e$ is a function of Γ_1/ω_1 and Γ_2/ω_2.

(d) Suppose a pump wave with frequency ω_3 propagates into a plasma with a monotonically increasing density. Show the relative locations of stimulated Brillouin and stimulated Raman backscattering in terms of a criterion for the local density. Assume the wavelengths of all modes are much shorter than the density gradient scale length.

(e) The National Ignition Facility at the Lawrence Livermore National Lab in the USA has an array of 351 nm wavelength lasers designed to apply 5×10^{14} W over the surface of a target having an initial radius of 1 mm. If these lasers are considered to pump electromagnetic waves with frequency ω_3, at what density does $\omega_3 = \omega_{pe}$ occur? Assuming $T_e = 300$ eV, what is the value of $\varepsilon_0 E^2/n\kappa T_e$ at the location where $\omega_3 = \omega_{pe}$? If $\Gamma_1/\omega_1 = \Gamma_2/\omega_2 = 0.1$ will this plasma be susceptible to the parametric decay instability?

16
Non-neutral plasmas

16.1 Introduction

Conventional plasmas, the main topic of this book, consist of a quasi-neutral collection of mutually interacting ions and electrons. This description applies to the vast majority of entities called plasma (e.g., fusion, industrial, propulsion, ionospheric, magnetospheric, interplanetary, solar, and astrophysical plasmas). Non-neutral plasmas are an exception to this taxonomy and, not surprisingly, have certain behaviors that differ from conventional plasmas (Malmberg 1992, Davidson 2001). Applications of non-neutral plasmas include microwave oscillators and amplifiers, simulating the vortex dynamics of ideal two-dimensional hydrodynamics, testing basic nonlinear concepts, and confining anti-particles.

16.2 Brillouin flow

Using a cylindrical coordinate system $\{r, \theta, z\}$ we first consider the forces acting on an infinitely long, azimuthally symmetric cylindrical cloud of cold charged particles all having the *same* polarity. Because of the mutual electrostatic repulsion of the same-sign charged particles, the cloud of particles will expand continuously in the r direction and so will not be in radial equilibrium. The cloud would be even further from equilibrium if, in addition, it were to rotate with azimuthal velocity $\mathbf{u} = u_\theta \hat{\theta}$, because rotation would provide a centrifugal force mu_θ^2/r, which would add to the radially outwards electrostatic force and so increase the rate of radially outward expansion.

However, if this rotating cloud of same-sign charged particles were immersed in a uniform axial magnetic field $\mathbf{B} = B\hat{z}$, then the magnetic force $\sim q\mathbf{u} \times \mathbf{B}$ would push the charged particles radially inwards, counteracting both the electrostatic repulsion and the centrifugal force. This situation is shown in Fig. 16.1 for the situation where the cloud consists of electrons (Malmberg and de Grassie 1975). Negatively biased electrodes are used at the ends to prevent cloud expansion in

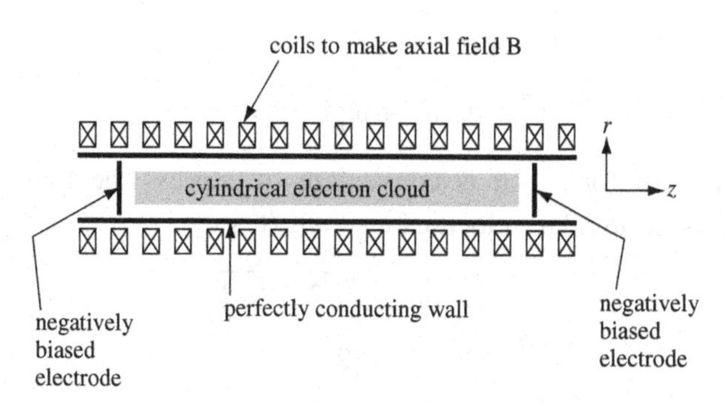

Fig. 16.1 Pure electron plasma configuration. Magnetic field $\mathbf{B} = B\hat{z}$ produced by coils, negatively biased electrodes on ends prevent axial expansion. Plasma radius is r_p, wall radius is a.

the z direction; if the cloud were made of ions, then the end electrodes would be biased positively.

Because there is only one charge species, there is no frictional drag due to collisions with a species of opposite polarity and, because the plasma is cold, the pressure is zero. The radial component of the fluid equation of motion Eq. (2.27) thus reduces to a simple competition between the electrostatic, magnetic, and centrifugal forces, namely

$$0 = q\left(E_r + u_\theta B_z\right) + \frac{mu_\theta^2}{r}. \tag{16.1}$$

Because of the assumed cylindrical and azimuthal symmetry, Poisson's equation reduces to

$$\frac{1}{r}\frac{\partial}{\partial r}(rE_r) = \frac{n(r)q}{\varepsilon_0}, \tag{16.2}$$

which can be integrated to give

$$E_r = \frac{q}{\varepsilon_0}\frac{1}{r}\int_0^r n(r')r'\mathrm{d}r'. \tag{16.3}$$

We will assume that the radius of the non-neutral plasma is r_p and that the non-neutral plasma is surrounded by a perfectly conducting wall of radius a, where a is larger than r_p so there is a vacuum region between the plasma and the wall. In the special case of uniform density up to the plasma radius r_p, Eq. (16.3) may be evaluated to give

$$E_r = \begin{cases} \dfrac{nq}{2\varepsilon_0} r \text{ for } r \leq r_p, \text{ plasma region} \\ \dfrac{nq}{2\varepsilon_0}\dfrac{r_p^2}{r} \text{ for } r_p \leq r \leq a, \text{ vacuum region.} \end{cases} \tag{16.4}$$

Thus, in the plasma Eq. (16.1) becomes

$$u_\theta^2 + u_\theta r\omega_c + \omega_p^2 r^2/2 = 0, \tag{16.5}$$

a quadratic equation for u_θ. It is convenient to express the two roots of this equation in terms of angular velocities $\omega_0 = u_\theta/r$ so

$$\omega_0 = \frac{-\omega_c \pm \sqrt{\omega_c^2 - 2\omega_p^2}}{2}. \tag{16.6}$$

Since ω_0 is independent of r, the non-neutral plasma rotates as a rigid body. This is a special case resulting from the assumption of a uniform density profile. In the more general case of a non-uniform density profile (to be discussed later), the rotation velocity is sheared so ω_0 varies with r. The two roots in Eq. (16.6) coalesce at $\omega_p^2 = \omega_c^2/2$. This point of coalescence is called the Brillouin limit and it is seen that real roots ω_0 exist only if the density is sufficiently low for ω_p^2 to be below this limit.

We will consider the situation typical for non-neutral plasma experiments where the density is well below the Brillouin limit so the two roots are well separated and given by

$$\omega_{0-} \simeq -\omega_c \left(1 + \omega_p^2/2\omega_c^2\right) \tag{16.7}$$

$$\omega_{0+} \simeq \omega_p^2/2\omega_c; \tag{16.8}$$

the plus and minus signs refer to the choice of signs in Eq. (16.6). The large root ω_{0-} is near the cyclotron frequency and the small root ω_{0+} is much smaller than the plasma frequency since $\omega_p \ll \omega_c$. The small root is called the diocotron frequency.

The rotation of the non-neutral plasma in the magnetic field provides an inward force balancing the radially outward electrostatic force. There also exist axial electrostatic forces due to the mutual electrostatic repulsion between the same-sign charges and these forces would cause the plasma to expand axially. Since these axial forces cannot be balanced magnetically, the axial forces are balanced by biased electrodes at the ends of the plasma. The electrodes have the same polarity as the plasma, thereby providing a potential well in the axial direction. Any particle attempting to escape axially is repelled by forces due to the repulsive bias on an end electrode and is reflected back into the main plasma cloud before it can reach the end electrode.

The fast motion of the charged particles in the axial direction smears out axial structure so, to first approximation, the non-neutral plasma can be considered axially uniform. The system then depends only on r and θ and a pair of deceptively simple-looking coupled equations relates the electrostatic potential ϕ and the density n. These two coupled equations govern not only the equilibrium but also the surprisingly rich low-frequency dynamics of a non-neutral plasma. Because the electric field is electrostatic and the magnetic field is a uniform vacuum field, the $\mathbf{E} \times \mathbf{B}$ drift

$$\mathbf{u} = \frac{-\nabla\phi \times \mathbf{B}}{B^2} \tag{16.9}$$

is incompressible, i.e., $\nabla \cdot \mathbf{u} = 0$. Because of this, the continuity equation, Eq. (2.19), reduces to a simple convection of density with flow, namely

$$\frac{\partial n}{\partial t} + \mathbf{u} \cdot \nabla n = 0. \tag{16.10}$$

Using Eq. (16.9) to substitute for $\mathbf{u}$, this can be written as

$$\frac{\partial n}{\partial t} = \frac{\nabla\phi \times \mathbf{B}}{B^2} \cdot \nabla n. \tag{16.11}$$

Equation (16.11) and Poisson's equation

$$\nabla^2 \phi = -\frac{nq}{\varepsilon_0} \tag{16.12}$$

provide two coupled equations in n and ϕ and are the governing equations for the configuration. Although Eqs. (16.11) and (16.12) appear simple, they actually allow quite complex behavior, which will be discussed in the remainder of this chapter. An interesting feature of these equations is that they have no mass dependence, a consequence of ignoring the centrifugal force term, which affects only the high-frequency branch, Eq. (16.7). Thus, the low-frequency dynamics described by Eqs. (16.11) and (16.12) can be thought of as the dynamics of a massless, incompressible fluid governed by a combination of $\mathbf{E} \times \mathbf{B}$ drifts and Poisson's equation.

16.3 Isomorphism to incompressible 2-D hydrodynamics

Let us put plasma physics aside for a moment and consider the equations governing an incompressible two-dimensional fluid. Incompressibility means that the mass density ρ is constant and uniform so the continuity equation reduces to

$$\nabla \cdot \mathbf{u} = 0. \tag{16.13}$$

Using the vector identity $\mathbf{u} \cdot \nabla \mathbf{u} = \nabla u^2/2 - \mathbf{u} \times \nabla \times \mathbf{u}$, the fluid equation of motion can be written as

$$\rho \left(\frac{\partial \mathbf{u}}{\partial t} + \frac{1}{2} \nabla u^2 - \mathbf{u} \times \nabla \times \mathbf{u} \right) = -\nabla P. \tag{16.14}$$

Taking the curl of this equation and defining the vorticity vector $\mathbf{\Omega} = \nabla \times \mathbf{u}$ gives

$$\frac{\partial \mathbf{\Omega}}{\partial t} = \nabla \times (\mathbf{u} \times \mathbf{\Omega}), \tag{16.15}$$

which has the same form as the ideal MHD induction equation, Eq. (2.82), and so can be interpreted as indicating that the vorticity is frozen into the convecting fluid (Kelvin vorticity theorem). Using cylindrical coordinates to express the velocity as $\mathbf{u} = u_r(r, \theta)\hat{r} + u_\theta(r, \theta)\hat{\theta}$ it is seen that

$$\nabla \times \mathbf{u} = \hat{z}\left(\frac{1}{r}\frac{\partial}{\partial r}(ru_\theta) - \frac{1}{r}\frac{\partial u_r}{\partial \theta}\right). \tag{16.16}$$

The vorticity vector therefore lies in the z direction and may be written as

$$\mathbf{\Omega} = \Omega \hat{z}. \tag{16.17}$$

The z component of Eq. (16.15) is

$$\begin{aligned}\frac{\partial \Omega}{\partial t} &= \hat{z} \cdot \nabla \times (\mathbf{u} \times \Omega \hat{z}) \\ &= \nabla \cdot ((\mathbf{u} \times \Omega \hat{z}) \times \hat{z}) \\ &= -\nabla \cdot (\mathbf{u}\Omega) \\ &= -\mathbf{u} \cdot \nabla \Omega. \end{aligned} \tag{16.18}$$

The incompressibility condition prescribed by Eq. (16.13) allows the velocity to be written in terms of a stream-function ψ,

$$\mathbf{u} = -\nabla\psi \times \hat{z}, \tag{16.19}$$

in which case Eq. (16.18) becomes

$$\frac{\partial \Omega}{\partial t} = \nabla\psi \times \hat{z} \cdot \nabla \Omega. \tag{16.20}$$

However, the z component of the curl of Eq. (16.19) can also be expressed in terms of ψ since

$$\begin{aligned}\Omega &= \hat{z} \cdot \nabla \times \mathbf{u} \\ &= -\hat{z} \cdot \nabla \times (\nabla\psi \times \hat{z}) \\ &= -\nabla \cdot ((\nabla\psi \times \hat{z}) \times \hat{z}) \\ &= \nabla^2 \psi. \end{aligned} \tag{16.21}$$

Table 16.1

Non-neutral plasma	Incompressible 2-D fluid
$\frac{\partial n}{\partial t} = \frac{\nabla\phi\times\hat{z}}{B}\cdot\nabla n$	$\frac{\partial\Omega}{\partial t} = \nabla\psi\times\hat{z}\cdot\nabla\Omega$
$\nabla^2\phi = -\frac{nq}{\varepsilon_0}$	$\nabla^2\psi = \Omega$

There is thus an exact isomorphism between the non-neutral plasma equations and the equations describing a 2-D incompressible fluid; this isomorphism is given in Table 16.1

These sets of equations can be made identical by setting

$$\psi = \frac{\phi}{B}, \quad \Omega = -\frac{nq}{\varepsilon_0 B}. \tag{16.22}$$

Thus, density corresponds to vorticity and electrostatic potential corresponds to stream-function.

A non-neutral plasma can therefore be used as an analog computer for investigating the behavior of an inviscid incompressible 2-D fluid (Driscoll and Fine 1990). This is quite useful because it is difficult to make a real fluid act in a truly two-dimensional inviscid fashion whereas it is relatively easy to make a non-neutral plasma do so. While real fluids are three dimensional, it is nevertheless very useful to develop an understanding for two-dimensional dynamics since this understanding can be of considerable help for understanding three-dimensional dynamics. The plasma analog contains all the nonlinear vortex interactions intrinsic to the 2-D fluid problem. Besides serving as an analog computer for investigations of 2-D fluid dynamics, non-neutral plasmas have also been successfully used as a method for trapping antimatter (Surko, Leventhal, and Passner 1989).

16.4 Near-perfect confinement

A curious feature of non-neutral plasmas is that collisions do not degrade confinement so long as axisymmetry is maintained (O'Neil 1995). Thus, in a pure electron plasma, electron–electron collisions do not cause leakage of the plasma out of the trap (loss of confinement comes only from collisions with neutrals and this can be minimized by using good vacuum techniques). To see this, consider the canonical angular momentum of the ith particle

$$P_{\theta i} = mr_i v_{\theta i} + qr_i A_{\theta i}. \tag{16.23}$$

A collision between two identical charged particles will conserve the total angular momentum of the two particles and so the total angular momentum of all the particles is

$$\sum_{i=1}^{N} P_{\theta i} = \sum_{i=1}^{N} \left(m r_i v_{\theta_i} + q r_i A_{\theta i} \right) = const. \tag{16.24}$$

even when there are electron–electron collisions. The vector potential for the magnetic field $\mathbf{B} = B\hat{z}$ is $\mathbf{A} = \hat{\theta} Br/2$, and if the magnetic field is sufficiently strong, the inertial term in Eq. (16.24) can be dropped, so that conservation of total canonical angular momentum reduces to the simple relationship

$$P_\theta = \sum_{i=1}^{N} P_{\theta i} \simeq \frac{q_e B}{2} \sum_{i=1}^{N} r_i^2 = const. \tag{16.25}$$

Equation (16.25) constrains the plasma from moving radially outwards in an axisymmetric fashion. Thus, interparticle collisions cannot make the plasma diffuse to the wall and so a collisional plasma is perfectly confined so long as axisymmetry is maintained.

An alternative interpretation can be developed by considering how collisions cause axisymmetric outward diffusion in a conventional (i.e., electron–ion) plasma. This was discussed in Section 2.8 and will be briefly reviewed here emphasizing the difference between a conventional plasma and a non-neutral plasma. The effect of collisions in a conventional plasma can be seen by considering the steady-state azimuthal component of the resistive MHD Ohm's law. This azimuthal component is

$$-U_r B = \eta J_\theta = \left(\frac{m_e \nu_{ei}}{n e^2} \right) n e (u_{i\theta} - u_{e\theta}) = \frac{m_e \nu_{ei}}{e} (u_{i\theta} - u_{e\theta}) \tag{16.26}$$

and shows that axisymmetric radial flow to the wall results from collisions between *unlike* particles since

$$U_r = -\frac{m_e \nu_{ei}}{eB} (u_{i\theta} - u_{e\theta}). \tag{16.27}$$

Thus, radial transport requires momentum exchange between unlike species and this exchange occurs at a rate dictated by the collision frequency ν_{ei} and by the difference between electron and ion azimuthal velocities. If only one species exists, it is clearly impossible for momentum to be exchanged between unlike species and so there cannot be a net radial motion. A practical consequence of this result is that particle confinement in single species plasmas is orders of magnitude better than confinement in conventional quasi-neutral plasmas (hours/weeks compared to microseconds/seconds).

16.5 Diocotron modes

Non-neutral plasmas support waves that differ significantly from the waves in a conventional, quasi-neutral plasma (Gould 1995). The theory of low-frequency linear waves in a cylindrical non-neutral plasma can be developed by linearizing the continuity equation Eq. (16.10) to obtain

$$\frac{\partial n_1}{\partial t} + \mathbf{u}_0 \cdot \nabla n_1 + \mathbf{u}_1 \cdot \nabla n_0 = 0. \tag{16.28}$$

Using Eq. (16.9) to give $\mathbf{u}_0, \mathbf{u}_1$, and using Poisson's equation, Eq. (16.12), to give n_0, n_1, results in the linear wave equation

$$\frac{\partial \nabla^2 \phi_1}{\partial t} - \frac{\nabla \phi_0 \times \mathbf{B}}{B^2} \cdot \nabla \nabla^2 \phi_1 - \frac{\nabla \phi_1 \times \mathbf{B}}{B^2} \cdot \nabla \nabla^2 \phi_0 = 0. \tag{16.29}$$

Because of the cylindrical geometry, it is convenient to decompose the potential perturbation ϕ_1 into azimuthal Fourier modes so

$$\phi_1(r, \theta, t) = \sum_{l=-\infty}^{\infty} \tilde{\phi}_l(r, t) \mathrm{e}^{\mathrm{i}l\theta}, \tag{16.30}$$

where

$$\tilde{\phi}_l(r, t) = \frac{1}{2\pi} \int_0^{2\pi} \mathrm{d}\theta \phi_1(r, \theta, t) \mathrm{e}^{-\mathrm{i}l\theta}. \tag{16.31}$$

The lth azimuthal mode is assumed to have a time dependence $\exp(-\mathrm{i}\omega_l t)$ so

$$\phi_1(r, \theta, t) = \sum_l \tilde{\phi}_l(r) \mathrm{e}^{\mathrm{i}l\theta - \mathrm{i}\omega_l t} \tag{16.32}$$

and the Fourier coefficient is given by

$$\tilde{\phi}_l(r) = \frac{1}{2\pi} \int_0^{2\pi} \mathrm{d}\theta \phi_1(r, \theta, t) \mathrm{e}^{-\mathrm{i}l\theta + \mathrm{i}\omega_l t}. \tag{16.33}$$

Using the equilibrium azimuthal velocity

$$u_{\theta 0}(r) = \frac{-\nabla \phi_0 \times \mathbf{B}}{B^2} \cdot \hat{\theta} \tag{16.34}$$

the temporal-azimuthal Fourier transform of Eq. (16.29) can be written as

$$\left(\omega_l - \frac{l u_{\theta 0}(r)}{r} \right) \nabla^2 \tilde{\phi}_l + \frac{l \tilde{\phi}_l}{rB} \frac{d}{dr} \nabla^2 \phi_0 = 0. \tag{16.35}$$

Using the relations

$$\nabla^2\tilde{\phi}_l = \frac{1}{r}\frac{\mathrm{d}}{\mathrm{d}r}\left(r\frac{\mathrm{d}\tilde{\phi}_l}{\mathrm{d}r}\right) - \frac{l^2}{r^2}\tilde{\phi}_l, \tag{16.36}$$

$$\nabla^2\phi_0 = \frac{1}{r}\frac{\mathrm{d}}{\mathrm{d}r}\left(r\frac{\mathrm{d}\phi_0}{\mathrm{d}r}\right), \tag{16.37}$$

this temporal-azimuthal Fourier transform of Eq. (16.29) can be expanded as

$$\left(\omega_l - \frac{lu_{\theta 0}(r)}{r}\right)\left(\frac{1}{r}\frac{\mathrm{d}}{\mathrm{d}r}\left(r\frac{\mathrm{d}\tilde{\phi}_l}{\mathrm{d}r}\right) - \frac{l^2}{r^2}\tilde{\phi}_l\right) + \frac{l\tilde{\phi}_l}{rB}\frac{\mathrm{d}}{\mathrm{d}r}\left(\frac{1}{r}\frac{\mathrm{d}}{\mathrm{d}r}\left(r\frac{\mathrm{d}\phi_0}{\mathrm{d}r}\right)\right) = 0. \tag{16.38}$$

The equilibrium angular velocity is

$$\omega_0(r) = \frac{u_{\theta 0}(r)}{r} = \frac{1}{rB}\frac{\mathrm{d}\phi_0}{\mathrm{d}r} \tag{16.39}$$

and so Eq. (16.38) can be expressed as

$$(\omega - l\omega_0(r))\left(\frac{1}{r}\frac{\mathrm{d}}{\mathrm{d}r}\left(r\frac{\mathrm{d}\tilde{\phi}_l}{\mathrm{d}r}\right) - \frac{l^2}{r^2}\tilde{\phi}_l\right) + \frac{l\tilde{\phi}_l}{r}\frac{\mathrm{d}}{\mathrm{d}r}\left(\frac{1}{r}\frac{\mathrm{d}}{\mathrm{d}r}\left(r^2\omega_0(r)\right)\right) = 0, \tag{16.40}$$

where the l subscript has been dropped from ω.

Since Eq. (16.37) gives

$$\frac{\mathrm{d}\phi_0}{\mathrm{d}r} = -\frac{q}{\varepsilon_0 r}\int_0^r n_0(r')r'\mathrm{d}r', \tag{16.41}$$

the equilibrium angular velocity can be evaluated in terms of the equilibrium density to obtain

$$\omega_0(r) = -\frac{q}{\varepsilon_0 B r^2}\int_0^r n_0(r')r'\mathrm{d}r'. \tag{16.42}$$

Equation (16.42) reduces to the special case of rigid body rotation when the density is uniform, but in the more general case there is a shear in the angular velocity. Combining Eq. (16.39) and Poisson's equation shows that

$$\frac{1}{r}\frac{\mathrm{d}}{\mathrm{d}r}\left(r^2\omega_0(r)\right) = \frac{1}{Br}\frac{\mathrm{d}}{\mathrm{d}r}\left(r\frac{\mathrm{d}\phi_0}{\mathrm{d}r}\right) = -\frac{nq}{\varepsilon_0 B} = -\frac{\omega_p^2(r)}{\omega_c}. \tag{16.43}$$

Furthermore, since

$$\frac{\mathrm{d}}{\mathrm{d}r}\left(\frac{1}{r}\frac{\mathrm{d}r^2}{\mathrm{d}r}\right) = 0, \tag{16.44}$$

an arbitrary constant can be added to $\omega_0(r)$ in the last term in Eq. (16.40), which consequently can be written as

$$[\omega - l\omega_0(r)]\left(\frac{\mathrm{d}}{\mathrm{d}r}\left(r\frac{\mathrm{d}\tilde{\phi}_l}{\mathrm{d}r}\right) - \frac{l^2}{r}\tilde{\phi}_l\right) - \tilde{\phi}_l\frac{\mathrm{d}}{\mathrm{d}r}\left(\frac{1}{r}\frac{\mathrm{d}}{\mathrm{d}r}\left[r^2(\omega - l\omega_0)\right]\right) = 0; \tag{16.45}$$

this is called the diocotron wave equation.

The diocotron wave equation can be written in a more symmetric form by defining

$$\Phi = r\tilde{\phi}_l,$$
$$G = r^2(\omega - l\omega_0(r)), \tag{16.46}$$

and using

$$\frac{\mathrm{d}}{\mathrm{d}r}\left(r\frac{\mathrm{d}\tilde{\phi}_l}{\mathrm{d}r}\right) = r\frac{\mathrm{d}}{\mathrm{d}r}\left(\frac{1}{r}\frac{\mathrm{d}\Phi}{\mathrm{d}r}\right) + \frac{\Phi}{r^2} \tag{16.47}$$

to obtain

$$r\frac{\mathrm{d}}{\mathrm{d}r}\left(\frac{1}{r}\frac{\mathrm{d}\Phi}{\mathrm{d}r}\right) + (1 - l^2)\,\Phi - \Phi\frac{r}{G}\frac{\mathrm{d}}{\mathrm{d}r}\left(\frac{1}{r}\frac{\mathrm{d}G}{\mathrm{d}r}\right) = 0. \tag{16.48}$$

16.5.1 Wall boundary condition

Although the wall boundary condition does not affect the equilibrium of a non-neutral plasma, it plays a critical role for the diocotron modes because these modes are non-axisymmetric. Because the wall is perfectly conducting, the azimuthal electric field must vanish at the wall, i.e., $E_\theta(a) = 0$. This boundary condition is trivially satisfied for the equilibrium because the equilibrium is axisymmetric and $E_\theta = -r^{-1}\partial\phi/\partial\theta$ vanishes everywhere for an axisymmetric field. However, for a non-axisymmetric perturbation, the azimuthal electric field is $\tilde{E}_{\theta,l} = -\mathrm{i}l\tilde{\phi}_l/r$ and so could, in principle, be finite. In order to have the azimuthal electric field vanish at the wall, each finite l mode must therefore satisfy the wall boundary condition $\tilde{\phi}_l(a) = 0$.

16.5.2 The $l = 1$ diocotron mode: a special case

The $l = 1$ mode is a special case because the $1 - l^2$ term in Eq. (16.48) vanishes and the remaining terms have an extremely simple relationship. Specifically, when $l = 1$ Eq. (16.48) reduces to

$$\frac{r}{\Phi}\frac{\mathrm{d}}{\mathrm{d}r}\left(\frac{1}{r}\frac{\mathrm{d}\Phi}{\mathrm{d}r}\right) = \frac{r}{G}\frac{\mathrm{d}}{\mathrm{d}r}\left(\frac{1}{r}\frac{\mathrm{d}G}{\mathrm{d}r}\right), \tag{16.49}$$

which has the exact solution

$$\Phi = \lambda G, \tag{16.50}$$

where λ is an arbitrary constant. Using Eq. (16.46), this solution can be expressed in terms of the original variables as (White, Malmberg, and Driscoll 1982)

$$\tilde{\phi}_{l=1}(r) = \frac{\omega - \omega_0(r)}{2\omega} Sr, \tag{16.51}$$

where S is a constant. Since Eq. (16.48) is a second-order ordinary differential equation, it must have two independent solutions and Eq. (16.51) must be one of these. The other solution has a singularity at $r = 0$ and is discarded as non-physical (this other solution can be found by assuming it is of the form $\Phi = FG$ and substituting this assumed Φ into Eq. (16.49) to find F). It is now evident that rigid body rotation corresponds to having $G = 0$ for all r since $\omega_0(r)$ is a constant and $\omega = \omega_0(a)$ for a rigid body. If $G = 0$ then Eq. (16.48) shows that the form of Φ is unrestricted and so rigid body rotation can have any perturbation profile. On the other hand, if the rotation has even the slightest amount of shear, then G must be non-zero and the solution is restricted to be of the form specified by Eq. (16.51).

The perfectly conducting wall boundary condition $E_\theta = 0$ implies $\tilde{\phi}_{l=1}(a) = 0$ and so Eq. (16.51) gives the $l = 1$ mode frequency to be

$$\begin{aligned} \omega &= \omega_0(a) \\ &= -\frac{q}{\varepsilon_0 B a^2} \int_0^{r_p} n_0(r') r' \mathrm{d}r' \\ &= -\frac{E_r(a)}{Ba}; \end{aligned} \tag{16.52}$$

this has the interesting feature of depending only on $E_r(a)$, the equilibrium radial field at the wall. Since $E_r(a)$ depends only on the total charge per length, the $l = 1$ mode frequency depends only on the total charge per length and so is independent of the radial profile of the charge density. The coefficient S has been chosen so that the *perturbed* radial electric field at the wall is

$$\tilde{E}_r(a) = -\frac{Sa}{2\omega} \left(\frac{\mathrm{d}\omega_0(r)}{\mathrm{d}r} \right)_{r=a}. \tag{16.53}$$

Since $n_0(a) = 0$ by assumption, Eq. (16.42) shows

$$\begin{aligned} \left(\frac{\mathrm{d}\omega_0(r)}{\mathrm{d}r} \right)_{r=a} &= \left(-\frac{2}{r} \omega_0(r) \right)_{r=a} \\ &= -\frac{2\omega}{a} \end{aligned} \tag{16.54}$$

and so

$$S = \tilde{E}_r(a). \tag{16.55}$$

16.5.3 Energy analysis of diocotron modes

The linearized current density of a diocotron mode is

$$\begin{aligned}\mathbf{J}_1 &= n_1 q\mathbf{u}_0 + n_0 q\mathbf{u}_1 \\ &= n_1 q r\omega_0(r)\hat{\theta} + n_0 q\frac{\mathbf{E}_1\times\mathbf{B}}{B^2}\end{aligned} \tag{16.56}$$

and the $n_1 q r\omega_0(r)\hat{\theta}$ equilibrium flow term enables the wave energy to become negative. The wave is electrostatic so Eq. (7.56) gives the wave-energy density to be (Briggs, Daugherty, and Levy 1970)

$$\begin{aligned}\bar{w} &= \int_{-\infty}^{t} \mathrm{d}t \left\langle \mathbf{E}_1\cdot\left(\mathbf{J}_1 + \varepsilon_0\frac{\partial\mathbf{E}_1}{\partial t}\right)\right\rangle \\ &= \frac{\varepsilon_0}{2}E_1^2 + \int_{-\infty}^{t}\mathrm{d}t\,\langle\mathbf{E}_1\cdot\mathbf{J}_1\rangle .\end{aligned} \tag{16.57}$$

Unlike the uniform plasma analysis in Section 7.4, here both plasma non-uniformity and wall boundary conditions must be taken into account. The change in system energy $\bar{W}$ due to establishment of the diocotron wave is found by integrating $\bar{w}$ over volume and is

$$\begin{aligned}\bar{W} &= \int \mathrm{d}^3r\left(\frac{\varepsilon_0}{2}\left(\nabla\cdot\phi_1\nabla\phi_1 - \phi_1\nabla^2\phi_1\right) + \int_{-\infty}^{t}\mathrm{d}t\, n_1 q r\omega_0(r)E_{1\theta}\right) \\ &= q\int \mathrm{d}^3r\left(\frac{1}{2}n_1\phi_1 + r\omega_0(r)\int_{-\infty}^{t}\mathrm{d}t\, n_1 E_{1\theta}\right),\end{aligned} \tag{16.58}$$

where Eq. (16.56) and the perfectly conducting wall boundary condition $\phi_1(a)=0$ have been used. Using Eq. (7.62) and (16.32), the energy for a given l mode can be written as

$$W_l = \frac{q}{2}\operatorname{Re}\int \mathrm{d}^3r\left(\frac{1}{2}\tilde{n}_l\tilde{\phi}_l^* \mathrm{e}^{2\omega_i t} + r\omega_0(r)\int_{-\infty}^{t}\mathrm{d}t\tilde{n}_l\tilde{E}_{l\theta}^*\mathrm{e}^{2\omega_i t}\right). \tag{16.59}$$

In order to evaluate this expression, it is necessary to express all fluctuating quantities in terms of the oscillating potential $\tilde{\phi}_l$. Using Poisson's equation in Eq. (16.35) it is seen that

$$\tilde{n}_l = -\frac{l\tilde{\phi}_l}{rB(\omega - l\omega_0(r))}\frac{\mathrm{d}n_0}{\mathrm{d}r} \tag{16.60}$$

and the complex conjugate of the linearized azimuthal electric field is

$$\tilde{E}_{1\theta}^{*} = \left(-\mathrm{i}\frac{l\tilde{\phi}_l}{r}\right)^{*} = \mathrm{i}\frac{l\tilde{\phi}_1^{*}}{r}. \tag{16.61}$$

The wave energy is thus

$$\begin{aligned}
\delta W_l &= -\frac{ql}{2B}\mathrm{Re}\int \mathrm{d}^3r \left\{ \begin{array}{c} \dfrac{\left|\tilde{\phi}_l\right|^2}{2r(\omega - l\omega_0(r))}\dfrac{\mathrm{d}n_0}{\mathrm{d}r}\mathrm{e}^{2\omega_i t} \\ +r\omega_0(r)\int_{-\infty}^{t}\mathrm{d}t\dfrac{\mathrm{i}l\left|\tilde{\phi}_l\right|^2\mathrm{e}^{2\omega_i t}}{r^2(\omega - l\omega_0(r))}\dfrac{\mathrm{d}n_0}{\mathrm{d}r} \end{array} \right\} \\
&= -\frac{qL}{4B}\mathrm{Re}\int_0^{2\pi}\mathrm{d}\theta\int_0^a \mathrm{d}r\left\{\left(\begin{array}{c} \dfrac{\mathrm{e}^{2\omega_i t}}{(\omega - l\omega_0(r))} \\ +\omega_0(r)\int_{-\infty}^{t}\dfrac{2\mathrm{i}l\mathrm{e}^{2\omega_i t}\mathrm{d}t}{(\omega - l\omega_0(r))} \end{array} \right)\frac{\mathrm{d}n_0}{\mathrm{d}r}l\left|\tilde{\phi}_l\right|^2\right\},
\end{aligned} \tag{16.62}$$

where L is the axial length of the plasma. The second term above can be directly evaluated as

$$\begin{aligned}
\mathrm{Re}\int_{-\infty}^{t}\mathrm{d}t\frac{2\mathrm{i}l\left|\tilde{\phi}_l\right|^2\mathrm{e}^{2\omega_i t}}{(\omega - l\omega_0(r))} &= \mathrm{Re}\int_{-\infty}^{t}\mathrm{d}t\frac{2\mathrm{i}l(\omega_r - l\omega_0(r) - \mathrm{i}\omega_i)}{(\omega_r - l\omega_0(r))^2 + \omega_i^2}\left|\tilde{\phi}_l\right|^2\mathrm{e}^{2\omega_i t} \\
&= \int_{-\infty}^{t}\mathrm{d}t\frac{2l\omega_i}{(\omega_r - l\omega_0(r))^2 + \omega_i^2}\left|\tilde{\phi}_l(t=0)\right|^2\mathrm{e}^{2\omega_i t} \\
&= \frac{l}{(\omega_r - l\omega_0(r))^2 + \omega_i^2}\left|\tilde{\phi}_l\right|^2\mathrm{e}^{2\omega_i t} \\
&\to \frac{l}{(\omega_r - l\omega_0(r))^2}\left|\tilde{\phi}_l\right|^2 \quad \text{as } \omega_i \to 0.
\end{aligned} \tag{16.63}$$

Thus, in the limit $\omega_i \to 0$,

$$\begin{aligned}
\delta W_l &= -\frac{qL}{4B}\int_0^{2\pi}\mathrm{d}\theta\int_0^a \mathrm{d}r\left[\left(\frac{1}{(\omega - l\omega_0(r))} + \frac{l\omega_0(r)}{(\omega - l\omega_0(r))^2}\right)\frac{\mathrm{d}n_0}{\mathrm{d}r}l\left|\tilde{\phi}_l\right|^2\right] \\
&= -\frac{\pi qL}{2B}\int_0^a \mathrm{d}r\frac{\omega}{(\omega - l\omega_0(r))^2}\frac{\mathrm{d}n_0}{\mathrm{d}r}l\left|\tilde{\phi}_l\right|^2.
\end{aligned} \tag{16.64}$$

For the $l = 1$ mode this can be evaluated using Eq. (16.51) to give

$$\begin{aligned}\delta W_{l=1} &= -\frac{\pi qL}{2B}\int_0^a \mathrm{d}r\frac{\omega}{(\omega-\omega_0(r))^2}\frac{\mathrm{d}n_0}{\mathrm{d}r}\left[Sr\left(\frac{\omega-\omega_0(r)}{2\omega}\right)\right]^2 \\ &= -\frac{\pi qLS^2}{8\omega B}\int_0^a \mathrm{d}r\frac{\mathrm{d}n_0}{\mathrm{d}r}r^2 \\ &= -\frac{\pi qLS^2}{8\omega B}\int_0^a \mathrm{d}r\left(-2rn_0(r)+\frac{\mathrm{d}}{\mathrm{d}r}\left[r^2n_0(r)\right]\right) \\ &= \frac{\pi qLS^2}{4\omega B}\int_0^{r_p} rn_0(r)\mathrm{d}r,\end{aligned} \tag{16.65}$$

where $n_0(r) = 0$ for $r_p < r \le a$ has been used. Finally, using the middle line of Eq. (16.52) this becomes

$$\delta W_{l=1} = -\frac{\varepsilon_0\pi La^2S^2}{4} \tag{16.66}$$

and so the $l = 1$ diocotron mode is seen to be a negative energy mode. This result can also be derived by considering the change in system energy that results when the non-neutral plasma is attracted to a fictitious image charge chosen to maintain the wall as an equipotential when the plasma moves off axis (see Assignments 3 and 4).

For $l \ge 2$, the negative energy nature of the diocotron mode can be understood by considering the change in electrostatic energy stored in an isolated parallel plate capacitor when the distance between the parallel plates is varied. The stored energy is $W = CV^2/2 = Q^2/2C$ and Q is constant because the capacitor is isolated. Because increasing the capacitance at constant Q reduces W, bringing the plates together makes available free energy, which can then be used to bring the plates even closer together. The system is therefore unstable with respect to perturbations that tend to increase the capacitance. This is the electrostatic analog of the ideal magnetohydrodynamic flux-conserving situation discussed on p. 315 where the magnetic energy was shown to be $W = \Phi^2/2L$ and Φ was the magnetic flux linked by an isolated inductor L; it was shown that any flux-conserving perturbation that increases inductance makes available free energy for driving an instability.

If all the charge of a non-neutral plasma is assumed to be located at the radius r_p, then the combination of the non-neutral plasma and the wall at radius a effectively constitutes a coaxial capacitor. Gauss' law shows that the radial electric field is $2\pi r\varepsilon_0 E_r = \lambda$, where λ is the charge per length. The voltage difference between the plasma and the wall is therefore $V = -\int_{r_p}^a E_r\mathrm{d}r = \lambda(2\pi\varepsilon_0)^{-1}\ln(a/r_p)$ and the capacitance per length is $C' = 2\pi\varepsilon_0\left[\ln(a/r_p)\right]^{-1}$. Thus, increasing r_p increases the capacitance and decreases the electrostatic energy associated with the fixed amount of charge. However, such an azimuthally symmetric increase of r_p is forbidden because it would rarefy the plasma, an impossibility since the only allowed motion is an $\mathbf{E} \times \mathbf{B}$ drift, which is incompressible for an electrostatic electric field.

The plasma can circumvent this constraint by undergoing an azimuthally periodic incompressible motion that, on average, increases the capacitance. This could be arranged by splitting the system into an even number of equally spaced azimuthally periodic segments and arranging for an incompressible motion where even-numbered segments move towards the wall and odd-numbered segments move away from the wall. The volume between the wall and the plasma would thus be conserved so the motion would be incompressible but, because capacitance scales as the inverse of the distance between a segment and the wall, the increase in capacitance due to the segments moving towards the wall would exceed the decrease in capacitance due to the segments moving away from the wall. Thus, the electrostatic energy of the system would decrease and free energy would be available to drive an instability.

16.5.4 Resistive wall

Let us now assume that the conducting wall has an insulated segment of length L_s with angular extent Δ and that this insulated segment is connected to ground via a small resistor R. The remainder of the wall is directly connected to ground; this is sketched in Fig. 16.2. The wall is thus at or near ground potential and so there are no electric fields exterior to the wall, and in particular there is no radial electric field just outside the wall.

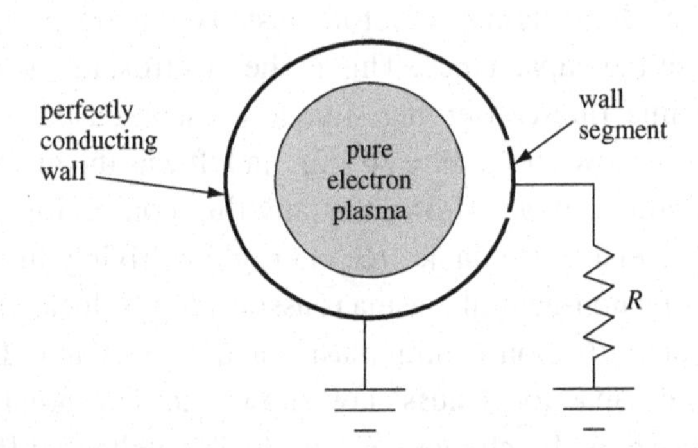

Fig. 16.2 Arrangement for resistive wall instability of a non-neutral plasma. A small resistor R is connected between ground and an isolated wall segment of azimuthal extent Δ and axial extent L_s. The remainder of the perfectly conducting wall is connected directly to ground.

In order for the radial electric field to vanish outside the wall, the wall must have a surface charge density that establishes a radial electric field equal and opposite to the wall radial electric field S produced by the plasma so

$$-S = \frac{\tilde{\sigma}_l}{\varepsilon_0}. \tag{16.67}$$

The surface charge flows back and forth between the wall segments and ground. The charge on the segment extending from $\theta = -\Delta/2$ to $\Delta/2$ is

$$\begin{aligned} Q_s &= L_s \int_{-\Delta/2}^{\Delta/2} a \mathrm{d}\theta \tilde{\sigma}_l \mathrm{e}^{il\theta} \\ &= L_s a \tilde{\sigma}_l \frac{\mathrm{e}^{il\Delta/2} - \mathrm{e}^{-il\Delta/2}}{il} \\ &= -\frac{2}{l} L_s a \varepsilon_0 S \sin\frac{\Delta}{2}. \end{aligned} \tag{16.68}$$

The electric current I flowing through the resistor is the rate of change of this surface charge, i.e., for the $l = 1$ mode this current is

$$I = 2i\omega L_s a \varepsilon_0 S \sin\frac{\Delta}{2}. \tag{16.69}$$

The rate at which energy is dissipated in the resistor is

$$\langle I^2 R \rangle = \frac{1}{2}\left(2\omega L_s a \varepsilon_0 S \sin\frac{\Delta}{2}\right)^2 R. \tag{16.70}$$

If the mode frequency is allowed to have a small imaginary part then the rate of change of electrostatic energy will be

$$P = 2\omega_i \delta W. \tag{16.71}$$

Since the sum of the thermal energy in the resistor and the electrostatic energy in the plasma constitutes the total energy in the system, conservation of this sum gives

$$2\omega_i \delta W + \langle I^2 R \rangle = 0. \tag{16.72}$$

or

$$\omega_i = -\frac{\langle I^2 R \rangle}{2\delta W}. \tag{16.73}$$

Because the mode has negative energy, ω_i is positive, i.e., the system is unstable.

Substitution into the right-hand side gives

$$\omega_i = \frac{4\omega^2 R L_s^2 \varepsilon_0 \sin^2\dfrac{\Delta}{2}}{\pi L}, \tag{16.74}$$

where L is the axial length of the wall and $L_s \leq L$ is the axial length of the segment. This dependence of the growth rate on resistance has been observed in experiments (White *et al.* 1982). If the resistor R is replaced by a parallel resonant circuit, then the instability will occur at the resonant frequency of this circuit because the effective resistance seen by the non-neutral plasma wall segment will be at a maximum at the resonant frequency. This is essentially the basis for the magnetron tube used in radar transmitters and microwave ovens (see Assignments 5 and 6).

16.5.5 Diocotron modes with $l \geq 2$

The analysis of $l \geq 2$ diocotron modes resembles the Landau analysis of electron plasma waves but is not exactly the same. In general, $l \geq 2$ diocotron modes must be considered using numerical methods because their behavior depends on the equilibrium density profile via coefficients of both the first and last terms in Eq. (16.38). However, some indication of the general behavior can be obtained analytically. An important condition can be obtained (Davidson 2001) by expressing the mode frequency as $\omega = \omega_r + \mathrm{i}\omega_i$ so Eq. (16.38) can be written as

$$\frac{1}{r}\frac{\mathrm{d}}{\mathrm{d}r}\left(r\frac{\mathrm{d}\tilde{\phi}_l}{\mathrm{d}r}\right) - \frac{l^2}{r^2}\tilde{\phi}_l - \frac{(\omega_r - l\omega_0(r)) - \mathrm{i}\omega_i}{(\omega_r - l\omega_0(r))^2 + \omega_i^2}\frac{l}{rB}\frac{n_0' q}{\varepsilon_0}\tilde{\phi}_l = 0. \tag{16.75}$$

After multiplying through by $r\tilde{\phi}_l^*$, integrating from $r = 0$ to $r = a$, and using the perfectly conducting boundary condition $\tilde{\phi}_l(a) = 0$ when integrating the first term by parts, the following integral relation is found:

$$\int_0^a \mathrm{d}r \left\{ r\left|\frac{\mathrm{d}\tilde{\phi}_l}{\mathrm{d}r}\right|^2 + \frac{l^2}{r}\left|\tilde{\phi}_l\right|^2 + \frac{(\omega_r - l\omega_0(r)) - \mathrm{i}\omega_i}{(\omega_r - l\omega_0(r))^2 + \omega_i^2}\left|\tilde{\phi}_l\right|^2 \frac{l}{rB}\frac{n_0' q}{\varepsilon_0} \right\} = 0. \tag{16.76}$$

The imaginary part of this expression is

$$-\omega_i \frac{lq}{\varepsilon_0 rB}\int_0^{r_p} \mathrm{d}r \left\{ \frac{\left|\tilde{\phi}_l\right|^2}{(\omega_r - l\omega_0(r))^2 + \omega_i^2}\frac{\mathrm{d}n_0}{\mathrm{d}r} \right\} = 0; \tag{16.77}$$

the upper limit has been changed to r_p because $n_0(r)$ and $\mathrm{d}n_0/\mathrm{d}r$ are by assumption zero in the region $r_p < r \leq a$. If $\mathrm{d}n_0/\mathrm{d}r$ has the same sign throughout the radial interval $0 \leq r \leq r_p$, then the integral would have to be non-zero since the integrand always has the same sign, and so ω_i would have to be zero. Thus, a necessary condition for ω_i to be finite is for $\mathrm{d}n_0/\mathrm{d}r$ to change signs in the interval $0 \leq r \leq r_p$. This necessary condition corresponds to $n_0(r)$ having a maximum in the interval $0 \leq r \leq r_p$. This sort of profile is commonly called hollow, because $n_0(r)$ starts at some finite value at $r=0$, increases to a maximum at some finite $r<r_p$, and then decreases to zero at $r=r_p$.

Further progress can be made by expressing the diocotron equations as a pair of coupled equations for the density and potential perturbations, i.e.,

$$\tilde{n}_l = -\frac{l\tilde{\phi}_l}{rB(\omega - l\omega_0(r))}\frac{dn_0}{dr} \tag{16.78}$$

$$\frac{1}{r}\frac{d}{dr}\left(r\frac{d\tilde{\phi}_l}{dr}\right) - \frac{l^2}{r^2}\tilde{\phi}_l = -\frac{\tilde{n}_l q}{\varepsilon_0}. \tag{16.79}$$

Rather than substitute for $\tilde{n}_l$ in order to obtain Eq. (16.38), instead Eq. (16.79) is solved for $\tilde{\phi}_l$ using a Green's function method (Schecter *et al.* 2000). This approach has the virtue of imposing the perfectly conducting wall boundary condition on $\tilde{\phi}_l$ at an earlier stage of the analysis before the entire wave equation is developed. The set of solutions to Eq. (16.79) is then effectively restricted to those satisfying the wall boundary condition, and only this restricted set is used when later combining Eqs. (16.79) and (16.78) to form a wave equation.

The Green's function solution to Eq. (16.79) is obtained by first recasting Eq. (16.79) as

$$\frac{1}{r}\frac{d}{dr}\left(r\frac{d\tilde{\phi}_l}{dr}\right) - \frac{l^2}{r^2}\tilde{\phi}_l = -\frac{q}{\varepsilon_0}\int_0^a ds\,\delta(r-s)\tilde{n}_l(s). \tag{16.80}$$

Thus, if $\psi(s,r)$ is the solution of

$$\frac{1}{r}\frac{d}{dr}\left(r\frac{d}{dr}\psi(r,s)\right) - \frac{l^2}{r^2}\psi(r,s) = -\delta(r-s), \tag{16.81}$$

where $0 \le s \le a$, then

$$\tilde{\phi}_l(r) = \frac{q}{\varepsilon_0}\int_0^a ds\psi(r,s)\tilde{n}_l(s) \tag{16.82}$$

is the solution of Eq. (16.79). The spatial boundary conditions are accounted for when solving Eq. (16.81) for the Green's function $\psi(r,s)$ and so are independent of the form of $\tilde{n}_l$.

Equation (16.81) is solved by finding separate solutions to its homogeneous counterpart

$$\frac{1}{r}\frac{d}{dr}\left(r\frac{d\psi}{dr}\right) - \frac{l^2}{r^2}\psi = 0 \tag{16.83}$$

for the inner interval $0 \le r < s$ and for the outer interval $s < r \le a$ and then appropriately matching these two distinct homogeneous solutions at $r=s$ where they meet. The inner solution must satisfy the regularity condition $\psi(0)=0$ and the outer solution must satisfy the wall boundary condition $\psi(a)=0$. Since the solutions of Eq. (16.83) are $\psi \sim r^{\pm l}$, the inner solution must be

$$\psi = \alpha\left(\frac{r}{a}\right)^l \quad \text{for } 0 \leq r < s \tag{16.84}$$

and the outer solution must be

$$\psi = \beta\left(\left(\frac{r}{a}\right)^l - \left(\frac{r}{a}\right)^{-l}\right) \text{for } s < r \leq a, \tag{16.85}$$

where the coefficients α and β are to be determined.

Integration of Eq. (16.81) across the delta function from $r=s_-$ to $r=s_+$ gives the jump condition

$$\left[\frac{\mathrm{d}}{\mathrm{d}r}\psi(r,s)\right]_{s_-}^{s_+} = -1 \tag{16.86}$$

and integrating a second time shows that ψ must be continuous at $r=s$. These jump and continuity conditions give two coupled equations in α and β,

$$\frac{\beta l}{a}\left(\left(\frac{s}{a}\right)^{l-1} + \left(\frac{s}{a}\right)^{-l-1}\right) - \frac{\alpha l}{a}\left(\frac{s}{a}\right)^{l-1} = -1 \tag{16.87}$$

$$\beta\left(\left(\frac{s}{a}\right)^l - \left(\frac{s}{a}\right)^{-l}\right) - \alpha\left(\frac{s}{a}\right)^l = 0. \tag{16.88}$$

Solving for α and β gives the Green's function,

$$\psi(r,s) = \begin{cases} -\dfrac{a}{2l}\left(\dfrac{s}{a}\right)^{l+1}\left(\left(\dfrac{r}{a}\right)^l - \left(\dfrac{r}{a}\right)^{-l}\right) & \text{for } r > s \\ -\dfrac{a}{2l}\left(\dfrac{r}{a}\right)^l\left(\left(\dfrac{s}{a}\right)^{l+1} - \left(\dfrac{s}{a}\right)^{-l+1}\right) & \text{for } r < s; \end{cases} \tag{16.89}$$

this satisfies Eq. (16.81) and also the boundary conditions at $r=0$ and $r=a$. Using the Green's function in Eq. (16.82) gives

$$\begin{aligned} \tilde{\phi}_l(r) &= \frac{q}{\varepsilon_0}\left[\int_0^r \mathrm{d}s\psi(r,s)\tilde{n}_l(s) + \int_r^a \mathrm{d}s\psi(r,s)\tilde{n}_l(s)\right] \\ &= -\frac{a}{2l}\frac{q}{\varepsilon_0}\left[\begin{array}{l} \displaystyle\int_0^r \mathrm{d}s\left(\left(\frac{s}{a}\right)^{l+1}\left(\left(\frac{r}{a}\right)^l - \left(\frac{r}{a}\right)^{-l}\right)\right)\tilde{n}_l(s) \\ \displaystyle+\int_r^a \mathrm{d}s\left(\left(\frac{r}{a}\right)^l\left(\left(\frac{s}{a}\right)^{l+1} - \left(\frac{s}{a}\right)^{-l+1}\right)\right)\tilde{n}_l(s) \end{array}\right]. \end{aligned} \tag{16.90}$$

Finally, using Eq. (16.78) to substitute for $\tilde{n}_l(s)$ gives

$$\tilde{\phi}_l(r) = \frac{q}{2\varepsilon_0 B} \left(\begin{array}{c} \displaystyle\int_0^r \mathrm{d}s \frac{s^l}{a^l}\left(\frac{r^l}{a^l} - \frac{a^l}{r^l}\right) \frac{\tilde{\phi}_l(s)}{\omega - l\omega_0(s)} \frac{\mathrm{d}n_0}{\mathrm{d}s} \\ \displaystyle + \int_r^a \mathrm{d}s \frac{r^l}{a^l}\left(\frac{s^l}{a^l} - \frac{a^l}{s^l}\right) \frac{\tilde{\phi}_l(s)}{\omega - l\omega_0(s)} \frac{\mathrm{d}n_0}{\mathrm{d}s} \end{array} \right). \tag{16.91}$$

This integral equation for $\tilde{\phi}_l(r)$ not only prescribes the mode dynamics, but also explicitly incorporates the spatial boundary conditions. The resonant denominators $\omega - l\omega_0(s)$ in the s integrals are reminiscent of the resonant denominators occurring in the velocity-space integrals of the Landau problem. Stability depends on the sign of $\mathrm{d}n_0/\mathrm{d}s$ at the location where $\omega = l\omega_0(s)$ in analogy to the dependence of Landau stability on the sign of $\mathrm{d}f_0/\mathrm{d}v$ where $v = \omega/k$. Because these equations are isomorphic to the 2-D incompressible hydrodynamic equations, it is seen that phenomena similar to Landau damping or instability can occur in 2-D incompressible hydrodynamics.

16.5.6 *Phase mixing and relation to Landau damping*

The dynamical evolution of an initially off-axis localized infinitesimal bump or patch of increased density provides insight into the vortex dynamics of a two-dimensional inviscid fluid since, as indicated by Eq. (16.22), a region of increased non-neutral plasma density is mathematically equivalent to a region of increased vorticity in a two-dimensional fluid.

If the non-neutral plasma equilibrium angular velocity is sheared, then the off-axis density patch will become sheared because the inner and outer portions of the patch will rotate at different angular velocities. Thus, the angular position of different radial positions of the patch will have trajectories scaling as $\theta(r) = t\omega_0(r)$. The angular separation between two points in the patch starting at respective radii r and $r+\delta r$ will scale as $\delta\theta = t\delta r \mathrm{d}\omega_0/\mathrm{d}r$ and this separation will eventually exceed 2π at sufficiently large t. This gives a sort of spatial phase mixing (Gould 1995) because a localized patch of increased density will eventually be stretched out to become a multi-turn thin spiral of increased density. The number of turns in the spiral increases linearly with time. The patch is thus smeared out over all angles and so is no longer azimuthally localized. Since the patch is stretched azimuthally and yet is incompressible, the thickness of each turn in the spiral must decrease as the number of turns in the spiral increases. Specifically, the length of the spiral increases with t and the radial thickness of each turn decreases as $1/t$ so that the area remains constant. A graphic demonstration of this stretching and thinning has been obtained by Bachman and Gould (1996) using direct numerical integration of the dynamical equations with an initial condition consisting of a prescribed density patch. Since the decay and eventual disappearance of the patch as it deforms into a nearly infinitely long and nearly infinitely thin spiral is a spatial analog to the velocity-space phase mixing underlying Landau damping, it might be expected that the nonlinear mixing of two such patches initiated at two successive times might give rise to a spatial echo effect (Gould 1995). This spatial echo effect involving the nonlinear beating of two spatial spirals has been seen experimentally (Yu and Driscoll 2002). In this experiment an $l = 2$ perturbation is applied and then decays as it is stretched into a nearly infinitely long, nearly infinitely thin spiral. Then an $l = 4$ perturbation is applied, which similarly decays. Finally, after both of the perturbations have decayed, a nonlinear $l = 2$ echo is observed because of nonlinear mixing of the two spirals associated with the respective initial perturbations.

If the patch has a finite amplitude, then its self-electric field will affect the dynamics. This is the nonlinear regime and will cause a sigmoidal curling of the patch since azimuthal electric fields associated with the patch will give radial motions in addition to the azimuthal motions.

16.6 Assignments

1. Relationship between non-neutral and conventional plasma equilibria. Compare the following two plasmas, both of which have cylindrical symmetry, axial uniformity, the same radial electron density profile $n_e(r)$, and are in equilibrium:
   (a) a cylindrical pure electron plasma immersed in a strong magnetic field $\mathbf{B} = B\hat{z}$, where $\omega_{pe}^2 \ll \omega_{ce}^2$;
   (b) an *unmagnetized* quasi-neutral electron–ion plasma, which has infinitely massive ions.

   What is the ion density in the quasi-neutral plasma and what electric field is produced by this ion density? How does the force associated with the electric field produced by the ions in the quasi-neutral plasma compare to the magnetic force associated with electron rotation (finite $u_{e\theta}$) in the pure electron plasma? What is the net equilibrium force on the electrons in the two cases? Assuming that the electrons have zero mass, which plasma has more free energy?
2. Free energy associated with sheared velocity profile in slab approximation. Suppose a non-neutral plasma does not rotate as a rigid body and instead has a sheared angular velocity. An observer rotating with the plasma at some radius r_{obs} would conclude that the plasma has positive angular velocity for $r > r_{obs}$ and a negative angular velocity for $r < r_{obs}$. Since the rotational velocity is due to an $\mathbf{E} \times \mathbf{B}$ drift, this means that the effective radial electric field measured in the observer frame must change sign at $r = r_{obs}$.
   (a) Show that this means that there is an effective charge sheet at $r = r_{obs}$.
   (b) Consider the limiting situation where the above situation is modeled using slab geometry so that $r \to x$ and $\theta \to y$, $\mathbf{B} = B\hat{z}$, and the non-neutral plasma initially centered about $x = 0$ has width w and a uniform density n for $|x| < w/2$. This

equilibrium is sketched in Fig. 16.3. There is vacuum in the region between the plasma and perfectly conducting walls at $x = \pm a$, where $a > w/2$. What are the electric field and the plasma flow velocity for $x > 0$ and for $x < 0$?

(c) Now suppose that the location of the non-neutral plasma charge sheet is perturbed so that, instead of being centered at $x = 0$, the location of the center becomes a function of y and is given by $x = \Delta(y) = \bar{\Delta}\cos ky$, where $kd \ll 1$ so that the wavelength in the y direction is very long. What is the boundary condition on E_y and hence on ϕ at the walls?

(d) Since there is vacuum between the plasma and the wall and $kd \ll 1$, what is the limiting form of Poisson's equation in the regions between the plasma and the walls?

(e) Assume that w is small, so that the non-neutral plasma can be approximated as being a thin sheet of charge. Being perfectly conducting, the walls must be equipotentials and so, without loss of generality, this potential can be defined to be zero. Show that the potential to the right of the charge sheet must be of the form

$$\phi_r = \alpha(a - x) \tag{16.92}$$

and the potential to the left of the charge sheet must be of the form

$$\phi_l = \beta(a + x), \tag{16.93}$$

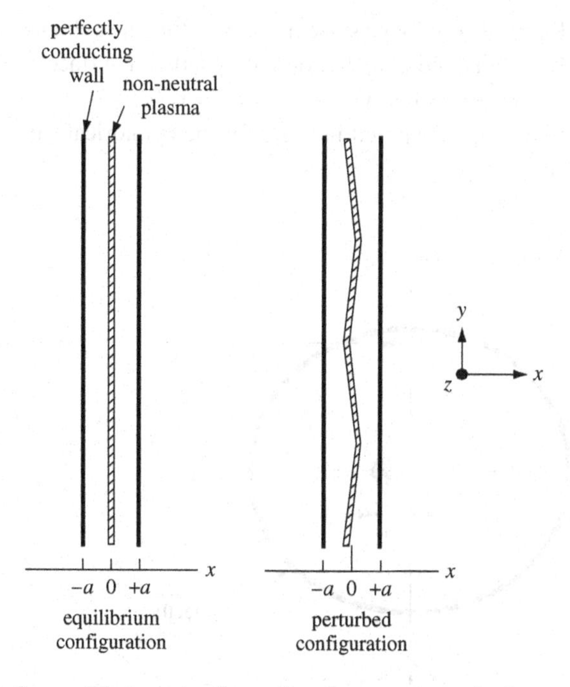

Fig. 16.3 Left: equilibrium configuration for non-neutral plasma of width w, centered between pair of perfectly conducting walls at $x = \pm a$. Right: perturbed configuration where position of x-midpoint of plasma is located at $\Delta = \bar{\Delta}\cos ky$.

where α and β are coefficients to be determined by considering the jump in electric field at the charge sheet.

(f) Taking into account the jump in the potential at the charge sheet and the continuity in the potential at the charge sheet, solve for α and β. Assume that $w \ll a$ consistent with the assumption that the non-neutral plasma can be considered as a thin charge sheet. Which of E_y^2, E_x^2 is larger?

(g) Suppose that the length of the plasma in the y direction is L, where L is an integral multiple of $2\pi/k$, the wavelength in the y direction. By calculating E_x to the right of the charge sheet and also to the left of the charge sheet, show that the energy stored in the electrostatic electric field is approximately

$$W = \frac{\varepsilon_0}{2} h \int_{-L/2}^{L/2} \mathrm{d}y \int_{-a}^{a} \mathrm{d}x E_x^2,$$

where h is the length in the z direction. How does W change when $\bar{\Delta}$ is increased? What does this suggest about the stability of sheared velocity profiles with respect to perturbations as prescribed in (c)?

3. Image charge in cylindrical geometry. Two line charges λ_1 and λ_2 are aligned along the z axis and located at $\mathbf{r}_1 = x_1\hat{x}$ and $\mathbf{r}_2 = x_2\hat{x}$ as shown in Fig. 16.4. What values of x_2 and λ_2 are required in order for the cylindrical surface $\sqrt{x^2+y^2} = a$ to be an equipotential as would be required in order to have a perfectly conducting wall at $r = a$? The following hints should be useful:

(a) Define a cylindrical coordinate system r, θ, z so that the cylindrical surface is given by $\mathbf{r} = a\hat{r}$. Determine the potential on the cylindrical surface $r = a$.

(b) Show that $|\hat{r} - \hat{x}x_1/a| = |\hat{r}a/x_2 - \hat{x}|$ if $a/x_2 = x_1/a$.

(c) Show that $\lambda_2 = -\lambda_1$ is required in order for the cylindrical surface $r = a$ to be an equipotential.

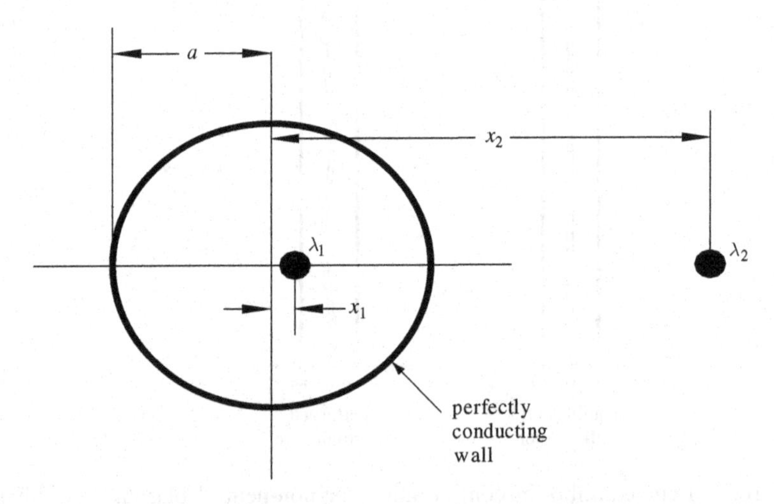

Fig. 16.4 Line charge λ_1 is displaced a distance x_1 from axis of perfectly conducting cylindrical wall of radius a. An image charge λ_2 is located at x_2.

4. Diocotron mode energy using image charge method. A cylindrical non-neutral plasma is translationally invariant in the z direction and is bounded by a perfectly conducting wall at $r = a$. The plasma has an equilibrium density profile $n = n(r)$ for $0 \leq r < r_p$ and $n = 0$ for $r_p < r \leq a$ so there is a finite-size vacuum region between the edge of the plasma r_p and the wall radius a. Calculate the change in system energy if the plasma axis is shifted off the axis and compare this result to Eq. (16.66). Hints are given below:

   (a) Show that the electric field outside the plasma is the same as the electric field of a line charge located on the plasma axis.
   (b) Suppose that x denotes the distance the plasma is shifted off the wall axis. What is the location and strength of the image charge required for the wall to be an equipotential? Is the force between the plasma and the image charge attractive or repulsive? How does the image charge move as x increases from zero to some finite value?
   (c) Calculate the force on the image charge. What is the force on the plasma? What work is done by the plasma in order to be displaced by a finite amount x? Express this work in terms of the radial electric field perturbation at the wall.

5. Negative energy in terms of two capacitors with fixed charge. Consider the situation where a charge Q is stored on two series-connected parallel plate capacitors that are deformed in such a way as to conserve the total volume of the regions between the plates. This situation is shown in Fig. 16.5(a) top and bottom, and in an electrically equivalent, but geometrically slightly different, way in Fig. 16.5(b) top and bottom.

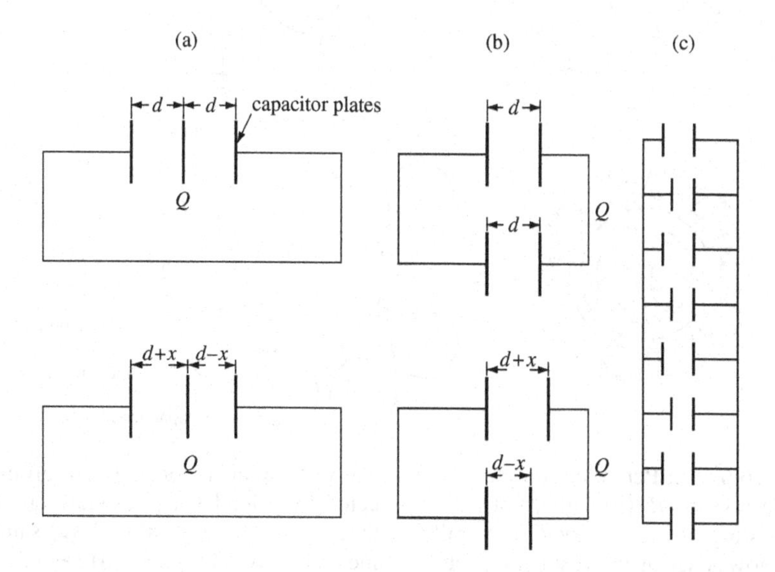

Fig. 16.5 (a) Two capacitors in series with moveable middle plate, (b) geometric rearrangement, (c) straightened-out model of azimuthal periodic incompressible perturbation.

(a) In Fig. 16.5(a) the center plate is displaced off center by an amount x so that the volume occupied by the two capacitors is conserved. What is the change in the system energy when the center plate is displaced? Assume that the value of each capacitor is $C_{1,2} = \varepsilon_0 A/d_{1,2}$, where $d_{1,2}$ are the distances between the parallel plates and initially (top of figure) $d_1 = d_2 = d$ but later (bottom of figure) $d_1 = d + x$ and $d_2 = d - x$.

(b) Show that the configuration in Fig. 16.5 (b) is electrically equivalent to the configuration in Fig. 16.5(a). Discuss how the system shown in Fig. 16.5(c) would relate to an $l = 4$ diocotron mode (recall that any perturbations to the plasma must be incompressible).

6. Magnetrons as a generalization of non-neutral plasmas undergoing a resistive wall instability. The magnetron vacuum tube used in radar transmitters and in microwave ovens can be considered as a non-neutral plasma undergoing a negative energy diocotron instability. These tubes are simple, rugged, and efficient. The relation between the non-neutral plasma resistive wall instability and the magnetron is shown in the sequence of sketches Figs. 16.6(a)–(d). The magnetron has a cylindrical geometry with an electron emitting filament (cathode) on the z axis, a segmented cylindrical

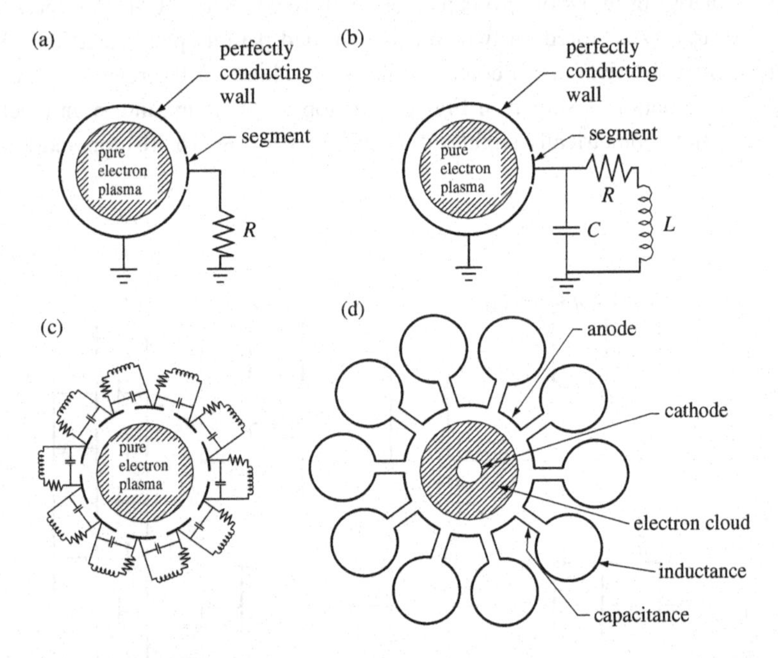

Fig. 16.6 (a) Pure electron plasma with one wall segment connected to ground via a resistor R, (b) same, but now connected to ground via a resonant circuit consisting of a capacitor C in parallel with a coil L and a resistor R, (c) same, but now a set of ten resonant circuits connected across ten gaps, (d) same, but now the coil is a single turn and the capacitance is in the leads connecting to the coil (the resistor is the loading of the coupler to the output circuit, typically a loop inserted into a coil).

wall (anode) and a z-directed magnetic field produced by permanent magnets. Instead of having a resistor R connected across the gap of a segmented wall as in Fig. 16.6(a), the magnetron has a set of cavities connected across the gap as in Fig. 16.6(d). The cavities function as a resonant circuit and the output circuit loading provides an effective resistance at the cavity resonant frequency. From the point of view of the plasma, the power appears to be dissipated in a resistor across the wall gap. However, since the cavity is coupled to an output waveguide, the power is actually transported away from the magnetron via a waveguide to some external location where the power is used to transmit a radar pulse or cook a meal. If the cavities in Fig. 16.6(d) are phased $0, \pi$ then the ten cavities provide five complete azimuthal wave periods or $l = 5$. If the electrons have near Brillouin flow, and $l = 5$, what axial magnetic field should be used in order to have an output frequency $f = 2450$ MHz, the frequency used in a home microwave oven? Why would the electron density increase to the point that the flow is nearly Brillouin, and why is the electron density not higher than this value? If the voltage drop between anode and cathode is 4 kV and the electron density is uniform, what nominal wall radius (distance between cathode and anode) should be used? Show that a parallel resonant circuit with a small resistance in series with the coil as in Fig. 16.6(b) has an effective resistance that peaks at the resonant frequency; why does one want the effective resistance to peak at the resonant frequency?

7. Spiral due to sheared rotation. Suppose that a non-neutral plasma has a sheared angular velocity so that $\omega_0 = \omega_0(r)$. Suppose that a radial line is painted onto the plasma at $t = 0$ extending from $x = r_p/4$, $y = 0$ to $x = r_p/2$, $y = 0$, where r_p is the plasma radius. Suppose that $\omega_0 = \bar{\omega}_0 r/r_p$. Calculate the trajectory of points on the painted line and plot this for successive times using a computer. If the initial line had finite thickness and hence finite area, what would happen to the thickness of the line with increasing time?

17

Dusty plasmas

17.1 Introduction

Non-neutral plasmas had one less species than a conventional plasma (one species instead of two); dusty plasmas have one more species (three species instead of two). Not surprisingly, the addition of another species provides new freedoms, which give rise to new behaviors.

The third species in a dusty plasma, electrically charged dust grains, typically have a charge to mass ratio quite different from that of electrons or ions. Several methods for charging dust grains are possible. Electron bombardment is the usual means for charging laboratory dusty plasmas, but photoionization or radioactive decay could also be operative mechanisms and may be important for certain space and astrophysical situations. Photoionization would make dust grains positive because photoionization causes electrons to leave dust grains. Radioactive decay of dust grains would make the dust grains develop a polarity opposite to that of the particle emitted in the decay process, e.g., alpha particle emission by dust grains would make the dust grains negative.

We shall consider here only the typical laboratory dusty plasma situation where the plasma is weakly ionized and dust grain charging is due to electron bombardment. Negative charging occurs because the electrons, being much lighter than the ions and usually much hotter, have a much larger thermal velocity than the ions. As a result, when a dust grain is inserted into the plasma it is initially subject to a greater flux of impacting electrons than impacting ions, thereby causing the dust grain to become negatively charged. The negative charge deposited on the dust grain eventually becomes sufficiently large to repel incident electrons and thus attenuate the incoming electron flux. On the other hand, the negative charge on the dust grain accelerates incident ions thereby increasing the ion flux to the dust grain. The net charge on the dust grain reaches equilibrium when the electron and ion fluxes intercepting the dust grain become equal. This is a dynamic equilibrium because it involves a continuous flow of plasma to the dust grain.

Some kind of external source is thus required to replenish the plasma electrons and ions continuously, because otherwise all the plasma electrons and ions would eventually accumulate on the dust grain. The dust grain charging process is quite similar to the development of a floating potential on an insulated probe immersed in a plasma (see discussion of Eq. (2.133)).

17.2 Electron and ion current flow to a dust grain

Because a dust grain is much heavier than an electron or an ion, a dust grain acts as an effectively infinitely massive scattering center for electrons and ions colliding with it. In typical dusty plasma laboratory experiments the collisional mean free path l_{mfp} greatly exceeds the Debye length. This situation is sketched in Fig. 17.1 where a dust grain with radius r_d is surrounded by an imaginary sphere with diameter l_{mfp} and the Debye radius is much smaller than l_{mfp}. The nominal distance from the dust grain to the location of the incident electron's or ion's previous collision is λ_{mfp}, the collision mean free path. Thus, incident electrons and ions can be considered collisionless inside a sphere of radius λ_{mfp} centered about the dust grain. Since electrons and ions are collisionless inside the l_{mfp} sphere they have Keplerian orbits associated with the electrostatic central force produced by the charge on the dust grain. Because the dust grain is shielded by other particles at distances greater than a Debye length, the electrons and ions experience this central force only when inside a sphere having the nominal Debye radius.

In order to develop a model for dust grain charging, it is first necessary to determine the effective collision cross-sections for electrons and ions colliding with the dust grain. A useful benchmark reference for this calculation is the

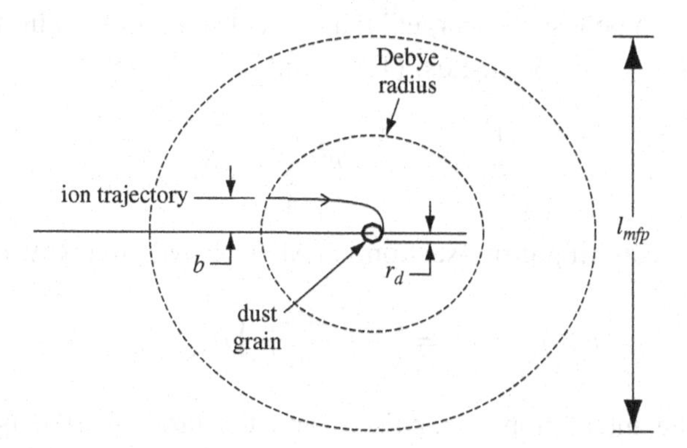

Fig. 17.1 Electrons and ions may be considered as being collisionless inside a sphere of diameter l_{mfp} surrounding a dust grain. The trajectory of an ion making a grazing collision is shown here; this ion has an impact parameter b.

cross-section of charged particles colliding with a neutral dust grain. Because charged particles incident on a neutral dust grain have straight line trajectories, the cross-section of a neutral dust grain is just the geometrical area it projects onto a plane, i.e., $\sigma_{geometric} = \pi r_d^2$. The cross-section of a *charged* dust grain differs from $\sigma_{geometric}$ because incident electrons and ions are deflected from straight-line trajectories by the electrostatic central force produced by the charge on the dust grain. The respective cross-sections for ions and electrons will now be calculated and related to $\sigma_{geometric}$.

Figure 17.1 shows the impact parameter b and trajectory of an ion colliding with a negatively charged dust grain; the corresponding trajectory of an electron would curl outwards instead of inwards and so the electron would require a much smaller impact parameter b in order to hit the dust grain. We define v to be the initial velocity of an incident particle having mass m and charge q, and define v_{impact} as the velocity of this particle at the instant it makes a grazing impact with the dust grain. Using these definitions it is seen that conservation of angular momentum imposes the requirement

$$vb = v_{impact} r_d \tag{17.1}$$

and conservation of energy imposes the requirement

$$\frac{1}{2}mv^2 = \frac{1}{2}mv_{impact}^2 + q\phi_d, \tag{17.2}$$

where ϕ_d is the potential on the surface of the dust grain. Figure 17.1 shows that b is larger than r_d for an ion. Because the repulsive force between the negatively charged dust grain and an electron causes the trajectory of an electron to swerve in the opposite sense from an ion, b is smaller than r_d for an electron; that is, the electron has to be more "on-target" than a neutral particle to hit the dust grain, whereas an ion can be less "on-target" than a neutral particle to hit the dust grain.

Eliminating v_{impact} between Eqs. (17.1) and (17.2) gives

$$\frac{1}{2}mv^2 = \frac{1}{2}mv^2\frac{b^2}{r_d^2} + q\phi_d \tag{17.3}$$

so the effective scattering cross-section is (Allen, Boyd, and Reynolds 1957)

$$\sigma(v) = \pi b^2 = \left(1 - \frac{2q\phi_d}{mv^2}\right)\sigma_{geometric}. \tag{17.4}$$

For $q\phi_d > 0$ the interaction is repulsive and the cross-section is smaller than $\sigma_{geometric}$, whereas for $q\phi_d < 0$ the interaction is attractive and the cross-section is larger than $\sigma_{geometric}$. The former case applies to electrons and the latter case applies to ions.

The total current flowing to the dust grain for attractive interactions is

$$I_{attractive} = q\int_0^\infty \sigma(v)vf(v)4\pi v^2 \mathrm{d}v. \tag{17.5}$$

However, because in the repulsive situation incident particles having $v < \sqrt{2q\phi_d/m}$ are reflected and do not hit the dust grain, the repulsive situation cross-section is zero for all particles having $v < \sqrt{2q\phi_d/m}$. Thus, the total current for repulsive interactions is

$$I_{repulsive} = q\int_{\sqrt{2q\phi_d/m}}^\infty \sigma(v)vf(v)4\pi v^2 \mathrm{d}v. \tag{17.6}$$

Since the region outside the l_{mfp} sphere sketched in Fig. 17.1 extends to infinity, particles can be considered to have made many collisions before entering the l_{mfp} sphere, and so the velocity distribution of particles incident upon the l_{mfp} sphere will be Maxwellian, i.e.,

$$f(v) = \frac{n_0}{\pi^{3/2}v_T^3}\mathrm{e}^{-v^2/v_T^2}, \tag{17.7}$$

where $v_T = \sqrt{2\kappa T/m}$ is the thermal velocity. The incident particles will be collisionless as they travel inside the l_{mfp} sphere. Integration of Eqs. (17.5) and (17.6) gives

$$\begin{aligned} I_{attractive} &= q\int_0^\infty \pi r_d^2\left(1-\frac{2q\phi_d}{mv^2}\right)vf(v)4\pi v^2\mathrm{d}v \\ &= 2\pi^{1/2}r_d^2 n_0 q v_T\int_0^\infty\left(1-\frac{2q\phi_d}{mv_T^2 x}\right)\mathrm{e}^{-x}x\mathrm{d}x \\ &= \frac{2}{\pi^{1/2}}\left(1-\frac{q\phi_d}{\kappa T}\right)n_0 q v_T\sigma_{geometric} \end{aligned} \tag{17.8}$$

and

$$\begin{aligned} I_{repulsive} &= 2\pi^{1/2}r_d^2 n_0 q v_T\int_{q\phi_d/\kappa T}^\infty\left(1-\frac{q\phi_d}{\kappa T x}\right)\mathrm{e}^{-x}x\mathrm{d}x \\ &= \frac{2}{\pi^{1/2}}\mathrm{e}^{-q\phi_d/\kappa T}n_0 q v_T\sigma_{geometric}. \end{aligned} \tag{17.9}$$

17.3 Dust charge

As discussed above, a neutral dust grain inserted into a plasma will become negatively charged because $v_{Te} \gg v_{Ti}$. The electron current I_e is thus a repulsive-type current and the ion current I_i is an attractive-type current. As the dust grain becomes more negatively charged, $|I_e|$ decreases and $|I_i|$ increases until $I_i + I_e = 0$, at which time dust grain charging ceases.

The presence of negatively charged dust grains affects the quasi-neutrality condition. Assuming singly charged ions and a charge of Z_d on the negatively charged dust grains, the quasi-neutrality condition generalizes to

$$n_{i0} = n_{e0} + Z_d n_{d0}. \tag{17.10}$$

Since the ion density is unaffected by the charging process, it is convenient to normalize all densities to the ion density and define

$$\alpha = \frac{Z_d n_{d0}}{n_{i0}} \tag{17.11}$$

so that

$$\frac{n_{e0}}{n_{i0}} = 1 - \alpha, \tag{17.12}$$

where n_{e0}, n_{i0}, and n_{d0} are the volume-averaged electron, ion, and dust grain densities. Furthermore, if r_d is small compared to the effective shielding length λ_d, then the dust grain charge Z_d is related to the dust grain surface potential and the dust grain radius r_d via the vacuum Coulomb relationship, i.e.,

$$\begin{aligned} \phi_d &= -\frac{Z_d e \exp(-r_d/\lambda_d)}{4\pi\varepsilon_0 r_d} \\ &\simeq -\frac{Z_d e}{4\pi\varepsilon_0 r_d} \quad \text{if } r_d \ll \lambda_d. \end{aligned} \tag{17.13}$$

Using Eq. (17.13) to give Z_d, Eq. (17.11) can then be written in terms of the dust grain surface potential as

$$\alpha = -\frac{4\pi\varepsilon_0 r_d n_{d0}}{n_{i0} e}\phi_d. \tag{17.14}$$

It is now convenient to introduce the dimensionless variable

$$\psi = -\frac{e\phi}{\kappa T_i} \tag{17.15}$$

so that α becomes

$$\alpha = 4\pi n_{d0} r_d \lambda_{di}^2 \psi_d, \tag{17.16}$$

where

$$\lambda_{di} = \sqrt{\frac{\varepsilon_0 \kappa T_i}{n_{i0} e^2}} \tag{17.17}$$

is the ion Debye length. Large ψ_d means that ions fall into a potential energy well much deeper than their thermal energy.

It is also useful to introduce the Wigner–Seitz radius a, a measure of the nominal spacing between adjacent dust grains. This spacing is defined by dividing the total volume of the system V by the total number of dust grains N to find a nominal volume surrounding each dust grain. It is then imagined that this nominal volume is spherical with radius a so that

$$N\frac{4\pi a^3}{3} = V, \tag{17.18}$$

in which case

$$n_{d0} = \frac{3}{4\pi a^3} \tag{17.19}$$

and

$$\alpha = \frac{3r_d\lambda_{di}^2}{a^3}\psi_d. \tag{17.20}$$

It is also convenient to normalize lengths to the ion Debye length so

$$\alpha = P\psi_d, \tag{17.21}$$

where

$$P = 3\frac{\bar{r}_d}{\bar{a}^3} \tag{17.22}$$

and the bar means normalized to an ion Debye length.

The dust charge can similarly be expressed in a non-dimensional fashion using Eq. (17.13) to give

$$\frac{Z_d}{4\pi n_{i0}\lambda_{di}^3} = \bar{r}_d\psi_d. \tag{17.23}$$

The floating condition $I_i + I_e = 0$ shows that the equilibrium dust potential must satisfy

$$0 = n_{i0}v_{T_i}\left(1 - \frac{e\phi_d}{\kappa T_i}\right) - n_{e0}v_{T_e}\exp\left(e\phi_d/\kappa T_e\right), \tag{17.24}$$

which can be rearranged as

$$(1+\psi_d)\sqrt{\frac{m_eT_i}{m_iT_e}}\exp\left(\psi_d T_i/T_e\right) = 1 - \alpha. \tag{17.25}$$

However, using Eq. (17.21) this becomes

$$(1+\psi_d)\sqrt{\frac{m_eT_i}{m_iT_e}}\exp\left(\psi_d T_i/T_e\right) = 1 - P\psi_d, \tag{17.26}$$

a transcendental equation relating ψ_d and P.

Equation (17.26) shows that, for a given temperature ratio T_i/T_e and a given mass ratio m_e/m_i, the normalized dust grain surface potential ψ_d is a function of

P, a dimensionless parameter incorporating the geometrical information characterizing the dust grains in the dusty plasma. Because ψ_d appears non-algebraically in Eq. (17.26), it is not possible to solve for $\psi_d(P)$ without resorting to numerical methods. However, solving the inverse relation, namely $P(\psi_d)$, is straightforward because P occurs only once in Eq. (17.26). Solving for P gives (Havnes *et al.* 1987)

$$P = \frac{1}{\psi_d} - \left(1 + \frac{1}{\psi_d}\right)\sqrt{\frac{m_e T_i}{m_i T_e}} \exp\left(\psi_d T_i / T_e\right). \tag{17.27}$$

The upper plot in Fig. 17.2 shows $\log P$ plotted versus log ψ_d for two cases corresponding to typical laboratory configurations where dusty plasma experiments have been conducted; the lower plot shows the corresponding dependence of α on ψ_d. The $T_e = T_i$ configuration corresponds to a potassium Q-machine plasma while the $T_e = 100T_i$ configuration corresponds to an argon rf discharge. An ion of 40 atomic mass units has been assumed for both cases. Figure 17.2 shows that P varies inversely with ψ_d up to some critical value and then, for ψ_d above this critical value, P heads sharply to zero for modest increases in ψ_d. The lower plot shows that α is near unity to the left of the knee and then drops sharply to zero to the right of the knee. Comparison of the solid and dashed curves shows that the saturated value of ψ_d is an increasing function of T_e/T_i. The saturation value of ψ_d is 2.5 for the Q-machine parameters whereas the saturated

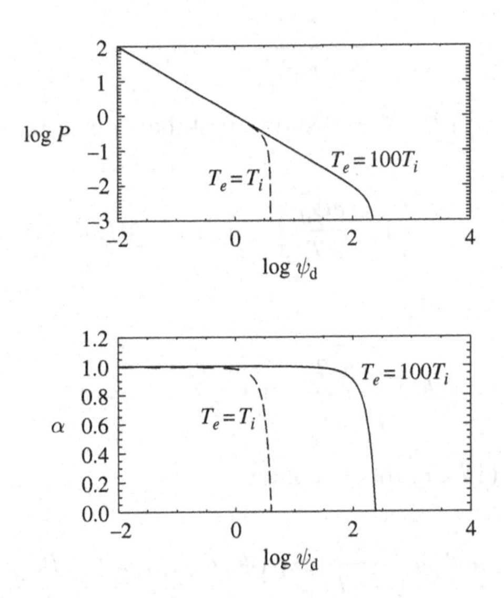

Fig. 17.2 Plots of $\log P$ and α vs. log ψ_d for $T_e = 100T_i$ and $T_e = T_i$. Logarithms are base 10 and ion mass is 40 amu. Because P is proportional to the dust grain density, the right-hand side of these curves, i.e., small P, corresponds to the limit of low dust grain density.

value of ψ_d for the rf discharge is $\sim 2 \times 10^2$. Since P is proportional to $\bar{a}^{-3}$ and hence to the dust grain density, the right-hand side of these plots (i.e., where P is small) corresponds to the limit of small dust grain density. Since Eq. (17.23) shows that Z_d is proportional to ψ_d, it is seen that Z_d also saturates on moving to the right in the plots and largest Z_d occurs at high T_e/T_i and small dust grain density.

The downward sloping segment on the left of Fig. 17.2 corresponds to the $\alpha \simeq 1$ regime, which is the situation where the dust grain density is large and nearly all the electrons are attached to the dust grains so the plasma has almost no free electrons. On the other hand, the saturated limit of ψ_d on the right of Fig. 17.2 corresponds to the $\alpha \ll 1$ regime, which is where there is minimal depletion of free electrons, the dust grain density is low, and there is a very high charge on each of the relatively small number of dust grains.

There are thus three regimes:

1. *The regime well to the left of the knee in the curves.* Here $\alpha \simeq 1$ and nearly all electrons are attached to the dust grains. To the extent that α approaches unity, the dust grains replace the electrons as the negative charge carriers.
2. *The regime to the right of the knee in the curves.* Here $\alpha \ll 1$, most electrons are free, and the potential of an individual dust grain is very high and near its saturation value (right-hand side of plots). This regime corresponds to small P and a very small dust grain density.
3. *The regime in the vicinity of the knee in the curves.* If $T_e \gg T_i$ then it is possible to have α of order unity and ψ_d large, but not quite at its saturation value. In this regime the majority of electrons reside on the dust grains, the dust grains are highly charged, and the dust grain density is appreciable. Crystallization of the dust grains can occur in this regime, since crystallization requires a combination of high dust grain charge and small separation between dust grains.

17.4 Dusty plasma parameter space

A dusty plasma is characterized by the normalized dust radius $\bar{r}_d$ and the normalized dust interparticle spacing $\bar{a}$. These quantities determine P and, given P, the normalized dust surface potential ψ_d is found by solving Eq. (17.27). Knowing P and ψ_d then gives α using Eq. (17.21). The consequence of this chain of argument is that α can be considered to be a function of $\bar{r}_d$ and $\bar{a}$.

A dusty plasma parameter space can thus be constructed where $\bar{a}$ is the horizontal component and $\bar{r}_d$ is the vertical component so that a given dusty plasma would correspond to a point in this parameter space. The quantities ψ_d, α, and $Z_d/4\pi n_{i0}\lambda_{di}^3$ are all functions of $\bar{a}$ and $\bar{r}_d$ and so contours of these three quantities can be drawn in the $\bar{a}, \bar{r}_d$ parameter space for specified m_e/m_i and T_e/T_i.

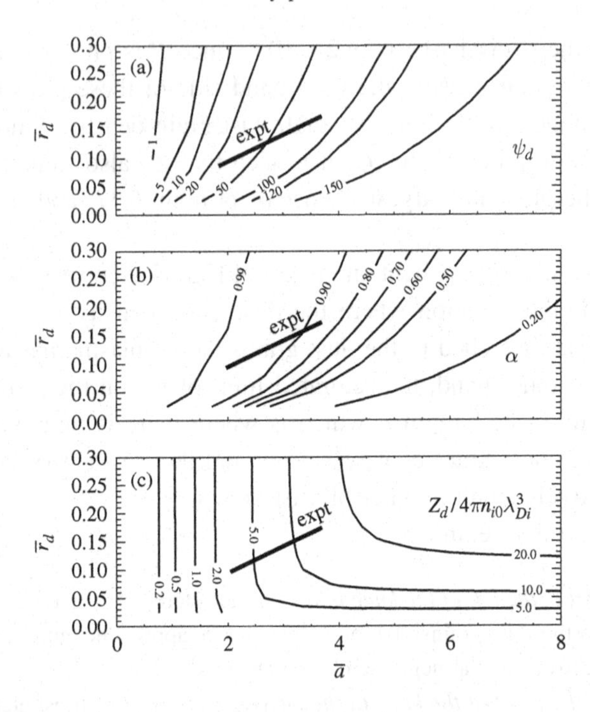

Fig. 17.3 Dusty plasma parameter space showing contours of constant ψ_d, α, and $Z_d/4\pi n_{i0}\lambda_{Di}^3$ for an argon plasma with $T_e/T_i = 100$. The experimental line corresponds to the density range of the Chu and I experiment (Chu and I 1994); these plots are from Bellan (2004b).

Examples of these contours are shown in Fig. 17.3 for an argon plasma with $T_e = 100T_i$. An actual experiment would have specific values of $\bar{a}$ and $\bar{r}_d$ and so would be represented as a point in this parameter space. Variation of the ion density in the experiment would change the value of λ_{di} while keeping the ratio $\bar{r}_d/\bar{a}$ fixed and so would correspond to moving along a sloped line in parameter space. The short sloped line in Fig. 17.3 represents the density in the dusty plasma experiment by Chu and I (1994) to be discussed later (the finite length of this line corresponds to the error bars for the density measurement).

17.5 Large *P* limit: dust acoustic waves

The large P limit (i.e., Regime 1 discussed above) has nearly all the electrons attached to the dust grains so that the plasma effectively consists of negatively charged dust grains and positive ions. A wave similar to the conventional ion acoustic wave can propagate in this regime, but the role played by positive and negative particles is reversed: here the ions are the light species and the dust grains are the heavy species. In order to appreciate the consequence of this role reversal, consider the conventional ion acoustic wave from the most simplistic point of view. The wave phase velocity ω/k is assumed to be much faster than the ion thermal velocity v_{Ti} (i.e., cold ion regime) but much slower than the electron thermal velocity v_{Te} (i.e., isothermal electron regime) so the respective approximations for the electron and ion equations of motion are

$$0 = -n_e e\mathbf{E} - \nabla(n_e \kappa T_e) \tag{17.28}$$

$$n_i m_i \frac{\mathrm{d}\mathbf{u}_i}{\mathrm{d}t} = n_i Ze\mathbf{E}. \tag{17.29}$$

Adding the electron and ion equations and assuming quasi-neutrality $n_e \simeq n_i Z$, gives

$$n_i m_i \frac{\mathrm{d}\mathbf{u}_i}{\mathrm{d}t} = -\nabla(n_e \kappa T_e) \tag{17.30}$$

showing that the effective force acting on the ions is the electron pressure gradient. This force is coupled to the ions via the electric field. In effect, the electron pressure gradient pushes against the electric field, which in turn pushes the ions. Linearizing Eq. (17.30) gives a result similar to a conventional sound wave, except that the system is isothermal with respect to the electron temperature (in a normal neutral gas sound wave, the gas temperature would appear in the right-hand side, the gas would be adiabatic, and so a γ would appear upon linearizing the gas pressure). Linearization of Eq. (17.30) and invoking the linearized quasi-neutrality relation $n_{e1} \simeq n_{i1} Z$, gives

$$n_{i0} m_i \frac{\partial \mathbf{u}_{i1}}{\partial t} = -Z\kappa T_e \nabla n_{i1}. \tag{17.31}$$

The linearized ion equation of continuity is

$$\frac{\partial n_{i1}}{\partial t} + n_{i0} \nabla \cdot \mathbf{u}_{i1} = 0. \tag{17.32}$$

The equations are combined by taking the divergence of Eq. (17.31) and then substituting Eq. (17.32) to obtain

$$\frac{\partial^2 n_{i1}}{\partial t^2} = \frac{Z\kappa T_e}{m_i} \nabla^2 n_{i1}, \tag{17.33}$$

which describes a wave with phase velocity $c_s^2 = Z\kappa T_e/m_i$.

This method can now be generalized to a plasma consisting of ions with charge $+e$, free electrons with charge $-e$, and dust grains with charge $-Z_d e$. The free electron density is assumed to be small so that $P \gg 1$, in which case the configuration is on the left of Fig. 17.2. The wave phase velocity is assumed to be much faster than the dust grain thermal velocity, but much slower than the ion/electron thermal velocities so the respective electron, ion, and dust grain equations of motion are

$$0 = -n_e e\mathbf{E} - \nabla(n_e \kappa T_e) \tag{17.34}$$

$$0 = +n_i e\mathbf{E} - \nabla(n_i \kappa T_i) \tag{17.35}$$

$$n_d m_d \frac{\mathrm{d}\mathbf{u}_d}{\mathrm{d}t} = -n_d Z_d e\mathbf{E}. \tag{17.36}$$

Adding the above three equations and invoking quasi-neutrality (i.e., $n_i = n_d Z_d + n_e$) results in

$$n_d m_d \frac{\mathrm{d}\mathbf{u}_d}{\mathrm{d}t} = -\nabla(n_i \kappa T_i + n_e \kappa T_e). \tag{17.37}$$

In analogy to the conventional ion acoustic wave, here the ion and electron pressures couple to the dust via the electric field. Linearization of Eq. (17.37) gives

$$n_{d0} m_d \frac{\partial \mathbf{u}_{d1}}{\partial t} = -\kappa T_i \nabla n_{i1} - \kappa T_e \nabla n_{e1}, \tag{17.38}$$

while linearization of Eqs. (17.34) and (17.35) gives

$$0 = -n_{e0} e\mathbf{E}_1 - \kappa T_e \nabla n_{e1} \tag{17.39}$$

$$0 = +n_{i0} e\mathbf{E}_1 - \kappa T_i \nabla n_{i1}. \tag{17.40}$$

Eliminating $\mathbf{E}_1$ between these last two equations shows that

$$\kappa T_e \nabla n_{e1} = -\frac{n_{e0}}{n_{i0}} \kappa T_i \nabla n_{i1}, \tag{17.41}$$

which can be integrated to give

$$n_{e1} = -\frac{n_{e0} T_i}{n_{i0} T_e} n_{i1}. \tag{17.42}$$

Inserting Eq. (17.42) into the linearized quasi-neutrality expression $n_{i1} = n_{d1} Z_d + n_{e1}$ gives

$$n_{i1} = \frac{1}{1 + \dfrac{n_{e0} T_i}{n_{i0} T_e}} Z_d n_{d1} \tag{17.43}$$

and hence

$$n_{e1} = -\frac{n_{e0} T_i / n_{i0} T_e}{1 + \dfrac{n_{e0} T_i}{n_{i0} T_e}} Z_d n_{d1}. \tag{17.44}$$

Substitution for n_{i1} and n_{e1} in Eq. (17.38) gives

$$n_{d0}m_d\frac{\partial \mathbf{u}_{d1}}{\partial t} = -\kappa T_i\frac{1-n_{e0}/n_{i0}}{1+\dfrac{n_{e0}T_i}{n_{i0}T_e}}Z_d\nabla n_{d1}$$

$$\simeq -\kappa T_i Z_d \nabla n_{d1} \tag{17.45}$$

since $n_{e0} \ll n_{i0}$ and typically $T_i/T_e \ll n_{i0}/n_{e0}$.

The linearized dust continuity equation is

$$\frac{\partial n_{d1}}{\partial t} + n_{d0}\nabla\cdot\mathbf{u}_{d1} = 0 \tag{17.46}$$

so taking the divergence of Eq. (17.45) and substituting Eq. (17.46) gives

$$\frac{\partial^2 n_{d1}}{\partial t^2} = \frac{Z_d\kappa T_i}{m_d}\nabla^2 n_{d1}, \tag{17.47}$$

which describes a wave with phase velocity

$$c_{da}^2 = \frac{Z_d\kappa T_i}{m_d}. \tag{17.48}$$

This wave is called the dust acoustic wave and its phase velocity is extremely low because of the large dust grain mass.

This analysis could be extended to include finite $k\lambda_D$ terms as obtained by using the full Poisson's equation instead of making the simplifying assumption of perfect neutrality. A Vlasov approach could also be invoked to demonstrate the effect of Landau damping. Rather than work through the details of these extensions, one can argue that the $\alpha \simeq 1$ dusty plasma is effectively a two-component plasma where the heavy particles are the negatively charged dust grains and the light particles are the positively charged ions. The previously derived results from both two-fluid theory and Vlasov theory could then be invoked by simply identifying the heavy and light particles in the manner stated above. Thus by making the identification shown in Table 17.1, taking into account finite $k^2\lambda_D^2$ terms will result in a dispersive dust acoustic wave

$$\omega^2 = \frac{k^2 c_{da}^2}{1+k^2\lambda_{Di}^2}. \tag{17.49}$$

Similarly, just as an electron flow velocity faster than the ion acoustic phase velocity would destabilize ion acoustic waves via inverse Landau damping, a Landau analysis would show that an ion flow velocity faster than the dust acoustic phase velocity would destabilize dust acoustic waves.

Destabilized dust acoustic waves have been observed in an experiment by Barkan, Merlino, and D'Angelo (1995) . The phase velocity of these waves was of

Table 17.1

	ion acoustic wave	dust acoustic wave
inertia provided by	ions	dust grains
restoring force provided by	electron pressure	ion pressure

the order of 10 cm s^{-1}, which is an order of magnitude less than a typical human walking speed.

17.6 Dust ion acoustic waves

The presence of dust grains can also substantially modify the propagation properties of conventional ion acoustic waves. In a conventional electron–ion plasma, ion acoustic waves can propagate only if $T_e \gg T_i$. This is because strong ion Landau damping occurs when the wave phase velocity is of the order of the ion thermal velocity. This damping is impossible to avoid unless $T_e \gg T_i$, since the conventional ion acoustic phase velocity scales as $c_s^2 = (\gamma\kappa T_i + \kappa T_e)/m_i$.

As in the conventional ion acoustic wave, the derivation of dust ion acoustic waves involves the linearized electron and ion equations

$$0 = -n_{e0}e\mathbf{E}_1 - \kappa T_e \nabla n_{e1} \tag{17.50}$$

$$n_{i0}m_i\frac{\partial \mathbf{u}_i}{\partial t} = +n_{i0}e\mathbf{E}_1 - \gamma\kappa T_i \nabla n_{i1}; \tag{17.51}$$

the possibility of finite T_i has been retained in order to allow consideration of the $T_e \sim T_i$ regime. The wave frequency is assumed to be sufficiently high so that the dust grains are unable to respond to the wave. The dust grains can thus be considered as being infinitely massive and therefore stationary. In this limit the dust grains contribute to the equilibrium quasi-neutrality condition $n_{i0} = n_{e0} + Z_d n_{d0}$ but, since the dust grains are assumed stationary, their density cannot change and so the linearized quasi-neutrality condition is $n_{i1} = n_{e1}$. Thus, infinitely massive dust grains affect the equilibrium electron density, but not the perturbed electron density.

Eliminating $\mathbf{E}_1$ between Eqs. (17.50) and (17.51) gives

$$n_{i0}m_i\frac{\partial \mathbf{u}_i}{\partial t} = -\frac{n_{i0}}{n_{e0}}\kappa T_e \nabla n_{e1} - \gamma\kappa T_i \nabla n_{i1}. \tag{17.52}$$

Since $n_{i1} = n_{e1}$ and $n_{i0}/n_{e0} = n_{i0}/(n_{i0} - Z_d n_{d0})$, Eq. (17.52) becomes

$$n_i m_i \frac{\partial \mathbf{u}_i}{\partial t} = -\frac{1}{1 - Z_d n_{d0}/n_{i0}}\kappa T_e \nabla n_i - \gamma\kappa T_i \nabla n_i. \tag{17.53}$$

Taking the divergence and using Eqs. (17.11) and (17.32) gives

$$m_i \frac{\partial^2 n_{i1}}{\partial t^2} = \left(\frac{1}{1-\alpha} \kappa T_e + \gamma \kappa T_i \right) \nabla^2 n_{i1}, \tag{17.54}$$

which gives the dust ion acoustic wave (Shukla and Silin 1992), a wave with phase velocity

$$c_{DIA}^2 = \frac{1}{1-\alpha} \frac{\kappa T_e}{m_i} + \gamma \frac{\kappa T_i}{m_i}. \tag{17.55}$$

As α approaches unity, the dust ion acoustic wave phase velocity becomes much greater than the ion thermal velocity even in a plasma having $T_i = T_e$. The dust ion acoustic wave thus propagates without being attenuated by ion Landau damping in a $T_i = T_e$ plasma having $\alpha \simeq 1$, which corresponds to the left side of Fig. 17.2. Thus, the presence of a large dust grain density enables the propagation of ion acoustic waves that normally would be damped in a $T_e = T_i$ plasma; this has been observed in experiments by Barkan, D'Angelo, and Merlino (1996).

17.7 The strongly coupled regime: crystallization of a dusty plasma

The mutual repulsive force between two negatively charged dust grains scales as Z_d^2 and so will be very large for highly charged dust grains. In the extreme limit, the electrostatic potential energy between dust grains might exceed their kinetic energy so that the grains would tend to form an ordered, crystallized state. The possibility that dust grains might crystallize was first suggested by Ikezi (1986) and has since been demonstrated in a number of experiments (Chu and I 1994, Melzer, Trottenberg, and Piel 1994, Thomas *et al.* 1994, Hayashi and Tachibana 1994, Nefedov *et al.* 2003, Morfill *et al.* 2002). The threshold criterion for crystallization will be discussed in this section following a model by Bellan (2004b). The threshold is determined by considering certain issues relating to the validity of the conventional Debye shielding model and the Boltzmann relation.

Large Z_d corresponds to large ψ_d, which in turn corresponds to operating towards the right of Fig. 17.2. Since the location of the saturation value of ψ_d increases with T_e/T_i, very large ψ_d can occur if $T_e \gg T_i$. In addition to the scaling with Z_d^2 the repulsive force also scales inversely with the square of the distance separating the two dust grains, i.e., the repulsive force also scales as $n_{d0}^{2/3}$. Since P is proportional to n_{d0}, the maximum repulsive force would be obtained around the knee in the $T_e \gg T_i$ curves in Fig. 17.2 since at this location it is possible to have large ψ_d without n_{d0} becoming infinitesimal.

The repulsive electrostatic force between two dust grains is attenuated by Debye shielding. This shielding can be calculated by considering a single dust grain to be a test particle immersed in a plasma consisting of electrons, ions, and other dust grains (these other dust grains will be referred to as "field" dust grains). The test particle will be completely shielded beyond some critical radius. A field dust grain located beyond this critical radius will experience no interaction with the test particle dust grain whereas if the field dust grain is located within the shielding radius, it will experience an enormous repulsive force. The test particle dust grain thus acts like a finite-radius hard sphere in its interactions with field particle dust grains.

A quantitative model for these interactions between dust grains can be developed by considering Poisson's equation for a dusty plasma,

$$\nabla^2\phi = -\frac{1}{\varepsilon_0}\left(n_i e - n_e e - Z_d n_d e\right). \tag{17.56}$$

The usual test-particle argument (see p. 10) involves linearization of the Boltzmann relation for each species to obtain a linearized density for each species. These linearized densities are then substituted into Poisson's equation resulting in the Yukawa-type solution,

$$\phi = \frac{q_t}{4\pi\varepsilon_0 r}\exp\left(-r/\lambda_D\right), \tag{17.57}$$

where $\lambda_D^{-2} = \sum \lambda_{D\sigma}^{-2}$. However, this linearization is based on the assumption $|q_t\phi/\kappa T_\sigma| \ll 1$, which is clearly not true in the vicinity of a highly charged dust grain. This inconsistency cannot be resolved in fluid theory and it is necessary to revert to the more fundamental Vlasov description.

According to the Vlasov description, the phase-space density of particles is characterized using a velocity distribution function $f_\sigma(\mathbf{r}, \mathbf{v}, t)$ for each species and the time evolution of f_σ is prescribed by the Vlasov equation. In order to determine the time-averaged potential in the vicinity of a test particle, a steady-state solution to the Vlasov equation must be found. To do this, we begin by letting the test particle location define the origin of a spherical coordinate system and then assume that all time-averaged quantities are spherically symmetric about this origin so $\phi = \phi(r)$ and $f_\sigma = f_\sigma(r, v)$. Since an incident particle can be considered as colliding with the test particle, the nominal distance from the test particle to the location of the incident particle's previous collision is λ_{mfp}, the collision mean free path. Thus, so far as collisions with particles other than the test particle are concerned, an incident particle can be considered as being collisionless inside a sphere of radius λ_{mfp} centered about the test particle, i.e., centered about the origin. Because a typical particle incident upon the test particle will have traveled many mean free paths, the velocity distribution of incident particles will be Maxwellian when these particles enter a sphere with radius of order l_{mfp} centered on the test particle. Since the incident particles are collisionless within the λ_{mfp}

sphere, their velocity distribution function must be a solution to the collisionless Vlasov equation inside this sphere. The boundary condition this inside solution must satisfy is that its large r limit should correspond to the collisional (i.e., Maxwellian) distribution outside the λ_{mfp} sphere.

Solutions to the collisionless Vlasov equation are functions of constants of the motion as shown in Section. 2.3 and the appropriate constant of the motion here is the particle energy $W = m_\sigma v^2/2 + q_\sigma \phi(r)$. Hence, the distribution function in the collisionless region near the test particle must be

$$f_\sigma(r, \mathbf{v}) = n_{\sigma 0}\left(\frac{m_\sigma}{2\pi\kappa T_\sigma}\right)^{3/2} \exp\left(-\frac{m_\sigma v^2/2 + q_\sigma \phi(r)}{\kappa T_\sigma}\right), \tag{17.58}$$

since this maps to a Maxwellian distribution at large distances r, where $\phi(r) \to 0$.

A negatively charged particle such as a dust grain or an electron experiences a repulsive force upon approaching the dust grain test particle and so slows down. Some approaching negatively charged particles reflect and so the minimum velocity of electrons or dust grains approaching the dust grain test particle is zero. The density of these particles will thus be

$$\begin{aligned} n_\sigma &= \int_0^\infty f_\sigma(r, \mathbf{v})\, \mathrm{d}^3 v \\ &= n_{\sigma 0} \exp\left(-\frac{q_\sigma \phi(r)}{\kappa T_\sigma}\right), \end{aligned} \tag{17.59}$$

which is the same as the fluid theory Boltzmann relation. It is useful at this point to change over to the non-dimensional scalar ψ defined in Eq. (17.15). Since the dust grains are negatively charged, ψ is large and positive in the vicinity of a dust grain. The respective normalized electron and field dust grain densities are thus

$$\frac{n_e}{n_{e0}} = \exp(-\psi T_i/T_e) \tag{17.60}$$

$$\frac{n_d}{n_{d0}} = \exp(-\bar{Z}\psi), \tag{17.61}$$

where

$$\bar{Z} = Z_d T_i/T_d. \tag{17.62}$$

Both the electron and field dust grain densities decrease in the vicinity of the dust grain test particle. However, because $\bar{Z}$ is typically very large and T_i/T_e is assumed to be very small, the field dust density scale length is much shorter than the electron density scale length.

Ion dynamics are qualitatively different because all ions approaching the dust grain are accelerated, leading to the situation that no zero velocity ions exist near the dust grain. In particular, an ion starting with infinitesimal inward velocity at infinity where $\psi = 0$ has nearly zero energy, i.e., $W \simeq 0$. The energy conservation equation for this slowest ion is

$$m_i v^2/2 + e\phi(r) = 0 \tag{17.63}$$

and so the velocity for this slowest ion has the spatial dependence

$$v_{\min} = \sqrt{-\frac{2e\phi(r)}{m_i}}. \tag{17.64}$$

The ion density is thus

$$\begin{aligned} n_i(r) &= \int_{v_{\min}}^{\infty} f_i(r, \mathbf{v})\, \mathrm{d}^3 v \\ &= n_{i0} \left(\frac{m_i}{2\pi\kappa T_i} \right)^{3/2} \exp\left(-\frac{q_i \phi(r)}{\kappa T_i} \right) \int_{v_{\min}}^{\infty} \exp\left(-\frac{mv^2/2}{\kappa T_i} \right) \mathrm{d}^3 v \\ &= n_{i0} \frac{4}{\sqrt{\pi}} \exp(\psi) \int_{\sqrt{\psi}}^{\infty} \exp(-\xi^2)\, \xi^2 \mathrm{d}\xi. \end{aligned} \tag{17.65}$$

This can be expressed in terms of the error function

$$\operatorname{erf} z = \frac{2}{\sqrt{\pi}} \int_0^z \exp(-\xi^2) \mathrm{d}\xi \tag{17.66}$$

and, in particular, using the identity

$$\frac{4}{\sqrt{\pi}} \int_z^{\infty} \exp(-\xi^2)\, \xi^2 \mathrm{d}\xi = \frac{2}{\sqrt{\pi}} \int_z^{\infty} \mathrm{d}\xi \exp(-\xi^2) - \frac{2}{\sqrt{\pi}} \int_z^{\infty} \mathrm{d}\xi \frac{\mathrm{d}}{\mathrm{d}\xi} (\xi \exp(-\xi^2)) \tag{17.67}$$

it is seen that (Laframboise and Parker 1973)

$$\frac{n_i}{n_{i0}} = \exp(\psi) \left(1 - \operatorname{erf} \sqrt{\psi} \right) + \frac{2}{\sqrt{\pi}} \sqrt{\psi}. \tag{17.68}$$

The error function has the small-argument limit

$$\lim_{z \to 0} \operatorname{erf} z \simeq \frac{2z}{\sqrt{\pi}} \tag{17.69}$$

and so for $\psi \ll 1$

$$\frac{n_i}{n_{i0}} = 1 + \psi, \tag{17.70}$$

which is identical to the Boltzmann result given by fluid theory.

However, because

$$\lim_{z \to \infty} \mathrm{e}^{-z^2} (1 - \operatorname{erf} z) = 0, \tag{17.71}$$

Table 17.2

	Region 1	Region 2	Region 3
Definition	$\psi_d > \psi > 1$	$1 > \psi > 1/\bar{Z}$	$1/\bar{Z} > \psi$
Location	adjacent to dust grain	middle spherical layer	outer layer
Ions	non-Boltzmann, fast	Boltzmann	Boltzmann
Electrons	Boltzmann	Boltzmann	Boltzmann
Dust	zero density	zero density	Boltzmann
Dependence	vacuum-like w/const.	growing/decaying Yukawa	decaying Yukawa

when $\psi \gg 1$ the ion density has a non-Boltzmann dependence

$$\frac{n_i}{n_{i0}} = \frac{2}{\sqrt{\pi}}\sqrt{\psi}, \tag{17.72}$$

which is much smaller than the $\exp(\psi)$ dependence predicted by the Boltzmann relation.

Poisson's equation, Eq. (17.56), can be written non-dimensionally as

$$\begin{aligned}\frac{1}{\bar{r}^2}\frac{\partial}{\partial \bar{r}}\left(\bar{r}^2\frac{\partial \psi}{\partial \bar{r}}\right) &= \frac{n_i}{n_{i0}} - \frac{n_e}{n_{i0}} - \frac{Z_d n_d}{n_{i0}} \\ &= \frac{n_i}{n_{i0}} - (1-\alpha)\frac{n_e}{n_{e0}} - \alpha\frac{n_d}{n_{d0}},\end{aligned} \tag{17.73}$$

where, following the normalization convention introduced earlier, $\bar{r} = r/\lambda_{di}$. Upon substituting for the normalized densities, Poisson's equation becomes

$$\begin{aligned}\underbrace{\frac{1}{\bar{r}^2}\frac{\partial}{\partial \bar{r}}\left(\bar{r}^2\frac{\partial \psi}{\partial \bar{r}}\right)}_{\text{vacuum}} &= \underbrace{e^{\psi}\left(1-\operatorname{erf}\left(\sqrt{\psi}\right)\right) + \frac{2}{\sqrt{\pi}}\sqrt{\psi}}_{\text{ions}} \\ &\quad \underbrace{-(1-\alpha)\exp\left(-\frac{\psi T_i}{T_e}\right)}_{\text{electrons}} \\ &\quad \underbrace{-\alpha\exp\left(-\bar{Z}\psi\right)}_{\text{dust}}.\end{aligned} \tag{17.74}$$

Because ψ becomes large near the dust grain, this equation is highly nonlinear and so the linearization technique used on p. 10 for the conventional Debye shielding derivation cannot be invoked. Instead, an approximate solution to Poisson's equation is obtained by separating the collisionless region around the dust grain test particle into three concentric layers (regions) according to the magnitude of ψ as shown in Table 17.2.

Poisson's equation has the following approximations in the three regions,

$$\text{Region 1}: \frac{1}{\bar{r}^2}\frac{\partial}{\partial \bar{r}}\left(\bar{r}^2\frac{\partial \psi}{\partial \bar{r}}\right) = 0 \tag{17.75}$$

$$\text{Region 2}: \frac{1}{\bar{r}^2}\frac{\partial}{\partial \bar{r}}\left(\bar{r}^2\frac{\partial \psi}{\partial \bar{r}}\right) = \psi + \alpha \tag{17.76}$$

$$\text{Region 3}: \frac{1}{\bar{r}^2}\frac{\partial}{\partial \bar{r}}\left(\bar{r}^2\frac{\partial \psi}{\partial \bar{r}}\right) = (1+\alpha\bar{Z})\psi. \tag{17.77}$$

The reasons for these approximations are as follows:

In Region 1, $\psi \gg 1$ so that the ion density is given in principle by Eq. (17.72) which would give a right-hand side scaling as $\sqrt{\psi}$. This ion term would dominate the electron and dust terms, since the latter are always less than unity in this region. Thus, keeping just the ion term on the right-hand side, Poisson's equation in Region 1 reduces to

$$\underbrace{\frac{1}{\bar{r}^2}\frac{\partial}{\partial \bar{r}}\left(\bar{r}^2\frac{\partial \psi}{\partial \bar{r}}\right)}_{\text{vacuum}} \simeq \underbrace{\frac{2}{\sqrt{\pi\psi}}}_{\text{ion}}\psi; \tag{17.78}$$

however, the ion term above can be neglected compared to the vacuum term because $\sqrt{\psi} \gg 1$ since $\psi \gg 1$. Thus Poisson's equation can be approximated in Region 1 by just the vacuum term. This assumption is quite good near the dust grain surface where ψ is indeed very large compared to unity, but becomes marginal when ψ approaches unity at the outer limit of Region 1.

In Region 2, defined by $1 > \psi > 1/\bar{Z}$, the ions have a Boltzmann distribution so $n_i/n_{e0} = 1+\psi$. The normalized electron density is $n_e/n_{e0} = \exp(-\psi T_i/T_e)$ and, since $T_i/T_e \ll 1$, this simplifies to $n_e/n_{e0} \simeq 1$. Also, because $\bar{Z}\psi \gg 1$ the dust density is nearly zero in this region.

In Region 3, defined by $1/\bar{Z} > \psi$, Eq. (17.74) approximates to

$$\underbrace{\frac{1}{\bar{r}^2}\frac{\partial}{\partial \bar{r}}\left(\bar{r}^2\frac{\partial \psi}{\partial \bar{r}}\right)}_{\text{vacuum}} \simeq \underbrace{1+\psi}_{\text{ions}} - \underbrace{(1-\alpha)\left(1-\frac{\psi T_i}{T_e}\right)}_{\text{electrons}} - \underbrace{\left(1-\bar{Z}\psi\right)\alpha}_{\text{dust}}$$

$$\simeq (1+\alpha\bar{Z})\psi, \tag{17.79}$$

where $T_i/T_e \ll 1$ has been used. For finite α, the dust term dominates because $\alpha\bar{Z} \gg 1$.

The system has two boundary conditions, one at the dust grain surface and the other at infinity. The former occurs in Region 1 and is set by the radial electric field at the dust grain surface. This boundary condition is obtained from Gauss' law, which relates $\partial\psi/\partial\bar{r}$ at the dust grain surface to the dust grain charge and radius. The boundary condition at infinity can be considered as the large $\bar{r}$ limit for Region 3 and requires ψ to vanish as $\bar{r}$ goes to infinity.

If the same form of Poisson's equation were to characterize Regions 1, 2, and 3, then the system would be governed by a single second-order ordinary differential equation and the two boundary conditions described in the previous paragraph would suffice to give a unique solution. However, the equations in Regions 1, 2, and 3 differ from each other and so must be solved separately with appropriate matching at the two interfaces between the three regions. The criteria for matching at the interfaces, determined by integrating Poisson's equation twice across each interface, are that both ψ and its radial derivative must be continuous at each interface. An important feature of this analysis is that the locations of the two interfaces are unknowns, which are to be determined by solving the matching problem. Since Regions 1 and 2 are of finite extent, both decaying and non-decaying ψ solutions are allowed in these regions. However, only a decaying solution is allowed in Region 3.

Region 1, which is governed by Eq. (17.75), has vacuum-like solutions of the form $\psi \sim c\bar{r}^{-1} + d$, where c and d are constants. The constant c is chosen to give the correct radial electric field at the surface of the dust grain test charge and d is chosen to set $\psi = 1$ at $\bar{r}_i$, which is the as yet undetermined location of the interface between Regions 1 and 2. Thus, using Gauss' law to prescribe the radial electric field at the dust grain surface and choosing d to set $\psi = 1$ at $\bar{r} = \bar{r}_i$ gives

$$\psi_1(\bar{r}) = \frac{\dfrac{\alpha\bar{a}^3}{3} + \left(1 - \dfrac{1}{\bar{r}_i}\dfrac{\alpha\bar{a}^3}{3}\right)\bar{r}}{\bar{r}}. \tag{17.80}$$

In Region 2, which is governed by Eq. (17.76), the effective dependent variable is $\psi + \alpha$, which has growing and decaying Yukawa-like solutions $\sim \exp(\pm\bar{r})/\bar{r}$. It is convenient to write the Region 2 solution as

$$\psi_2(\bar{r}) = \frac{\bar{r}_o\left(\dfrac{1}{\bar{Z}} + \alpha\right)\cosh(\bar{r} - \bar{r}_o) + B\sinh(\bar{r} - \bar{r}_o) - \alpha\bar{r}}{\bar{r}}, \tag{17.81}$$

where the coefficients have been chosen so that $\psi = 1/\bar{Z}$ at $\bar{r}_o$, the as yet undetermined location of the interface between Regions 2 and 3. The coefficient B in this expression is undetermined for now, and it should be noted that ψ is independent of B at $\bar{r} = \bar{r}_o$.

Region 3, governed by Eq. (17.77), has a solution that can be expressed as

$$\psi_3(\bar{r}) = \frac{\bar{r}_0}{\bar{Z}\bar{r}}\exp\left(-\sqrt{\alpha\bar{Z} + 1}\,(\bar{r} - \bar{r}_o)\right) \tag{17.82}$$

where the coefficients have been chosen to give $\psi = 1/\bar{Z}$ at $\bar{r}_o$, the interface between Regions 2 and 3.

The condition that $\partial\psi/\partial\bar{r}$ is continuous at $\bar{r}_o$ gives

$$B = \alpha - \frac{\bar{r}_o}{\bar{Z}}\sqrt{\alpha\bar{Z}+1} \tag{17.83}$$

and this result completes the matching process at $\bar{r}_o$ since both ψ and $\partial\psi/\partial\bar{r}$ have now been arranged to be continuous at $\bar{r}_o$. Since $\psi_1 = 1$ has already been arranged at $\bar{r}_i$, all that is required to have continuity of ψ at $\bar{r}_i$ is to set $\psi_2 = 1$ at $\bar{r}_i$, i.e., set

$$\bar{r}_i = \bar{r}_o\left(\frac{1}{\bar{Z}}+\alpha\right)\cosh\left(\bar{r}_i-\bar{r}_o\right) + B\sinh\left(\bar{r}_i-\bar{r}_o\right) - \alpha\bar{r}_i. \tag{17.84}$$

What remains to be done is arrange for continuity of $\partial\psi/\partial\bar{r}$ at $\bar{r} = \bar{r}_i$. Since ψ has already been arranged to be continuous at $\bar{r} = \bar{r}_i$, continuity of $\partial\left(\bar{r}\psi\right)/\partial\bar{r}$ at $\bar{r} = \bar{r}_i$ implies continuity of $\partial\psi/\partial\bar{r}$. This means that continuity of $\partial\psi/\partial\bar{r}$ at $\bar{r} = \bar{r}_i$ is attained by equating the derivatives of the numerators of the right-hand sides of Eqs. (17.80) and (17.81) at $\bar{r}_i$, i.e., continuity of $\partial\psi/\partial\bar{r}$ at $\bar{r} = \bar{r}_i$ is attained by setting

$$1 - \frac{1}{\bar{r}_i}\frac{\alpha\bar{a}^3}{3} = \bar{r}_o\left(\frac{1}{\bar{Z}}+\alpha\right)\sinh\left(\bar{r}_i-\bar{r}_o\right) + B\cosh\left(\bar{r}_i-\bar{r}_o\right) - \alpha. \tag{17.85}$$

Since $\bar{Z} \gg 1$ and α is of order unity, Eqs. (17.84) and (17.85) can be approximated as

$$(1+\alpha)\,\bar{r}_i = \alpha\bar{r}_o\cosh\left(\bar{r}_i-\bar{r}_o\right) + \alpha\sinh\left(\bar{r}_i-\bar{r}_o\right) \tag{17.86}$$

$$1+\alpha - \frac{1}{\bar{r}_i}\frac{\alpha\bar{a}^3}{3} = \alpha\bar{r}_o\sinh\left(\bar{r}_i-\bar{r}_o\right) + \alpha\cosh\left(\bar{r}_i-\bar{r}_o\right). \tag{17.87}$$

These constitute two coupled nonlinear equations in the unknowns $\bar{r}_i$ and $\bar{r}_o$ and can be solved numerically. Since the parameters α and $\bar{a}$ in these equations are functions of position in dusty plasma parameter space, Eqs. (17.86) and (17.87) can be solved for $\bar{r}_i$ and $\bar{r}_o$ at any point $\bar{a}, \bar{r}_d$ in dusty plasma parameter space. Because a different solution set $\{\bar{r}_i, \bar{r}_o\}$ exists at each point in dusty plasma parameter space, $\bar{r}_i$ and $\bar{r}_o$ may be considered as functions of position in dusty plasma parameter space, i.e., $\bar{r}_i = \bar{r}_i(\bar{a}, \bar{r}_d)$ and $\bar{r}_o = \bar{r}_o(\bar{a}, \bar{r}_d)$.

Now consider the density of the field dust grains in the vicinity of $\bar{r} = \bar{r}_o$. For $\bar{r} > \bar{r}_o$ the normalized potential is prescribed by Eq. (17.82) and decays at the dust Debye length, which is an extremely short length because $\bar{Z} \gg 1$. This precipitous spatial attenuation of ψ for $\bar{r} > \bar{r}_o$ means that dust grains are completely shielded from each other when their separation distance slightly exceeds $\bar{r}_o$. If this is the case, then the dust grains will not interact with each other; i.e., they can be considered as a gas of non-interacting particles. On the other hand, if $\bar{r} < \bar{r}_o$, the normalized potential ψ then exceeds $1/\bar{Z}$ and the dust grains will experience a strong mutual repulsion. This abrupt change-over is implicit in Eq. (17.61) which indicates that n_d/n_{d0} is near unity when $\psi \ll 1/\bar{Z}$ (i.e., just outside $\bar{r} = \bar{r}_o$) but n_d/n_{d0} is near zero when ψ becomes significantly larger than $1/\bar{Z}$ (i.e., just inside $\bar{r} = \bar{r}_o$). Hence, a dust grain test particle can be considered as a hard sphere of radius $\bar{r}_o$ when interacting with field dust grains.

Since the nominal separation between dust grains is $\bar{a}$, if $\bar{a} > \bar{r}_o$ then each dust grain is completely shielded from its neighbors, in which case the dust grains behave as an ideal gas of non-interacting particles. However, if $\bar{a} < \bar{r}_o$ is attempted, each dust grain experiences the full, unshielded repulsive force of its neighbors. This extreme repulsive force means that it is not possible for $\bar{a}$ to become less than $\bar{r}_o$. In effect, each dust grain sees its neighbor as a hard sphere of radius $\bar{r}_o$. Thus, when $\bar{a} = \bar{r}_o$ the collection of pressed-together dust grains should form a regularly spaced lattice structure with lattice spacing of order $\bar{a} = \bar{r}_o$. The dusty plasma has thus crystallized.

The condition for crystallization of a dusty plasma is thus

$$\bar{a} \leq \bar{r}_o(\bar{a}, \bar{r}_d), \tag{17.88}$$

which defines a curve in dusty plasma parameter space. Figure 17.4 shows a plot of contours of $\bar{r}_o/\bar{a}$ in dusty plasma parameter space; for reference, the region

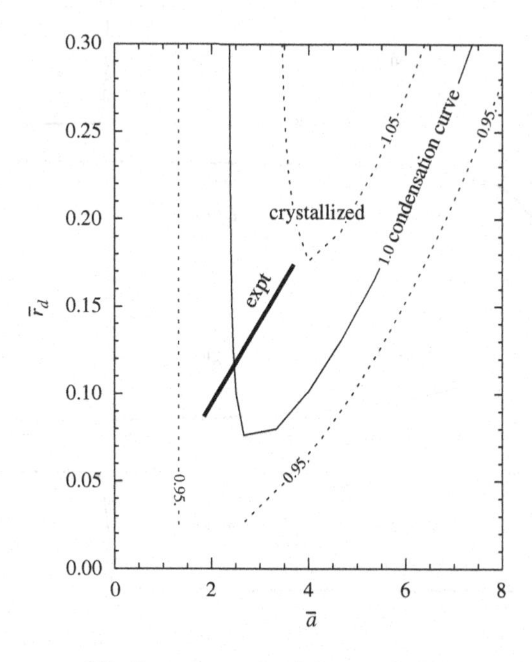

Fig. 17.4 Contours of $\bar{r}_o/\bar{a}$ as determined from solution to matching problem. Condensation (crystallization) is predicted to occur above the $\bar{r}_o/\bar{a} = 1$ line. Chu and I (1994) crystallization experiment, shown as bold line, is seen to lie in the condensation region.

in parameter space associated with the Chu and I dusty plasma crystallization experiment is indicated as a bold line (as mentioned before, the finite length of this line corresponds to the density measurement error bars). The contour $\bar{r}_o(\bar{a}, \bar{r}_d)/\bar{a} = 1$ gives the crystallization condition; above this contour the dust grains are crystallized. It is also necessary to have $\bar{r}_d$ small as was specified in Eq. (17.13) so that ψ_d is not attenuated from its vacuum value by the effect of the shielding cloud.

Figure 17.5 shows plots of $\log_{10}\psi$, ψ, $10^4\psi$, n_e/n_{e0}, n_i/n_{i0}, and n_d/n_{d0} as functions of $\bar{r}$ for a dusty plasma on the verge of condensation. The functional

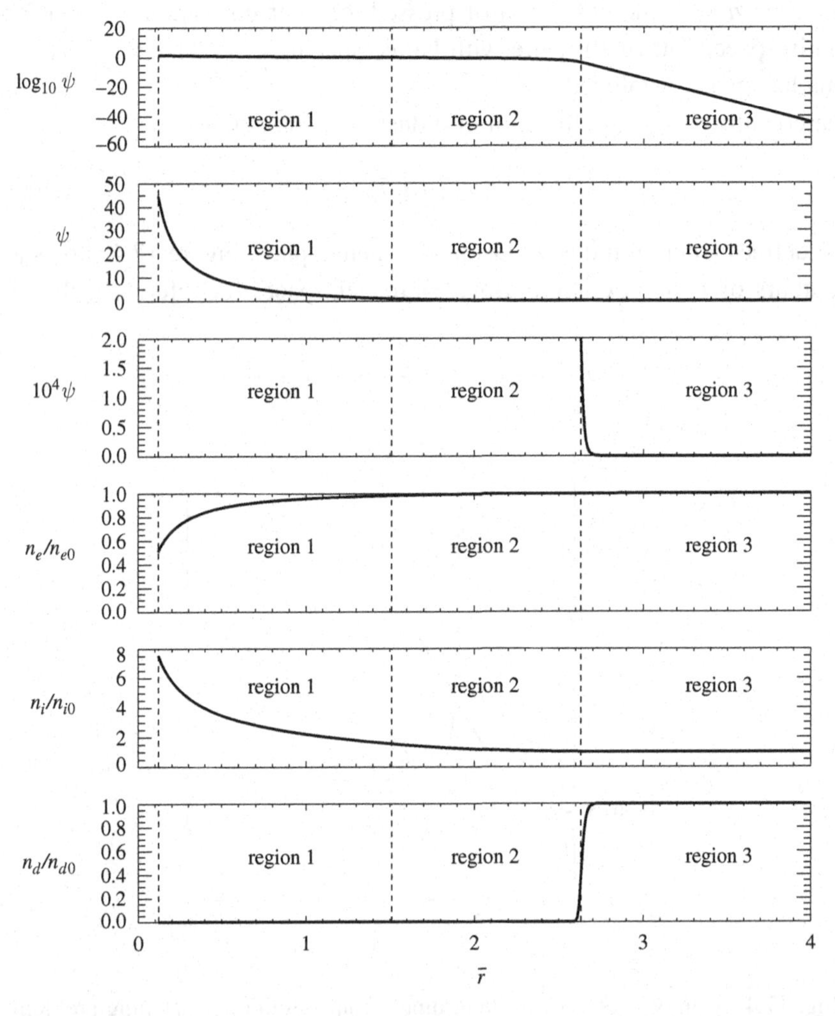

Fig. 17.5 Plots of $\log_{10}\psi$, ψ, $10^4\psi$, n_e/n_{e0}, n_i/n_{i0}, and n_d/n_{d0} as functions of $\bar{r}$ for a dusty plasma on the verge of condensation. Dust grain radius is at left of Region 1, parameters are from Bellan (2004b).

form of ψ is obtained from the $\bar{r}_i$ and $\bar{r}_o$ values determined by solving Eqs. (17.86) and (17.87). The extremely rapid cutoff of the potential at $\bar{r}_o$, the interface between Regions 2 and 3, is evident and the characteristic scale of this cutoff is the dust Debye length. The normalized densities are plotted using Eqs. (17.60), (17.68), and (17.61). The field dust grain density vanishes for $\bar{r}$ inside $\bar{r}_o$, consistent with the hard-sphere behavior of the dust grain test particle. Although ψ is tiny at the interface between Regions 2 and 3, it nevertheless affects the average dust grain density profile at this location because of the extremely large charge on the dust grains.

17.8 Assignments

1. Assuming $T_e = 3\,\text{eV}$, $T_i = 0.03\,\text{eV}$, and $n_i = 10^9\ \text{cm}^{-3}$, estimate to within an order of magnitude how long it would take for $r = 5\,\mu\text{m}$ dust grains with density $n_d = 10^5$ cm^{-3} to become fully charged. Hint: estimate the initial electron current impinging on a dust grain and compare this to the final charge on the dust grain. Use Fig. 17.3 to estimate Z_d.
2. Orbital motion limit and angular momentum. The OML approximation ignores the effective potential resulting from centrifugal force. The validity of this approximation is examined here for the cases of an algebraically decaying central force and an exponentially decaying central force.

   (a) Show that when centrifugal force is taken into account, the radial equation of motion for a particle in a spherically symmetric electrostatic potential $\phi(r)$ is

   $$m\ddot{r} = -\frac{\partial \chi}{\partial r},$$

   where the effective potential χ is given by

   $$\chi = q\phi(r) + \frac{mb^2v^2}{2r^2}$$

   and m, b, and v are defined as in Section 17.2.

   (b) Show that if $q\phi$ is negative, a local minimum of the effective potential exists at some radial position r_1; compare this situation to planetary motion in the solar system.

   (c) Show that if $q\phi \sim -r^{-p}$ there can also be a local maximum at a radius r_2, where $r_2 > r_1$, providing p satisfies a certain condition. What is this condition for p? Plot $\chi(r)$ for the situation where there is a local maximum. Can a particle incident from infinity reach r_1 if its energy is less than $\chi(r_2)$? What sort of condition does this place on m, v, and b? Under what circumstances is the OML approximation valid? Plot trajectories for the situations where the OML approximation is valid and situations where it fails. Comment on whether OML is a reasonable approximation for dust charging.

   (d) Repeat (c) above for the situation where $q\phi \sim -\exp(-r/\lambda)$.

3. Assume the OML approximation is valid so that the effective potential can be considered to have a minimum in the vicinity of a negatively charged dust particle. Show that, if an incident ion does not hit the dust grain, it will reflect back to infinity. Show that a collision could cause an incident ion to be trapped in the effective potential well at r_1 as discussed in Assignment 1(b). Once an ion becomes trapped how long would it stay trapped (see Goree 1992)?
4. Dust Alfvén waves. Show that for waves with frequency below the dust grain cyclotron frequency, a dusty plasma will support MHD Alfvén waves with dispersion $\omega^2 = k_z^2 v_A^2$, where $v_A^2 = B^2/n_d m_d$.
5. Dust whistler waves. Show that a dusty plasma with $\alpha \simeq 1$ will support whistler-like waves with dispersion $\omega \simeq \omega_{ci} \cos\theta$, where $\omega_{cd} \ll \omega \ll \omega_{ci}$.
6. Make a plot like Fig. 17.3 for a dusty plasma with $T_e = T_i$ and discuss whether such a plasma could condense and form crystals.
7. Consider a methane plasma consisting of CH_4 ions and electrons. Assume that there exist some infinitesimal dust nuclei at $t = 0$ and that these become negatively charged as discussed in Section 17.2. Further assume that any methane ion hitting the dust grain sticks to the dust grain so that the mass of the dust grain increases with time. Plot the dust grain radius as a function of time. How long would it take for the dust grains to become sufficiently large to form a crystal (assume that $n_d = 10^5$ cm^{-3}, $T_e/T_i = 100$, $\lambda_{di} = 100\,\mu$m)?
8. Consider a weakly ionized dusty plasma with 5 μm diameter dust grains, 3 eV electrons and room temperature (0.025 eV) argon ions. The dusty plasma is located above a horizontal metal plate lying in the $z = 0$ plane. Because electrons move faster than ions and dust grains, the electron flux to the metal plate is initially much larger than the ion or dust grain flux. The faster rate of loss for electrons causes the dusty plasma to become positively charged so that an electric field develops, which tends to retard the electrons and accelerate the ions towards the metal plate. Thus, the plasma potential is positive with respect to the metal plate and if the plasma potential is defined to be zero, then the metal plate has a negative potential, say ϕ_{plate}. The electron flux to the plate will thus be

$$\Gamma_e = n v_{Te} \exp\left(-q_e \phi_{plate}/\kappa T_e\right)$$

and, using a sheath analysis as in Section 2.9, the ion flux to the plate is

$$\Gamma_i = n c_s,$$

where $c_s = \sqrt{\kappa T_e/m_i}$ is the ion acoustic velocity. In equilibrium, the electron and ion fluxes will balance so that

$$\exp\left(-q_e \phi_{plate}/\kappa T_e\right) = \sqrt{\frac{m_e}{m_i}}$$

or

$$\phi_{plate} = -\frac{T_e}{2} \ln \frac{m_i}{m_e},$$

where the temperature is expressed in electron volts. The change in potential will occur over a distance of the order of the electron Debye length going from the plate to the bulk plasma so that

$$\phi(z) = -\frac{T_e}{2} \exp(-z/\lambda_{De}) \ln \frac{m_i}{m_e},$$

where z is the distance above the metal plate. By considering the combined effect of (i) the vertical electric field associated with this potential and (ii) gravity acting on a dust grain, show that dust grains will tend to levitate above the metal plate. Plot the potential energy of dust grains in the combination of the electrostatic and gravitational potential fields. Assume that $Z_d = 10^4$ and that the dust grains are made of a material that has a mass density of 1 gm cc^{-1}.

9. Suppose a negatively charged dust grain has a charge Z_d and a radius r_d. Show that the potential on the surface of the dust grain is much less than $Z_d/4\pi\varepsilon_0 r_d$ if $r_d \gg \lambda_D$, where λ_D is the nominal shielding length for the shielding cloud around the dust grain. What implication does this have for dust charging theory and why do interesting dusty plasmas typically have $r_d \ll \lambda_D$?

Appendix A
Intuitive method for vector calculus identities

Instead of providing the traditional "back of the book" list of vector calculus identities, an intuitive method[1] for deriving these identities will now be presented.

This method is based on combining the product rule of calculus with the vector algebra triple product and dot-cross product rules. The two vector rules will first be reviewed and then the method for combining these vector rules with the product rule of calculus will be presented.

Vector algebra triple product

There are two forms for the vector triple product, depending on the location of the parenthesis on the left-hand side, namely:

$$\mathbf{A}\times\left(\underbrace{\mathbf{B}}_{\text{middle}}\times\underbrace{\mathbf{C}}_{\text{outer}}\right)=\underbrace{\mathbf{B}}_{\text{middle}}\left(\underbrace{\mathbf{A}\cdot\mathbf{C}}_{\substack{\text{other two}\\ \text{dotted together}}}\right)-\underbrace{\mathbf{C}}_{\text{outer}}\left(\underbrace{\mathbf{A}\cdot\mathbf{B}}_{\substack{\text{other two}\\ \text{dotted together}}}\right) \tag{A.1}$$

$$\left(\underbrace{\mathbf{A}}_{\text{outer}}\times\underbrace{\mathbf{B}}_{\text{middle}}\right)\times\mathbf{C}=\underbrace{\mathbf{B}}_{\text{middle}}\left(\underbrace{\mathbf{A}\cdot\mathbf{C}}_{\substack{\text{other two}\\ \text{dotted together}}}\right)-\underbrace{\mathbf{A}}_{\text{outer}}\left(\underbrace{\mathbf{B}\cdot\mathbf{C}}_{\substack{\text{other two}\\ \text{dotted together}}}\right). \tag{A.2}$$

Both of these distinct forms can be remembered by the two-word mnemonic "middle-outer." The words middle and outer are defined with reference to the left-hand side of both equations; middle refers to the middle vector in the group

[1] This method was explained to the author by the late C. Oberman.

of three and outer refers to the outer vector in the parentheses. The first term on the right-hand side is in the vector direction of the "middle" vector with the other two vectors dotted together; the second term on the right-hand side is in the direction of the "outer" vector with the other two vectors dotted together.

Dot-Cross product

Now consider the combination of dot and cross, $\mathbf{A}\cdot\mathbf{B}\times\mathbf{C}$. Here the rules are that the dot and cross can be interchanged without changing the result and the order can be cyclically permuted without changing the result, but if the cyclic order is changed then the sign changes. Thus,

$$\mathbf{A}\cdot\mathbf{B}\times\mathbf{C}=\mathbf{A}\times\mathbf{B}\cdot\mathbf{C} \quad \text{interchange dot and cross,} \tag{A.3}$$

$$\mathbf{A}\cdot\mathbf{B}\times\mathbf{C}=\mathbf{B}\times\mathbf{C}\cdot\mathbf{A} \quad \text{permutation maintaining cyclic order,} \tag{A.4}$$

$$\mathbf{A}\cdot\mathbf{B}\times\mathbf{C}=-\mathbf{A}\cdot\mathbf{C}\times\mathbf{B} \quad \text{permutation changing cyclic order.} \tag{A.5}$$

Derivation of the vector calculus identities

The basic idea is to replace one of the vectors by the ∇ operator and then rearrange terms and if necessary add terms. The criterion for these maneuvers is that, just like getting the right piece placed in a jigsaw puzzle, here all applicable rules of vector algebra and of calculus must be satisfied simultaneously. The simple example

$$\nabla\cdot(\psi\mathbf{A})=\psi\nabla\cdot\mathbf{A}+\mathbf{A}\cdot\nabla\psi \tag{A.6}$$

illustrates this principle of satisfying the vector algebra and the calculus rules simultaneously. Here the dot always goes between the ∇ and $\mathbf{A}$ in order to satisfy the rules of vector algebra and the ∇ operates once on $\mathbf{A}$ and once on ψ in order to satisfy the product rule $(ab)'=ab'+a'b$ of calculus.

A less trivial example is $\nabla\cdot(\mathbf{B}\times\mathbf{C})$. Here the ∇ must operate on both $\mathbf{B}$ and $\mathbf{C}$ according to the calculus product rule so, neglecting vector issues for now, the result must be of the form $\mathbf{B}\nabla\mathbf{C}+\mathbf{C}\nabla\mathbf{B}$. This basic product rule result is then adjusted to satisfy the vector dot-cross rules. In particular, the dots and crosses may be interchanged at will and the sign is plus if the cyclic order is $\nabla\mathbf{BC}$, $\mathbf{BC}\nabla$, or $\mathbf{C}\nabla\mathbf{B}$ and the sign is minus if the cyclic order is $\nabla\mathbf{CB}$, $\mathbf{CB}\nabla$, or $\mathbf{B}\nabla\mathbf{C}$. Since the ∇ is supposed to operate on only one vector at a time, we pick the arrangement where the ∇ is in the middle and either $\mathbf{B}$ or $\mathbf{C}$ is to its right so that each of $\mathbf{B}$ or $\mathbf{C}$ has a turn at being operated on by the ∇. The arrangements of interest are thus $\mathbf{C}\cdot\nabla\times\mathbf{B}$ and $-\mathbf{B}\cdot\nabla\times\mathbf{C}$, where the minus sign has been inserted to account for the change in cyclic order. The result is

$$\nabla\cdot(\mathbf{B}\times C)=\mathbf{C}\cdot\nabla\times\mathbf{B}-\mathbf{B}\cdot\nabla\times\mathbf{C}, \tag{A.7}$$

which satisfies both the vector algebra dot-cross rule and the product rule of calculus.

Next consider the triple product $\mathbf{A}\times(\nabla\times\mathbf{C})$. Here the ∇ operates only on the $\mathbf{C}$ and the vector triple product generates two terms as indicated by Eq. (A.1). Expanding the triple product according to Eq. (A.1) while arranging an ordering of terms where the ∇ operates only on $\mathbf{C}$ gives

$$\mathbf{A}\times\left(\underbrace{\nabla}_{\text{middle}}\times\underbrace{\mathbf{C}}_{\text{outer}}\right)=\left(\underbrace{\nabla}_{\text{middle}}\underbrace{\mathbf{C}}_{\text{outer}}\right)\cdot\mathbf{A}-\left(\mathbf{A}\cdot\underbrace{\nabla}_{\text{middle}}\right)\underbrace{\mathbf{C}}_{\text{outer}}. \tag{A.8}$$

Thus, the first term on the right-hand side has its vector direction determined by ∇ with the other two terms dotted together, and the parenthesis indicates that the ∇ operates only on $\mathbf{C}$. The second term on the right-hand side has its vector direction determined by $\mathbf{C}$ with the other two terms dotted together and again the ∇ operates only on $\mathbf{C}$.

This can be rearranged as

$$(\nabla\mathbf{C})\cdot\mathbf{A}=\mathbf{A}\times(\nabla\times\mathbf{C})+\mathbf{A}\cdot\nabla\mathbf{C}. \tag{A.9}$$

Interchanging $\mathbf{A}$ and $\mathbf{C}$ gives

$$(\nabla\mathbf{A})\cdot\mathbf{C}=\mathbf{C}\times(\nabla\times\mathbf{A})+\mathbf{C}\cdot\nabla\mathbf{A}. \tag{A.10}$$

Adding these last two expressions gives

$$(\nabla\mathbf{C})\cdot\mathbf{A}+(\nabla\mathbf{A})\cdot\mathbf{C}=\mathbf{A}\times(\nabla\times\mathbf{C})+\mathbf{A}\cdot\nabla\mathbf{C}+\mathbf{C}\times(\nabla\times\mathbf{A})+\mathbf{C}\cdot\nabla\mathbf{A}. \tag{A.11}$$

However, using the same arguments about maintaining the dot and satisfying the product rule shows that

$$\begin{aligned}\nabla(\mathbf{A}\cdot\mathbf{C})&=\nabla(\mathbf{C}\cdot\mathbf{A})\\&=(\nabla\mathbf{A})\cdot\mathbf{C}+(\nabla\mathbf{C})\cdot\mathbf{A}.\end{aligned} \tag{A.12}$$

Here the ∇ operates once on the $\mathbf{A}$ and once on the $\mathbf{C}$ according to the product rule and the dot is always between the $\mathbf{A}$ and the $\mathbf{C}$. Combining Eqs. (A.11) and (A.12) gives the standard vector identity

$$\nabla(\mathbf{A}\cdot\mathbf{C})=\mathbf{A}\times(\nabla\times\mathbf{C})+\mathbf{A}\cdot\nabla\mathbf{C}+\mathbf{C}\times(\nabla\times\mathbf{A})+\mathbf{C}\cdot\nabla\mathbf{A}. \tag{A.13}$$

Finally, the triple product with two ∇'s can be expressed as

$$\nabla\times\left(\underbrace{\nabla}_{\text{middle}}\times\underbrace{\mathbf{A}}_{\text{outer}}\right)=\underbrace{\nabla}_{\text{middle}}\left(\underbrace{\nabla\cdot\mathbf{A}}_{\substack{\text{other two}\\\text{dotted together}}}\right)-\underbrace{\nabla^2}_{\substack{\text{other two}\\\text{dotted together}}}\underbrace{\mathbf{A}}_{\text{outer}} \tag{A.14}$$

or, without the labeling, as

$$\nabla\times\nabla\times\mathbf{A}=\nabla\nabla\cdot\mathbf{A}-\nabla^2\mathbf{A}. \tag{A.15}$$

The relationships $\nabla\cdot\nabla\times\mathbf{A}=0$ and $\nabla\times\nabla\psi=0$ can be proved by direct evaluation using Cartesian coordinates.

The expression $\nabla\times(\mathbf{B}\times\mathbf{C})$ can be evaluated using the vector cross-product rule in conjunction with the product rule. Thus, the rule $\mathbf{A}\times(\mathbf{B}\times\mathbf{C})=\mathbf{B}(\mathbf{A}\cdot\mathbf{C})-\mathbf{C}(\mathbf{A}\cdot\mathbf{B})$ imposed by Eq. (A.1) must be satisfied with the ∇ operator playing the role of $\mathbf{A}$, and also the product rule requirement $(\psi\chi)'=\chi'\psi+\chi\psi'$ must be satisfied. Since $\mathbf{B}(\mathbf{A}\cdot\mathbf{C})=(\mathbf{C}\cdot\mathbf{A})\mathbf{B}$ we must have ∇ operate on both $\mathbf{B}$ and on $\mathbf{C}$ as a derivative operator and in addition satisfy the vector requirement. Similarly, we can use $\mathbf{C}(\mathbf{A}\cdot\mathbf{B})=(\mathbf{B}\cdot\mathbf{A})\mathbf{C}$ in order to give ∇ represented by $\mathbf{A}$ a chance to operate on both $\mathbf{B}$ and on $\mathbf{A}$ according to the product rule. Thus, using the vector rule $\mathbf{A}\times(\mathbf{B}\times\mathbf{C})=\mathbf{B}(\mathbf{A}\cdot\mathbf{C})-\mathbf{C}(\mathbf{A}\cdot\mathbf{B})$, letting ∇ replace $\mathbf{A}$, and taking into account the conditions imposed by the product rule of calculus gives

$$\nabla\times(\mathbf{B}\times\mathbf{C})=\mathbf{B}(\nabla\cdot\mathbf{C})+\mathbf{C}\cdot\nabla\mathbf{B}-\mathbf{C}(\nabla\cdot\mathbf{B})-\mathbf{B}\cdot\nabla\mathbf{C}.$$

Summary of vector identities

$$\mathbf{A}\times(\mathbf{B}\times\mathbf{C})=\mathbf{B}(\mathbf{A}\cdot\mathbf{C})-\mathbf{C}(\mathbf{A}\cdot\mathbf{B})$$

$$(\mathbf{A}\times\mathbf{B})\times\mathbf{C}=\mathbf{B}(\mathbf{A}\cdot\mathbf{C})-\mathbf{A}(\mathbf{B}\cdot\mathbf{C})$$

$$\mathbf{A}\cdot\mathbf{B}\times\mathbf{C}=\mathbf{A}\times\mathbf{B}\cdot\mathbf{C},\ \text{interchange dot and cross}$$

$$\mathbf{A}\cdot\mathbf{B}\times\mathbf{C}=\mathbf{B}\times\mathbf{C}\cdot\mathbf{A},\ \text{cyclic permutation, cyclic order maintained}$$

$$\mathbf{A}\cdot\mathbf{B}\times\mathbf{C}=-\mathbf{A}\cdot\mathbf{C}\times\mathbf{B},\ \text{cyclic permutation, cyclic order changed}$$

$$\nabla\cdot(\psi\mathbf{A})=\psi\nabla\cdot\mathbf{A}+\mathbf{A}\cdot\nabla\psi$$

$$\nabla\cdot(\mathbf{A}\times\mathbf{B})=\mathbf{B}\cdot\nabla\times\mathbf{A}-\mathbf{A}\cdot\nabla\times\mathbf{B}$$

$$(\nabla\mathbf{B})\cdot\mathbf{A}=\mathbf{A}\times(\nabla\times\mathbf{B})+\mathbf{A}\cdot\nabla\mathbf{B}$$

$$\nabla(\mathbf{A}\cdot\mathbf{B})=\mathbf{A}\times(\nabla\times\mathbf{B})+\mathbf{A}\cdot\nabla\mathbf{B}+\mathbf{B}\times(\nabla\times\mathbf{A})+\mathbf{B}\cdot\nabla\mathbf{A}$$

$$\nabla\times\nabla\times\mathbf{A}=\nabla(\nabla\cdot\mathbf{A})-\nabla^2\mathbf{A}$$

$$\nabla\times(\mathbf{B}\times\mathbf{C})=\mathbf{B}(\nabla\cdot\mathbf{C})+\mathbf{C}\cdot\nabla\mathbf{B}-\mathbf{C}(\nabla\cdot\mathbf{B})-\mathbf{B}\cdot\nabla\mathbf{C}$$

$$\nabla\cdot\nabla\times\mathbf{A}=0$$

$$\nabla\times\nabla\psi=0$$

Appendix B

Vector calculus in orthogonal curvilinear coordinates

Derivation of generalized relations

Let x_1, x_2, x_3 be a right-handed set of orthogonal coordinates. Let h_i be the scale factor relating an increment in the ith coordinate to an increment in length in that direction so

$$dl_i = h_i dx_i; \tag{B.1}$$

an example is the ϕ direction of cylindrical coordinates where $dl_\phi = rd\phi$. Since the coordinates are assumed to be orthogonal, the increment in distance along a curve in three-dimensional space is

$$(ds)^2 = h_1^2 (dx_1)^2 + h_2^2 (dx_2)^2 + h_3^2 (dx_3)^2 . \tag{B.2}$$

The $\{x_1, x_2, x_3\}$ coordinates and corresponding $\{h_1, h_2, h_3\}$ scale factors for Cartesian, cylindrical, and spherical coordinate systems are listed in Table B.1.

The differential of the scalar ψ for a displacement by dl_1 of the coordinate x_1 is

$$d\psi = dl_1 \hat{x}_1 \cdot \nabla\psi \tag{B.3}$$

so the component of the gradient operator in the direction of $\hat{x}_1$ is

$$\hat{x}_1 \cdot \nabla\psi = \frac{d\psi}{dl_1} = \frac{1}{h_1}\frac{\partial\psi}{\partial x_1}. \tag{B.4}$$

Equation (B.1) has been used to obtain the right-most expression and partial derivative notation is invoked for this expression because the displacement is just in the direction of x_1.

Generalized gradient operator

Because the coordinates are independent, a displacement in the direction of one coordinate does not affect the dependence on the other coordinates and so the

Table B.1 *Coordinate systems and their scale factors.*

Coordinate system	Distance along a curve	$\{x_1, x_2, x_3\}$	$\{h_1, h_2, h_3\}$
Cartesian	$(dx)^2 + (dy)^2 + (dz)^2$	$\{x, y, z\}$	$\{1, 1, 1\}$
cylindrical	$(dr)^2 + (rd\phi)^2 + (dz)^2$	$\{r, \phi, z\}$	$\{1, r, 1\}$
spherical	$(dr)^2 + (rd\theta)^2 + (r\sin\theta d\phi)^2$	$\{r, \theta, \phi\}$	$\{1, r, r\sin\theta\}$

gradient operator is just the sum of its components in the three orthogonal directions, i.e.,

$$\nabla = \frac{\hat{x}_1}{h_1}\frac{\partial}{\partial x_1} + \frac{\hat{x}_2}{h_2}\frac{\partial}{\partial x_2} + \frac{\hat{x}_3}{h_3}\frac{\partial}{\partial x_3}. \tag{B.5}$$

Since $\nabla x_1 = \hat{x}_1/h_1$, the unit vector in the x_1 direction is

$$\hat{x}_1 = h_1 \nabla x_1. \tag{B.6}$$

Also, because the coordinates form a right-handed orthogonal system the unit vectors are related by

$$\hat{x}_1 \times \hat{x}_2 = \hat{x}_3, \;\; \hat{x}_2 \times \hat{x}_3 = \hat{x}_1, \;\; \hat{x}_3 \times \hat{x}_1 = \hat{x}_2. \tag{B.7}$$

Generalized divergence and curl

Let $\mathbf{V}$ be an arbitrary vector

$$\mathbf{V} = V_1\hat{x}_1 + V_2\hat{x}_2 + V_3\hat{x}_3 \tag{B.8}$$

and consider the divergence of the first term,

$$\begin{aligned}\nabla\cdot(V_1\hat{x}_1) &= \nabla\cdot(V_1\hat{x}_2\times\hat{x}_3)\\ &= \nabla\cdot(V_1 h_2\nabla x_2 \times h_3\nabla x_3)\\ &= \nabla(h_2h_3V_1)\cdot\nabla x_2\times\nabla x_3\\ &= \nabla(h_2h_3V_1)\cdot\frac{\hat{x}_2\times\hat{x}_3}{h_2h_3}\\ &= \frac{\hat{x}_1}{h_2h_3}\cdot\nabla(h_2h_3V_1)\\ &= \frac{1}{h_1h_2h_3}\frac{\partial}{\partial x_1}(h_2h_3V_1).\end{aligned} \tag{B.9}$$

Extending this to all three terms gives the general form for the divergence to be

$$\nabla\cdot\mathbf{V} = \frac{1}{h_1h_2h_3}\left(\frac{\partial}{\partial x_1}(h_2h_3V_1) + \frac{\partial}{\partial x_2}(h_1h_3V_2) + \frac{\partial}{\partial x_3}(h_1h_2V_3)\right). \tag{B.10}$$

Now consider the curl of the first term of the arbitrary vector, namely

$$\begin{aligned}\nabla\times(V_1\hat{x}_1) &= \nabla\times(V_1h_1\nabla x_1)\\ &= \nabla(V_1h_1)\times\nabla x_1\\ &= \frac{1}{h_1}\nabla(V_1h_1)\times\hat{x}_1\end{aligned} \tag{B.11}$$

and so

$$\nabla\times\mathbf{V} = \frac{1}{h_1}\nabla(V_1h_1)\times\hat{x}_1 + \frac{1}{h_2}\nabla(V_2h_2)\times\hat{x}_2 + \frac{1}{h_3}\nabla(V_3h_3)\times\hat{x}_3. \tag{B.12}$$

The component in the direction of $\hat{x}_1$ is

$$\begin{aligned}\hat{x}_1\cdot\nabla\times\mathbf{V} &= \frac{1}{h_2}\nabla(V_2h_2)\times\hat{x}_2\cdot\hat{x}_1 + \frac{1}{h_3}\nabla(V_3h_3)\times\hat{x}_3\cdot\hat{x}_1\\ &= \frac{1}{h_2}\nabla(V_2h_2)\cdot\hat{x}_2\times\hat{x}_1 + \frac{1}{h_3}\nabla(V_3h_3)\cdot\hat{x}_3\times\hat{x}_1\\ &= \frac{1}{h_3}\hat{x}_2\cdot\nabla(V_3h_3) - \frac{1}{h_2}\hat{x}_3\cdot\nabla(V_2h_2)\\ &= \frac{1}{h_2h_3}\frac{\partial}{\partial x_2}(V_3h_3) - \frac{1}{h_2h_3}\frac{\partial}{\partial x_3}(V_2h_2).\end{aligned} \tag{B.13}$$

Thus, the general form of the curl of the arbitrary vector is

$$\begin{aligned}\nabla\times\mathbf{V} &= \frac{1}{h_2h_3}\left(\frac{\partial}{\partial x_2}(V_3h_3) - \frac{\partial}{\partial x_3}(V_2h_2)\right)\hat{x}_1\\ &\quad + \frac{1}{h_1h_3}\left(\frac{\partial}{\partial x_3}(V_1h_1) - \frac{\partial}{\partial x_1}(V_3h_3)\right)\hat{x}_2\\ &\quad + \frac{1}{h_1h_2}\left(\frac{\partial}{\partial x_1}(V_2h_2) - \frac{\partial}{\partial x_2}(V_1h_1)\right)\hat{x}_3\\ &= \frac{1}{h_1h_2h_3}\begin{vmatrix} h_1\hat{x}_1 & h_2\hat{x}_2 & h_3\hat{x}_3\\ \partial/\partial x_1 & \partial/\partial x_2 & \partial/\partial x_3\\ h_1V_1 & h_2V_2 & h_3V_3\end{vmatrix}.\end{aligned} \tag{B.14}$$

Generalized Laplacian of a scalar

The Laplacian of a scalar is

$$\nabla^2\psi = \nabla\cdot\left(\frac{\hat{x}_1}{h_1}\frac{\partial\psi}{\partial x_1} + \frac{\hat{x}_2}{h_2}\frac{\partial\psi}{\partial x_2} + \frac{\hat{x}_3}{h_3}\frac{\partial\psi}{\partial x_3}\right). \tag{B.15}$$

Using Eq. (B.10) this becomes

$$\nabla^2\psi = \frac{1}{h_1h_2h_3}\left(\frac{\partial}{\partial x_1}\left(\frac{h_2h_3}{h_1}\frac{\partial\psi}{\partial x_1}\right)+\frac{\partial}{\partial x_2}\left(\frac{h_3h_1}{h_2}\frac{\partial\psi}{\partial x_2}\right)+\frac{\partial}{\partial x_3}\left(\frac{h_1h_2}{h_3}\frac{\partial\psi}{\partial x_3}\right)\right). \tag{B.16}$$

The Laplacian of a vector in general differs from the Laplacian of a scalar because the Laplacian of a vector can contain derivatives of unit vectors. However, the general formula is overly complex and so the Laplacian of a vector will only be calculated for cylindrical coordinates since these are of most interest here.

Application to Cartesian coordinates

$\{x_1, x_2, x_3\} = \{x, y, z\}; \quad \{h_1, h_2, h_3\} = \{1, 1, 1\}$

$$\nabla\psi = \hat{x}\frac{\partial\psi}{\partial x}+\hat{y}\frac{\partial\psi}{\partial y}+\hat{z}\frac{\partial\psi}{\partial z} \tag{B.17}$$

$$\nabla\cdot\mathbf{V} = \frac{\partial V_x}{\partial x}+\frac{\partial V_y}{\partial y}+\frac{\partial V_z}{\partial z} \tag{B.18}$$

$$\nabla\times\mathbf{V} = \left(\frac{\partial V_z}{\partial y}-\frac{\partial V_y}{\partial z}\right)\hat{x}+\left(\frac{\partial V_x}{\partial z}-\frac{\partial V_z}{\partial x}\right)\hat{y} + \left(\frac{\partial V_y}{\partial x}-\frac{\partial V_x}{\partial y}\right)\hat{z} \tag{B.19}$$

$$\nabla^2\psi = \frac{\partial^2\psi}{\partial x^2}+\frac{\partial^2\psi}{\partial y^2}+\frac{\partial^2\psi}{\partial z^2} \tag{B.20}$$

$$\nabla^2\mathbf{V} = \frac{\partial^2\mathbf{V}}{\partial x^2}+\frac{\partial^2\mathbf{V}}{\partial y^2}+\frac{\partial^2\mathbf{V}}{\partial z^2} \tag{B.21}$$

Application to cylindrical coordinates

$\{x_1, x_2, x_3\} = \{r, \phi, z\}; \quad \{h_1, h_2, h_3\} = \{1, r, 1\}$

$$\nabla\psi = \hat{r}\frac{\partial\psi}{\partial r}+\frac{\hat{\phi}}{r}\frac{\partial\psi}{\partial\phi}+\hat{z}\frac{\partial\psi}{\partial z} \tag{B.22}$$

$$\nabla\cdot\mathbf{V} = \frac{1}{r}\frac{\partial}{\partial r}(rV_r)+\frac{1}{r}\frac{\partial V_\phi}{\partial\phi}+\frac{\partial V_z}{\partial z} \tag{B.23}$$

$$\nabla\times\mathbf{V} = \left(\frac{1}{r}\frac{\partial V_z}{\partial\phi}-\frac{\partial V_\phi}{\partial z}\right)\hat{r}+\left(\frac{\partial V_r}{\partial z}-\frac{\partial V_z}{\partial r}\right)\hat{\phi} + \frac{1}{r}\left(\frac{\partial}{\partial r}(rV_\phi)-\frac{\partial V_r}{\partial\phi}\right)\hat{z} \tag{B.24}$$

$$\nabla^2\psi = \frac{1}{r}\frac{\partial}{\partial r}\left(r\frac{\partial\psi}{\partial r}\right)+\frac{1}{r^2}\frac{\partial^2\psi}{\partial\phi^2}+\frac{\partial^2\psi}{\partial z^2} \tag{B.25}$$

Laplacian of a vector in cylindrical coordinates

Before proceeding, note that the cylindrical coordinate unit vectors can be expressed in terms of Cartesian unit vectors as

$$\hat{r} = \hat{x}\cos\phi + \hat{y}\sin\phi \tag{B.26}$$

$$\hat{\phi} = -\hat{x}\sin\phi + \hat{y}\cos\phi \tag{B.27}$$

so

$$\frac{\partial}{\partial\phi}\hat{r} = \hat{\phi}, \quad \frac{\partial}{\partial\phi}\hat{\phi} = -\hat{r}. \tag{B.28}$$

Thus,

$$\nabla\hat{r} = \left(\hat{r}\frac{\partial}{\partial r} + \frac{\hat{\phi}}{r}\frac{\partial}{\partial\phi} + \hat{z}\frac{\partial}{\partial z}\right)\hat{r} = \frac{\hat{\phi}\hat{\phi}}{r} \tag{B.29}$$

$$\nabla\hat{\phi} = \left(\hat{r}\frac{\partial}{\partial r} + \frac{\hat{\phi}}{r}\frac{\partial}{\partial\phi} + \hat{z}\frac{\partial}{\partial z}\right)\hat{\phi} = -\frac{\hat{\phi}\hat{r}}{r} \tag{B.30}$$

and

$$\nabla^2\hat{r} = -\frac{1}{r^2}\hat{r} \tag{B.31}$$

$$\nabla^2\hat{\phi} = -\frac{1}{r^2}\hat{\phi}. \tag{B.32}$$

These results can now be used to calculate $\nabla^2(V_r\hat{r})$ and $\nabla^2\left(V_\phi\hat{\phi}\right)$, which can then be used to construct the full Laplacian. The first calculation gives

$$\begin{aligned}
\nabla^2(V_r\hat{r}) &= \nabla\cdot\nabla(V_r\hat{r}) \\
&= \nabla\cdot((\nabla V_r)\,\hat{r} + V_r\nabla\hat{r}) \\
&= \hat{r}\nabla^2 V_r + (\nabla V_r)\cdot\nabla\hat{r} + \nabla\cdot(V_r\nabla\hat{r}) \\
&= \hat{r}\nabla^2 V_r + 2\nabla V_r\cdot\nabla\hat{r} + V_r\nabla^2\hat{r} \\
&= \hat{r}\nabla^2 V_r + \frac{2\hat{\phi}\hat{\phi}}{r}\cdot\nabla V_r + \frac{V_r}{r^2}\frac{\partial^2}{\partial\phi^2}\hat{r} \\
&= \hat{r}\nabla^2 V_r + \frac{2\hat{\phi}}{r^2}\frac{\partial V_r}{\partial\phi} - \frac{V_r}{r^2}\hat{r}
\end{aligned} \tag{B.33}$$

while the second calculation gives

$$\begin{aligned}\nabla^2\left(V_\phi\hat{\phi}\right) &= \nabla\cdot\nabla\left(V_\phi\hat{\phi}\right)\\ &= \nabla\cdot\left(\left(\nabla V_\phi\right)\hat{\phi}+V_\phi\nabla\hat{\phi}\right)\\ &= \hat{\phi}\nabla^2 V_\phi+2\nabla V_\phi\cdot\nabla\hat{\phi}+V_\phi\nabla^2\hat{\phi}\\ &= \hat{\phi}\nabla^2 V_\phi-\frac{2\hat{r}}{r^2}\frac{\partial V_\phi}{\partial\phi}-\frac{V_\phi}{r^2}\hat{\phi}.\end{aligned} \tag{B.34}$$

Since $\nabla^2\left(V_z\hat{z}\right)=\hat{z}\nabla^2 V_z$, it is seen that the Laplacian of a vector in cylindrical coordinates is

$$\begin{aligned}\nabla^2\mathbf{V} = &\ \hat{r}\left(\nabla^2 V_r-\frac{2}{r^2}\frac{\partial V_\phi}{\partial\phi}-\frac{V_r}{r^2}\right)\\ &+\hat{\phi}\left(\nabla^2 V_\phi+\frac{2}{r^2}\frac{\partial V_r}{\partial\phi}-\frac{V_\phi}{r^2}\right)\\ &+\hat{z}\nabla^2 V_z.\end{aligned} \tag{B.35}$$

Equation (B.28) can also be used to calculate $\mathbf{V}\cdot\nabla\mathbf{V}$ giving

$$\begin{aligned}\mathbf{V}\cdot\nabla\mathbf{V} &= \left(V_r\frac{\partial}{\partial r}+\frac{V_\phi}{r}\frac{\partial}{\partial\phi}+V_z\frac{\partial}{\partial z}\right)\left(V_r\hat{r}+V_\phi\hat{\phi}+V_z\hat{z}\right)\\ &= \hat{r}\left(V_r\frac{\partial V_r}{\partial r}+\frac{V_\phi}{r}\frac{\partial V_r}{\partial\phi}+V_z\frac{\partial V_r}{\partial z}-\frac{V_\phi^2}{r}\right)\\ &\quad+\hat{\phi}\left(V_r\frac{\partial V_\phi}{\partial r}+\frac{V_\phi}{r}\frac{\partial V_\phi}{\partial\phi}+V_z\frac{\partial V_\phi}{\partial z}+\frac{V_\phi V_r}{r}\right)\\ &\quad+\hat{z}\left(V_r\frac{\partial V_z}{\partial r}+\frac{V_\phi}{r}\frac{\partial V_z}{\partial\phi}+V_z\frac{\partial V_z}{\partial z}\right).\end{aligned} \tag{B.36}$$

Application to spherical coordinates

$\{x_1, x_2, x_3\}=\{r,\theta,\phi\};\quad \{h_1,h_2,h_3\}=\{1,r,r\sin\theta\}$

$$\nabla\psi=\hat{r}\frac{\partial\psi}{\partial r}+\frac{\hat{\theta}}{r}\frac{\partial\psi}{\partial\theta}+\frac{\hat{\phi}}{r\sin\theta}\frac{\partial\psi}{\partial\phi} \tag{B.37}$$

$$\nabla\cdot\mathbf{V}=\frac{1}{r^2}\frac{\partial}{\partial r}\left(r^2V_r\right)+\frac{1}{r\sin\theta}\frac{\partial}{\partial\theta}\left(\sin\theta\, V_\theta\right)+\frac{1}{r\sin\theta}\frac{\partial}{\partial\phi}\left(V_\phi\right) \tag{B.38}$$

$$\nabla \times \mathbf{V} = \frac{1}{r \sin\theta}\left(\frac{\partial}{\partial\theta}\left(V_\phi \sin\theta\right) - \frac{\partial V_\theta}{\partial\phi}\right)\hat{r} + \frac{1}{r\sin\theta}\left(\frac{\partial V_r}{\partial\phi} - \frac{\partial}{\partial r}\left(V_\phi r \sin\theta\right)\right)\hat{\theta} + \frac{1}{r}\left(\frac{\partial}{\partial r}\left(V_\theta r\right) - \frac{\partial V_r}{\partial\theta}\right)\hat{\phi} \tag{B.39}$$

$$\nabla^2\psi = \frac{1}{r^2}\frac{\partial}{\partial r}\left(r^2\frac{\partial\psi}{\partial r}\right) + \frac{1}{r^2\sin\theta}\frac{\partial}{\partial\theta}\left(\sin\theta\frac{\partial\psi}{\partial\theta}\right) + \frac{1}{r^2\sin^2\theta}\frac{\partial^2\psi}{\partial\phi^2} \tag{B.40}$$

Appendix C

Frequently used physical constants and formulae

Physical constants (to 3 significant figures)

	symbol	value	unit
electron mass	m_e	9.11×10^{-31}	kg
proton mass	m_p	1.67×10^{-27}	kg
vacuum permeability	μ_0	$4\pi \times 10^{-7}$	$\mathrm{N\ A^{-2}}$
vacuum permittivity	ε_0	8.85×10^{-12}	$\mathrm{F\ m^{-1}}$
speed of light	c	3.00×10^{8}	$\mathrm{m\ s^{-1}}$
electron charge	e	1.60×10^{-19}	C
Avogadro's number	N_A	6.02×10^{23}	$\mathrm{mol^{-1}}$
Boltzmann constant	κ	1.60×10^{-19}	$\mathrm{J\ eV^{-1}}$

Formulae (all quantities in SI units, temperatures in eV, A is ion atomic mass number)

Lengths

Debye length (p. 10)

$$\frac{1}{\lambda_D^2} = \sum_\sigma \frac{1}{\lambda_\sigma^2},$$

where σ is over all species participating in shielding and

$$\lambda_\sigma = \sqrt{\frac{\varepsilon_0 \kappa T_\sigma}{n_{0\sigma} q_\sigma^2}} = 7.4 \times 10^3 \sqrt{\frac{T_\sigma}{n_{0_\sigma}}}\ \mathrm{m}.$$

Electron Larmor radius (p. 271)

$$r_{Le} = \frac{\sqrt{\kappa T_e / m_e}}{|\omega_{ce}|} = 2.4 \times 10^{-6} \frac{\sqrt{T_e}}{B}\ \mathrm{m}.$$

Ion Larmor radius (p. 271)

$$r_{Li} = \frac{\sqrt{\kappa T_i/m_i}}{|\omega_{ci}|} = 1.0 \times 10^{-4} \frac{\sqrt{AT_i}}{B\sqrt{Z}} \text{ m},$$

where A is atomic mass and Z is ion charge.
Electron collisionless skin depth (p. 170)

$$\frac{c}{\omega_{pe}} = \frac{5.3 \times 10^6}{\sqrt{n_e}} \text{ m}.$$

Ion collisionless skin depth (assuming quasi-neutral plasma so $n_i Z = n_e$)

$$\frac{c}{\omega_{pi}} = 2.3 \times 10^8 \sqrt{\frac{A}{Zn_e}} \text{ m}.$$

Frequencies

Electron plasma frequency (p. 150)

$$f_{pe} = \frac{\omega_{pe}}{2\pi} = \frac{1}{2\pi}\sqrt{\frac{n_e e^2}{\varepsilon_0 m_e}} = 9\sqrt{n_e} \text{ Hz}.$$

Ion plasma frequency (p. 150)

$$f_{pi} = \omega_{pi} = \frac{1}{2\pi}\sqrt{\frac{n_i q_i^2}{\varepsilon_0 m_i}} = 0.21\sqrt{\frac{Zn_e}{A}} \text{ Hz}.$$

Electron cyclotron frequency (p. 94)

$$f_{ce} = \frac{1}{2\pi}|\omega_{ce}| = \frac{eB}{2\pi m_e} = 2.8 \times 10^{10} B \text{ Hz}.$$

Ion cyclotron frequency (p. 94)

$$f_{ci} = \frac{1}{2\pi}|\omega_{ci}| = \frac{ZeB}{2\pi m_i} = 1.52 \times 10^7 \frac{ZB}{A} \text{ Hz}.$$

Upper hybrid frequency (p. 215)

$$f_{uh} = \sqrt{f_{pe}^2 + f_{ce}^2}.$$

Lower hybrid frequency (p. 215)

$$f_{lh} = \sqrt{f_{ci}^2 + \frac{f_{pi}^2}{1 + \frac{f_{pe}^2}{f_{ce}^2}}}.$$

Diocotron frequency (pure electron plasma, p. 532)

$$f_{dioc} \simeq f_{pe}^2/2f_{ce}.$$

Velocities

Electron thermal velocity (p. 181)

$$v_{Te} = \sqrt{\frac{2\kappa T_e}{m_e}} = 5.9 \times 10^5 \sqrt{T_e} \text{ m s}^{-1}.$$

Ion thermal velocity (p. 181)

$$v_{Ti} = \sqrt{\frac{2\kappa T_i}{m_i}} = 1.4 \times 10^4 \sqrt{\frac{T_i}{A}} \text{ m s}^{-1}.$$

Ion acoustic velocity (p. 152)

$$c_s = \sqrt{\frac{\kappa T_e}{m_i}} = 9.8 \times 10^3 \sqrt{\frac{T_e}{A}} \text{ m s}^{-1}.$$

Electron diamagnetic drift velocity (p. 289)

$$u_{d,e} = \frac{\kappa T_e}{eB} \left| \frac{1}{n} \nabla n \right| = \frac{T_e}{B} \left| \frac{1}{n} \nabla n \right| \text{ m s}^{-1}.$$

Ion diamagnetic drift velocity (flow in opposite direction from electrons, p. 289)

$$u_{d,i} = \frac{\kappa T_i}{q_i B} \left| \frac{1}{n} \nabla n \right| = \frac{T_i}{ZB} \left| \frac{1}{n} \nabla n \right| \text{ m s}^{-1}.$$

Alfvén velocity (p. 156)

$$v_A = \frac{B}{\sqrt{\mu_0 n_i m_i}} = \frac{B}{\sqrt{\mu_0 n_e A m_p / Z}} = 2.2 \times 10^{16} B \sqrt{\frac{Z}{n_e A}} \text{ m s}^{-1}.$$

Wave phase velocity (p. 234)

$$v_{ph} = \frac{\omega}{k} = f\lambda.$$

Dimensionless

Plasma beta (p. 62, p. 366)

$$\beta = \frac{2\mu_0 n \kappa T}{B^2} = \frac{4.03 \times 10^{-25}}{B^2} nT.$$

Lundquist number (p. 435)

$$S = \frac{\mu_0 v_A L}{\eta}.$$

Collisions, resistivity, and runaways

Electron–electron collision rate (p. 24)

$$\nu_{ee} = 4 \times 10^{-12} \frac{n \ln \Lambda}{T_{eV}^{3/2}} \text{ s}^{-1},$$

where $\ln \Lambda \sim 10$, electron–ion rate same magnitude, ion–ion slower by $\sqrt{m_e/m_i}$.
Spitzer resistivity (p. 451)

$$\eta = 1.03 \times 10^{-4} \frac{Z \ln \Lambda}{T_e^{3/2}} \text{ Ohm m}.$$

Dreicer runaway electric field (p. 452)

$$E_{Dreicer} = 5.6 \times 10^{-18} n_e Z \frac{\ln \Lambda}{T_e} \text{ V m}^{-1}.$$

Typical neutral cross-section (p. 19)

$$\sigma_{\text{neut}} \sim 3 \times 10^{-20} \text{ m}^2.$$

Warm plasma waves

Electrostatic susceptibility (p. 192)

$$\chi_\sigma = \frac{1}{k^2 \lambda_{D\sigma}^2} \left[1 + \alpha Z(\alpha)\right],$$

where $\alpha = \omega / k v_{T\sigma}$.
Plasma dispersion function (p. 192)

$$Z(\alpha) = \frac{1}{\pi^{1/2}} \int_{-\infty}^{\infty} \mathrm{d}\xi \frac{\exp(-\xi^2)}{(\xi - \alpha)}$$

$$\lim_{\alpha<<1} Z(\alpha) = -2\alpha \left(1 - \frac{2\alpha^2}{3} + \ldots\right) + \mathrm{i}\pi^{1/2} \exp(-\alpha^2)$$

$$\lim_{\alpha>>1} Z(\alpha) = -\frac{1}{\alpha} \left[1 + \frac{1}{2\alpha^2} + \frac{3}{4\alpha^4} + \ldots\right] + \mathrm{i}\pi^{1/2} \exp(-\alpha^2).$$

Bibliography and suggested reading

Bateman, G., *MHD Instabilities*, Cambridge, Mass.: MIT Press, 1978.
Bellan, P. M., *Spheromaks: A Practical Application of Magnetohydrodynamic Dynamos and Plasma Self-Organization*, London: Imperial College Press, 2000.
Birdsall, C. K. and Langdon, A. B., *Plasma Physics via Computer Simulation*, New York: McGraw Hill, 1985.
Boyd, T. J. M. and Sanderson, J. J., *The Physics of Plasmas*, Cambridge: Cambridge University Press, 2003.
Chen, F. F., *Introduction to Plasma Physics and Controlled Fusion*, 2nd edition, New York: Plenum Press, 1984.
Cramer, N. F., *The Physics of Alfvén Waves*, Berlin: Wiley-VCH, 2001.
Davidson, R. C., *Methods in Nonlinear Plasma Theory*, New York: Academic Press, 1972.
Davidson, R. C., *Physics of Nonneutral Plasmas*, London: Imperial College Press, 2001.
Drazin, P. and Johns, R., *Solitons: An Introduction*, Cambridge: Cambridge University Press, 1989.
Freidberg, J. P., *Ideal Magnetohydrodynamics*, New York: Plenum Press, 1987.
Goldston, R. J. and Rutherford, P. H., *Introduction to Plasma Physics*, Bristol: Institute of Physics Publishing, 1995.
Hasegawa, A. and Uberoi, C., *The Alfvén Wave*, Oak Ridge: National Technical Information Service, US Department of Commerce, 1982.
Hutchison, I. H., *Principles of Plasma Diagnostics*, 2nd edn., Cambridge: Cambridge University Press, 2002.
Kadomtsev, B. B., *Plasma Turbulence*, London: Academic Press, 1965.
Kivelson, M. G. and Russell, C. T., *Introduction to Space Physics*, Cambridge: Cambridge University Press, 1995.
Krall, N. A. and Trivelpiece, A. W., *Principles of Plasma Physics*, New York: McGraw-Hill, 1973.
Landau, L. D. and Lifshitz, E. M., *Mechanics*, Oxford: Pergamon Press, 1989.
Mathews, J. and Walker, R. L., *Mathematical Methods of Physics*, New York: W.A. Benjamin, 1965.
Melrose, D. B., *Instabilities in Space and Laboratory Plasmas*, Cambridge: Cambridge University Press, 1986.
Miyamoto, K., *Plasma Physics for Nuclear Fusion*, Cambridge, Mass: MIT Press, 1989.
Moffatt, H. K., *Magnetic Field Generation in Electrically Conducting Fluids*, Cambridge: Cambridge University Press, 1978.

Nicholson, D. R., *Introduction to Plasma Theory*, Malabar, Florida: Krieger Pub. Co., 1992.
Sagdeev, R. Z. and Galeev, A., *Nonlinear Plasma Theory*, New York: W. A. Benjamin, 1969.
Schmidt, G., *Physics of High Temperature Plasmas*, New York: Academic Press, 1979.
Shukla, P. K. and Mamun, A. A., *Introduction to Dusty Plasma Physics*, Bristol: Institute of Physics Publishing, 2002.
Stix, T. H., *Waves in Plasmas*, New York: American Institute of Physics, 1992.
Storey, L. R. O. (1956) Whistlers, *Scientific American* **194**, 34.
Sturrock, P. A., *Plasma Physics: An Introduction to the Theory of Astrophysical, Geophysical, and Laboratory Plasmas*, Cambridge: Cambridge University Press, 1994.
Swanson, D. G., *Plasma Waves*, 2nd edn., Bristol: Institute of Physics Publishing, 2003.

References

Alfvén, H. (1943). On the existence of electromagnetic-hydrodynamic waves, *Arkiv Mat. Astron. Fysik*, **B29**, 2.

Allen, J. E., Boyd, R. L. F., and Reynolds, P. (1957). The collection of positive ions by a probe immersed in a plasma, *P. Phys. Soc. Lon. B*, **70**, 297.

Allis, W. P. (1955). *Waves in a Plasma*, MIT Res. Lab. Electronics Quarterly Report, Vol. 54, p. 5.

Appleton, E. V. (1932). Wireless studies of the ionosphere, *J. Inst. Elec. Engrs.*, **71**, 642.

Bachman, D. A. and Gould, R. W. (1996). Landau-like damping in rotating pure electron plasmas, *IEEE Trans. Plasma Sci.*, **24**, 14.

Barkan, A., D'Angelo, N., and Merlino, R. L. (1996). Experiments on ion acoustic waves in dusty plasmas, *Planet. Space Sci.*, **44**, 239.

Barkan, A., Merlino, R. L., and D'Angelo, N. (1995). Laboratory observation of the dust acoustic wave mode, *Phys. Plasmas*, **2**, 3563.

Barkhausen, H. (1930). Whistling tones from the Earth, *Proc. Inst. Radio Engrs.*, **18**, 1115.

Barnes, A. (1977). Geometrical meaning of curl operation when $\mathbf{A} \cdot \text{curl}\ \mathbf{A} \neq 0$, *Am. J. Phys.*, **45**, 371.

Bateman, G. (1978). *MHD Instabilities*, Cambridge, Mass.: MIT Press.

Bellan, P. M. (1992), Vorticity model of flow driven by purely poloidal currents, *Phys. Rev. Lett.*, **69**, 3515.

(2000). *Spheromaks: A Practical Application of Magnetohydrodynamic Dynamos and Plasma Self-Organization*, London: Imperial College Press, pp. 61–64.

(2003). Why current-carrying magnetic flux tubes gobble up plasma and become thin as a result, *Phys. Plasmas*, **10**, 1999.

(2004a). A microscopic, mechanical derivation of the adiabatic gas relation, *Am. J. Phys.*, **72**, 679.

(2004b). A model for the condensation of a dusty plasma, *Phys. Plasmas*, **11**, 3368.

Bellan, P. M. and Porkolab, M. (1975). Excitation of lower-hybrid waves by a slow-wave structure, *Phys. Rev. Lett.*, **34**, 124.

Bennett,W. H. (1934). Magnetically self-focussing streams, *Phys. Rev.*, **45**, 890.

Berger, M. A. and Field, G. B. (1984). The topological properties of magnetic helicity, *J. Fluid Mech.*, **147**, 133.

Bernstein, I. B. (1958). Waves in a plasma in a magnetic field, *Phys. Rev.*, **109**, 10.

Bernstein, I. B. and Engelmann, F. (1966). Quasi-linear theory of plasma waves, *Phys. Fluids*, **9**, 937.

Bernstein, I. B., Frieman, E. A., Kruskal, M. D., and Kulsrud, R. M. (1958). An energy principle for hydromagnetic stability problems, *Proc. R. Soc. Lon. Ser. A*, **244**, 17.

Birdsall, C. K. and Langdon, A. B. (1985). *Plasma Physics via Computer Simulation*, New York: McGraw Hill.
Bohm, D. and Gross, E. P. (1949). Theory of plasma oscillations. A. Origin of medium-like behavior, *Phys. Rev.*, **75**, 1851.
Braginskii, S. I. (1965). Transport processes in a plasma, *Rev. Plasma Phys.*, **1**, 205.
Briggs, R. J., Daugherty, J. D., and Levy, R. H. (1970). Role of Landau damping in crossed-field electron beams and inviscid shear flow, *Phys. Fluids*, **13**, 421.
Briggs, R. J. and Parker, R. R. (1972). Transport of rf energy to lower hybrid resonance in an inhomogeneous plasma, *Phys. Rev. Lett.*, **29**, 852.
Brillouin, L. (1926). La mécanique ondulatoire de Schrödinger, une méthode générale de résolution par approximations successives, *Compt. Rend.*, **183**, 24.
Buchsbaum, S. J. (1960). Resonance in a plasma with two ion species, *Phys. Fluids*, **3**, 418.
Calugareanu, G. (1959). L'intégrale de Gauss et l'analyse des nuds tridimensionnels, *Rev. Math. Pures Appl.*, **4**, 5.
Chew, G. F., Goldberger, M., and Low, F. E. (1956). The Boltzmann equation and the one-fluid hydromagnetic equations in the absence of particle collisions, *Proc. R. Soc. Lon. Ser. A*, **236**, 112.
Chu, J. H. and I, L. (1994). Direct observation of Coulomb crystals and liquids in strongly coupled rf dusty plasmas, *Phys. Rev. Lett.*, **72**, 4009.
Clemmow, P. C. and Mullaly, R. F. (1955). Dependence of the refractive index in magneto-ionic theory on the direction of the wave normal, *Physics of the Ionosphere: Report of Phys. Soc. Conf. Cavendish Lab.*, London: Physical Society, 340.
Crawford, F. W. (1965). Cyclotron harmonic waves in warm plasmas, *J. Res. N. B. S. D. Rad. Sci.*, **D 69**, 789.
Davidson, R. C. (2001). *Physics of Nonneutral Plasmas*, London: Imperial College Press; World Scientific.
Debye, P. and Hückel, E. (1923). Zur Theorie der Elektrolyte. I. Gefrierpunktserniedrigung und verwandte Erscheinungen, *Physik. Zeits.*, **24**, 185.
Drazin, P. G. and Johns, R. S. (1989). *Solitons: An Introduction*, Cambridge: Cambridge University Press.
Dreicer, H. (1959). Electron and ion runaway in a fully ionized gas. 1., *Phys. Rev.*, **115**, 238.
Driscoll, C. F. and Fine, K. S. (1990). Experiments on vortex dynamics in pure electron plasmas, *Phys. Fluids B – Plasma*, **2**, part 2, 1359.
Drummond, W. E. and Pines, D. (1962). Non-linear stability of plasma oscillations, *Nucl. Fusion*, 1049.
Eckersley, T. L. (1935). Musical atmospherics, *Nature*, **135**, 104.
Fermi, E. (1954), Galactic magnetic fields and the origin of cosmic radiation, *Astrophys. J.*, **119**, 1.
Fisch, N. J. (1978). Confining a tokamak plasma with rf-driven currents, *Phys. Rev. Lett.*, **41**, 873.
Fisher, R. K. and Gould, R. W. (1969). Resonance cones in the field pattern of a short antenna in an anisotropic plasma, *Phys. Rev. Lett.*, **22**, 1093.
(1971). Resonance cones in the field pattern of a radio frequency probe in a warm anisotropic plasma, *Phys. Fluids*, **14**, 857.
Furth, H. P., Killeen, J., and Rosenbluth, M. N. (1963). Finite-resistivity instabilities of a sheet pinch, *Phys. Fluids*, **6**, 459.
Gardner, C. S., Greene, J. M., Kruskal, M. D., and Miura, R. M. (1967). Method for solving Korteweg–deVries equation, *Phys. Rev. Lett.*, **19**, 1095.
Goree, J. (1992). Ion trapping by a charged dust grain in a plasma, *Phys. Rev. Lett.*, **69**, 277.
Gould, R. W. (1995). Dynamics of nonneutral plasmas, *Phys. Plasmas*, **2**, 2151.

Gould, R. W., O'Neil, T. M., and Malmberg, J. H. (1967). Plasma wave echo, *Phys. Rev. Lett.*, **19**, 219.

Grad, H. and Rubin, H. (1958). *MHD Equilibrium in an Axisymmetric Toroid*, Proceedings of the 2nd UN Conf. on the Peaceful Uses of Atomic Energy, Vol. 31, Vienna: IAEA p. 190.

Hansen, J. F. and Bellan, P. M. (2001). Experimental demonstration of how strapping fields can inhibit solar prominence eruptions, *Astrophys. J.*, **563**, L183.

Havnes, O., Goertz, C. K., Morfill, G. E., Grun, E., and Ip, W. (1987). Dust charges, cloud potential, and instabilities in a dust cloud embedded in a plasma, *J. Geophys. Res. – Space*, **92**, 2281.

Hayashi, Y. and Tachibana, K. (1994). Observation of Coulomb crystal formation from carbon particles grown in a methane plasma, *Jpn. J. Appl. Phys. 2*, **33**, L804.

Hudgings, D. W., Meger, R. A., Striffler, C. D., *et al.* (1978). Trapping of cusp-injected, nonneutral electron rings with resistive walls and static mirror coils, *Phys. Rev. Lett.*, **40**, 764.

Ikezi, H. (1986). Coulomb solid of small particles in plasmas, *Phys. Fluids*, **29**, 1764.

Jackson, J. D. (1998). *Classical Electrodynamics*, 3rd edn. New York: Wiley.

Jeans, J. H. (1915). On the theory of star-streaming and the structure of the universe, *Mon. Not. R. Astron. Soc.*, **76**, 70.

Kim, H. C., Stenzel, R. L., and Wong, A. Y. (1974). Development of cavitons and trapping of rf field, *Phys. Rev. Lett.*, **33**, 886.

Klimchuk, J. (2000). Cross-sectional properties of coronal loops, *Sol. Phys.*, **193**, 53.

Korteweg, D. J. and de Vries, F. (1895). On the change of form of long waves advancing in a rectangular canal and on a new type of long stationary waves, *Philos. Mag.*, **39**, 422.

Kramers, H. A. (1926). Wellenmechanik und halbzahlige Quantisierung, *Z. Physik.*, **39**, 828.

Kribel, R. E., Shinksky, K., Phelps, D. A., and Fleischmann, H. H. (1974). Generation of field-reversing electron rings by neutralized cusp injection into a magnetic-mirror trap, *Plasma Phys. Cont. F.*, **16**, 113.

Kruskal, M. D., Johnson, J. L., Gottlieb, M. B., and Goldman, L. M. (1958). Hydromagnetic instability in a stellarator, *Phys. Fluids*, **1**, 421.

Kruskal, M. D. and Schwarzschild, M. (1954). Some instabilities of a completely ionized plasma, *Proc. R. Soc. Lon. Ser. A*, **223**, 348.

Laframboise, J. G. and Parker, L. W. (1973). Probe design for orbit-limited current collection, *Phys. Fluids*, **16**, 629.

Landau, L. D. (1946). On the vibrations of the electronic plasma, *J. Physics (USSR)*, **10**, 25.

Langmuir, I. (1928). Oscillations in ionized gases, *P. Natl. Acad. Sci.*, **14**, 627.

Leuterer, F. (1969). Forward and backward Bernstein waves, *Plasma Phys.*, **11**, 615.

Livio, M. (1999). Astrophysical jets: a phenomenological examination of acceleration and collimation, *Phys. Rep.*, **311**, 225.

Lundquist, S. (1950). Magneto-hydrostatic fields, *Arkiv for Fysik*, **B2**, 361.

Maecker, H. (1955). Plasmastromungen in Lichtbogen infolge eigenmagnetischer Kompression, *Z. Physik*, **141**, 198.

Malmberg, J. H. (1992). Some recent results with nonneutral plasmas, *Plasma Phys. Cont. F.*, **34**, 1767.

Malmberg, J. H. and de Grassie, J. S. (1975). Properties of nonneutral plasma, *Phys. Rev. Lett.*, **35**, 577.

Malmberg, J. H. and Wharton, C. B. (1964). Collisionless damping of electrostatic plasma waves, *Phys. Rev. Lett.*, **13**, 184.

Malmberg, J. H., Wharton, C. B., Gould, R. W., and O'Neil, T. M. (1968). Plasma wave echo experiment, *Phys. Rev. Lett.*, **20**, 95.

Manley, J. M. and Rowe, H. E. (1956). Some general properties of nonlinear elements 1. General energy relations, *P. I. R. E.*, **44**, 904.
Marshall, J. (1960). Performance of a hydromagnetic plasma gun, *Phys. Fluids*, **3**, 134.
McChesney, J. M., Stern, R. A., and Bellan, P. M. (1987). Observations of fast stochastic ion heating by drift waves, *Phys. Rev. Lett.*, **59**, 1436.
Melzer, A., Trottenberg, T., and Piel, A. (1994) Experimental determination of the charge on dust particles forming Coulomb lattices, *Phys. Lett. A*, **191**, 301.
Moffatt, H. K. (1978). *Magnetic Field Generation in Electrically Conducting Fluids*, Cambridge: Cambridge University Press.
Moffatt, H. K. and Ricca, R. L. (1992). Helicity and the Calugareanu invariant, *P. R. Soc. Lon. Ser. A*, **439**, 411.
Moncuquet, M., Meyervernet, N., and Hoang, S. (1995). Dispersion of electrostaticwaves in the Io plasma torus and derived electron temperature, *J. Geophys. Res. – Space*, **100**, 21697.
Morfill, G. E., Annaratone, B. M., Bryant, P., *et al.* (2002). A review of liquid and crystalline plasmas – new physical states of matter?, *Plasma Phys. Cont. F.*, **44**, B263.
Nefedov, A. P., Morfill, G. E., Fortov, V. E., *et al.* (2003). PKE-Nefedov: plasma crystal experiments on the International Space Station, *New J. Phys.*, **5**, art. no. 33.
Nishikawa, K. (1968a). Parametric excitation of coupled waves 1. General formulation, *J. Phys. Soc. Jpn.*, **24**, 916.
(1968b). Parametric excitation of coupled waves 2. Parametric plasmon–photon interaction, *J. Phys. Soc. Jpn.*, **24**, 1152.
O'Neil, T. M. (1995). Plasmas with a single sign of charge (an overview), *Phys. Scripta*, **T59**, 341.
Ono, M. (1979). Cold, electrostatic, ion cyclotron waves and ion–ion hybrid resonances, *Phys. Rev. Lett*, **42**, 1267.
Parker, E. N. (1957). Sweet's mechanism for merging magnetic fields in conducting fluids, *J. Geophys. Res.*, **62**, 509.
Penrose, O. (1960). Electrostatic instabilities of a uniform non-Maxwellian plasma, *Phys. Fluids*, **3**, 258.
Pfister, H. and Gekelman, W. (1991). Demonstration of helicity conservation during magnetic reconnection using Christmas ribbons, *Am. J. Phys.*, **59**, 497.
Rosenbluth, M. N., Macdonald, W. M., and Judd, D. (1957). Fokker–Planck equation for an inverse-square force, *Phys. Rev.*, **107**, 1.
Rutherford, P. H. (1973). Nonlinear growth of tearing mode, *Phys. Fluids*, **16**, 1903.
Sagdeev, R. Z. and Galeev, A. (1969). *Nonlinear plasma theory*, New York: W. A. Benjamin.
Schecter, D. A., Dubin, D. H. E., Cass, A. C., Driscoll, C. F., Lansky, I. M., and O'Neil, T. M. (2000), Inviscid damping of asymmetries on a two-dimensional vortex, *Phys. Fluids*, **12**, 2397.
Schmitt, J. M. (1973). Dispersion and cyclotron damping of pure ion Bernstein waves, *Phys. Rev. Lett.*, **31**, 982.
Shafranov, V. D. (1958). On magnetohydrodynamical equilibrium configurations, *Sov. Phys. – JETP*, **6**, 545.
(1966). Plasma equilibrium in a magnetic field, *Reviews of Plasma Physics*, Vol. 2, New York: Consultants Bureau, p. 103.
Shukla, P. K. and Silin, V. P. (1992). Dust ion acoustic wave, *Phys. Scripta*, **45**, 508.
Silin, V. P. (1965). Parametric resonance in a plasma, *Sov. Phys. JETP*, **21**, 1127.
Solov'ev, L. S. (1976). Hydromagnetic stability of closed plasma configurations, *Rev. Plasma Phys.*, **6**, 239.
Spitzer, L. and Harm, R. (1953). Transport phenomena in a completely ionized gas, *Phys. Rev.*, **89**, 977.

Stasiewicz, K., Bellan, P., Chaston, C., *et al.* (2000). Small scale Alfvénic structure in the aurora, *Space Sci. Rev.*, **92**, 423.

Stasiewicz, K., Lundin, R., and Marklund, G. (2000). Stochastic ion heating by orbit chaotization on electrostatic waves and nonlinear structures, *Phys. Scripta*, **T84**, 60.

Stenzel, R. and Wong, A. Y. (1972). Threshold and saturation of parametric decay instability, *Phys. Rev. Lett.*, **28**, 274.

Stix, T. H. (1962). *The Theory of Plasma Waves*, New York: McGraw-Hill.

(1965). Radiation and absorption via mode conversion in an inhomogeneous collision-free plasma, *Phys. Rev. Lett.*, **15**, 878.

(1992). *Waves in Plasmas*, New York: American Institute of Physics.

Storey, L. R. O. (1953). An investigation of whistling atmospherics, *Philos. T. R. Soc. Lon.*, **246**, 113.

Surko, C. M., Leventhal, M., and Passner, A. (1989). Positron plasma in the laboratory, *Phys. Rev. Lett.*, **62**, 901.

Swanson, D. G. (1989). *Plasma Waves*, San Diego: Academic Press. See p. 24 for discussion regarding Altar.

(2003). *Plasma waves*, 2nd edn. Bristol Institute of Physics Publishing.

Sweet, P. A. (1958). The neutral point theory of solar flares. In *Electromagnetic Phenomena in Cosmical Physics*, ed. B. Lehnert. Cambridge: Cambridge University Press.

Taylor, J. B. (1974). Relaxation of toroidal plasma and generation of reverse magnetic fields, *Phys. Rev. Lett.*, **33**, 1139.

Thomas, H., Morfill, G. E., Demmel, V., Goree, J., Feuerbacher, B., and Mohlmann, D. (1994). Plasma crystal – Coulomb crystallization in a dusty plasma, *Phys. Rev. Lett.*, **73**, 652.

Trintchouk, F., Yamada, M., Ji, H., Kulsrud, R. M., and Carter, T. A. (2003). Measurement of the transverse Spitzer resistivity during collisional magnetic reconnection, *Phys. Plasmas*, **10**, 319.

Trivelpiece, A. W. and Gould, R. W. (1959). Space charge waves in cylindrical plasma columns, *J. Applied Phys.*, **30**, 1784.

Vedenov, A. A., Velikhov, E. P., and Sagdeev, R. (1962). Quasi-linear theory of plasma oscillations, *Nucl. Fusion Suppl.* 2, p. 465.

Wang, C. L., Joyce, G., and Nicholson, D. R. (1981). Debye shielding of a moving test charge in plasma, *J. Plasma Phy.*, **25**, 225.

Watson, G. N. (1922). *A Treatise on the Theory of Bessel Functions*, Cambridge: Cambridge University Press.

Watson, K. M. (1956). Use of the Boltzmann equation for the study of ionized gases of low density 1, *Phys. Rev.*, **102**, 12.

Wentzel, G. (1926). Eine Verallgemeinerung der Quantenbedingungen für die Zwecke der Wellenmechanik, *Z. Physik.*, **38**, 518.

White, R., Chen, L., and Lin, Z. H. (2002). Resonant plasma heating below the cyclotron frequency, *Phys. Plasmas*, **9**, 1890.

White, W. D., Malmberg, J. H., and Driscoll, C. F. (1982). Resistive-wall destabilization of diocotron waves, *Phys. Rev. Lett.*, **49**, 1822.

Woltjer, L. (1958). A theorem on force-free magnetic fields, *P. Natl. Acad. Sci.*, **44**, 489.

Wong, A. Y., Motley, R. W., and D'Angelo, N. (1964). Landau damping of ion acoustic waves in highly ionized plasmas, *Phys. Rev. A*, **133**, 436.

You, S., Yun, G. S., and Bellan, P. M. (2005). Dynamic and stagnating plasma flow leading to magnetic-flux-tube collimation, *Phys. Rev. Lett.*, **95**, 045002.

Yu, J. H. and Driscoll, C. F. (2002). Diocotron wave echoes in a pure electron plasma, *IEEE Trans. Plasma Sci.*, **30**, 24.

Index

Printed in the United States
By Bookmasters